Macmillan Education
LaunchPad

ADDITIONAL CHAPTERS AVAILABLE ONLINE IN LAUNCHPAD!

BRIEF CONTENTS

Publisher: Katherine Parker
Senior Acquisitions Editor: Bill Minick
Developmental Editor: Andrea Gawrylewski
Executive Marketing Manager: John Britch
Marketing Assistant: Bailey James
Media and Supplements Editor: Amanda Dunning
Editorial Assistants: Tue Tran & Shannon Moloney
Art Director: Diana Blume
Cover Design and Text Illustrations: MGMT. design
Text Design: Dirk Kaufman
Photo Editor: Sheena Goldstein
Photo Researcher: Bianca Moscatelli
Art Manager: Matthew McAdams
Senior Production Supervisor: Susan Wein
Printing and Binding: RR Donnelley
Cover Photo: Annie Marie Musselman

Library of Congress Control Number: 2014957697
ISBN-13: 978-1-4641-6220-6
ISBN-10: 1-4641-6220-4

Printed in the United States of America

First printing

W. H. Freeman and Company
41 Madison Avenue
New York, NY 10010
Houndmills, Basingstoke RG21 6XS, England
www.whfreeman.com

ITY PLEDGE
ommitted to lessening our
ct on the environment.
mily of publishing houses
ur 2020 CO2 emissions
09 baseline.

JAMES P. BLAIR/National Geographic
Creative

ENVIRONMENTAL SCIENCE

FOR A CHANGING WORLD

SECOND EDITION

SUSAN KARR
Carson-Newman University

JENEEN INTERLANDI
Science Writer

ANNE HOUTMAN
California State University, Bakersfield

· A PARTNERSHIP BETWEEN ·

macmillan education & SCIENTIFIC AMERICAN

BRIEF CONTENTS

DETAILED CONTENTS

INTRODUCTION TO ENVIRONMENTAL, SCIENCE, AND INFORMATION LITERACY

Paul Souders/WorldFoto/Aurora Photos

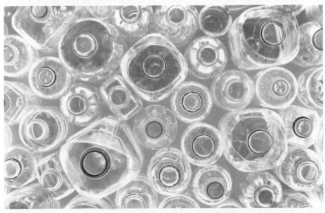

ULTRA.F/Digital Vision/Getty Images

HUMANS POPULATIONS AND ENVIRONMENTAL HEALTH

Christian Kober/Robert Harding/Newscom

CHAPTER 4 HUMAN POPULATIONS   62

ONE-CHILD CHINA GROWS UP

A country faces the outcomes of radical population control

Vanessa Vick/The New York Times/Redux

CHAPTER 5 ENVIRONMENTAL HEALTH   80

ERADICATING A PARASITIC NIGHTMARE

Human health is intricately linked to the environment

CONSUMPTION AND THE ENVIRONMENTAL FOOTPRINT

CHAPTER 6 ECOLOGICAL ECONOMICS AND CONSUMPTION   98

WALL TO WALL, CRADLE TO CRADLE

A leading carpet company takes a chance on going green

jacus/iStockphoto/Thinkstock

ECOLOGY

© ClassicStock/CAMERIQUE/Alamy

© Stephen Vincent/Alamy

EVOLUTION AND BIODIVERSITY

James Balog/The Image Bank/Getty Images

JAMES P. BLAIR/National Geographic Creative

WATER RESOURCES

Constantinos Z/iStock/360/Getty Images

FOOD RESOURCES

YASUYOSHI CHIBA/AFP/Getty Images

CHAPTER 16 FEEDING THE WORLD   298
A GENE REVOLUTION
Can genetically engineered food help end hunger?

Moment/Getty Images

CHAPTER 17 SUSTAINABLE AGRICULTURE: RAISING CROPS   316
FARMING LIKE AN ECOSYSTEM
Creative solutions to feeding the world

CONVENTIONAL ENERGY: FOSSIL FUELS

George Steinmetz/Corbis

AIR POLLUTION:
CONSEQUENCES OF USING FOSSIL FUELS

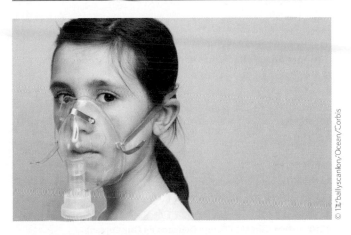

© 1.2'ballyscanlon/Ocean/Corbis

ALTERNATIVES TO FOSSIL FUELS

AP Photo/David Guttenfelder

Alessandro Grassani/Invision/Aurora Photos

SUSTAINABLE LIVING IN COMMUNITY

TAO Images Limited/Getty Images

CHAPTER 24 ENVIRONMENTAL POLICY   470
COUNTERFEIT COOLING
*In the global efforts to thwart climate change, some lessons are
learned after the fact*

Nina Berman/Noor/Redux

CHAPTER 25 URBANIZATION AND SUSTAINABLE
COMMUNITIES   490
THE GHETTO GOES GREEN
Building a better backyard in the Bronx

AVAILABLE ONLINE IN Macmillan Education LaunchPad

http://www.macmillanhighered.com/launchpad/saes2e

Can some populations adapt to ocean acidification?

Coral reefs are complex communities with lots of interspecific interactions.

INFOGRAPHIC
29.5   Coral Biology
29.6   Coral Bleaching

The world's oceans face many other threats.

INFOGRAPHIC
29.7   Threats to Oceans

Reducing the threats to oceans requires a multi-pronged approach.

TABLE
29.1   Reducing the Threats to Ocean Ecosystems

ONLINE CHAPTER 30 AGRICULTURE: RAISING LIVESTOCK

A CARNIVORE'S CONUNDRUM

Disease, pollution, and the true costs of meat

The way we raise livestock may jeopardize the safety of food products.

INFOGRAPHIC
30.1   *E. Coli*—Just the Tip of the Iceberg

Affluence influences diet.

INFOGRAPHIC
30.2   Affluence Affects Diet and Health

CAFOs can raise a large number of animals quickly, but incur a huge environmental cost.

INFOGRAPHIC
30.3   Growing Livestock: Feed and Water Needs
30.4   From Farm to You

A variety of methods can reduce *E. coli* contamination.

INFOGRAPHIC
30.5   *E. Coli* 0157:H7 Infections are Decreasing in the United States

There are more sustainable ways to grow livestock.

U.S. food policies support industrial agriculture.

INFOGRAPHIC
30.6   Agricultural Policy Must Consider Trade-Offs

Consumer choices can increase food supply.

INFOGRAPHIC
30.7   Diet and Carrying Capacity

ONLINE CHAPTER 31 FISHERIES AND AQUACULTURE

FISH IN A WAREHOUSE?

How one Baltimore fish scientist could change the way we eat

Industrial fishing is impacting fisheries worldwide.

INFOGRAPHIC
31.1   Meet the Cod
31.2   Bottom Trawling

Humans rely on protein from fish but overfishing of wild stocks makes it harder for fish populations to recover.

INFOGRAPHIC
31.3   Fishing Down the Food Chain
31.4   Status of Marine Fisheries

Laws exist to protect and manage fisheries.

INFOGRAPHIC
31.5   Protection for Marine Areas

Scientists study the possibility of growing marine fish indoors.

Aquaculture presents environmental challenges.

TABLE
31.1   Net Pen and Pond Aquaculture: Problems and Possible Solutions

Indoor fish farming may provide a solution.

INFOGRAPHIC
31.6   Biomimicry in the Pool

ONLINE CHAPTER 32 BIOFUELS

GAS FROM GRASS

Will an ordinary prairie grass become the next biofuel?

Biofuels are a potentially important alternative to fossil fuels.

INFOGRAPHIC
32.1   Biofuel Sources

Biofuels can come from unexpected sources.

INFOGRAPHIC
32.2   Waste to Energy

Turning grass into gas is less environmentally friendly than it sounds.

TABLE
32.1   Biofuel Trade-Offs

Tilman's experiments showed the importance of biodiversity.

INFOGRAPHIC
32.3   LIHD Crops Offer Advantages Over Traditional Monoculture Biofuel Crops

There is another rising biofuel star: Algae.

INFOGRAPHIC
32.4   Biofuels from Algae

There are many reasons why biofuels have not solved our dependence on fossil fuels.

INFOGRAPHIC
32.5   Bioethanol Production

Multiple solutions will be needed to help replace fossil fuels.

INFOGRAPHIC
32.6   Energy Efficiency and Conservation are Part of the Solution

Despite ongoing controversies and setbacks, the future of biofuels looks bright.

ABOUT THE AUTHORS

SUSAN KARR, MS, is an Instructor in the biology department of Carson-Newman University in Jefferson City, Tennessee, and has been teaching for more than 20 years. She has served on campus and community environmental sustainability groups and helps produce an annual "State of the Environment" report on the environmental health of her county. In addition to teaching non-majors courses in environmental science and human biology, she teaches an upper-level course in animal behavior where she and her students train dogs from the local animal shelter in a program that improves the animals' chances of adoption. She received degrees in animal behavior and forestry from the University of Georgia.

Stephen Karr

© 2011 Macmillan. Photo by Andrea Gawrylewski

JENEEN INTERLANDI, MA, MS, is a science writer who contributes to *Scientific American* and *The New York Times Magazine*. Previously, she spent four years as a staff writer for *Newsweek*, where she covered health, science, and the environment. Jeneen has worked as a researcher at both Harvard Medical School and Lamont Doherty Earth Observatory. She was a 2013 Nieman Fellow. In 2014 she received a grant from the Pulitzer Center for Crisis Reporting, to cover the struggles of Roma ascension in Hungary. Jeneen holds Master's degrees in environmental science and journalism, both from Columbia University in New York.

Will Prouty

ANNE HOUTMAN, DPHIL, is Dean of the School of Natural Sciences, Mathematics and Engineering, and Professor of Biological Sciences at California State University, Bakersfield. Her research interests are in the behavioral ecology of birds. She is strongly committed to evidence-based, experiential education and has been an active participant in the national dialogue on science education—how best to teach science to future scientists and future science "consumers"—for almost 20 years. Anne received her doctorate in zoology from the University of Oxford and conducted postdoctoral research at the University of Toronto.

Dear Reader,

For more than 20 years as an environmental science and biology instructor I've found that "stories" capture the imagination of my students. Students are genuinely interested in environmental issues—and using stories to teach these issues makes the science more relevant and meaningful to them. Many leave the class with an understanding that what they do really matters, and they feel a willingness to act on that knowledge. This is why I am enthusiastic about our textbook, *Environmental Science for a Changing World*.

Each chapter will keep students engaged and reading to find out "what happens next." At the same time, explanations of science are woven into the narrative and illustrated in vivid infographics that give additional detail without slowing down the story. We've heard from instructors using the book (and our own students) and we know students are actually reading the textbook and being drawn into the stories, making it is easier for them to make connections between the environmental science concepts being taught and the "bigger picture" of why it matters.

In this book we've broken some topics down into multiple chapters—for instance, we present separate, short chapters on coal, petroleum and natural gas, and nuclear power rather than the traditional single chapter on conventional energy. This gives instructors the flexibility to focus on discrete topics if they choose.

In this second edition, we have added some popular chapters (originally in the extended edition only): Environmental Health, Environmental Policy, World Hunger, and Biodiversity Preservation. We also moved some of the chapters online to LaunchPad, the textbook's web platform, as a way to make all of our chapters available to every adopter, giving even more flexibility in choosing which chapters and topics to cover. (LaunchPad chapters new to the second edition include a chapter on Mineral Resources and Geology as well as an additional agriculture chapter on Raising Livestock in CAFOs. Other chapters found in the first edition that now reside on LaunchPad include chapters on Forest Resources, Grasslands and Soils, Marine Resources, Fisheries, and Biofuels.) Every chapter has been updated to include the latest information.

We also listened to instructor feedback and in this second edition we have added pedagogical aids such as key concepts and end-of-chapter study aids that are tied to the guiding questions that open each chapter. This will help students focus on the important environmental science concepts being presented in the chapter. We've also added questions to each infographic to better engage the student with the figures and diagrams.

As with the first edition, the text focuses on building core competencies for the non-major: environmental literacy, science literacy, and information literacy. End-of-chapter and online exercises provide further opportunities to develop these competencies, as well as critical thinking skills.

Environmental Literacy: The scientific, social, political, and economic facets of contemporary environmental issues are examined with a focus on the scientific concepts and drivers underlying issues. Material is presented in a balanced way, especially for controversial topics. Sustainable solutions are presented.

Science Literacy: Each chapter includes experimental evidence and graphical data representation, and describes the day-to-day work of scientists, giving students many opportunities to evaluate evidence and understand the process of science.

Information Literacy: Students must be able to both find information and assess its quality. We explain how to effectively search for and find scientific information, and how to critically analyze that information.

Every person involved in this book—the writers, illustrators, editors, and fellow instructors—has one sincere objective: to help students become informed citizens who are able to analyze issues, evaluate arguments, discuss solutions, and recognize trade-offs as they make up their own minds about our most pressing environmental challenges.

Sincerely,

Susan Karr

Susan Karr

CAPTIVATING STORIES

STUDENTS FOLLOW ONE RIVETING STORY THROUGHOUT THE ENTIRE CHAPTER

FROM THE OPENING PAGE TO THE END-OF CHAPTER QUESTIONS, AND EVERYWHERE IN BETWEEN, EACH CHAPTER FOLLOWS AN ENGAGING REAL STORY THAT ILLUSTRATES AND MOTIVATES CORE SCIENCE CONCEPTS. HERE ARE FOUR OF MANY PLACES IN THIS CHAPTER WHERE THE STORY UNFOLDS.

SEE PAGE 220

Laurel Sutherlin of the Rainforest Action Network went to Sumatra to gauge the impact of the expanding palm oil industry on the island.

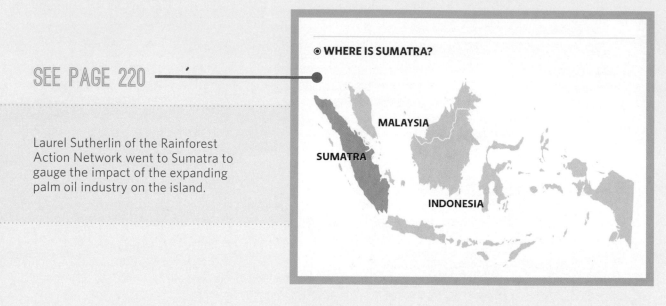

⊙ **WHERE IS SUMATRA?**

MALAYSIA

SUMATRA

INDONESIA

SEE PAGES 223, 235, 238 & 242

The orangutan is among the many species being driven to the edge of extinction by the pursuit of palm oil. So much of the Sumatran rain forest land has been laid low that Laurel Sutherlin and his team had to travel overnight and through the morning to reach an untouched section. But once there, they witnessed firsthand its rich biodiversity.

Vier Pfoten/Four Paws/RHOI/Rex Features via AP Images

CHAPTER 12:

PALM PLANET: Can we
have tropical forests and
our palm oil too?

> " *Sustainable palm is the key
> to protecting our biodiversity,
> and our heritage.* "
> —Raviga Sambanthamurthi.

Virtually all of the forest's inhabitants are facing
annihilation. "The remaining populations of endemic
Sumatran rhinos are widely considered to be the living
dead," says Sutherlin. "Their habitat is too sparse, too
fragmented and too disturbed, their numbers too few."

SEE PAGE 233

In 2012, Malaysian geneticist Raviga
Sambanthamurthi and her team of
researchers discovered the SHELL gene,
which is responsible for the most productive
palm oil fruit. Farmers now can identify the
seeds for this fruit before planting, and
produce more oil per hectare.

BENEFITS OF PALM OIL	DISADVANTAGES OF PALM OIL
Palm oil is the preferred dietary replacement for trans fats.	Though probably better than trans fats for one's health, it is still a fat whose consumption should be limited.
Palm oil is already widely used in many products.	The high demand for palm oil in products increases the need for more oil palm plantations.
Palm oil is inexpensive.	If palm oil is priced in a way that reflects its true costs, the prices of goods that contain it will increase.
Oil palm plantations can produce much more oil per acre than can other oil crops.	Oil palm plantations reduce biodiversity and decrease the ability of local ecosystems to provide ecosystem services.
ADDRESSING THE TRADE-OFFS	
Cultivation of oil palms should be done in a way to minimize damage to local ecosystems and biodiversity, leaving some areas uncultivated as refuges for local biodiversity. The most productive palm varieties should be used to increase production per acre.	
Manufacturers should only source palm oil that is certified as sustainably grown, and this information should be readily available to consumers.	
Consumers should reduce their use of palm oil products in general and opt for products that contain certified sustainable palm oil to support companies that use it.	
Consumers should be educated so that they understand that consumption of palm oil (in foods), like consumption of any other fat, must be moderate.	

SEE PAGES 244-245

Consumer demands are leading
to changes in the palm oil
industry. Thanks to the "Snack
Food 20" Campaign by Laurel
Sutherlin's Rainforest Action
Network, some of the world's
largest snack food companies
have pledged to use only
sustainably produced palm oil.

EMPOWERING SCIENCE

ENVIRONMENTAL SCIENCE FOR A CHANGING WORLD OFFERS A CONSISTENT METHODOLOGY FOR TEACHING THE FIELD'S ESSENTIAL SCIENTIFIC CONCEPTS, WITH EACH CHAPTER CENTERED AROUND FIVE GUIDING QUESTIONS.

THESE QUESTIONS ESTABLISH A CLEAR, STEP-BY-STEP PATHWAY THROUGH THE CHAPTER FROM THE OPENING STORY; THROUGH THE NARRATIVE, KEY CONCEPT CALLOUTS, PHOTOS, INFOGRAPHICS; TO THE END-OF-CHAPTER ASSESSMENT. THE QUESTIONS BRING THE SCIENCE IN THE CHAPTER TO THE FOREFRONT, SO STUDENTS NEVER LOSE SIGHT OF FUNDAMENTAL CONCEPTS WHILE READING THE STORY.

CHAPTER 12 BIODIVERSITY

PALM PLANET

Can we have tropical forests and our palm oil too?

CORE MESSAGE
The variety of life on Earth is tremendous. This biodiversity provides important ecological services to ecosystems; we depend on these services for things like food, medicine, and economic development. The decline of biodiversity has serious ramifications for other species as well as human well-being, so we should evaluate actions that threaten biodiversity and take steps to reduce the impact when possible.

AFTER READING THIS CHAPTER, YOU SHOULD BE ABLE TO ANSWER THE FOLLOWING **GUIDING QUESTIONS**

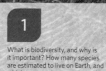

1 What is biodiversity, and why is it important? How many species are estimated to live on Earth, and which taxonomic groups have the most species?

2 How do genetic, species, and ecosystem diversity each contribute to ecosystem function and services?

Tropical forests tend to be particularly flush with both ecological and species diversity, thanks largely to the abundant sunlight and climatic conditions conducive to growth. The forests that are currently being laid low in Sumatra are no exception. They have been left unhampered for so many millennia that these steamy amphibious ecosystems swarm with a cornucopia of life: elephants, orangutans, tapirs, tigers, and every manner of bird and beetle the human imagination can fathom. "The truth is, no one has any idea how many species used to live here," Sutherlin says. "Half the species in these forests have yet to be described to science."

KEY CONCEPT 12.4

Protecting biodiversity hotspots, areas with high numbers of endangered endemic species, can be a cost-effective way to protect a large number of endangered species.

pg 235

KEY CONCEPT 12.5

Isolated populations are especially vulnerable to detrimental environmental changes because they cannot freely breed with other populations and thereby increase their genetic diversity and chances of survival.

Endemism increases with isolation, as does extinction risk.

Species come to islands such as Sumatra in various ways: They may be blown in on storms, they may arrive as lost migrators, or the island may have broken off from a larger landmass at some point. Because these are rare events, once a species arrives, it is unlikely to be joined by other members of its species. The founding population is therefore isolated, and as it adapts over time to its new island home, it may eventually evolve into a new species (see Chapter 11 for more on the founder effect). For this reason, the number of unique species (the degree of endemism) generally increases with isolation. On the Hawaiian Islands, almost 3 (2,400 miles) from the nearest mainland, all species are endemic.

pg 236

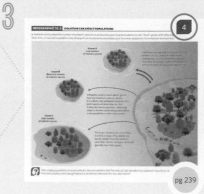

pg 239

· A PARTNERSHIP BETWEEN ·

macmillan education & SCIENTIFIC AMERICAN

environmental science FOR A CHANGING WORLD

SUSAN KARR

JENEEN INTERLANDI

ANNE HOUTMAN

SECOND EDITION

3

What are biodiversity [hotspots and]
why are they importa[nt?]

4

What role does isolation play in a
species' vulnerability to extinction?
How do habitat destruction and
fragmentation threaten species?

5

How can we acquire the food, fiber, fuel,
and pharmaceutical resources we need
without damaging the ecosystems that
provide those resources?

In the past two decades
more than 20 million acres of
rain forest have been cleared
and planted as oil palm
plantations.

1 Each chapter begins with a
Core Message and a series of
Guiding Questions that focus
students on the chapter's
central scientific content.

2 The new Key Concepts
correspond to the Guiding
Questions, re-emphasizing the
chapter's essential scientific
ideas.

3 Moving through the chapter,
students encounter icons
that connect the Guiding
Questions to specific sections
and infographics, where they'll
find the information they need
to think critically about the
question.

4 Guiding Questions provide
the framework for the end-
of-chapter pedagogy, which
includes references back to
relevant infographics to help
students answer the questions
and prepare for exams.

pg 240

INFOGRAPHIC 12.7   GLOBAL FOREST CHANGE

◆ Deforestation has been highest in tropical areas in recent years. Afforestation projects, including forests planted and managed for timber
harvesting, have actually led to an increase in forest cover in North America and China.

Net change in forest area by country, 2005–2010

Net Change (hectare/year)

Net loss
- More than 500,000
- 250,000–500,000
- 50,000–250,000

Small change (gain or loss)
- Less than 50,000

Net gain
- 50,000–250,000
- 250,000–500,000
- More than 500,000

 What areas show a net gain in forest cover on this map? Which country has experienced the highest net gain? What are
contributing to these gains?

pg 242

ENVIRONMENTAL LITERACY UNDERSTANDING THE ISSUE

1 What is biodiversity, and why is it
important? How many species are
estimated to live on Earth, and which
taxonomic groups have the most species?
INFOGRAPHICS 12.1–12.2

1. Give a basic definition of biodiversity.

2. The total number of different species on Earth is
   a. is unknown, but insects are the most numerous.
   b. is a few million, mostly bacteria and fungi.
   c. is more than 20 million, with half of them insects.
   d. is less than 5 million, mostly vertebrates.

3. Why is biodiversity into a concern?

3 What are biodiversity hotspots, and why are
they important?
INFOGRAPHIC 12.4

7. True or False: Biodiversity hotspots are areas with many

4 What role does isolation play in a species'
vulnerability to extinction? How do habitat
destruction and fragmentation threaten
species?
INFOGRAPHICS 12.5 AND 12.6

9. The leading human cause of species endangerment is

10. Why are many of the biodiversity hotspots around the world on
    islands?
    a. Islands accumulate species from many different areas.
    b. Populations of island species are isolated.
    c. Islands have more diverse habitats.
    d. There are more niches on islands.

11. Why are isolated populations more vulnerable to extinction than
    populations that are not isolated from each other?

2 How do genetic, species, and
diversity each contribute to
ecosystem function and services?
INFOGRAPHIC 12.3

pg 246

WHAT'S NEW

ONE UNIFIED BOOK THAT COVERS THE MOST ESSENTIAL CONTENT IN THE COURSE

REVISED, LEARNING-FRIENDLY **TABLE OF CONTENTS**

THE MOST CURRENT STORIES IN ENVIRONMENTAL SCIENCE. **BRAND NEW STORIES** IN THIS EDITION:

Dimas Ardian/Bloomberg via Getty Images

CHAPTER 12

In the past two decades, more than 20 million acres of rainforest—primarily in Indonesia—have been cleared and planted as palm oil plantations. **Chapter 12** dives into the Trade-Offs associated with this new, ubiquitous ingredient, and how we might use sustainable practices to keep our palm oil, and rainforests too.

VICTORIA LOE/KRT/Newscom

CHAPTER 15

A patch of the Gulf of Mexico the size of Connecticut is plagued by low-oxygen waters—conditions that threaten wildlife and the local economy. Researchers profiled in **Chapter 15** have been working tirelessly to pinpoint the oxygen-depleting culprit and come up with ways we might help the Gulf breathe again.

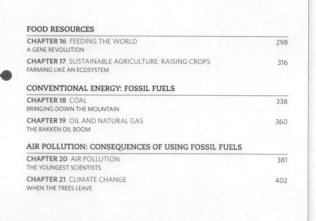

© Jim West/Alamy

CHAPTER 19

An oil boom is transforming sleepy towns into overcrowded, bustling, urban hubs. To access the oil, the industry is using a new kind of technology called fracking. **Chapter 19** delves into the impacts, both social and environmental, of this new way of harvesting oil.

NEW WAYS OF EMPOWERING SCIENCE LEARNING

CHAPTER 12 BIODIVERSITY

PALM PLANET

Can we have tropical forests and our palm oil too?

CORE MESSAGE
The variety of life on Earth is tremendous. This biodiversity provides important ecological services to ecosystems; we depend on these services for things like food, medicine, and economic development. The decline of biodiversity has serious ramifications for other species as well as human well-being. So we should evaluate actions that threaten biodiversity and take steps to reduce the impact when possible.

AFTER READING THIS
CHAPTER, YOU SHOULD
BE ABLE TO ANSWER
THE FOLLOWING
GUIDING QUESTIONS

1. What is biodiversity, and why is it important? How many species are estimated to live on Earth, and which taxonomic groups have the most species?

2. How do genetic, species, and ecosystem diversity each contribute to ecosystem function and services?

3. What are biodiversity hotspots, and why are they important?

4. What role does isolation play in a species' vulnerability to extinction? How do habitat destruction and fragmentation threaten species?

5. How can we acquire the food, fiber, fuel, and pharmaceutical resources we need without disrupting the ecosystems that provide those resources?

A LEARNING PATH AT THE BEGINNING OF EACH CHAPTER HOMES STUDENT ATTENTION TO THE CORE MESSAGE AND GUIDING QUESTIONS.

---

cal forests tend to be particularly flush with both
...tical and species diversity, thanks largely to the
...ant sunlight and climatic conditions conducive
...wth. The forests that are currently being laid
... Sumatra are no
...tion. They have
...eft unhampered for
...ny millennia that
...steamy amphibious
...tems swarm with
...ucopia of life:
...nts, orangutans,
..., tigers, and every
...er of bird and beetle
...man imagination
...thom. "The truth
...one has any idea
...many species used to
...ere," Sutherlin says.
...the species in these
...s have yet to be described
...ence."

...stunning example
...ecological diversity
...biodiversity

KEY CONCEPT 12.4

Protecting biodiversity hotspots, areas with high numbers of endangered endemic species, can be a cost-effective way to protect a large number of endangered species.

instrumental value An object's or species' worth, based on its

KEY CONCEPTS ALERT STUDENTS TO SALIENT TAKE-HOME MESSAGES WITHIN THE CHAPTER TEXT AND ENFORCE THE GUIDING QUESTIONS.

INFOGRAPHIC 12.4 BIODIVERSITY HOTSPOTS

3

Biodiversity hotspots, areas with high numbers of endemic but endangered species, cover a small percentage of land and water but hold more than 40%–50% of all plant and vertebrate endemic animal species. Most hotspots are located in tropical biomes or in forested terrestrial ecosystems, such as mountains or islands. Even small disturbances, such as a small farm plot or road that cuts through the area, can threaten endemic species that populate specialized niches in these hotspots.

Malaysia
Sumatra

TERRESTRIAL
aquatic

TROPICAL ANDES 15,000 endemic plants, 487 threatened species, 2 extinct

GUINEAN FORESTS 1,800 endemic plants, 115 threatened species, 0 extinct

POLYNESIA-MICRONESIA 3,074 endemic plants, 99 threatened species, 43 extinct

INDO-MALAYAN ARCHIPELAGO 16,000 endemic plants, 162 threatened species, 4 extinct

? Many hotspots are in tropical areas, but some are also found in areas further... non-tropical coastal areas are biodiversity hotspots?

EACH INFOGRAPHIC IS TAGGED TO A GUIDING QUESTION—THE CONTENT IN THE FIGURE HELPS STUDENTS ANSWER THE QUESTION. NEW **CRITICAL THINKING QUESTIONS** HAVE BEEN ADDED TO EACH FIGURE IN THE BOOK.

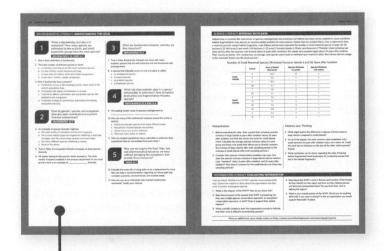

NEW PRACTICE QUESTIONS IN THE END-OF-CHAPTER MATERIAL TEACH EACH GUIDING QUESTION.

RESOURCES TARGET THE MOST CHALLENGING CONCEPTS AND SKILLS IN THE COURSE

Classroom activities, animations, tutorials, and assessment materials are all built around the concepts and skills that are most difficult for students to master

INSTRUCTOR RESOURCES

Macmillan Education LaunchPad

LaunchPad A new standard in online course management, LaunchPad makes it easier than ever to create interactive assignments, track online homework, and access a wealth of extraordinary teaching and learning tools. Fully loaded with our customizable e-Book and all student and instructor resources, the LaunchPad is organized around a series of prebuilt LaunchPad units— carefully curated, ready-to-use collections of material for each chapter of *Environmental Science for a Changing World.*

Scientific American newsfeed Available on the LaunchPad welcome page, the *Scientific American* newsfeed allows you and your students to access current and relevant content from *Scientific American.* You can also choose to assign the articles or bookmark them for later.

LECTURE TOOLS

Optimized Art (JPEGs and layered PowerPoint slides) Infographics are optimized for projection in large lecture halls and split apart for effective presentation.

Layered or Active PowerPoint Slides PowerPoint slides for select figures deconstruct key concepts, sequences, and processes in a step-by-step format, allowing instructors to present complex ideas in clear, manageable parts.

Lecture Outlines for PowerPoint Adjunct professors and instructors who are new to the discipline will appreciate these detailed companion lectures, perfect for walking students through the key ideas in each chapter. These rich, prebuilt lectures make it easy for instructors to transition to the book.

Team-Based Learning Activities Developed by author Susan Karr, these classroom activities use proven active-learning techniques to engage students in the material and inspire critical thinking. These activities are intended for an instructor who is interested in taking an active learning approach to the course.

Clicker Questions Designed as interactive in-class exercises, these questions reinforce core concepts and uncover misconceptions.

Story Abstracts The abstracts offer a brief story synopsis, providing interesting details relevant to the chapter and to the online resources not found in the book.

ASSESSMENT

Learning Curve A game-like interface to guide students through a series of questions tailored to their individual level of understanding.

Videos A new suite of videos for the 2nd edition, including videos on Biodiversity, Climate Change, Human Populations, Energy (including Fracking), and more. Videos bring the stories of environmental science to life and make the materials meaningful to students. Each video is assignable and includes assessment questions to gauge student understanding.

Test Bank A collection of over a thousand questions, organized by chapter and Guiding Question, presented in a sortable, searchable platform. The Test Bank features multiple-choice and short-answer questions, and uses infographics and graphs from the book.

Post-Chapter Quizzes Post-Chapter quizzes are brief 5-10 question quizzes that test the students' basic knowledge of the chapter. They are aligned to the new 2nd edition chapters and Guiding Questions.

Course Management System e-packs available for Blackboard, WebCT, and other course management platforms.

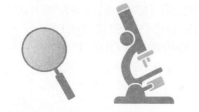

ORGANIZED BY GUIDING QUESTIONS

Student and instructor resources are arranged to support the learning goals of each chapter

SUPPORT ENVIRONMENTAL, SCIENCE, AND INFORMATION LITERACY

Supplementary materials explore timely environmental issues, develop crucial science literacy skills such as data analysis and graph interpretation, and provide practice in evaluating sources of information

STUDENT RESOURCES

Student resources reinforce chapter concepts and give students the tools they need to succeed in the course. All student resources are organized by Guiding Questions and can be found in the LaunchPad.

LaunchPad Students have access to a variety of study tools in the LaunchPad along with a complete online version of the textbook. Carefully curated LaunchPad Units provide suggested learning paths for each chapter in the text.

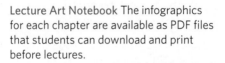

Learning Curve This set of formative assessment activities uses a game-like interface to guide students through a series of questions tailored to their individual level of understanding. A personalized study plan is generated based upon their quiz results. LearningCurve is available to students in the LaunchPad.

GraphingTutorials let students build and analyze graphs, using their critical thinking skills to predict trends, identify bias, and make cause-and-effect connections.

Video Case Studies Videos from an array of trusted sources bring the stories of the book to life and allow students to apply their environmental, scientific, and information literacy skills. Each video includes questions that engage students in the critical thinking process.

Interactive Animated Infographics Most infographics in the text include an animated interactive tutorial or an infographic activity.

Key Term Flashcards Students can drill and learn the most important terms in each chapter using interactive flashcards.

Lecture Art Notebook The infographics for each chapter are available as PDF files that students can download and print before lectures.

ACKNOWLEDGEMENTS

From Susan Karr...

I've discovered that an undertaking this big is truly a collaborative effort. It is amazing what you can accomplish when you work with talented and highly skilled people. I want to thank Jerry Correa for his original vision for the book and personal support and W. H. Freeman acquisitions editor Bill Minick for his continued support and guidance; and developmental editor Andrea Gawrylewski for her patience, insights, and outstanding editorial skills in crafting each chapter. I also want to thank Jeneen Interlandi, the accomplished and gifted writer who has made these chapters such a pleasure to read. Outstanding writing was also done by Alison McCook and Melinda Wenner Moyer. I gratefully acknowledge the entire team at MGMT. design for their skills, vision, and patience, and the folks at Cenveo who expertly coordinated the production of the book. Thanks also go to John Britch, the executive marketing manager, and the sales force, who work tirelessly to see that the book is a success in the marketplace

I've had tremendous support from my biology department colleagues at Carson-Newman University and from focus group participants, who offered advice, answered questions, and helped track down elusive information—thanks for sharing your expertise. In particular, long-time environmental science instructors Teri Balser, Shamili Sandiford, and Lissa Leege helped develop the original roadmap for the book and Michelle Cawthorn's insightful advice guided much of the revision of this second edition. I also owe a debt of gratitude to my environmental science students over the years for their questions, interests, demands, and passion for learning that have always challenged and inspired me. Finally, I want to thank my husband, Steve, for supporting me in so many different ways it is impossible to count them all.

From Jeneen Interlandi...

Each of these chapters is a story—of scientists and everyday people, often doing extraordinary things. It has been my great pleasure to tell those stories here. For that, I thank each and every one of my sources. Their time and patience are what made this book possible.

I would also like to thank Susan, with whom it has been an honor to work, and the entire W. H. Freeman team for their tireless efforts.

REVIEWERS

We would like to extend our deep appreciation to the following instructors who reviewed, tested, and advised on the book manuscript at various stages of development.

Chapter Reviewers

Matthew Abbott, *Des Moines Area Community College*

David Aborn, *University of Tennessee at Chattanooga*

Michael Adams, *Pasco-Hernando Community College*

Shamim Ahsan, *Metropolitan State College of Denver*

John Anderson, *Georgia Perimeter College*

Walter Arenstein, *San Jose State University*

Deniz Ballero, *Georgia Perimeter College*

Marcin Baranowski, *Passaic County Community College*

Brad Basehore, *Harrisburg Area Community College*

Sean Beckmann, *Rockford College*

Ray Beiersdorfer, *Youngstown State University*

Tracy Benning, *University of San Francisco*

David Berg, *Miami University*

Bert Berquist, *University of Northern Iowa*

Joe Beuchel, *Triton College*

Aaron Binns, *Florida State University*

Karen Blair, *Pennsylvania State University*

Barbara Blonder, *Flagler College*

Steve Blumenshine, *California State University, Fresno*

Ralph Bonati, *Pima Community College*

Polly Bouker, *Georgia Perimeter College*

Richard Bowden, *Allegheny College*

Jennifer Boyd, *University of Tennessee at Chattanooga*

Scott Brame, *Clemson University*

Allison Breedveld, *Shasta College*

Mary Brown, *Lansing Community College*

Brett Burkett, *Collin College*

Alan Cady, *Miami University*

Elena Cainas, *Broward College*

Deborah Carr, *Texas Tech University*

Daniel Capuano, *Hudson Valley Community College*

Mary Kay Cassani, *Florida Gulf Coast University*

Michelle Cawthorn, *Georgia Southern University*

Niccole Cerveny, *Mesa Community College*

Lu Anne Clark, *Lansing Community College*

Jennifer Cole, *Northeastern University*

Eric Compas, *University of Wisconsin, Whitewater*

Jason Crean, *Saint Xavier University & Moraine Valley Community College*

Michael Dann, *Pennsylvania State University*

James Dauray, *College of Lake County*

Michael Denniston, *Georgia Perimeter College*

Robert Dill, *Bergen Community College*

Craig Dilley, *Des Moines Area Community College*

Michael Draney, *University of Wisconsin, Green Bay*

JodyLee Estrada Duek, *Pima Community College, Desert Vista*

Don Duke, *Florida Gulf Coast University*

James Dunn, *Grand Valley State University*

John Dunning, *Purdue University*

Kathy Evans, *Reading Area Community College*

Brad Fiero, *Pima County Community College*

Linda Fitzhugh, *Gulf Coast Community College*

Stephan Fitzpatrick, *Georgia Perimeter College*

April Ann Fong, *Portland Community College, Sylvania Campus*

Steven Forman, *University of Illinois at Chicago*

Nicholas Frankovits, *University of Akron*

Phil Gibson, *University of Oklahoma*

Michael Golden, *Grossmont College*

Sherri Graves, *Sacramento City College*

Michelle Groves, *Oakton Community College*

Myra Carmen Hall, *Georgia Perimeter College*

Sally Harms, *Wayne State College*

Stephanie Hart, *Lansing Community College*

Wendy Hartman, *Palm Beach State College*

Alan Harvey, *Georgia Southern University*

Keith Hench, *Kirkwood Community College*

Robert Hollister, *Grand Valley State University*

Tara Holmberg, *Northwestern Connecticut Community College*

Jodee Hunt, *Grand Valley State University*

Meshagae Hunte-Brown, *Drexel University*

Kristin Jacobson, *Illinois Central College*

Jason Janke, *Metropolitan State College of Denver*

David Jeffrey, *Georgia Perimeter College*

Thomas Jurik, *Iowa State University*

Charles Kaminski, *Middlesex Community College*

Michael Kaplan, *College of Lake County*

John Keller, *College of Southern Nevada*

Myung-Hoon Kim, *Georgia Perimeter College*

Elroy Klaviter, *Lansing Community College*

Paul Klerks, *University of Louisiana at Lafayette*

Janet Kotash, *Moraine Valley Community College*

Jean Kowal, *University of Wisconsin, Whitewater*

John Krolak, *Georgia Perimeter College*

James Kubicki, *Pennsylvania State University*

Diane LaCole, *Georgia Perimeter College*

Katherine LaCommare, *Lansing Community College*

Andrew Lapinski, *Reading Area Community College*

Jennifer Latimer, *Indiana State University*

Stephen Lewis, *California State University, Fresno*

Jedidiah Lobos, *Antelope Valley College*

Eric Lovely, *Arkansas Tech University*

Marvin Lowery, *Lone Star College System*

Steve Luzkow, *Lansing Community College*

Steve Mackie, *Pima Community College*

Nilo Marin, *Broward College*

Eric Maurer, *University of Cincinnati*

Costa Mazidji, *Collin College*

DeWayne McAllister, *Johnson County Community College*

Vicki Medland, *University of Wisconsin, Green Bay*

Alberto Mestas-Nunez, *Texas A&M University, Corpus Christi*

Chris Migliaccio, *Miami Dade College*

Jessica Miles, *Palm Beach State College*

Dale Miller, *University of Colorado, Boulder*

Kiran Misra, *Edinboro University of Pennsylvania*

Scott Mittman, *Essex County College*

Edward Mondor, *Georgia Southern University*

Zia Nisani, *Antelope Valley College*

Ken Nolte, *Shasta College*

Kathleen Nuckolls, *University of Kansas*

Segun Ogunjemiyo, *California State University*

Bruce Olszewski, *San José State University*

Jeff Onsted, *Florida International University*

Nancy Ostiguy, *Pennsylvania State University*

Daniel Pavuk, *Bowling Green State University*

Barry Perlmutter, *College of Southern Nevada*

Chris Petrie, *Brevard College*

Craig Phelps, *Rutgers, The State University of New Jersey*

Neal Phillip, *Bronx Community College*

Greg Pillar, *Queens University of Charlotte*

Keith Putirka, *California State University, Fresno*

Bob Remedi, *College of Lake County*

Erin Rempala, *San Diego City College*

Eric Ribbens, *Western Illinois University*

Angel Rodriquez, *Broward College*

Dennis Ruez, *University of Illinois at Springfield*

Robert Ruliffson, *Minneapolis Community and Technical College*

Melanie Sadeghpour, *Des Moines Area Community College*

Seema Sah, *Florida International University*

Jay Sah, *Florida International University*

Shamili Sandiford, *College of DuPage*

Judith Scherff, *Washburn University*

Waweise Schmidt, *Palm Beach State College*

Jeffery Schneider, *State University of New York at Oswego*

Joe Shaw, *Indiana University*

William Shockner, *Community College of Baltimore County*

David Serrano, *Broward College*

Patricia Smith, *Valencia College*

Dale Splinter, *University of Wisconsin, Whitewater*

Jacob Spuck, *Florida State University*

Craig Steele, *Edinboro University*

Rich Stevens, *Monroe Community College*

Michelle Pulich Stewart, *Mesa Community College*

John Sulik, *Florida State University*

Donald Thieme, *Valdosta State University*

Jamey Thompson, *Hudson Valley Community College*

Susanna Tong, *University of Cincinnati*

Michelle Tremblay, *Georgia Southern University*

Karen Troncalli, *Georgia Perimeter College*

Caryl Waggett, *Allegheny College*

Meredith Wagner, *Lansing Community College*

Deena Wassenberg, *University of Minnesota*

Kelly Watson, *Eastern Kentucky University*

Jennifer Willing, *College of Lake County*

Danielle Wirth, *Des Moines Area Community College*

Lorne Wolfe, *Georgia Southern University*

Janet Wolkenstein, *Hudson Valley Community College*

James Yount, *Eastern Florida State College*

Douglas Zook, *Boston University*

Focus Group Participants

John Anderson, *Georgia Perimeter College*

Teri Balser, *University of Wisconsin, Madison*

Elena Cainas, *Broward College*

Mary Kay Cassani, *Florida Gulf Coast University*

Kelly Cartwright, *College of Lake County*

Michelle Cawthorn, *Georgia Southern University*

Mark Coykendall, *College of Lake County*

JodyLee Estrada Duek, *Pima Community College*

Jason Janke, *Metropolitan State College of Denver*

Janet Kotash, *Moraine Valley Community College*

Jean Kowal, *University of Wisconsin, Whitewater*

Nilo Marin, *Broward College*

Edward Mondor, *Georgia Southern University*

Brian Mooney, *Johnson & Wales University, North Carolina*

Barry Perlmutter, *College of Southern Nevada*

Matthew Rowe, *Sam Houston State University*

Shamili Sandiford, *College of DuPage*

Ryan Tainsh, *Johnson & Wales University*

Michelle Tremblay, *Georgia Southern University*

Kelly Watson, *Eastern Kentucky University*

Comparative Reviewers

Buffany DeBoer, *Wayne State College*

Dani DuCharme, *Waubonsee Community College*

James Eames, *Loyola University Chicago*

Bob East, *Washington & Jefferson College*

Matthew Eick, *Virginia Polytechnic Institute and State University*

Kevin Glaeske, *Wisconsin Lutheran College*

Rachel Goodman, *Hamdpen-Sydney College*

Melissa Hobbs, *Williams Baptist College*

David Hoferer, *Judson University*

Paul Klerks, *University of Louisiana at Lafayette*

Troy Ladine, *East Texas Baptist University*

Jennifer Latimer, *Indiana State University*

Kurt Leuschner, *College of the Desert*

Quent Lupton, *Craven Community College*

Jay Mager, *Ohio Northern University*

Steven Manis, *Mississippi Gulf Coast Community College*

Nancy Mann, *Cuesta College*

Heidi Marcum, *Baylor University*

John McCarty, *University of Nebraska at Omaha*

Chris Poulsen, *University of Michigan*

Mary Puglia, *Central Arizona College*

Michael Tarrant, *University of Georgia*

Melissa Terlecki, *Cabrini College*

Jody Terrell, *Texas Woman's University*

Class Test Participants

Mary Kay Cassani, *Florida Gulf Coast College*

Ron Cisar, *Iowa Western Community College*

Reggie Cobb, *Nash Community College*

Randi Darling, *Westfield State University*

JodyLee Estrada Duek, *Pima Community College*

Catherine Hurlbut, *Florida State College at Jacksonville*

James Hutcherson, *Blue Ridge Community College*

Janet Kotash, *Moraine Valley*

Offiong Mkpong, *Palm Beach State College*

Edward Mondor, *Georgia Southern University*

Anthony Overton, *East Carolina University*

Shamilli Sandiford, *College of DuPage*

Keith Summerville, *Drake University*

ENVIRONMENTAL SCIENCE

FOR A CHANGING WORLD

ON THE ROAD TO COLLAPSE

What lessons can we learn from a vanished Viking society?

CORE MESSAGE
Humans are a part of the natural world and are dependent on a healthy, functioning planet. We put pressure on the planet in a variety of ways, but our choices can help us move toward sustainability.

AFTER READING THIS CHAPTER, YOU SHOULD BE ABLE TO ANSWER THE FOLLOWING **GUIDING QUESTIONS**

1

What constitutes the "environment," and what fields of study collaborate under the umbrella of environmental science?

2

What are some of the environmental dilemmas that humans face, and why are many of them considered "wicked problems"?

The remains of Hvalsey, a
Viking settlement church,
in southern Greenland.

3

What does it mean to be sustainable,
and what are the characteristics of a
sustainable ecosystem?

4

What can human societies and
individuals do to pursue sustainability?

5

What challenges does humanity face in
dealing with environmental issues, and
how can environmental literacy help us
make more informed decisions?

Although not much of a tourist destination, Greenland offers some spectacular sights—colossal ice sheets, a lively seascape, rare and precious wildlife (whales, seals, polar bears, eagles). But on his umpteenth trip to the island, Thomas McGovern was not interested in any of that. What he wanted to see was the garbage—specifically, the ancient, fossilized garbage that Viking settlers had left behind some seven centuries before.

McGovern, an archaeologist at the City University of New York, had been on countless expeditions to Greenland over the preceding 40 years. Digging through layers of peat and permafrost, he and his team had unearthed a museum's worth of artifacts that, when pieced together, told the story of the Greenland Vikings. But as thorough as their expeditions had been, that story was still maddeningly incomplete.

> ### KEY CONCEPT 1.1
>
> Environmental science draws from science and non-science disciplines to understand and address environmental problems.

Here's what they knew so far: A thousand or so years ago, an infamous Viking by the name of Erik the Red led a small group of followers across the ocean from Norway, to a vast expanse of snow and ice that he had dubbed Greenland. Most of Greenland was not green. In fact, it was a forbidding place marked by harsh winds and sparse vegetation. But tucked between two fjords along the southwestern coast, protected from the elements by jagged, imposing cliffs, the Vikings found a string of verdant meadows, brimming with wildflowers. They quickly set up camp here and proceeded to build a society similar to the one they had left behind in Norway. They farmed, hunted, and raised livestock. They also built barns and churches as elaborate as the ones back home. They established an economy and a legal system, traded goods with mainland Europe, and, at their peak, reached a population of 5,000 (a large number in those days).

And then, after 450 years of prosperity, they disappeared—seemingly into thin air—leaving little more than the beautiful, tragic ruins of a handful of barns and churches in their wake.

The how and why of this vanishing act remained a tantalizing mystery, one that has drawn hundreds of scientists—McGovern among them—to Greenland each summer. Recently, some of McGovern's colleagues had begun to suspect that disturbances in the natural environment—a cooling climate, loss of soil, problems with the food supply—may have been the deciding factors.

While other researchers probed ice sheets and soil deposits in search of clues, McGovern stuck to the garbage heaps, or *middens*, as Vikings called them. Every farmstead had one, and every generation of the farmstead's owners threw their waste into it. The result was an archaeological treasure trove: fine-grain details about what people ate, how they dressed, and the kinds of objects they filled their homes with. It gave McGovern and his team a clear picture of how they lived.

If they dug deep enough, McGovern thought, it might also explain how they died.

Environmental science is all encompassing.

From a modern developed society like the United States, it can be difficult to imagine a time and place when the natural world held such sway over our fate.

⊙ **WHERE IS THE VIKING SETTLEMENT IN GREENLAND?**

CANADA

GREENLAND

NORSAQ

ICELAND

INFOGRAPHIC 1.1   ENVIRONMENTAL SCIENCE IS HIGHLY INTERDISCIPLINARY

Environmental science studies the natural world and how humans interact with and impact it. We must look to the natural and social sciences as well as to the humanities to help us understand our world and effectively address environmental issues and environmental questions such as, "Why did the Vikings disappear from this region in Greenland, and how do humans live now in such a harsh environment?"

NATURAL SCIENCES

- What is the climate like?
- Which plants and animals live here?
- Which crops or animals can be raised here?
- How can soil erosion be prevented?
- What energy sources are available, and how do they impact the environment?

HUMANITIES

- How do religion and tradition influence choices?
- How can people express their love, fears, and hopes for their homeland (literature, theatre, music)?

SOCIAL SCIENCES

- How have indigenous people lived here?
- What environmental policies would best fit this culture and place?
- Will residents accept changes to their lifestyle that might benefit the environment?
- Which energy sources are most cost-effective?

? How would you use your particular college major to help address an environmental problem?

PETE RYAN/National Geographic Creative

Our food comes from a grocery store, our water from a tap; even our air is artificially heated and cooled to our liking. These days, it seems more logical to consider societal conflict, or even collapse, through the lens of politics or economics. But, as we will see time and again throughout this book, the natural environment—and how we interact with it—plays a leading role in the sagas that shape human history; this is as true today as it was in the time of the Vikings.

Environment is a broad term that describes the surroundings or conditions (including living and nonliving components) in which any given organism exists. **Environmental science**—a field of research that is used to understand the natural world and our relationship to it—is extremely interdisciplinary. It relies on a range of natural and applied sciences (such as ecology, geology, chemistry, and engineering) to unlock the myster of the natural world, and to look at the role and impact of humans in the world. It also draws on social sciences (such as anthropology, psychology, and economics) and the humanities (such as art, literature, and music) to understand the ways that humans interact with, and thus impact, the ecosystems around them. **INFOGRAPHIC 1.1**

environment The biological and physical surroundings in which any given living organism exists.

environmental science An interdisciplinary field of research that draws on the natural and social sciences and the humanities in order to understand the natural world and our relationship to it.

Environmental science is an **empirical science**: It scientifically investigates the natural world through systematic observation and experimentation. It is also an **applied science**: We use its findings to inform our actions and, in the best cases, to bring about positive change. **INFOGRAPHIC 1.2**

The ability to understand environmental problems is referred to as **environmental literacy**. Such literacy is crucial to helping us become better stewards of Earth. Environmental problems can be extremely complicated and tend to have multiple causes, each one difficult to address. We must also understand that because of their complexity, any given response to an environmental problem involves significant **trade-offs**, and no one response is likely to present the ultimate solution. Scientists refer to such problems as "wicked problems." In confronting them, we must consider not only their environmental but also their economic and social causes and consequences. Scientists refer to this trifecta as the **triple bottom line**. **INFOGRAPHIC 1.3**

In his book *Collapse*, University of California at Los Angeles biologist Jared Diamond details how wicked problems can lead to a society's ultimate demise. He identifies five factors in particular that determine whether any given society will succeed or fail: natural climate change, failure to properly respond to environmental changes, self-inflicted environmental damage, hostile neighbors, and loss of friendly neighbors. According to Diamond, the relative impact of each factor varies by society.

The situation of the Greenland Vikings was a rare case. It turns out that, as in a perfect storm, all five of these factors conspired together.

The Greenland Vikings' demise was caused by natural events and human choices.

Greenland's interior is covered by vast ice sheets that stretch toward the horizon—3,000 meters (10,000 feet) thick and more than 250,000 years old. To residents of the hard land, these ice sheets are not good for much— they create harsh winds and brutal cold—but to climate

KEY CONCEPT 1.2

Environmental problems are difficult to solve because there are multiple causes and consequences and because potential solutions come with trade-offs.

INFOGRAPHIC 1.2   **DIFFERENT APPROACHES TO SCIENCE HAVE DIFFERENT GOALS AND OUTCOMES**   1

Environmental science is used to systematically collect and analyze data to draw conclusions and use these conclusions to propose reasonable courses of action.

EMPIRICAL SCIENCE IS USED TO INVESTIGATE THE NATURAL WORLD

↑ Through observation, glaciologists study and record the rate of glacier melt in Greenland; it is increasing dramatically in some far flung places.

IN APPLIED SCIENCE, KNOWLEDGE IS USED TO ADDRESS PROBLEMS OR NEEDS

↑ Engineers use their understanding of flowing water's potential energy to harness its power; glacial meltwater can be diverted to produce hydroelectric power.

 If we don't have an application in mind for an empirical research topic, is it worth pursuing?

2

INFOGRAPHIC 1.3 WICKED PROBLEMS

Wicked problems are difficult to address because, in many cases, each stakeholder hopes for a different solution. Solutions that address wicked problems usually involve trade-offs, so there is no clear "winner." One example of a wicked problem is climate change. There are many causes of the current climate change we are experiencing, both natural and anthropogenic (caused by human actions), and the effects of climate change will be varied for different species and people, depending on where they live and their ability to adapt to the changes.

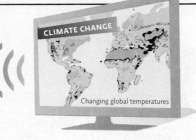

CLIMATE CHANGE

Changing global temperatures

CLIMATE CHANGE

SOME OF THE CAUSES:

- Burning fossil fuels
- Deforestation
- Methane from agriculture
- Overconsumption by modern society

SOME OF THE CONSEQUENCES:

- Sea level rise
- Habitat loss and species endangerment
- Spread of tropical disease
- Agriculture: worse in some areas, better in others

POSSIBLE ACTIONS (AND POTENTIAL TRADE-OFFS) INCLUDE:

- Alternative energy sources (less pollution but can be costly)
- Irrigation (increases crop yields but can cause water shortages and soil problems)
- Reforestation projects (lessen CO_2 in atmosphere and increase habitats but may take land needed for agriculture or other uses)
- Protecting flood-prone areas with levees or sea walls (may protect cities and farms but may fragment aquatic habitats and isolate species' populations)

? What are some other environmental "wicked problems" we face?

scientists, they're a treasure trove. As snow falls, it absorbs various particles from the atmosphere and lands on the ice sheets. As time passes, the snow and particles compact into ice, freezing in time perfect samples of the atmosphere as it existed when that snow first fell. By analyzing those ice-trapped particles—dust, gases, chemicals, even the water molecules—scientists can get a pretty good idea of what was happening to the climate at any given time. "It's like perfectly preserved slices of atmosphere from the past," says Lisa Barlow, a geologist and climate researcher at the University of Colorado at Boulder. "It gives us additional clues as to what was going on."

To uncover those clues, a team of scientists and engineers picked an accessible segment

of ice sheet, not far from the Viking settlements, drilled from the surface all the way down to the bedrock below, and extracted a 12-centimeter-wide (5-inch-wide), 3,000-meter-long cylinder of ice, which they then divided and dispersed among a handful of labs around the globe, including Jim White's light stable isotope lab—also at the University of Colorado at Boulder. Analysis of thin sequential segments of the ice core showed that when the Vikings first arrived in Greenland, the temperature was anomalously higher than the average over the preceding 1,000 years. (Atmospheric temperature can be deduced from the amount of oxygen-18 or deuterium—heavy hydrogen—present in the sample.) By the time the Vikings had vanished, temperatures had lowered so

empirical science A scientific approach that investigates the natural world through systematic observation and experimentation.

applied science Research whose findings are used to help solve practical problems.

environmental literacy A basic understanding of how ecosystems function and of the impact of our choices on the environment.

trade-offs The imperfect and sometimes problematic responses that we must at times choose between when addressing complex problems.

triple bottom line The combination of the environmental, social, and economic impacts of our choices.

NSIDC courtesy Ted Scambos and Rob Bauer.

↑ Scientist from the National Snow and Ice Data Center at the University of Colorado at Boulder, working with an ice core drill.

much that scientists call the period the Little Ice Age—a time when all the seasons were cooler than normal and winters were exceptionally cold. "It's no wonder they didn't make it," says Barlow. "With lower temperatures, livestock would have starved for lack of hay over the long winter, and self-inflicted environmental damage made the situation worse."

Indeed, in addition to natural climate change (Diamond's first factor), the Vikings also suffered from self-inflicted environmental damage (Diamond's second factor).

In addition to that single ice core, scientists have analyzed hundreds of mud cores taken from lake beds around the Viking settlements. These mud samples—which contain large amounts of soil that was blown into the lakes during Viking times—indicate that soil erosion had become a significant problem long before the region descended into a mini ice age. "This wasn't a climate problem," says Bent Fredskild, a Danish scientist who extracted and studied many of the mud cores. "This was self-inflicted. It happened the same way that soil erosion happens today—they overgrazed the land, and once it was denuded, there was nothing to anchor the soil in place. So the wind carried it away."

Overgrazing wasn't their only mistake. The Greenland Vikings also used grassland to insulate their houses against the cold of winter; typical insulation consisted of 2-meter-thick (about 6 feet) slabs of turf, and a typical home took about 4 hectares (10 acres) of grassland to insulate. On top of that, they chopped down the forests, harvesting enough timber to not only provide fuel and build houses but also to make the innumerable wooden objects to which they had become accustomed back in Norway.

Greenland's ecosystem was far too fragile to endure such pressure, especially as the settlement grew from a few hundred to a few thousand. The short, cool growing season meant that plants developed slowly, which in turn meant that the land could not recover quickly enough from the various assaults to protect the soil.

As climate cooling and overharvesting conspired to destroy pasturelands, summer hay yields shrank. When scientists counted the fossilized remains of insects that lived in the fields and haylofts of Viking Greenland, they found that their numbers fell dramatically in the settlement's final years. "The falloff in insects tells us that hay production dwindled to the point of crisis," said the late Peter Skidmore, a Sheffield University entomologist. Without hay, livestock could not survive the ever-colder, ever-longer winters. And without livestock, the Vikings themselves went hungry. As scientists have discovered, they needn't have.

Responding to environmental problems and working with neighbors help a society cope with changes.

Back in his Manhattan lab, McGovern sorts through hundreds of animal bones collected from various Greenland middens. By examining the bones and making careful note of which layers they were retrieved from, McGovern can tell what the people ate and how their diets changed over time. "This is a pretty typical set of remains for these people from this region and time period," he says, leaning over a shiny metal tray of neatly arranged bone fragments. Some are the bones of cattle imported from Europe. Others are the remains of sheep and goats; still others of local wildlife such as caribou. Conspicuously absent, McGovern says, are fish of any kind.

"If we look at a comparable pile of bones from [Norwegian settlers of] the same time period, from Iceland, we see something very different," McGovern explains. "We have fish bones and bird bones and little fragments of whale bones. Most of it, in fact, is fish—including a lot of cod." It turns out that while the Greenland Vikings were guilty of Diamond's third factor, failure to respond to the natural environment, their Icelandic cousins were not.

Like their cousins in Greenland, the Vikings who settled Iceland at about the same time were initially fooled into thinking that their newly discovered land could sustain their cow-farming, wood-dependent ways: Plants and animals looked similar to those back in Norway, and grasslands seemed lush and abundant. They cleared about 80% of Iceland's forests and allowed their cows, sheep,

KEY CONCEPT 1.3

Living sustainably means living within the means of one's environment in a way that does not diminish the environment's ability to support life in the future.

and goats to chew the region's grasslands down to nothing before finally noticing how profound the differences between Iceland and Norway actually were: Growing seasons in Iceland were shorter, both soil and vegetation were much more fragile, and because the land could not rebound quickly, cow farming was unsustainable.

But once they saw that their old ways would not work in this new country, the Icelandic Vikings made changes. Not only did they switch from beef to fish, they also began conserving their wood and abandoned the highlands, where soil was especially fragile. And, as a result of these and other adaptations, they survived and prospered. The Icelanders responded to the limitations of their natural environment in a way that allowed them to meet present needs without compromising the ability of future generations to do the same—an approach known today as **sustainable development**.

Some of the most telling clues to the mystery of the Greenland Vikings' demise come not from the Viking colonies but from another group of people who lived nearby: the Inuit. The Inuit arrived in the Arctic centuries before the Vikings. They were expert hunters of ringed seal—an exceedingly difficult to-catch but very abundant food source. They knew how to heat and light their homes with seal blubber (instead of firewood). And they were experts at fishing.

Fishing is not nearly as labor intensive as raising cattle, and in the lakes and fjords of Greenland, it provides an easy, reliable source of protein. A comparison of Inuit and Viking middens shows that even as the Greenland Vikings were scraping off every last bit of meat and marrow from their cattle bones, the Inuit had more food than they could eat.

The Vikings might have learned from their Inuit neighbors; by adapting some of their customs, they might have survived the Little Ice Age and gone on to prosper as the Icelandic Vikings did. But excavations show that virtually no Inuit artifacts made their way into Viking settlements. And according to written records, the Norse detested the Inuit, who, on at least one occasion, attacked the Greenland colony; they called the Inuit *skraelings*, which is Norse for "wretches," considered them inferior, and refused to seek their friendship or their counsel.

In addition to these hostile neighbors (Diamond's fourth factor), the Greenland Vikings also suffered a loss of friendly neighbors (Diamond's fifth factor). As the productivity of the Viking colonies declined, so did visits from European ships. As time wore on, it became apparent that the Greenland Vikings could expect very little in the way of trade; royal and private ships that had visited every year came less and less often. After a while, they did not come at all. For the Greenland Vikings, who depended on the Europeans for iron, timber, and other essential supplies, this loss proved devastating. Among other things, it meant that, as the weather grew colder, and food supplies dwindled, they had no one to turn to for help.

Humans are an environmental force that impacts Earth's ecosystems.

When it comes to the environment, modern societies are not as different from the Vikings as one might assume. Vikings chose livestock and farming methods that were ill suited to Greenland's climate and natural environment. We, too, use farming practices that strip away topsoil and diminish the land's fertility. We have overharvested our forests and in so doing have triggered a cascade of environmental consequences: loss of vital habitat and biodiversity, soil erosion, and water pollution. We have overfished and overhunted and have allowed invasive species to devastate some of our most valuable ecosystems.

KEY CONCEPT 1.4

Modern humans inflict tremendous environmental impact by virtue of our sheer numbers and the high per-person impact of some societies.

In part, these problems stem from a disconnect in our understanding of the relationship between our actions and their environmental consequences. For example, unless they live close by, many people in the United States don't realize that entire mountains are being leveled to produce their electricity; thousands of acres of habitat and miles of streams and rivers have been destroyed to access coal seams deep beneath the surface of West Virginia and Wyoming. We are slow to make the connection between the burning of that coal and mercury-contaminated fish or increased asthma rates.

We also face a suite of new problems that did not trouble the Vikings. Chief among them

sustainable development Development that meets present needs without compromising the ability of future generations to do the same.

In part, these problems stem from a disconnect in our understanding of the relationship between our actions and their environmental consequences.

is population growth; as we will discuss in subsequent chapters, global population is poised to top 9 billion come 2050. The sheer volume of people will strain Earth's resources like never before. This is relevant because every environment has a **carrying capacity**—the population size that an area can support indefinitely—but some of our actions are decreasing carrying capacity, even as our population swells. (See Chapter 4 for more on human populations and carrying capacity.) In addition, we generate more pollution than the Vikings did, and much of what we generate is more toxic.

Environmental scientists evaluate the impact any population has on its environment—due to the resources it takes and the waste it produces—by calculating its **ecological footprint** (see Chapter 5). Some analysts feel we have already surpassed the carrying capacity of Earth; our collective footprint already surpasses what Earth can support over a long period of time.

Anthropogenic climate change is another serious consequence of larger populations, increasing affluence, and more sophisticated technology. While the Vikings

had to contend with periodic warming and cooling periods that were part of the natural climate cycle, the vast majority of scientists today conclude that modern humans are faced with rapidly warming temperatures caused largely by our own use of greenhouse gas—emitting fossil fuels. **INFOGRAPHIC 1.4**

We have something else in common with the Vikings of Greenland: Our attitudes frequently prevent us from responding effectively to environmental changes. According to Diamond, the Vikings were stymied by their own sense of superiority. They considered themselves masters of their surroundings, and so they did not notice signs that they were causing irreparable harm to their environment, nor did they bother to learn the ways of their Inuit neighbors, who had managed to survive in the same region for centuries before their arrival.

The United Nations' Millennium Ecosystem Assessment is a scientific appraisal of current research that evaluates what is currently known about the ways in which environmental problems affect humans and makes recommendations about addressing those problems. According to its 2005 assessment, a consensus report of more than 1,500 scientists, human actions are so straining the environment that the ability of the planet's ecosystems to sustain future generations is gravely imperiled. But there is hope: If we act now, the report's authors write, we can still reverse much of the damage. Some of the best lessons about how we can do this come from the natural environment itself.

INFOGRAPHIC 1.4   2

MANY ENVIRONMENTAL PROBLEMS CAN BE TRACED TO THREE UNDERLYING CAUSES

Mads Nissen/Panos

↑ Population size: There are more than 7 billion people on the planet. Though some have more impact than others, our sheer numbers contribute to many environmental problems.

© Oleksiy Maksymenko Photography/Alamy

↑ Resource use: We use resources faster than they are replaced and convert matter into forms that don't readily decompose.

Martin Roemers/Panos

↑ Pollution: We generate pollution that compromises our own health and that of ecosystems.

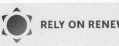 **INFOGRAPHIC 1.5** **FOUR CHARACTERISTICS OF A SUSTAINABLE ECOSYSTEM** 3

Ecosystems found on Earth today have the capacity to be naturally sustainable—those that were not died out long ago. They all share characteristics that allow the capture of energy and use of matter in a way that allows them to persist over time, all without degrading the environment itself.

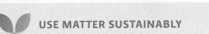

SUSTAINABLE ECOSYSTEMS

RELY ON RENEWABLE ENERGY

Ecosystems must rely on sources of energy that are replenished daily because energy that is used by one organism is "used up" and cannot be used by another; energy is NOT recycled. This means new inputs are constantly needed.

USE MATTER SUSTAINABLY

No new matter arrives on Earth, so ecosystems must make do with what they have. Fortunately, matter can be recycled, and organisms in ecosystems use matter resources over and over again; the waste of one becomes resource for the next.

HAVE POPULATION CONTROL

The sizes of the various populations in an ecosystem are kept in check by disease, predators, and competitors. This prevents a population from getting too large and damaging the ecosystem it, and others, depends on.

DEPEND ON LOCAL BIODIVERSITY

Ecosystems access energy, recycle matter, and control population sizes largely through the actions of their resident species. Higher biodiversity (greater number of species and more variation of individuals within a species) generally means more energy can be captured, more matter can be recycled and at a faster pace, and population sizes can be better controlled.

 Identify an example of each of these sustainability characteristics from the ecosystem where you live.

Ecosystems are naturally sustainable and a good model for human societies hoping to become more sustainable.

Unlike their counterparts in Greenland, the Icelandic Vikings responded to their environment and adopted more **sustainable** practices. They didn't have to look far for a model of sustainability: Natural ecosystems are sustainable. This means they use resources—namely, energy and matter—in a way that ensures that those resources continue to be available.

To survive, organisms need a constant, dependable source of energy. But as energy passes from one part of an ecosystem to the next, the usable amount declines; therefore, new inputs of energy are always needed. A sustainable ecosystem is one that makes the most of **renewable energy**—energy that comes from an infinitely available or easily replenished source. For almost all natural ecosystems, that energy source is the Sun. Photosynthetic organisms such as plants trap solar energy and convert it to a form that they can readily use or that can be passed up the food chain to other organisms.

Unlike energy, matter (anything that has mass and takes up space) can be recycled and reused indefinitely; the key is not using it faster than it is recycled. Naturally sustainable ecosystems waste nothing; they recycle matter so that the waste from one organism ultimately becomes a resource for another. They also keep populations in check so that the resources are not overused and there is enough food, water, and shelter for all.

Finally, sustainable ecosystems depend on local **biodiversity** (the variety of species present) to perform many of the jobs just mentioned; different species have different ways of trapping and using energy and matter, the net result of which boosts productivity and efficiency (see Chapter 8). And predators, parasites, and competitors serve as natural population checks.
INFOGRAPHIC 1.5

KEY CONCEPT 1.5

Natural ecosystems are sustainable because of the way they acquire energy, use matter, control population sizes, and depend on local biodiversity to meet these needs.

carrying capacity The population size that a particular environment can support indefinitely.

ecological footprint The land needed to provide the resources and assimilate the waste of a person or population.

anthropogenic Caused by or related to human action.

sustainable A method of using resources in such a way that we can continue to use them indefinitely.

renewable energy Energy that comes from an infinitely available or easily replenished source.

biodiversity The variety of species on Earth.

INFOGRAPHIC 1.6 | **SUSTAINABLE ECOSYSTEMS CAN BE A USEFUL MODEL FOR HUMAN SOCIETIES**   4

SUSTAINABLE ECOSYSTEMS:	WE CAN MIMIC THIS BY:	
RELY ON RENEWABLE ENERGY	**USING SUSTAINABLE ENERGY SOURCES** We can move away from nonrenewable fuels such as fossil fuels by turning to sustainable energy sources such as solar, wind, geothermal, and biomass (harvested at sustainable rates).	
USE MATTER SUSTAINABLY	**USING MATTER CONSERVATIVELY AND SUSTAINABLY** We can reduce our waste by reducing our use of resources and by recovering, reusing, and recycling matter that we do use; we would also benefit from minimizing the toxins we create or release into the environment that degrade our natural resources.	
HAVE POPULATION CONTROL	**GETTING HUMAN POPULATION GROWTH UNDER CONTROL** While predation controls many natural populations, there are many ways to reduce human birth rates without increasing the death rate through war or disease.	
DEPEND ON LOCAL BIODIVERSITY TO MEET THE FIRST THREE REQUIREMENTS	**DEPENDING ON LOCAL HUMAN CONTRIBUTIONS AND BIODIVERSITY** Protecting biodiversity will help us achieve the three above goals; we can emulate nature by using a variety of local energy sources, building materials, and crops and by exploring the many ideas and innovations that come from a diverse human community.	

What obstacles stand in society's way in pursuing the actions mentioned in this graphic?

Thus, natural ecosystems live within their means, and each organism contributes to the ecosystem's overall function. This is not to imply that natural ecosystems are perfect places of total harmony, but those that are sustainable meet all four of these characteristics.

KEY CONCEPT 1.6

Human societies can become more sustainable by mimicking the way natural ecosystems operate.

Human ecosystems are another story. Humans tend to rely on **nonrenewable resources**—those whose supply is finite or is not replenished in a timely fashion. The most obvious example of this is our reliance on fossil fuels such as coal, natural gas, and petroleum, culled from deep within Earth, to power our society. Fossil fuels are replenished only over vast geologic time—far too slowly to keep pace with our rampant consumption of them. On top of that, we have a hard time keeping our population under control, despite (or maybe because of)

nonrenewable resources Resources whose supply is finite or not replenished in a timely fashion.

all the advances of modern technology. We also generate volumes of waste, much of it toxic, and have yet to fully master the art of recycling.

Increasingly, however, we humans are looking to nature to help us learn how to change our ways. In Janine Benyus' book *Biomimicry* (the term used to describe the strategy of mimicking nature), Benyus explains that biomimicry involves using nature as a *model*, *mentor*, and *measure* for our own systems. Emulating nature (nature as model) gives us an example of what to do; it can also teach us how to do it (nature as mentor) and the level of response that is appropriate (nature as measure). As we'll see in later chapters, scientists are using biomimicry to design more sustainable methods of growing crops and livestock for human consumption, and of trapping and using energy. **INFOGRAPHIC 1.6**

Despite such efforts, some concerned scientists and environmentalists say that modern global societies are not acting nearly as quickly as they could or should. Once again, the charge echoes those that archaeologists have leveled at the Vikings of Greenland. To understand what prevents us from changing our ways even in the face of brewing calamity, it helps to understand why the Greenland Vikings failed to do the same.

Humanity faces some challenges in dealing with environmental issues.

Experts agree that if the Greenland Vikings had simply switched from cows to fish, they might have avoided their own demise. Such a switch would have saved at least some of the pastureland, not to mention the tremendous amount of time and labor it took to raise cattle in such an ill-suited environment. But cows were a status symbol in Europe, beef was a coveted delicacy, and for reasons that still elude researchers, the Vikings had a cultural taboo against fish. It was a taboo they clung to, even as they starved to death.

<div class="key-concept">

KEY CONCEPT 1.7

Environmental literacy helps us understand and respond to environmental problems. Impediments to solving environmental problems include social traps, wealth inequities, and conflicting worldviews.

</div>

↑ A replica of the first church built in Brattahlid, Greenland—a Viking settlement founded by Erik the Red more than 1,000 years ago.

Scott Warren/Getty Images

"It's clear that the Vikings' decisions made them especially vulnerable to the climatic and environmental changes that descended upon them," says McGovern. "When you're building wooden cathedrals in a land without trees, when you create a society reliant on imports in a situation where it's difficult to travel back and forth between the homeland, when you absolutely refuse to collaborate with your neighbors in such a harsh, unforgiving environment, you are setting yourself up for trouble." Ultimately, he says, the decisions this society made played as much of a role in its demise as the actual environmental changes did.

The problem, McGovern says, is not that the Greenland Vikings made so many mistakes in their early days but that they were unwilling to change later on. "Their conservatism and rigidity, which we can see in many different aspects, seems to have kept them on the same path, maybe even prodded them to try even harder—build bigger churches, etc.—instead of trying to adapt."

Decisions by individuals or groups that seem good at the time and produce a short-term benefit but that hurt society in the long run are called **social traps**. The

> **❝** *It's clear that the Vikings' decisions made them especially vulnerable to the climatic and environmental changes that descended upon them.* **❞**
> —Thomas McGovern.

tragedy of the commons is a social trap, first described by ecologist Garrett Hardin, that often emerges when many people are using a commonly held resource, such as water or public land. Each person will act in a way to maximize his or her own benefit, but as everyone does this, the resource becomes overused or damaged. Herders might put more animals on a common pasture because they are driven by the idea that "if I don't use it, someone else will." We do the same thing today as we overharvest forests and oceans and as we release toxins into the air and water. Other social traps include the **time delay** and **sliding reinforcer** traps—actions that, like the tragedy of the commons, have a negative effect later on.

INFOGRAPHIC 1.7

Education is our best hope for avoiding such traps. When people are aware of the consequences of their decisions, they are more likely to examine the trade-offs to determine whether long-term costs are worth the short-term gains and, hopefully, to make different choices when necessary. In the United States, environmental laws, such as the Clean Air Act, Clean Water Act, and Endangered Species Act, have gone a long way toward protecting our natural

social traps Decisions by individuals or groups that seem good at the time and produce a short-term benefit but that hurt society in the long run.

tragedy of the commons The tendency of an individual to abuse commonly held resources in order to maximize his or her own personal interest.

time delay Actions that produce a benefit today and set into motion events that cause problems later on.

sliding reinforcer Actions that are beneficial at first but that change conditions such that their benefit declines over time.

INFOGRAPHIC 1.7   SOCIAL TRAPS

Social traps are decisions that seem good at the time and produce a short-term benefit but that hurt society (usually in the long run).

 Why do you suppose humans are so prone to being caught in these social traps?

↘ **TRAGEDY OF THE COMMONS** When resources aren't "owned" by anyone (they are commonly held), individuals who try to maximize their own benefit end up harming the resource itself.

This common resource (pasture) can support four animals sustainably. Each of the four farmers benefits equally.

If one farmer adds two more cows, the commons becomes degraded. All four farmers share in the degradation (less milk produced per cow), but the farmer with three cows gets all the benefit from adding the extra cows.

The other farmers also must add cows to return to the same level of production as before. **Over time, the commons degrades even more, and at some point it will no longer support any animals.**

↘ **TIME DELAY** Actions that produce a benefit today set into motion events that cause problems later on.

Modern fishing techniques that use giant nets can harvest large numbers of fish. Fishers try to get the most fish possible in the short term.

If more fish are taken over a number of years than are replaced naturally through fish reproduction, the population decreases. Other populations also decline as the fish they depend on are depleted.

After a decade or so, overfishing may so deplete the population that fishers cannot catch enough to meet needs. **The effects of decisions about how many fish to harvest are not felt until later.**

↘ **SLIDING REINFORCER** Actions that are beneficial at first may change conditions such that their benefit declines over time.

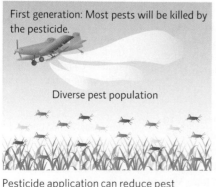  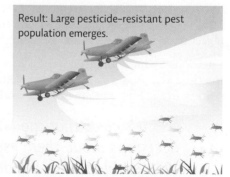

Pesticide application can reduce pest numbers on a crop, but a few pests might survive.

The surviving pests reproduce. The pesticide is only helpful if more is applied or a more toxic pesticide is used. This can be harmful to other organisms and to humans.

In addition to becoming resistant to the pesticide, the pest population may even become bigger if the pest's predators are also killed. **What was helpful at first is no longer helpful and is even harmful over time.**

environment from various social traps, but enforcing those laws is a constant challenge.

Social traps are not the only challenge societies face. Another obstacle to sustainable growth is wealth inequality—the unequal distribution of wealth (and power) in a community or society. In the Greenland Viking colony, wealth at first insulated the people in power from the environmental problems, and they didn't feel the strain of the decline (as other, less powerful colony members undoubtedly did) until it was too late. In the same way, wealthier nations today are less affected by resource availability, while 2 billion or more people lack adequate resources to meet their needs. In fact, just 20% of the population controls roughly 80% of all the world's resources. On one hand, the affluent minority (of which the United States is a part) uses more than its share. Deep pockets allow us to exploit resources for wants, not just needs, and to exploit them all over the world, so that we can spare our own natural environments at the expense of someone else's. For example, our demand for mahogany furniture drives deforestation in Central and South America, where the tree flourishes. Because we are far away from the trees, the environmental fallout is easy to ignore at the moment, but as it did for the Vikings, it may reach a point where it affects everyone.

On the other hand, the underprivileged also exploit the environment in an unsustainable way. With limited access to external resources, they are often forced to overexploit their immediate surroundings just to survive. The above example applies here, too. For an impoverished landowner in Costa Rica or Brazil, chopping down trees may be the only way she can feed her family; to her, harvesting the forest is more valuable than preserving it.

There are societal costs, as well, to this inequality. A growing gap between the haves and have-nots exacerbates tension and strife all over the world. In fact, fighting over resources has long been one of the contributing factors to societal decline and collapse. **INFOGRAPHIC 1.8**

Conflicting worldviews are another challenge to sustainable living. Because our **worldviews**—the windows through which we view our world and existence—are influenced by cultural, religious, and personal experiences, they vary across countries and geographic regions, even within a society. People's worldviews determine their **environmental ethic**, or how they interact with their natural environment; worldviews also impact how people respond to environmental problems. When different people or groups, with different worldviews,

worldview The window through which one views one's world and existence.

environmental ethic The personal philosophy that influences how a person interacts with his or her natural environment and thus affects how one responds to environmental problems.

INFOGRAPHIC 1.8   **WEALTH INEQUALITY**

5

Wealth and power was not evenly distributed in the Greenland Viking colony; a few powerful leaders made decisions that impacted everyone. The same is true today in many regions of the world. The World Bank estimates that about 40% of the world's population lives on less than $2 a day; almost half of the 2.2 billion children worldwide live in poverty. A small percentage of our population (perhaps 20%) controls 80% of all resources. Actions by the wealthy today that degrade the environment degrade it for all sooner or later.

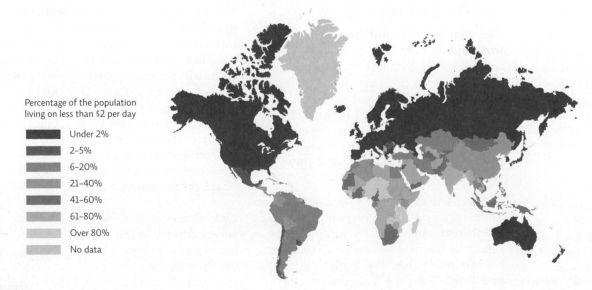

Percentage of the population living on less than $2 per day

- Under 2%
- 2–5%
- 6–20%
- 21–40%
- 41–60%
- 61–80%
- Over 80%
- No data

 How might environmental problems differ between wealthy and poorer nations?

INFOGRAPHIC 1.9   **WORLDVIEWS AND ENVIRONMENTAL ETHICS**

People's environmental worldviews describe how they see themselves in relation to the world around them. Their worldview influences their environmental ethic, which in turn influences how they interact with the natural world. We present three common worldviews here (there are others).

What is your own environmental worldview? Does it fall squarely into one of these camps, or is it a combination of two or more?

ECOCENTRISM
System-centered: Value is given to the importance of the ecosystem as a whole, including interactions such as those between wind and soil and between species (predators and their prey, for instance), as well as natural processes (such as the water cycle).

BIOCENTRISM
Life-centered: Humans and other species have a right to exist and are worthy of protection.

ANTHROPOCENTRISM
Human-centered: Only humans have intrinsic value, and resources are here to meet human needs and wants.

---

anthropocentric worldview A human-centered view that assigns intrinsic value only to humans.

instrumental value The value or worth of an object, an organism, or a species, based on its usefulness to humans.

biocentric worldview A life-centered approach that views all life as having intrinsic value, regardless of its usefulness to humans.

intrinsic value The value or worth of an object, an organism, or a species, based on its mere existence.

ecocentric worldview A system-centered view that values intact ecosystems, not just the individual parts.

approach environmental problems, they are bound to draw very different conclusions about how best to proceed.

The Vikings had an **anthropocentric worldview**—one where only human lives and interests are important. Meanwhile, the Vikings saw other species as having **instrumental value**—meaning the Vikings valued them only for what they could get out of them. Forests were nothing more than a source of timber; grasslands a source of home insulation and a feeding ground for cattle.

A **biocentric worldview** values all life. From a biocentric standpoint, every organism has an inherent right to exist, regardless of its benefit (or harm) to humans; each organism has **intrinsic value**. This worldview would lead us to be mindful of our choices and

avoid actions that indiscriminately harm other organisms or put entire species in danger of extinction. An **ecocentric worldview** values the ecosystem as an intact whole, including all of the ecosystem's organisms and the nonliving processes that occur within the ecosystem. Considering the same forests and grassland from an ecocentric worldview, the Vikings might have decided to protect both, not just for the resources they could harvest but to protect the complex processes that can produce those resources only when they remain intact. **INFOGRAPHIC 1.9**

Each of these worldviews may be found among citizens of the United States today, but we can also trace a historical and gradual change in the way a modern nation such as the United States has viewed the natural

An ecocentric worldview values all living creatures and nonliving processes of an ecosystem, from animals and wildflowers to nutrient cycling in the soil and water flow from snowmelt. Sassalb mountain range, Switzerland.

INFOGRAPHIC 1.10   U.S. ENVIRONMENTAL HISTORY   5

The predominant worldview of a society shapes its actions and the choices made by its citizens at any given point in time. Throughout history, notable individuals with different worldviews have greatly influenced the evolution of our relationship with the natural world. The gradual shift from unrestrained use to focus first on conservation and later on sustainability that was seen in United States and some other developed countries has now become an international movement.

How might differing worldviews influence international efforts to address environmental problems?

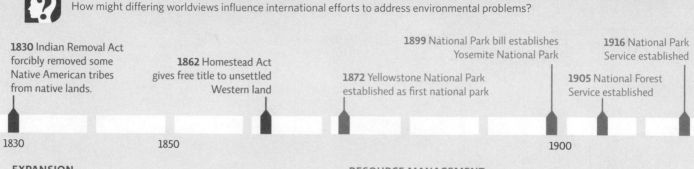

1830 Indian Removal Act forcibly removed some Native American tribes from native lands.

1862 Homestead Act gives free title to unsettled Western land

1872 Yellowstone National Park established as first national park

1899 National Park bill establishes Yosemite National Park

1905 National Forest Service established

1916 National Park Service established

1830   1850   1900

EXPANSION

When Europeans first settled in what is now the United States, the vast land was seen as an unending supply of natural resources and our actions focused on extracting those resources and spreading out, for the most part farther west.

RESOURCE MANAGEMENT

In the late 1800s it became apparent that we actually could deplete natural resources found in the United States; they were not, in fact, inexhaustible. The first pieces of restrictive environmental legislation were passed at that time, creating national parks, forests, and reserves to limit use. Key leaders in this conservation movement were President Theodore Roosevelt, Gifford Pinchot, and John Muir.

John Parrot/Stocktrek Images/Getty Images

Thomas Jefferson penned the General Land Ordinance of 1785, encouraging settlement of Western regions; later, the Homestead Act of 1862 gave free title to unsettled Western land.

akg-images/Newscom

John Muir (right), an ardent preservationist, worked tirelessly for the preservation of Yosemite Valley and the passage of the National Park bill. He was sometimes in conflict with President Theodore Roosevelt (left) and Gifford Pinchot (center), the first head of the National Forest Service who tried to maximize the commercial and recreational value of nature.

---

world. In fact, we can see these worldviews expressed in the ethical positions of many prominent environmental scientists, politicians, and citizens whose actions are associated with landmark events in environmental history. **INFOGRAPHIC 1.10**

Back in Greenland, in the silt-covered ruins of a Viking farmhouse, archaeologists found the bones of a hunting dog and a newborn calf, dating back several centuries. The knife marks covering both indicate that the animals were butchered and eaten. "It shows how desperate they were," says McGovern. "They would not have eaten a baby calf, or a hunting dog, unless they were starving." By then, it was probably far too late for the Greenland Vikings to adapt in any meaningful way. Their tale had already been

written—in ice cores and mud cores and in hundreds of years' worth of midden heaps.

Our own story is being written now, in much the same way. But unlike our forebears, we still have enough time to shape our own narrative. Will it be dug up, 1,000 years hence, from the ruins of what we leave behind? Or will it be passed down by the voices of our successors, who continue to thrive long after we are gone? Ultimately, the answer is up to us.

Select References:

Benyus, J. (1997). *Biomimicry*. New York: William Morrow and Company.
Diamond, J. (2005). *Collapse*. New York: Viking Press.
McGovern, T. H. (1980). Cows, harp seals, and churchbells: Adaptation and extinction in Norse Greenland. *Human Ecology*, *8*(3): 245–275.

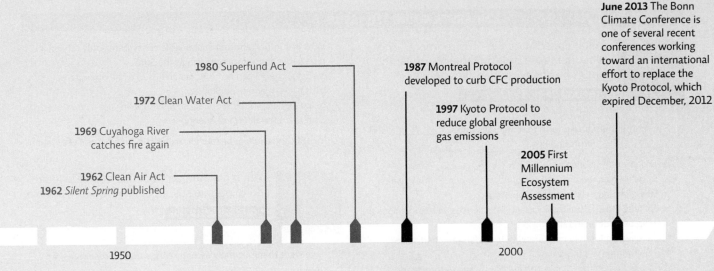

1962 Clean Air Act
1962 *Silent Spring* published

1969 Cuyahoga River
catches fire again

1972 Clean Water Act

1980 Superfund Act

1987 Montreal Protocol
developed to curb CFC production

1997 Kyoto Protocol to
reduce global greenhouse
gas emissions

2005 First
Millennium
Ecosystem
Assessment

June 2013 The Bonn
Climate Conference is
one of several recent
conferences working
toward an international
effort to replace the
Kyoto Protocol, which
expired December, 2012

1950

2000

POLLUTION CONCERNS

In the mid–1900s, pollution became a prominent national issue. Extreme events such as the Cuyahoga River catching fire due to all the flammable liquids discharged by industry caught the attention of the nation and led to the passage of some of our landmark environmental laws.

INTERNATIONAL EFFORTS

Today, improving the environment requires international cooperation and coordinated efforts. International groups such as the Intergovernmental Panel on Climate Change work to understand climate change; many nonprofit organizations address environmental issues, not just in their own region or nation, but around the world.

Pittsburgh Post- Gazette/ ZUMAPRESS/Newscom

Rachel Carson's 1962 book *Silent Spring* opened a new dialogue on the dangers of toxins in our environment. She provided evidence that these chemicals could have unexpected effects on wildlife and that we too were vulnerable to these poisons (see Chapter 3).

*/Kyodo/Newscom

Wangari Maathai began the Greenbelt Movement in Kenya in 1977 to reforest severely damaged areas of her country. The movement has spread to more than 15 African nations, and to the United States and Europe, and has planted more than 30 million trees (see LaunchPad Chapter 28).

BRING IT HOME

PERSONAL CHOICES THAT HELP

The concept of sustainability unites three main goals: environmental health, economic profitability, and social and economic equity. All sorts of people, philosophies, policies, and practices contribute to these goals; concepts of sustainable living apply on every level, from the individual to the society as a whole. In other words, every one of us participates. Throughout this book, you will have the opportunity to learn about personal actions that can help address environmental issues, but a good starting place is to learn about your own environment and the place you call home.

Individual Steps

• Discover your local environment. What parks or natural areas are close by? Does your campus have any natural areas? Visit one and spend a little time observing nature and your own reactions. Write down your thoughts or share you experiences with a friend.

• Are there restaurants, grocery stores, or other retail venues accessible through public transportation or within walking distance of your campus or home? For a week or two, try walking or riding a bike or bus to these businesses instead of driving to others farther away. Is this a reasonable option for you? Why or why not?

Group Action

• Discover your own interests. There is a group for every interest—from outdoor recreation, wildlife viewing and preservation, and environmental education, to transportation and air-quality issues. Get involved with organizations working to improve environmental issues or address social change and human rights. One person can make a difference, but a group of people can cause a sea of change.

Policy Change

• Discover what's happening in your community. Read the newspapers and monitor blogs covering environmental and quality-of-life issues. Alert your local, regional, and national representatives about the issues you care about and vote for government officials who support the causes you support.

ENVIRONMENTAL LITERACY UNDERSTANDING THE ISSUE

1 What constitutes the "environment," and what fields of study collaborate under the umbrella of environmental science?
INFOGRAPHICS 1.1 AND 1.2

1. Environment is a broad term that includes one's _____ and _____ surroundings.

2. Environmental science:
   a. is defined as the physical surroundings or conditions in which any given organism exists or operates.
   b. includes the natural sciences, social sciences, and humanities to help us understand the world and our place in it.
   c. focuses exclusively on our ability to understand and solve environmental problems.
   d. is the study of the plants and animals of the natural world and how they interact with each other.

3. Some scientists study peatlands to look for clues about how ancient civilizations lived and died. Others work with peatlands to manage these ecosystems for their water purification services, or as sources of fuel or soil additives (peat moss). Which of the activities above describes empirical science, and which describes applied science? Explain.

2 What are some of the environmental dilemmas that humans face, and why are many of them considered "wicked problems"?
INFOGRAPHICS 1.3 AND 1.4

4. The *triple bottom line* is an approach to problem analysis that includes consideration of the _____, _____, and _____ aspects of any potential solution.

5. Why is it that many problems are considered to be *wicked problems*?
   a. Their potential solutions almost always come with trade-offs.
   b. They arise out of the greed and malice of other people.
   c. These problems cannot be solved or even effectively addressed.
   d. They are hard to study using empirical methods, and so science can't help find solutions.

6. Consider the wicked problem of air pollution. Identify some of the causes and consequences of air pollution. What are some possible solutions, and what are the trade-offs that come with those solutions?

3 What does it mean to be sustainable, and what are the characteristics of a sustainable ecosystem?
INFOGRAPHIC 1.5

7. Which of these is an example of acting sustainably?
   a. Using available resources to support our current lifestyle
   b. Using resources now as we need them but also researching new ways to meet future needs
   c. Dramatically reducing our use of resources, even if it means a lower quality of life for most people
   d. Meeting today's needs without compromising the ability of future generations to meet theirs

8. Why is it important that sustainable ecosystems only rely on a renewable energy source such as the Sun?
   a. Solar energy can be captured and used by all organisms.
   b. It is clean and nonpolluting.
   c. Energy cannot be recycled, so new supplies are always needed.
   d. It is a natural form of energy.

9. Identify and explain the four characteristics of a natural ecosystem.

4 What can human societies and individuals do to pursue sustainability?
INFOGRAPHIC 1.6

10. If we were to mimic the way nature uses matter resources, we would:
    a. reduce the waste we produce by reusing and recycling the matter resources we do use.
    b. use the most abundant resources first before turning to less abundant ones.
    c. focus on reducing what we "take" from the environment but not need to worry about what we put back into the environment.
    d. rely on one or two ways to use and recycle material globally rather than using a variety of methods that might differ in different places.

11. Identify at least one action we could take as a society to follow each of the four characteristics of a sustainable ecosystem.

5 What challenges does humanity face in dealing with environmental issues, and how can environmental literacy help us make more informed decisions?
INFOGRAPHICS 1.7, 1.8, 1.9, AND 1.10

12. When you drive your car, it releases a small amount of air pollution. You know air pollution is bad, but if you stop driving, you'll be forced to find another way to get around, and the air will still be polluted because everyone else will still keep driving. So you keep driving, too. This is an example of the social trap known as _____.

13. Jack is opposed to the selective killing of deer that have vastly overpopulated the local forest. Jill argues that the ecosystem will be destroyed if some are not removed. While Jack is _____ as he sees deer as having intrinsic value, Jill is _____ as she values not just the species but the ecosystem processes as well.
    a. biocentric; ecocentric
    b. ecocentric; biocentric
    c. anthropocentric; ecocentric
    d. biocentric; anthropocentric

14. Compare the responses of the Vikings who settled in Greenland and Iceland. Why did one civilization persist, while the other died out? How is this an example of the importance of environmental literacy?

SCIENCE LITERACY WORKING WITH DATA

The following data show the number of different types of livestock grazed in the world and in Nigeria from 1961 to 2008. Use these graphs and Table 1.1 to answer the next five questions.

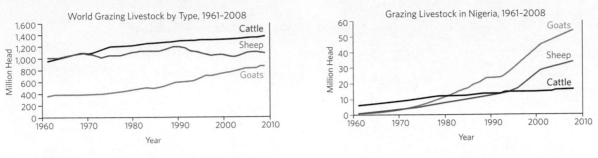

Table 1.1

Grazing Livestock Population (in millions)

Type of livestock	World in 1961	World in 2008	% change	Nigeria in 1961	Nigeria in 2008	% change
Cattle	942	1,372		6	16.3	
Goats	349	864		0.6	53.8	
Sheep	994	1,086		1	33.9	

Interpretation

1. How many different livestock types are included in the graphs, and what are the trends in the numbers?

2. Which animals constituted the bulk of the grazing herds around 2008 on a worldwide basis? In Nigeria? Was this true in 1961 (the first year for which data are reported)? Provide data to explain your responses.

3. Using the data from the table, calculate the percentage change for each type of animal. According to your calculations, which animal changed the most in each case (i.e., worldwide and in Nigeria)?

   Hint: percentage change $= \dfrac{(\text{2nd value} - \text{1st value})}{\text{1st value}}$

Advance Your Thinking

4. Unlike cattle and sheep, goats are much more flexible in what they eat, but their sharp hooves also pulverize soil more easily. The data show that the growth in goat populations is particularly dramatic in a developing country like Nigeria. What might explain this pattern? What might be some potential consequences?

5. How might the increase in the size of the goat herd be a potential social trap? What are some ways that we could avoid this trap?

INFORMATION LITERACY EVALUATING INFORMATION

The Lorax, a children's book by Dr. Seuss, tells the story of the Lorax, a fictional character who speaks for the trees against the greedy Once-ler who represents industry. Written in 1971, *The Lorax* was banned in parts of the United States for being an allegorical political commentary. Today the book is used for educating children about environmental concerns (see www.seussville.com/loraxproject/). Even so, some people consider the book inappropriate for young children due to its "doom and gloom" environmentalism.

The book *The Truax*, by Terri Birkett, involves a forest industry representative offering a logging-friendly perspective to an anthropomorphic tree, known as the Guardbark. This story was criticized for containing skewed arguments, and in particular a nonchalant attitude toward endangered species. About 400,000 copies of the book have been distributed to elementary schools nationwide.

Read both books. You can find *The Lorax* at your local public library, and you can download *The Truax* as a PDF from http://woodfloors.org/truax.pdf.

Evaluate the stories and work with the information to answer the following questions:

1. What are the credentials of the author of each book? In each case, do the person's credentials make him or her reliable/unreliable as a storyteller? Explain.

2. Connect each story to the key concepts in the chapter:
   a. What are the underlying attitudes and worldviews of each story?
   b. Does each story reflect social traps and, if so, in what way?
   c. How might each story contribute to environmental literacy? Explain.
   d. What does each story have to say about sustainability? Explain.

3. What supporting evidence can you find for the main message in each story? In the story itself? From doing some research?

4. What is your response to each story? What do you agree and disagree with in each case? Explain.

Find an additional case study online at http://www.macmillanhighered.com/launchpad/saes2e

SCIENCE AND THE SKY

Solving the mystery of disappearing ozone

CORE MESSAGE

Depletion of the stratospheric ozone caused by synthetic chemicals allows more dangerous solar radiation to reach Earth's surface, threatening the health and well-being of many plants and animals, including humans. Researchers employed the scientific process to collect physical evidence, analyze it, and report their findings to the scientific community. This helped us understand what was causing the depletion and guided wise policy decisions at the national and international levels to address this tragedy of the commons.

AFTER READING THIS CHAPTER, YOU SHOULD BE ABLE TO ANSWER THE FOLLOWING **GUIDING QUESTIONS**

How do scientists study the natural world? Why do we say science is a "process" and that conclusions are always open to further study?

What is stratospheric ozone, and why are scientists worried about its depletion?

Setup for launch of the high-altitude research balloon for ozone testing, McMurdo Station, Antarctica. *Courtesy Linnea Avallone/Concordiasi Team*

3

What lines of evidence suggest that CFCs are causing the depletion of stratospheric ozone?

4

Why are both observational and experimental studies needed to investigate the natural world? How does statistics help researchers analyze the data they gather from these studies?

5

How do approaches that use the precautionary principle and adaptive management help policy makers address environmental problems?

There's a point of no return halfway into the 9-hour flight from New Zealand to the Antarctic. Once that 4.5-hour mark passes, if something goes wrong with the plane, there's nowhere to stop in the Southern Ocean for repairs.

Susan Solomon is very familiar with that trip's all-important midway point. During her very first flight to the southernmost continent, the pilot told the passengers that their plane was not working properly: The front ski at the nose of their C-130 was frozen and couldn't be lowered into position, so landing on the packed snow at the Antarctic research station would be impossible. They had to turn around.

It was August 1986—late winter in the Antarctic—and the atmospheric chemist was on her way to the southernmost continent to investigate a mystery: Why was the ozone layer above the South Pole disappearing? Suddenly, the remoteness of where she was going hit home.

"I remember, as we were flying back to New Zealand, thinking, 'Wow, I really am going to the Antarctic,'" Solomon says.

The next night, Solomon and her team from various research institutions, including the National Oceanic and Atmospheric Administration (NOAA), managed to make it to Antarctica, landing at McMurdo Station. On this, her very first excursion to the Antarctic, Solomon and her colleagues collected the initial data that would eventually grab the world's attention and settle a long-standing, hard-fought scientific debate that was taking place on an international stage.

Science gives us tools to observe and make sense of the natural world.

Solomon became a scientist because she was curious about the natural world around her. **Science** is both a body of knowledge (facts and explanations) and the process used to get that knowledge. Understanding the process is more important than the "facts," since facts may change as more information is collected through the scientific process. This process is a powerful tool that allows us to gather evidence to test our ideas *and* to evaluate the quality of that evidence.

Science, however, is limited to asking questions about the *natural* world; not all questions are open to science. Scientific investigation, in both the natural and social sciences, is based on data gathered through **empirical evidence**, or observations. Only physical phenomena that can be objectively observed—meaning data that could be collected by *anyone* in the same place, using the same equipment, etc.—are fair game for science. Scientists can gather empirical evidence about the environment and living things using a wide variety of tools, including natural tools, such as their eyes, ears, and other senses. Phenomena that are not objectively observable (What is my dog thinking? Do ghosts exist?) and ethical or religious questions (Is the death penalty wrong? What is the meaning of life?) cannot be empirically studied and, therefore, are not within the purview of science.

⊙ WHERE IS McMURDO STATION?

KEY CONCEPT 2.1

Science is best viewed not just as a body of facts but as the process used to gain that knowledge.

GEORGE STEINMETZ/National Geographic Creative

↑ McMurdo Station, the largest human settlement in the Antarctic.

Solomon and the team had decided to fly to the other end of the world after reading a scientific paper published the year before, in which Joe Farman of the British Antarctic Survey and his colleagues showed that, since the late 1970s, the ozone layer had thinned by about one-third during the Antarctic spring.

The British team had collected nearly three decades of data in Antarctica, starting in 1957 with on-the-ground

KEY CONCEPT 2.2

Scientists collect evidence (observations) and use this evidence to draw conclusions (inferences) in an effort to understand the natural world.

instruments. Like all other good scientists, Farman and his team depended on **observations** (information detected with the senses or with equipment that extends our senses) of the natural world. His team collected data on the atmosphere's composition (lower-than-normal ozone levels) and then used these observations to draw conclusions or make **inferences**—explanations of what else might be true or what might have caused the observed phenomenon.

science A body of knowledge (facts and explanations) about the natural world and the process used to get that knowledge.

empirical evidence Information gathered via observation of physical phenomena.

observations Information detected with the senses—or with equipment that extends our senses.

inferences Conclusions we draw based on observations.

Courtesy: Dr. Susan Solomon

↑ Susan Solomon at McMurdo Station in 1987.

The ozone layer protects life on Earth

It was a serious proposition: Without ozone, the world as we know it would not exist. Ozone is a key element of the **atmosphere**, the blanket of gases surrounding our planet that is made up of discernable layers, which differ in temperature, density, and gas composition. The lowest level, the **troposphere**, extends about 11 km (7 miles) up from Earth's surface. This level is familiar to us: It is the air we breathe and where our weather occurs. The next level in the atmospheric blanket, the **stratosphere**, rises to 50 km (31 miles) above Earth's surface. The stratosphere is much less dense than the troposphere but contains a "layer" of **ozone** (abbreviated as O_3 because it contains three oxygen atoms), a region where most of the atmosphere's ozone is found.

Without ozone, the world as we know it would not exist.

Farman's group also connected their results to studies by other researchers that had shown higher concentrations of an important humanmade compound: chlorofluorocarbons (CFCs), which in turn produce atmospheric chlorine (Cl). The concentrations of these chemicals seemed to increase at a rate matching the disappearance of ozone. The observation of a decrease in ozone did not come from just a few readings but represented data collected at two different sites over more than a dozen seasons. *Replication* within a study (multiple test subjects or measurements) and among studies (independent tests that collect the same data, preferably conducted by other researchers) increase the reliability of the data. When multiple studies produce similar results, it is less likely that the original data was "a fluke," or an unusual result. Farman's research exemplified this hallmark of good science—in this case, multiple data points at two different testing sites. Since then, other researchers have confirmed his team's results.

Farman's team inferred that the ozone depletion in the Antarctic was somehow connected to the increased presence of chlorine compounds in the atmosphere.

atmosphere The blanket of gases that surrounds Earth and other planets.

troposphere The region of the atmosphere that starts at ground level and extends upward about 7 miles.

stratosphere The region of the atmosphere that starts at the top of the troposphere and extends up to about 31 miles; contains the ozone layer.

ozone A molecule with three oxygen atoms that absorbs UV radiation in the stratosphere.

ultraviolet (UV) radiation Short-wavelength electromagnetic energy emitted by the Sun.

Solar radiation enters the atmosphere every day, including three forms of **ultraviolet (UV) radiation**: UV-A, UV-B, and UV-C. Each form of radiation travels at a different wavelength, and each differs in its ability to penetrate materials, like skin. In humans, exposure to UV-B radiation increases the risk of cataracts, skin damage, and cancer. Fortunately, ozone prevents most of the UV-B radiation from reaching Earth's surface, minimizing the damage it can cause to living things. About 50% of incoming UV-A radiation still reaches Earth's surface, where it can damage skin, cause sunburn, prematurely age the skin, and cause skin cancer. A suntan is actually a response by the skin to the damaging effects of UV radiation and will not necessarily protect an individual from further Sun damage. Stratospheric (good) ozone should not be confused with ground-level (bad) ozone found in the troposphere. Ground-level ozone is a component of smog and is harmful to living things (see Chapter 21).
INFOGRAPHIC 2.1

KEY CONCEPT 2.3

Human actions can significantly alter the natural world we depend on. The stratospheric ozone layer is critical for life on Earth, but human impact is causing its depletion in some areas.

Arriving at McMurdo Station, Solomon knew she had to apply the scientific process to understand why the ozone

INFOGRAPHIC 2.1 THE ATMOSPHERE AND UV RADIATION

↓ The atmosphere is composed of layers that differ in temperature and chemical composition. There is not much ozone in the atmosphere, but most of what ozone there is occurs in a layer in the stratosphere. Ozone (O_3) is important because it prevents some UV radiation from reaching Earth's surface.

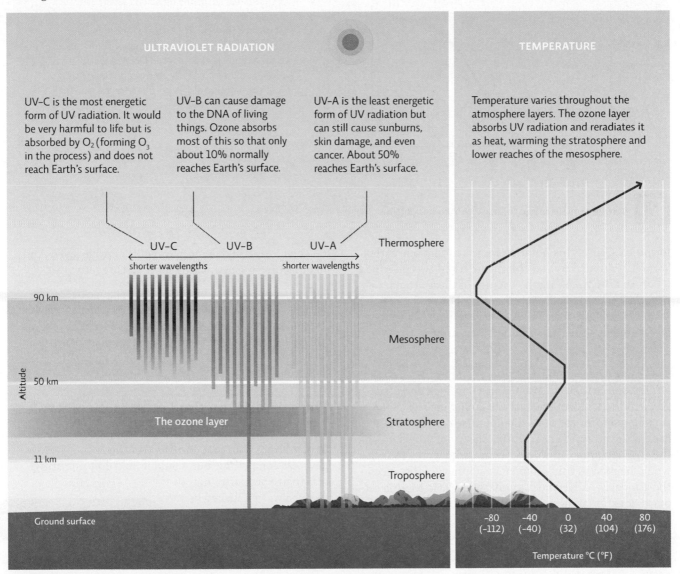

ULTRAVIOLET RADIATION

UV-C is the most energetic form of UV radiation. It would be very harmful to life but is absorbed by O_2 (forming O_3 in the process) and does not reach Earth's surface.

UV-B can cause damage to the DNA of living things. Ozone absorbs most of this so that only about 10% normally reaches Earth's surface.

UV-A is the least energetic form of UV radiation but can still cause sunburns, skin damage, and even cancer. About 50% reaches Earth's surface.

TEMPERATURE

Temperature varies throughout the atmosphere layers. The ozone layer absorbs UV radiation and reradiates it as heat, warming the stratosphere and lower reaches of the mesosphere.

UV-C   UV-B   UV-A   Thermosphere

shorter wavelengths   shorter wavelengths

90 km   Mesosphere

Altitude

50 km

The ozone layer   Stratosphere

11 km

Troposphere

Ground surface

-80 (-112)   -40 (-40)   0 (32)   40 (104)   80 (176)

Temperature °C (°F)

? What problems could emerge if the ozone layer lost some of its ozone?

layer was thinning in the Antarctic. And she already had a culprit in mind.

Scientific views rarely change overnight.

Throughout her childhood, Solomon was surrounded by items that contained the synthetic molecules known as CFCs. The compounds were first developed in the 1930s as a commercial coolant, to replace more toxic ammonia and sulfur dioxides, and were included in refrigerators, air conditioners, and other household and industrial items. By the time Solomon reached graduate school at the University of California, Berkeley, in the late 1970s, CFCs were being used in everything from hairspray to foam containers for fast food.

CFCs contain atoms of carbon, fluorine, and chlorine. For example, CFC_{12}, a common refrigerant, has the molecular

INFOGRAPHIC 2.2   OZONE DEPLETION AND CFC LEVELS

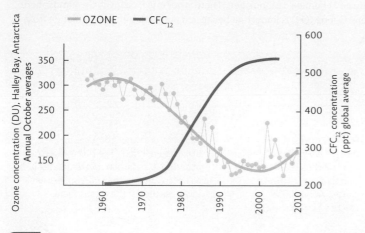

← This graph shows a correlation between the levels of CFC_{12} and ozone over Antarctica. As CFC_{12} increased, ozone declined (a negative correlation). When CFC_{12} amounts levelled off, around the year 2000, the ozone hole over Antarctica started to recover. (Ozone concentration is expressed in Dobson units, DU, a measure of how thick the air sample would be if it were compressed at 1 atmosphere of pressure at 0°C; 1 DU = 1 mm thick.)

 Why is it unreasonable to conclude from this data that the rise in CFCs caused the drop in ozone levels?

formula CCl_2Fl_2; this notation indicates that one carbon atom (C) is bound to two chlorine (Cl) and two fluorine (Fl) atoms. By tweaking the balance of these atoms, chemists synthesized CFCs that they believed were stable, nonflammable, and harmless to people and the environment—qualities that led to their broad acceptance.

Over time, these items emitted CFC molecules into the air; in 1971, British scientist James Lovelock detected CFCs in the atmosphere over England. Since CFCs were considered safe, the news did not raise concern. But it caught the attention of scientist Sherwood Rowland of the University of California, Irvine, who wondered what would happen to this completely synthetic substance in the atmosphere. He and his colleague Mario Molina set out to look for answers.

In 1974, Rowland and Molina proposed that CFCs were not entirely harmless in the atmosphere. Chemists had specifically designed these compounds to be stable, and Molina and Rowland realized that CFCs would stay aloft for a remarkably long time in the atmosphere, residing there and accumulating for 100 years or more. And once in the stratosphere, the molecules would be exposed to UV light so intense that it would break them apart. The process would release solitary chlorine atoms, which previous research had shown could chemically react with—and destroy—ozone, aligning with Farman's observations from Antarctica. Molina and Rowland calculated that at the rate CFCs were being produced in 1972,

the chemicals could destroy 6% of the ozone layer. And manufacturers were making more CFCs every year.

But the scientific view of CFCs didn't change overnight. Because all conclusions in science are considered open to revision (since our understanding of a concept or process will change as scientists learn more), more evidence was needed to overturn the prevailing conclusion that CFCs were safe.

The scientific method systematically rules out explanations.

Farman's 1985 data that one-third of the ozone layer was gone was compelling, but would it hold up to scrutiny? Less than a year after his research was published in the journal *Nature*, NASA scientists published their own report, verifying what Farman had observed. This corroboration strengthened Farman's conclusions. This faster-than-anticipated ozone depletion set off alarm bells. These data also provided a **correlation** between the presence of CFCs and ozone depletion: Both occurred together. However, although the correlation suggested that CFCs might be related to ozone decline, it did not establish a **cause-and-effect relationship**. The two trends could occur together by coincidence, or something else entirely could be causing both to occur. **INFOGRAPHIC 2.2**

KEY CONCEPT 2.4

Scientists use the scientific method to propose hypotheses and systematically test predictions; they then base their conclusions on the evidence gained.

correlation Two things occurring together but not necessarily having a cause-and-effect relationship.

cause-and-effect relationship An association between two variables that identifies one (the effect) occurring as a result of or in response to the other (the cause).

To get a better picture of what was going on, scientists needed to further apply the **scientific method**, in which they would work logically and systematically to design studies specific to the question being asked. The fact that the ozone layer was thinning above the Antarctic triggered great debate among scientists, resulting in more than one possible explanation, or **hypothesis**, for what was occurring. Some researchers thought that Antarctic air was mixing and lifting lower, low-ozone air into the stratosphere, changing the concentration. Other researchers believed that solar activity was creating nitrogen oxides (NO_x), which could be destroying ozone, as the amount of NO_x fluctuated with sunlight in the Antarctic. But Susan Solomon had a different idea.

Solomon was a young scientist when she visited Antarctica in 1986 and 1987, the beginning of her long relationship with the southern, icy world. While trying to understand why ozone was disappearing over the region, Solomon kept thinking about temperature. October is spring in Antarctica, and scientists knew that cold spring winds would swirl around, producing a cyclone of air in the atmosphere (a polar vortex), keeping cold air in place over the poles and leading to the formation of polar clouds in the stratosphere. Solomon proposed the hypothesis that particles in the polar stratospheric clouds were providing surfaces for the reactions that would free chlorine molecules (Cl_2) from CFCs. In sunlight, the chlorine molecules would then break up into chlorine atoms. These isolated chlorine atoms destroyed ozone—particularly in the Antarctic spring, when sunlight streamed in.

Scientific hypotheses like Solomon's must be **testable** (that is, generate **predictions** about what we could objectively observe if we conducted a test). In turn, these predictions must be **falsifiable**, meaning that it would be possible to produce evidence to show that the prediction is wrong. (Predictions based on untestable ideas—such as "reincarnation exists"—are not falsifiable and therefore are not considered suitable for science.) If a prediction is falsifiable and the falsifying evidence does not appear, it may be reasonable to conclude that the evidence supports the prediction.

Once a study is conducted and data are gathered, the data are evaluated to determine whether they confirm or fail to confirm the hypothesis. But the scientific process doesn't end there. If a hypothesis is rejected, alternative hypotheses can be tested. If a hypothesis is confirmed, the researchers should repeat the study to validate the data. They can also generate new predictions that test the same hypothesis from different angles. As evidence mounts from replicate studies and from multiple predictions, we become more confident in our data and conclusions.

Before publishing, scientific reports are subjected to **peer review**, meaning that they are reviewed by a group of third-party experts. Studies that are not well designed or well conducted are not accepted for publication. Therefore, peer-reviewed published research represents high-quality scholarship in the field. **INFOGRAPHIC 2.3**

Notice we do not claim that the hypothesis is *proven*, only that it is supported (or confirmed). This is a hallmark of the tentative nature of science. "Proven" suggests that we have the final answer; science, however, is open ended, and no matter how much evidence accumulates, there are always new questions to ask and new studies to conduct that could alter our conclusions. But this does not mean that all possible explanations are equally valid. This is precisely the reason hypotheses must be tested again and again and in different ways. As evidence mounts in support of a hypothesis, the probability that it is wrong lessens, and it becomes unreasonable to reject the hypothesis in favor of another, less-supported explanation.

A hypothesis with a tremendous amount of support from multiple lines of evidence may eventually be considered to be a scientific **theory**—a widely accepted explanation that has been extensively and rigorously tested. But, in keeping with the tentative nature of science, even well-substantiated theories are always open to further study and will be revised or even abandoned if new data strongly support a new conclusion. This differs from the casual

KEY CONCEPT 2.5

Scientific evidence from different types of studies supporting the same conclusion increases our confidence that a hypothesis is valid.

scientific method The procedure scientists use to empirically test a hypothesis.

hypothesis A possible explanation for what we have observed that is based on some previous knowledge.

testable Having a possible explanation that generates predictions for which empirical evidence can be collected to verify or refute the hypothesis.

prediction A statement that identifies what is expected to happen in a given situation.

falsifiable Being capable of being proved wrong by evidence.

peer review A process whereby researchers submit a report of their work to outside experts who evaluate the study's design and results to determine whether it is of a high enough quality to publish.

theory A widely accepted explanation of a natural phenomenon that has been extensively and rigorously tested scientifically.

INFOGRAPHIC 2.3 **SCIENTIFIC PROCESS**

↓ Scientists work from previous knowledge and observation to ask new questions and pose possible explanations (hypotheses) for what they observe. They then design a study to gather evidence to test predictions made from their hypotheses. The scientific method is not necessarily a linear sequence. Rather, it is more of a cycle that scientists move through in whatever order of "steps" best suits their needs.

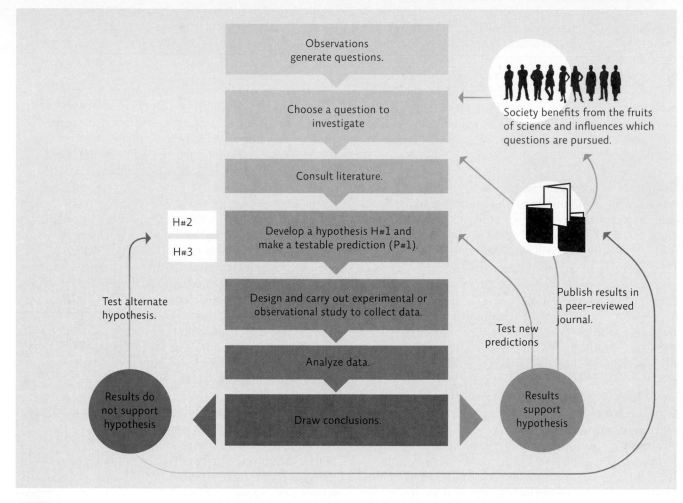

? If the data confirm a hypothesis, new predictions are still tested. Why is this useful?

meaning of *theory*, which suggests a speculative idea that has little substance. To discount any scientific theory as "just a theory" represents a serious flaw in one's understanding of what a scientific theory really is. **INFOGRAPHIC 2.4**

Different types of studies amass a body of evidence.

Solomon's hypothesis generated the prediction that the stratosphere would contain high levels of chlorine monoxide, or ClO. If polar clouds and sunlight were causing chlorine to react with ozone, then the atmosphere should contain many molecules of chlorine

bound to individual oxygen atoms. This prediction was falsifiable, since Solomon might not find high levels of ClO.

On both of her trips to the Antarctic, her team raised balloons into the air to measure the composition of the atmosphere where the ozone hole was found. They came back with measurements of ClO that would turn out to be their so-called smoking gun.

At the time, NASA ozone modeler Paul Newman believed that the loss of ozone in the Antarctic spring was due to excess solar activity. But when the ClO measurements from Solomon's team streamed out of a fax machine in NASA's Goddard Research Center, he

INFOGRAPHIC 2.4   CERTAINTY IN SCIENCE

↓ There are degrees of certainty in science; we know some ideas are better than others. The more evidence we have in support of an idea, especially when the evidence comes from different lines of inquiry, the more certain we are that we are on the right track. But since all scientific information is open to further evaluation, we do not expect or require "absolute" proof.

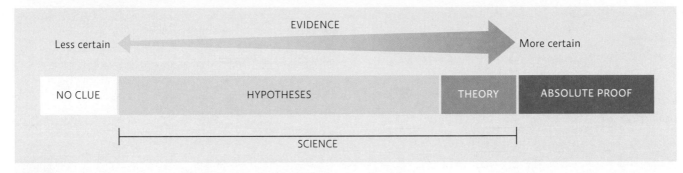

EVIDENCE

Less certain ←————————————————————————————→ More certain

| NO CLUE | HYPOTHESES | THEORY | ABSOLUTE PROOF |

SCIENCE

? Why do scientists say that a hypothesis can be disproved but never proven?

David Hay Jones/Science Source

↑ Researchers at the Finnish Ultraviolet International Research Center in the Arctic Circle research the effects of increased UV-B radiation on plant growth and biology. UV-B radiation at high latitudes damages living organisms and can alter their DNA (genetic material).

INFOGRAPHIC 2.5   **THE CHEMISTRY OF OZONE FORMATION AND BREAKDOWN**

↓ Ozone is naturally formed and broken down in the stratosphere.

↓ **Susan Solomon's proposed mechanism for ozone destruction**
Solomon's research provided evidence that CFCs catalyze additional ozone breakdown, leading to ozone depletion.

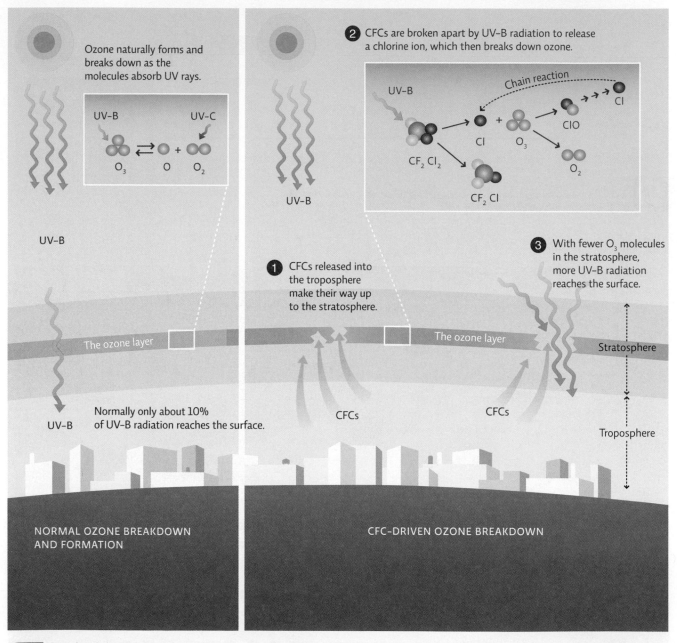

Ozone naturally forms and breaks down as the molecules absorb UV rays.

UV–B   UV–C

O_3 O + O_2

NORMALLY only about 10% of UV–B radiation reaches the surface.

NORMAL OZONE BREAKDOWN AND FORMATION

2 CFCs are broken apart by UV–B radiation to release a chlorine ion, which then breaks down ozone.

UV–B   Chain reaction
                                                    Cl
                              Cl +   → ClO
$CF_2 Cl_2$ O_3
                              $CF_2 Cl$ O_2

1 CFCs released into the troposphere make their way up to the stratosphere.

3 With fewer O_3 molecules in the stratosphere, more UV–B radiation reaches the surface.

The ozone layer   Stratosphere

CFCs   CFCs   Troposphere

CFC-DRIVEN OZONE BREAKDOWN

Based on Solomon's proposed mechanism, which molecule or atom would you look for in the atmosphere to determine whether CFCs were contributing to ozone depletion: ClO, O_2, CF_2Cl or Cl? Explain why this, and only this, molecule or atom would identify CFCs as the culprit.

knew the evidence supported Solomon's polar cloud hypothesis. **INFOGRAPHIC 2.5**

Solomon's experiment is an example of an **observational study**, which involves collecting data in the real world without intentionally manipulating the subject of study. In these types of studies, researchers may simply be gathering data to learn about a system or phenomenon, or they may be comparing different groups or conditions found in nature. Often, researchers can conduct observational studies that take advantage of natural changes in the environment, such as collecting "before and after" data

Both observational and experimental studies are needed to help us make sense of the natural world. Each supplies different lines of evidence for analysis.

in an area to examine the effect of a natural perturbation such as a flood or volcanic eruption or, in this case, comparing areas with different levels of ozone depletion. These types of opportunities represent valuable "natural experiments" that can allow us to test cause-and-effect hypotheses, just as a controlled **experimental study**, in which the researcher intentionally manipulates the conditions of the experiment in a lab or field setting, allows us to do.

Much research is under way on how ozone depletion and elevated UV exposures are affecting living organisms. Previous studies have shown that UV exposure can increase the incidence of skin cancer. A reasonable question would be: Is ozone depletion causing more skin cancer? This could lead to an experimental hypothesis such as: Lower ozone levels will lead to more cases of skin cancer. This hypothesis generates many predictions; some would be tested with observational studies and others experimentally.

In 2002, Chilean researchers Jaime Abarca and Claudio Casiccia investigated this very hypothesis by evaluating skin cancer incidence in residents of Punta Arenas, Chile, a region exposed to higher-than-typical UV-B radiation due to its proximity to the ozone hole. Because of the ozone hole, global air circulation patterns have actually shifted; the Southern Hemisphere jet stream now circulates at a latitude closer to the South Pole, taking the ozone-depleted air over more populated regions of the Southern Hemisphere, such as southern Chile. The researchers predicted that more cases of skin cancer would be seen there in years when ozone depletion was high. To test this, they calculated skin cancer rates of the population between 1987 and 1993 and compared those to rates from the same population between 1994 and 2000. The results of their study showed that nonmelanoma skin cancer rates were significantly higher in times when ozone depletion was higher. This evidence supports their hypothesis and correlates living close to the Antarctic ozone hole with one's chance of developing skin cancer. This observational study does not manipulate any variable—no people were exposed to higher or lower levels of UV-B or intentionally relocated to areas of higher or lower ozone depletion. The researchers simply collected data that were available in the population and then evaluated the data to see if the two variables—ozone depletion and skin cancer incidence—were correlated.

We can't ethically manipulate people to further test this hypothesis, but we can conduct an experimental study on cells or model organisms such as mice. Australian researcher Scott Menzies and his colleagues did just that in the early 1990s: They tested the prediction that mice exposed to high UV-B radiation would develop more skin cancer than mice exposed to normal levels of UV-B radiation. This experiment compared two groups—the **control group** of mice, which was exposed to normal levels of UV-B radiation, and the **test group**, which was exposed to the same amount of UV-B radiation received in areas where ozone depletion has been observed. The two groups are identical in every way except for the test variable (in this case, the amount of UV-B radiation exposure). The inclusion of a control group is key because it allows researchers to attribute any differences seen between the two groups to the single test variable that was altered. If researchers looked only at the test group, they would have no way to determine whether the incidence of skin cancer was higher than normal.

In an experimental study, we have both an **independent variable** and a **dependent variable**. We manipulate the independent variable (in this case, the amount of UV-B radiation), and measure the dependent variable (the incidence of skin cancer) to see if it is affected. In other words, if development of skin cancer is *dependent* on UV-B radiation, then we should see skin cancer incidence change as UV-B exposure changes. Scientists often represent their data on a graph, on which the x-axis (horizontal axis) displays the independent variable and the y-axis (vertical axis) shows the response (dependent variable). Menzies' data showed a clear difference between his control group (none developed skin cancer) and his test group (100% developed skin cancer). It is rare to see such an absolute difference and, as with any study, we would like to see these results replicated. Regardless, a well-designed and well-conducted study increases our confidence that the results are valid.

INFOGRAPHIC 2.6

Both observational and experimental studies gather data systematically to produce

observational study Research that gathers data in a real-world setting without intentionally manipulating any variable.

experimental study Research that manipulates a variable in a test group and compares the response to that of a control group that was not exposed to the same variable.

control group The group in an experimental study that the test group's results are compared to; ideally, the control group will differ from the test group in only one way.

test group The group in an experimental study that is manipulated such that it differs from the control group in only one way.

independent variable The variable in an experiment that a researcher manipulates or changes to see if the change produces an effect.

dependent variable The variable in an experiment that is evaluated to see if it changes due to the conditions of the experiment.

BACKGROUND KNOWLEDGE
UV light can cause
skin cancer.

← Scientists collect evidence to test ideas. Experimental studies are used when the test subjects can be intentionally manipulated; observational studies allow scientists to look at entire ecosystems or other complex systems.

QUESTION
Is ozone depletion causing
more skin cancer?

HYPOTHESIS
Lower ozone levels will lead
to more cases of skin cancer.

Scientifically test the hypothesis.

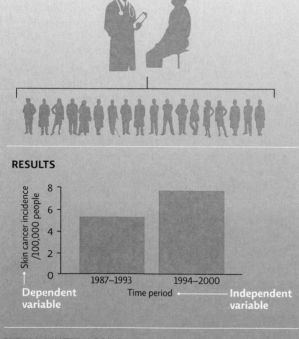

OBSERVATIONAL STUDY (Abarca, et al.)
Prediction: The incidence of skin cancer in an area where ozone depletion is occurring will be higher than it was in times of less ozone depletion.
Procedure: Collect data on skin cancer incidence in southern Chile.

RESULTS

Skin cancer incidence /100,000 people (Dependent variable) vs Time period / Independent variable

- 1987–1993
- 1994–2000

EXPERIMENTERS' CONCLUSION: Living close to the ozone hole increases one's chance of developing skin cancer.

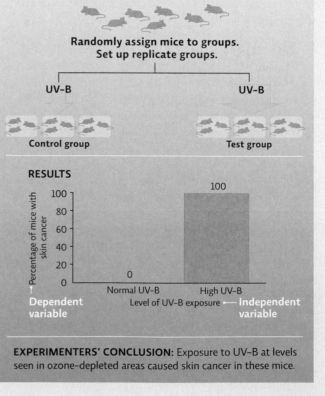

EXPERIMENTAL STUDY (Menzies, et al.)
Prediction: Mice exposed to UV–B levels comparable to those seen in areas where ozone depletion is observed will develop more skin cancer than mice exposed to normal levels of UV–B.
Procedure: Expose mice to different UV–B levels and monitor them for skin cancer.

**Randomly assign mice to groups.
Set up replicate groups.**

UV–B   UV–B

Control group   **Test group**

RESULTS

Percentage of mice with skin cancer (Dependent variable) vs Level of UV–B exposure / Independent variable

- Normal UV–B: 0
- High UV–B: 100

EXPERIMENTERS' CONCLUSION: Exposure to UV–B at levels seen in ozone-depleted areas caused skin cancer in these mice.

 What else could have caused the increase in skin cancer seen in the Abarca study?

 Do the results from the Menzies study tell us anything about how other species would respond to these levels of UV–B radiation exposure? Explain.

scientifically valid evidence. In contrast, anecdotal accounts (individual occurrences or observations) represent data that was not systematically collected or tested and cannot be compared to any control (because many uncontrolled variables exist). While anecdotes are not considered acceptable scientific evidence, a claim based on anecdotal accounts could be tested to see if a correlation or cause-and-effect relationship exists.

Multiple ozone depletion hypotheses were tested but only the CFC hypothesis was confirmed.

Solomon and her team reported their evidence to the international scientific community by publishing their results in the peer-reviewed journals *Nature* and the *Journal of Geophysical Research* in 1987. Prior studies had already presented evidence that CFCs were present in the stratosphere, and lab studies showed that they were capable of destroying ozone. Now, Solomon and her team were presenting evidence that, at the time of year when ozone was dropping, the Antarctic stratosphere contained high levels of ClO, demonstrating that free chlorine atoms were reacting with ozone.

The evidence was amassing that CFCs were contributing to ozone depletion, but that didn't mean the other hypotheses would be immediately abandoned. Research on those alternative hypotheses continued, but scientists were not finding evidence to support the predictions of the other hypotheses. If lower, ozone-poor air was lifting and mixing with the stratosphere, then researchers should observe gases moving upward in the atmosphere. Instead, they saw the opposite: Air seemed to be flowing downward. If solar activity was creating nitrogen oxides (NO_x) that were destroying ozone, as NASA modeler Newman had once believed, then scientists should have observed increases in NO_x at the South Pole. But when they took measurements, they found that NO_x levels were actually decreasing. This observation provided further support for Solomon's hypothesis, which predicted that NO_x levels should be low, not high—because otherwise NO_x would combine with ClO molecules and prevent them from interacting with ozone. The multiple lines of evidence collected for more than 30 years are so compelling that they have elevated the "CFC hypothesis" to the status of theory.

KEY CONCEPT 2.7

Scientists statistically evaluate data to assign a degree of certainty (known as a *p*-value) to their conclusions.

↑ In 1975, Representative Perry Bullard of Michigan introduced legislation to outlaw aerosol cans, 12 years before the Montreal Protocol was signed. This is an example of applying the precautionary principle.

As mentioned earlier, in science there are *degrees of certainty*; we know some things better than others. The more evidence we have in support of an idea, especially from different types of experiments, the more certain we are that we are on the right track. These degrees of certainty are expressed mathematically in terms of probabilities, using **statistics**. A mathematical

statistics The mathematical evaluation of experimental data to determine how likely it is that any difference observed is due to the variable being tested.

© Bettmann/CORBIS

analysis of the data is done to determine the probability that the occurrence of a phenomenon (in this case, the experiment's result) is a random event rather than being caused by the variable being investigated.

This type of data analysis allows us to quantitatively assign a level of certainty to our conclusions. In statistics, this probability is expressed as a *p-value* that represents the likelihood that our conclusions are wrong. Scientists generally require a high probability (at least 95%) that their conclusions are correct. $p \leq 0.05$ represents a level of certainty of 95%, and it means that there is only a 5% chance we have incorrectly accepted or rejected our hypothesis. The more evidence that accumulates in favor of a particular conclusion, the more certain we are that it is likely to be correct and can be used to inform our decisions. Just because we don't know exactly what will happen if we continue to release CFCs into the atmosphere (including how fast the ozone layer will deplete or when it could jeopardize human health) doesn't mean we don't know enough to take action. (For more on statistics and experimental design, see Appendix 3.)

Indeed, the loss of ozone was starting to have effects in other places besides regions close to the Antarctic ozone hole. Other observational studies revealed that UV-B radiation levels were increasing in high- and midlatitude regions. Research by NASA scientists showed that the amount of UV-B radiation reaching the ground in the midlatitudes in the mid-1990s had increased over 1979 levels. Other research showed that UV-B radiation levels were 45% higher than normal in the spring of 1990 at the southern tip of Argentina.

In the meantime, skin cancer incidence has increased in human populations (as have other skin disorders and eye problems, such as cataracts), and other organisms are also affected by extra UV-B exposure. Photosynthesis rates in marine organisms such as the tiny phytoplankton of the Antarctic sea are lower than normal in the spring because of increased UV-B exposure—a troublesome fact since phytoplankton form the base of the Antarctic food chain. Decreases in photosynthesis are also troubling because they could lead to lower agricultural productivity and

---

Montreal Protocol An international treaty that laid out plans to phase out ozone-depleting chemicals such as CFCs.

policy A formalized plan that addresses a desired outcome or goal.

precautionary principle A principle that encourages acting in a way that leaves a margin of safety when there is a potential for serious harm but uncertainty about the form or magnitude of that harm.

---

reduce the value of these areas as "carbon sinks"—areas where carbon is stored and kept out of the atmosphere. (See Chapter 21, on climate change.)

The international community got together to meet the problem head on.

Even while scientists were still trying to understand why the ozone layer was depleting, they knew they had to do something about it. Concerns raised by researchers in the 1970s had already led to a ban on CFCs for use as a propellant in hairsprays and other products in some countries, including the United States.

In 1985, a group of experts from around the world met in Vienna, Austria, to discuss ways to research and solve the problem. This set the stage for the international community to come together in Montreal, Canada, in 1987, to produce a plan to deal with the problem of ozone depletion. The plan, called the **Montreal Protocol**, would involve sacrifices—notably, phasing out dangerous chemicals including CFCs. At that first meeting in September 1987, only two dozen governments signed on to the protocol.

As the evidence mounted, industry, the public, and governments began to realize the seriousness of the ozone problem. By 2009, the protocol was eventually ratified by all 196 countries in the world when East Timor signed on.

The Montreal Protocol, administered by the United Nations, outlined a series of deadlines over the next decade for cutting back production of CFCs. Governments would have to put in place their own plans for achieving a desired outcome, or **policy**, for reducing CFCs.

Policy has been described as *translating our values into action*. Science provides information we can use to make informed decisions to protect our health, safety, or environment. Society then decides what "ought" to be addressed (Do we address ozone depletion?) and hopefully uses the best scientific information available to set reasonable policies that take into account the ecological, economic, social, and political issues at stake.

KEY CONCEPT 2.8

Policy makers should base decisions on the best scientific evidence available, and they should employ the *precautionary principle* and *adaptive management* when appropriate.

INFOGRAPHIC 2.7   **THE MONTREAL PROTOCOL AND ITS AMENDMENTS HAVE BEEN EFFECTIVE**   5

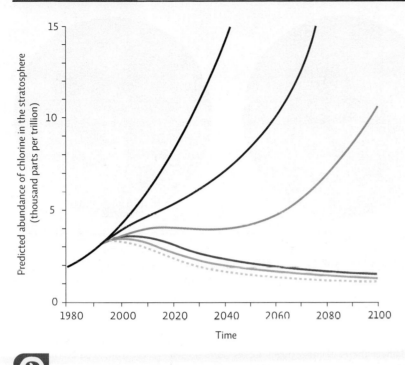

← Actual and projected change over time for total global emissions of ozone-depleting substances (ODS) with and without the Montreal Protocol and its amendments. Adjustments to the phase-out schedule of various ODSs in the form of amendments represent the success of adaptive management in dealing with complex environmental issues.

— No protocol
— Montreal Protocol, 1987
— London Amendment, 1990
— Copenhagen Amendment, 1992
— Beijing Amendment, 1999
---- Zero emissions

? What type of new information might have led policy makers to amend the Montreal Protocol?

To achieve the goals of the Montreal Protocol, the U.S. government required that CFCs be phased out starting in the early 1990s, through regulations from the U.S. Environmental Protection Agency, the federal agency responsible for seeing that environmental laws are followed.

Interestingly, the 1987 Montreal Protocol and the international commitment to address CFCs and ozone depletion came before Susan Solomon's definitive studies were published in 1988. This is an example of applying the **precautionary principle**—acting in the face of uncertainty when there is a chance that serious consequences might occur. By 1987, though we didn't know all the details, we knew enough to take action. As more information poured in, it quickly became apparent that the Montreal Protocol targets would not be sufficient to stop ozone depletion. Amendments to the protocol are still proposed and negotiated in annual meetings that strengthen the response and adjust the target dates to phase out harmful compounds. This is an ongoing process, and it is an example of **adaptive management**—allowing room for altering strategies as new information comes in or the situation itself changes.
INFOGRAPHIC 2.7

This same approach is needed to address other global environmental issues, such as climate change. We cannot know the exact details of future climate change impacts before they occur, but we do have a good idea of what we are facing. By acting now, we can lessen the severity of those impacts and adjust our response as we go along.

Fortunately, companies that produced CFCs, such as DuPont, were already researching alternative chemicals in the lab that could serve the same purposes with less damage. If companies could sell a CFC alternative just as easily, then cutting back on CFCs would not hurt them financially. Still, some of the replacements, particularly some hydrochlorofluorocarbons (HCFCs) used in refrigerants, would eventually also prove to be detrimental to ozone. In addition, illegal stockpiles and old refrigerators continue to release some CFCs, though that amount is decreasing rapidly, thanks to the Montreal Protocol. Pockets of CFCs remain because of continued legal "essential" uses, such as for medicines and nuclear power.

adaptive management A plan that allows room for altering strategies as new information becomes available or as the situation itself changes.

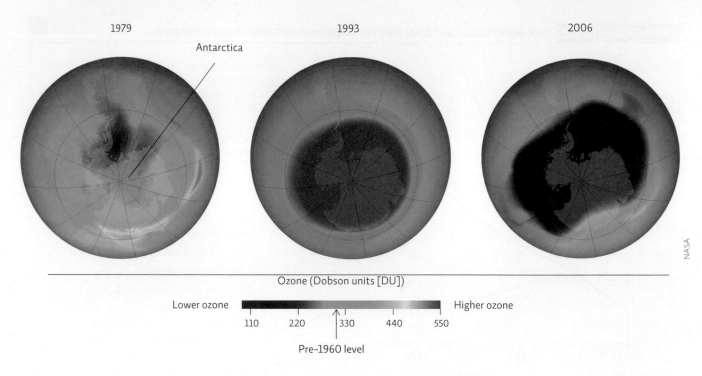

1979   1993   2006

Antarctica

Ozone (Dobson units [DU])

Lower ozone   Higher ozone

110   220   330   440   550

Pre–1960 level

↑ These images show the size of the ozone hole over Antarctica in 1979 (the first year ozone satellite images were available), 1993 (a "mid" year), and 2006 (the worst ozone hole ever recorded). Though the 1979 image might not be "normal," since CFC-driven depletion had already begun, ozone levels are considerably higher than in subsequent years. The 1993 image shows a much larger ozone hole than that in 1979, but the record ozone hole of 2006 is wider and deeper (indicated by the purple color—lowest ozone levels ever recorded).

Today, ozone-depleting compounds in the stratosphere are decreasing at a rate consistent with the policy changes of the Montreal Protocol and all its subsequent amendments. Outside polar regions, ozone levels declined through the 1990s but seem to be holding steady more recently. Still, the Montreal Protocol has been hailed as the most successful international environmental agreement in history.

There is still much we don't understand about the atmospheric chemistry of ozone depletion. Just when it seemed that ozone was beginning to recover, the 2004–2005 Arctic winter was unusually cold, and ozone loss was very high that year. This was eclipsed by a record 40% loss of ozone over the Arctic in spring of 2011. And Antarctica set two new records for ozone depletion in 2006—the widest hole ever observed and the deepest hole ever observed, with some areas near 90% depletion. This tells us that even though CFCs are declining—the 2012 Antarctic hole was smaller than it had been in 21 years,

thanks in part to a warmer-than-average Antarctic stratosphere—very cold winters will still give us years of higher-than-expected ozone depletion.

Projections estimate that midlatitude areas should be back to pre-1980 ozone levels by 2050, and polar regions should be back by 2075; however, the next 15 years should show periodic large declines and produce large "ozone holes" in very cold years like those seen in 2006 and 2011. Although she retired from NOAA in 2011, Solomon continues her research on CFCs and is both hopeful and realistic. "It's clear that ozone will ultimately recover, but it's also clear that it will take many decades to do so," she says. "It has been a real

By acting now, we can lessen the severity of those impacts and adjust our response as we go along.

privilege to work on such an interesting problem and to feel that the world found it useful in making choices about the Montreal Protocol."

Select References:

Abarca, J. F., & C. C. Casiccia. (2002). Skin cancer and ultraviolet-B radiation under the Antarctic ozone hole: Southern Chile. *Photodermatology, Photoimmunology & Photomedicine, 18*(6): 294−302.

Farman, J., et al. (1985). Large losses of total ozone in Antarctica reveal seasonal ClOx/NOx interaction. *Nature, 315*(6016): 207−210.

Menzies, S. W., et al. (1991). Ultraviolet radiation-induced murine tumors produced in the absence of ultraviolet radiation-induced systemic tumor immunosuppression. *Cancer Research, 51*(11): 2773−2779.

Molina, M. J., & F. S. Rowland. (1974). Stratospheric sink for chlorofluoromethanes: Chlorine atom-catalysed destruction of ozone. *Nature, 249*(5460): 810−812.

NASA. "Ozone Hole Watch," http://ozonewatch.gsfc.nasa.gov.

Solomon, S., et al. (1987). Visible spectroscopy at McMurdo station, **Antarctica: 2.** Observations of OClO, *Journal of Geophysical Research, 92*(D7): 8329−8338.

Solomon, S., et al. (1988). Observation of the nighttime abundance of OClO in the winter stratosphere above Thule, Greenland. *Science, 242*(4878): 550−555.

BRING IT HOME

PERSONAL CHOICES THAT HELP

©iStock.com/BartCo

The depletion of the ozone layer is a great example of how science documented a problem and its cause, and public action confronted the problem. All the environmental changes we face, from rising levels of greenhouse gases to loss of biodiversity, can be addressed using science. How scientific information is or is not put into action has far-reaching consequences, making science literacy a matter of importance for every individual.

Individual Steps
• Practice thinking like a scientist. Go outside for 10 minutes and observe the world around you. Make observations of what you see or hear. What predictions could you make from your observations? How could you test them?
• Stay informed. Read or watch a science-related article or show once a month.

Group Action
• Demonstrate the importance of scientific literacy to your friends and family. Develop three additional questions from the material in the chapter and discuss them over dinner.
• Support science education: Find out about public lectures and programs in your area and attend one with your friends.

Policy Change
• Attend a city council or county board meeting to see how policy issues are addressed in your area.
• Make knowledgeable voting decisions on ballot initiatives.
• Serve on local civic committees that address environmental issues in your community.

ENVIRONMENTAL LITERACY UNDERSTANDING THE ISSUE

1 How do scientists study the natural world? Why do we say science is a "process" and that conclusions are always open to further study?
INFOGRAPHICS 2.3 AND 2.4

1. Why is the statement "It is wrong to inflict pain on animals" not scientifically testable?

2. Let's say that you are going to test whether exposure to UV-B radiation causes cells to become cancerous. Why is it necessary to have a control group in this experimental study?
   a. You need to feel like you have some kind of control over your experiment.
   b. It is not actually necessary in this type of experiment.
   c. You need to make sure cells not exposed to UV-B radiation don't turn cancerous.
   d. You need to determine how likely it is that cells not exposed to UV-B radiation will turn cancerous.

3. Some people reject the explanation that the hole in the ozone layer is caused by CFCs on the basis that it is "just a theory." These people:
   a. are correct if they are using a different line of evidence to support their conclusions.
   b. are equating the term *scientific theory* with an untested hypothesis or a guess.
   c. do not understand the chemistry involved in ozone depletion.
   d. are requiring that the explanation gain the status of hypothesis before they will accept it as probably true.

2 What is stratospheric ozone, and why are scientists worried about its depletion?
INFOGRAPHIC 2.1

4. Tanning beds use light that is mostly UV-A radiation and only 1%–3% UV-B radiation. Does this mean they are safe to use?
   a. Yes. They deliver less UV-B radiation than you would get if you laid out in the Sun for a tan.
   b. No. Even though exposure to UV-A radiation is safe, tanning beds still deliver some dangerous UV-B radiation.
   c. Yes. Having a tan means you won't get a sunburn or other skin damage.
   d. No. UV-A radiation is also damaging, so this exposure is not without risk.

5. Ozone, a component of smog, is a dangerous substance that can damage sensitive tissue such as lungs. So why is it okay, and even necessary, to have ozone up in the stratosphere?

3 What lines of evidence suggest that CFCs are causing the depletion of stratospheric ozone?
INFOGRAPHICS 2.2 AND 2.5

6. What is the relationship between CFCs and ozone in the stratosphere?
   a. Ozone naturally breaks down in the stratosphere, but substances like CFCs regulate its re-formation so that less-harmful UV radiation reaches Earth's surface.
   b. Ozone from the stratosphere migrates down to the troposphere, where it reacts with chemicals like CFCs to produce more oxygen.
   c. Ozone in the stratosphere is broken down by chemicals like CFCs, but as CFCs themselves are broken apart by UV radiation, ozone depletion slows.
   d. Ozone is naturally formed and broken down in the stratosphere, but substances like CFCs catalyze additional ozone breakdown, which results in increased UV radiation reaching Earth's surface.

7. What is a *correlation* in science? How is it different from a *cause-and-effect relationship*? Explain how both types of relationships are used to understand the connection between CFCs and stratospheric ozone.

4 Why are both observational and experimental studies needed to investigate the natural world? How does statistics help researchers analyze the data they gather from these studies?
INFOGRAPHIC 2.6

8. Scientists want to know whether ocean water is becoming more acidic due to the extra CO_2 that is released by the burning of fossil fuels. They have set up monitoring sites in the water to collect daily data on the water's pH. This is an _____ study.

9. Other scientists decide to add extra CO_2 to tanks of ocean water that contain coral to see if the lower pH harms the animals by weakening their skeletons. This is an _____ study.

10. Identify at least three circumstances in which it would be more acceptable or reasonable to conduct an observational study than to conduct an experimental one.

5 How do approaches that use the precautionary principle and adaptive management help policy makers address environmental problems?
INFOGRAPHIC 2.7

11. **True or False:** *Adaptive management* focuses on finding a solution that seems best and sticking with it.

12. When is it reasonable to invoke the *precautionary principle* when setting policy?

SCIENCE LITERACY WORKING WITH DATA

The graph below comes from a study exploring the correlation between chlorine monoxide (CIO) and ozone. The data were collected by placing the measurement instrument on the wing of a plane and flying through the Antarctic ozone hole.

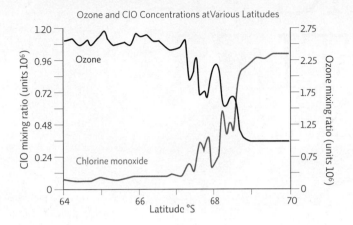

Interpretation

1. What does this graph show? Describe the pattern that you see.

2. Looking at the levels of ozone relative to the levels of chlorine monoxide, which side of the graph do you think most likely represents the inside of the ozone hole? On what basis do you think this?

3. Compare this graph to the graph in Infographic 2.2 (on ozone depletion and CFC levels). Although both graphs show a similar pattern, they are about two different aspects of the CFC–ozone story. Compare and contrast the information presented in the two graphs.

Advance Your Thinking

4. Do the data in this graph show a correlation or a cause-and-effect relationship? Do they support the hypothesis that chlorine is the cause of ozone depletion? Explain your responses.

5. How are these data similar to and/or different from the data collected by Susan Solomon's team? What do these data mean for the hypothesis that Solomon proposed?

6. How does this research study support the idea that science is a community enterprise?

INFORMATION LITERACY EVALUATING INFORMATION

You can use your understanding of the nature of science to evaluate ongoing environmental issues. For example, the Montreal Protocol's phase-out of CFCs was made possible by the availability of working alternatives. But do these alternatives come with unacceptable trade-offs?

The hydrocholorfluorocharbons (HCFCs) and hydrofluorocarbons (HFCs) that have largely replaced CFCs for industrial purposes don't damage stratospheric ozone, but it turns out they do have a negative impact on the environment. Should they now be phased out, too?

Search the library or Internet for information about the drawbacks of HCFCs and HFCs. (As a starting point, read the *ScienceDaily* article at www.sciencedaily.com/releases/2012/02/120224110737.htm.)

Consult at least two other sources of information (three sources total) and write an essay that addresses these questions:

1. Are HCFCs and HFCs good alternatives to CFCs with regard to stratospheric ozone depletion?

2. What environmental problems are associated with the use of HCFCs and HFCs?

3. What groups are advocating the ban of HCFCs and HFCs?

4. What is the current status of HCFCs and HFCs regarding a ban according to the Montreal Protocol?

5. What replacements are on tap for HCFCs and HFCs if they are banned?

6. What is your position on a possible ban of both of these chemicals? Support your answer. Cite your sources of information and provide a bibliography for your essay.

Find an additional case study online at www.macmillanhighered.com/launchpad/saes2e

TOXIC BOTTLES?

On the trail of chemicals in our everyday lives

CORE MESSAGE

Our understanding of the natural world changes almost daily as new evidence is gathered. Unfortunately, misinformation about science abounds in the popular press and on the Internet. For example, we live in an environment full of natural and synthetic chemicals, some of which are toxic, but knowing how to respond to the latest toxic scare is often difficult as we receive conflicting messages about the safety or risks of various chemicals. Developing information literacy skills enables us to better evaluate the usefulness and trustworthiness of various sources of information and then use the highest-quality information we can find to make reasoned decisions about how to respond.

AFTER READING THIS CHAPTER, YOU SHOULD BE ABLE TO ANSWER THE FOLLOWING **GUIDING QUESTIONS**

1

What are *toxic substances*, and how do we make decisions regarding the risks of exposure to these chemicals? Why is the solubility of a toxic substance important to individuals and to ecosystems?

2

What is information literacy, and why is it important?

Plastic bottles come in all shapes and sizes.
ULTRA.F/Digital Vision/Getty Images

3

What factors influence a chemical's toxicity? How is toxicity determined?

4

What are endocrine disruptors, and why is it often harder to determine a "safe dose" of them than it is to determine safe doses of other types of toxic substances?

5

How can we use critical thinking skills to logically evaluate the quality of information and its source? What common logical fallacies are used in presenting arguments?

In 2008, after a decade of study and contentious debate, a U.S. government-appointed panel of scientists known as the National Toxicology Program (NTP) finally arrived at a tentative consensus: Based upon data they had accumulated up to that point, they wrote in a report that would splash across headlines, they had "some concern for effects on the brain, behavior, and prostate gland in fetuses, infants, and children at current human exposures to bisphenol A."

Bisphenol A, or *BPA,* as it is more commonly known, is a synthetic chemical. Since the late 1940s, it has been a staple ingredient in the linings of metal food cans and plastic products of every kind, including food and beverage containers and plastic baby bottles. But in the two preceding decades, a mountain of scientific studies had implicated the unassuming compound in a rash of serious medical conditions, from impaired neurological and sexual development to cancer. At that time, both Canada and the European Union were considering banning the use of BPA in baby bottles and baby formula cans. Most of the NTP panel's scientists felt that the data were still too uncertain to warrant such a drastic step. But, they thought, it would be prudent for industries that used the chemical to start looking for a replacement. "The panel raised important research questions and public health concerns," says Sarah Vogel, a public health historian with the Johnson Family Foundation who has closely tracked the case of BPA. "For the first time ever, the U.S. government was suggesting that BPA might not be safe." The report incited a frenzy.

The plastic industry decried the report's conclusions. In a firestorm of press releases, in newspapers across the country and on cable news, industry spokespeople insisted that their own data showed BPA to be perfectly safe.

Parents everywhere were torn. On the one hand, it was hard to believe that something as commonplace as BPA could be so dangerous; if at least some studies were showing it to be safe, maybe it was. On the other hand, if there was even a chance that this chemical could harm their children, shouldn't the government take every possible precaution? BPA-free products were exceedingly difficult to find, and without some sort of federal mandate, that was unlikely to change. What were average consumers to do?

We live in an environment full of toxic substances.

Toxic substances (toxics) are chemicals that can harm living organisms. They fall into two broad categories: synthetic and natural. Natural toxics are not to be taken lightly: Arsenic, a basic element that sometimes leaches into groundwater, can cause cancer and nervous system damage in humans.

But synthetic toxic substances are a particular problem because there are quite a lot of them, and many are **persistent chemicals** (meaning they don't readily degrade over time). According to the U.S. **Environmental Protection Agency (EPA)**, more than 80,000 chemicals are used in the United States alone. And some 1,000–2,000 new chemicals enter the consumer market each year.

The debate over how to regulate these chemicals—how to determine what quantity of any particular compound is safe for humans or the environment and then how to ensure that exposure levels stay well below those

KEY CONCEPT 3.1

The environment is full of natural and synthetic toxic substances that have the potential to harm living things, and most of them have not been tested to determine safe exposures.

ASSOCIATED PRESS

↑ Rachel Carson, testifying before a Senate subcommittee in 1963 on the effects of pesticides.

↑ A group of men from Todd Shipyards Corporation run their first public test of an insecticidal fogging machine at Jones Beach State Park, New York, in July 1945. As part of the testing, a 6.5-kilometer (4 mile) area was blanketed with the DDT fog.

quantities—began in 1962, with a book called *Silent Spring*.

In this book, legendary environmental activist Rachel Carson asked her readers to imagine a world without the sounds of spring, a world in which the birds, frogs, and crickets had all been poisoned to death by toxic chemicals. Just 20 years had passed since the widespread introduction of herbicides and pesticides (like DDT), she explained, but in that relatively short time, they had thoroughly permeated our society.

These chemicals were obviously great for killing off weeds and pests; they had done an amazing job conquering mosquito-borne diseases like malaria during World War II and combating world hunger by boosting global food production. But no one seemed terribly concerned about the effects they might have on nontarget species or on their (or our) ecosystems. After all, they were designed to kill living things. Wasn't it at least possible that what killed one organism might also kill others?

Carson went on to identify three specific concerns that were being overlooked at the time: Some chemicals can have large effects at small doses, certain stages of human development are especially vulnerable to these effects, and mixtures of different chemicals can have unexpected impacts. The book created an uproar, which led to much stricter regulations for chemical pesticides in general,

and, in the United States, a complete ban on DDT in particular. But half a century later, we are still struggling to effectively regulate the chemicals in our world.

Regulation happens even in the face of change.

Regulation begins with **risk assessment**—a careful weighing of the risks and benefits associated with any given chemical. In an ideal world, unbiased, professional regulators would assess the safety of every new chemical before it entered our lives. They would discern all the potential consequences of excessive or continued long-term exposure and, in so doing, would protect us from any slow, unwitting poisoning. In reality, of course, a variety of factors—practicality, economic forces, sheer need—makes implementing such thorough precautions nearly impossible.

Federal agencies (namely the Food and Drug Administration [FDA] and EPA) are mandated to protect us from harmful chemicals; they have the

toxic substances/toxics Chemicals that cause damage to living organisms through immediate or long-term exposure.

persistent chemicals Chemicals that don't readily degrade over time.

Environmental Protection Agency (EPA) The federal agency responsible for setting policy and enforcing U.S. environmental laws.

risk assessment The process of weighing the risks and benefits of a particular action in order to decide how to proceed.

KEY CONCEPT 3.2

Risk assessment helps us determine the best way to deal with dangerous chemicals. If there are uncertainties, we can apply the precautionary principle and leave a wider margin of safety when setting exposure limits.

precautionary principle
A principle that encourages acting in a way that leaves a margin of safety when there is a potential for serious harm but uncertainty about the form or magnitude of that harm.

information literacy The ability to find and evaluate the quality of information.

primary sources Sources that present new and original data or information, including novel scientific experiments or observations and firsthand accounts of any given event.

peer review A process whereby researchers submit a report of their work to outside experts who evaluate the study's design and results to determine whether it is of a high enough quality to publish.

secondary sources Sources that present and interpret information solely from primary sources.

authority to heavily regulate or ban outright chemicals deemed to be dangerous. For substances where the data are uncertain and where the substances may cause unexpected or unpredictable effects, these agencies can employ a "better safe than sorry" strategy known as the **precautionary principle**. This rule of thumb calls for leaving a wide safety margin when setting the *exposure limit*—the maximum quantity humans can safely be exposed to. The width of that margin depends on the severity of the potential health effects and environmental damage.

The precautionary principle is becoming a favored tactic in the European Union. But in the United States, for the vast majority of chemicals, we take a different approach: "innocent until proven guilty." Rather than thoroughly testing each individual compound, regulators make educated guesses about safety, based on how other, similar compounds have fared. As a result, toxic products are often discovered only after (sometimes long after) reaching the marketplace—usually when some person or group of people suffer the effects. Rather than preventing these products from reaching store shelves in the first place, we recall them after the fact. This ad hoc regulation puts on the public the burden of proving that a chemical is actually more dangerous than expected.

As the case of BPA shows, even when concerns about safety emerge, deciding which precautions to take can seem like an impossible task. Part of the problem is that, even in our era of warp-speed communication, information is a slippery thing; this is especially true when it comes to science. We are constantly uncovering

In fact, much of what we learn in science class today will be outdated 5 years from now—not because we are wholly ignorant in the present but because we will know so much more in the future.

new information and gleaning new insights about the environment and our relationship to it. And as our understanding grows and changes, existing information often becomes obsolete. In fact, much of what we learn in science class today will be outdated 5 years from now—not because we are wholly ignorant in the present but because we will know so much more in the future.

In this rapidly moving current of knowledge floats a seemingly endless array of information sources: newspapers and magazines, websites, scientific journals, and so on. Not all of these sources are equal. While some are carefully vetted for accuracy, others are incomplete or deliberately misleading.

Information sources vary in their reliability.

The ability to distinguish between reliable and unreliable sources of information is referred to as **information literacy**. It's the key to drawing reasonable, evidence-based conclusions about any given issue or topic, and it is especially important in cases like that of BPA because when it comes to scientific issues, hyperbole and misinformation abound.

Primary sources are sources that present new and original data or information, including novel scientific experiments and firsthand accounts of any given observation. Scientific journals are primary sources; they contain original reports of scientific studies. Almost all of these reports, or papers, are rigorously evaluated through **peer review**—a process whereby experts in the field (a panel of the author's or authors' "peers") assess the quality of the study's design, data, and statistical analysis, as well as the soundness of the paper's conclusions. Good studies are published; bad ones are rejected. There has been a recent, rapid increase of *open-access online journals* (articles available without a paid subscription), but the quality of peer review varies greatly among these journals.

Secondary sources present and interpret information solely from primary sources. **Tertiary sources** present and interpret at least some information from secondary

Published information about toxics (and other scientific topics) abounds in our modern world. It is up to an individual, using information literacy skills, to determine the reliability of that information.

Because the popular (nonscientific) press runs on catchy sound bites and easily digestible bits of information, news outlets (both secondary and tertiary) tend to oversimplify the results of individual studies or present them as if they provided definitive answers. But there are rarely easy answers to environmental questions, and science is almost never as straightforward as we would like it to be. In fact, by its very nature, science is incremental; each study is just one small piece of a much larger puzzle, and existing hypotheses are subject to endless revision and qualification as new bits of data trickle in.

sources. Because they usually use secondary sources, even if they also use some primary sources of information, most books, including textbooks, and reports from the popular press are tertiary sources. Most blogs, websites, and even news shows also qualify as tertiary sources: They provide additional commentary on, or foster debate over, reports from the popular press. **INFOGRAPHIC 3.1**

What are the dangers presented by toxics, and how do we determine safe exposure levels?

In the wake of the National Toxicology Program's 2008 report, a team of scientists at the Centers for Disease Control and Prevention (CDC) analyzed more than 2,000 urine samples collected from a statistically representative cross section of the American population. More than 90% of those samples tested positive for BPA; the average concentration was 2.6 parts per billion (ppb), though the top 5% of samples had an average concentration of almost 16 ppb. That's well past the amount of BPA known to cause harm in rodents. Concentrations seemed to decrease with age, so that children had more BPA in their systems than adolescents, and adolescents had more than adults. "This study really laid to rest any doubts about whether or not BPA was in fact leaching from our food containers into our bodies," says Vogel. But did that necessarily mean that BPA was dangerous to humans?

Any given toxic substance has several characteristics we must consider when evaluating its safety. One such characteristic is its **persistence**—how long

tertiary sources Sources that present and interpret information from secondary sources.

persistence The ability of a substance to remain in its original form; often expressed as the length of time it takes a substance to break down in the environment.

INFOGRAPHIC 3.1   INFORMATION SOURCES

PRIMARY SOURCES

Primary sources are firsthand accounts of research and observations; they also include interviews or personal diaries. In science they are usually peer-reviewed articles in scholarly journals that provide methods and data but can include other kinds of reports and books based on primary research.

SECONDARY SOURCES

- Scholarly reviews and editorials
- Nonprofit agency publications
- Books
- Magazine/newspaper articles
- Websites (including some social media)
- Government and international agencies (EPA, FDA, World Health Organization, etc.)

TERTIARY SOURCES

Tertiary sources draw from, or summarize, a secondary source and can include any of the publications shown as secondary sources. Many blogs, websites, and news shows qualify as tertiary sources. They provide additional commentary on, or foster debate over, reports from the popular press. They may be accurate; however, they may introduce errors because they do not rely on the original source for facts. They also may perpetuate any errors that appeared in a secondary source.

 Look at the references cited at the end of this chapter. Does this chapter qualify as a primary, secondary, or tertiary resource?

KEY CONCEPT 3.4

How hazardous a substance is depends on its persistence, solubility, and toxicity to cells. Interactions with other chemicals may alter toxicity, making it difficult to determine safe exposure limits.

↑ University of Alberta scientists take water and vegetation samples near a fossil fuel mining site to test for mercury and other contaminants.

Peter Essick/Aurora Photos

it takes the substance to break down in the environment. Chemicals with low persistence tend to break down quickly, especially in the presence of sunlight. Chemicals with high persistence tend to linger for a long time and can affect ecosystems well after their initial release.

Another trait that must be considered is a toxic's **solubility**—its ability to dissolve in liquid, particularly water. In some cases, *water-soluble* chemicals are safe for humans but not so good for the environment. Because we can excrete them in our urine, they don't linger in our bodies for very long. (Of course, at high enough doses, these chemicals can still prove toxic, and at low but continual doses, they can cause kidney damage.) But because water-soluble chemicals are easily taken up by aquatic organisms, they can wreak slow havoc on aquatic environments and, by extension, on the ecosystems that surround them.

solubility The ability of a substance to dissolve in a liquid or gas.

bioaccumulation The buildup of substances in the tissue of an organism over the course of its lifetime.

Fat-soluble chemicals present an extra level of complexity. Because they pass easily through cell membranes, our cells can readily absorb these chemicals. Once they're inside, our bodies have a hard time expelling fat-soluble chemicals. In some cases, the liver can convert a fat-soluble molecule into a water-soluble one, so that it can be broken down and excreted in urine. But when our livers can't work this magic, fat-soluble chemicals are stored in our fatty tissue, where they can pile up in a process known as bioaccumulation.

Bioaccumulation refers to the buildup of fat-soluble substances in the tissue of an organism over the

---

INFOGRAPHIC 3.2   BIOACCUMULATION AND BIOMAGNIFICATION   1

Animals can acquire fat-soluble toxic substances through air, water, or food sources. The substances build up in the tissue of the animal over its lifetime if it has continued exposure; the lifetime accumulation is stored in fatty tissue.

BIOACCUMULATION

Mercury in the water or food

A little more mercury accumulates in the fish each day.

Even more mercury accumulates in the fish—maybe enough to make the fish sick.

TIME

 Why don't animals bioaccumulate or biomagnify water-soluble substances?

KEY CONCEPT 3.5

Potential toxics are evaluated using *in vitro, in vivo,* and epidemiological studies. Each type of study asks different kinds of questions, and contributes to a fuller understanding.

do those lower on the chain. The best examples in the human diet are tuna and swordfish. These fish are large predators and thus are high up on the ocean food chain. So when we eat them, we consume all the toxic substances—of special concern is mercury—that they have picked up from preying on smaller fish. **INFOGRAPHIC 3.2**

BPA has a low persistence, meaning that it breaks down rapidly in the environment. Although it is fat soluble, liver and gut cells can readily convert it to a water-soluble form, so that it's easily excreted in urine. This means it should not bioaccumulate or biomagnify.

So how then was it present in more than 90% of the 2,000 human urine samples analyzed by the CDC? BPA

course of its lifetime. **Biomagnification** describes a consequence of bioaccumulation; it's what happens when animals that are higher up on the food chain eat other animals that have bioaccumulated toxic substances: They consume their prey's entire lifetime dose of those toxics. Biomagnification means that animals higher on the food chain accumulate far more toxics than

is so commonplace and people are exposed to it so continuously that it remains ever present in our systems. Even as we are breaking down and excreting some bits of BPA, we are ingesting more. Scientists have spent the past decade trying to determine what such exposure might mean for human health.

Figuring out the cause-and-effect relationships between our bodies and the chemicals that enter them is tricky work. **Epidemiologists**—the

biomagnification The increased levels of substances in the tissue of predatory (higher levels of the food chain) animals that have consumed organisms that contain bioaccumulated toxic substances.

epidemiologist A scientist who studies the causes and patterns of disease in human populations.

↑ A caution sign warns people not to harvest or eat shellfish at a contaminated beach.

Top predators living in ecosystems contaminated with persistent, fat-soluble toxic substances such as mercury or DDT will have much higher levels in their flesh than organisms lower on the food chain. Fish advisories often reflect this danger.

BIOMAGNIFICATION

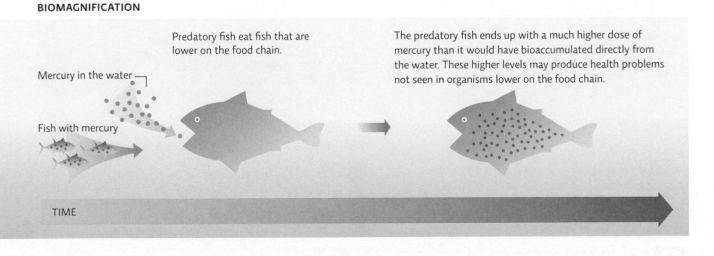

Predatory fish eat fish that are lower on the food chain.

The predatory fish ends up with a much higher dose of mercury than it would have bioaccumulated directly from the water. These higher levels may produce health problems not seen in organisms lower on the food chain.

Mercury in the water

Fish with mercury

TIME

Both *in vivo* and *in vitro* studies are used to experimentally study the effects of BPA. Note that none of the researchers discussed below concludes that they have definitive evidence that BPA is harmful, but each bit of research presents a piece of the puzzle. It is the accumulation of evidence that will be most helpful in drawing conclusions about the danger or safety of BPA in our food containers and other plastic goods.

TOXICOLOGICAL STUDIES

In vivo studies

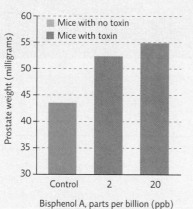

EXPERIMENTAL DESIGN

Researchers Nagel and vom Saal studied the effect of low-dose BPA on the prostate size of male mice pups whose mothers were fed one of two doses of BPA when pregnant, compared to males whose mothers were not fed BPA.

RESULT

At both doses, male mice had significantly larger (30%–35% larger) prostate glands (p < 0.01) but did not vary from controls in overall body size.

RESEARCHERS' CONCLUSIONS

Prenatal exposure to low doses of BPA alters prostate growth in male mice and is thus biologically active at doses seen in nature. The effect was greater than predicted by *in vitro* studies. Perhaps BPA acts synergistically with naturally occuring estrogens in the animal feed or is converted to a more active product in the body. Further studies on the processing of BPA in the body are needed.

 Why was a control group used in the Nagel and vom Saal prostate study?

In vitro studies

EXPERIMENTAL DESIGN

Researchers Ishido and Suzuki placed droplets of rat neural stem cells onto a dish treated with various doses of BPA. They then monitored the cells with a microscope. As the stem cells migrated out of the spherical droplets across the dish, the researchers measured, on average, how far the cells migrated in both the test and control groups.

RESULT

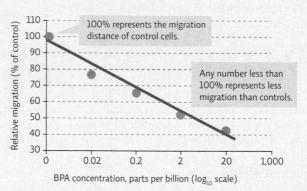

BPA had a negative effect on cell migration compared to control cells; as BPA concentration went up, cell migration (and survival) went down.

RESEARCHERS' CONCLUSIONS

BPA negatively affects migration ability of rat brain cells *in vitro*. Since successful migration is necessary for proper brain development, low doses of BPA may impact brain development and function in these animals. Further testing is needed to determine if the same effect is seen *in vivo*.

 Could the researchers have seen cells with y-axis migration values at 110% or greater? Explain.

Epidemiological studies look at human populations to see if any BPA effects emerge in particular groups, such as those with a diagnosis of disease. The Lang et al. study shown here compared the amount of BPA in the urine of subjects with and without various common health conditions to see if high BPA correlated with any of the illnesses.

EPIDEMIOLOGICAL STUDIES

EXPERIMENTAL DESIGN

The Lang et al. study evaluated 1,455 subjects' urine BPA levels. Subjects were interviewed about whether they had been diagnosed with various health conditions. The data were adjusted for age and sex and then statistically analyzed to see if any groups (those with or without a certain health condition) differed in their levels of BPA.

RESULTS

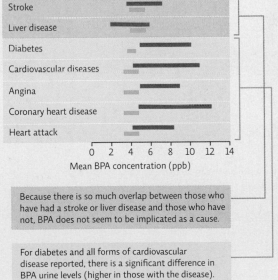

- ■ Subjects diagnosed with the condition
- ■ Subjects who have not been diagnosed with the condition

Stroke
Liver disease
Diabetes
Cardiovascular diseases
Angina
Coronary heart disease
Heart attack

0 2 4 6 8 10 12 14
Mean BPA concentration (ppb)

Because there is so much overlap between those who have had a stroke or liver disease and those who have not, BPA does not seem to be implicated as a cause.

For diabetes and all forms of cardiovascular disease reported, there is a significant difference in BPA urine levels (higher in those with the disease). Notice that the bars overlap very little, if at all.

RESEARCHERS' CONCLUSIONS

Concentrations of BPA were associated with an increased prevalence of coronary heart disease, cardiovascular disease, and diabetes. These findings add to the evidence that suggests there are adverse effects of low–dose BPA in animals. Independent replication and follow–up studies are needed to confirm these findings and to provide evidence about whether the associations are causal.

 Why is it unlikely that researchers will attempt to do an *in vivo* study in humans to verify the results obtained in the Lang et al. epidemiology study?

scientists charged with this type of research—can't just give a test group of humans a toxic substance to see what effects it has. They must do a bit of detective work. They can start by looking for health problems in specific populations and work their way backward to find the culprit. Or they can look at groups that have been exposed to a given toxic and see if any common health problems emerge or have already emerged. This latter approach is the one researchers took with BPA. Epidemiologists looked at the health profiles of hundreds of individuals who had BPA in their urine; in one study of 1,455 such people, they found a correlation between BPA concentrations and cardiovascular disease.

The task of determining exactly how BPA might go about wreaking havoc inside actual human bodies falls to toxicologists. **Toxicologists** concern themselves with determining the specific properties of potentially toxic substances and how they affect cells or tissues. They do this by testing lab animals through *in vivo* (*in vivo* means "in the body") studies, or by testing cells in Petri dishes in what scientists call *in vitro* studies (*in vitro* means "in glass"). Toxicologists use these data to determine how toxic a substance is and what effects it has on living organisms. **INFOGRAPHIC 3.3**

Toxicity can be affected by a host of factors. Individual susceptibility varies with genetics, age, and underlying health status. When a person is exposed to a toxic substance, the type and amount of chemicals already in the person's system are important. Some chemicals in the body might combine to increase overall toxicity (**additive effects**); other chemicals may reduce toxicity due to interactions between the toxins that "cancel each other out" or at least lessen the effect (**antagonistic effects**). Still other chemicals may work together to produce an even bigger effect than expected (**synergistic effects**). Route of exposure (for example, inhalation, injection, or skin contact) and the dose at the time of exposure also play a role. (Large doses can cause

toxicologist A scientist who studies the specific properties of potentially toxic substances.

***in vivo* study** Research that studies the effects of an experimental treatment in intact organisms.

***in vitro* study** Research that studies the effects of experimental treatment on cells in culture dishes rather than in intact organisms.

additive effects Exposure to two or more chemicals that has an effect equivalent to the sum of their individual effects.

antagonistic effects Exposure to two or more chemicals that has a lesser effect than the sum of their individual effects would predict.

synergistic effects Exposure to two or more chemicals that has a greater effect than the sum of their individual effects would predict.

immediate effects that differ from those caused by lower doses acquired continually over a longer time period.) **INFOGRAPHIC 3.4**

But in general, toxicologists like to say that "the dose makes the poison." This means that almost anything can be tolerated in low enough doses; conversely, anything—even water—can be toxic if the dose is big enough. And in most cases, as the dose increases, so does the severity of the effect. This idea—that higher doses of something harmful are worse for you than lower doses—makes obvious sense. It guides both regulatory efforts and modern medicine. And it applies to almost every chemical you can think of—except for the class of which BPA is a part: endocrine disruptors.

Endocrine disruptors cause big problems at small doses.

As their name suggests, **endocrine disruptors** interfere with the endocrine system, typically by mimicking a **hormone** or preventing a hormone from having an effect. Estrogen is a hormone that plays many roles in the body; its main task is to guide reproduction and development in both males and females. BPA is an estrogen mimic; it binds to the body's cellular estrogen

---

INFOGRAPHIC 3.4   FACTORS THAT AFFECT TOXICITY   3

↘ Some chemicals are more toxic than others due to their mode of action. Other factors also affect how toxic a particular chemical will be for an individual.

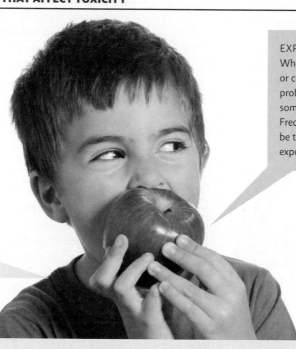

EXPOSURE
Whether a toxic substance is inhaled, ingested, or contacts the skin may affect how much of a problem it causes. Dose is also important, since some substances have a threshold of toxicity. Frequency also matters: A single exposure may be tolerable at a given dose, but repeated exposure may cause problems.

INDIVIDUAL FACTORS
Factors related to the individual may affect how toxic a chemical is. Some chemicals are more of a problem for the very young or very old, or for those who are ill. In some cases, genetic differences make a person more or less vulnerable to a given chemical.

CHEMICAL INTERACTIONS
We are never exposed to just one chemical. The fact that chemicals can interact in ways that increase or decrease their toxic effects complicates our efforts to determine a "safe dose". For instance, suppose there are two chemicals (A and B); each raises one's temperature 2° at a given dose. If we are exposed to both chemicals at the same time, one of three interactions is possible:

Temperature increase

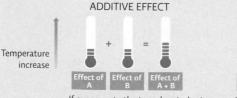

ADDITIVE EFFECT

Effect of A + Effect of B = Effect of A + B

If exposure to the two chemicals gives the effect we'd expect (2 + 2 = 4), then the effect is additive.

ANTAGONISTIC EFFECT

Effect of A + Effect of B = Effect of A + B

If exposure to both actually lessens the effect we would expect, then the interaction is antagonistic (2 + 2 < 4).

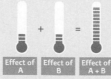

SYNERGISTIC EFFECT

Effect of A + Effect of B = Effect of A + B

If exposure to both produces an effect much greater than the sum of the two individual effects, the chemicals are working synergistically (2 + 2 > 4).

 Based on your own individual factors (age, health, etc.), do you predict you are more or less vulnerable to toxics than the average person?

INFOGRAPHIC 3.5 **HOW HORMONES WORK**

↓ Steroid hormones like estrogen (and its mimics) pass into cells, where they bind to DNA and actually "turn on" genes. These effects can be far-reaching since the activated genes may affect other genes and many cell processes, having many later effects.

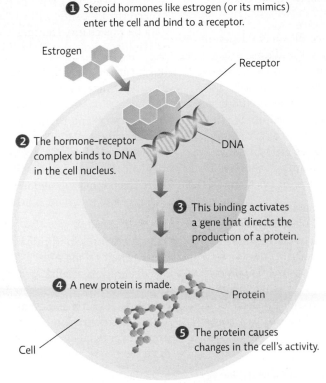

❶ Steroid hormones like estrogen (or its mimics) enter the cell and bind to a receptor.

Estrogen

Receptor

❷ The hormone–receptor complex binds to DNA in the cell nucleus.

DNA

❸ This binding activates a gene that directs the production of a protein.

❹ A new protein is made.

Protein

❺ The protein causes changes in the cell's activity.

Cell

Do you think that the change in cellular activity that results when estrogen or an estrogen mimic binds to its receptor is immediate, or does it take a while? Explain.

receptors and triggers the same effects that actual estrogen would trigger. In the wild, chemicals like this have been shown to cause feminization of males (even sex changes), as evidenced by lower sperm counts and the production of egg proteins normally produced only by females. **INFOGRAPHIC 3.5**

In humans, we know that sperm counts are down among men and that puberty onset is earlier in both boys and girls than it has ever been before. We also know that 95% of the U.S. population has trace amounts of BPA in their urine, and since there are no known natural sources of BPA, this suggests that it is leaching from our food and beverage containers into our bodies. No one can say for sure whether one fact (changes in sexual development) is related to the other (BPA in our systems), but because hormones control the development of body organs, scientists are especially concerned about the exposure of developing fetuses, newborns, and infants to endocrine disruptors.

Endocrine disruptors are curiously different from most other chemicals, where the relationship between dose and effect is linear (the more you ingest, the sicker you get—"the dose makes the poison"). Endocrine disruptors can have one set of effects at a very low dose and no effects (or much different effects) at higher doses.

KEY CONCEPT 3.6

The threat posed by endocrine disruptors is difficult to determine because they may have different effects at different times in life and because they can have greater effects at very small doses than at larger doses.

endocrine disruptor A substance that interferes with the endocrine system, typically by mimicking a hormone or preventing a hormone from having an effect.

hormone A chemical released by organisms that directs cellular activity and produces changes in how their bodies function.

receptor A structure on or inside a cell that binds to a particular molecule, thus allowing the molecule to affect the cell.

To track the effects of a dose of a chemical, toxicologists use data from *in vitro* and *in vivo* studies to create a **dose–response curve**, from which they can calculate an **LD50** (lethal dose 50%), the dose that would kill 50% of the population. The lower the LD50, the more toxic the substance. **INFOGRAPHIC 3.6**

Because endocrine disruptors like BPA have different effects at low and high doses, LD50s and dose–response curves are much trickier to calculate, and a "safe dose" is much tougher to determine. "We can't just find the threshold dose, where effects are first seen, and test higher doses from there to see what the impact will be," says Frederick vom Saal, a biologist at the University of Missouri who has spent two decades studying low-dose effects of BPA. "For endocrine disruptors, we have to test the effects of high doses and low doses separately."

Early claims of BPA safety did not take this detail into account. In the late 1980s, when EPA and FDA scientists were testing the chemical, they started with very high doses, given to rats, lowered the dose until they saw no adverse effects, and then stopped. As is the usual practice, a "safe dose" was established by dividing that "no effect" level by 1,000, to account for the possibility that humans might be more sensitive than lab animals or that some people, such as children and elderly or sick individuals, might be more sensitive than others.

This high-dose assessment is the basis for statements like the one Steven Hentges, spokesperson for the

American Plastics Council, made to the press when the 2008 NTP report came out. "At the rate of migration into food," he told hordes of reporters, "a consumer would have to ingest more than 1,300 pounds of food and beverage in contact with BPA every day for a lifetime to exceed the EPA safety limit." Like the scientists whose data he was citing, Hentges was not considering the possibility of separate low-dose effects.

But it turns out that, thanks to one unfortunate chapter in our chemical history, we already have a very good idea of what low doses of estrogen mimics can do to humans. Between 1938 and 1971, doctors used the estrogen mimic diethylstilbestrol (DES) to prevent premature labor in expectant mothers. Thirty years later, when a higher incidence of reproductive abnormalities started showing up in the population, epidemiologists traced it back to DES given to pregnant mothers.

The DES research gave scientists an important tool for future estrogen mimic studies. In the course of confirming the chemical's toxicity, toxicologists discovered a particular strain of mouse that, when given DES—or any other estrogen mimic, for that matter—responded exactly as humans did. "It's a rare human–animal corroboration," says vom Saal. "It gives us an unprecedented degree of certainty about how well these mice studies relate to humans."

Back in 1997, when vom Saal tested low doses of BPA on these rodents, he found a litany of effects: increased

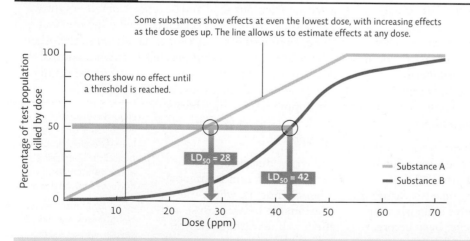

Some substances show effects at even the lowest dose, with increasing effects as the dose goes up. The line allows us to estimate effects at any dose.

Others show no effect until a threshold is reached.

LD50 = 28

LD50 = 42

— Substance A
— Substance B

← Dose-response studies evaluate the effect of a toxic substance at various doses (on animals or cells). Knowing the LD50 allows us to compare toxics directly— that is, which are more toxic than others. Charting the actual change in effect as the dose increases (the "dose response") allows us to predict effects at doses other than those tested. We can also determine if there is a threshold of exposure that must be reached before harmful effects are seen.

"SAFE DOSE"
Because there is uncertainty in the determination of "safe dose," regulatory agencies factor in a safety margin of 100 or even 1000 to be on the safe side. Example: If we have evidence indicating that a chemical is safe at 100 ppm, we might set the environmental standard (what is allowed) at 1 ppm (or even 0.1 ppm if children are at risk).

 Consider substance C that has an LD50 of 50 ppm. Is substance C more or less toxic than substances A and B?

Karen Sherlock/Milwaukee Journal Sentinel/MCT/Newscom

↑ Frederick vom Saal conducts a chemical audit for a homeowner, explaining some of the potential chemical hazards in her home.

postnatal growth in both males and females, early onset of sexual maturation in females, decreased testosterone and increased prostate size in males, altered immune function and increased mortality of embryos. Over the years, as other scientists replicated these data, more and more of them began to worry that, at low doses—such as the quantities that leach from plastic containers into food and drinks—BPA might be particularly dangerous for pregnant mothers, developing fetuses, and young children.

An equal number of researchers thought that developing fetuses were safe from any potential effects. They reasoned that BPA would be broken down in the mother's gut and excreted in her urine long before it had a chance to reach the womb. Even if the gut failed, they thought, BPA would never get past the placenta.

According to one 2002 study of 37 pregnant women, they were mistaken. In that study, researchers found BPA not only in maternal blood but also in fetal blood and the placenta. Although no one could say for certain whether the quantities that reached the fetus were enough to cause harm, this study and others like it were enough to convince some scientists that BPA ought to be more tightly regulated.

But when the 2008 report came out saying as much, a firestorm erupted.

Critical thinking gives us the tools to uncover logical fallacies in arguments or claims.

A wide range of media outlets jumped on the NTP report, producing scores of articles that read a lot like this one:

Tests performed on liquid baby formula found that they all contained bisphenol A (BPA). This leaching, hormone-mimicking chemical is used by all major baby formula manufacturers in the linings of the metal cans in which baby formula is sold. BPA has been found to cause hyperactivity, reproductive abnormalities and pediatric brain cancer in lab animals. Increasingly, scientists suspect that BPA might be linked to several medical problems in humans, including breast and testicular cancer.

While the article is factually correct, it commits several errors of omission. By failing to

dose–response curve A graph of the effects of a substance at different concentrations or levels of exposures.

LD$_{50}$ (lethal dose 50%) The dose of a substance that would kill 50% of the test population.

explain the high degree of uncertainty (similar studies reached different conclusions, and scientists had yet to reach any sort of consensus about what the risks might be), the author creates the impression that the risks are far more certain, and dire, than they actually are. Hundreds of articles in dozens of publications followed a similar tack, and before long, the story was shortened, in the public's mind, to one simple statement: BPA causes cancer.

Almost immediately, environmental and consumer groups began calling for a nationwide ban on the chemical. These groups frequently attacked the plastics industry, suggesting that because large, powerful companies profited handsomely from BPA-made

products, they had an incentive to downplay or even suppress troubling data. Such *red herring* and *ad hominem* attacks succeeded in stirring up the public's fears but told them nothing about whether BPA was actually safe. **TABLE 3.1**

The plastics industry countered by repeatedly reminding the public that neither the EPA nor the FDA had moved to ban BPA from use and that the recent NTP report cited only "some concern" about safety. This was also true. But it ignored the fact that a small army of National Institutes of Health–funded scientists had long expressed concerns about the purported dangers of low-dose BPA exposure and were now urging regulatory agencies to at least set a new, lower exposure limit.

TABLE 3.1

↓ Logical fallacies are arguments used to confuse or sway a reader or listener to accept a claim/position in the absence of evidence.

COMMON LOGICAL FALLACIES	EXAMPLES
Hasty generalization: Drawing a broad conclusion on too little evidence.	A study might show that BPA is present in the urine of babies who drink from plastic bottles; however, this is not evidence that babies are being poisoned by their bottles. All we could conclude would be that babies who drink from plastic bottles have more BPA in their urine than babies who do not.
Red herring: Presents extraneous information that does not directly support the claim but that might confuse the reader/listener.	An argument that the buildup of toxic chemicals in modern humans is significant and that many of these buildups have led to problems may be true but tells us nothing about the safety of BPA.
Ad hominem attack: Attacks the person/group presenting the opposite view rather than addressing the evidence.	Opposing the use of BPA on the grounds that the chemical industry is untrustworthy (because it profits from plastics) does not address the safety of BPA.
Appeal to authority: Does not present evidence directly but instead makes the case that "experts" agree with the position/claim.	Claiming that BPA is a health hazard (or not) because a noted toxicologist or scientific group has drawn that conclusion is not evidence in itself. If the experts agree, there should be a reason, and that evidence should be presented.
Appeal to ignorance: A statement or an implication that the issue is too complex, and the reader/listener is not capable of understanding it.	In order to justify doing nothing about the use of BPA, one might claim that there is no way to know its effect since humans are exposed to so many toxic substances. We may not know everything about the effects of BPA, but we do know some things, and it is on that evidence that we should draw our conclusions.
False dichotomy: The argument sets up an either/or choice that is not valid. Issues in environmental science are rarely black and white, so easy answers (it is "this" or "that") are rarely accurate.	The claim that BPA must either be *completely avoided* or is *totally safe for everyone* is a false dichotomy. The evidence so far suggests that the effects of BPA may be negligible for healthy adults but problematic for fetuses and very young children. Recommendations differ for different groups.

 Which logical fallacy do you feel is the hardest to recognize and might slip by the average person? Why?

Critical thinking allows us to logically evaluate the evidence behind a claim or conclusion to determine whether it is indeed supported by the evidence.

Critical thinking is the antidote to **logical fallacies**; it enables individuals to logically assess and reflect on information and reach their own conclusions about it. The skill set can be broken down into a handful of tenets, or measures.

Be skeptical. Just like a good scientist, a nonscientist should not accept claims without evidence, even from an expert. This doesn't mean refusing to believe anything; it simply means requiring evidence before accepting a claim as reasonable. For example, a *Science News* article, "Receipts a Large and Little-Known Source of BPA," quoted a researcher as saying that he believed BPA from paper cash register receipts "penetrated deeply into the skin, perhaps as far as the bloodstream." The actual research documented the amount of BPA in the receipts and recovered from the surface of fingers that held a receipt, not the amount in the bloodstream of individuals who handled the receipts. A skeptical reader would therefore question the inference that BPA from receipts enters the body since no evidence was presented regarding its uptake, only its presence on the skin's surface. Perhaps BPA does enter the bloodstream through receipts, but we need evidence in order to view this claim as reasonable.

Evaluate the evidence. Is the claim being made derived from anecdotal evidence (unscientific observations, usually relayed as secondhand stories) or from actual scientific studies? If it is based on actual studies, how relevant are those studies to the claim? Were they done on primates? Rodents? Cells in a Petri dish? Or did the researchers look at human populations? Most of the BPA studies were done on rodents. As we'll see, some of those rodents were very good models for human health effects, and some were not.

Be open-minded. Try to identify your own biases or preconceived notions (most chemicals are dangerous, most people overreact to things like this, etc.) and be willing to follow the evidence wherever it takes you.

Average consumers were left to figure out for themselves which side to trust: Were environmental activists exaggerating, or was BPA truly dangerous? Why did some scientific studies show deleterious effects from the chemical, while others did not? More importantly, what precautions could, or should, individuals take?

Watch out for author biases. Is the author of the study or person making the claim trying to promote a position? Is that person financially tied to one conclusion or another? Is he or she trying to use evidence to support a predetermined conclusion?

In defense of BPA, the plastics industry frequently cites a 2004 report by the Harvard Center for Risk Analysis, which had concluded that the evidence for adverse effects from low-dose BPA was weak. Vom Saal criticized this report because it looked at only 19 of the 47 published studies available at the time. The report was funded by the American Plastics Council, and vom Saal charged that the authors chose to examine only studies that gave the desired "no effect" results. When he himself evaluated those 47 published studies, and an additional 68 new studies, he found that results (effects versus no effects) were strongly correlated with the source of funding. None of the 11 industry-funded studies found adverse effects from BPA, whereas 90% (94 out of 104) of the government-funded studies showed significant effects of one kind or another.

The report's authors argued that they excluded studies where rats had been injected with BPA rather than fed food pellets laden with the chemical. They reasoned that because ingestion was the most common route of exposure for humans, it made more sense to feed the rodents than to inject them. They felt that effects seen in the injection studies were not relevant to humans because in humans, BPA would not enter directly into the bloodstream. Vom Saal and others disagreed. "Fetuses and newborns—the ones we are most concerned about—haven't yet acquired the gut microbes that adults have," he said. "So we can't expect their digestive systems to break BPA down as efficiently as an adult digestive system might. That's as true for rodents as it is for humans."

As it turned out, there was another technical reason that some of the industry studies did not find low-dose effects. It had to do with the type of rodent those researchers were using. Rather than use the strain of mice that had been so instrumental in cracking the DES case, industry scientists used a strain of rat known to be highly resistant to estrogen. This was akin to stacking the deck in favor of BPA: If the rats weren't sensitive to estrogen, there was virtually no chance that they'd be affected by an estrogen mimic like BPA. The panel that composed the 2008 report chose not to include these studies in its assessment.

critical thinking Skills that enable individuals to logically assess information, reflect on that information, and reach their own conclusions.

logical fallacies Arguments that attempt to sway the reader without using reasonable evidence.

Risk assessments help determine safe exposure levels.

By the time the NTP panel drew its conclusions, BPA was already a $3 billion industry unto itself. BPA was used in an infinite variety of products, from baby bottles and food and beverage cans to dental sealants and football helmets. And, in some cases at least, it had no obvious substitute. That made employing the precautionary principle or resetting the exposure limits almost impossible. "Makers of plastic bottles had other things they could substitute BPA with," says Aaron L. Brody, a food scientist at the University of Georgia. "But what about cans?" Food manufacturers have long used BPA to create epoxy linings for steel cans. This coating makes the cans extremely durable; it protects the contents of canned goods without affecting taste, and, most importantly, the coatings are integral to reducing the incidence of bacterial contamination and foodborne illness. "BPA-based plastic is among the few materials so far that can withstand the high temperatures and pressures used to kill bacteria," says Brody. "It's not a perfect plastic, but without it, you'd probably have a lot more food preservation and safety issues."

As scientists debated and regulatory agencies struggled, people were left to decide for themselves whether the risk posed by continued exposure to BPA warranted the hassle of trying to purge it from their lives or at least from their baby bottles. "Sometimes, if there's enough momentum from the public, regulatory change will follow," says Vogel. "But more often than not, because of the way the system works, it really has to start with individual consumers deciding for themselves that they won't purchase or use certain items."

To make such decisions, we need to know two things: whether a given chemical has the potential to harm us and how great that harm might be. It's crucial to consider both. Say, for example, a new study reports that BPA exposure doubles one's chances of developing a particular disease. Before hitting the panic button, we must first ask what the chances are of getting the disease in the first place, without BPA exposure. If it's a rare condition, with a 1% probability, then BPA increases our chances to a mere 2%. If it is more common, with, say, a 20% probability, then BPA doubles our chances to 40%—a significantly greater risk. How important that increase is, and thus what we should do in light of it, depends on the seriousness of the disease.

So, what to do about BPA? Let's take a look at what we've learned so far: We know that chemicals like BPA can have effects at low doses. We know that the fetuses and young offspring of many mammal species, including humans, are particularly susceptible to these low-dose effects. We know that BPA is, in fact, leaching into our systems from food and beverage containers and that it can indeed cross the human placental barrier.

But we also know that BPA is water soluble and can be excreted in our urine. This means that, in most cases, it may not be building up, or bioaccumulating, in our tissues.

What we don't know—and what we may never know—is how well the effects seen in mice correspond to the risks faced by humans. And because of all the additive and synergistic effects BPA is likely to have with all the countless other chemicals we encounter, it will be difficult to tell, even in the future, how much of any given health condition can be specifically attributed to BPA. The bane of risk assessment is that we can't really wait for those facts to come in.

Over the past few years, the public has employed a de facto precautionary principle. With the scientific jury still out, consumer and environmental groups took matters into their own hands. They issued dozens of public statements, advising consumers to avoid using food and beverage containers made with BPA. They also lobbied Congress for a full, nationwide ban. The public response to those campaigns led Walmart to say that it would no longer stock its shelves with any bottles or containers made with BPA. Facing hundreds of millions of dollars in lost sales, Nalgene, a leading plastic bottle maker, finally conceded, promising that it would phase out BPA over the next 5 years. Other companies quickly followed suit.

By 2011, both Canada and the European Union banned BPA; the United States followed in late 2012, with an official ban on BPA in baby bottles and "sippy" cups. Even before the U.S. ban, more and more companies had stopped using the chemical. In fact, many plastic bottles now tout their BPA-free status. What replaced it? A very similar chemical known as BPS (bisphenol S).

A century hence, such drastic measures may prove unwarranted. Even vom Saal concedes that it is possible that the long-term studies on BPA will not yield conclusions nearly as horrendous as those on, say, DES. It's possible that BPA will prove to be relatively harmless. But in the meantime, he insists, we're much better safe than sorry.

KEY CONCEPT 3.8

Risk assessment requires an equal evaluation of the potential for harm of a given action and of any alternatives. This allows us to avoid the mistake of replacing a potentially harmful action with one that is more harmful.

Select References:
Bohannon, J. (2013). Who's afraid of peer review? *Science* 4(6154): 66—65.
Ishido, M., & J. Suzuki. (2010). Quantitative analyses of inhibitory effects of bisphenol A on neural stem-cell migration using a neurosphere assay in vitro. *Journal of Health Science*, 56(2): 175—181.
Lang, I. A., et al. (2008). Association of urinary bisphenol A concentration with medical disorders and laboratory abnormalities in adults. *Journal of the American Medical Association*, 300(11): 1303—1310.

Nagel, S. C., et al. (1997). Relative binding affinity-serum modified access (RBA-SMA) assay predicts the relative *in vivo* bioactivity of the xenoestrogens bisphenol A and octylphenol. *Environmental Health Perspectives*, 105(1): 70—76.
Schönfelder, G., et al. (2002). Parent bisphenol A accumulation in the human maternal—fetal—placental unit. *Environmental Health Perspectives*, 110: 703—707.
vom Saal, F. S., & C. Hughes. (2005). An extensive new literature concerning low-dose effects of bisphenol a shows the need for a new risk assessment. *Environmental Health Perspectives*, 113(8): 926—933.

BRING IT HOME

PERSONAL CHOICES THAT HELP

Margaret Edwards/iStock/360/Thinkstock

What do some air fresheners, nail polishes, and plastic storage containers have in common? They are all potential sources of chemicals that people are exposed to every day. Many of the products designed to improve our lives actually contain chemicals that may harm us in the long run. With just a few changes, you can dramatically reduce your long-term chemical exposure.

Individual Steps
• Check to see if the body products you use contain potentially harmful chemicals at www.ewg.org/skindeep.
• Avoid microwaving food in plastic containers; chemicals such as BPA can leach into your food when heated. A safer bet is to use microwave-safe glass or ceramic containers when microwaving food.
• Check with your city's solid waste agency for guidelines about how to correctly dispose of household wastes, including paint, medication, cleaning products, and yard chemicals.

Group Action
• Talk to your friends and family members, especially those you live with, to see if you can switch to products with fewer harmful chemicals.

Policy Change
• Research the policy points of the Safe Chemical Acts of 2013, the proposed overhaul of the 35-year-old Toxic Substances Control Act. Chronicle its progress through Congress and contact your legislator to voice an opinion about the policy.

ENVIRONMENTAL LITERACY UNDERSTANDING THE ISSUE

1 What are *toxic substances* and how do we make decisions regarding the risks of exposure to these chemicals? Why is the solubility of a toxic substance important to individuals and to ecosystems?

INFOGRAPHIC 3.2

1. Which of the following is true of all toxic substances?
   a. They don't readily break down through natural means.
   b. They are synthetic chemicals; no natural substances are considered toxic.
   c. They cause damage through immediate or long-term exposure.
   d. They biomagnify up the food chain.

2. At low doses that don't hurt test mice, a new pesticide can effectively control an insect pest that's been ravaging local trees. However, when that pesticide is combined with another commonly used pesticide, it is very toxic to mice. In deciding to approve the pesticide for sale anyway, policy makers were applying:
   a. the precautionary principle.
   b. logical fallacies.
   c. information literacy.
   d. a risk assessment.

3. In what way does the story of BPA reflect the precautionary principle?

2 What is information literacy, and why is it important?

INFOGRAPHIC 3.1

4. Which of the following is a secondary source of information?
   a. A blog in which the author enters personal observations on his day-to-day life
   b. A magazine article that uses primary and secondary sources
   c. A scholarly review article that references only peer-reviewed original research articles
   d. A government website that cites original research and scholarly review articles

5. Why are tertiary information sources considered less reliable than primary or secondary sources?

6. What is peer review, and how does it increase the reliability of an information source?

3 What factors influence a chemical's toxicity? How is toxicity determined?

INFOGRAPHICS 3.3, 3.4, AND 3.6

7. The toxicity of a chemical is evaluated using cells in culture or live animal subjects by creating a(n) _____; a high LD50 indicates _____.
   a. *in vitro* study; a safe dose
   b. dose analysis; a threshold dose
   c. *in vivo* study; high toxicity
   d. dose–response curve; low toxicity

8. Pesticides such as DDT are known to biomagnify in food chains. This means that:
   a. organisms in lower food chain levels accumulate lethal doses of toxic substances.
   b. organisms at higher food chain levels have more concentrated levels of toxic substances.
   c. toxic substances build up in the tissue of an organism over the course of its lifetime.
   d. the environment has higher concentrations of toxic substances than organisms in the food chain.

9. Why is it so hard to determine a *safe exposure* for a given chemical? How do regulators setting safe exposure standards deal with the uncertainty associated with these factors?

4 What are endocrine disruptors, and why is it often harder to determine a "safe dose" of them than it is to determine safe doses of other types of toxic substances?

INFOGRAPHIC 3.5

10. Why are toxicologists more concerned about exposure to BPA in children and infants than in adults?
    a. BPA does not bind to receptors in adult cells.
    b. BPA might interfere with normal development in the young.
    c. Adults can break down and excrete BPA, whereas children cannot.
    d. BPA is only found in products that children are exposed to, such as formula cans and baby bottles.

11. Why must a *hormone* and a *hormone receptor* be present for a hormone to have an effect?

5 How can we use critical thinking skills to logically evaluate the quality of information and its source? What common logical fallacies are used in presenting arguments?

TABLE 3.1

12. Pesticides help increase agricultural yields, but some people oppose their use because of their inherent toxicity. An argument against pesticide use that attacks the pesticide maker on the grounds that he or she is simply profit driven is:
    a. an ad hominem attack.
    b. an appeal to authority.
    c. an appeal to ignorance.
    d. a false dichotomy.

13. Identify and explain four tenets in critical thinking that can help you logically access and reflect on information in order to draw your own conclusions.

SCIENCE LITERACY **WORKING WITH DATA**

The graphs below come from a 2005 study by the Toxic-Free Legacy Coalition and the Washington Toxics Coalition to identify chemical residues in the human body (http://pollutioninpeople.org/results). Samples of hair, blood, and urine of 10 Washingtonians were tested for a variety of toxic substances, including mercury and DDT. Data for each individual (identified by his or her initials) are shown for p,p'-DDE (a breakdown product of DDT) and for mercury.

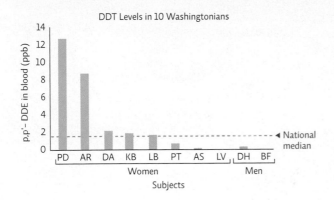

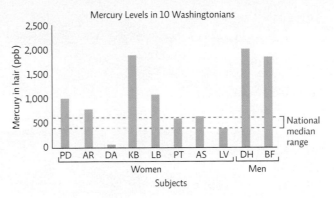

Interpretation

1. Identify the people with the highest and lowest levels of mercury and DDT.

2. Compare the levels of mercury and DDT in each study participant to the national median. Which study participants have levels of DDT and mercury that are at least twice the national median?

3. The EPA "safe dose" for mercury is 1,100 ppb (parts per billion) for women in their childbearing years, as mercury levels above this value may impair neurological development in the fetus. Which study participants are above the safe level, and by how much?

Advance Your Thinking

4. Who conducted this study, and for what purpose? What type of study do these data represent? Are the data reliable? Explain your responses.

5. The pesticide DDT has been banned in the United States since 1972. How do you explain the presence of DDT in 8 of the 10 study participants?

INFORMATION LITERACY **EVALUATING INFORMATION**

Some sources estimate that more than 12,000 chemical ingredients are used in cosmetics, of which fewer than 20% have been reviewed for safety. There are no laws in the United States to regulate chemicals in cosmetic products; it is up to us individually to decide on the safety of a particular product. The Environmental Working Group's Skin Deep cosmetics database (www.ewg.org/skindeep) has ratings for more than 68,000 products from approximately 3,000 brands. Each product is rated on a scale of 0 to 10 for the level of hazard (low, moderate, high) and data availability for the ingredients.

Evaluate the website and work with the information to answer the following questions:

1. Is this a reliable information source? Does it have a clear and transparent agenda?
   a. Who runs this website? Do the credentials of the organization running the site make the information presented reliable or unreliable? Explain.
   b. What is the mission of this website? What are its underlying values? How do you know this?
   c. What data sources does EWG rely on, and what methodology does the organization employ in constructing its databases? Are the sources EWG uses reliable?

   d. Identify a claim made on this website. Is there sufficient evidence provided in support of this claim? Are there any logical fallacies used? Explain.
   e. Do you agree with EWG's assessment of the problems with and concerns about cosmetics and other personal care products? What about its solutions? Explain.

2. Select one of your favorite personal care products that is listed on the EWG website:
   a. Check out the company website for that product. What sort of information about the product and its ingredients does the company website offer? Does the company employ any logical fallacies in talking about the product or its ingredients? Is the information provided useful to you as a consumer? Explain your responses.
   b. How do the two websites (Skin Deep and the company website) compare in providing useful and reliable information about your particular product? Explain.
   c. Search the Skin Deep website for alternative brands that have a safer rating than your personal care product. Are there other choices? Would you consider switching your brand? Explain.

Find an additional case study online at http://www.macmillanhighered.com/launchpad/saes2e

ONE-CHILD CHINA GROWS UP

A country faces the outcomes of radical population control

CORE MESSAGE

A variety of factors influence whether and how fast a population grows. Many human populations have grown explosively in the recent past. The human population is still growing, especially in developing nations, which can lead to overpopulation in those areas. We can pursue a variety of approaches to reduce population growth and stabilize population size; many of these approaches focus on issues of social justice. While population size has increased our impact on the environment, that rising impact is also being caused by an increasing per capita use of resources and generation of waste.

AFTER READING THIS CHAPTER, YOU SHOULD BE ABLE TO ANSWER THE FOLLOWING **GUIDING QUESTIONS**

1

How and why have human population size and growth rate changed over time? How many people live on Earth today?

2

What cultural and demographic factors influence population growth in a given country? How do they differ between more and less developed countries?

Crowded streets of China.
Christian Kober/Robert Harding/
Newscom

3

What is the demographic transition, and why is it important? What is the current trend for global population growth?

4

What are some strategies for achieving zero population growth?

5

What determines Earth's carrying capacity for humans? Is Earth's carrying capacity enough to support the current or future (projected) human population?

It's not hard to imagine a life without siblings; plenty of families have just one child. But what about a life without cousins or aunts and uncles? Imagine not just one family with a single child but an entire country of single children, and you'll begin to get a sense of what China is like for the current generation of young adults.

Their elders call them the "Little Emperors"—a title that is meant to reflect the spoiled life and haughty temperament we sometimes associate with only children. These young adults are the result of a colossal social experiment the Chinese government set in motion some 30 years ago, in an effort to curb population growth: one child per family.

From its very inception, the policy has been fraught with controversy. Critics say that enforcing it—often through legally mandated abortions—has amounted to gross human rights violations. Proponents say such measures have been essential to stave off an even greater catastrophe of human suffering that would have resulted from unchecked population growth. In recent years, as the policy has been revised, its success remains a subject of intense debate.

population growth rate The change in population size over time that takes into account the number of births and deaths as well as immigration and emigration numbers.

life expectancy The number of years an individual is expected to live.

crude death rate The number of deaths per 1,000 individuals per year.

crude birth rate The number of offspring born per 1,000 individuals per year.

Human populations grew slowly at first and then at a much faster rate in recent years.

At least twice in the course of human history, global human population has hit a dramatic growth spurt. The first came around 10,000 years ago, when the Agricultural Revolution dramatically increased the amount of food we could obtain and, thus, the number of mouths we could feed. As more food translated into healthier people, lower death rates, and longer life spans, the **population growth rate** increased, and global population raced toward the 100 million mark. The next, much bigger, growth spurt

INFOGRAPHIC 4.1   HUMAN POPULATION THROUGH HISTORY   1

↓ For most of human history, there have been fewer than 1 billion people on the planet.

Agricultural Revolution
10,000–8000 BCE

| OLD STONE AGE | NEW STONE AGE | BRONZE AGE |

| 9000 | 8000 | 3000 | 2000 |

began in the 1700s, when the Industrial Revolution ushered in a rapid succession of advances in agriculture as well as sanitation and health care—including vaccines, cleaner water, and better nutrition. Once again, death rates fell, life expectancy rose, and the population swelled.

KEY CONCEPT 4.1

The human population grew very slowly for most of human history before the Industrial Revolution. Since then, a growth spurt has sent our population soaring past 7 billion.

INFOGRAPHIC 4.1

In China, this cycle played out most dramatically during the latter half of the 20th century, when the country experienced a spectacular improvement in its standard of living. Between the 1950s and 1970s, **life expectancy** increased from 45 to 60. But as the **crude death rate** fell, the **crude birth rate** held steady. By 1949, China was home to 500 million people—triple the U.S. population at that time—making it the most populous country on Earth. By 1970, the population of China had swelled to almost 900 million: roughly one-quarter of the world's people living on just 7% of the planet's arable land.

Today, China is the most populous nation, with more than 1.3 billion people, but it will soon be surpassed by India, which is projected to have 1.5 billion by 2030. Like

What kinds of factors might result in a population of 16 billion by 2100 (the high estimate), and what might prevent our population from growing this large? For the medium and low estimates, what circumstances might lead to populations of those sizes?

→ It took 130 years to go from 1 billion to 2 billion people. Since 1974, it has taken about 12 years to add each additional billion. In 2011 we passed the 7 billion mark.

→ Population size in 2050 and beyond will largely depend on future growth rates and may stabilize around 9 billion or, if we maintain our current growth rate, will exceed 16 billion by 2100.

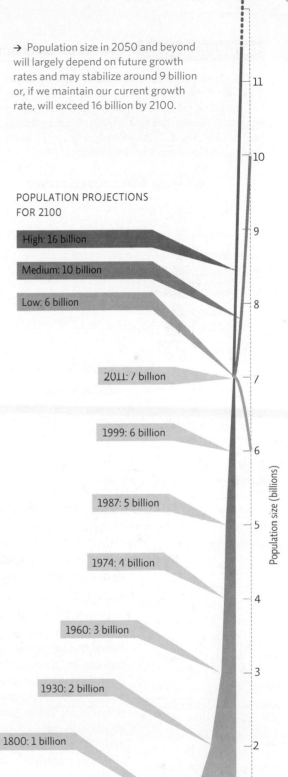

POPULATION PROJECTIONS FOR 2100

High: 16 billion

Medium: 10 billion

Low: 6 billion

2011: 7 billion

1999: 6 billion

1987: 5 billion

1974: 4 billion

1960: 3 billion

1930: 2 billion

1800: 1 billion

Industrial Revolution

Black Death in Europe

Population size (billions)

IRON AGE

MIDDLE AGES

MODERN TIMES

1000 — BCE (before common era) — CE (common era) — 1000 — 2000 — 2100

INFOGRAPHIC 4.2 **POPULATION DISTRIBUTION**   `1`

↓ The 10 most populous countries are not necessarily those with the largest land mass. China and the United States have roughly the same area—so China has a higher population density than the United States. Bangladesh has one of the highest population densities in the world, at almost 1,000 people per square kilometer.

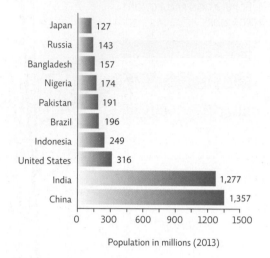

TEN MOST POPULOUS COUNTRIES

Country	Population
Japan	127
Russia	143
Bangladesh	157
Nigeria	174
Pakistan	191
Brazil	196
Indonesia	249
United States	316
India	1,277
China	1,357

Population in millions (2013)

? What are some of the advantages and disadvantages of living in areas with high population density?

population density The number of people per unit area.

immigration The movement of people into a given population.

emigration The movement of people out of a given population.

age structure The percentage of the population that is distributed into various age groups.

sex ratio The relative number of males to females in a population; calculated by dividing the number of males by the number of females.

age structure diagram A graphic that displays the relative sizes of various age groups, with males shown on one side of the graphic and females on the other.

many other nations around the world, more and more of China's people live in big cities, with many urban areas reaching record **population densities**. **INFOGRAPHIC 4.2**

Migration into or out of a population also affects the size of a *local* population. **Immigration** refers to the movement of people into a given population; **emigration** refers to their movement out of a given population. Taken together, annual migration data and birth and death rates can tell us how quickly a population is changing in size.

Demographers (social scientists who statistically analyze human populations and how they change over time) create a diagram that displays a population's size as well as its **age structure** and **sex ratio**. These **age structure diagrams** give insight into the future growth potential of a population. **INFOGRAPHIC 4.3**

In China, rapid population growth became a political issue. In the late 1950s, famine claimed 30 million lives. During the 1970s, a severe shortage in consumer goods—soap, eggs, sugar, and cotton—led to strict rationing. The government blamed the country's woes on **overpopulation**. Today, experts say that the real culprit was botched state planning. "China was not unlike other parts of the world at that time, in terms of higher life expectancies and growing population," says Wang Feng, Director of the Brookings-Tsinghua Center for Public Policy. "We could have managed, as other countries did, with better agricultural and economic policies."

Maybe so. But in the late 1970s, with two-thirds of the population under the age of 30, and the baby boomers of the 1950s and 1960s just entering their reproductive years, China's leaders had good reason to panic. When the majority of the people in a population are young, the age structure diagram looks very much like a pyramid; there are nearly equal portions of men and women, and there are fewer and fewer people at each higher age group. However, when many people are young and have not yet reached reproductive age, the population will continue to grow rapidly over time. A population like this has **population momentum**; it will continue to increase for another generation, even if each couple has only two children—just enough to replace themselves when they die (called *replacement fertility*). If the Chinese government could not feed the mouths it already had, how in the world would it feed any more?

In 1979, China's leaders issued a mandatory decree: No family could have more than one child. The government vowed to deny state-funded health care and education to all but the firstborn child. Parents who didn't comply also risked losing their jobs and faced severe fines (often several times their annual income) and penalties (including a loss of government subsidies for baby formula and other foodstuffs).

Dramatic as it sounded, the idea of strictly enforced population control was not entirely new. In fact, modern societies have fretted over swelling ranks since the Industrial Revolution (that second growth spurt); that's when an English priest by the name of

KEY CONCEPT 4.2

Populations with a lot of young people have a great deal of population momentum and will continue to grow even at replacement fertility.

INFOGRAPHIC 4.3 **AGE STRUCTURE AFFECTS FUTURE POPULATION GROWTH**

↓ The fastest-growing regions are those with a youthful or very young population.

Very young
Youthful
Transitional
Mature
No data

↓ Age structure diagrams show the distribution of males and females of a population in various age classes. The width of a bar shows the percentage of the total population that is in each gender and age class. The more young people in a population, the more population momentum it has; it will continue to grow for some time. The more people of reproductive age, the higher the growth rate, but this measure is also influenced by income; growth rates decline as income increases.

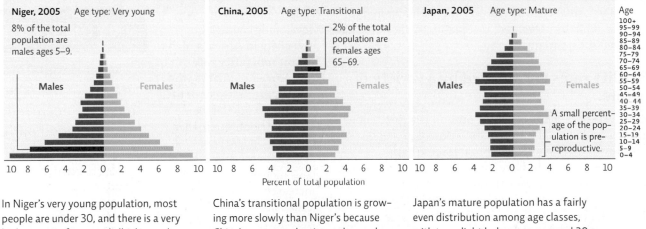

Niger, 2005   Age type: Very young

8% of the total population are males ages 5–9.

Males   Females

China, 2005   Age type: Transitional

2% of the total population are females ages 65–69.

Males   Females

Japan, 2005   Age type: Mature

A small percentage of the population is pre-reproductive.

Males   Females

Age
100+
95–99
90–94
85–89
80–84
75–79
70–74
65–69
60–64
55–59
50–54
45–49
40–44
35–39
30–34
25–29
20–24
15–19
10–14
5–9
0–4

10 8 6 4 2 0 2 4 6 8 10   10 8 6 4 2 0 2 4 6 8 10   10 8 6 4 2 0 2 4 6 8 10

Percent of total population

In Niger's very young population, most people are under 30, and there is a very high capacity for growth (high population momentum).

China's transitional population is growing more slowly than Niger's because China's pre-reproductive and reproductive age cohorts are smaller. The slight skew toward males is noticeable.

Japan's mature population has a fairly even distribution among age classes, with two slight bulges seen around 30 and 60. This population is fairly stable or may even be decreasing slowly as deaths start to outnumber births.

? What percentage of the population is between 0 and 4 years of age in each of the three countries shown here (Niger, China, and Japan)?

Thomas Malthus first noticed that food supplies were not increasing in tandem with population. Unless something changed, he warned, we would soon have more mouths to feed than food to feed them with. And when that happened, disease, famine, and war would surely follow. According to Susan Greenhalgh, Harvard University anthropologist and author of the book *Just*

One Child: Science and Policy in Deng's China, such catastrophic thinking was a big part of how the Chinese government arrived at its one-child policy.

When China's leaders enacted the policy, they promised that the policy would be a short-term one, set to expire in 30 years. By then, they said, Chinese society would have

overpopulation More people living in an area than its natural and human resources can support.

population momentum The tendency of a young population to continue to grow even after birth rates drop to "replacement fertility" (two children per couple).

↑ An overcrowded train approaches as other passengers wait to board at a railway station in Dhaka, Bangladesh.

adapted to a small-family culture, and with hundreds of millions of births prevented, the quality of life would have improved substantially.

Fertility rates are affected by a variety of factors.

China was once a culture built around large, extended families and shaped by a constellation of *pronatalist pressures*—cultural and economic forces that encourage families to have more children. Parents and grandparents lived under the same roof; aunts, uncles, and cousins remained close by, usually in the same village. And couples had as many children as possible so that there would be enough hands to work the family farm, tend to household chores, and, most importantly, care for parents as they aged.

Desired family size (how many children a couple wants) is one of the best predictors of actual fertility (how many children a couple has); therefore, any factors that increase or decrease one's desire to have children will significantly impact population growth rates.

The need for labor is a common pronatalist pressure in agrarian societies—as are the status and prestige associated with large families and a lack of other options for women. (In many countries women's rights and freedoms are still strictly limited.) In many developing countries, especially in Africa, a high **infant mortality rate**— the number of infants who die in their first year of life, per every 1,000 births— also contributes to higher fertility. Under these conditions, couples tend to have more children, which increases the odds that at least some will survive to adulthood.

KEY CONCEPT 4.3

Pronatalist pressures strongly affect desired family size, one of the best predictors of total fertility.

KEY CONCEPT 4.4

Most population growth in the recent past has occurred in less developed nations; projections for future growth continue that trend.

For centuries—right up until the 1970s, in fact—China's **total fertility rate (TFR)**, the average number of children a woman has in the course of her lifetime, hovered between five and six. Ironically, the most dramatic decline in that rate took place in the decade before China implemented its "one-child" policy. Between 1970 and 1979, a largely voluntary mandate known as "late, few, long" managed to cut the total fertility rate in half—from 5.9 to 2.9—just by encouraging couples to marry later, delay childbearing, and space conceptions as far apart as possible.

Around the world, we see big differences in the kinds of factors that influence how a population changes in size and composition. **Demographic factors** such as health (especially that of children), education, economic conditions, and cultural influences are, on average, quite different for **more developed** than they are for **less developed countries**. INFOGRAPHIC 4.4

INFOGRAPHIC 4.4   WE LIVE IN TWO DEMOGRAPHIC WORLDS

↓ Values for demographic factors differ tremendously between developing nations like many African nations and developed nations like the United States. Most of the world's population, as well as most population growth, occurs in developing nations, but most wealth is in developed nations. The higher death rate in developed nations is due to an aging population; the much higher infant mortality rate in developing nations reveal the differences in quality of life and health care between the two categories.

Demographic Factor (2013 data)	World	More Developed Countries	Less Developed Countries	Niger (the fastest growing country)
Population size	7 billion	1.3 billion	5.9 billion	16.9 million
% growth rate	1.2	0.1	1.40	3.8
Crude birth rate (births/ 1,000 individuals)	20	11	22	50
Crude death rate (deaths/ 1,000 individuals)	8	10	7	12
Total fertility rate	2.5	1.6	2.6	7.6
Infant mortality rate (deaths/ 1,000 children under one year old)	40	5	44	51
Life expectancy	70 yrs	78 yrs	69 yrs	57 yrs
Wealth (per capita GNP; 2012 U.S.$)	$11,690	$35,800	$6,600	$650

Projected growth (UN medium projection)

Most projected growth is expected to continue to be in less developed nations.

Population size (billions) — Year

■ More developed nations   ■ Less developed nations

What demographic factors do you think most strongly affect the high population growth rate of Niger, the world's fastest, growing country?

In China, cutting the average family size from six children down to three seemed easy. But making the leap from three children down to one proved considerably more difficult. One child meant no siblings, and in subsequent generations, no cousins or aunts and uncles, either. Who would help tend the farms? If a couple's child died or was disabled, who would care for them when they aged? "Many Chinese families in the countryside found it unthinkable at the time to have just one child," says Greenhalgh. Even with the penalties they faced, many resisted.

The policy's most horrendous consequences remain sources of great speculation and worry. Many experts have said that in the early years, enforcers carried out numerous sterilizations and abortions—some estimates range as high as the tens of millions—many of

them by force. But because official numbers are very difficult to come by, it's nearly impossible to say whether such figures grossly overestimate or underestimate the problem. Ultimately, the true numbers may be unknowable.

In 1998, however, a former Chinese family planning official shed some light on the inner workings of the one-child policy, when she testified before the U.S. Congress that women as far along as 8.5 months were forced to abort by injection of saline solution, and women in their ninth month of pregnancy, or even women in labor, were having their babies killed in the birth canal or immediately after birth. The policy's defenders say that such practices have abated, but as recently as 2001, the British newspaper *The Telegraph*

infant mortality rate The number of infants who die in their first year of life per every 1,000 live births in that year.

total fertility rate (TFR) The number of children the average woman has in her lifetime.

demographic factors Population characteristics such as birth rate or life expectancy that influence how a population changes in size and composition.

more developed country A country that has a moderate to high standard of living on average and an established market economy.

less developed country A country that has a lower standard of living than a developed country and has a weak economy; may have high poverty.

↑ A billboard promoting China's one-child policy.

reported that a single county in Guangdong Province was tasked with an annual quota of 20,000 abortions and sterilizations after some residents disregarded the one-child policy.

Chinese officials have argued that their focus on population control has actually improved women's access to health care—both contraception and prenatal classes are free to all—and dramatically reduced the incidence of childbirth-related deaths and injuries.

But critics say that the human rights violations engendered by the one-child policy far outweigh any possible health gains and that, at any rate, health care is wildly imbalanced. One study, conducted by public health researchers from Oregon State University in 1990 in the rural Sichuan Province, found that women whose pregnancies were unapproved were twice as likely to die in childbirth as those whose pregnancies were sanctioned by the government. And while nearly 90% of all married women in China use contraception—compared with roughly one-third in most other developing countries—most women are not given a choice about which type of contraceptive to use; in a 2001 survey conducted by the Chinese family planning commission, 80% of women said they were compelled to accept whatever method their family planning worker chose for them. Overall, IUDs and sterilizations account for more than 90% of contraceptive methods used since the mid-1980s.

Still, there are signs of improvement. The number of sterilizations has declined considerably since the early 1990s. And in 2002, facing strong international pressure, the Chinese government finally outlawed the use of physical force to compel a woman to have an abortion or be sterilized. (Experts say that prohibition is not entirely enforced. The policy is largely exercised at the local level. And in many provinces, local governments still demand abortions.)

In any case, by 2011, China's fertility rate had plummeted to 1.54, spurring China's leaders to proclaim that the one-child policy had succeeded in preventing some 400 million births, and all the calamities those births would have brought—disease, more famine, overtaxed social services, and so on.

But while the drop in fertility was plain, not everyone would agree that the one-child policy deserved the credit. Critics argue that by the time the first generation of "Little Emperors" reached adulthood, many other forces had reshaped their society and the world.

Factors that decrease the death rate can also decrease overall population growth rates.

When Qin Yijao's mom was pregnant with him back in 1981, the one-child policy was being strictly enforced, and she already had a son. "People from the Communist Brigade said I couldn't have a second baby," the mother told National Public Radio (NPR) in a 2005 interview. "But I was determined. I needed a son to work our farm." So taking care to evade the birth police, she snuck off and

KEY CONCEPT 4.5

Helping low-income countries develop economically may spur a demographic transition similar to that experienced by many industrialized countries.

INFOGRAPHIC 4.5   **DEMOGRAPHIC TRANSITION**

↓ Some industrialized nations have gone through the demographic transition from high birth and death rates to low birth and death rates, giving them a stable population. It is unclear whether industrialization will produce the same pattern in all developing countries. Steps to improve the quality of life and decrease death rates are important worldwide, even if other measures are needed to slow birth rates.

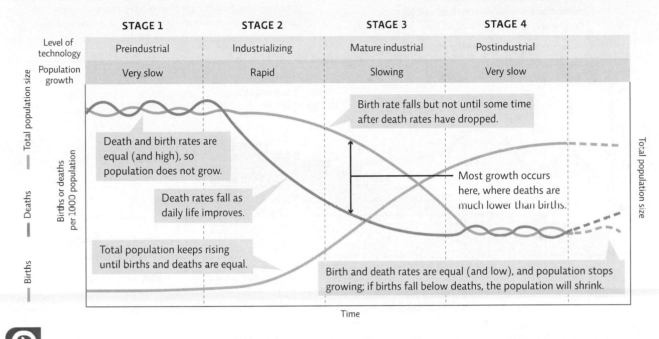

	STAGE 1	STAGE 2	STAGE 3	STAGE 4
Level of technology	Preindustrial	Industrializing	Mature industrial	Postindustrial
Population growth	Very slow	Rapid	Slowing	Very slow

Death and birth rates are equal (and high), so population does not grow.

Death rates fall as daily life improves.

Total population keeps rising until births and deaths are equal.

Birth rate falls but not until some time after death rates have dropped.

Most growth occurs here, where deaths are much lower than births.

Birth and death rates are equal (and low), and population stops growing; if births fall below deaths, the population will shrink.

Total population size

Births or deaths per 1000 population

Total population size

— Deaths

— Births

Time

? Why might industrialization *not* lead to a demographic transition in all countries?

delivered her son in a cave. When she was found out, she was penalized. "I had to pay double the price for grain, and they confiscated our television and sewing machine." In the end, she got to keep and raise her son. But at 30, he has never worked a single day on the farm. "I don't even know where my mom's farm plot is," Qin also told NPR, adding that while he appreciates the risks his mother took, he and his wife have no desire for a second child themselves. The anecdote underscores a key point: Chinese culture has changed dramatically in the years since the one-child policy was enacted, becoming much more urban and thus lessening the need for farmhand children. Many demographers say that such social changes may be more responsible for smaller families than the policy itself.

The **demographic transition** holds that as a country's economy changes from preindustrial to postindustrial, low birth and death rates replace high birth and death rates. The reasons are cultural as much as economic. As health care improves, infant mortality declines, and with it, the need for couples to have many children just to ensure that some make it to adulthood. As cities grow and jobs materialize, the need for children to work the farm decreases. Meanwhile, as women find themselves with more opportunities, better health care, and greater access to education, many come to want fewer children.

INFOGRAPHIC 4.5

More developed countries, including the United States and countries of the European Union, have already undergone the demographic transition. Demographers are split over whether other countries are likely to do the same. But that uncertainty has not stopped world leaders from trying to cull lessons from the demographic transition model. Around the world, countries from Bangladesh to Brazil are working to reduce pronatalist pressures, increase gender equality, and combat poverty and infant mortality, all in an effort to improve standards of living, which in turn should lower fertility rates. The ultimate goal is to achieve **zero population growth**, when population size is stable, neither increasing nor decreasing. This occurs when the number of people born equals the number of people dying—in other words, when the population reaches the **replacement fertility rate**.

In the United States, the replacement rate is 2.1 (rather than the expected 2.0), since a few children die before maturity and not every female has children. In countries with high infant mortality rates, the replacement rate is higher.

demographic transition A theoretical model that describes the expected drop in once-high population growth rates as economic conditions improve the quality of life in a population.

zero population growth The absence of population growth; occurs when birth rates equal death rates.

replacement fertility rate The rate at which children must be born to replace those dying in the population.

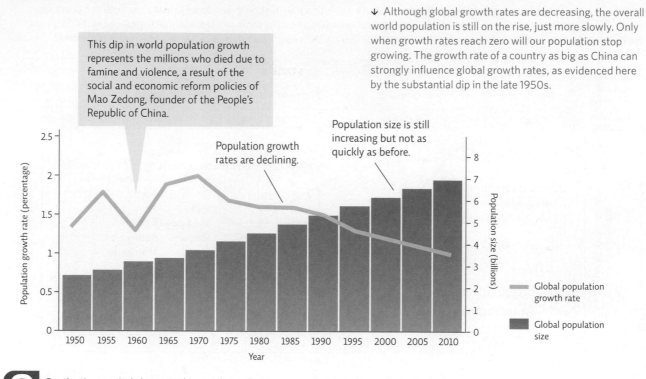

This dip in world population growth represents the millions who died due to famine and violence, a result of the social and economic reform policies of Mao Zedong, founder of the People's Republic of China.

↓ Although global growth rates are decreasing, the overall world population is still on the rise, just more slowly. Only when growth rates reach zero will our population stop growing. The growth rate of a country as big as China can strongly influence global growth rates, as evidenced here by the substantial dip in the late 1950s.

Population growth rates are declining.

Population size is still increasing but not as quickly as before.

Global population growth rate

Global population size

For the time period shown in this graph, in what year was the human population growth rate the highest? When was the human population size the greatest?

At least a handful of clues suggest that the demographic transition in China is as responsible for reductions in total fertility rates as is the one-child policy. Clue number one: The most significant fertility declines in the policy's history coincided with the largest gains in the Chinese economy, not with periods of strict enforcement. Clue number two: Many other countries have had substantial declines in fertility during the past 25 years without the use of strict population-control policies, including Japan, Singapore, and Thailand, whose fertility rates mirror those of China's almost exactly—falling from 5 children per family in 1970 to 1.8 children per family in 2011.

"The parallels are hard to ignore," says Therese Hesketh, a global health professor at University College London. "[They suggest] that China could have expected a continued reduction in its fertility rate just from continued economic development." Feng puts it more bluntly. "The claim that one-child prevented 400 million births is bogus for sure,"

he says. "Many other countries experienced nearly identical fertility declines over the same time frame. So unless you say Chinese people are necessarily different from other populations, it's very hard to argue the decline is due to government policy."

Worldwide, population growth rates are declining, but overall the number is still positive, so we are still increasing. The United Nations estimates future population size leveling off somewhere between 6 and 16 billion people by 2100, depending on how quickly we reduce growth rates (see Infographic 4.1). **INFOGRAPHIC 4.6**

But if the predominant causes of declining fertility remain a source of debate, the consequences are becoming clearer and more stark by the day. Because the reasons for population growth are likely to be different in different regions, reaching zero population growth requires that the pronatalist pressures in each region be addressed. A variety of measures that don't include government-restricted family size have proven successful.

Programs that address the needs of a given population and work within the cultural and religious traditions are likely to be most successful. Many demographers believe that addressing the social justice issues associated with overpopulation (poverty, high infant

KEY CONCEPT 4.6

Currently, the global population growth rate is decreasing, but it is not yet zero, so the world's population is still increasing.

mortality, lack of education and opportunities for women) will prove the most instrumental in enabling countries with high TFRs to confront what lies ahead. For example, programs that improve health care also increase the survival of infants and children; education and job opportunities for women have been shown to reduce the desired family size. Targeting such health, education, and family planning programs in specific regions of the country where there is an unmet need for such services can lead to significant reductions in TFR. Targeting areas of unmet need has been particularly effective in other fast-growing regions of the world, such as many African nations, where as many as 39% of the woman in rural areas want birth control but have no access to it. But such efforts can work only when paired with cultural and societal shifts that reduce the desired family size. **INFOGRAPHIC 4.7**

The age and gender composition of a population affects more than just its potential for growth.

Currently, many industrialized nations—including the United States and many European countries—have top-heavy age structure diagrams with many older people. In many cases, the struggle to care for these rapidly aging populations has turned prickly. In 2010, in order to relieve retirement age problems (and thus government pension payouts), France raised the retirement age from 60 to 62, sparking riots throughout the nation. In the United States, the prospect of baby boomers' retirement bankrupting Social Security has spurred intense and vitriolic debate.

China's age structure diagram has lost the bottom-heavy shape, reflecting its changing population age structure. Falling birth rates and rising life expectancy have tipped the balance between old and young. In 1982, just 5% of China's population was older than 65. In 2004, that figure had climbed to 7.5%; by 2050, it is expected to exceed 15%. These figures are lower than those in most industrialized countries (especially Japan, where the proportion of people over age 65 years is 20%). But because elderly people are still dependent on their children for support, this situation could spell disaster.

> ### KEY CONCEPT 4.7
>
> In many cases, addressing social justice issues, especially improving the health and well-being of women and children, is the key to reducing total fertility and reaching zero population growth.

Demographers call it the 4—2—1 conundrum: As they settle into middle age, the members of each successive generation of only children will find themselves responsible for two aging parents and four grandparents. If that single child fails, family elders would be left with virtually no options—no second children, or nieces and nephews, and not even friends and neighbors who could help out, since most other families would be in an equally precarious situation. In a country without an extensive pension program, this prospect has the government especially worried.

There's another problem with China's evolving age structure: the workforce. Economists predict that between 2010 and 2020, the annual size of the labor force aged 20—24 will shrink by 50%. This forecast has global implications. "The deep fertility declines have plunged China into a demographic watershed," says Feng. "It means the days of cheap, abundant Chinese labor are over." For countries like the United States, which have come to depend on a steady influx of cheap goods from China, that's not such good news. But for the only children of China, it almost certainly means higher wages, better working conditions, and more jobs to choose from in the coming decades.

An out-of-whack age structure isn't the only unintended consequence of the one-child policy. The sex ratio of men to women has also grown alarmingly high. It's true that almost everywhere in the world, there are slightly more males than females. (However, this differs according to age. Because they tend to live longer, there are more women than men in older age groups.) But in most industrialized countries, sex ratio of men to women tends to hover close to 1.05 (which, coincidentally, is the exact sex ratio of the United States). In China, the ratio soared from 1.06 in 1979 to 1.17 in 2011; in some rural provinces, it was as high as 1.3 in 2011. What does this mean in actual numbers? The State Population and Family Planning Commission estimates that come 2020, there will be 30 million more men than women in China. How did this happen?

Like most Asian countries, and many African ones, China has a long tradition of preference for sons, based on the belief that boys are better suited to the rigors of manual labor that drive rural, and even industrial, economies. This means that they are better able to provide for retired and aging parents than daughters are.

A recent study by Chinese researchers from Zhejiang Normal University shows that the sex ratio is high (favoring males) in urban areas where only one child is allowed. However, in rural areas where a second child is allowed, if the first is a girl, sex ratios are even higher for the second child—as high as 160 males

↓ Reaching zero population growth takes two steps: identifying why birth rates are high (what are the pronatalist pressures?) and then taking steps (education, birth control, social campaigns) to address those pronatalist pressures and reduce birth rates.

STEP 1: IDENTIFY PRONATALIST PRESSURES THAT INCREASE TOTAL FERTILITY RATE (TFR)

INFANT MORTALITY RATE IMPACTS BIRTH RATE

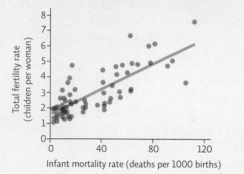

High infant mortality is closely correlated with poverty: TFR goes up as infant mortality and poverty increase.

DESIRED FERTILITY AND ACTUAL TOTAL FERTILITY

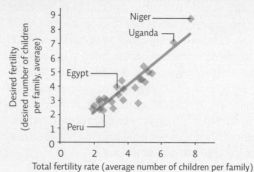

TFR closely aligns with the desired fertility of families—families usually have the number of children they would like to have.

STEP 2: EMPLOY STRATEGIES TO REDUCE TFR

AS EDUCATION OF WOMEN INCREASES, TFR DECREASES

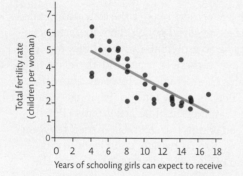

Fertility declines as educational opportunities for girls and women increase. This means that funding for education and job opportunities for women may be more effective at lowering TFR than other approaches. It also decreases infant mortality rates, which helps decrease birth rates.

AS USE OF MODERN CONTACEPTIVES IN AFRICA GOES UP, TFR GOES DOWN

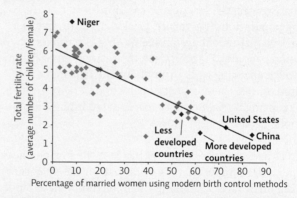

Family planning programs have been effective in many areas of the world. However, in countries with high desired fertility, such as Niger, providing contraceptives may have little impact on TFR.

DIFFERENT SOLUTIONS FOR DIFFERENT REGIONS

Thailand's very successful family planning program started with better maternal and infant health care, family planning programs, and access to contraceptives, but also reflected the culture's love of humor—from condom distribution at public gatherings to free vasectomies on the king's birthday, to "ads" painted on the sides of water buffalos.

Kerala, India, a state in the world's second most populous country, adopted a population-control plan centered around the three "e's": education, employment, and equality. As a result, Kerala's school system boasts more than a 90% literacy rate, which is almost identical between boys and girls. Educated women join the work-force before deciding to have children, and some 63% (compared to 48% in India as a whole) of women there use contraceptives. The state's birth rate has stabilized at around 2.0.

In Brazil, the dropping birth rate over the past 40 years is attributed in part to pop culture. Families in popular TV shows are small; this coupled with rising opportunities for women and crowded living spaces in large cities, has produced a generation of young women whose desired family size is only one or two children.

? Which of the graphs shown here depict positive correlations? Which are negative correlations? Are positive correlations "good" and negative correlations "bad"? Explain.

> " *The deep fertility declines have plunged China into a demographic watershed.* " —Wang Feng

for every 100 females. The researchers attribute this high male bias to sex-selective abortion—deliberately terminating a pregnancy based on the gender of the fetus (in this case, aborting females). The practice has been outlawed in China, but most experts say it is still common, thanks in large part to private-sector health care and the ability of a growing number of Chinese citizens to pay the high cost of a gender-determining ultrasound. Although most demographers agree that outright infanticide is increasingly rare, subtler forms of gendercide—the systematic killing of a specific gender—are known to occur. If a female infant falls ill, for example, she might be treated less aggressively than a male infant would be.

These days, however, it is much more common for girl babies to be put up for adoption than boys. Officially registered adoptions increased 30-fold, from about 2,000 in 1992 to 55,000 in 2001. The trend was especially pronounced in the United States, where families adopted nearly 8,000 Chinese babies in 2005 alone (compared with just 200 or so in 1992). The practice is not necessarily as benign as it sounds. In 2009, the *Los Angeles Times* reported that many babies put up for adoption had not been abandoned by their parents but confiscated by Chinese family planning officials.

The result of the skewed sex ratio, 30 years out, seems to be lots of prospective husbands in want of wives. Recent census data show that in a growing portion of rural Chinese provinces, one in four men are still single at 40. "The marriage market is already getting more intense and competitive," says Feng. "Men of lower social ranks—who make up a very significant chunk of the population—are being left out. And it's only going to get worse."

A growing number of social scientists say that the dearth of women will ultimately threaten the country's very stability. Experts worry that such large numbers of young men who can't find partners will prove a recipe for disaster. "The scarcity of females has resulted in kidnapping and trafficking of women for marriage and increased numbers of commercial sex workers," writes Therese Hesketh, "with a potential rise in human immunodeficiency virus infection and other sexually transmitted diseases."

Carrying capacity: Is zero population growth enough?

Every given environment has a **carrying capacity**—the maximum population size that the area can support. How large that population can be is determined by a range of forces, including the supply of nonrenewable resources, the rate of replenishment for renewable resources on which we depend, and the impact each individual person has on the environment. (A degraded ecosystem can sustain far fewer people than a well-maintained, healthy one.)

There are roughly 7 billion people on the planet today. Whether we stabilize at 9 or 10 billion or more depends on how quickly we lower TFR to achieve replacement fertility worldwide. Even at 7 billion people, we may have already exceeded the carrying capacity of Earth. One problem we face is our dependence on nonrenewable energy sources; they will not

carrying capacity The population size that an area can support for the long term; depends on resource availability and the rate of per capita resource use by the population.

↓ Multiple generations of a Chinese family pose at their home in an undated photograph from the 20th century. Historically, sons were so highly valued that a second son might be sent to join the family of a close relative who had no sons.

© Bettmann/CORBIS

INFOGRAPHIC 4.8   HOW MANY PEOPLE CAN EARTH SUPPORT?

↓ Many ecologists and economists think that, at our current rate of consumption, human population size has already surpassed what the Earth can support in the long term (reduced its "carrying capacity"). This reduces the abillity of the planet to meet our current needs, much less the needs of an ever growing population.

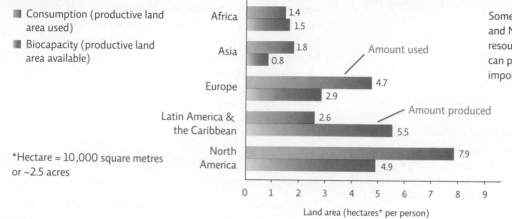

■ Consumption (productive land area used)

■ Biocapacity (productive land area available)

Some areas, like Asia, Europe, and North America, use more resources than their own region can produce and depend on imports from others.

Africa — 1.4 / 1.5
Asia — 1.8 / 0.8
Europe — 4.7 / 2.9
Latin America & the Caribbean — 2.6 / 5.5
North America — 7.9 / 4.9

Amount used
Amount produced

*Hectare = 10,000 square metres or ~2.5 acres

Land area (hectares* per person)

How could the human population increase the ability of the productive land area, a finite, though renewable resource base, to support it?

last indefinitely. But we are also overusing our biological resources. There are differences in how well different regions of the world live within the *biocapacity* (the ability of the ecosystem's living components to produce and recycle resources and assimilate wastes) of their own environments. North America, and the United States in particular, uses far more resources than its land area can provide, whereas Latin America and the Caribbean use less than half of the biocapacity of their regions. This means nations like the United States and many others are importing resources from other regions. **INFOGRAPHIC 4.8**

The problem, then, is twofold: On one hand, we are increasing, rapidly, in sheer numbers. On the other, we are consuming more resources per person than ever before.

In general, population increase is a problem of the developing world. The highest fertility rates tend to be in the least developed countries—those that have yet to complete the demographic transition. Overconsumption, meanwhile, is a problem of more developed countries like the United States, where a single person might consume more food, fuel, and other resources (and thus place more strain on the environment) than a whole group of people in a less developed country like Sudan or Bangladesh. The impact humans have on Earth is thus due to a combination of factors: population size, affluence, and how we use resources, especially our use of technology, which tends to increase our overall use of resources. This is discussed more fully in Chapter 6.

In the years since the one-child policy was enacted, China has found itself careening from one end of this spectrum to the other. The problems that come

with overpopulation—such as slums, epidemics, overwhelmed social services, and the production of high volumes of waste—have been mitigated by a reduction in the fertility rate. But as fertility has declined, relative affluence has increased, and the Chinese are now straining the environment in a different way: by overconsuming.

It's important to understand that overconsumption in one region or country can strain carrying capacity in other, disparate regions. In Bangladesh, for example, rising sea levels, spurred by global climate change, are pushing that nation's carrying capacity to the brink. Most experts agree that a major driver of current climate change is rampant fossil fuel consumption by nations all over the world, many very far from Bangladesh. China now leads the way in the release of fossil fuel-based carbon emissions linked to climate change—because of its huge population and its growing affluence.

Ultimately, the question of how many people the planet can support may not be the right one. Trying to pinpoint whether Earth's carrying capacity is 7 billion or 9 billion or even 16 billion is meaningless unless we identify what an acceptable quality of life is. And while overpopulation may be an issue that has

KEY CONCEPT 4.8

The number of people Earth can support depends on how many resources each uses. We may have already exceeded our planet's carrying capacity by using more resources than it can replace in the long term.

global significance and causes, it is one that ultimately plays out at the regional level: As land becomes degraded and water supplies become depleted or polluted, people (and other organisms) live in an increasingly impoverished environment.

What awaits China's generation of Little Emperors?

Evidence suggests that despite the plight of the latest generation, China is indeed becoming a small-family culture. The National Family Planning and Reproductive Health Survey found that 35% of women prefer having only one child; 57% prefer having two children; and only 5.8% want more than two. True to the demographic transition's predictions, educated, urban women wanted fewer children than their rural, farm-dwelling counterparts. That's good news for those striving to bring China toward zero population growth: In the past three decades, the federal government has allowed hundreds of millions of Chinese people to move to cities in search of work. Those families no longer need lots of children to work farms.

In recent years, the Chinese government has softened its stance. In 2008, it added several significant exceptions to the one-child rule, permitting couples made up of two only children, rural families with land to farm, and several groups of ethnic minorities to have more than one child if they so choose. In 2013, the government went a step further, declaring that couples in which only one spouse (rather than both) is an only child could now have two children without breaking the law.

Still, it's unlikely that the one-child policy will officially expire any time soon. In a 2008 interview with the state newspaper, *China Daily*, the country's minister of State Population and Family Planning, Zhang Weiqing, said

that loosening the one-child policy would unleash a new baby boom. "Given such a large population base, there would be major fluctuations in population growth if we abandoned the one-child rule now. It would cause serious problems and add extra pressure on social and economic development." Implementation of the latest amendments has largely been left to individual provincial governments. While some areas are moving quickly to standardize the rule change, in other areas, it's as if no policy change has been made.

For better or worse, China is now a country of small families and of ever-fewer children. That may mean more resources per child, but it also means fewer workers, taxpayers, and innovators. Ultimately, it means that even as they bask in the spoils of the one-child policy, the Little Emperors are burdened with the hopes and fears of an entire country.

Select References:

Ewing, B., et al. (2010). *Ecological Footprint Atlas 2010.* Oakland, CA: Global Footprint Network.

Greenhalgh, S. (2008). *Just One Child: Science and Policy in Deng's China.* Berkeley: University of California Press.

Hesketh, T., et al. (2005). The effect of China's one-child family policy after 25 years. *New England Journal of Medicine,* 353(11): 1171–1176.

Ni, H., & A. M. Rossignol. (1994). Maternal deaths among women with pregnancies outside of family planning in Sichuan, China. *Epidemiology,* 5(5): 490–494.

Pritchett, L. (1994). Desired fertility and the impact of population policies. *Population and Development Review,* 20(1): 1–55.

Sedgh, G., et al. (2007). *Women with an Unmet Need for Contraception in Developing Countries and Their Reasons for Not Using a Method.* New York, New York: Guttmacher Institute.

Yin, Q. (2003). *Theses Collection of 2001 National Family Planning and Reproductive Health Survey.* Beijing: China Population Publishing House.

Zhu, W., et al. (2009). China's excess males, sex selective abortion, and one child policy: Analysis of data from 2005 national intercensus survey. *British Medical Journal,* 338: 1211.

BRING IT HOME

PERSONAL CHOICES THAT HELP

The impact of humans on the planet is created by a combination of population size and resource use. The issue of population and carrying capacity is complex. We cannot have a truly sustainable society until key components such as poverty, lack of education, and basic human rights are addressed.

Individual Steps

• Buy Fair Trade Certified products. These products provide a livable wage to workers and are often linked to education and community development.

• Research the products you buy to make sure that you are not supporting child slave labor or sweatshop facilities.

Group Action

• Raise money and invest it in a socially responsible project. Kiva is a nonprofit organization that provides microloans to help people start small businesses in less developed countries (www.kiva.org).

• Join an organization such as Habitat for Humanity, which builds housing for low income families.

Policy Change

• We all know that our tendency as humans is to build up and out; however, revitalizing our current older downtown areas is important to prevent new habitat destruction as well as to make use of current infrastructure. Urge your local government to keep shopping and dining establishments in historic downtowns. If needed, work to develop a community partnership that starts clean-up programs and community gardens in abandoned lots.

ENVIRONMENTAL LITERACY **UNDERSTANDING THE ISSUE**

1 How and why have human population size and growth rate changed over time? How many people live on Earth today?

INFOGRAPHICS 4.1 AND 4.2

1. A graph that shows human population size over the course of human history:
   a. is a line that has steadily increased at a constant rate.
   b. produces an undulating curve that rises and falls repeatedly, with our current population similar in size to past peaks.
   c. resembles a J-shaped curve—flat at first, with a rapid upward rise since about 1800.
   d. is an S-shaped curve—flat at first, then increasing rapidly for a few hundred years and then leveling off in the past 50 years.

2. Approximately _____ people live on Earth today.

3. What factors led to the increase in population growth rate during and after the Industrial Revolution?

2 What cultural and demographic factors influence population growth in a given country? How do they differ between more and less developed countries?

INFOGRAPHICS 4.3 AND 4.4

4. Reducing infant mortality generally:
   a. increases the rate at which a population grows.
   b. decreases the rate at which a population grows.
   c. has little effect on the rate at which a population grows.

5. An age structure diagram can be used to:
   a. measure potential population migration to cities.
   b. see the historical growth of a population.
   c. predict the future growth of a population.
   d. measure the quality of life of a population.

6. True or False: A population with more older people than young people has a lot of population momentum.

7. Identify several factors that contribute to the high growth rate in developing nations like Niger.

3 What is the demographic transition, and why is it important? What is the current trend for global population growth?

INFOGRAPHICS 4.5 AND 4.6

8. The term *demographic transition* refers to:
   a. slower growth as the population size approaches carrying capacity.
   b. the decline in death rates and then birth rates as a country becomes industrialized.
   c. the requirement for a population to reach a specific size before it becomes stable.
   d. migration from the overpopulated countryside to urban centers.

9. In recent years, the world population's growth rate has declined, but it is still a positive number. This means human population size:
   a. has stabilized (neither increasing nor decreasing).
   b. has started to fall.
   c. is still increasing but more slowly than before.
   d. will peak soon and then fall slightly.

10. Why might a modern-day developing nation in an intermediate stage of the demographic transition (experiencing lower death rates) *not* complete the transition to lower birth rates?

4 What are some strategies for achieving zero population growth?

INFOGRAPHIC 4.7

11. All of the following are pronatalist pressures *except*:
    a. A child is part of the family labor pool.
    b. There is a high infant mortality rate.
    c. Contraceptives are not available.
    d. Women have many opportunities to participate in the workforce.

12. What generally happens to a population's total fertility rate (TFR) when the education that women receive increases?
    a. TFR increases
    b. TFR decreases
    c. TFR is not affected by educational opportunities for women

13. In terms of reducing population growth rates, how important is reducing desired family size?

5 What determines Earth's carrying capacity for humans? Is Earth's carrying capacity enough to support the current or future (projected) human population?

INFOGRAPHIC 4.8

14. Which statement illustrates human overconsumption?
    a. Japan's fertility rate of 1.4 is below replacement-level fertility.
    b. Average per capita daily water usage in the United States is 575 liters, while the amount needed for a reasonable quality of life is 80 liters.
    c. At 176 deaths per 1,000 live births, Angola has the highest infant mortality rate in the world.
    d. Per capita use of fossil fuels is decreasing in Denmark as a result of the carbon tax.

15. How could it be possible that we are already living beyond the long-term carrying capacity of Earth?

SCIENCE LITERACY WORKING WITH DATA

The United Nations Population Division projects future population growth. The four projections in Graph A are each based on a different future fertility rate. The medium fertility variant assumes that the global fertility rate will drop to replacement (2.1), the constant fertility variant assumes no change, and the high and low fertility variants assume that the global fertility rate will be 2.6, and 1.6, respectively.

GRAPH A Estimated and Projected World Population According to Different Variants

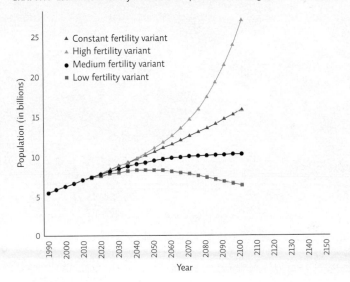

GRAPH B Projected Population Change between 2010 and 2100 by Major Region

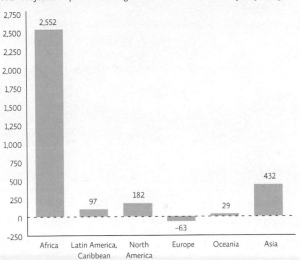

Interpretation

1. What are the population projections for each fertility variant in 2050? In 2100? What do you predict the population will be in 2150 for the constant fertility variant?

2. What is the trend in the numbers in Graph A for each line?

3. How does the information in Graph A relate to that in Graph B? Explain.

Advance Your Thinking

4. Explain why the UN might have four population projections based on different fertility rates.

5. Why is Africa projected to have such large growth if population projections assume that global fertility rates will be at replacement? Use information from the chapter to support your conclusion.

INFORMATION LITERACY EVALUATING INFORMATION

Many people consider providing education and job opportunities for women to be important, not just for their effects on fertility but also as a matter of social justice. One organization that works on this is the Foundation for International Community Assistance (FINCA), which provides microfinance services to low-income people, mostly women. Explore the FINCA website (www.finca.org).

Evaluate the website and work with the information to answer the following questions:

1. Is FINCA a trustworthy organization? Does it have a clear and transparent agenda?
   a. Who runs this organization? Do his or her credentials make the work of the organization and the information presented reliable or unreliable? Explain.
   b. What are the mission and vision of this organization? What are its underlying values? How do you know this?

   c. Do you agree with FINCA's assessment of the problems and concerns about poverty? Explain.

2. Explore the "About FINCA" links.
   a. What are *microfinance* and *village banking*? Does the website provide supporting evidence for FINCA's claims about how these programs can help the poor? Is the evidence reliable?
   b. What is the Smart Microfinance Campaign, and why is it important to microfinance clients?
   c. Do you think FINCA's business model is valid and effective? Explain.

3. Select the "Take Action" link.
   a. Who can be a part of FINCA's solution? How can they get involved?
   b. Identify a specific solution and the strategy to accomplish it. Are the solution sufficient and the strategy reasonable? Explain.

4. How might the FINCA model influence cultural, economic, and demographic factors that influence population growth?

Find an additional case study online at www.macmillanhighered.com/launchpad/saes2e

CORE MESSAGE

Human health is impacted by the environment. Our choices can alter the environment to either facilitate disease or to reduce its transmission. Infectious diseases are the leading cause of death in less developed nations, while water contamination, land and waterway alteration, deforestation, and climate change are contributing to these and other health problems. Health can be improved significantly with better sanitation, access to clean air and water, and public health programs that reach people in affected areas.

AFTER READING THIS CHAPTER, YOU SHOULD BE ABLE TO ANSWER THE FOLLOWING **GUIDING QUESTIONS**

1

What types of environmental hazards impact the health of people?

2

How do the fields of public health and environmental health help improve the health of human populations? What environmental factors contribute to the global burden of disease?

ERADICATING A PARASITIC NIGHTMARE

Human health is intricately linked to the environment

Women gather water from Ogi, a sacred pond in Nigeria. Previously contaminated with water fleas that carried Guinea worm larvae, the pond was successfully decontaminated by local health officials. Guinea worm disease is on track to become the first disease since smallpox to be totally eradicated. Vanessa Vick/The New York Times/Redux

3

What types of pathogens cause disease? Specifically, what factors facilitate the spread of Guinea worm disease, and what steps are needed to eradicate it?

4

How do factors that affect human health differ between more and less developed nations?

5

What can be done to reduce environmentally mediated health problems? How close are we to eradicating GWD?

Ernesto Ruiz-Tiben shook the tube of water and held it up to the sunlight so that the women who were gathered around him could see the tiny black flecks that had settled out. There was a soft, collective gasp at the spectacle. The black flecks—tiny water fleas known as copepods—offered the Nigerian women the first visible proof of what Ruiz-Tiben, director of the Carter Center's Guinea Worm Eradication Program, had been trying to explain to them: The water they drank and bathed in was contaminated with tiny bugs, and these bugs were solely responsible for the searing worm infections that seemed to sweep through the village every year or so, usually right around harvest time.

For those unlucky villagers, the infection began when a person ingested contaminated water that contained copepods infected with Guinea worm larvae. Once digested, the copepods released the larvae, which burrowed into the victim's abdominal tissue. There the larvae matured into adult males and females who then mated. The males died, but the females continued to grow—up to 100 centimeters (3 feet) long—all the while migrating through the victim's tissue. About a year after infection, the full-grown females forced their way out by releasing acid just beneath the skin, which in turn created a blister. When the fiery pain of that blister drove the victim to plunge into water, the female squirted out a dense cloud of milky white larvae, starting the cycle over again.

The disease is not fatal, but recovery is both very slow and very debilitating. Even after she releases her larvae, the mother worm can take as long as 3 months to fully emerge from the skin, during which time the victim is often completely laid up—unable to work for the entire harvest—or even to walk or move much, depending on which limbs are infected and with how many worms. Worse yet, there are no medications or vaccines for dracunculiasis, or Guinea worm disease (GWD) as it is more commonly known, and the infection itself does not confer immunity; that means the same people can fall prey to the worms over and over again.

KEY CONCEPT 5.1

Water- and vector-borne infectious diseases threaten many human populations. Human actions that increase habitat for pathogens or their vectors can facilitate the spread of these diseases.

Efforts to reduce the incidence of GWD began as part of an international program to provide safe drinking water to all people. According to the World Health Organization (WHO), some 1 billion people per year fall victim to **waterborne diseases**—those acquired by consuming contaminated water. In fact, water- and **vector-borne diseases** are the main **infectious disease** threats to human health. (Vectors are organisms that transmit a **pathogen** from one host to another; in the case of GWD, copepods serve as the vector.) GWD is just one of countless such diseases, but in the course of eliminating it, Ruiz-Tiben and his colleagues hoped that they might also eliminate others.

◉ **WHERE IS NIGERIA?**

↑ Guinea worm disease is a major impediment to a farmer's ability to work. A health volunteer in Ghana educates children on how to use pipe filters when they go to the fields with their families. Pipe filters, individual filtration devices worn around the neck, work similarly to a straw, allowing people to filter their water to avoid contracting Guinea worm disease while away from home.

Human manipulation of the environment can increase our exposure to pathogens.

Worldwide, people face a variety of health hazards. One of the biggest influences on the seriousness of any given hazard is the health of the natural environment, something we've had unprecedented influence over in recent centuries. "We humans have been remarkably effective at rearranging the natural world to meet our own needs," says Sam Myers, a scientist at the Harvard University School of Public Health, in a recent article in *Annual Review of Environment and Resources*. In the article, he says that between one-third and one-half of global resources produced by ecosystem functions are now diverted to human uses. In the past 300 years alone, we have completely deforested between 7 and 11 million square kilometers (2.7–4.2 million square miles) of land—an area the size of the continental United States—and converted some 40% of the planet's ice-free land surface to cropland or pasture. We have converted an additional 2 million square kilometers (0.8 million square miles) of forest into

" We humans have been remarkably effective at rearranging the natural world to meet our own needs. "
—Sam Myers

highly managed plantations with significantly less biodiversity. We are already using roughly half of the planet's accessible surface freshwater and fishing three-quarters of monitored fisheries at or beyond their sustainable limits. And with population rising, these numbers are only likely to increase.

Though this type of manipulation has made life better for many, Myers and others worry that by

waterborne disease An infectious disease acquired through contact with contaminated water.

vector-borne disease An infectious disease acquired from organisms that transmit a pathogen from one host to another.

infectious disease An illness caused by an invading pathogen such as a bacterium or virus.

pathogen An infectious agent that causes illness or disease.

so dramatically altering the ecosystems around us, we have not only imperiled our access to some of the most basic components of human health—namely adequate nutrition, safe water, and clean air—but have also increased our exposure and vulnerability to both natural disasters (see LaunchPad Chapter 28: Forest Resources) and infectious diseases.

In the northeastern United States, for example, a cascade of interrelated environmental changes have helped drive the incidence of Lyme disease upward. Habitat fragmentation (a consequence of land development) has increased the edge habitats where field mice thrive. The mice are the main host reservoir for the bacterium that causes Lyme disease: It thrives in them and can be passed from host to host via ticks that bite the mice and then bite other animls. Biodiversity loss has also played a role. Unlike field mice, squirrels and possums are usually able to kill and remove the ticks before becoming infected. As biodiversity decreases, and populations of these less vulnerable hosts diminish, the ticks increasingly end up on the mice that more readily transmit the disease, building the infected tick population and increasing the chance that one will bite a person and transmit the disease.

Meanwhile, in the Amazon basin, deforestation has increased the breeding habitat for mosquitoes that transmit malaria. In Cameroon, it has altered aquatic habitats in ways that favor a schistosomiasis-carrying snail. In Asia and South America, monsoon rains and agricultural runoff have conspired to alter the salt and nutrient levels of coastal waters in ways that favor cholera population explosions. Floods that washed away latrines, emptying sewage into nearby waters, also sparked a 2012 cholera outbreak in Malawi.

Throughout the developing world, both dam building and urbanization have increased the incidence of a wide range of waterborne pathogens, including but not limited to GWD. In cities, not only do people live in much closer quarters, but an infinite variety of humanmade objects—namely old tires and discarded plastic food containers—find second life as vessels for rainwater that collects during wet seasons, bakes in sunlight, and grows dank and stagnant over time. This water provides an excellent habitat for a whole suite of vectors that transmit a host of diseases: mosquitoes that carry the dengue virus and the *Plasmodium* protozoan that causes malaria, black flies and snails that carry worms that cause debilitating diseases like elephantiasis, river blindness (onchocerciasis), and schistosomiasis. Dams—especially those in tropical regions—do something similar: They create large bodies of standing water that have been associated with an uptick in the same cadre of diseases.

In general, these hazards can be divided into three broad categories: physical hazards (see Chapter 2), chemical hazards (see Chapter 3), and biological hazards (the focus of this chapter). *Biological hazard* typically refers to infectious diseases—illnesses caused by an invading pathogen such as a bacterium, virus, or parasite. However, as illustrated above, we are never exposed to any hazard—be it physical, chemical, or biological—in isolation. Rather, different types of hazards interact with one another and with other elements of the human environment in ways that can make it tricky for health-care workers to map cause-and-effect relationships. For example, someone negatively affected by a chemical hazard such as air pollution may be more susceptible to a biological hazard such as a lung infection. **INFOGRAPHIC 5.1**

Public health programs seek to improve community health.

GWD has been around for centuries but persists today only in remote populations whose only drinking water comes from ponds or other standing water sources—ideal habitat for the copepod vectors. But Ruiz-Tiben and his staff at the Carter Center were aided in their quest by a simple fact: Unlike some of those other infectious organisms, Guinea worms are utterly dependent on humans to complete their life cycle; humans are the only known reservoir for adult worms (larvae can only survive in copepods, and only for a few weeks). That meant Ruiz-Tiben and his colleagues could break the Guinea worm's life cycle—and thus obliterate the disease—simply by changing human behavior.

Convincing villagers to filter their water before drinking it would stop new infections. Getting the villagers to apply a mild pesticide would decontaminate area ponds. And teaching people to avoid communal swimming or bathing while worms were emerging from the skin, and to treat new infections as soon as blisters emerge, would break the worm's life cycle once and for all. Their goal was total eradication of GWD; if they succeeded, it would be only the second disease in human history (after smallpox) to be completely wiped off the face of Earth.

KEY CONCEPT 5.2

Environmental hazards can be biological, chemical, or physical in nature, and any given hazard can interact with others, making these hazards more difficult to address.

INFOGRAPHIC 5.1   TYPES OF ENVIRONMENTAL HAZARDS

↓ The WHO recognizes that modifiable environmental factors contribute significantly to disease, injury, and death worldwide. Environmental hazard categories can be broken down in a variety of ways, but here is one simple breakdown: physical, chemical, and biological hazards. All of these hazards can interact, sometimes in unpredictable ways, which increases the complexity of public health issues. Of these three groups, biological hazards remain the major risk in low-income nations, the areas where death rates are the highest.

BIOLOGICAL HAZARDS
These types of hazards include infectious agents (pathogens), which can be bacteria, viruses, protozoa, fungi, and worms. Water- and vector-borne diseases are the major infectious disease threat worldwide, though diseases may also be spread via contaminated food or direct person-to-person contact.

Someone whose health is impacted by pollution may be more vulnerable to infection by a biological hazard; and those weakened by infection may be more affected by air pollution.

Floodwaters are likely to be contaminated with pathogens or provide a habitat for mosquitoes (biological hazards). Additionally, people who are ill or malnourished may be less able to cope with such natural disasters.

CHEMICAL HAZARDS
This type of hazard includes environmental pollution and exposure to hazardous chemicals in the home or workplace or that occur naturally in the environment (such as arsenic in water). The largest contributor in this group is smoke from burning solid fuels indoors and smoking, including secondhand smoke.

Natural disasters can increase the risk of exposure to toxic chemicals. For example, harmful chemicals can be flushed from industrial or agricultural sites into residential areas during a flood.

PHYSICAL HAZARDS
Natural disasters, extreme weather events, exposure to ultraviolet radiation, and even traffic accidents are examples of physical hazards.

? How might biological and chemical hazards be affected during a natural disaster such as a flood?

As straightforward as it all sounded, Ruiz Tiben had not had much luck so far. Despite his careful detailing of the science, most of the villagers he had spoken with still believed—rather fiercely—that the sickness was delivered by angry gods, as punishment for various misdeeds. And if that explanation seemed ludicrous to Ruiz-Tiben, well, his counter-explanation—that the worms actually came from water the villagers had consumed a year prior—seemed equally ludicrous to them. He thought he might make some headway by showing them the dead bugs in the water.

Public health is a field that deals with the health of human populations as a whole. Public health **epidemiologists** like Ruiz-Tiben work to gauge the overall health status of a population or even of a nation. They use statistical analysis (e.g., rates of infant mortality, incidence of various diseases) to identify specific health threats to

public health The science that deals with the health of human populations.

epidemiologist A scientist who studies the causes and patterns of disease in human populations.

KEY CONCEPT 5.3

Public health officials track the health of a population, identifying health threats and implementing or recommending ways to mitigate those threats.

INFOGRAPHIC 5.2

Environmental health is a branch of public health that focuses on potential health hazards in the natural world and the human-built environment; such hazards include not only things like contaminated water, air,

groups of people; they then recommend ways to mitigate those threats. Devising a plan of action is often the trickiest part; it requires the study of a whole host of interacting variables—from cultural and social forces (which influence things like diet and smoking habits) to economic stability or instability (which determines a given population's access to resources) to environmental factors like water cleanliness or changes in habitat that affect disease transmission.

and soil but also human behaviors—hand washing and water drinking, for example—that help determine whether those natural factors become hazards. According to the WHO, environmental hazards are responsible for about 25% of disease and deaths worldwide.

Fortunately, many environmental hazards are *modifiable*—that is, we can take action to change them. Like GWD, they can be mitigated (indeed, some 13 million deaths could be prevented each year) through reasonable measures. It is on these modifiable hazards—like contaminated drinking water—that environmental public health workers like Ruiz-Tiben focus their efforts.

To be sure, infectious diseases account for less of the global burden of disease than **noncommunicable diseases (NCDs)** like cardiovascular illness and diabetes. (Also known as lifestyle diseases, NCDs are largely determined by choices about things like diet and exercise.) In fact, NCDs cause the most deaths globally.

INFOGRAPHIC 5.2   PUBLIC HEALTH PROGRAMS SEEK TO IMPROVE HEALTH OF THE POPULATION AS A WHOLE   2

↓ The goal of public health programs is to improve the health of human populations through prevention and treatment of disease at the community level.

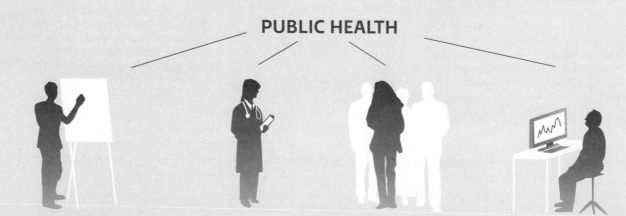

PUBLIC HEALTH

EDUCATES
Public health professionals provide information and health care advice to communities. Changing behaviors can be a critical part of improving public health.

PROVIDES HEALTH CARE
Public health care workers provide needed preventive medical care and treatment.

PROPOSES ACTIONS
Once risks are identified, public health professionals make recommendations to improve health in specific groups and in the population as a whole.

CONDUCTS RISK ASSESSMENTS
Epidemiologists analyze statistics related to a population's health to determine risk for various groups in the population (e.g., children, the elderly, the chronically ill) to a variety of environmental factors (e.g., sanitation, pollution, climate change).

? How does the U.S. childhood vaccination program protect the health of children too young or ill to receive the vaccinations?

INFOGRAPHIC 5.3   ENVIRONMENTAL FACTORS CONTRIBUTE TO THE GLOBAL BURDEN OF DISEASE

2

↳ Modifiable environmental factors contribute to some diseases more than others. This is a result of how a particular disease is transmitted (e.g., if it is a waterborne disease) and how well a particular region is able to address environmental conditions (e.g., whether there is clean water).

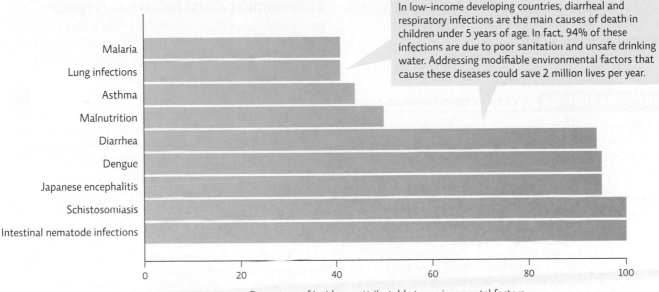

DISEASES WITH A HIGH ENVIRONMENTAL BURDEN

In low-income developing countries, diarrheal and respiratory infections are the main causes of death in children under 5 years of age. In fact, 94% of these infections are due to poor sanitation and unsafe drinking water. Addressing modifiable environmental factors that cause these diseases could save 2 million lives per year.

Percentage of incidence attributable to environmental factors

? Other than pathogens, what environmental factors might be contributing to asthma and lung infections?

But infectious diseases are still a major problem. They account for about 26% of deaths worldwide each year, the vast majority of them in the developing world.

The main environmentally mediated infectious diseases are diarrheal diseases (due to environmental factors like unsafe drinking water or poor hygiene and sanitation), lung infections (linked to air pollution), and mosquito-transmitted diseases like malaria and dengue fever. **INFOGRAPHIC 5.3**

KEY CONCEPT 5.4

Many parasitic infections and other health threats are linked to contaminated water and air; modifications to the environment can reduce the risk of these infections.

Unlike GWD, for which humans are the only host for the worm, many of these diseases are **zoonotic**—that is, they can spread between infected animals and humans. In fact, 75% of all **emerging infectious diseases**—those that are new to humans or have rapidly increased

their range or incidence in recent years—are zoonotic; this includes new strains of influenza, one of the most common zoonotic diseases. About 72% of zoonotic diseases come from wildlife, a number that is increasing due to human encroachment into formerly wild areas and the *bushmeat trade*—killing wild animals such as primates for food. Many pathogens that infect wild primates can also infect humans, and the hunting and consumption of those animals increases the chances that individuals will be exposed to these pathogens.

Environmental changes are believed to play a role in many of these emerging infectious diseases. In the United States,

environmental health The branch of public health that focuses on factors in the natural world and the human-built environment that impact the health of populations.

noncommunicable diseases (NCDs) Illnesses that are not transmissible between people; not infectious.

zoonotic disease A disease that is spread to humans from infected animals (not merely a vector that transmits the pathogen but another host that harbors the pathogen through its life cycle).

emerging infectious diseases Infectious diseases that are new to humans or that have recently increased significantly in incidence, in some cases by spreading to new ranges.

recent increases in cases of West Nile virus may be linked to climate change—specifically extreme heat and drought. West Nile virus first showed up in the United States in 1999 in New York City; since then it has rapidly spread from coast to coast. High temperatures increase the ability of the mosquito vector to pick up the virus from its bird hosts. Though mosquitoes require water to breed, droughts tend to increase urban mosquito populations by increasing standing water in drains that would normally be flushed out by rains. With the hottest summer on record in Texas and other areas in the United States, 2012 broke the record for deaths due to

West Nile virus (286 deaths out of 5,674 reported cases). Such outbreaks may become more common as a warming global climate allows vectors like mosquitoes to expand their range and thrive in areas where they were formerly scarce or nonexistent. **INFOGRAPHIC 5.4**

Addressing biological hazards requires environmental and behavioral changes.

In the Nigerian village where Ruiz-Tiben was working, however, the main problem was not urbanization, or habitat fragmentation, or even deforestation. Rather,

---

INFOGRAPHIC 5.4   **A VARIETY OF PATHOGENS CAUSE DISEASE**   3

↓ A wide variety of infectious agents can cause disease in humans and other organisms. The main pathogenic threats to human health are waterborne and vector-borne diseases, with the mosquito being the number-one vector of pathogens. Public health officials follow emerging infectious diseases closely; many of these diseases are zoonotic, and their recent rise is usually related to environmental factors that increase the spread of the pathogen or its vector.

PATHOGEN CLASS	VIRUSES	BACTERIA	PROTOZOA	FUNGI	WORMS
	Dengue, Japanese encephalitis, and West Nile virus are emerging viral diseases that are spread by mosquitoes. Increased habitat, international human travel, and climate change are all implicated in these diseases. Other viruses are spread by direct contact or by contaminated water.	Bacterial pathogens can be spread by contaminated water (typhoid fever) or through vectors (Lyme disease). The 2010–2011 cholera outbreak in Haiti (caused by the *Vibrio* bacterium) killed more than 7,000 people.	Protozoa are single-celled organisms. Some are pathogenic and cause diseases such as giardiasis, contracted by drinking contaminated water, and African sleeping sickness. Malaria, caused by the *Plasmodium* protozoan, kills as many as 1 million people annually, mostly African children.	Many fungal infections plague humans. Ringworm is not a worm at all but is caused by a fungus. It is most common in areas with inadequate water for washing. Candidiasis is another common fungal infection; outbreaks in hospitals are linked to the overuse of antibiotics. (The fungus can thrive when bacteria are killed.)	A variety of worms can infect humans and other animals via ingestion or direct contact with the worm or its vector. The Guinea worm is a microscopic worm known as a nematode. Other nematode species cause trichinosis, elephantiasis, and hookworm.

EXAMPLES					
Pathogen	Dengue virus	*Vibrio cholerae*	*Plasmodium*	*Tinea corporis*	*Dracunculus medinenisis*
Transmission	Vector: Mosquito	Contaminated water	Vector: Mosquito	Contaminated water or direct contact	Vector: Copepods
Disease	Dengue fever	Cholera	Malaria	Ringworm	Guinea worm disease

[?] How has human travel contributed to the spread of vector-borne diseases normally found only in tropical areas?

KEY CONCEPT 5.5

A wide variety of pathogens cause disease and many diseases can pass from one species to another (zoonotic). The range of some diseases is expanding, further threatening human populations

The main problem was a normal human trait: resistance to change. Even after the villagers saw the bugs in the water (which he had killed with a dash of the pesticide Abate), they continued to resist using both the water filters and pesticides. When Ruiz-Tiben discovered a hidden pond infested with copepods, the women of the village formed a human shield around it so that his team could not treat it with Abate. The pond was a sacred ancestral pool, they insisted. To douse it with chemicals would invite the wrath of their gods.

The belief was anchored in an ancient history. In fact, GWD is itself ancient; it has been traced all the way back through the Old Testament and to the Egyptian mummies. Some say the very symbol of modern medicine—a serpent coiled around a staff (known as a caduceus)—derives from our treatment of GWD: To prevent the worm from breaking off at the blister, it is wound slowly around a stick as it emerges from the body—a practice that has not changed for thousands of years.

Both the worms and the copepods that carry them are native to Africa and Asia; both evolved across human history to exploit human hosts and human water sources. The copepods thrive in open, stagnant water sources—like ponds and pools formed by dams—whose availability varies by season and region. In the Sahelian zone, transmission generally occurs in the rainy season (from May to August), when shallow ponds grow deep enough to bathe in. In the humid savanna and forest zones, infections peak during the dry season (from September to January),

KEY CONCEPT 5.6

Guinea worm disease (GWD) is contracted by drinking water contaminated with worm larvae. Because humans are needed for the worm to complete its life cycle, if we can prevent exposure, GWD could be eradicated.

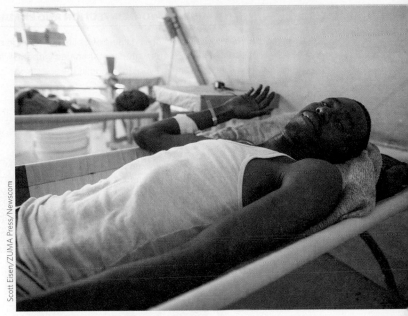

Scott Eisen/ZUMA Press/Newscom

↑ A man who is recovering from a severe case of cholera at the Doctors Without Borders clinic in Haiti.

when water holes shrivel and grow still, inviting copepods to multiply.

The only humans who come in contact with copepods are those who drink unpurified water from these sources. Like most other such people, the villagers Ruiz-Tiben was working with were poor and lived in a naturally dry area; that meant that for generations, they had had no choice but to drink and bathe in whatever water they could find. Consistent water sources—those that do not dry up when the rains pass, or evaporate during particularly hot years—were sacred. And the women were understandably leery of polluting their most treasured supply with a foreign chemical that had been made in a distant land.

It was not until a revered general from the region intervened on Ruiz-Tiben's behalf that the women finally relented. The general, a former president of Nigeria, assured the women that the pesticide would not harm their fish but would instead keep their families from getting sick. Their ancestors, he said, would not want them to be sick. Reluctantly, the human shield broke up, and Ruiz-Tiben and his team were able to treat the pond with Abate, ridding it of copepods. **INFOGRAPHIC 5.5**

But as the Carter Center's army of public health workers would soon discover, Nigeria's challenges paled in comparison to the obstacles faced by other Guinea worm–infested countries.

INFOGRAPHIC 5.5 **GUINEA WORM INFECTION AND ERADICATION PROGRAMS**   3

↓ The Guinea worm is a parasite that spends part of its life cycle inside copepods (water fleas) and part in a human host. Humans are exposed to the parasite when they drink water contaminated with the water fleas. Because there is no other animal host in which the worm can complete its life cycle, if we can prevent infection at the source, we can eradicate this parasite.

THE LIFE CYCLE OF THE GUINEA WORM

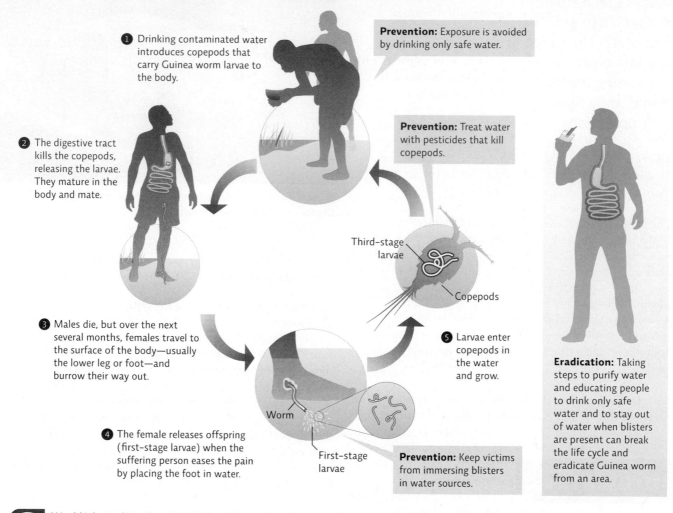

❶ Drinking contaminated water introduces copepods that carry Guinea worm larvae to the body.

Prevention: Exposure is avoided by drinking only safe water.

❷ The digestive tract kills the copepods, releasing the larvae. They mature in the body and mate.

Prevention: Treat water with pesticides that kill copepods.

Third-stage larvae

Copepods

❸ Males die, but over the next several months, females travel to the surface of the body—usually the lower leg or foot—and burrow their way out.

❺ Larvae enter copepods in the water and grow.

Worm

Eradication: Taking steps to purify water and educating people to drink only safe water and to stay out of water when blisters are present can break the life cycle and eradicate Guinea worm from an area.

❹ The female releases offspring (first-stage larvae) when the suffering person eases the pain by placing the foot in water.

First-stage larvae

Prevention: Keep victims from immersing blisters in water sources.

Would it be technically possible to eradicate GWD if only step 4 of the Guinea worm's life cycle (delivery of eggs into water) were eliminated?

The factors that affect human health differ significantly between more and less developed nations.

Back in 1995, when Sudan was still one nation with two warring factions—north versus south—Nabil Aziz Mikhail bore witness to a quiet sort of miracle.

The country's Ministry of Health had long reported Guinea worm cases in the low thousands. But Mikhail,

who had just assumed the role of Guinea Worm Eradication Coordinator, had quickly discovered that the number was much, much higher than that: A better estimate was 100,000-plus cases. Mikhail knew that simple measures, such as those the Carter Center was employing elsewhere, could stop the disease in its tracks. But he also knew that no such measures could be employed in Sudan, marred as it was by poverty and extreme violence. The most afflicted areas were simply too dangerous to venture into. Even if they could be

KEY CONCEPT 5.7

People in wealthier nations are more likely to die from life-style diseases, whereas in low-income countries, the leading causes of death are infectious diseases, many of which are linked to poor environmental conditions.

reached, eradication programs—community education, latrine building, even pesticide application—would be impossible to implement under the circumstances.

Here's where the miracle comes in: Desperate to make a dent in the problem, Mikhail and his colleagues called former U.S. President Jimmy Carter and invited him to host a conference on GWD in their war-torn home. Carter went a step further: Not only did he come to Sudan, but he quickly negotiated what would later be called the "Guinea Worm Cease Fire"—a laying down of arms that lasted 6 full months and allowed public health workers to secure unprecedented gains in the region. Infected water sources were detected and decontaminated, filters were distributed, and active infections were treated—on a scale the country had never seen before. "It was a dream," Mikhail says now. "I've never heard of a health activity that brought any sort of cease-fire."

It was also a lesson, Mikhail says: If there's one human behavior that favors the Guinea worm even more than bathing in infested water, it's war. Plain and simple.

In fact, war is just one reason that the death rate from environmentally mediated diseases is much greater (12 times greater, in 2006) in less developed nations than it is in more developed ones. Poverty is another; poor basic nutrition is another still.

As we discuss elsewhere in this book, the differences between more developed and less developed countries are vast. In more developed countries, cardiovascular disease and cancer represent the bulk of the disease burden. (These are caused by lifestyle choices and industry and vehicle pollution.) In less developed countries, while NCDs are ticking upward, infectious diseases, caused by all the environmental factors we've already discussed (e.g., lack of clean water, poor sanitation, burning of solid fuels indoors for heat and energy), are still the leading cause of death. **INFOGRAPHIC 5.6**

The Sudan cease-fire gave health workers a fighting chance to address the environmental conditions that favored Guinea worms. "Once the violence stopped,

progress was imminent," says Ruiz-Tiben. "The Sudanese health workers and foreign nongovernment organizations surprised themselves with what they were able to accomplish in those 6 months. But when the cease fire ended, the infection rates climbed right back up." Today, though South Sudan became an independent nation in 2011, violence persists and has left the country one of the few still plagued by GWD. The number of infections in South Sudan is beginning to fall once again, from 521 cases in 2012 down to 113 cases in 2013. Those 113 South Sudan cases represented 76% of the world's GWD cases in 2013; the remaining 35 cases were reported in Chad, Ethiopia, and Mali.

Environmentally mediated diseases can be mitigated with funding, support, and education.

In Ezza Nkwubor, in southeastern Nigeria, a team of elderly men—all local villagers—stand guard over a large, silent pond. The water has been treated with pesticides, and the men's job is to ensure that nobody with emerging worms comes into contact with it. Elsewhere, the sick have been quarantined—their wounds carefully tended and water and free food brought to them for the entire month that it takes the worm to emerge. Young boys—travelers and hunters—carry whistle-shaped cylinders tied to strings around their necks. The cylinders serve as portable filters so that the boys can drink directly from an environmental water source and still protect themselves from infection when they're out hunting. And season after season, women teach their protégés the importance of filtering water before giving it to their families.

KEY CONCEPT 5.8

Steps that improve air quality, sanitation, and access to clean water can reduce environmental health hazards but require education and effective public policy for success.

It's the picture of success that Ruiz-Tiben and his colleagues have envisioned for decades. "It just shows what you can accomplish when the support is there," Ruiz-Tiben says, underscoring what has been a key lesson of the GWD eradication campaign: Implementing even the simplest technologies requires financial and political support—not only from the developed world or the international community but from the countries themselves.

INFOGRAPHIC 5.6   DEATH RATES AND LEADING CAUSES OF DEATH DIFFER AMONG NATIONS   4

ENVIRONMENTAL DISEASE BURDEN, 2004

Deaths per 100,000 population

- < 150
- 151–200
- 201–350
- 351–500
- 501–1,000
- No data

→ Death rates due to modifiable environmental factors are highest in developing countries; almost half of those deaths are in children. Poverty that restricts access to medical care, clean water, and an adequate diet is perhaps the leading "health risk" worldwide. This is nowhere more evident than in the low-income nations of Africa. Over 70% of people in high-income countries will live to at least age 70; in middle-income nations, that number drops to 40%; and in low-income countries, only 17% will make it to age 70.

→ The main causes of death differ from country to country, depending on income level. A World Health Organization comparison of low-, lower-middle-, and high-income countries shows that infectious disease is more of a problem in low-income countries. As income increases, so does the burden of disease caused by lifestyle diseases related to diet and exercise. Upper-middle-income countries (not shown here) continue the transition toward lifestyle-related causes of mortality that is begun in lower-middle income countries.

 Why do you think middle-income nations suffer both from lifestyle factors and from infectious diseases like diarrheal diseases and tuberculosis?

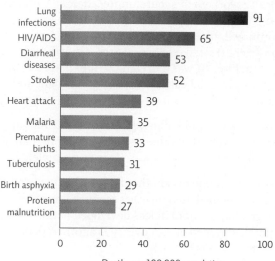

Top 10 causes of death in low-income countries 2012

Cause	Deaths per 100,000
Lung infections	91
HIV/AIDS	65
Diarrheal diseases	53
Stroke	52
Heart attack	39
Malaria	35
Premature births	33
Tuberculosis	31
Birth asphyxia	29
Protein malnutrition	27

Deaths per 100,000 population

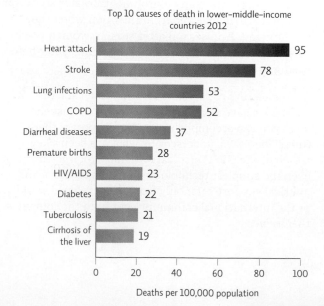

Top 10 causes of death in lower–middle-income countries 2012

Cause	Deaths per 100,000
Heart attack	95
Stroke	78
Lung infections	53
COPD	52
Diarrheal diseases	37
Premature births	28
HIV/AIDS	23
Diabetes	22
Tuberculosis	21
Cirrhosis of the liver	19

Deaths per 100,000 population

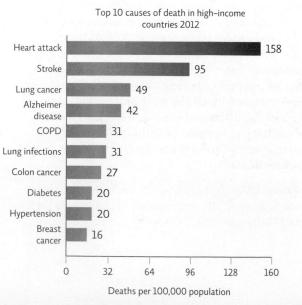

Top 10 causes of death in high-income countries 2012

Cause	Deaths per 100,000
Heart attack	158
Stroke	95
Lung cancer	49
Alzheimer disease	42
COPD	31
Lung infections	31
Colon cancer	27
Diabetes	20
Hypertension	20
Breast cancer	16

Deaths per 100,000 population

TABLE 5.1   REDUCING ENVIRONMENTAL HEALTH HAZARDS

Provide access to clean water	Digging wells and filtering surface water, including using personal filters such as the LifeStraw, are two common methods to purify water, but these methods cost money, and many low-income countries need financial assistance to make access to clean water a reality.
Improve sanitation and hygiene	Proper disposal of human and animal fecal waste includes methods to keep sewage out of surface waters such as building latrines, planting streamside vegetation to reduce runoff, and keeping animals out of water sources.
Reduce vector exposure	Removing vector habitat (like standing water for mosquitoes), providing barriers like mosquito netting, and applying pesticides can significantly reduce infection. Vaccinating pets can also reduce exposure to zoonotic diseases.
Reduce air pollution	Using cleaner-burning fuels and better-constructed and better-ventilated indoor stoves reduces indoor air pollution. Solar ovens can eliminate smoke and are inexpensive and effective. Adopting and enforcing air quality standards can reduce outdoor air pollution.
Education	Public education programs are vital in areas where simple behavioral steps can reduce exposure to hazards. Educational campaigns can teach people how to avoid exposure to pathogens and how to protect themselves from infection and hazardous chemicals.
Effective public policy	Government support is needed to reduce environmental hazards and improve health care. (Governments can accept foreign aid if they cannot provide support directly.) Unfortunately, not all areas are politically stable, and armed conflict or disinterested leaders can disrupt progress.

What do you think will be the biggest impediment to the eradication of GWD in Africa—the implementation of technical solutions to provide safe water or the education of people in local communities on how to reduce their exposure to contaminated water?

Yes, it is relatively cheap and technologically simple to build pit latrines and septic systems that make proper waste disposal possible, or to build fences that can keep animals out of human water supplies, or to plant vegetative buffers that can soak up runoff before it pollutes area streams. And yes, when combined, such a roster of straightforward measures might dramatically reduce the incidence of any number of life-threatening diseases (including, for example, the diarrheal diseases that kill so many children each year). But without the money to buy wood (for fences) or plants (for buffer zones), and without know-how and "local buy-in," such projects would never get off the ground.

Even education campaigns that help people understand how to avoid exposure to certain infectious agents—a measure that requires almost no material support—still take a concerted effort, and thus a well-trained workforce. "It's a lot of work to overcome preconceived notions," says Ruiz-Tiben. "You have to present the information in a way they can relate and respond to." **TABLE 5.1**

To generate those things—international support, local support, and so on—environmental health workers from the developed world need, urgently, to consider the perspectives of their developing-world brethren.

INFOGRAPHIC 5.7   ERADICATING GUINEA WORM DISEASE   5

↓ Eradication programs have been in effect since 1980 and have made great progress. Prevention methods include purifying water and education about how to acquire safe water sources and handle infections to avoid reintroducing larvae into the water. Donald Hopkins, the Carter Center Vice President of Health Programs, wrote in a 2011 report that the last cases are always the hardest to wipe out, but doing so is not a matter of *if* but *when*.

1986

3.5 million cases in 20 countries

2013

148 cases in 4 countries

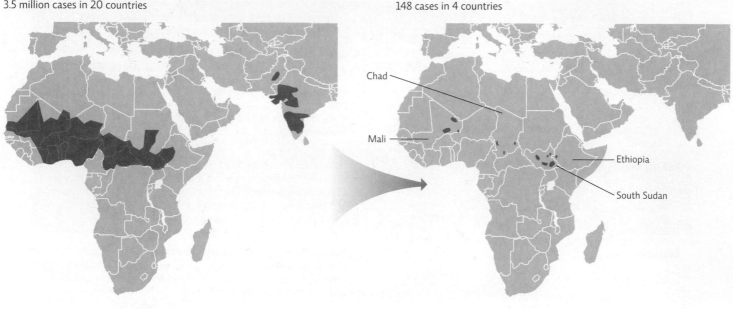

Chad

Mali

Ethiopia

South Sudan

Climate change is projected to reduce rainfall in some areas where GWD still persists. Will this help or hurt the fight to eradicate GWD?

Consider the case of DDT: The pesticide can go a long way toward keeping insect vectors like mosquitoes from human hosts. But when scientists in the developed world linked it to a roster of poor health and poor environmental outcomes (see Chapter 3), the chemical was banned in many regions, including places where malaria is common. Leaders in those countries were, by many accounts, responding to pressure from the developed world. Today, world leaders are reconsidering: In some malaria hotspots—certainly in places where mosquito-borne diseases kill tens of thousands of people every year—it turns out that judiciously applied DDT still provides the best mitigation strategy and may yet prove to be worth those risks.

As with most other environmental problems, there are no easy solutions. To conquer GWD, environmental health workers like Ruiz-Tiben have had to battle indifference, poverty, human stubbornness, and more. In the end, those battles have paid off: In 2009, Nigeria became the 15th African country to rid itself of the ancient worm. At that time, it was estimated that just 3,500 or so cases remained throughout the entire continent, and those numbers were dwindling rapidly. By 2012, only 542 cases were reported in all of Africa; in 2013, that number had dropped to 148. Indeed, some three decades after beginning its quest, the Carter Center is finally closing in on its ultimate goal: eradicating Guinea worm disease, everywhere, once and for all. **INFOGRAPHIC 5.7**

Select References:
The Carter Center. "Health Programs," www.cartercenter.org/index.html.
Guimarãe, R. M., et al. (2007). DDT reintroduction for malaria control: The cost–benefit debate for public health. *Cadernos de Saúde Pública*, 23(12): 2835–2844.

LoGiudice, K., et al. (2003). The ecology of infectious disease: Effects of host diversity and community composition on Lyme disease risk. *Proceedings of the National Academy of Sciences*, 100(2): 567–571.
Myers, S. S., & J. A. Patz. (2009). Emerging threats to human health from global environmental change. *Annual Review of Environment and Resources*, 34: 223–252.

BRING IT HOME

PERSONAL CHOICES THAT HELP

People in developed countries typically do not experience the same prevalence of infectious disease as do those in the developing world. However, outbreaks of illnesses like whooping cough, West Nile virus, bacterial food poisoning, and antibiotic-resistant bacterial infections do occur in the developed world and are largely preventable.

Individual Steps
• Many diseases are spread by contaminated hands. The most effective way to remove infectious bacteria and viruses is with 20 seconds or more of washing with soap and water. This is even more effective than using hand sanitizer.
• Reduce the likelihood that antibiotic-resistant bacteria will emerge by taking the entire prescription of any antibiotic you are prescribed.

Group Action
• Mosquitoes are responsible for spreading many diseases. Organize your neighbors to take preventive steps to reduce mosquito breeding, including removing containers that might trap rainwater, draining areas of standing water, and cleaning out rain gutters.
• Organize a fund-raising campaign to help purchase pipe filters such as the LifeStraw or finance well digging in areas that need access to clean water.

Policy Change
• While the Safe Drinking Water Act requires the Environmental Protection Agency to test public water supplies for contaminants and bacteria, the Food and Drug Administration is not empowered to require the same level of testing of bottled water. Ask your legislators what steps could be taken to hold bottled water to similar standards.

ENVIRONMENTAL LITERACY UNDERSTANDING THE ISSUE

 1 What types of environmental hazards impact the health of people?
INFOGRAPHIC 5.1

1. Infectious agents that cause disease are called _____ .

2. What route of exposure represents the main infectious disease threat to humans?
   a. Water- and vector-borne diseases
   b. Disease caused by malnutrition
   c. Direct transmission from human to human
   d. Disease caused by eating contaminated food

3. Environmental hazards can be divided into physical, chemical, and biological hazards. Give an example of each of these hazards. Describe a scenario in which exposure to one of the hazards could make a person more vulnerable to another type of hazard.

 2 How do the fields of public health and environmental health help improve the health of human populations? What environmental factors contribute to the global burden of disease?
INFOGRAPHICS 5.2 AND 5.3

4. Which of the following is NOT true of environmental health?
   a. It is a branch of public health.
   b. It focuses only on how natural disasters impact health.
   c. It focuses on health risks from the natural and human-built environment.
   d. It focuses on human health, not on the health of other species in the environment.

5. Zoonotic diseases:
   a. are increasing in frequency worldwide.
   b. come mainly from domesticated animal species.
   c. are diseases that spread between humans and plant species.
   d. All of the above:

6. How do the focus and goal of public health programs differ from those of individual health care providers?

 3 What types of pathogens cause disease? Specifically, what factors facilitate the spread of Guinea worm disease, and what steps are needed to eradicate it?
INFOGRAPHICS 5.4 AND 5.5

7. Which of the following can be pathogens?
   a. Bacteria and viruses
   b. Fungi
   c. Protozoa and worms
   d. All of the above

8. How is the increase in standing water in cities linked to increased human health hazards?
   a. It increases habitat for vectors such as mosquitoes.
   b. It reduces the amount of water available for irrigation.
   c. It increases the chance for flooding.
   d. None of the above; standing water is not a health hazard.

9. Guinea worm disease is still found in South Sudan. What can be done to eradicate it, and why has this not already been done?

 4 How do factors that affect human health differ between more and less developed nations?
INFOGRAPHIC 5.6

10. Compared to those in high-income nations, people in low-income developing nations:
    a. are more likely to die from infectious diseases.
    b. develop more genetically based diseases.
    c. do not develop "lifestyle" diseases.
    d. are less affected by environmental hazards.

11. Compare the leading causes of death in low-, middle-, and high-income countries. Why do these differences exist?

 5 What can be done to reduce environmentally mediated health problems? How close are we to eradicating GWD?
TABLE 5.1 AND INFOGRAPHIC 5.7

12. Which one of the following factors reduces our ability to address many environmental health problems?
    a. Scientists don't understand how most infectious diseases are spread.
    b. Most environmental health problems are in rural areas, far from health professionals.
    c. People in developed countries don't care about the health of those in developing countries.
    d. Affected regions may not have the political will or financial means to address environmental problems.

13. Which action below would help reduce the health problem that has the largest percentage of its occurrence attributed to environmental conditions?
    a. Provide solar ovens to low-income areas.
    b. Provide better sanitation and clean water.
    c. Improve diets to help people lose weight.
    d. Use cleaner-burning fuels indoors.

14. Identify possible interventions that could decrease health problems associated with air pollution. (Smoking, pollutants due to burning solid fuels indoors without adequate ventilation, and vehicle exhaust are the most common air pollutants that impair health.)

SCIENCE LITERACY WORKING WITH DATA

Accurate estimates of deaths due to malaria are important for many reasons, including decisions regarding vector control and health interventions, and also to direct charitable donations effectively. A recent analysis estimated much higher annual death rates from malaria than did previous studies, with the majority of the increase coming from victims over age 5. The table below shows the annual number of deaths from malaria in different world regions, estimated by the Institute for Health Metrics and Evaluation (IHME) and the World Health Organization (WHO).

Comparison of IHME and WHO Estimates of Malaria Deaths by WHO Region

WHO Region	Number of Malaria Deaths		
	IHME	WHO	Difference
Africa	1,098,818	596,000	502,818
Americas	986	1,000	–14
Eastern Mediterranean	47,499	15,000	32,499
Europe	3	—	3
Southeast Asia	84,573	38,000	46,573
Western Pacific	5,596	5,000	596
Total	1,237,475	655,000	582,475
% malaria deaths under age 5	58%	86%	

Interpretation

1. Describe in one sentence what the table shows about the total number of deaths due to malaria.

2. What region of the world has the highest mortality due to malaria? What proportion of the world total comes from this region?

3. What percentage and what number of malaria deaths do IHME and WHO estimate are of children under age 5?

Advance Your Thinking

4. In which region are the new mortality estimates most enlarged? Why do you think this might be the case?

5. It is often extremely difficult to estimate the cause of death in a developing country; health workers rely on a "verbal autopsy," in which they ask surviving family members a series of questions about the deceased's symptoms. If it were possible to confirm that a person who died was or was not infected by malaria, rather than relying on verbal autopsies, would you predict that the estimates of mortality from malaria would increase or decrease? Explain your reasoning.

INFORMATION LITERACY EVALUATING INFORMATION

Humanitarian organizations are increasingly interested in ensuring that their time and energy are invested in the most efficient and effective manner possible. The Bill and Melinda Gates Foundation has led the way with this approach and has donated a large amount of money to the Institute for Health Metrics and Evaluation (IHME) with this goal in mind.

Go to the IHME website (www.healthdata.org). Evaluate the website and work with the information to answer the following questions:

1. **What is the IHME's mission?**
   a. Review the topics listed under the "About" tab. What is the overall mission of IHME? Do you believe this mission is reasonable? Explain.
   b. How is IHME funded? Could the funding source(s) influence the IHME's mission? Explain.

2. Select the "News" link under the "News & Events" tab and look over the titles shown. Choose two of the articles and read each. Identify each article by name and complete the following evaluation for each:
   a. Is this a primary, secondary, or tertiary information source? Justify your answer.
   b. Identify the authors of the article. Do their credentials qualify them as suitable information sources?

   c. Identify a claim made in the article. Does the article give supporting evidence for this claim? If so, identify the evidence.
   d. Based on your evaluation of the article you examined, do you feel that the news coverage is consistent with the IHME's stated goals? Explain.

3. Select the "Research Articles" link under the "Results" tab. Look over the titles and choose two or more articles to evaluate. For each article, answer a–g:
   a. What type of article is this—primary research, a review article, or an opinion piece (editorial or blog)? How do you know?
   b. How does this differ from the news article you read?
   c. Is this a primary, secondary, or tertiary information source? Justify your answer.
   d. Identify the authors of the article. Do their credentials qualify them as suitable information sources?
   e. Try to access the actual article: Click on the "Read the article" link above the list of authors. (You may then have to click on another link to access the "Full Text" or a PDF of the article.) Is the full article available for you to read?
   f. The full article may or may not be available for you to read. Explain what steps you would take to access the article if a full copy is not available from this website. (Do this even if the full article is available for the article you chose.)
   g. How useful are these research articles? Do they help fulfill the mission of the IHME? Explain.

Find an additional case study online at http://www.macmillanhighered.com/launchpad/saes2e

CORE MESSAGE

Human impact on Earth can be measured in terms of our ecological footprint, which is closely tied to the way we use resources. Our economic choices, both corporate and individual, tend to focus on short-term gain rather than long-term sustainability, but we can make better and more informed decisions by taking all the costs—economic, social, and environmental—of a given action into account. Using nature as a model can help us make more sustainable choices while still supporting a viable economy.

AFTER READING THIS CHAPTER, YOU SHOULD BE ABLE TO ANSWER THE FOLLOWING **GUIDING QUESTIONS**

1

What are ecosystem services, and why are they important to ecosystems and human populations?

2

What is an ecological footprint, and how does it relate to our use of natural interest and natural capital?

WALL TO WALL, CRADLE TO CRADLE

A leading carpet company takes a chance on going green

Carpet tiles are paving the way for a more sustainable way to produce and market carpet. © Sophie James/Alamy

3

What factors influence how much human actions impact the environment, and how can we reduce that impact?

4

What are externalities and internalities in the business world, and how do they relate to true costs?

5

What concerns do ecological economists have with mainstream economics, and what suggestions do they offer to help businesses and consumers make better choices?

It was the summer of 1994, and Ray Anderson was feeling pretty good about things. His Atlanta-based company, Interface Carpet, was the world's leading seller of carpet tiles—small, square pieces of carpet that are easier to install and replace than rolled carpet—and it was raking in more than $1 billion per year. One day, though, an associate from Anderson's research division approached him with a question. Some customers apparently wanted to know what Interface was doing for the environment. One potential customer had told Interface's West Coast sales manager that, environmentally speaking, Interface "just didn't get it."

Anderson was dumbfounded. The carpet industry was not generally an eco-conscious industry; after all, synthetic carpet is made from petroleum in a toxic process that releases significant amounts of air and water pollution, along with solid waste. Indeed, Interface used more than 1 billion pounds of oil-derived raw materials each year, and its plant in LaGrange, Georgia, released 6 tons of carpet trimming waste to landfills each day. "I could not think of what to say, other than 'we obey the law, we comply,'" he recalled—in other words, his company did things by the book, in terms of the environment. Wasn't that enough? His research associate suggested that the company launch a task force to create a companywide environmental vision. Anderson agreed, albeit reluctantly.

Desperate for inspiration, Anderson began leafing through *The Ecology of Commerce*, a book by environmental activist, entrepreneur, and writer Paul Hawken, which one of his sales managers had lent him. The book told the story of a small island in Alaska, on which the U.S. Fish and Wildlife Service had introduced a population of reindeer during

⊙ **WHERE IS LAGRANGE, GEORGIA?**

Jorgen Caris/Hollandse Hoogte/Redux

↑ Ray Anderson, founder of Interface. The background displays sample pieces of his carpet tiles.

World War II. Although the reindeer thrived for a time on the available plants, eventually the population exploded beyond what the environment could support. The reindeer ultimately died out because, as Anderson explained, "you can't go on consuming more than your environment is able to renew." Yet that, he suddenly realized, was precisely what Interface was doing—using more resources than it could possibly renew. "As I read the book, it became clear that, God almighty, we're on the wrong side of history, and we've got to do something."

Anderson realized that he had to make changes to Interface; he needed to build it into a **sustainable**, environmentally sound business. "I didn't know what it

INFOGRAPHIC 6.1   VALUE OF ECOSYSTEM SERVICES

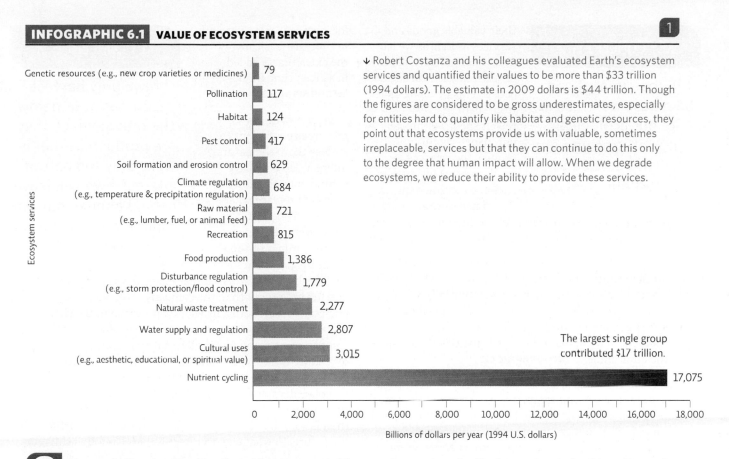

↓ Robert Costanza and his colleagues evaluated Earth's ecosystem services and quantified their values to be more than $33 trillion (1994 dollars). The estimate in 2009 dollars is $44 trillion. Though the figures are considered to be gross underestimates, especially for entities hard to quantify like habitat and genetic resources, they point out that ecosystems provide us with valuable, sometimes irreplaceable, services but that they can continue to do this only to the degree that human impact will allow. When we degrade ecosystems, we reduce their ability to provide these services.

Ecosystem service	Billions of dollars per year (1994 U.S. dollars)
Genetic resources (e.g., new crop varieties or medicines)	79
Pollination	117
Habitat	124
Pest control	417
Soil formation and erosion control	629
Climate regulation (e.g., temperature & precipitation regulation)	684
Raw material (e.g., lumber, fuel, or animal feed)	721
Recreation	815
Food production	1,386
Disturbance regulation (e.g., storm protection/flood control)	1,779
Natural waste treatment	2,277
Water supply and regulation	2,807
Cultural uses (e.g., aesthetic, educational, or spiritual value)	3,015
Nutrient cycling	17,075

The largest single group contributed $17 trillion.

? How would the price of food be affected if we incorporated the ecosystem services of pollination, pest control, soil formation, and water supply into that price?

would cost, and I didn't know what our customers would pay, so it was a leap of faith," Anderson recalled. "I knew we had to do this, but it was like stepping off of a cliff and not knowing where your foot was going to come down."

Businesses and individuals impact the environment with their economic decisions.

The choices businesses (and, by extension, consumers) make have tremendous impacts on the environment. The amount and type of energy and water they use, the way they handle the waste they produce, the raw materials they use—these decisions affect not only business operations themselves but also Earth as a whole, especially considering the magnitude of the resources and waste that some large businesses use and produce.

Businesses that are environmentally mindful aren't limited to simply trying to minimize their impact on nature; they can actually look to nature as an economic model from which to learn and model their choices. After all, **economics**—the social science that deals with how we allocate scarce resources—is not just about money. Most of the resources we depend on actually come from the environment. Environmental resources like timber and water can be considered ecosystem goods. And ecological processes like water purification, pollination, climate regulation, and nutrient cycling are essential and economically valuable **ecosystem services**, according to work published by Robert Costanza, an ecological economist currently at Australian National University. We take many of these services and resources for granted, but some are priceless because there are no substitutes—such as the oxygen produced by green plants, which we need in order to survive. **INFOGRAPHIC 6.1**

When ecosystems are intact, they are naturally sustainable: They rely on renewable resources and also provide services that help relplenish and recycle these resources. But ecosystems will only be able to provide us with

KEY CONCEPT 6.1

All species on Earth, including humans, depend on ecosystem services provided by nature. Recognizing the value of these services may motivate us to protect them.

sustainable Capable of being continued indefinitely.

economics The social science that deals with the production, distribution, and consumption of goods and services.

ecosystem services Essential ecological processes that make life on Earth possible.

KEY CONCEPT 6.2

Human impact can be measured in terms of our ecological footprint—the amount of land needed to support our lifestyle.

Like many other businesses, Interface Carpet has a large **ecological footprint**—that is, the land needed to provide its resources and assimilate its waste (typically expressed as hectares [ha] or acres [ac] per person or population). The ecological footprint is a value that businesses, individuals, and populations use to quantify their impact on the environment. **INFOGRAPHIC 6.2**

The United States, for instance, has a particularly high per capita (per person) footprint, in that it requires much more land area to support each person than it actually possesses. The country is forced to import resources from other countries and even to export some waste. In fact, if the 7 billion people who populate the planet all lived like the average person in the United States, we would need the landmass of six or more Earths to sustain everyone. According to the World Wildlife Federation's 2012 *Living Planet Report*, humans use 50% more resources than is ultimately sustainable. This means that unless we stop using them so quickly, we are going to run out.

What kinds of essential resources does Earth provide us? Considered in financial terms, our **natural capital**

their valuable goods and services as long as we let them. As Anderson came to realize, when we degrade ecosystems by using more from them than can be replenished, we threaten our planet's ability to provide the services we need, and this ultimately threatens our own future. By using nature as a model, businesses can lessen their impact on the environment and still make choices that support a viable industrial economy.

includes the natural resources we consume, like oxygen, trees, and fish, as well as the natural systems—forests, wetlands, and oceans—that produce some of these resources. Our **natural interest** is what is produced from this capital, over time—more trees and oxygen, for example—much like the interest you earn with a bank account. Natural interest represents the amount of readily produced resources that we *could* use and still leave enough behind to, in time, replace what we took. Natural interest might be represented by an increase in a fish population,

for instance, or new growth in a forest—basically, the extra that is added in a given time frame. **INFOGRAPHIC 6.3**

If we only withdraw resources equivalent to (or less than) the natural interest, we will leave behind enough natural capital to replace what we took. When Anderson spoke to his employees in the summer of 1994 about his new plan for sustainability, he stressed that his goal was to begin putting back more than the company took from the planet; in other words, he wanted Interface to be what he called a "restorative enterprise." Up to that point, the company was using up far more natural capital in the form of resources like petroleum and water than was ultimately sustainable. Anderson realized that if we take more than is replaced, capital will shrink and therefore produce less the next year. Essentially, by taking 50% more resources than is sustainable, we are taking resources away from the future, in what eco-architect Bill McDonough calls *intergenerational tyranny*. When we liquidate our natural capital more quickly than it can be replaced and call that "income," the question becomes this: Where will future income come from?

This can be an especially big problem with commonly held resources like water: Once we remove it from wells or rivers, we have to wait for the next rainfall to replenish it. When many users are accessing the resource, it can quickly become degraded if they do not work together to manage it—a tragedy of the commons (see Chapter 1).

From 1994 to 2006, Interface made major changes in order to achieve its new goals. The company cut the amount of energy it derived from fossil fuels by 55% and reduced its total energy use by 43%. It did this in part by maximizing energy efficiency in its facilities, installing skylights and solar tubes to replace artificial, electricity-dependent lighting, and installing more energy-efficient heating, ventilation, and air conditioning systems. In one of its factories, Interface also installed a real-time energy tracker that displays energy use prominently for its employees to see, inspiring them to think of new ways to conserve energy. Although Interface declined to reveal how much money it invested in such improvements and technology, the company has ultimately recouped its costs in energy savings, according to a company spokesperson.

Researchers often use the **IPAT model** to estimate the size of a population's ecological footprint, or impact (I),

KEY CONCEPT 6.3

If we only harvest resources at or below the rate at which they are produced—that is, take only the natural interest—we will leave behind enough natural capital to replace what we took

---

ecological footprint The land area needed to provide the resources for, and assimilate the waste of, a person or population.

natural capital The wealth of resources on Earth.

natural interest Readily produced resources that we could use and still leave enough natural capital behind to replace what we took.

IPAT model An equation (I = P x A x T) that measures human impact (I), based on three factors: population (P), affluence (A), and technology (T).

INFOGRAPHIC 6.2   ECOLOGICAL FOOTPRINT

↓ The ecological footprint is the land area needed to provide the resources for and assimilate the waste of a person or population and may extend far beyond the actual land occupied by the person or population; it is usually expressed as a per capita value (hectares or acres/person). The current world footprint would require about 1.5 Earths to maintain, but obviously we just have one to work with.

Raw materials used in the city are imported from elsewhere.

Some waste is assimilated by areas outside the city.

Physical footprint of the city.

Ecological footprint.

BOTH PER CAPITA IMPACT AND POPULATION SIZE AFFECT HOW MUCH ENVIRONMENTAL IMPACT A NATION HAS

↓ In 2008, the United States had the fifth-largest ecological footprint of all the countries in the world at 7 global hectares per person. (A global hectare represents the average productivity per hectare worldwide.) Fossil fuel use accounts for more than half of that footprint.

ECOLOGICAL FOOTPRINTS: THE TOP TEN COUNTRIES (2008)

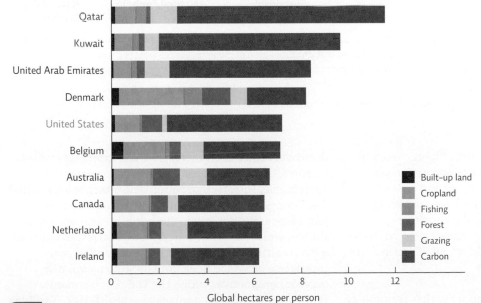

Qatar
Kuwait
United Arab Emirates
Denmark
United States
Belgium
Australia
Canada
Netherlands
Ireland

0   2   4   6   8   10   12

Global hectares per person

Legend:
- Built–up land
- Cropland
- Fishing
- Forest
- Grazing
- Carbon

If everyone on Earth had a footprint like the United States, we would need over 6 Earths.

If everyone on Earth had a footprint like China, we would have land and resources to spare.

Identify some personal choices you could make to reduce your own ecological footprint.

INFOGRAPHIC 6.3 **CAPITAL AND INTEREST**

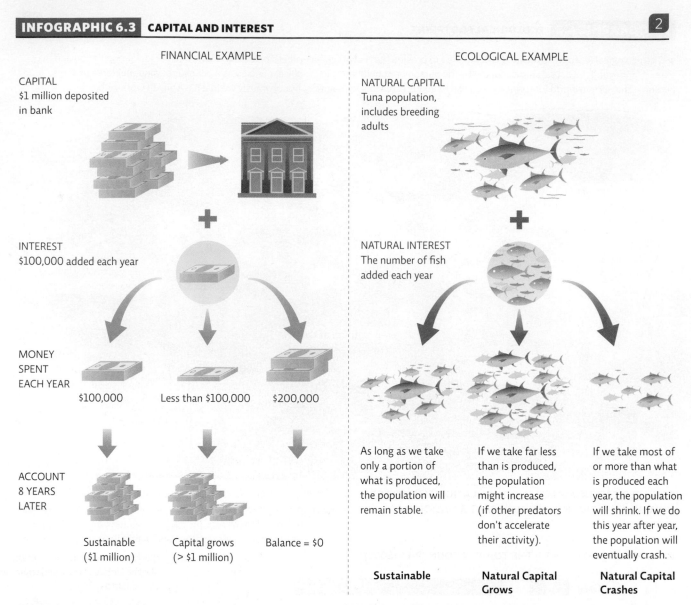

FINANCIAL EXAMPLE

CAPITAL
$1 million deposited
in bank

+

INTEREST
$100,000 added each year

**MONEY
SPENT
EACH YEAR**

$100,000 | Less than $100,000 | $200,000

**ACCOUNT
8 YEARS
LATER**

Sustainable
($1 million) | Capital grows
(> $1 million) | Balance = $0

ECOLOGICAL EXAMPLE

NATURAL CAPITAL
Tuna population,
includes breeding
adults

+

NATURAL INTEREST
The number of fish
added each year

As long as we take only a portion of what is produced, the population will remain stable. | If we take far less than is produced, the population might increase (if other predators don't accelerate their activity). | If we take most of or more than what is produced each year, the population will shrink. If we do this year after year, the population will eventually crash.

Sustainable | **Natural Capital Grows** | **Natural Capital Crashes**

↑ Natural resources can be compared to the financial concepts of capital and interest. Natural capital is the wealth of resources on Earth and includes all the natural resources we use as well as the natural systems that produce some of those resources (forests, wetlands, oceans, etc.). Natural interest is the amount produced regularly that we could use and still leave enough natural capital behind to replace what we took.

 Why might it be difficult to harvest a natural resource like a tuna population sustainably, even if we set that as our goal?

based on three factors: population (P), affluence (A), and technology (T). The premise is that as population size increases, so does impact. More affluent and technology-dependent populations use more resources and generate more waste than do less affluent and technology-dependent populations; technology allows us to build more things, dig deeper, and fly higher, all of which drain the environment. **INFOGRAPHIC 6.4**

One caveat with regard to this model is that technology can have the opposite effect: Some technologies can

decrease, rather than increase, environmental impact. In 2006, for instance, after deciding to become sustainable, Interface invented a new technology called TacTiles—2.5 × 2.5-inch squares of adhesive tape that join carpet tiles together. The adhesive is made from the same plastic used to make soda bottles. In contrast with traditional "spread on the floor" adhesives, Interface's new tape does not contain any volatile organic compounds, which the U.S. Environmental Protection Agency recognizes as a health risk. TacTiles also make it possible for customers to replace single

INFOGRAPHIC 6.4   **THE IPAT EQUATION**

↓ Population (P), affluence (A), and technology (T) all affect how much of an impact an individual or a population has on the environment. As any or each of these factors increase, so does the population's overall impact, as indicated by the model's equation: I = P × A × T. The right kind of technology can actually lower overall impact, in which case the equation becomes I= (P × A)/T.

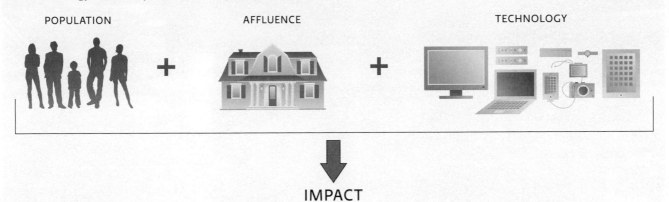

POPULATION   +   AFFLUENCE   +   TECHNOLOGY

IMPACT

 Identify some technologies that you use that increase your impact. Are there alternative technologies you could use (or propose be developed) that would decrease that impact?

KEY CONCEPT 6.4

Human impact on the environment generally increases as a population's size, affluence, and use of technology increase. However, the right technology can reduce resource use and pollution generation, thus helping to decrease impact.

carpet tiles easily, when, for instance, there has been a spill. Although resources are still required to make TacTiles, they present fewer health risks and produce less waste than other approaches. When technologies such as TacTiles reduce the environmental footprint rather than increase it, the equation used to describe their impact changes to I = (P × A)/T.

In 2007, to further reduce its impact, Interface launched a major carpet-recycling initiative called ReEntry 2.0. More than 2 million metric tons (2.5 million U.S. tons) of carpet are pulled up and discarded globally each year, and less than 5% of that has historically been reused or recycled. With ReEntry 2.0, Interface developed a way to recycle carpets—both its own and those made by its competitors—to make new carpet, using only a small amount of virgin materials to do so. With ReEntry 2.0, Interface has diverted about 90,000 metric tons (100,000 U.S. tons) of material from landfills. Interface has promised to eliminate *any* negative impact it has on the environment by 2020, in a plan it calls "Mission Zero."

Mainstream economics supports some actions that are not sustainable

Current-day, or "mainstream," economics allows managers to evaluate possible resource-use choices and make the most profitable decisions, often by seeking to *maximize value*—that is, achieving the greatest benefit at the lowest cost. However, one of the complaints against mainstream economics is that it doesn't take into account *all* potential costs when trying to maximize value. For instance, a carpet tile might require a certain amount of material that has a particular monetary cost; but what about the environmental costs associated with drilling enough oil to make that material in the first place, or the costs associated with cleaning up the pollution it creates?

The direct cost of the material is an **internal cost**—a cost that is accounted for when a product or service is priced—but it is often incomplete. There can also be **external costs**, such as the health costs associated with the waste produced by making the carpet tiles or the environmental damage caused by pollution. Historically, economists have regarded these as external to the business (the business doesn't pay for them), and they aren't reflected in the price the consumer pays for the good or service. But if the business doesn't pay for the costs or pass those costs on to

internal cost A cost—such as for raw materials, manufacturing costs, labor, taxes, utilities, insurance, or rent—that is accounted for when a product or service is evaluated for pricing.

external cost A cost associated with a product or service that is not taken into account when a price is assigned to that product or service but rather is passed on to a third party who does not benefit from the transaction.

Michiel Wijnbergh/Hollandse Hoogte/Redux

↑ Interface produces carpet made from yarns that contain up to 100% recycled content. The company has procedures in place to retrieve thread and carpet trimming from the production floor and recycle those into new product. They also reclaim used fishing nets and used carpet and recycle the components into new carpet.

the consumer, who does pay? Other people, present and future, and other species do. They pay in the form of degraded health, ecosystems, or opportunities.

An assessment of the cost of a good or service (or any of our choices) should include more than just the economic costs; it should also include the social and environmental costs—the **triple bottom line**. By ignoring the external costs, economies create a false idea of the true and complete costs of particular choices. A customer may pay $8 for every 50-square-centimeter (about 8 square inches) carpet tile, but the **true cost** for that piece of carpet would be much higher if it included the cost of greenhouse gas emissions and, for instance, the cost of treating people for asthma if they have fallen ill as a result of the particulate matter released during the carpet's production. The inadequate valuation of a product could eventually lead to the exploitation or overuse of resources needed to produce it—an example of market failure. When external costs are internalized, on the other hand, people (or species) who don't benefit from the transaction do not pay for it. In this case, the product or service can be more appropriately priced or valued; this new price more accurately

triple bottom line The combination of the environmental, social, and economic impacts of our choices.

true cost The sum of both external and internal costs of a good or service.

reflects the true cost of the product or service.

Because we are so accustomed to not paying true costs, we would most likely be appalled at how much goods and services would really cost if all externalities were internalized. Although it sounds discouraging, any time we purchase products that were made in a more environmentally or socially sound manner, we come a little closer to bearing the consequences of our choices. We also create a demand for these products in the marketplace. And, if businesses are forced to internalize external costs, it then becomes profitable for them to take steps to lower those costs—for example, by installing pollution prevention technologies—a benefit that could lower the environmental and societal costs overall.

INFOGRAPHIC 6.5

KEY CONCEPT 6.5

The price of a product usually reflects only the internal costs of doing business, not the social and environmental external costs. This means those costs are not passed on to the consumer but rather are paid by others. Internalizing these external costs better reflects the true cost of a product.

INFOGRAPHIC 6.5   TRUE COST ACCOUNTING

↓ Many environmental and health costs of our goods and services are externalized (not included in the price the consumer pays). But if consumers don't pay all the costs to produce a product, such as paper, who does?

What does it take to produce the paper you use every day?

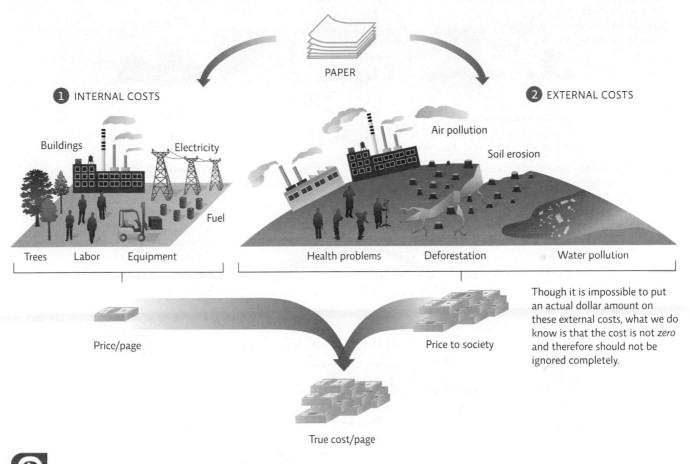

PAPER

1 INTERNAL COSTS

Buildings   Electricity

Fuel

Trees   Labor   Equipment

2 EXTERNAL COSTS

Air pollution

Soil erosion

Health problems   Deforestation   Water pollution

Price/page

Price to society

Though it is impossible to put an actual dollar amount on these external costs, what we do know is that the cost is not *zero* and therefore should not be ignored completely.

True cost/page

? What could a paper company do to reduce the external costs of harvesting trees to make paper?

KEY CONCEPT 6.6

Ecological economists argue that mainstream economics will fail in the long run because it makes some assumptions that are inconsistent with the way nature operates.

Ecologically minded economists are considering how to incorporate environmental considerations into economic decisions. **Ecological economics** is a discipline that considers the long-term impact of our choices on human society and the environment. With this focus, ecological economists argue that mainstream economic theory will fail because it is based on several erroneous assumptions.

One inaccurate assumption of mainstream economics is that natural and human resources are either infinite or that substitutes can be found if needed. This is true for some but not all resources. For instance, fossil fuels are finite and will run out, even with technological advances that allow us to access more of the fuel that is left. It remains to be seen if we can replace fossil fuels with sustainable alternatives at current levels of use. In addition, our actions can degrade air and water resources faster than nature can restore them; also, crop productivity has limits.

Mainstream economics also assumes that economic growth will go on forever. Since there are inherent limits to what Earth can provide, unlimited economic growth (at least that which

ecological economics New theory of economics that considers the long-term impact of our choices on people and the environment.

INFOGRAPHIC 6.6 | **ECONOMIC MODELS**

↓ Mainstream economics assumes that resources will always be available and that waste can be disposed of in a linear (one-way) system.

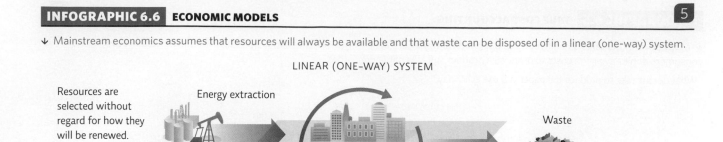

LINEAR (ONE-WAY) SYSTEM

Resources are selected without regard for how they will be renewed.

Energy extraction

Business & consumer

Waste

Resource extraction

Human economy—exchange of goods and services

↓ Environmental economics recognizes that natural ecosystems provide our resources and assimilate our wastes. If companies could fold "waste" back into production or make sure it can be decomposed by nature, we could reduce our extraction costs and be operating in a sustainable closed-loop system.

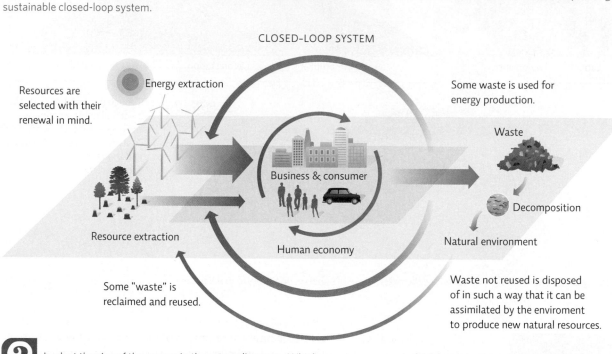

CLOSED-LOOP SYSTEM

Resources are selected with their renewal in mind.

Energy extraction

Some waste is used for energy production.

Business & consumer

Waste

Decomposition

Resource extraction

Human economy

Natural environment

Some "waste" is reclaimed and reused.

Waste not reused is disposed of in such a way that it can be assimilated by the enviroment to produce new natural resources.

? Look at the size of the arrows in these two diagrams. Which ones increase and which ones decrease in size? Why?

depends on finite resources) is not, in fact, possible. We have to work within the limits of available resources in ways that allow essential ecosystem services to continue.

These assumptions lead to yet another misconception— that models of production follow a linear sequence: Raw materials come in, humans transform those materials into some kind of product, and then they discard the waste generated in the process. But because some resources are finite—and waste in the form of pollution can damage natural capital such as air, water, and soil—linear models of production will eventually fail. For instance, most traditionally produced carpet tiles are made from fossil

fuels, a practice that is not sustainable. Old unwanted tiles are then discarded, and some are eventually burned, releasing toxic pollutants and greenhouse gases. A sustainable approach would be more cyclical, where "waste" becomes the raw material once again and can be used to make new products. Interface's ReEntry 2.0 program uses old carpet tiles to make new ones, and it uses old carpet backings to make new carpet backings, in an effort to make the production process more cyclical. This is an example of a **closed-loop system**, where the product is folded back into the resource stream when consumers are finished with it or is disposed of in such a way that nature can decompose it. **INFOGRAPHIC 6.6**

lumen-digital/Shutterstock

↑ More and more industries are investing in wind and solar panels as a way to decrease the ecological footprint of their operations.

Interface seeks to operate as a closed-loop system by managing its product in a **cradle-to-cradle** fashion: It considers the entire *life cycle* of the product, from the beginning (acquisition of raw materials) to the end of its useful life (disposal), and is responsible for the impact of its use at every stage of the process. This can lead to better material choices (less toxic, more sustainable) and better process choices (reusable materials, less waste and pollution). **INFOGRAPHIC 6.7**

Another problem with mainstream economics is that it **discounts future value**: It tends to give more weight to short-term benefits and costs than it does long-term ones. In other words, mainstream economics considers something that benefits or harms us today more important than something that might do so tomorrow. For instance, we value the tuna we can harvest today more highly than tuna we might harvest 10 years from now, so the value of taking a large harvest of tuna today outweighs the benefits of taking less now to ensure that there is still some later. If the money we could earn by using the resource now is higher than sustainable harvesting yields, modern economics tells us it is more profitable to use it now and invest the resulting money in another venture. But this investment approach doesn't take into account where those other ventures might come from or whether they are in any way diminished

by the elimination of the first resource. How might the loss of tuna affect the ecosystem and other populations? Will there always be another fish population to harvest?

What about consumers? We can all decrease our impact by making more sustainable choices and by consuming less. This doesn't necessarily mean "doing without," but it does mean being mindful of our choices and opting for sustainable or low-impact choices whenever possible. However, this requires transparency from the industries that produce and sell us goods and services. That is hard to come by with current business models, often because the businesses themselves don't know all the external costs associated with their products. In order for its new plan to be successful, Interface was counting on its customers to make more sustainable choices as well. And make them they did. ReEntry 2.0 drew many new customers to Interface, including the Georgia state legislature, which purchased 13,000 yards of new carpets.

closed-loop system A production system in which the product is returned to the resource stream when consumers are finished with it or is disposed of in such a way that nature can decompose it.

cradle-to-cradle Management of a resource that considers the impact of its use at every stage, from raw material extraction to final disposal or recycling.

discounting future value Giving more weight to short-term benefits and costs than to long-term ones.

↓ In cradle-to-cradle management, the manufacturer is responsible for the product from its production (cradle) to its final disposition after the consumer is finished with it. If the item were merely disposed of, it would be sent to its "grave" and those resources wasted. If, however, the item is disassembled and the parts reused, these parts become raw material again for a new product—a new "cradle." This provides an incentive to produce the product in a way that uses durable, reusable parts and that minimizes toxicity, since the manufacturer is responsible for dealing with those toxins.

Factory

Products

Some parts are reprocessed into new products.

Some parts are reused in their original forms.

Products are disassembled.

Explain why this flowchart is an example of a cradle-to-cradle management system.

Businesses can learn a great deal about how to be sustainable from nature.

Anderson vowed in 1994 that Interface Carpet would become the world's first sustainable business. But exactly what does that mean? By definition, **sustainable development** meets present needs without preventing future generations from meeting their needs. It enhances quality of life without damaging the environment that helps meet those needs. In short, Anderson says, sustainability means "Take nothing. Do no harm." Some people have said that the term *sustainable development* is an oxymoron because development implies constant upward progress and, at some point, resource restriction will prevent further development. However, this would be true if development could only be considered a physical process, dependent on resource extraction. In reality, development can also be abstract: Some, for instance, consider improved quality of life and happiness to be development, even if it is not tied to physical resource use.

To combat the erroneous assumptions of mainstream economics, ecological economists support actions such as improving technology to increase production efficiency and reduce waste; valuing resources as realistically as possible; moving away from dependence on nonrenewable resources; and shifting away from a product-oriented economy. They look to natural ecosystems as models for how to efficiently use resources and live within the limits of nature. But ecological economists feel that our ingenuity will take us only so far and that economic growth has limits; therefore, we must significantly change the way we do things in order to become sustainable as a society.

> ### KEY CONCEPT 6.7
>
> We can become more sustainable by using nature as a model—reducing waste with increased efficiency, making choices that allow waste to be used as a resource, and relying on sustainable energy sources.

When he first vowed to make Interface sustainable, Anderson did not know whether his business would thrive or suffer as a result. "I was very apprehensive about it," he recalls. But he, along with other entrepreneurs who have followed suit, are finding that **green business**—doing business in a way that is good for people and the environment—is also profitable. It can provide a competitive advantage either

sustainable development Economic and social development that meets present needs without preventing future generations from meeting their needs.

green business Doing business in a way that is good for people and the environment.

Consumers buy products.

Consumers use products.

Consumers return products to the factory when finished with them.

Waste is minimized because many components are reclaimed.

because the consumer is willing to support the company's efforts or because green actions end up saving money.

Between 1996 and 2013 Interface Carpet has reduced greenhouse gas emissions per unit of production from its manufacturing facilities by 71%, slashed total energy use by 39%, and now relies on recycled or bio-based ingredients for 49% of its raw materials. During this same time, the company has increased sales by two-thirds and doubled its earnings. Some of this extra money directly resulted from its efforts; by reducing the amount of waste it produces, for instance, the company has saved $438 million in waste elimination costs since 1994. But Interface has also won many new customer contracts as a result of the changes it has made. At one point, for instance, Interface was in competition with two other carpet companies over a $20 million contract at the University of California. After Interface filled out a 200-page questionnaire about how the company was addressing various environmental issues, one of the university's representatives turned to a colleague of Anderson's and exclaimed, "This is *real*."

How does a company find inspiration to become sustainable? One way is to look to natural ecosystems—a perfect example of sustainable resource use and waste minimization (biomimicry; see Chapter 1).

At Interface, Anderson was strongly inspired by biomimicry. For instance, the TacTiles technology that the company developed to replace glue was based on the physics that explains how a gecko lizard clings to walls and ceilings. The microscopic hairs on a gecko's foot bond to the molecular layer of water that's present on nearly every surface, allowing its feet to cling. Interface used this information to develop tiles that bond to one another rather than to the floor, making a kind of "floating" carpet that stays in place due to gravity rather than being glued to the floor. This makes carpet installation and removal much faster and easier.

Interface also revolutionized its operations by considering itself part of a **service economy**; it focuses on selling a *service* rather than a *product*. The idea is simple: A customer pays for the service, and the vendor makes sure that the service is always available. The service might be the ability to photocopy pages, walk on comfortable carpet, or keep refrigerated food cold. Interface, for instance, sells the service of carpet—its color, texture, durability, and comfort—rather than the product itself. The customer pays a monthly fee to "lease" the carpet, and Interface maintains

service economy A business model whose focus is on leasing and caring for a product in the customer's possession rather than on selling the product itself (that is, selling the *service* that the product provides).

it and replaces it as needed. This encourages Interface to produce carpet that is durable and recyclable and also easily replaceable. **INFOGRAPHIC 6.8**

Another sustainable business practice involves *take-back programs*, particularly for products with a defined life span, such as electronics: Customers return the product to the producer when they are finished with it or when they need an upgrade. This provides an incentive to the producer to make a durable, high-quality product that can be reused or recycled.

There are many tactics for achieving sustainability.

Changing the way we do business is not going to be easy. Even Interface has progressed slowly, despite its strong desire to become sustainable. Start-up or upgrade costs can be substantial, and even though improvements may pay for themselves in the long run, many businesses simply do not have the funds to pay for them. Plus, they may find themselves at a competitive disadvantage with businesses that are not trying to internalize costs. Consumers also have a role to play. For example, recycled paper's higher cost may more closely reflect the true cost of paper, but if consumers are not willing to put their buying dollars behind their environmental ideals, businesses that make and sell paper from trees will still be more successful. In other words, it will take changes from both consumers and producers to put business and industry on the path to sustainability. But there are things that can be done to level the playing field.

Governments can encourage sustainability by providing incentives for businesses to account for true costs rather than just internal costs. This could be accomplished by taxing companies based on how much pollution they generate, subsidizing environmentally friendly processes, or giving out pollution "permits" that companies could sell if they release less pollution than they are allowed (*cap-and-trade*; see Chapter 20). For instance, if there is a pollution tax, it will be passed on to the consumer, who then decides whether to buy the product. The manufacturer that minimizes waste production and relies on lower fossil fuel inputs than its competitors would be able to meet the regulations at the lowest cost, offer lower prices, and, as a result, have a major market advantage. While promising, these policies can be complex to implement because external costs are hard to quantify (How is a pollution tax fairly assessed?) and because they may put a burden on smaller manufacturers that have less ability to absorb the cost of upgrades; passing these costs on to the consumer also stresses low-income households.

INFOGRAPHIC 6.8 **PRODUCT VERSUS SERVICE ECONOMY**

⬇ A service economy that focuses on providing the consumer the service desired, rather than a product, decreases resource drain and lessens waste while still potentially providing a profit for the seller.

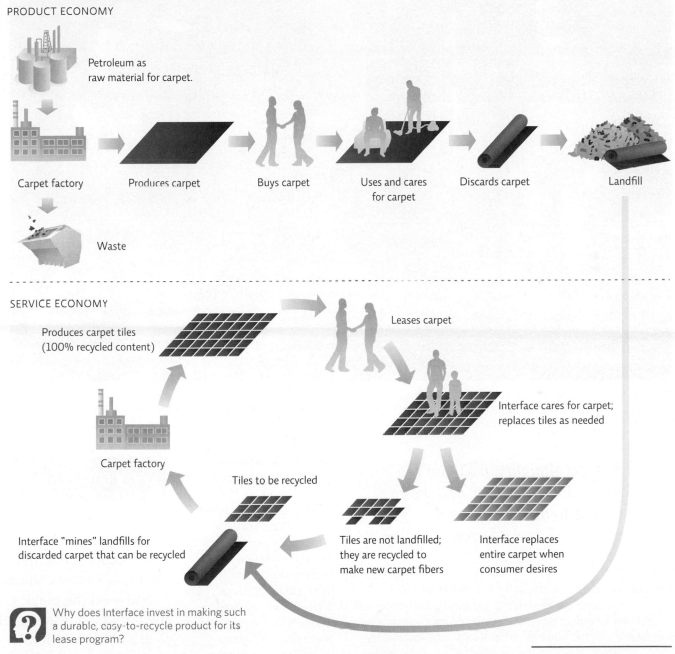

PRODUCT ECONOMY

Petroleum as raw material for carpet.

Carpet factory

Produces carpet

Buys carpet

Uses and cares for carpet

Discards carpet

Landfill

Waste

SERVICE ECONOMY

Produces carpet tiles (100% recycled content)

Leases carpet

Carpet factory

Interface cares for carpet; replaces tiles as needed

Tiles to be recycled

Interface "mines" landfills for discarded carpet that can be recycled

Tiles are not landfilled; they are recycled to make new carpet fibers

Interface replaces entire carpet when consumer desires

Why does Interface invest in making such a durable, easy-to-recycle product for its lease program?

According to Anderson, the consumer needs to know how a good or service is made—what the cradle-to-grave environmental impact of that product is. "You lay out for the consumer everything that goes into that product, and you lay it out for your competitors too—it's a totally transparent revelation of how you made that product, and what that footprint is at every step," he explained. This is hard to achieve when we buy many products made in faraway places and shipped over long distances.

One way to communicate this information is through **ecolabeling**. But consumers have to be wary of labels because as "green"

ecolabeling Providing information about how a product is made and where it comes from. Allows consumers to make more sustainable choices and support sustainable products and the businesses that produce them.

© Andy Ryan/Corbis Outline

↑ Craig Martineau, Brandon Sargent, and Dan Blake (from left) dropped out of Brigham Young University to pursue their green business, EcoScraps. Founded in 2010, the company collects roughly 18 metric tons of food waste a day from more than 70 grocers, produce wholesalers, and Costco stores across Utah and Arizona. Then it composts the waste into potting soil, which retails for up to $8.50 a bag in nurseries.

❝ *God almighty, we're on the wrong side of history, and we've got to do something.* ❞
—*Ray Anderson*

products become more attractive to consumers, more companies engage in *greenwashing*—claiming environmental benefits for a product when they are minor or nonexistent. *Fair trade* items are, however, more likely to be sustainably produced. For a product to be certified as fair trade, workers must be paid a fair wage and work in reasonable conditions to produce the goods or services. *Share programs* are another useful option for items that people need infrequently, such as a car for those who live in a large city. Rather than buying, owning, and then storing the product for a large part of the time, consumers share ownership and use the product only when they need it.

Although Interface has come a long way since 1994, it is still working hard to achieve its sustainability goals. In June 2011, the company began producing its first 100% non-virgin fiber carpet tiles, made from reclaimed carpet, fiber derived from salvaged commercial fishnets, and postindustrial waste. In addition to enjoying substantial energy savings, Interface has reduced waste some 94% since 1996 and has several LEED-certified facilities. (LEED is an internationally recognized green building certification system; the acronym stands for Leadership in Energy & Environmental Design.) Practices like intercepting industrial waste destined for landfills have a positive effect, while other

KEY CONCEPT 6.8

Transparency in how a business operates will allow consumers to make better choices and level the playing field for businesses that are trying to operate more sustainably.

Andrew Hetherington/Redux

↑ Steve Ells, the founder, Co-CEO, and Chairman of the Chipotle Mexican Grill, is committed to doing business in a way that is good for people and the environment, the definition of a Green Business. The company's environmental efforts include using organic vegetables and ethically raised meat as much as possible and constructing new restaurant buildings to be energy and water efficient.

efforts lessen the company's overall negative impact: less toxic glues, less carpet waste. All the while, Interface is still the world's leading manufacturer of commercial carpet tiles, and its 2010 operating income increased 47% over 2009. "I think we're on the right track, and we'll keep on going," says Anderson, who stepped down as the company's CEO in 2001 but still played the role of the company's conscience until his death in 2011 at the age of 77. "We'll get to the top of that mountain."

Select References:

Anderson, R. (1998). *Mid-course Correction: Toward a Sustainable Enterprise: The Interface Model.* Atlanta: Peregrinzilla Press.

Anderson, R. (2009). "The Business Logic of Sustainability," www.ted.com/talks/ray_anderson_on_the_business_logic_of_sustainability.

Costanza, R. et al. (1997). The value of the world's ecosystem services and natural capital. *Nature,* 387: 253–260.

World Wildlife Federation. (2012). "Living Planet Report, 2012," http://wwf.panda.org/about_our_earth/all_publications/living_planet_report/.

BRING IT HOME

PERSONAL CHOICES THAT HELP

You have an impact on creating a sustainable society. Every time you buy a product or service, you are telling the manufacturer that you agree with the principles behind the product. You can use your purchasing power to show companies that people are interested in good-quality products that support environmental and social values.

Individual Steps

• Reduce the amount of stuff you accumulate by buying fewer items and by choosing products that are well made and last longer.

• Use the GoodGuide app on your smartphone to scan product bar codes and see how the products rank on different scales of environmental impact, social responsibility, and health.

• Ask your local food store or pharmacy to stock fair trade–certified products if it doesn't already.

• Instead of buying a new or used car, join a car-share program like Zipcar.

Group Action

• Get together with family and friends and write a letter to your favorite companies, asking them to reduce their ecological footprint. You can ask them to become more transparent by publishing how their business practices impact the environment.

Policy Change

• Start a blog or Facebook page to chronicle the changes you make in your buying habits and encourage others to do the same. Discuss the companies whose environmental policies you agree with.

Tara Walton/The Toronto Star/ZUMA-RESS.com/Newscom

ENVIRONMENTAL LITERACY UNDERSTANDING THE ISSUE

1 What are ecosystem services, and why are they important to ecosystems and human populations?
INFOGRAPHIC 6.1

1. The water cycle is an example of:
   a. natural interest.
   b. an external cost.
   c. an internal cost.
   d. an ecosystem service.

2. Which ecosystem service is estimated to have the highest annual monetary value?
   a. Nutrient recycling
   b. Food production
   c. Pollination
   d. Waste treatment

3. How can it be useful to place a monetary value on ecosystem services, even if we know it will not be accurate?

2 What is an ecological footprint, and how does it relate to our use of natural interest and natural capital?
INFOGRAPHICS 6.2 AND 6.3

4. The land needed to provide the resources for and assimilate the waste of a person or population is referred to as:
   a. sustainable development.
   b. natural interest.
   c. an ecological footprint.
   d. true cost accounting.

5. Readily produced resources that can be used and still leave enough natural capital behind to replace what we took are known as:
   a. renewable resources.
   b. natural interest.
   c. investment capital.
   d. scarce resources.

6. The sap of maple trees (sap is "food" for the tree) can be tapped to make maple syrup, but taking too much will kill the tree. In this example, what would constitute the natural capital, and what would be the natural interest?

3 What factors influence how much human actions impact the environment, and how can we reduce that impact?
INFOGRAPHIC 6.4

7. True or False: Researchers use the IPAT model to estimate the size of an individual's ecological footprint.

8. What is the IPAT model? How is the equation I = P × A × T similar to and/or different from the equation I = (P × A)/T?

4 What are externalities and internalities in the business world, and how do they relate to true costs?
INFOGRAPHIC 6.5

9. Which of the following is an internal cost of coal mining?
   a. Pollution to nearby communities
   b. Long-term health effects suffered by miners
   c. Wages paid to workers
   d. The loss of wildlife close to the coal mine

10. If we included external costs in the cost of a good or service, we would expect the price to:
    a. go up as the users become responsible for paying all the costs.
    b. go down because the environmental impact decreases.
    c. go down because it is shared by more people.
    d. stay the same because external costs don't affect pricing.

11. What is true cost accounting, and why would it be good for the environment if businesses internalized all external costs?

5 What concerns do ecological economists have with mainstream economics, and what suggestions do they offer to help businesses and consumers make better choices?
INFOGRAPHICS 6.6, 6.7, AND 6.8

12. What does the term *cradle-to-cradle* mean when talking about product management?
    a. Product materials must be tracked from production to disposal.
    b. The product is potentially more dangerous to children.
    c. Current legislation is too restrictive on new product development and causes the early demise of new businesses.
    d. Production is cyclical: "Waste" becomes the raw material once again and can be reused.

13. Which of the following statements about sustainability is false?
    a. The ecological footprint of industrial processes could be reduced by transforming linear production processes into circular ones.
    b. The U.S. rate of consumption is not sustainable; if the world population consumed as much as the average U.S. citizen, we would need over six Earths.
    c. In its current form, mainstream economics is the optimal model for building sustainability because external costs are built in.
    d. Technology can be used to promote sustainability and decrease human environmental impact.

14. In addition to failure to consider true costs, identify and explain four erroneous assumptions that mainstream economics makes with regard to the environment.

15. What actions did Interface Carpet use to become more sustainable?

SCIENCE LITERACY WORKING WITH DATA

China and the United States are two of the world's largest consumers of resources due to sheer population size and per capita consumption rates. Look at graphs A and B which come from the Earth Policy Institute's article "Learning from China: Why the Existing Economic Model Will Fail."

GRAPH A Coal Consumption in the United States and China

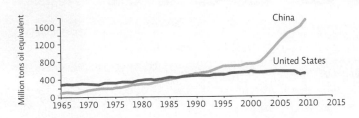

GRAPH B Oil Consumption in the United States and China. 2010, with Projections for 2035

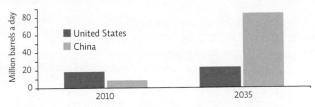

Interpretation

1. What do these figures show? How many different fuel types are included in the figures?

2. How did China and the United States compare in their use of coal and oil in 2010? Use the data to explain your response.

3. What is the projected oil consumption for China in 2035, and how does this compare to what is projected for the United States? What is the potential percentage change in oil consumption for each country? How does this trend compare to coal consumption for the two countries?

Advance Your Thinking

4. China's ecological footprint is much smaller than that of the United States. What explains the trend in coal consumption in China and why it exceeds that of the United States today? Using the IPAT model, how might China and the United States compare?

5. The world produces 86 million barrels of oil a day. What percentage of this oil production did the United States consume in 2010? China? How does this translate into per capita consumption, if the current population of China is 1.3 billion and that of the United States is 300 million? Assuming that we hold to this level of oil production, what proportion of it will China need in 2035? What about the United States? How does this translate in per capita terms, assuming that the population sizes in China and in the United States do not change significantly? Discuss the environmental consequences in terms of the IPAT model and the ecological footprint.

INFORMATION LITERACY EVALUATING INFORMATION

As consumers, we can rely on ecolabeling to some extent, but we need to remain vigilant about greenwashing, given that there are no clear labeling guidelines. One helpful source is the GoodGuide, which has ratings for more than 120,000 consumer products. Each product is given a summary score on a scale of 0 to 10 that is based on three subscores that address the product's health, environmental, and social impacts.

Explore the GoodGuide website (www.goodguide.com).

Evaluate the website and work with the information to answer the following questions:

1. Is this a reliable information source? Does it have a clear and transparent agenda?
   a. Who runs this website? Do the credentials of the individual or organization make the information presented reliable/unreliable? Explain.
   b. What is the mission of this website? What are its underlying values? How do you know this?
   c. What data sources does GoodGuide rely on, and what methodology does it employ in calculating its rating? Are its sources reliable?
   d. Do you agree with GoodGuide's assessment of the problems with and concerns about consumer products? Explain.
   e. Do you agree with GoodGuide's solutions (e.g., its rating system)? Do you think the criteria it uses for rating products are sufficient and reasonable? Which criteria are most important to you as a consumer? Explain.

2. Select one of your favorite products that is rated on the GoodGuide website.
   a. How is your product rated by the GoodGuide? Discuss both the overall score as well as the details of the three subscores.
   b. Check out the company website for your product. What sort of information about the product does the company website offer? How does it compare to the information on the GoodGuide site? Which information source is more useful to you as a consumer? Explain your responses.
   c. On the GoodGuide website, search for alternative brands that have a better rating than your preferred product. Are there other choices? Would you consider switching your brand? Explain.

Find an additional case study online at http://www.macmillanhighered.com/launchpad/saes2e

CHAPTER 7 MANAGING SOLID WASTE

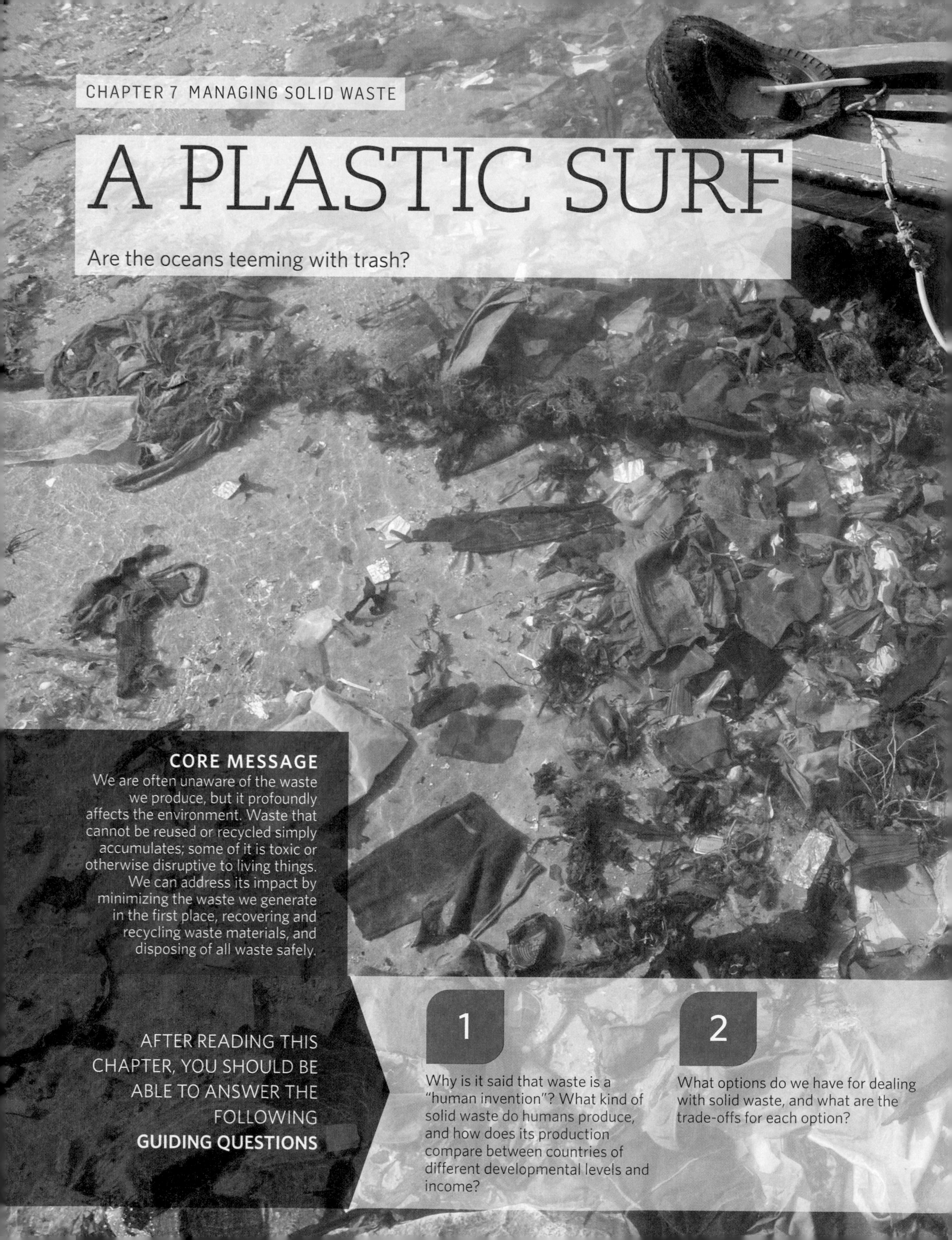

A PLASTIC SURF

Are the oceans teeming with trash?

CORE MESSAGE

We are often unaware of the waste we produce, but it profoundly affects the environment. Waste that cannot be reused or recycled simply accumulates; some of it is toxic or otherwise disruptive to living things. We can address its impact by minimizing the waste we generate in the first place, recovering and recycling waste materials, and disposing of all waste safely.

AFTER READING THIS CHAPTER, YOU SHOULD BE ABLE TO ANSWER THE FOLLOWING **GUIDING QUESTIONS**

1. Why is it said that waste is a "human invention"? What kind of solid waste do humans produce, and how does its production compare between countries of different developmental levels and income?

2. What options do we have for dealing with solid waste, and what are the trade-offs for each option?

Trash litters coastal waters near the Vietnamese island of Phu Quoc. Millions of tons of solid waste find their way to the world's oceans every year. jacus/iStockphoto/Thinkstock

3

What are some of the negative consequences of solid waste pollution?

4

What is hazardous waste, and what can the average person do to reduce his or her production of it?

5

How can industry and individuals reduce the amount of waste they produce?

From the deck of the *SSV Corwith Cramer,* the surface waters of the North Atlantic looked like smooth, dark glass. The 134-foot oceanographic research vessel had set out from Bermuda just 24 hours before on a month-long expedition aimed at tracking human garbage across the deep sea.

It was a cool mid-June evening, the Sun was setting, and the crew had just launched its first "neuston tow" of the trip. Giora Proskurowski, the expedition's chief scientist, stood watching as the tangle of mesh skidded along the water's surface, or *neuston layer,* keeping pace with the ship as it went. It was hard to imagine that any garbage would be found in such a flat, serene seascape.

After 30 minutes, crew members pulled the net—dripping with seagrass and stained red with jellyfish—from the water. Sure enough, when Proskurowski got closer, he could make out scores of tiny bits of plastic glistening in the mesh. The crew members counted 110 pieces, each one smaller than a pencil eraser and no heavier than a paper clip. Based on the size of the net and the area they had dragged it over, 110 pieces came out to nearly 100,000 bits of plastic per square kilometer (about 260,000 bits per square mile).

Despite the water's pristine look, the yield was hardly surprising. In the past 20 years, undergraduate students on expeditions like this one had handpicked, counted, and measured more than 64,000 pieces of plastic from some 6,000 net tows. That might not sound like much, given the vastness of the Atlantic and the smallness of the plastic. But the tiny bits were gathering in very specific areas, known as *gyres.* Gyres are regions of the world's oceans where strong currents circle around areas with very weak, or even no, currents. Lightweight material, like plastic, that is delivered to a gyre by ocean currents becomes trapped and cannot escape the stronger circling currents. When scientists first evaluated all the data from the Atlantic Gyre, they discovered surprisingly dense patches of plastic—more than 100,000 pieces per square kilometer—across a surprisingly large "high-concentration zone." The press and general public would come to know this region as the "Great Atlantic Garbage Patch," cousin to a "Great Pacific Garbage Patch" discovered around the same time, and, they suspected, to several other patches scattered about the ocean.

Scientists knew where the plastic was coming from (seagoing vessels, open landfills, and litter-polluted gutters around the world). And they understood why it was being trapped in the gyres (wind patterns and ocean currents). But other questions remained. No one could say how big the Atlantic Patch actually was, or how all that plastic was affecting ocean ecosystems. Were the toxic substances found in plastic accumulating up the food chain, making their way into fish, bird, or even mammal diets? Were all those tiny bits of solid surface providing transport for invasive species? And why wasn't there more of it? Despite a fivefold increase in global plastic production and a fourfold increase in the amount of plastic discarded by the United States, the concentration of plastic in North Atlantic surface waters had remained fairly steady across the 22 years for which data existed.

To answer these questions, Proskurowski, his team, and their captain would sail the *Corwith Cramer* all the way out to the Mid-Atlantic Ridge, an underwater mountain range that lies about 1,600 km (1,000 miles) farther east than any previous plastics expedition had ever gone. "That's far enough from Bermuda that getting back will be a challenge," he wrote on the ship's blog as land faded from view. "The same forces that drew plastic into this particular part of the ocean—namely low and variable winds and currents—will make operating a tall ship sailing vessel tricky, to say the least." But if all went according to plan, Proskurowski thought, he and his colleagues would be the first to reach the so-called garbage patch's easternmost edge.

◉ WHERE IS BERMUDA?

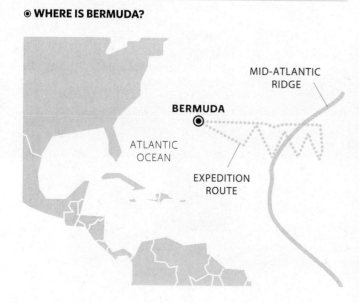

↑ Team members ready a neuston net for deployment.

↓ Giora Proskurowski on the *Thomas G. Thompson.*

Waste is a uniquely human invention, generated by uniquely human activities.

In natural ecosystems, there is no such thing as waste. Matter expelled by one organism is taken up by another organism and used again. This natural recycling is consistent with the *law of conservation of matter*, which states that matter is never created or destroyed; it only changes form. Forms of matter that are dangerous to living things (think arsenic and mercury) tend to stay buried, deep underground, and are released only during extreme events like volcanic eruptions.

Human ecosystems are another story. By taking

KEY CONCEPT 7.1

Solid waste is unique to human societies because in nature there is no "waste"; the discarded matter of one organism becomes a resource for another. Some of the waste we produce is not readily degradable and can persist for a long time.

matter out of the reach of organisms that can use it, we continually disrupt this natural cycle. We do this by converting usable matter into synthetic chemicals that can't easily be broken down and by burying readily degradable things in places and under conditions where natural processes can't run their course.

For example, paper and cardboard break down easily in a compost bin, thanks to the microbes that feed on them. But we typically keep this type of waste in landfills, where the lack of water, oxygen, and microbes force it to decompose much more slowly than it otherwise might.

Thus *matter*—the physical substance of which the universe is made—becomes **waste**—a uniquely human term used to describe all the things we throw away. Waste that can be broken down by chemical and physical processes is considered *degradable*, even if that degradation takes a long time. Waste that can be broken down by living organisms such as microbes is considered **biodegradable**. Some waste—mostly synthetic molecules like the pesticide DDT and the CFCs once found in aerosols—is considered **nondegradable**. These molecules are chemically stable and don't degrade in normal atmospheric conditions. And because they haven't been around for very long, no organism has yet developed (through mutation or genetic recombination) the enzymes to use them as food.

This explains why plastics—a human creation that uses petroleum or natural gas as the starting material—take so long to fall apart and break down; nothing "eats" them in their original form. Different types of plastics have different degradation rates, but for all it is generally quite slow. After all, one of the perks of plastics is their durability. Exposure to sunlight in an oxygen-rich environment causes plastics to chemically change and become brittle, breaking into smaller pieces and making some (but probably not all) components of the plastic accessible and digestible by microbes. This is an incredibly slow process when plastic is buried in a landfill (with no sunlight and little oxygen), but on beaches and the surface of the ocean, it can occur more quickly due to the exposure to sunlight and the additional weathering action of wind and waves, which produces the tiny plastic bits found in the ocean garbage patches.

Almost any human activity you can think of generates some form of waste. Processes that produce food and consumer goods generate *agricultural* and *industrial waste*, which in the United States account for about 54% of all garbage. The harvesting of coal and precious metals like gold and copper generates *mining waste*, which can pollute air, water, and soil and makes up an additional 33% of U.S. waste. And the increasingly complicated act of living—in houses, apartments, dormitories, and small businesses around the world—produces its own steady stream of trash, referred to as **municipal solid waste (MSW)**, or at the community level, an MSW stream.
INFOGRAPHIC 7.1

Currently, urban dwellers produce about 1.3 billion metric tons (0.9 billion U.S. tons) of solid waste each year and this amount is expected to almost double by 2025. In the United States, MSW streams make up just 2% of total waste. But because Americans create more garbage per person than just about any other country in the world, 2% is still a lot of trash. In 2012 alone, each American produced about 1.99 kilograms (4.38 pounds) of solid waste per day, up from about 1.22 kilograms (2.7 pounds) per day in the 1960s. With more than 300 million people living in the United States, this adds up to about 251 million tons of household trash per year; that's twice the per capita amount produced by the European Union and as much as 10 times the amount produced by most less developed countries.

The vast majority of this garbage comes from a familiar array of goods: paper, wood, glass, rubber, leather, textiles, and of course plastic—cheap enough to have become a staple of both advanced and developing societies, light enough to float, and durable enough to persist for hundreds of years across thousands of miles of ocean.

KEY CONCEPT 7.2

While agricultural, industrial, and mining wastes represent the majority of our waste, municipal solid waste (MSW) volumes are still significant. Urbanites generate the most MSW per person; waste generation also increases as income increases.

---

waste Any material that humans discard as unwanted.

biodegradable Capable of being broken down by living organisms.

nondegradable Incapable of being broken down under normal conditions.

municipal solid waste (MSW) Everyday garbage or trash (solid waste) produced by individuals or small businesses.

In 2012 alone, each American produced about 1.99 kilograms (4.38 pounds) of solid waste per day.

INFOGRAPHIC 7.1   **U.S. MUNICIPAL SOLID WASTE STREAM**

↓ Solid waste represents a major and growing environmental issue, especially in urban areas. In just 10 years (2002–2012) urban solid waste generation per year roughly doubled to reach 1.3 billion metric tons. Urban areas in more developed nations such as those in the Organisation for Economic Co-operation and Development (OECD)—which includes the United States and other free-economy countries—generate the most solid waste per person. Though municipal solid waste makes up a small part of the total waste produced (about 2% in the United States), it is the type of waste we can most directly impact with our day-to-day choices.

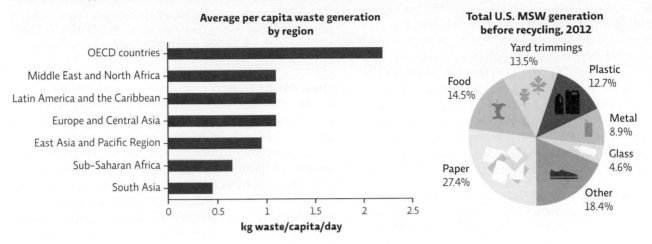

Average per capita waste generation by region

- OECD countries
- Middle East and North Africa
- Latin America and the Caribbean
- Europe and Central Asia
- East Asia and Pacific Region
- Sub-Saharan Africa
- South Asia

kg waste/capita/day

Total U.S. MSW generation before recycling, 2012

- Yard trimmings 13.5%
- Plastic 12.7%
- Food 14.5%
- Metal 8.9%
- Glass 4.6%
- Paper 27.4%
- Other 18.4%

❓ Which of the categories of waste shown in the pie chart are naturally degradable? Which could be recycled?

How big is the Atlantic Garbage Patch, and is it growing?

The first garbage patch was discovered in 1997 by Captain Charles Moore, a veteran seafarer who was crossing the Pacific on his way back from an international yacht race.

As the press caught wind of what sounded like a giant plastic island in the middle of the ocean, a media frenzy ensued. Some reports said the patch was twice the size of Texas; others claimed that the patch was growing exponentially. But these widely reported claims were misleading, if not downright wrong.

Part of the trouble was semantic. "An oceanographer understands the term 'patch' to mean 'an uneven distribution,'" says Kara Law, a scientist at the Woods Hole Oceanographic Institute who has led several expeditions on the *Corwith Cramer*. "We say the upper ocean is 'patchy' because organisms are often observed

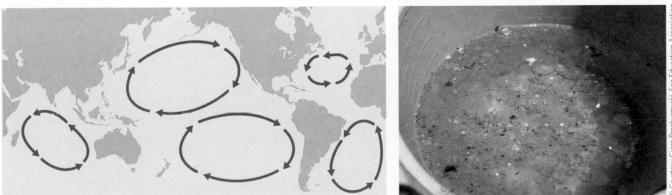

↑ Five major gyres are found in the world's oceans, and there are floating bits of plastic in all of them. These "garbage patches" are not floating islands and aren't even necessarily visible when gazing at the water. Much of the debris is very small and lies just below the surface. The highest density captured in one of the 6,100 sampling tows was approximately 200,000 pieces per square kilometer. It is hard to estimate the size of any garbage patch since the material is so spread out and may reach down to depths of 20 meters.

in clumps, separated by regions with sparse populations. But when reporters and laypeople hear 'patch,' they incorrectly think of an 'island' or a 'continent' of trash."

A bigger problem was the data itself. MSW and global plastic production data indicated that more plastic was being produced, used, and discarded throughout the world. According to one scientific study, the amount of plastic being ingested by Subarctic seabirds had nearly doubled between 1975 and 1985. But it was impossible to tell whether the patches themselves were growing. Data from the Pacific Patch suggested that the concentration of plastic had risen by an order of magnitude between the 1980s and 1990s. But that comparison was of limited value because sampling methods differed widely from one expedition to the next. In the Atlantic, where sampling methods had remained consistent across the years, no such increase could be detected.

Some scientists had suggested that the nets used on most plastics expeditions were too porous to trap the smallest of particles and that, with finer nets, much more plastic would be found at the surface. Others thought that plastic was hiding below the surface, throughout the mixed layer—a layer of uniform density at the top of the water column that is saturated with sunlight, low in nutrients, and easily mixed by wind. Part of the *Corwith Cramer*'s mission was to try and figure out which hypothesis might be correct.

By July 3, Proskurowski's team had gleaned at least part of the answer. In the ship's daily blog, he wrote:

The results from the past several days of tucker trawls [which collect water samples from discrete depths within the mixed layer] have been very interesting. While we've seen fairly low numbers of plastic at the surface, we have observed an almost equal amount of plastic from one meter's water depth, slightly less at ten meters and no plastic below the mixed layer. These tows show that plastic is undoubtedly being mixed down into the water column, and what we measure at the surface is, in many cases, not the sum total of what is out here. Like many environmental problems, the closer you look at the system, the worse it appears.

open dumps Places where trash, both hazardous and nonhazardous, is simply piled up.

leachate Water that carries dissolved substances (often contaminated) that can percolate through soil.

sanitary landfills Disposal sites that seal in trash at the top and bottom to prevent its release into the atmosphere; the sites are lined on the bottom, and trash is dumped in and covered with soil daily.

It would be a few more weeks before they found out just how much worse.

How we handle waste determines where it ends up.

"It is with startling accuracy that so many tiny particles of plastic end up in such obscure but well-defined stretches of ocean," Proskurowski wrote one

morning, as the *Corwith Cramer* floated under the early summer sun near the Mid-Atlantic Ridge. "Especially given how long and convoluted the journey they took to get here was."

In cities of the United States and elsewhere around the world, collected solid waste makes a similar journey, one that begins when we toss something we no longer need into a trash bag that is then carried from building to curb to garbage truck, before making its way to one of several kinds of waste facilities.

Open dumps are places where trash is simply piled up. Because they are one of the cheapest ways to get rid of human trash, they are common in less developed countries, where entire communities often spring up around the dumps and people survive on what they scavenge from the waste piles. Open dumps attract pests such as flies and rats, which can be human health hazards. Open dumps also contribute to water pollution: Rain either washes pollutants away from the dump to surrounding areas or pulls it along as it soaks into the ground. If this contaminated water, called **leachate**, continues to travel downward, it can contaminate the soil and groundwater.

Sanitary landfills, more common in developed countries, seal in trash at the top and bottom in an attempt to prevent its release into the environment. Several protective layers of gravel, soil, and thick plastic prevent leachate from delivering toxic substances to groundwater below the landfill. The trash is covered regularly with a layer of soil that reduces unpleasant odors, thus attracting fewer pests.

> ### KEY CONCEPT 7.3
>
> There are a variety of solid waste disposal methods ranging from problematic open dumps to the modern methods of landfilling or incineration; all of these come with trade-offs.

But there is a downside to these, too. The compacting of trash under a layer of soil excludes oxygen and water so well that the aerobic bacteria (those that require oxygen to live) and other organisms that normally decompose at least some of the waste can't survive. Newspaper that would degrade in a matter of weeks is preserved in landfills for decades. Anaerobic bacteria (those that live in oxygen-poor environments) pick up some of the slack. But they are much slower and produce lots of methane, a combustible greenhouse gas that is 20 times more potent than CO_2 (see Chapter 21). As a result, landfills are a significant anthropogenic contributor of methane in developed countries like the United States. **INFOGRAPHIC 7.2**

INFOGRAPHIC 7.2   **MUNICIPAL SOLID WASTE DISPOSAL**

↓ As indicated in the Environmental Protection Agency's solid waste hierarchy, reducing waste at its source (homes and small businesses) is the top choice in waste management, with landfilling as the last choice. However, more than 50% of our solid waste ends up in a landfill.

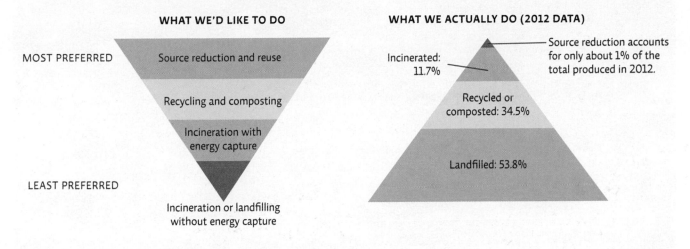

WHAT WE'D LIKE TO DO

MOST PREFERRED

Source reduction and reuse

Recycling and composting

Incineration with energy capture

LEAST PREFERRED

Incineration or landfilling without energy capture

WHAT WE ACTUALLY DO (2012 DATA)

Incinerated: 11.7%

Source reduction accounts for only about 1% of the total produced in 2012.

Recycled or composted: 34.5%

Landfilled: 53.8%

↓ In a sanitary landfill, an area is dug out and lined to prevent groundwater contamination from leachate; trash is dumped and covered with soil frequently. (This soil may take up to 20% of the landfill area.) Newer landfills have a leachate-collection system built in; older landfills can be retrofitted to collect leachate. Leachate from holding ponds is treated before being released into the environment.

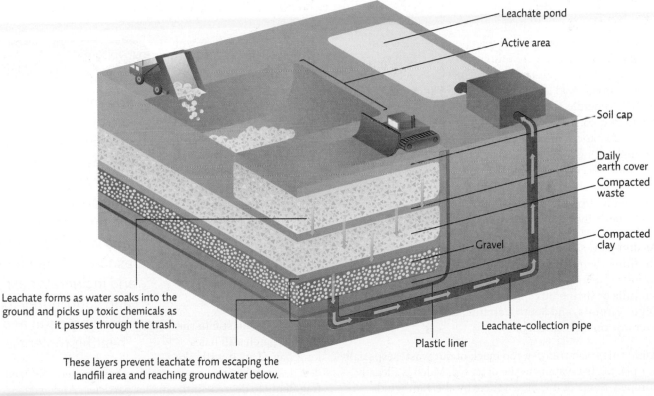

Leachate pond

Active area

Soil cap

Daily earth cover

Compacted waste

Compacted clay

Gravel

Leachate-collection pipe

Plastic liner

Leachate forms as water soaks into the ground and picks up toxic chemicals as it passes through the trash.

These layers prevent leachate from escaping the landfill area and reaching groundwater below.

How is solid waste disposed of in your community? How does this compare to EPA's waste hierarchy of preferred methods?

↑ At Phnom Pen, Cambodia's municipal garbage dump, people work around the clock collecting plastic, metals, wood, cloth, and paper, which they sell to recyclers.

A lot of trash—thousands of tons per day—also winds up in specially designed **incinerators**. Burning waste in this way reduces its volume dramatically—by about 80% to 90%. And the heat produced by burning can be used to generate steam for heat or electricity. But burning waste that contains plastics and other chemicals releases toxic substances into the air, polluting air and water and producing toxic ash, which must be disposed of in a separate, specially designed landfill. Incinerators are also extraordinarily expensive to build, and tipping fees (fees charged to drop off trash) are usually much higher at an incinerator than at a landfill. **INFOGRAPHIC 7.3**

As dumpsters and landfills fill up, cities and towns begin shuffling their waste from state to state and country to country. New York City, for example, sends its trash to landfills or incinerators in New Jersey, Pennsylvania, Ohio, Virginia, and South Carolina. Other cities ship trash overseas on garbage barges.

Amid this trash transfer, too much of our waste, especially our plastic, is escaping to the open sea. Much is illegally dumped. Some of it is blown there by aberrant winds from the tops of open landfills. Some is carried through faulty sewage systems or in the trickling currents of litter-polluted gutters. To be sure, not all of it is plastic, but other types of waste—textiles, glass,

incinerators Facilities that burn trash at high temperatures.

wood, and rubber—sink or degrade relatively quickly. Plastic just floats along. And as pieces accumulate organisms, they become heavier and can sink. Eventually, time, saltwater, and sunlight break it down from its recognizable, everyday forms—combs, candy bar wrappers, CD cases—into fragments so tiny that even thousands of them together can't be seen by a naked eye trained on a calm sea.

And as far as it has traveled, the plastic in the garbage patch still has a long way to go. It will take decades, maybe even centuries, for those fingernail-sized fragments to degrade into smaller particles. And even then, they will not cease to pollute. In one study of the North Pacific Gyre, virtually all water samples, even those that were free of plastic debris, contained traces of polystyrene, a common plastic used in a wide range of consumer goods.

KEY CONCEPT 7.4

Uncollected solid waste contributes to flooding and air and water pollution. Even MSW which is disposed of using modern techniques traps valuable matter resources in landfills and incinerator ash, and contributes to environmental and health problems.

INFOGRAPHIC 7.3   HOW IT WORKS: AN INCINERATOR

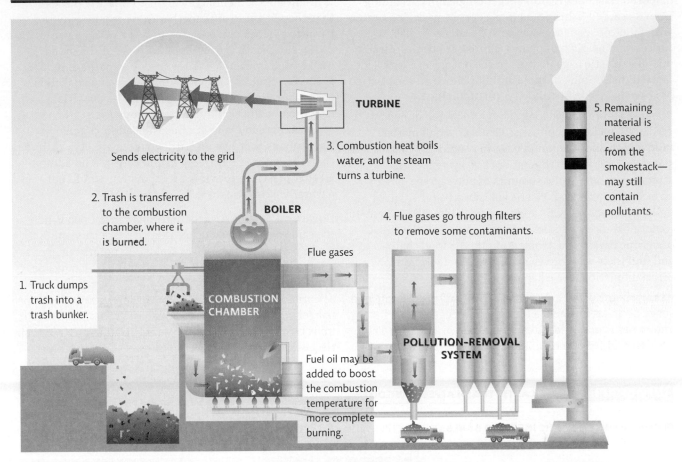

Sends electricity to the grid

TURBINE

3. Combustion heat boils water, and the steam turns a turbine.

2. Trash is transferred to the combustion chamber, where it is burned.

BOILER

4. Flue gases go through filters to remove some contaminants.

5. Remaining material is released from the smokestack— may still contain pollutants.

1. Truck dumps trash into a trash bunker.

Flue gases

COMBUSTION CHAMBER

Fuel oil may be added to boost the combustion temperature for more complete burning.

POLLUTION-REMOVAL SYSTEM

↑ Trash can be burned at very high temperatures in incinerators (some of which are designed to also generate electricity), but in some facilities, fuel oil must be added to the mix for more complete combustion. In modern facilities, cleaning systems remove particulates, sulfur, and nitrogen pollutants, as well as toxic pollutants like mercury and dioxins. The ash is considered toxic waste and must be buried in the hazardous waste landfills. Municipal solid waste, medical waste, and some hazardous waste are incinerated in the United States.

 What could be done to reduce the toxicity of incinerator ash?

Solid waste pollution threatens all living things.

The consequences of mismanaging our trash are manifold. Deciding how best to dispose of our trash is a problem that certainly requires our attention, but the bigger, often overlooked issue is the fact that disposal removes valuable resources: Landfilled or incinerated products cannot be readily reclaimed or recycled to make new products, nor can biodegradable materials readily decompose to return usable matter to natural ecosystems.

In urban areas of lower-income countries, *uncollected* solid waste is a major problem. Environmental impacts include flooding and water and air pollution, all of which can lead to health problems. Blocked storm drains can cause flooding, which can lead to water pollution and offer stagnant-water habitat to mosquitoes, flies, and rodents—vector organisms that can spread pathogens for diseases such as dengue fever, diarrhea, and parasitic infections.

The modern disposal methods used in more developed countries can also contribute to health and environmental problems; incinerators create small-particle air pollution, and landfills produce methane. When disposed of improperly, chemical waste can wreak havoc on plant and animal life. Even some household trash is considered hazardous to health and should not be disposed of in a landfill or municipal incinerator.

Aquatic life is especially vulnerable to improperly disposed trash. Sea mammals get tangled in everything from discarded fishing nets to plastic six-pack rings, often with fatal consequences. On top of that, many fish, sea turtles, and nearly half of all seabirds eat plastic, often by mistake (plastic bags floating in the open ocean look a lot like jellyfish). Some of these animals choke on the plastic or are poisoned by its toxicity. **INFOGRAPHIC 7.4**

Many sea animals live long enough to be consumed by predators, including humans. That's no small matter. BPA, an organic compound used in plastic, has been shown to interfere with reproductive systems, and styrene monomers, the subunits of polystyrene, are a suspected carcinogen. Plastic also absorbs fat-soluble pollutants such as PCBs and pesticides like DDT. These toxic substances are known to accumulate in the tissue of marine organisms, biomagnify up the food chain, and find their way into the foods we eat. (For more on bioaccumulation, see Chapter 3.)

Researchers suspect that floating bits of plastic can also serve as an attachment point for fish eggs, barnacles, and many types of larval and juvenile organisms. Thus each tiny bit of plastic could potentially transport

harmful, invasive, or exotic species to new locales. It might also add enough weight to floating pieces that they begin to sink (perhaps one reason why the garbage patches don't appear to be growing, even as our use of plastics grow). "I think one of the most underrated impacts of these so-called garbage patches is the introduction of hard surfaces to an ecosystem that naturally has very few of them," says Miriam Goldstein, a PhD candidate at Scripps Institute of Oceanography who studies the Pacific patch. "Organisms that live on hard surfaces are very different than those that float freely in the ocean. And adding all that plastic is providing habitat that would not naturally exist out there."

However, knowing that these things can happen is not the same as finding evidence that they are happening in the patches. In general, as Goldstein points out, gyres tend to be areas of very low productivity, which means there are very few large fish there. It is not yet clear to scientists whether large numbers or important species of fish (or sea birds) are ingesting plastic from the gyre, or if toxic chemicals from the plastic are accumulating in their tissues.

INFOGRAPHIC 7.4   **PLASTIC TRASH AFFECTS WILDLIFE**   3

COMPARISON OF PLASTIC INGESTED BY ALBATROSS CHICKS IN TWO REGIONS OF THE PACIFIC OCEAN

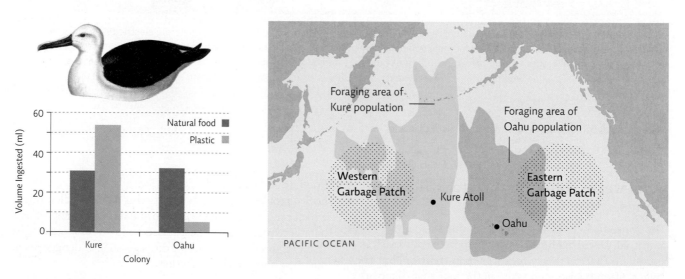

↑ Research by Lindsay Young, at the University of Hawaii, and her colleagues compared the food ingested by Laysan albatross chicks in two populations, more than 1,900 kilometers apart in the Pacific Ocean. Their data show that while both populations consumed roughly the same amount of actual food, chicks in the western Pacific near Kure ingested 10 times more plastic than chicks near Oahu. In addition, the Kure chicks had 4 times as many plastic pieces and these pieces were, on average, twice as large as those of the Oahu chicks.

 Why do you think the birds of the Kure population ingest more plastic than the birds of the Oahu population?

On June 21, 3 weeks into their journey, Proskurowski wrote on the ship's blog:

At 0930 this morning, we sampled what I predict will be the largest amount of plastic [the program] has ever recorded in 25 years of sampling. When the tow started, I could distinctly see a few pieces in the upper layer of water and it looked like it was going to be a "good one"—one where we got enough plastic to keep the lab busy. A couple minutes into the tow, suddenly we started to see more and more macro debris—a toilet seat, white plastic bags, oil jugs, a few bread bag fasteners, Styrofoam cups, several shoes, a few foot insoles, a loofah sponge, sunscreen, and liquor bottles—sometimes appearing to form loosely organized windrows [indicative of a special type of turbulence in the upper ocean called Langmuir circulation]. While everyone was commenting on all the debris, a red 5-gallon bucket drifted by the port bow with a school of about 20 fish underneath it. [The bucket got caught in the net.] Fearing it would tear the net, we ended the tow early and pulled the bucket with two of its associated schools of grey triggerfish and thousands of tiny fragments of plastic on deck. [Team scientist] Skye Moret quickly dissected one fish and found 46 pieces of plastic in its guts. This school of fish, a coastal species that typically lives on reefs, was thousands of miles from land with plastic-filled stomachs.

© Liang Xu/Xinhua/ZUMAPRESS.com

↑ In some areas of China, the process of recovering electronic waste is not done safely and exposes workers and the community at large to toxic substances.

He was right. At 23,000 pieces of plastic—more than 26 million pieces per square kilometer—the June 21 tow was the largest in the research program's 25-year history. It would take two lab members (rotating every hour or so) more than 14 hours of continuous work to process all of it.

KEY CONCEPT 7.5

Some solid waste, including e-waste, is considered hazardous and must be disposed of properly to reduce the risk to workers, avoid environmental contamination, and reclaim valuable mineral resources.

Some waste is hazardous and must be handled carefully.

A large part of our solid waste stream includes **hazardous waste** (waste that presents a health risk of some kind). Modern households often contain a bevy of dangerous chemicals and products, ranging from household cleaners, paints, and automotive supplies to fluorescent light bulbs and batteries that should not be disposed of with the regular trash.

A new, modern category of hazardous waste is **e-waste**, discarded electronic devices such as computers, cell phones, and televisions—any device that contains a circuit board. These devices contain significant amounts of precious and rare-earth metals (see LaunchPad Chapter 26); reclaiming these materials from e-waste can reduce our need to extract them from mines, a process with a huge environmental impact. But many countries (including the United States) send some of their e-waste to countries in Asia and Africa, where impoverished villagers do their best to extract the precious metals within. It's dangerous work. In addition to gold, copper, and zinc, e-waste contains a suite of toxic metals such as lead, mercury, and chromium. When they are released by unsafe and poor extraction methods, these substances cause a wide range of medical conditions in workers or community members—from birth defects to brain, lung, and kidney damage to cancer. The soil, air, and water

hazardous waste Waste that is toxic, flammable, corrosive, explosive, or radioactive.

e-waste Unwanted computers and other electronic devices such as discarded televisions and cell phones.

surrounding e-waste disposal sites don't fare much better: Toxic chemicals are released into all three during these unsafe extraction processes. **INFOGRAPHIC 7.5**

When it comes to managing waste, the best solutions mimic nature.

While nondegradable trash like plastics presents a challenge for disposal, much of our waste is biodegradable, and we can apply the concept of biomimicry—emulating nature—to better deal with this part of our waste stream. **Composting**—allowing waste to biologically decompose in the presence of oxygen and water—can turn some forms of trash into a soil-like mulch that nourishes the soil and can be used for gardening and landscaping. Composting can be done on a small scale (in homes, schools, and small offices) or on a large one (in municipal "digesters"). But this works only for organic waste, such as paper, kitchen scraps, and yard debris. **INFOGRAPHIC 7.6**

Another promising solution to some of our waste woes involves converting garbage or its by-products into usable energy. As mentioned before, the heat produced during incineration might be converted into steam energy or electricity. And the methane produced during landfill anaerobic decomposition can also be harnessed as an energy source. Engineers and scientists are working out ways to use methane-producing microorganisms to process trash in vats or digesters.

composting Allowing waste to biologically decompose in the presence of oxygen and water, producing a soil-like mulch.

In fact, the concept of spinning waste into energy that could then heat or light our homes or fuel our cars is so appealing that it has even extended to waste found in the open ocean. Not long after the Atlantic and Pacific patches were found, a San Francisco–based environmental group tried devising ways to turn the plastic bits into diesel fuel. That plan proved impossible. "There is no way to remove all those millions of tiny pieces without also filtering out all the millions of tiny creatures that inhabit the same ecosystem," says Kara Law. "So once it's out there, all we can really do is hope that nature eventually breaks it all the way down." That could take hundreds to thousands of years. In the meantime, she says, we need to stop adding to the patches that already exist.

Life-cycle analysis and better design can help reduce waste.

The real trick to keeping garbage out of the ocean is to go back to the beginning, before the things we throw away are even made. Environmentally friendly products

> ### KEY CONCEPT 7.6
>
> Composting organic trash is a waste disposal method that mimics nature: It allows waste to decompose and produces a mulch-like product that can be used to return nutrients to soil.

INFOGRAPHIC 7.5 | **HOUSEHOLD HAZARDOUS WASTES**

4

↓ Hazardous wastes are those that are toxic, flammable, explosive, or corrosive (like acids). Many chemicals that enter your home are actually hazardous and should not be discarded in the regular household trash. The EPA recommends that you contact your local solid waste agency for information on disposing hazardous materials. You can also call 1-877-EARTH-911.

To protect your health and that of your family and the environment, avoid or reduce your use of hazardous chemicals such as:

Drain openers	Glue
Oven cleaners	Pesticides and insecticides
Engine oil and fuel additives	Mold and mildew removers
Grease and rust removers	Paint thinners, strippers, and removers

Other materials that are also considered hazardous and may need to be disposed of as hazardous waste include:

Batteries
Fluorescent light bulbs
Mercury thermometers
E-waste (cell phones, computers, printers, televisions, video game consoles, etc.)

 How can you safely dispose of hazardous waste in your community?

INFOGRAPHIC 7.6   COMPOSTING

↓ Composting can reduce the amount of a household's trash tremendously. A simple compost pile can be started in the backyard, or a compost bin can be built or purchased.

HOME COMPOSTING

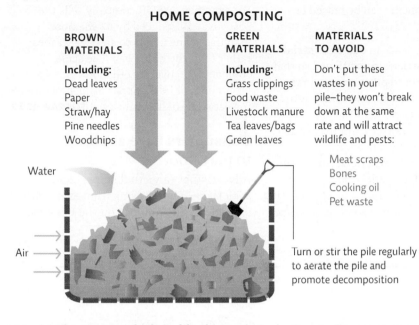

BROWN MATERIALS

Including:
Dead leaves
Paper
Straw/hay
Pine needles
Woodchips

GREEN MATERIALS

Including:
Grass clippings
Food waste
Livestock manure
Tea leaves/bags
Green leaves

MATERIALS TO AVOID

Don't put these wastes in your pile—they won't break down at the same rate and will attract wildlife and pests:

Meat scraps
Bones
Cooking oil
Pet waste

Water

Air →

Turn or stir the pile regularly to aerate the pile and promote decomposition

↑ A variety of compost bins can be used. Small tubs and countertop green bins can be used indoors.

© Alan Marsh/Wave/Corbis

↓ Municipal composting facilities like this one in Sevier County, Tennessee, use large digesters to process household waste after recyclables have been removed. The digesters produce a mulch that residents can pick up at no cost.

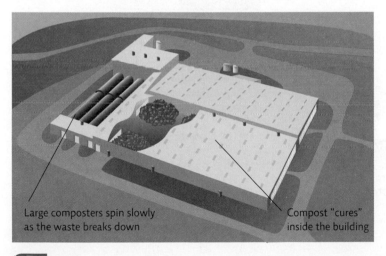

Large composters spin slowly as the waste breaks down

Compost "cures" inside the building

↑ Household plant-based food scraps can be added, but meat products should not be included.

© Chris Price/iStockphoto

↑ The end product is a rich, soil-like mulch that can be used in gardens.

© Mode/MW/Aurora Photos

? Does your school cafeteria have a composting program? If not, what would it take to implement one?

continue to grow in popularity, and manufacturers have given a name to the tactics used to create these products: life-cycle analysis. By assessing the environmental impact of every stage of a product's life—from production, to use, to disposal—an increasing number of companies are trying to reduce the amount of waste generated by the things they design, make, and sell. Cradle-to-cradle analysis takes this even further as it tries to increase reuse potential and turn *waste* back into *resource* (see Chapter 6).

Part of this shift has been spurred by legislation. Some European countries and at least 19 U.S. states have implemented "take-back laws," which require manufacturers to take back some of their products— namely, computers—after consumers are finished with them. This creates an incentive for manufacturers to design products from which components can easily be salvaged and reused. Indeed, many companies are trying to cut down on waste by *de-manufacturing*—or disassembling

KEY CONCEPT 7.7

We can reduce industrial waste by identifying ways to use resources more efficiently so there is less waste to begin with, and then identifying ways to reuse and recycle waste that is produced.

equipment, machinery, and appliances into the component parts so that those parts can be reused. BMW, for example, designs its vehicles so that at least 97% of the components can be reused or recycled.

On a grander scale, in **eco-industrial parks**, industries are positioned so that they can use each other's waste in the same industrial area; one company's trash becomes another's raw material. This type of strategic thinking is known as *industrial ecology*; in addition to "waste-to-feed" exchanges, participants in eco-industrial parks also decrease their ecological footprint by coordinating activities and sharing some spaces, like

warehouses and shipping facilities. This form of waste reduction can occur within a single company, between two companies, or between a company and a community. For example, one UK company sells the soapy water produced by its shampoo manufacturing plants to area car washes rather than throw it away. The same company also uses its ethanol waste—a by-product of its manufacturing processes— to heat some of its facilities. **INFOGRAPHIC 7.7**

Consumers have a role to play, too.

Advertising—a virtual staple of advanced societies—bombards us from every corner of modern life: not just on the televisions

---

INFOGRAPHIC 7.7   **INDUSTRIAL ECOLOGY**   5

↓ The industrial park in Kalundborg, Denmark, is a prime example of industrial ecology. Here, by design, there are 24 different connections between various industries and local farms, such that the waste of one becomes the resource for another.

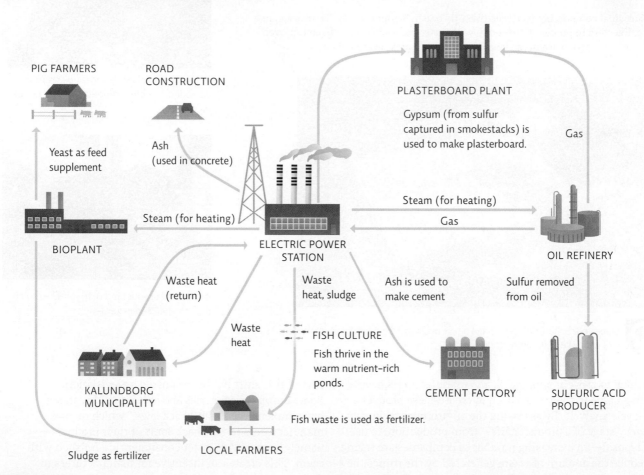

 Suppose a community wanted to transition an industrial park to the eco-industrial model. Explain how a life-cycle analysis of the industry products already made there can help determine which other industries to "invite" to join their park.

↑ The Foster family of Stowe, Ohio, with their polymer-based possessions.

© SARAH LEEN/National Geographic Creative

harmful to the environment. This may be as minor as declining to take a plastic bag for a few items purchased at the drugstore or as major as biking or walking to work rather than taking the car. The logic is simple: When we save a resource by refusing to use it, that resource lasts longer, which in turn means that less pollution will be generated disposing of it and producing replacements, and more will be available for future uses. "Refusal doesn't mean never using the resource," says David Bruno, founder of the 100 Thing Challenge, a popular movement to pare down our worldly possession to 100 items or fewer. "It just means using it at a more sustainable rate."

If we can't completely refuse a given commodity, we can still try to **reduce** our consumption of it or minimize our overall ecological footprint by making careful purchases. People who must drive to work can minimize their fossil fuel consumption by choosing a more fuel-efficient vehicle. Those who choose not to drink tap water might purchase a specialized faucet or pitcher filter instead of relying on bottled water. And all consumers can greatly reduce the amount of waste they generate by paying special attention to packaging, which accounts for one-third of all U.S. trash and roughly half of all paper used.

And if we can't avoid using a product, our next best choice is to **reuse**—the third R— something consumers can do with just a little effort, by choosing durable products over disposable ones. "Products produced for limited use are really just made to be trash," Bruno says. "They pull resources out of the environment and produce pollution at every step of production, shipping, and disposal." He advises considering use and reuse each time we head to the store. Whether you are purchasing clothing, razors, cups,

in our living rooms but on taxicab computer screens, billboard-laden subway cars, and the pop-up ads that invade our laptops. The message is surprisingly uniform: To live a happy, more fulfilling life, we simply must have more "stuff." A good deal of that stuff—though certainly not all of it—is made of plastic. "Plastics are incredibly useful and have made possible many of the greatest breakthroughs in technology and standard of living," Proskurowski says. "But they have also made our lives lazier. We now buy a bottle of water rather than refill a canteen. We buy individually wrapped bags of mini carrots, instead of buying carrots that are straight from the ground and have to be washed. There are countless other examples, and as we learn with every net tow, there are significant costs to the planet for those choices."

So how do we start making different choices? As any good environmentalist will tell you, it comes down to the "four Rs": refuse, reduce, reuse, recycle.

The first thing we can do is simply **refuse** to use things that we don't really need, especially if they are

KEY CONCEPT 7.8

Consumers can follow the four *Rs* to reduce their use of resources and generation of waste. This requires that we consider the reuse and recycling potential of any product before we decide to buy it and that we help "close the loop" by purchasing recycled products.

eco-industrial parks Industrial parks in which industries are physically positioned near each other for "waste-to-feed" exchanges; the waste of one becomes the raw material for another.

refuse The first of the waste-reduction four *Rs*: Choose not to use or buy a product if you can do without it.

reduce The second of the waste-reduction four *Rs*: Make choices that allow you to use less of a resource by, for instance, purchasing durable goods that will last or can be repaired.

reuse The third of the waste-reduction four *Rs*: Use a product more than once for its original purpose or for another purpose.

INFOGRAPHIC 7.8 | **THE FOUR *R*s: Help You Reduce Waste**

↓ There are ways to reduce the amount of waste we generate, but for the waste we do have, there are better options than simply throwing away many products. An item like a plastic bottle can be recovered and reused, as the innovative company TerraCycle does, or the bottle may be recycled into another product.

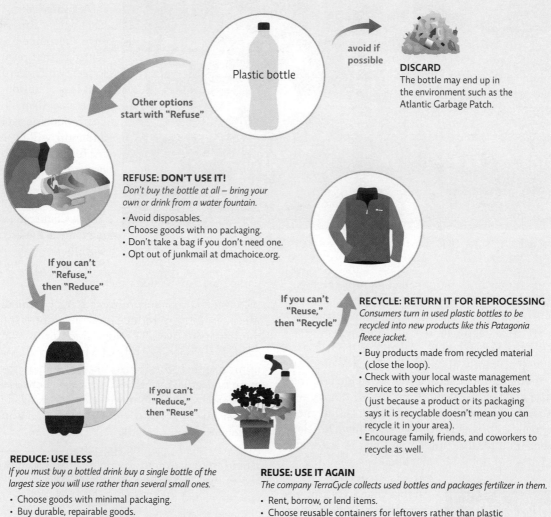

Plastic bottle

avoid if possible

DISCARD
The bottle may end up in the environment such as the Atlantic Garbage Patch.

Other options start with "Refuse"

REFUSE: DON'T USE IT!
Don't buy the bottle at all – bring your own or drink from a water fountain.

- Avoid disposables.
- Choose goods with no packaging.
- Don't take a bag if you don't need one.
- Opt out of junkmail at dmachoice.org.

If you can't "Refuse," then "Reduce"

If you can't "Reuse," then "Recycle"

RECYCLE: RETURN IT FOR REPROCESSING
Consumers turn in used plastic bottles to be recycled into new products like this Patagonia fleece jacket.

- Buy products made from recycled material (close the loop).
- Check with your local waste management service to see which recyclables it takes (just because a product or its packaging says it is recyclable doesn't mean you can recycle it in your area).
- Encourage family, friends, and coworkers to recycle as well.

If you can't "Reduce," then "Reuse"

REDUCE: USE LESS
If you must buy a bottled drink buy a single bottle of the largest size you will use rather than several small ones.

- Choose goods with minimal packaging.
- Buy durable, repairable goods.
- Buy local fresh food; it comes with less packaging.
- Use both sides of paper.
- Post notices on a bulletin board or via email to reduce copies.

REUSE: USE IT AGAIN
The company TerraCycle collects used bottles and packages fertilizer in them.

- Rent, borrow, or lend items.
- Choose reusable containers for leftovers rather than plastic bags or wrap.
- Buy and sell old clothes or donate to charity.
- Reuse products in different ways: use yogurt containers to hold screws; scrap paper for a note pad, etc.

Collection

Purchase   Production

RECYCLING REQUIRES 3 STEPS: Consumers and industry must turn in or collect materials for recycling, the material must be used to make new products, and the products must be bought by consumers.

1 PETE 2 HDPE 3 V 4 LDPE 5 PP 6 PS 7 OTHER

RECYCLING PLASTIC: The number code on a plastic item indicates the type of resin it is made with. In many areas, #1 and #2 are the only plastics that can be recycled; in others, all types are taken by recyclers. Check with your community recyclers to see what is taken in your area.

? Why is "recycling" considered the fourth choice of the four *R*s?

or plates, ask yourself: How long will this last, and for what other purpose might it be used?

Reusing also applies to industry. TerraCycle is a U.S. company that produces liquid worm-compost fertilizer and packages it in used soda and water bottles. The company also collects hard-to-recycle packaging like candy bar wrappers and juice pouches and turns them into new products like backpacks.

Once we've refused, reduced, and/or reused a given commodity as much as possible, we are left with the final R, **recycling**, the reprocessing of waste into new products. Recycling has several advantages. It certainly reduces the trash we generate, but it does more than that. By reclaiming raw materials from an item that we can no longer use, we limit the amount of raw materials that must be harvested, mined, or cut down to make new items. In most cases, this helps conserve limited resources—not only trees and precious metals but also energy and water. To execute this step properly, we must first have purchased items that can be recycled. We must also *close the loop* by purchasing items that are made of recycled materials to encourage manufacturers to make products from recycled materials. **INFOGRAPHIC 7.8**

Of course, even recycling comes with trade-offs. For example, while less energy and water are used by making paper from paper than are used by making paper from trees, there are also energy and environmental costs in collecting used paper, storing it, and shipping it to recycling plants. Once these costs are factored in, sustainably harvesting a forest of fast-growing trees may prove to be more environmentally friendly than recycling used paper.

By journey's end, the crew of the *Corwith Cramer* had spent a total of 34 days at sea, traveled 3,817 nautical miles (that's about 7,000 km or 4,400 miles), conducted 128 net tows, and counted 48,571 pieces of plastic along the way. Crew members found the northern and southern boundaries of the Atlantic Patch, at latitudes near Virginia and Cuba, respectively. But despite making it as far as the Mid-Atlantic Ridge, they never found the eastern edge.

As the coast of Bermuda came back into view on July 14, 2010, Proskurowski wrote his last blog post:

It is easy to brush off the topic of plastic pollution in the ocean. It occurs thousands of miles from land, in regions of the planet that are rarely visited by humans, and relatively sparsely populated by marine life. It is also easy to pass off responsibility, to say "I recycle," or "It must come from some other (developing) country," or "It is all fishing or marine industry waste." While there may be kernels of truth in all those arguments, the reality of the situation is that open ocean plastic pollution occurs over incredibly large regions of the Earth, has widely distributed point sources, and—because the oceans connect the whole globe—has far-reaching consequences.

recycle The fourth of the waste-reduction four *Rs*: Return items for reprocessing into new products.

Select References:

Environmental Protection Agency. (2014). *Municipal Solid Waste Generation, Recycling, and Disposal in the United State: Facts and Figures for 2012* (EPA 530-R-14-001).

Hoornweg, D., & P. Bhada-Tata. (2012). "What a Waste: A Global Review of Solid Waste Management," Urban Development Knowledge Papers, World Bank, http://go.worldbank.org/BCQEPOTMOO.

United Nations Environmental Programme. (2011). Plastic debris in the ocean. *UNEP Yearbook 2011*, www.unep.org/yearbook/2011/.

Webb, H. K., et al. (2012). Plastic degradation and its environmental implications with special reference to poly (ethylene terephthalate). *Polymers*, 5(1): 1—18.

Young, L. C., et al. (2009). Bringing home the trash: Do colony-based differences in foraging distribution lead to increased plastic ingestion in Laysan Albatrosses? *PLoS ONE*, 4(10): e7623.

BRING IT HOME

PERSONAL CHOICES THAT HELP

How much solid waste you produce is under your control. By reducing the amount of waste you produce, you reduce how much money we spend on waste disposal as a whole and at the same time place less pressure on the resources used to produce consumer goods. Reducing your solid waste is very easy and can save you money in the process.

Individual Steps

• Track your trash. Record what you throw out for a week by category and weight. How could you reduce your total trash weight by one-quarter? By half?

• Use the information in Infographic 7.8 (strategies to refuse, reduce, reuse, and recycle) to identify five changes you can make to reduce your solid waste.

Group Action

• Start recycling unusual items in your community. TerraCycle is a company that takes items that usually end up in the garbage, like candy wrappers, corks, and chip bags, and recycles them into new products, like purses or backpacks.

• Talk to friends and family about having "no gift" or "low gift" celebrations. Instead of buying lots of presents, treat friends to a dinner or a fun activity. For large families, use a grab bag or draw names and buy or upcycle (give used goods) for only specific people.

Policy Change

• Talk to community leaders to discuss the possibility of starting a communitywide composting program.

• Research rates of recycling participation in your community. Advocate for recycling education and curbside recycling programs.

ENVIRONMENTAL LITERACY UNDERSTANDING THE ISSUE

Why is it said that waste is a "human invention"? What kind of solid waste do humans produce, and how does its production compare between countries of different developmental levels and income?

INFOGRAPHIC 7.1

1. All of the following are consistent with the law of conservation of matter *except*:
   a. there is no such thing as waste.
   b. matter is neither created nor destroyed, but it can change form.
   c. if we encounter dangerous matter that we don't want around, we can destroy it.
   d. one organism's waste matter is taken up by another organism and reused.

2. Distinguish between degradable, biodegradable, and nondegradable waste. How well do plastics degrade?

3. How does waste generation relate to developmental status of countries? Why do you suppose that the regions with the highest per capita waste generation produce so much more waste than other regions?

What options do we have for dealing with solid waste, and what are the trade-offs for each option?

INFOGRAPHICS 7.2, 7.3 AND 7.6

4. True or False: Garbage decomposes more quickly in a sanitary landfill than it would in an open dump.

5. Which of the following statements about incinerators is *false*?
   a. Incinerators reduce the volume of material that needs to be landfilled.
   b. Incinerators are expensive to build and operate.
   c. The heat released during burning reduces the toxic and hazardous materials in the waste stream.
   d. Burning solid waste in incinerators can produce energy.

6. Compare and contrast landfilling and composting. What are the trade-offs for each option?

What are some of the negative consequences of solid waste pollution?

INFOGRAPHIC 7.4

7. Which of the following is a consequence of uncollected waste?
   a. Air pollution
   b. Water pollution
   c. Flooding
   d. A and B
   e. A, B, and C

8. In what ways does the plastic trash in oceans harm ocean life?

9. From an economic point of view, why is solid waste considered a mismanagement of resources?

What is hazardous waste, and what can the average person do to reduce his or her production of it?

INFOGRAPHIC 7.5

10. True or False: Hazardous waste is an industrial waste issue and is not a problem for MSW.

11. What is e-waste, and why is it a concern?

12. Identify some hazardous substances in your home.

How can industry and individuals reduce the amount of waste they produce?

INFOGRAPHICS 7.6, 7.7, AND 7.8

13. True or False: Any vegetable based kitchen or yard scraps can be added to a compost pile but you should avoid adding meat and bones.

14. The best solutions to managing waste include which one of the following?
   a. Source reduction and application of natural decomposition processes
   b. Supporting the siting of a landfill in your community to reduce waste transportation costs
   c. Disposing of hazardous waste in oceans, as the water dilutes the waste
   d. Shipping electronic waste to developing countries to provide jobs there in extracting precious metals from the electronic devices

15. An industrial ecologist's goals would *not* include:
   a. increasing resource-use efficiency by altering manufacturing processes.
   b. designing products that are durable and can be de-manufactured.
   c. identifying points in the product life cycle at which waste products can be used in other processes.
   d. identifying points in the product life cycle at which waste products can be landfilled.

16. Explain what is meant by the four *R*s, in order of preference, and give an example of each.

SCIENCE LITERACY **WORKING WITH DATA**

The data in the following table come from a report by Greenpeace, based on research published on plastic debris in the world's oceans between 1990 and 2005.

Number and Percentage of Marine Species Worldwide with Documented Entanglement and Ingestion Records

Species Group	Total Number of Species Worldwide	Number and Percentage of Species with Entanglement Records	Number and Percentage of Species with Ingestion Records
Sea turtles	7	6 (86%)	6 (86%)
Seabirds	312	51 (16%)	111 (36%)
Penguins	16	6 (38%)	1 (6%)
Grebes	19	2 (10%)	0
Albatrosses	99	10 (10%)	62 (63%)
Pelicans and cormorants	51	11 (22%)	8 (16%)
Gulls and terns	122	22 (18%)	40 (33%)
Marine mammals	115	32 (28%)	26 (23%)
Baleen whales	10	6 (60%)	2 (20%)
Toothed whales	65	5 (8%)	21 (32%)
Fur seals and sea lions	14	11 (79%)	1 (7%)
True seals	19	8 (42%)	1 (5%)
Manatees and dugongs	4	1 (25%)	1 (25%)
Sea otters	1	1 (100%)	0

Interpretation

1. How many different species are included in these data, and how many are affected by entanglement or ingestion of trash?

2. For seabirds:
   a. What percentage of seabird species had entanglement records? Which two groups had the highest rates of entanglements?
   b. What percentage of seabird species had ingestion records? Which two groups of seabirds had the highest rates of ingestions?

3. For marine mammals:
   a. What percentage of marine mammal species had entanglement records? Which two groups had the highest rates of entanglements?
   b. What percentage of marine mammal species had ingestion records? Which two groups had the highest rates of ingestions?

Advance Your Thinking

4. Scientists suspect that entanglement is a significant cause of population decline for many species, but they consider the reported entanglement rates to be conservative.
   a. Based on the data in the table, which group (sea turtles, seabirds, or marine mammals) is likely to be most affected by entanglement? Why?
   b. Why might reported entanglement rates underestimate the real problem?

5. Ingestion refers to animals eating plastic. While many species of marine mammals, seabirds, and sea turtles ingest plastic, some groups ingest more than others. What might explain these differences in ingestion rates among species?

6. Not much is known about the specific consequences of aquatic species ingesting plastics. In the process of science, observation leads to more questions. List three questions about ingestion of plastics that should be studied. How would you go about answering one of these questions?

INFORMATION LITERACY **EVALUATING INFORMATION**

Curbside recycling programs are a very convenient system for recycling metals, paper, glass, and plastic. However, they are available to only about half the population in the United States today, and what can be recycled is limited. So where do you go if you do not have curbside recycling or if you want to recycle or safely dispose of such things as electronics, batteries, or books? One source of help is the Earth911 website, which provides information on how to recycle a vast range of items as well as information on the latest recycling laws, ideas for living a green lifestyle, and stories on people and companies who are working to make a difference.

Go to the Earth911 website (http://earth911.com).

Evaluate the website and work with the information to answer the following questions:

1. Is this a reliable information source? Does the organization have a clear and transparent agenda?
   a. Who runs this website? Do the organization's credentials make the information presented reliable or unreliable? Explain.
   b. What is the mission of this website? What are its underlying values? How do you know this?
   c. What data sources does Earth911 rely on for its information, and what is its policy on what it puts on its website? Are the sources it uses reliable?
   d. Do you agree with the organization's assessment of the problems with and concerns about waste and recycling? What about its solutions? Explain.

2. On the home page, select the link "Recycling Search" and click the "Electronics" icon. Enter your location in the box provided (city and state or zip code).
   a. What options exist in your community for recycling electronics?
   b. Where is the nearest location where you could recycle a desktop computer? A cell phone? A video game console?

3. Click on any of the tabs at the top of the home page and investigate one topic thoroughly. Identify which tab you chose and answer the following questions:
   a. What types of articles are offered?
   b. Who writes these articles?
   c. Does the content seem useful and credible? Explain your response.

4. How useful is a website like Earth911? How can a website like this influence societal understanding of waste issues and facilitate a change in behavior on the part of individuals and businesses?

Find an additional case study online at http://www.macmillanhighered.com/launchpad/saes2e

ENGINEERING EARTH

An ambitious attempt to replicate Earth's life support systems falls short

CORE MESSAGE

Ecosystems are complex assemblages of many interacting living and nonliving components. Living organisms play irreplaceable roles in nature, supporting life and allowing ecosystems to function over the long term. It is important that we protect and work to restore ecosystems in nature to keep these connections intact so that we and other species can continue to live and thrive on this planet.

AFTER READING THIS CHAPTER, YOU SHOULD BE ABLE TO ANSWER THE FOLLOWING **GUIDING QUESTIONS**

1

What is the hierarchy of organization recognized by ecologists, and why might it be useful to recognize such distinctions?

2

Why do ecosystems need a constant input of energy? How do they deal with the fact that Earth does not receive appreciable new inputs of matter?

Biosphere 2 from afar.
© ClassicStock/CAMERIQUE/Alamy

3

What are biomes, and how do environmental factors affect their distribution and makeup?

4

What is a population's range of tolerance, and how does it affect the distribution of a population within its ecosystem or its ability to adapt to changing conditions?

5

How do carbon, nitrogen, and phosphorus cycle through ecosystems? How are these cycles being disrupted, and what problems can this disruption cause?

On September 26, 1993, with their first mission complete, four men and four women emerged from Biosphere 2—a hulking dome of custom-made glass and steel—back into the Arizona desert, where throngs of spectators stood cheering. They had been sealed inside the facility, along with 3,000 other plant and animal species, for exactly 2 years and 20 minutes; it was the longest anyone had ever survived in an enclosed structure.

The feat was part of a grand experiment, the goals of which were twofold. First, scientists wanted to prove that an entirely self-contained, humanmade system—the kind they might one day use to colonize the Moon or Mars—could sustain life. Second, they hoped that by studying this mini-Earth, which could be controlled and manipulated in ways the real Earth could not, they might better understand our own planet's delicate balance and how best to protect it.

Despite the fanfare surrounding the biospherians' emergence, it was tough to say whether the mission had been a success or a failure. More than one-third of the flora and fauna had become extinct, including most of the vertebrates and all of the pollinating insects. Morning glory vines had overrun other plants, including food crops. Cockroaches and "crazy ants" were thriving. Too little wind had prevented trees from developing stress wood—wood that grows in response to mechanical stress and helps trunks and branches shift into an optimal position; without stress wood, the trees were brittle and prone to collapse. And eating too many sweet potatoes had turned the biospherians themselves bright orange. (A string of plant diseases had decimated other crops.)

On top of that, nitrous oxide (laughing gas) had grown concentrated enough to "reduce vitamin B12 synthesis to a level that could impair or damage the brain," according to one interim report. And oxygen levels had plummeted from 21% (roughly the same as Earth's atmosphere) to 14% (just barely enough to sustain human life). To fix this, project engineers had been forced to pump in outside air, violating the facility's sanctity as a closed system.

Worst of all, missteps and course corrections had been mired in secrecy—each one leaked to the press only months after the fact. Rumors had begun to circulate that the eight people sealed inside—not to mention the ones they took their orders from—were more interested in creating a futuristic utopia than in conducting rigorous scientific research. As evidence for this rumor mounted, the scientific community grew suspicious. Was Biosphere 2 legitimate science, a publicity stunt, or some bizarre mix of the two?

species   A group of plants or animals that have a high degree of similarity and can generally only interbreed among themselves.

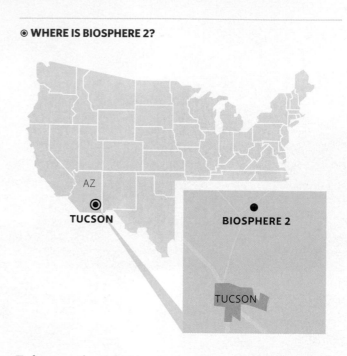

WHERE IS BIOSPHERE 2?

AZ

TUCSON

BIOSPHERE 2

TUCSON

To be sure, the eight biospherians had survived, and many experts agreed that, in principle at least, the facility still held enormous potential as a scientific tool. But before that potential could be realized, the scientific community and the public at large would need to know exactly what had happened inside the desert dome.

To answer that question, we need to answer a few others first: What exactly is a biosphere, and just how did Biosphere 2's creators set about building one?

Organisms and their habitats form complex systems.

The field of ecology focuses on how **species** interact with other components in their environment. In other words, it is about relationships. These

KEY CONCEPT 8.1

Ecologists identify a nesting organization of life from individuals to the biosphere, often focusing on the levels of population, community, and ecosystem to examine how species respond to and affect the natural world.

← The eight biospherians emerge from Biosphere 2 after living there for two years.

↓ Visitors can tour the facility. Here they view the desert biome.

relationships can be examined at different levels. The term **biosphere** refers to the total area on Earth where living things are found—the sum total of all its ecosystems. An **ecosystem** includes all the organisms in a given area plus the nonliving components of the physical environment in which they interact. In the natural world, ecosystems assume a range of shapes and sizes—a single, simple tide pool qualifies as an ecosystem; so does the entire Mojave Desert.

But having the entire biosphere (or even a large ecosystem) as the focus of study is usually too expansive to manage,

so ecologists study how ecosystems function by focusing on species' interactions with their physical surroundings (their **habitat**) and with other species in their **community**. They also study interactions between individuals of the same species within a **population**. These relationships define the unique ecological **niche** of each individual species. **INFOGRAPHIC 8.1** and **INFOGRAPHIC 8.2**

All ecosystems function through two fundamental processes that are collectively referred to as ecosystem processes: nutrient cycling and **energy flow**. **Nutrient cycles** are biogeochemical cycles that refer specifically

---

INFOGRAPHIC 8.1   ORGANIZATION OF LIFE: FROM BIOSPHERE TO INDIVIDUAL

ECOSYSTEM
A specific portion of a biome consisting of the living (biotic) and nonliving (abiotic) environmental components that interact.

BIOSPHERE
The total area on Earth (air, land, or water) where living things are found.

BIOME
A portion of the biosphere characterized by a distinct climate and a particular assemblage of plants and animals adapted to it.

COMMUNITY
All the populations (plants, animals, and other species) living and interacting in an area. Communities represent the "living" portion of the ecosystem.

← Ecologists recognize a nesting hierarchy of organization from the biosphere down to the individual organism. Each category is made up of the smaller ones. Ecologists often focus their study on populations, communities, and ecosystems and the interactions between organisms and their surroundings. In contrast, a zoologist or botanist might focus on individual animals or plants.

POPULATION
A group of individuals of the same species living and interacting in the same region.

INDIVIDUAL
A single member of a population.

 You are studying a pocket mouse that lives in an underground borrow where it stores seeds. Make a list of biotic and abiotic things that might be important to its survival.

INFOGRAPHIC 8.2   HABITAT AND NICHE

↓ Species depend on suitable habitats in which to live. Species fill a specific niche in their community.

SPECIES
A group of plants or animals that have a high degree of similarity and can generally only interbreed among themselves.

HABITAT
The physical environment in which individuals of a particular species can be found.

NICHE
The role a species plays in its community, including how it gets its energy and nutrients, its habitat requirements, and what other species and parts of the ecosystem it interacts with.

 Why do ecologists usually study populations rather than entire species? When might they focus on an entire species?

to the movement of life's essential chemicals or nutrients through an ecosystem. Energy, on the other hand, enters ecosystems as solar radiation and is passed along from organism to organism, some released as heat, until there is no more usable energy left. Therefore, we can say that matter *cycles* but energy *flows* in a one-way trip.

Earth—or "Biosphere 1," as the creators of Biosphere 2 liked to call it—is materially closed but energetically open. In other words, the plants and other organic material that make up an ecosystem, called *biomass*, cannot enter or leave the system, but energy can: Some energy leaves as heat or light, and new energy is absorbed from outside. In fact, plant biomass is produced with energy from the Sun through photosynthesis. **INFOGRAPHIC 8.3**

Biomes are specific portions of the biosphere determined by climate and identified by the predominant vegetation and organisms adapted to live there. Biomes can be divided into three broad categories—terrestrial, marine, and freshwater. Within those three categories are several narrower groups, and within those are a variety of subgroups. An entire biome itself may be considered an ecosystem, as are the smaller groups and subgroups. For example, forests, deserts, and grasslands are the three main types of terrestrial biomes. Within the forest biome category are different types of forests, such as tropical, temperate, and boreal forests, and within each of those groups are subgroups (for example, dry tropical forest and tropical rain forest). **INFOGRAPHIC 8.4**

INFOGRAPHIC 8.3   EARTH IS A CLOSED SYSTEM FOR MATTER BUT NOT FOR ENERGY

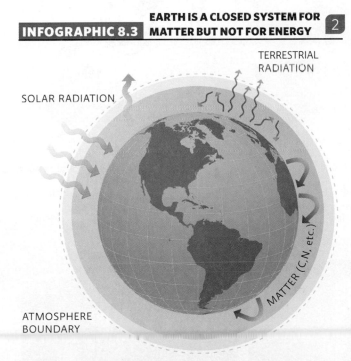

TERRESTRIAL RADIATION

SOLAR RADIATION

MATTER (C, N, etc.)

ATMOSPHERE BOUNDARY

↑ Energy can enter and leave Earth as light (solar radiation) and heat (terrestrial radiation), but matter stays in the biosphere, cycling in and out of organisms (biomass) and environmental components.

 Why is it so important that species recycle matter and that they depend on a source of energy that is readily replenished (renewable)?

biosphere The sum total of all of Earth's ecosystems.

ecosystem All of the organisms in a given area plus the physical environment in which, and with which, they interact.

habitat The physical environment in which individuals of a particular species can be found.

community All the populations (plants, animals, and other species) living and interacting in an area.

population All the individuals of a species that live in the same geographic area and are able to interact and interbreed.

niche The role a species plays in its community, including how it gets its energy and nutrients, what habitat requirements it has, and what other species and parts of the ecosystem it interacts with.

energy flow The one-way passage of energy through an ecosystem.

nutrient cycles Movement of life's essential chemicals or nutrients through an ecosystem.

biome One of many distinctive types of ecosystems determined by climate and identified by the predominant vegetation and organisms that have adapted to live there.

INFOGRAPHIC 8.4   **GLOBAL TERRESTRIAL BIOMES**

3

What biome do you live in? Identify an area on another continent where you could travel and visit that same biome.

↘ Terrestrial biomes are specific types of terrestrial ecosystems with characteristic temperature and precipitation conditions. Temperature varies with latitude (decreasing as one moves away from the equator) and altitude (decreasing as elevation increases); thus a cold climate can be found above 60° north and south latitudes, as well as on an equatorial mountaintop. Latitude also affects precipitation, with wet areas occurring at the equator and at 30° north and south and dry areas occurring at 60° north and south (due to global air circulation patterns).

30° Tropic of Cancer

0° Equator

30° Tropic of Capricorn

- ● Ice
- ● Tundra
- ● Desert
- ● Mountain
- ● Boreal forest (taiga)
- ● Temperate forest
- ● Grassland
- ● Mediterranean scrub
- ● Tropical rain forest
- ● Tropical seasonal forest
- ● Savannah
- ● Tropical scrub

TUNDRA

George Burba/Shutterstock

TROPICAL RAIN FOREST

Dr. Morley Read/Shutterstock

BOREAL FOREST

IDAK/Shutterstock

↓ This biome climograph shows the approximate distribution of terrestrial biomes with regard to annual precipitation and temperature. As precipitation increases and more plant life is able to be supported, deserts, scrublands, or grasslands give way to forests. The temperature also influences what type of desert, grassland, or forest is present (for example, temperate forest versus boreal forest).

Which biome exists across the widest range of temperatures? Precipitation?

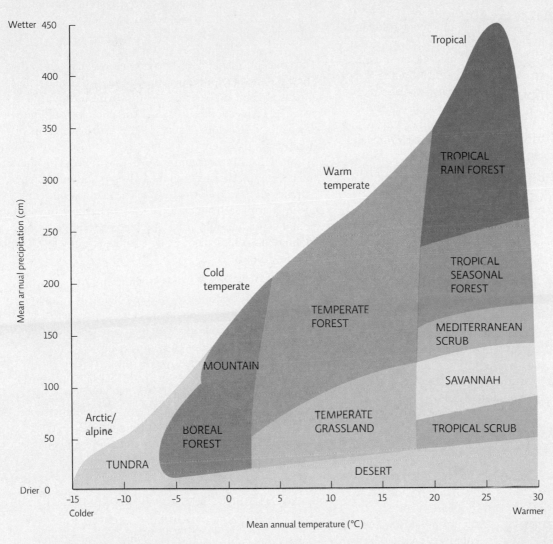

Mediterranean Scrub

MEDITERRANEAN SCRUB
DESERT
SAVANNA

Sergey Borisov/iStock/360/Thinkstock/Getty Images

© Norma Jean Gargasz/AGE fotostock

Vladimir Kondrachov/Shutterstock

KEY CONCEPT 8.2

All life on Earth depends on a constant input of new energy because once it is used, energy is degraded and no longer useful to organisms. New inputs of matter to Earth, however, are negligible, so life also depends on the constant cycling of matter resources.

KEY CONCEPT 8.3

The biome that is present in a given area is influenced by the physical and climatic characteristics of its environment, particularly precipitation amount and temperature for terrestrial biomes.

When ecologists study entire ecosystems, they are limited to making observations and trying to discern cause and effect from those observations. This is no small challenge; precise measurement of each and every relevant ecological factor is simply not possible, and even the simplest factor (for example, a change in rainfall, the loss of a single species, variations in solar radiation reaching the ground) can impact many of the other factors and affect the ecosystem as a whole. But this doesn't mean we can't gain insights from ecosystem-scale studies. Scientists measure important parameters of natural ecosystems to gather evidence to try to discern cause and effect; they often compare those that are affected by natural disasters or human impact to less disturbed ecosystems in a kind of "natural experiment." From this, ecologists make their best estimations of how multiple factors affect each another (often via mathematical modeling). The more parameters that are properly measured and linked to one another, the better, but because all parameters and relationships cannot possibly be included, there may be a lot of room for interpretation when it comes to understanding what is happing at the ecosystem level.

Biosphere 2 offered ecologists an unprecedented research tool: a mini-planet where a variety of environmental variables—from temperature and water availability to the relative proportions of oxygen and carbon dioxide (CO_2) at any given moment—could be tightly controlled and precisely measured. "Manipulating these variables and tracking the outcomes could greatly advance our understanding of natural ecosystems and all the minute, complex interactions that make them work," says Kevin Griffin, a Columbia University plant ecologist who conducted research at the Biosphere 2 facility. "The plan was to use that knowledge to figure out how to repair degraded ecosystems in the real world, so that they continue to provide the services so essential to our survival."

On top of all that, proving that humans could survive in a completely enclosed, manufactured system would take us one giant step closer to colonizing space.

But would it work?

The concept of enclosed ecosystems was not a new one. Since before they put a person on the Moon, astronauts and engineers had been tinkering with their own artificial ecosystems—systems they hoped could one day be used to colonize space. The earliest versions of this technology were developed in the 1960s and 1970s by Russian and American scientists who, in the spirit of their times, had pitted themselves against one another in a mad dash to the finish line. The ecosystems they designed ranged from small, crude structures in which a single person might last a single day, to larger, more sophisticated enterprises that could sustain a few people for a few months.

Biosphere 2 is by far the most elaborate. At just over 1 hectare (3 acres or about 2 football fields in diameter), it remains the largest enclosed ecosystem ever created and the only one to house several biomes under one roof. From a mountain under the dome's 28-meter (91-foot) zenith, a stream rushes down through a tropical rain forest before snaking southward into a savanna. From there, the stream wends its way through a mangrove swamp into a million-gallon ocean, complete with a coral reef. On the other side of the ocean lies a desert. Biosphere 2 also includes a human habitat and an agricultural biome. **INFOGRAPHIC 8.5**

Living things survive within a specific range of environmental conditions.

Each biome required a mind-boggling array of considerations—not only how diverse plant and animal species would interact within and across biomes but also the nutrient requirements of each organism they planned to include. Termites, for example, would need enough dead wood at the beginning of closure to sustain them until some of the larger plants began dying off. Termites live in the soil, stirring it and allowing air to penetrate soil particles. If the termites ran out of dead wood and starved to death, organisms living in the soil would not get enough oxygen, and the entire desert would be jeopardized. Hummingbirds, on the other hand, would need nectar-filled flowers. "Try figuring out how many flowers a day a hummingbird needs," says Tony Burgess, a University of Arizona ecologist who helped design

3

INFOGRAPHIC 8.5   **MAP OF BIOSPHERE 2**

↓ Biosphere 2 houses several biomes under one roof, each contributing to overall function. One of the challenges faced by designers was how to include a variety of biomes in the close quarters of the 3-acre Biosphere 2 structure. For example, in nature, a tropical rain forest would not be next to an arid desert. To deal with this, an ocean was placed between the desert and rain forest to serve as a temperature buffer.

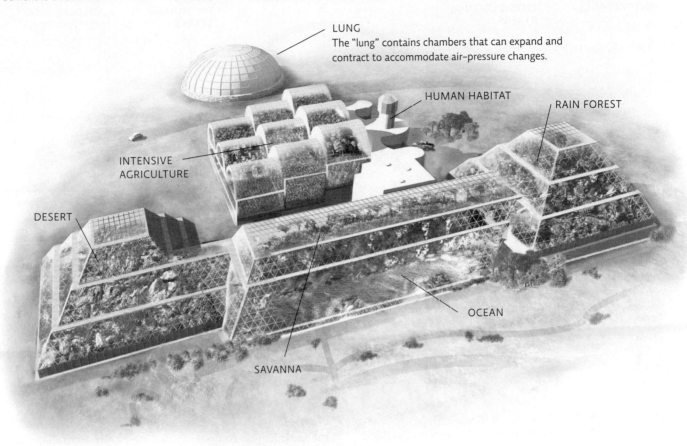

LUNG
The "lung" contains chambers that can expand and contract to accommodate air-pressure changes.

HUMAN HABITAT

RAIN FOREST

INTENSIVE AGRICULTURE

DESERT

OCEAN

SAVANNA

Why was it important for the desert and rain forest biome sections of Biosphere 2 to be separated by another biome such as the ocean?

the biomes in Biosphere 2 and remained involved until 2004. "From there you need to know what the blooming season is, and then what the nectar load per flower is. And then you have to translate all of that into units of hummingbird support. Now imagine doing that sort of thing about 3,000 times."

Biome-level planning was equally complicated. A rain forest would need consistently warm temperatures, but most desert biomes fluctuate wildly between day and night. In the summer, the grassland biome might literally starve the rain forest by gobbling up all the CO_2 needed for photosynthesis.

In this context, dead wood, flowers, and CO_2 are all examples of **limiting factors**—resources so critical that their availability controls the distribution of species and thus of biomes. The principle of limiting factors states that the critical resource in least

supply is what determines the survival, growth, and reproduction of a given species in a given biome. Living things can only survive and reproduce within a certain range (between the upper and lower limits that they can tolerate) for a given critical resource or environmental condition, referred to as their **range of tolerance**. **INFOGRAPHIC 8.6**

Ecologists routinely monitor a variety of limiting factors as a way of assessing ecosystem health, but anticipating what each individual organism would need to survive before the fact proved daunting.

An all-star team of scientists—oceanographers, forest ecologists, and plant physiologists—spent

limiting factor The critical resource whose supply determines the population size of a given species in a given ecosystem.

range of tolerance The range, within upper and lower limits, of a limiting factor that allows a species to survive and reproduce.

INFOGRAPHIC 8.6   RANGE OF TOLERANCE FOR LIFE   4

| ZONE OF INTOLERANCE | ZONE OF PHYSIOLOGICAL STRESS | ZONE OF OPTIMUM RANGE | ZONE OF PHYSIOLOGICAL STRESS | ZONE OF INTOLERANCE |

(POPULATION SIZE axis: High / Low)

Organisms abundant

Organisms absent

Organisms absent

Low temperature   ENVIRONMENTAL GRADIENT   High temperature

↑ Populations have a range of tolerance for any given environmental factor (such as temperature). Every species has an upper and a lower limit beyond which it cannot survive (in this example, temperatures that are too cool or too hot). Most individuals, like the butterflies in this population, can be found around the optimum temperature, though what is "optimum" for each individual may differ slightly because of genetic variability. Some individuals may find themselves in areas of the habitat that are warmer or colder than the optimum. They can tolerate these conditions but may be physiologically stressed and not grow well or successfully breed. Genetic differences that allow some individuals to tolerate or even thrive at the edges of the population's tolerance offer the population a chance to adapt to changing conditions (such as a warmer climate) if needed. The more narrow the range of tolerance and the less genetically diverse the population, the less likely it will survive a change in conditions.

 Which population would have the greatest chance of adapting to an increase or decrease in the average temperature—one with a broad range of tolerance for temperature or one with a narrow range of tolerance?

2 years sorting through these challenges. Drawing on their combined expertise, they set about choosing the combination of soils, plants, and animals that seemed most capable of working together to re-create the delicate balances that had made Biosphere 1 such a spectacular success. A summer-dormant desert, like the ones found in Baja California, was chosen because it would reduce the desert's CO_2 demands when the savanna's productivity was at its highest. The ocean was situated between the desert and rain forest so that it could serve as a temperature buffer between the two. And each biome was created from a carefully selected array of species: The marsh biome was composed of intact chunks of swampland harvested from the Florida Everglades, and the savanna was composed of grasses from Australia, South America, and Africa. Well water mixed with aquarium salt filled the ocean, to which were added coral reef sections culled from the Caribbean.

But it wasn't long before the rigor and pragmatism of good science began to clash with the idealism of Biosphere 2 financiers. And when that happened, critics say, science lost out.

Some scientists worried that the ocean wouldn't get enough sunlight to support plant and animal life. Others opposed the use of soil high in organic matter; soil microbes decompose the soil's organic carbon and release it into the air as CO_2. Although this organic soil might eliminate the need for chemical fertilizers, there were concerns that it would provide too much fuel for the soil microbes and would thus send atmospheric CO_2 concentrations through the roof. Despite these concerns, scientific advisors to the project were overruled.

The first few months of Mission 1 went smoothly enough, but eventually, plants and animals started dying. Humans grew hungry and mysteriously sleepy. And before long, tempers flared.

KEY CONCEPT 8.4

Populations are able to survive and thrive in habitats with the needed resources and environmental conditions. Individual variability increases a population's range of tolerance, expanding its distribution within the habitat as well as increasing the chance that a population will be able to adapt to changing conditions.

KEY CONCEPT 8.5

The matter cycles that move nutrients though ecosystems depend on living organisms and abiotic reservoirs of those resources.

Like Earth, Biosphere 2 was designed as a materially closed and energetically open system: Plants would conduct photosynthesis fueled by sunlight that streamed through the glass, but no biomass would enter or leave. Temperature, wind, rain, and ocean waves would be controlled mechanically. "But the facility would have to be self-sustaining," says Burgess. "Everything would die, unless the biota met its most fundamental purpose—using energy flow for biomass production." Humans would have to survive exclusively on what they could grow or catch under the dome. The system as a whole would have to continuously recycle every last bit of nutrient that was in the soil on day one.

At first, the carefully constructed agricultural biome seemed well suited to the challenge. Carrots, broccoli, peanuts, kale, lettuce, and sweet potatoes were grown on broad half-acre terraces that sat adjacent to the sprawling six-story human habitat. A bevy of domestic animals also provided sustenance—goats for milk, chickens for eggs, and pigs for pork. Indeed, eating only what they could grow made the biospherians healthier. Bad cholesterol and blood pressure went down; so did white blood cell counts. And slowly, the biospherians say, their relationship to food changed. "Inside, I knew exactly where my food came from, and I totally understood my place in the biosphere and how it impacted the food I ate," says Jayne Poynter, the biospherian responsible for tending the farm. "When I breathed out, my CO_2 fed the sweet potatoes that I was

growing. When I first got out, I lost sense of that. I would stand for hours in the aisles of shops, reading all the names on all the food things and think 'where does this stuff come from?' People must have thought I was nuts."

But the glazed glass of the dome admitted less sunlight than had been anticipated. Less sunlight meant less biomass production. And that meant less food. Mites and disease also cut crop production. Biosphere 2's size precluded the use of pesticides and herbicides: In that small an atmosphere, the toxins would build up rapidly and could have had a deleterious impact on air quality and human health.

Now, a few months in, food was starting to run out. And that wasn't the only problem.

Nutrients such as carbon cycle through ecosystems.

After several months under the dome, the humans grew so tired they couldn't work. Nobody knew why, but scientists on the outside suspected that it had something to do with nutrient cycles.

On Earth, nutrients cycle through both **biotic** and **abiotic** components of an ecosystem—organisms, air, land, and water. They are stored in abiotic or biotic parts of the environment called **reservoirs**, or **sinks**, and linger in each for various lengths of time, known as *residence times*. Organisms acquire nutrients from the reservoir, and

biotic The living (organic) components of an ecosystem, such as the plants and animals and their waste (dead leaves, feces).

abiotic The nonliving components of an ecosystem, such as rainfall and mineral composition of the soil.

reservoirs (or sinks) Abiotic or biotic components of the environment that serve as storage places for cycling nutrients.

↓ Decomposers like these mushrooms (a fungus) growing on a decaying log are important consumers that help recycle nutrients in an ecosystem.

C. Kurt Holter/Shutterstock

INFOGRAPHIC 8.7   **CARBON CYCLES VIA PHOTOSYNTHESIS AND CELLULAR RESPIRATION**   5

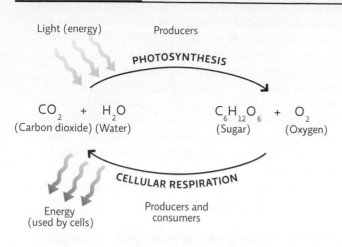

← In photosynthesis, producers use solar energy to combine CO_2 and H_2O to make sugar, releasing O_2 in the process. When producers (or any consumer who eats another organism) need energy, they break apart the sugar via the reverse reaction, cellular respiration. Oxygen is required for this reaction (which is why it is called "respiration").

? Identify some producers and consumers in your own ecosystem.

those nutrients are then cycled through the food chain and eventually are returned to the reservoir.

For carbon, the atmosphere—where carbon is stored as CO_2—is the most important reservoir. (Oceans and soil are also abiotic reservoirs for carbon. Oceans absorb CO_2 directly from the atmosphere, and soils accumulate it during decomposition.) Plants and other photosynthesizers use carbon molecules from atmospheric CO_2 to build sugar, and they release oxygen in the process. Because they "produce" sugar (an organic molecule) from inorganic atmospheric CO_2, they are called **producers**.

This sugar molecule represents stored chemical energy that the producer can use. A **consumer**, the organism that eats the plant (or that eats the organism that eats the plant), also uses the chemical energy of sugar. This energy is released to the cell via the process of **cellular respiration**. All organisms—producers and consumers—perform cellular respiration. (For more information on producers, consumers, and the food chain, see Chapter 10.) **INFOGRAPHIC 8.7**

From its initial incorporation into living tissue via photosynthesis, to its ultimate return to the atmosphere through respiration or through the burning of carbon-based fuels, carbon cycles in and out of various molecular forms and in and out of living things as it moves through the **carbon cycle**. **INFOGRAPHIC 8.8**

producer An organism that converts solar energy to chemical energy via photosynthesis.

consumer An organism that obtains energy and nutrients by feeding on another organism.

cellular respiration The process in which all organisms break down sugar to release its energy, using oxygen and giving off CO_2 as a waste product.

carbon cycle Movement of carbon through biotic and abiotic parts of an ecosystem. Carbon cycles via photosynthesis and cellular respiration as well as in and out of other reservoirs, such as oceans, soil, rock, and atmosphere. It is also released by human actions such as the burning of fossil fuels.

Even though a single carbon atom might be captured by photosynthesis and returned to the atmosphere via cellular respiration in a few weeks or years, decades can pass before changes in a Brazilian rain forest impact a farm in Iowa. Inside Biosphere 2, the same cycle took approximately 3 days, which meant that changes in one biome could be felt in another biome much more quickly than on Earth.

Still, Biosphere 2's carbon cycle was not that different from Earth's; carbon moved from living tissue to the atmosphere and back in the same predictable manner. Or at least it should have. As the biospherians' energy waned, it became clear that something had gone terribly wrong.

It turned out that oxygen levels had fallen steadily—from 21% down to 14%. At such low concentrations, the biospherians were unable to convert the food they consumed into usable energy. "We were just dragging ourselves around the place," Poynter says. "And we had sleep apnea at night. So we'd wake up gasping for air because our blood chemistry had changed."

In just a few months, some 7 metric tons of oxygen—enough to keep six people breathing for 6 months—had gone missing. As scientists from Columbia University later discovered, soil microbes were gobbling up all that O_2 and converting it into CO_2 as they decomposed the organic matter in the soil.

The biospherians responded by filling all unused planting areas with morning glory

KEY CONCEPT 8.6

Carbon cycles through the environment via photosynthesis and cellular respiration. This cycle is becoming unbalanced due to human actions that increase the amount of atmospheric carbon dioxide.

vines, a pretty and fast-growing (but, as it turned out, invasive) species they hoped would maximize the amount of CO_2 converted back into O_2 by photosynthesis. But even with an abundance of plants and enough CO_2, photosynthesis was still limited by the availability of sunlight; even the morning glories couldn't keep up with the soil microbes in their warm, well-watered, highly organic soil.

Biosphere 2 is not alone with regard to a disrupted carbon cycle. Human activity has greatly altered carbon amounts in Earth's atmosphere. Many of our actions (like burning fossil fuels) increase the amount of carbon normally released into the atmosphere or degrade natural ecosystems so that less carbon is removed from the atmosphere (as in the case of deforestation). Just as with Biosphere 2, this extra atmospheric carbon causes problems such as global climate change, acidification of oceans, and alterations of communities worldwide.

Adding to the confusion, concrete used to build parts of Biosphere 2 was absorbing some of the CO_2 and converting it into calcium carbonate, trapping some of the carbon and oxygen in this unexpected sink.

Besides carbon, other chemicals essential for life, such as nitrogen and phosphorus, cycle through ecosystems. Nitrogen, the most abundant element in Earth's atmosphere, is needed to make proteins and nucleic acids, but plants cannot utilize nitrogen in its atmospheric form (N_2). All plant life, and ultimately all animal life, too,

INFOGRAPHIC 8.8   THE CARBON CYCLE

5

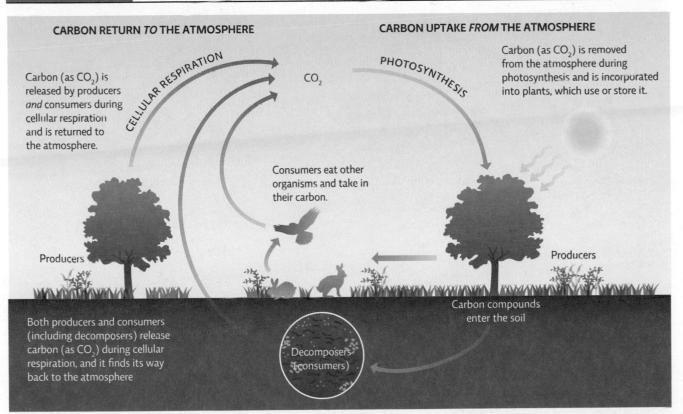

CARBON RETURN *TO* THE ATMOSPHERE

CARBON UPTAKE *FROM* THE ATMOSPHERE

Carbon (as CO_2) is released by producers *and* consumers during cellular respiration and is returned to the atmosphere.

CELLULAR RESPIRATION

CO_2

PHOTOSYNTHESIS

Carbon (as CO_2) is removed from the atmosphere during photosynthesis and is incorporated into plants, which use or store it.

Consumers eat other organisms and take in their carbon.

Producers

Producers

Carbon compounds enter the soil

Both producers and consumers (including decomposers) release carbon (as CO_2) during cellular respiration, and it finds its way back to the atmosphere

Decomposers (consumers)

↑ Carbon cycles in and out of living things during photosynthesis and cellular respiration. As consumers (including decomposers) eat other organisms, carbon is transferred. Most of Earth's carbon is actually stored in rocks or dissolved in the planet's oceans, but some carbon is stored in the bodies of organisms and in soil. Without human interference, over the long term the carbon cycle is balanced between photosynthesis and respiration.

Humans unbalance the carbon cycle via activities that increase the amount of CO_2 in the atmosphere.

Burning fossil fuels

Forest fire

Deforestation

Explain how each of these human impacts (burning fossil fuels, forest fires, and deforestation) can result in a net increase in atmospheric carbon.

depends on microbes (bacteria) to convert atmospheric nitrogen into usable forms as part of the **nitrogen cycle**.

---

nitrogen cycle A continuous series of natural processes by which nitrogen passes from the air to the soil, to organisms, and then returns back to the air or soil.

nitrogen fixation Conversion of atmospheric nitrogen into a biologically usable form, carried out by bacteria found in soil or via lightning.

In a process called **nitrogen fixation**, bacteria convert atmospheric nitrogen (N_2) into ammonia (NH_3), which plants subsequently take up through their roots; consumers take in nitrogen via their diet. A small amount of N_2 is fixed by lightning, producing nitrate (NO_3). In other steps of the nitrogen cycle (decomposition, nitrification, and denitrification), various types of bacteria feed on nitrogen compounds in organic matter or the soil, eventually returning it to the atmosphere as N_2.

INFOGRAPHIC 8.9

In Biosphere 2, the nitrogen cycle was disrupted. Thanks to an overabundance of soil microbes, nitrous oxide (N_2O), or laughing gas, a normal by-product of denitrification, reached levels high enough to interfere with the metabolism of vitamin B12, which is essential to the brain and nervous system.

The phosphorous cycle was also disrupted in Biosphere 2. Unlike nitrogen and carbon, phosphorus—which is needed to make DNA and RNA—is found only in solid or

INFOGRAPHIC 8.9   **THE NITROGEN CYCLE**   5

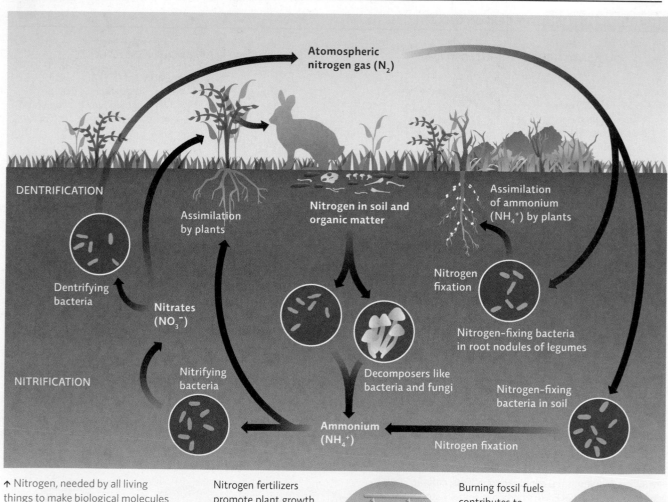

 ↑ Nitrogen, needed by all living things to make biological molecules like protein and DNA, continuously moves in and out of organisms and the atmosphere in a cycle absolutely dependent on a variety of bacteria.

Nitrogen fertilizers promote plant growth, but this depletes other soil nutrients; they can also leach out of soils and pollute aquatic ecosystems.

Fertilizing

Burning fossil fuels contributes to nitrogen pollution such as smog and acid rain.

Burning fossil fuels

Look closely at the nitrogen cycle. How many different types of microbes are needed to complete the entire cycle?

INFOGRAPHIC 8.10 **THE PHOSPHORUS CYCLE**

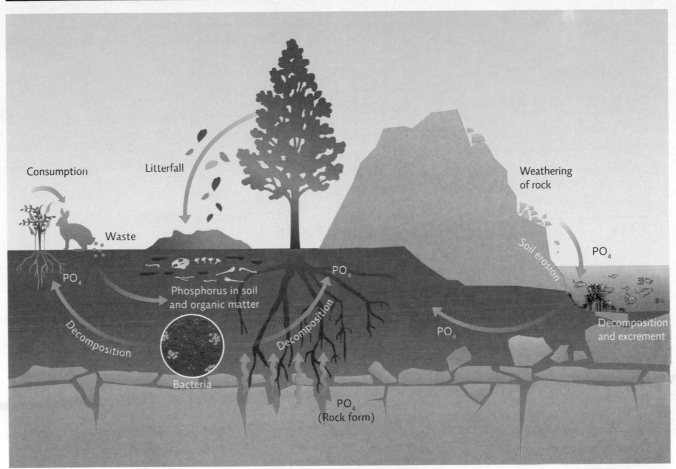

Consumption

Litterfall

Weathering of rock

Waste

PO_4

Soil erosion

PO_4

Phosphorus in soil and organic matter

PO_4

Decomposition

Decomposition

PO_4

Decomposition and excrement

Bacteria

PO_4 (Rock form)

↑ Phosphorus, needed by all organisms to make DNA, cycles very slowly. It has no atmospheric component but instead depends on the weathering of rock to release new supplies of phosphate (PO_4) into bodies of water or the soil, where it dissolves in water and can be taken up by organisms. Microbes also play a role when they break down organic material and release the phosphate to the soil.

Dust released through mining or in eroded areas can introduce phosphorus into the environment much more quickly than it would normally enter.

Open-pit mining

Fertilizers and animal waste (including sewage) can alter plant growth and nutrient cycling, especially in aquatic ecosystems where phosphorus is usually a limiting nutrient.

Animal waste

? How might phosphorus from farms enter aquatic ecosystems (rivers, streams, lakes, and oceans)?

liquid form on Earth, so the **phosphorus cycle** does not move through the atmosphere but passes from inorganic to organic form through a series of interactions with water and organisms. **INFOGRAPHIC 8.10**

In Biosphere 2 phosphorus got trapped in the water system, polluting aquatic habitats, because the underwater and terrestrial plants there were dying off too quickly to complete this cycle. Biospherians removed excess nutrients from their water supply by passing the water over algal mats that would absorb the phosphorus and could then be harvested, dried, and stored.

As food reserves dwindled, the eight biospherians split into two factions. One group felt that scientific research was the top priority and wanted to import food so that they would have enough energy to continue with their experiments. The other group felt that maintaining a truly closed system—one where no biomass was allowed to enter or leave—was the project's most important goal; proving that humans could survive exclusively on what the dome provided would be essential to

phosphorus cycle A series of natural processes by which the nutrient phosphorus moves from rock to soil or water, to living organisms, and back to soil.

↑ Justin Peterson, a Biosphere 2 undergraduate intern who assists PhD candidate Henry Adams with his Pinon Pine Tree Drought Experiment. The experiment's goal is to predict the effects of climate shifts on the trees.

one day colonizing the Moon or Mars. To them, importing food would amount to a mission failure. "It was a heartbreaking split," Poynter says. "Just 6 months into the mission, and two people on the other side of the divide had been my closest friends going in." Eventually, Poynter snuck in food. That wasn't the only breech. To solve the various nutrient cycle conundrums, the project's engineers installed a CO_2 scrubber and pumped in 17,000 cubic meters (600,000 cubic feet or about 4 million gallons) of oxygen.

Biosphere officials hid these actions from all but a few key people. When one reporter finally broke the story, the public was outraged. Spectators of every ilk—seasoned scientists, skeptical reporters, and casual observers alike—became convinced that other data were also being fudged. "Secrets are like kryptonite to the scientific process," says Griffin. "Once you find out something has been deliberately overhyped or downplayed, or just plain lied about, all the data from that research becomes suspect. And data that can't be trusted has no scientific value."

Three years after the first mission was completed, the editors of the respected journal *Science* deemed the entire project a failure. "Isolating small pieces of large biomes and juxtaposing them in an artificial enclosure changed their functioning and interactions, rather than creating a small working Earth as originally intended," they wrote. For the $200 million dome to survive as a scientific enterprise, they concluded, it would need dramatic retooling.

KEY CONCEPT 8.7

Nitrogen and phosphorus move through the environment in cycles that depend on physical and biological processes and are heavily dependent on a wide variety of bacteria. Human impact has unbalanced these cycles.

Three years after the first mission was completed, the editors of the respected journal Science deemed the entire project a failure.

Ecosystems are irreplaceable, but learning how they function may help us restore degraded ones.

Biosphere 2 taught scientists that Earth may be far too complex, that ecosystem components intertwine in far too many complicated ways, for humans to re-create. Each is governed by a countless array of interacting factors, and a change in one can set off a whole chain of events that degrade the system's capacity to sustain life.

From a scientific point of view, the fact that Biosphere 2 was not able to sustain life as hoped does not mean the project was of no value. Negative results can be just as informative as positive ones—in some cases even more so because they uncover gaps in our knowledge and help us decide how to move forward. In fact, Biosphere 2's greatest liability—its skyrocketing CO_2 levels—proved to be its most valuable asset. "Now it's like a time machine," says Griffin who points out that it is allowing us a look at the consequences of elevated atmospheric CO_2 levels, the main contributor to climate change today. Recent research by Griffin has uncovered some of the complexities of carbon cycling. His group saw unexpected fluctuations in carbon release at various levels in the tree canopy, telling him there is much we still don't know about how carbon cycles—data that could only be gathered in an enclosed forest such as that found in Biosphere 2. Today, scientists from all over the world still use the facility to study the effects of an atmosphere loaded with carbon dioxide. Ultimately, though, the most valuable lesson Biosphere 2 has provided is how irreplaceable Biosphere 1 is.

Select References:

Cohen, J. E., & D. Tilman. (1996). Biosphere 2 and biodiversity: The lessons so far. *Science*, 274(5290): 1150–1151.

Griffin, K. L., et al. (2002). Canopy position affects the temperature response of leaf respiration in Populus deltoides. *New Phytologist*, 154(3): 609–619.

Marino, B., et al. (1999). The agricultural biome of Biosphere 2: Structure, composition and function. *Ecological Engineering*, 13(1): 199–234.

BRING IT HOME

PERSONAL CHOICES THAT HELP

Nutrient cycling is critical for maintaining Earth's ecosystems, but we interfere with nutrient cycles through our daily activities. Driving a car interferes with the carbon cycle by releasing carbon from fossil fuel reservoirs. Applying synthetic chemical fertilizers to food crops interferes with the nitrogen cycle by adding soluble nitrogen compounds to aquatic ecosystems through runoff. The challenge is to figure out ways to work with nutrient cycles rather than against them—in other words, to return nutrients to the reservoirs from which they come. How might this be done in our daily lives and in our own communities?

Individual Steps

• Reduce your fossil fuel use to curtail carbon, nitrogen, and sulfur emissions. Take public transportation, walk, ride a bike, and drive a fuel-efficient vehicle.

• Compost food and yard waste. Then use this material to fertilize flowerbeds, trees, and garden plots. Composting will reduce or eliminate the need for inorganic fertilizers in your yard.

Group Action

• Participate in or organize an event to plant trees or native grasses. By doing so, you can help recapture the carbon put into the atmosphere by driving a car.

• Many urban areas are grateful for individuals being willing to plant native trees, shrubs, and wildflowers along roadways or in parks and other public spaces.

Policy Change

• Public policy currently prevents large-scale composting of municipal wastes in most areas. Working to change these policies will extend the life of our landfills and make use of valuable nutrient-rich materials.

• Support legislation to increase fuel efficiency of vehicles and subsidies for clean, renewable energy. More fuel-efficient cars and cleaner energy sources will reduce our carbon outputs from fossil fuel use.

Peter Essick/Aurora Photos

ENVIRONMENTAL LITERACY UNDERSTANDING THE ISSUE

1 What is the hierarchy of organization recognized by ecologists, and why might it be useful to recognize such distinctions?

INFOGRAPHICS 8.1 AND 8.2

1. Which of the following is the correct heirarchy of life?
   a. Individual, community, biome, ecosystem, population, biosphere
   b. Individual, population, community, ecosystem, biome, biosphere
   c. Biosphere, community, biome, ecosystem, population, individual
   d. Biome, biosphere, population, ecosystem, community, individual

2. The word *niche* refers to an organism's _____, while its *habitat* refers to its _____.
   a. role and relationships in its community; environmental surroundings
   b. life cycle; role and relationships in its community
   c. environmental surroundings; tolerance limits
   d. "address"; "job description"

3. Why do ecologists focus mainly on the study of populations, communities, and ecosystems?

2 Why do ecosystems need a constant input of energy? How do they deal with the fact that Earth does not receive appreciable new inputs of matter?

INFOGRAPHIC 8.3

4. What does it mean for the Earth to be energetically open but a closed system with regard to matter?
   a. Energy can be used again and again, but matter can be used only once.
   b. Organisms can create new energy but not matter.
   c. Earth receives new energy inputs but not significant new matter inputs.
   d. Energy inputs exceed matter outputs.

5. Why is it essential that sustainable ecosystems rely on an energy source that is readily replenished, like sunlight, rather than nonrenewable sources, such as fossil fuels?

3 What are biomes, and how do environmental factors affect their distribution and makeup?

INFOGRAPHICS 8.4 AND 8.5

6. Which biome description is correct?
   a. Grasslands receive less rainfall than forests but more than deserts.
   b. Forests have freezing temperatures regularly.
   c. Grasslands are much warmer annually than forests.
   d. Deserts are hot year round—much hotter than forests.

7. Look at the location of temperate deciduous forests on Earth in the biome climograph in Infographic 8.4. What can you surmise about the climatic conditions in North America and Europe/Asia where these forests are found?

4 What is a population's range of tolerance, and how does it affect the distribution of a population within its ecosystem or its ability to adapt to changing conditions?

INFOGRAPHIC 8.6

8. Which of these is an example of a population's *range of tolerance*?
   a. The season with the most rainfall
   b. The number of competing species in the area
   c. The hottest and coldest temperatures it can survive
   d. The quality of nesting sites in the area

9. The population that would have the best chance of surviving an environmental change would be the one with:
   a. the narrowest range of tolerance for temperature.
   b. the greatest genetic diversity.
   c. the largest population size.
   d. the least variation among individuals.

10. Using the example of spring wildflowers and the critical factor of rainfall, explain the term *environmental gradient* (for rainfall) and the term *range of tolerance* (in terms of the distribution of wildflowers within their range).

5 How do carbon, nitrogen, and phosphorus cycle through ecosystems? How are these cycles being disrupted, and what problems can this disruption cause?

INFOGRAPHICS 8.7, 8.8, 8.9, AND 8.10

11. True or False: Plants and other producers are the only types of species that perform photosynthesis, whereas all species perform cellular respiration.

12. Of the three matter cycles discussed here, the only one that does not have an atmospheric component is the _____ cycle.

13. What is the purpose of photosynthesis?
    a. It produces a form of chemical energy that the plant can use as needed.
    b. It allows plants to remove excess CO_2 from the atmosphere.
    c. Plants do it to produce oxygen so animals (and humans) can breathe.
    d. It is a way for producers and consumers to release the energy stored in sugar molecules.

14. List the steps involved in the nitrogen cycle. What would happen if a wildfire burned so hot that it sterilized the soil, killing all the microbes?

15. Consider the ways in which human impact is affecting the carbon, nitrogen, and phosphorus cycles. What do these various human actions all have in common? Why might this be a concern?

SCIENCE LITERACY WORKING WITH DATA

The following table shows six tropical forests, arranged from least to most fertile soils, estimated from total soil nitrogen:

Comparison of Fertility Parameters in Different Forest Types

Parameter	Lower Montane Rain Forest, Puerto Rico	Amazon Caatinga, Venezuela	Oxisol Forest, Venezuela	Evergreen Forest, Ivory Coast	Dipterocarp Forest, Malaysia	Lowland Rain Forest, Costa Rica
Above-ground biomass (metric tons/ha)	228	268	264	513	475	382
Root biomass (metric tons/ha)	72.3	132	56	49	20.5	14.4
Total soil nitrogen (kg/ha)	—	785	1,697	6,500	6,752	20,000
Turnover time of leaves (years)	2	2.2	1.7	—	1.3	—

Interpretation

1. Which two areas have the poorest soils, in terms of nitrogen availability?

2. Look at the root biomass in the two areas from Question 1. What is the relationship between root biomass and poor soils? Propose an explanation.

3. Can you determine soil nutrient levels by looking at the above-ground biomass? Explain.

Advance Your Thinking

4. Notice how long it takes leaves to break down ("Turnover time of leaves"). Speculate as to why trees in poorer soils produce leaves that last longer.

5. If you removed the trees through logging or burning, would these areas be good for agriculture? (Hint: Cleared areas are usually used for small farms that do not have funds to purchase farm equipment, fertilizers, or other chemicals.)

INFORMATION LITERACY EVALUATING INFORMATION

Biosphere 2 covers a mere 3 acres and has finite resources. Although Earth is vastly larger, its resources are also finite. People have to decide, for each biome, whether to leave some places untouched or use all of an area and its resources for human purposes. One example would be plowing under an entire prairie and growing wheat, displacing all of the native plants and animals. The rain forests are another such biome. Left alone, they produce huge amounts of oxygen for the whole planet and also support millions of species. Many humans want to use the trees for lumber and the land to grow crops or raise cattle.

Some groups advocate using rain forests in sustainable ways: Manufacturing herbal products or raising shade-grown coffee under the canopy are two examples. Will these tactics help preserve the biome?

Go to the Rainforest Alliance's website (www.rainforest-alliance.org) and read about sustainable agriculture and Rainforest Alliance–certified products such as coffee, tea, and cocoa.

Evaluate the website and work with the information to answer the following questions:

1. Is this a reliable information source?
   a. Does the organization give supporting evidence for its claims?
   b. Does it give sources for its evidence?
   c. What is the mission of this organization?

2. Note the range of publications available on the Rainforest Alliance website (www.rainforest-alliance.org/publications) and view the white paper entitled *Cocoa Certification* (search for "cocoa certification," using the "search publications" option).
   a. Do you agree that certification is an adequate tool to ensure sustainability in cocoa farming?
   b. Identify a claim the organization makes and the evidence it gives in support of this claim. Is it convincing?

3. The Global Canopy Programme (www.globalcanopy.org) is a group based in the United Kingdom that has joined with the Carbon Disclosure Project to ask companies to disclose their "forest footprint" (www.cdproject.net/en-US/Programmes/Pages/forests.aspx). Should companies be required to participate?
   a. Should some intergovernmental group investigate companies and estimate impacts?
   b. Is it important to allow a "free market" for goods and services, assuming that businesses will take care of the lands they use but do not own?

THE WOLF WATCHERS

Endangered gray wolves return to the American West

CORE MESSAGE

Ecologists study populations to better understand what makes them survive and thrive. The size and distribution of populations in an ecosystem is influenced by a variety of factors, such as the availability of resources and the presence of other species, like predators. Species vary in their ability to adapt to changes in their environment, but many species are facing environmental disturbances that are threatening their populations. Determining the factors that affect a given population is an important part of managing it, especially for populations that are endangered.

AFTER READING THIS CHAPTER, YOU SHOULD BE ABLE TO ANSWER THE FOLLOWING **GUIDING QUESTIONS**

What ecological significance do a population's size, density, and distribution have?

What determines a population's growth rate? Under what conditions do we see exponential population growth? Logistic growth?

Biologists track wolves from a helicopter. Once they spot a wolf, they dart it with a tranquilizer so that they can fit it with a radio collar. The wolf is unharmed. Barrett Hedges/National Geographic

3

How do density-dependent and density-independent factors affect population growth?

4

What are the life-history strategies of *r*- and *K*-selected species, and how do they relate to population growth patterns and potential?

5

What are top-down and bottom-up regulation, and which is most important in determining the size of a population?

At least a half a dozen times each winter, Doug Smith climbs into a helicopter, gun in tow, and hunts wolves (*Canis lupus*) in Yellowstone National Park. He's not looking to kill them—just to put them to sleep for a little while so that he can outfit them with radio collars to track the sizes of their packs, what they eat, and where they go over the course of the following year. Smith, a biologist, spends about 200 hours per year on these "hunting" expeditions, which are part of the Yellowstone Gray Wolf Restoration Project that he leads. The project has been responsible for reintroducing a total of 41 wolves to Yellowstone since 1995, after their disappearance as a result of predator control programs implemented by the U.S. government in the early 20th century.

Sometimes, though, things go awry on Smith's radio collar missions. For instance, the tranquilizer dart doesn't fully sedate big wolves—some of which can reach 80 kilograms (175 pounds). Smith is forced to approach the animals while they're still awake. "I have to grab them on the scruff of the neck and manhandle them until I get the collar on," he explains. "They're typically not dangerous then—they've had enough drug to be kind of out of it—but they're still able to walk around. It's a wild experience." Sometimes the wolves—who are typically frightened of the helicopter—try to attack it while it's hovering with the doors open a few feet above the ground. "I've had two females turn and run at the helicopter, teeth gnashing, jumping up trying to get me," Smith recalls. "I'm hovering above it, going back and forth, and I can't get a shot because all I'm seeing is face and teeth."

Despite these adventures, Smith says a wolf has never actually bitten him—and he has tranquilized and collared more than 300 of them.

Ecologists monitor **populations** of organisms in ecosystems around the world for a variety of reasons, whether to protect endangered species, to manage economically valuable species such as commercial fisheries or timber, or to control pests. In Yellowstone, Smith and others monitor wolf populations. Elk (*Cervus elaphus*) populations, a popular

◉ WHERE IS YELLOWSTONE?

MONTANA

YELLOWSTONE NATIONAL PARK

WY

IDAHO

WYOMING

game species, are also closely monitored, as are aspen, willow, and cottonwood trees, the most important winter food for elk. In some cases, elk numbers and aspen growth are closely tied to the size of the wolf population—but not in all areas. Other factors play a role in the sizes of these populations.

Understanding how any population interacts with biotic and abiotic forces in its environment, through programs like the Wolf Restoration Project, is key to preventing species from disappearing from their ecosystem forever. That's because many factors influence the livelihood of a species and its ability to survive and reproduce. In the early 20th century, humans not only threatened wolves by hunting them, they also destroyed the animals' natural habitat (and that of their prey) when they cut down forests to build farms and ranches. They starved the animals by hunting elk, deer, and bison, wolves' usual sources of food. At the time, people didn't think that this combination of changes would very nearly cause wolves to go extinct or result in exploding elk and deer populations. But now

KEY CONCEPT 9.1

Ecologists study populations to better understand what makes them thrive, decline, or become overpopulated in an effort to manage them and the other populations they impact.

---

population All the individuals of a species that live in the same geographic area and are able to interact and interbreed.

National Park Service

↑ Doug Smith and Yellowstone Delta wolf 487M, the largest wolf collared in the winter of 2005.

that scientists like Smith have spent years watching how the animals live, they have a much better understanding of what needs to be done to keep them alive.

Populations fluctuate in size and have varied distributions.

Before humans started killing wolves in the early 1900s as part of a government-sponsored program, the exact numbers of gray wolves living in Yellowstone were unknown, but estimates range from 300 to 400. Elk populations at that time hovered around 10,000 animals. The size of a population in a given geographic area is determined by the interplay of factors that simultaneously increase the number of individuals in a population (births and immigration) and those that decrease numbers (deaths and emigration). An understanding of these factors helps us predict population size at any given time. Ecologists who study changes in population size and makeup (**population dynamics**) find that the population size

of some species increases and decreases rather predictably (barring a catastrophic event), while others tend to fluctuate more randomly, affected by a variety of factors.

In the early 20th century, Congress allocated $125,000 for the Predator Control Program, which employed poison—and later hunting—to eradicate predators and rodents that might harm crops or livestock. Wolves were one of the targeted species because wolves preyed upon livestock. Wolf eradication was also supported because it boosted deer and elk populations for human hunting. "It was park policy to kill all predators, and wolves were their biggest objective," Smith explains. Between 1914 and 1926, at least 136 wolves were killed in Yellowstone. In 1944, the last known wolf in the Yellowstone area was killed.

Every population has a **minimum viable population**, or the smallest number of individuals that would still allow a population to be able to persist or grow, ensuring long-term survival. This is an

KEY CONCEPT 9.2

Populations require minimal sizes and densities to reproduce successfully and maintain social ties, but high population density can lead to problems such as disease and overuse of resources.

population dynamics   Changes over time in population size and composition.

minimum viable population   The smallest number of individuals that would still allow a population to be able to persist or grow, ensuring long-term survival.

important concept when considering how to conserve endangered or threatened species. A population that is too small may fail to recover for a variety of reasons. For example, some species' courtship rituals require a minimum number of individuals for success. Other activities that depend on numbers—like flocking, schooling, and foraging—fail below certain population sizes. Genetic diversity (inherited variety between individuals in a population) is also important: A population with little genetic diversity is less able to adapt to changes and is therefore more vulnerable to environmental change. A small population is also subject to inbreeding, which allows harmful genetic traits to spread and weaken the population. Ecologically, however, the minimum viable population may not be as significant as what is known as an *ecologically effective population*—one large enough to perform the important ecosystem services it normally contributes to its community. One small pack of wolves in Yellowstone National Park probably has little impact on elk populations.

Critics began raising concerns about the Predator Control Program in the 1920s, and wolves in Montana and Wyoming became protected under the Endangered Species Act in 1973, but it wasn't until 1987 that the U.S. Fish and Wildlife Service proposed reintroducing an "experimental population" of wolves into Yellowstone, in its Northern Rocky Mountain Wolf Recovery Plan.

Not everyone was in favor of reintroducing wolves to the landscape of the American West, and opposition remains today. Ranchers in particular were and still are wary of the damage wolves might do to livestock. In some years, wolves were responsible for thousands of herd deaths—although not nearly as many as are killed by coyotes, says the U.S. Fish and Wildlife Service. Current programs compensate ranchers for lost livestock, but they must prove that their animal was killed by a wolf, which is not always easy to do.

After the Fish and Wildlife Service unveiled its reintroduction plan, Congress funded the organization to prepare an *environmental impact statement* (an evaluation of the positive and negative impacts of a proposed action) on what was likely to happen if wolves were reintroduced to Yellowstone. By considering all aspects of wolf ecology, scientists were able to organize the reintroduction to maximize survival of the species. For instance, scientists believed that it would be best to "softly" release wolves in the park by holding them temporarily in packs in areas that would be suitable for them to live instead of doing a "hard" release in which the wolves could immediately disperse anywhere in the park. The soft release curtailed the wolves' movement and helped them survive and acclimate to the move.

After many years of tracking wolves in Yellowstone, Smith has a good idea of where to look for the wolf packs. And the helicopter hovers a mere 15 meters (50 feet) off the ground, making it easy to spot roaming groups of wolves.

The number of wolves distributed throughout Yellowstone National Park is their **population density**—the number of individuals per unit area. Population density is an important feature that varies enormously among species, or even among populations of the same species in different ecosystems. If a population's density is too low, individuals may have difficulty finding mates, or the only potential mates may be closely related individuals, which can lead to inbreeding, loss of genetic variability, and, ultimately, extinction. Density that is too high can also cause problems, such as increased competition, fighting, and spread of disease. Deer, elk, and moose populations in the United States, whose density has increased in recent years because of exploding numbers combined with shrinking habitats, now frequently suffer from an infectious disease known as chronic wasting disease.

In addition to size and density, another important feature is **population distribution**, or the location and spacing of individuals within their range. A number of factors affect distribution, including species characteristics, topography, and habitat makeup. Ecologists typically speak of three types of distribution. In a **clumped distribution**, individuals are found in groups or patches within the habitat. Yellowstone examples include social species like prairie dogs or beaver that are clustered around a necessary resource, like water. Wolves travel in packs and therefore have a clumped distribution. Elk, one of their prey species, congregate as well; living in herds offers some protection against the wolves.

In a **random distribution**, individuals are spread out in the environment irregularly, with no discernible

population density The number of individuals per unit area.

population distribution The location and spacing of individuals within their range.

clumped distribution A distribution in which individuals are found in groups or patches within the habitat.

random distribution A distribution in which individuals are spread out over the environment irregularly, with no discernable pattern.

uniform distribution A distribution in which individuals are spaced evenly, perhaps due to territorial behavior or mechanisms for suppressing the growth of nearby individuals.

KEY CONCEPT 9.3

The distribution of a population may be influenced by behavioral and/or ecological factors. Understanding a population's distribution is important in managing it.

INFOGRAPHIC 9.1   POPULATION DISTRIBUTION PATTERNS

↓ The distribution of individuals in populations varies from species to species and is influenced by biotic and/or abiotic factors.

CLUMPED

↑ Elk stay in herds, which offers some protection against predators.

RANDOM

↑ The seeds of many Yellowstone flower species are distributed randomly and germinate where they fall.

UNIFORM

↑ The creosote bush of the desert Southwest produces toxins that prevent other bushes from growing close by.

? In which population distribution pattern would individuals within the population experience the most competition with other individuals in their population? Why, then, is this distribution pattern ever seen?

pattern. Random distributions are sometimes seen in homogeneous environments, in part because no particular spot is considered better than another. Species that rely on wind and water to disperse their offspring—like wind-blown seeds or the free-floating larvae of coral—also often have a random distribution. **Uniform distributions**, rare in nature, include individuals that are spaced evenly, perhaps due to territorial behavior or mechanisms for suppressing growth of nearby individuals (seen in some plant species). **INFOGRAPHIC 9.1**

Populations display various patterns of growth.

The researchers in the wolf restoration project could observe these factors—distribution, genetic diversity, initial population size, and density—as well as factors like prey availability and habitat structure and see how they influenced the size and dynamics of a population in Yellowstone. However, in order to predict population dynamics, scientists use some simple mathematical models that describe population growth over time. The annual **population growth rate** is determined by *birth rate* (for example, the number of births per 1,000 individuals per year) minus the population *death rate* (for example, number of deaths per 1,000 individuals per year) or by taking a simple census of population size at the two time points in question. For example, between 2009 and 2010, the wolf population went from 320 to 343, an increase of 23 animals. This is a growth rate of 7% (23/320 = 0.07). In that same time frame, the elk population decreased from 6,070 to 4,635, giving this population a *negative* growth rate (4,635 - 6,070 = -1, 435; -1,435/6,070= -0.24, or -24%).

Population growth is dependent on the presence of **growth factors** (resources individuals need to survive or reproduce). Conversely, **resistance factors** (things that reduce population size by directly or indirectly killing individuals or prompting emigration), such as predators, competitors, or diseases, will decrease population size. When there are no environmental limits to survival or reproduction, a population will reach its maximum per capita rate of increase (*r*), called its **biotic potential**. This occurs, theoretically, when every female reproduces to her maximal potential and every offspring survives. A population increasing in this manner will quickly grow to fill its environment. This period of growth, which can't go on indefinitely, is referred to as **exponential growth**, named for the mathematical function it represents.

Populations that have a high biotic potential have high *fecundity* (females typically produce lots of offspring, reach reproductive maturity quickly, and produce many "clutches"

per year). The higher the biotic potential, the faster the population of a given species will grow under ideal conditions. Yellowstone species such as deer mice and the problematic non-native weed known as spotted knapweed have higher biotic potential than species such as grizzly bears and spruce trees.

In nature, exponential growth is typically seen when a species first enters a new environment or when there is an influx of new resources. The population must have a high birthrate—most individuals must have access to enough food, water, and habitat in which to reproduce—and a low death rate. The loss of predators can also lead to exponential growth among their prey species. For instance, without wolves to thin their ranks, elk numbers in Yellowstone doubled between 1914 and 1932, after the Predator Control Program had been implemented. This led to the need to cull the herd (in spite of increased hunting pressure), a standard practice for many years that reduced the herd from 16,000 in 1932 to around 6,000 in 1968, when this practice ended; elk population sizes climbed to a high of around 19,000 in the 1980s before wolves were reintroduced.

A population that is growing exponentially will have a *J-shaped curve* if plotted on a graph with time on the x-axis and population size on the y-axis. The J curve shows a slight lag at first and then a rapid increase. This is due to the fact that the larger the population, the faster it grows, even at the same growth rate. Think of it this way: Doubling a small number yields a number that is still small. Doubling a large number, on the other hand, produces a very large number. **INFOGRAPHIC 9.2**

As an example of how profound exponential growth can be, imagine if someone offered to give you a penny one day and then, each subsequent day for a month, doubled the amount given to you the previous day. On day 1 you would have 1 cent; on day 2 you would be given 2 cents; on day 3 you'd be given 4 cents, and on day 4, you would be given 8 cents. By day 31, you would have more than $21 million. Exponential growth can create large populations quickly.

---

population growth rate The change in population size over time that takes into account the number of births and deaths as well as immigration and emigration numbers.

growth factors Resources individuals need to survive and reproduce that allow a population to grow in number.

resistance factors Things that directly (predators, disease) or indirectly (competitors) reduce population size.

biotic potential (*r*) The maximum rate at which the population can grow due to births if each member of the population survives and reproduces.

exponential growth The kind of growth in which a population becomes progressively larger each breeding cycle; produces a J curve when plotted over time.

INFOGRAPHIC 9.2   **EXPONENTIAL GROWTH OCCURS WHEN THERE ARE NO LIMITS TO GROWTH**

2

BIOTIC POTENTIAL OF DEER MICE

Assume each pair produces 10 pups/litter and none die: $r = 5$ (r is expressed as surviving offspring per adult; we divide the litter size by 2 since there are 2 parents per litter).

← Because deer mice have a high biotic potential, even a single pair could produce more than 31,000 descendants in their lifetime.

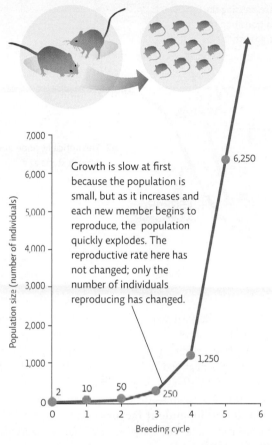

Growth is slow at first because the population is small, but as it increases and each new member begins to reproduce, the population quickly explodes. The reproductive rate here has not changed; only the number of individuals reproducing has changed.

Y-axis: Population size (number of individuals) — 0, 1,000, 2,000, 3,000, 4,000, 5,000, 6,000, 7,000

X-axis: Breeding cycle — 0, 1, 2, 3, 4, 5, 6

Data points: 2, 10, 50, 250, 1,250, 6,250

Cockroaches thrive in some homes and apartments. What could trigger exponential growth in a cockroach population? What could prevent it?

Exponential growth can't last forever, however; as a population begins to fill its environment, its growth rate decreases because as more individuals use the available resources, the resources become scarce. Some individuals starve or are unable to find habitat in which to reproduce, and crowding may also bring about an increase in disease and aggression. Predation pressure may increase as the more numerous prey is easier to track and capture or simply because the predator population itself has increased. This kind of growth—in which as population size increases, growth rate decreases—is called **logistic growth**.

A population that grows logistically will produce an *S-shaped* curve if plotted on a graph with time on the x-axis and population size on the y-axis. The S is created by the initial growth of the J-shaped curve of exponential growth, followed by decelerating growth as the species approaches its maximum sustainable population size, where it levels off.

The population size that a particular environment can support indefinitely—without long-term damage to the environment—is called its **carrying capacity**, signified as K in population mathematical models. Carrying capacity is determined by the presence of growth factors and varies between species; the same

KEY CONCEPT 9.5

As a population's size approaches carrying capacity, exponential growth may transition to logistic growth, slowing population growth rates.

logistic growth The kind of growth in which population size increases rapidly at first but then slows down as the population becomes larger; produces an S curve when plotted over time.

carrying capacity (K) The maximum population size that a particular environment can support indefinitely.

INFOGRAPHIC 9.3   LOGISTIC POPULATION GROWTH   2

→ Exponential growth turns into logistic growth (S curve) as population size approaches carrying capacity (*K*) and resistance factors begin to limit survival.

What could increase or decrease the carrying capacity for a particular population in an environment?

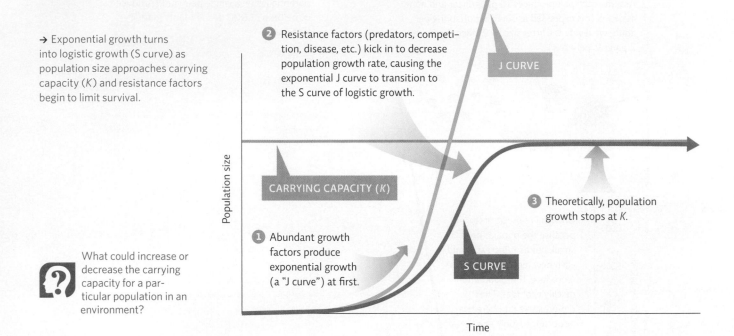

2 Resistance factors (predators, competition, disease, etc.) kick in to decrease population growth rate, causing the exponential J curve to transition to the S curve of logistic growth.

J CURVE

Population size

CARRYING CAPACITY (*K*)

1 Abundant growth factors produce exponential growth (a "J curve") at first.

S CURVE

3 Theoretically, population growth stops at *K*.

Time

---

environment can support many more elk than wolves, for example. Over time, a population's carrying capacity can change. If resources are diminished at a faster rate than they are replenished, the carrying capacity will drop. If, on the other hand, new resources are added or become available, perhaps due to the loss of a competitor, the carrying capacity for a given species will rise. **INFOGRAPHIC 9.3**

A variety of factors affect population growth.

Populations will grow as long as growth factors are available, but as the population gets larger, resources start to decline. The *limiting factor*, the resource that is most scarce, tends to determine carrying capacity. The effects that predators (a resistance factor) have on populations can vary widely, in part because predators, along with disease and competition, are **density-dependent factors**—their effects all increase as the prey's population size goes up. Wolves and other predators are density-dependent factors

affecting elk herd size. In the same way, elk are density-dependent factors on the ability of young aspens to grow after sprouting (aspen is an important winter food for elk; they eat the young shoots). On the other hand, some factors affect a population no matter how large or small it is, such as droughts, storms, and fire. These **density-independent factors** don't necessarily regulate population size, but they can increase or decrease it. **INFOGRAPHIC 9.4**

The biology of a species (which reflects its adaptations for growth, reproduction, and survival) also affects its populations' growth potential. For instance, ecologists recognize a continuum of **life-history strategies** among species. Species whose members mature early, have high fecundity, and have relatively short life spans are known as **r-selected species**—so named because of their high rate of population increase (*r*). Yellowstone r-selected species, such as deer mice and spotted knapweed, are well adapted to exploit unpredictable

---

density-dependent factors Factors, such as predation or disease, whose impact on a population increases as population size goes up.

density-independent factors Factors, such as a storm or an avalanche, whose impact on a population is not related to population size.

life-history strategies Biological characteristics of a species (for example, life span, fecundity, maturity rate) that influence how quickly a population can potentially increase in number.

r-selected species Species that have a high biotic potential and that share other characteristics, such as short life span, early maturity, and high fecundity.

KEY CONCEPT 9.6

Some factors have an increased effect on population size when a population's size is large (density-dependent), and others have an effect that is not related to population size (density-independent).

INFOGRAPHIC 9.4 **DENSITY-DEPENDENT AND DENSITY-INDEPENDENT FACTORS AFFECT POPULATION SIZE**

3

Density-dependent factors exert more of an effect as population size increases. On the other hand, density-independent factors have the same effect regardless of population size.

DENSITY DEPENDENT

COMPETITION

↑ Mule deer compete with elk for food; the larger the deer population, the greater the competition.

DISEASE

↑ Infectious diseases, such as chronic wasting disease, which weakens and eventually kills the animal, spread more easily in large populations of elk, deer, or moose.

PREDATORS

↑ Predators are more successful at capturing prey from larger populations than from smaller ones.

DENSITY INDEPENDENT

FIRE/FLOOD

↑ Forest fire.

STORMS

↑ Row of trees flooded by a river.

AVALANCHE AND OTHER NATURAL DISASTERS

↑ Snow avalanche.

? Identify the following as either density-dependent or density-independent factors for an elk population: a tick infestation, building a dam that floods a valley in elk habitat, drought, bison, and a blizzard.

INFOGRAPHIC 9.5 LIFE-HISTORY STRATEGIES

↓ Different species have different potentials for population growth, known as life-history strategies. A species' biology may place it anywhere along a continuum between two extremes—the *r*- and *K*-selected species.

CHARACTERISTICS OF *r*-SELECTED SPECIES

1. Short life
2. Rapid growth of individual
3. Early maturity
4. Many, small offspring
5. Little parental care
6. Adapted to unstable environment
7. Prey
8. Uses many habitats and resources

CHARACTERISTICS OF *K*-SELECTED SPECIES

1. Long life
2. Slower growth of individual
3. Late maturity
4. Few, large offspring
5. High parental care
6. Adapted to stable environment
7. Predators
8. Needs specific habitat and resources

r-SELECTED SPECIES — DANDELIONS — DEER MICE — ELK — BEARS — *K*-SELECTED SPECIES — SPOTTED KNAPWEED — SPRUCE TREES

From left: Norbert Rosing/National Geographic Stock; © Wayne Lynch/AGE fotostock; © Juniors Bildarchiv GmbH/Alamy; Tom Reichner/Shutterstock; John W. Bova/Science Source; Taylor S. Kennedy/National Geographic Stock

? Consider an aquatic ecosystem. Where would you place the following organisms on a life-history continuum: tuna, sperm whale, plankton, and jellyfish? (Hint: Identify the most extreme *r* and *K* species and then place the others in between those on the continuum.)

KEY CONCEPT 9.7

The life-history strategy of a species influences the growth potential of its populations. Population size of *r*-selected species can increase or decrease quickly in response to environmental changes. *K*-selected species are more likely to have stable population sizes close to carrying capacity but are less adaptable in the face of environmental change.

environments and are able to increase quickly if resources suddenly become available.

On the other hand, **K-selected species**, which in Yellowstone include bears, wolves, and slow-growing trees like spruce, are found at the other end of the continuum. Individuals in these species have longer life spans, are slow to mature, and have lower fecundity. Because of this, their reproductive rates are lower, but this means their population growth rates are responsive to environmental conditions; they decrease or increase slowly if resources decrease or increase in availability. This responsiveness tends to keep population sizes close to carrying capacity (*K*).

INFOGRAPHIC 9.5

Some species, like elk and deer, have characteristics of both *r* and *K* species;

they fall somewhere in the middle of the continuum. They are large organisms that have one or two offspring per year and provide parental care (*K* characteristics), but their population sizes can increase rapidly if conditions are favorable for growth and survival (*r*-characteristics).

K-species and *r*-species often experience different types of population change. For instance, population sizes tend to be stable, especially for *K*-species, in undisturbed, mature areas. On the other hand, *r*-species with rapid reproductive potential sometimes have sudden, rapid population growth, characterized by occasional surges to very high population numbers, which may overshoot carrying capacity, followed by sudden crashes, especially in response to seasonal availability of food or temperature changes; their high rate of reproduction does not allow the population the time to adjust and produce fewer offspring as resources become scarce. When this occurs, the population that exceeds carrying capacity will drop below carrying capacity and then increase again; some populations will eventually level off close to carrying capacity, while others continue to overshoot and crash.

Predator and prey also respond to each other; predator populations increase as their prey population increases. But more predators eventually reduce the prey population. Fewer prey means less food and eventually decreases the number of predators, allowing the prey species to recover. In some cases, the fluctuations in population size are large enough to result in **boom-and-bust cycles**, with the peaks of the predator population size lagging behind those of its prey or food source. A classic example is that of the snowshoe hare and Canada lynx. Long-term data based on the fur trade and going back as far as the 1800s show that snowshoe hare populations undergo cyclic oscillations. According to recent studies by University of British Columbia researcher Charles Krebs, these mirror the size of predator populations, such as lynx and great horned owls. **INFOGRAPHIC 9.6**

The loss of the wolf emphasized the importance of an ecosystem's top predator.

Thanks in part to Smith's determination, the Yellowstone Gray Wolf Restoration project is going strong. There are approximately 80 wolves in Yellowstone National Park (and more than 5,000 living in the lower 48 states). But one chief lesson hammered home by observing wolves in Yellowstone is that populations do not exist in isolation. "Work in central Yellowstone clearly demonstrated that the addition of a keystone species, such as wolves, can result in observable changes in the behavior of its prey," says Claire Gower, a wildlife biologist at Montana Fish, Wildlife, and Parks. "These consequences may not stop at the prey individual, but may have cascading effects on other community-level processes."

K-selected species Species that have a low biotic potential and that share characteristics such as long life span, late maturity, and low fecundity; generally show logistic population growth.

boom-and-bust cycles Fluctuations in population size that produce a very large population followed by a crash that lowers the population size drastically, followed again by an increase to a large size and a subsequent crash.

INFOGRAPHIC 9.6   SOME POPULATIONS FLUCTUATE IN SIZE OVER TIME   4

OVERSHOOT AND CRASH

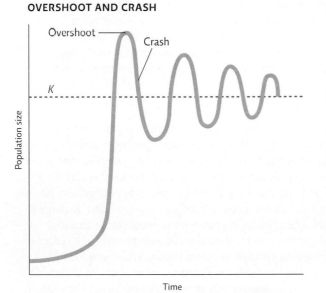

↑ Not all populations show logistic growth that levels off nicely at carrying capacity. Some populations might overshoot carrying capacity, drop below it, and increase and overshoot it again until they settle down close to carrying capacity.

SNOWSHOE HARE AND CANADA LYNX POPULATION SIZE ESTIMATES AT KLUANE LAKE, YUKON

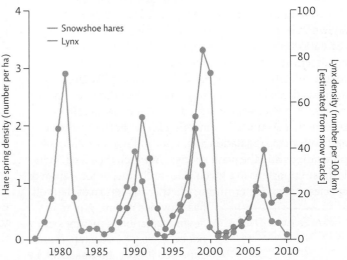

↑ Some predator-prey populations, like that of the snowshoe hare and Canada lynx, regularly go through boom-and-bust cycles as part of their natural population cycle.

 Why don't *K*-selected species usually go through overshoot and crash cycles?

KEY CONCEPT 9.8

Population size is influenced by top-down and bottom-up factors, but which one exerts the greatest effect varies from population to population.

After wolves were reintroduced to Yellowstone, project scientists documented that coyote populations in the area shrank, in part because the wolves were killing them, usually when the coyotes would approach a fresh wolf kill. Coyotes, which were running in packs in Yellowstone before the reintroduction of the wolf (unusual for coyotes), have begun reverting to traveling in pairs, more common for this species. Wolf reintroduction also indirectly impacted grizzly bear diets. The foraging of elk on berry bushes declined significantly, allowing these bushes to produce more summer fruit. Grizzly bear consumption of berries increased 20 and 2 fold in July and August, respectively, when compared to berry consumption before wolves were reintroduced (1968–1987). The increased consumption of this late summer food source may be especially important for breeding females.

Elk populations dropped after the wolf introduction. The ability of a predator (or other resistance factor) to control the size of its prey populations is known as **top-down regulation**. However, ecologists have debated for years the relative importance of top-down regulation versus the regulation by growth factors that *boost* numbers such as the availability of water and food or sunlight (**bottom-up regulation**). The lynx–hare relationship shown in Infographic 9.6 is an example of top-down regulation because, according to studies, hare populations are determined by predation pressures rather than food availability. Even physical disturbances like fire are considered bottom-up regulators when they free up nutrients, boosting plant growth. In Yellowstone, the picture that seems to be emerging is that both top-down and bottom-up regulation are important in controlling elk populations and the plant species on which they feed. Which is most important in a given place and time, or for a given or species, varies. Determining whether the return of the wolf has provided meaningful top-down regulation is turning out to not be as easy as researchers once expected it to be. **INFOGRAPHIC 9.7**

For example, after the *extirpation* (local extinction) of the wolf and the increase of the elk population that followed, willow and aspen trees were overgrazed. Elk like to eat the tender young shoots, but they removed an

top-down regulation The control of population size by factors that reduce population size (resistance factors) such as predation, competition, or disease.

bottom-up regulation The control of population size by factors that enhance growth and survival (growth factors) such as nutrients, water, sunlight, and habitat.

> *" Work in central Yellowstone clearly demonstrated that the addition of a keystone species, such as wolves, can result in observable changes in the behavior of its prey. "* —Claire Gower.

important resource for other animals like birds (habitat) and beaver (food and building material) and also made it difficult for the new trees to grow and become part of the mature stand of trees. The decline of the beaver population was especially significant because beaver dams change the flow of water, creating lakes that support a wide variety of fish, amphibian, and plant species that would not frequent the faster-flowing stream. The loss of these dams allowed streams to return to their original flow, changing the community that lived in the area.

After the wolves' return to Yellowstone, willow stands appear to be recovering in some areas. In others, the ecological changes that resulted from the earlier overgrazing are proving difficult to overcome. A study by Kristen Marshall, then at Colorado State University, showed that even in areas where elk are physically blocked by fencing, willows are not growing tall enough to be useful to beaver unless simulated beaver dams (flooded areas) are included in the area. Increased feeding by elk over time contributed to changes in the ability of the soil to hold water; without vegetation to prevent soil erosion, less water is in the soil and available to plants. If the willows cannot grow tall enough to lure back beaver (due to the very lack of beaver dams), these streamside ecosystems may not return to their former configuration, even with fewer elk. An altered ecosystem may yet emerge in these areas.

Aspen are also regrowing in some areas but not all. Elk are less likely to browse in aspen groves if wolves are in the area, but wolf presence does not always lead to aspen recovery. Oregon State University researcher Cristina Eisenberg discovered that in areas where there are lots of wolves, elk returned to feeding on aspen. She surmised that when predation pressure is predictable and high everywhere (field, forest, and aspen groves), elk returned to feeding on their preferred food. Eisenberg found that aspen recovery was only seen in areas of high wolf population density where wildfire had swept through. Wildfires clear out competing plants (including adult aspen trees), spurring a burst of growth of young aspens. Fire that leaves behind fallen trees creates a habitat that elk generally stay away from (escaping the wolves would be difficult in these deadfall littered areas). Without elk present, young aspen shoots in these areas grow tall enough to escape elk browsing when the elk eventually return, allowing the aspen stand to recover.

INFOGRAPHIC 9.7 | **TOP-DOWN AND BOTTOM-UP REGULATION**

↓ Population size is affected by both the presence of resistance factors (things that reduce population size) and growth factors (the availability of resources that allow the population to grow), but ecologists have long debated which one has the greatest influence on population size. In most cases, both impact population size, though which plays the greater role may vary from species to species or even within a species, depending on a wide variety of factors, such as relative population sizes of a population versus its predators, competitors or prey, or seasonal climate changes.

TOP-DOWN REGULATION

BOTTOM-UP REGULATION

CONTROL IS FROM PREDATORS HIGH ON THE FOOD CHAIN

CONTROL IS FROM THE BOTTOM OF THE FOOD CHAIN

(REMOVAL OF WOLVES)

(POOR GROWTH CONDITIONS FOR PLANTS)

When wolves were removed from Yellowstone, the elk population increased, and the abundance of aspen, cottonwood, and willow decreased due to overgrazing by the elk.

↑ According to this model, population size is primarily determined by resistance factors such as the control exerted by top predators that eat herbivores; this then limits herbivore consumption of plants, increasing the population size of plants in the ecosystem.

In Yellowstone, willow shoots do not grow well in dry areas; this could limit the elk population, which in turn could limit the wolf population.

↑ Conversely, growth factors could be the most important determinant of population size. The availability of needed resources—nutrients, sunlight, and water—influences the growth of plants, which determines the population size of animals that eat plants (herbivores), which in turn determines the population size of animals that eat the herbivores.

? In most unprotected areas of the United States, wolf population densities are much lower than they are inside the protected Yellowstone National Park. Do you think wolves exert top-down control on elk populations in these unprotected areas? Explain.

As with any complex system, it is proving difficult to tease apart the impacts of several contributing factors on elk, aspen, and willow populations. Are wolves facilitating a recovery of these trees by reducing elk populations and reducing the amount of time elk spend foraging on these plants? As Eisenberg puts in her book *The Carnivore's Way*, the answer seems to be "It depends." It depends on the size of the given elk population and the abundance of food; it depends on climate and season; it depends on fire and drought; it depends on wolf populations and perhaps other, as-yet-unidentified factors.

The overall effects of returning wolves to Yellowstone is just beginning to unfold and be uncovered by researchers. The fact that the wolves were removed for so long may have changed the ecosystem for good. This is reason to pause

↑ The bare trunks of these aspen trees show the winter browse line from herbivores such as elk and deer, which eat leaves as high as they can reach during winter. As forage becomes more and more scarce, these animals will even strip off the bark, damaging the trees.

and consider human actions today that threaten other species. Many populations of animals other than wolves are declining worldwide due to human impact, especially from habitat loss, the introduction of non-native species, and predator removals. In fact, the number-one reason that species become endangered today is habitat destruction (see Chapter 13). People damage habitats, remove resources, and break needed connections within ecosystems, and populations respond: Some species may benefit from the change and their population could increase (spotted knapweed, a non-native plant in Yellowstone, spreads quickly in disturbed areas), while other species may decline in number because they are displaced by others or because needed conditions for growth are no longer present. (Songbirds may lose habitat if the woodland patches they nest in are cut down.) Understanding community interactions and population dynamics helps managers monitor, protect, and restore populations.

The success of the reintroduction program has led to the wolf being "delisted" in Montana, Idaho, and Wyoming (a delisted species is no longer protected by the ESA; management authority is returned to the state), though not without opposition from some conservation groups. This delisting has allowed wolf hunts with quotas set by the Fish and Wildlife Service. To Smith, the management of wolves includes the protection of some packs but policies that allow hunting of others. He reasons that hunting wolves only in areas where they conflict with humans may

be one of the best ways to protect wolves in wild places like Yellowstone. If people know they can protect their animals and livelihoods, they may be more amenable to allowing the wolves to remain in wilderness areas.

To ensure that the wolves reintroduced to Yellowstone are given a chance to really flourish, Smith and his colleagues diligently stay on the wolves' trail, studying their population dynamics. "We want to know their population size, what they are eating, where they are denning, how many pups they have and how many survive, and how the wolves interact with each other," he explains. Why is it so important to ensure that the wolves do well? Simply put: They were here first, he says. "We want to restore the original inhabitants to the Park."

Select References:
Beschta, R. L., & W. J. Ripple. (2014). Divergent patterns of riparian cottonwood recovery after the return of wolves in Yellowstone, USA. *Ecohydrology.* doi: 10.1002/eco.1487.
Eisenberg, C., et al. (2013). Wolf, elk, and aspen food web relationships: Context and complexity. *Forest Ecology and Management,* 299: 70–80.
Krebs, C. J. (2011). Of lemmings and snowshoe hares: The ecology of northern Canada. *Proceedings of the Royal Society B: Biological Sciences,* 278(1705): 481–489.
Marshall, K. N., et al. (2013). Stream hydrology limits recovery of riparian ecosystems after wolf reintroduction. *Proceedings of the Royal Society B: Biological Sciences,* 280(1756): 20122977.
Ripple, W. J., et al. (2014). Trophic cascades from wolves to grizzly bears in Yellowstone. *Journal of Animal Ecology,* 83(1): 223-233.
Smith, D., et al. (2013). *Yellowstone Wolf Project: Annual Report, 2012.* Yellowstone National Park, WY: National Park Service, Yellowstone Center for Resources.

Wild Horizon/Universal Images Group/Getty Images

↑ A grey wolf on the prowl at Yellowstone National Park.

BRING IT HOME

PERSONAL CHOICES THAT HELP

Understanding the factors that influence how populations change can help us manage species that are facing extinction or help us control (or even eliminate) non-native species that are causing problems. How people view species and their connection to our world has a large impact on how management plays out.

Individual Steps

• Learn more about wolves at the International Wolf Center (www.wolf.org).
• Use the Internet and books on wildlife to research what your area might have been like prior to human settlement. Which species have been extirpated, and which ones have been introduced? How have wildlife populations changed as a result of human action?
• See if you can recognize distribution patterns in the wild. Do some flowers or trees grow in clumps or patches? Can you find species that appear to have a random or uniform distribution?

Group Action

• Explore organizations that support predator preservation, such as Defenders of Wildlife and Keystone Conservation, for suggestions on how you can help educate others about the importance of predators.
• Join a local, regional, or national group that works to monitor, protect, and restore wildlife habitats, such as the Defenders of Wildlife Volunteer Corps.
• Investigate predator compensation funds such as the Defenders of Wildlife Wolf Compensation Trust and the Maasailand Preservation Trust (for livestock losses due to lion predation). Do you feel this is a worthwhile approach?

Policy Change

• Check the U.S. Fish and Wildlife Service website, www.fws.gov, for updates on wolf management and protection status.
• Write a letter to the editor of your newspaper in support of or in opposition to the decision to delist the wolf in parts of the American West.

ENVIRONMENTAL LITERACY UNDERSTANDING THE ISSUE

What ecological significance do a population's size, density, and distribution have?

INFOGRAPHIC 9.1

1. The number of individuals in a given area, such as an acre or a square mile, is a measurement known as _____ _____.

2. The concept of minimum viable population:
   a. predicts how many individuals can fit into a habitat.
   b. describes the potential number of individuals if there are no predators.
   c. describes the potential number of individuals if resources are limited.
   d. describes the smallest number of individuals needed to ensure the long-term continuation of a particular population.

3. Why is it important for an ecologist to understand how a species she is studying is distributed within its ecosystem?

What determines a population's growth rate? Under what conditions do we see exponential population growth? Logistic growth?

INFOGRAPHICS 9.2 AND 9.3

4. Exponential growth of a population:
   a. is often seen if a population reaches a new environment that is favorable.
   b. is a J-shaped curve on a population graph.
   c. occurs when most or all females reproduce at every opportunity and most or all offspring survive.
   d. All of the above are true.

5. Consider the roach. Why does its population size never reach its biotic potential?
   a. Resistance factors limit population size.
   b. Females produce few offspring at a time.
   c. Its tolerance limits are too broad.
   d. Too many growth factors are present.

6 Kangaroo rats eat seeds and are eaten by coyotes. Under what conditions might the kangaroo rat population increase exponentially? Logistically?

How do density-dependent and density-independent factors affect population growth?

INFOGRAPHIC 9.4

7. True or False: Fire is an example of a density-dependent factor.

8. Lesser goldfinches are small, seed-eating birds. In cities, both wild hawks and domestic cats eat these birds. Discuss several density-dependent and density-independent factors, including both growth and resistance factors, that could affect their carrying capacity.

What are the life-history strategies of *r*- and *K*-selected species, and how do they relate to population growth patterns and potential?

INFOGRAPHICS 9.5 AND 9.6

9. What type of species are more likely to have a population overshoot carrying capacity and then die back?
   a. Species with low biotic potential
   b. Species with high parental care
   c. *r*-selected species
   d. *K*-selected species

10. Compare the life-history strategy of a deer mouse with that of a bear and identify each as either an *r*- or *K*-selected species.

11. Beavers in some areas of Michigan and Minnesota live almost exclusively on slow-growing aspen trees and can harvest them faster than the trees can reproduce. Describe how these two populations—beaver and aspen trees—undergo population fluctuations, or boom-and-bust cycles.

What are top-down and bottom-up regulation, and which is most important in determining the size of a population?

INFOGRAPHIC 9.7

12. True or False: If predators are present in a community, top-down regulation will always trump bottom-up regulation.

13. In a bottom-up regulation scenario, the size of an elk herd is determined by:
   a. the number of wolves in the area.
   b. the number of wolf predators in the area.
   c. the amount of grass, aspen, and other food sources.
   d. All of these are examples of bottom-up regulation.

14. How can it be that a single population might sometimes be controlled from the top down and other times be controlled from the bottom up? Give an example of an instance in which an elk population would be controlled by top-down regulation and another example of its control by bottom-up regulation.

SCIENCE LITERACY WORKING WITH DATA

The following graph shows the numbers of Kaibab deer on the isolated Kaibab Plateau in northern Arizona. It was declared a National Game Preserve in 1906, and predator removal was encouraged to protect the deer. Between 1907 and 1937, more than 800 mountain lions were removed. Wolves were exterminated by 1926, and more than 7,000 coyotes and 500 bobcats were also removed. The deer population increased, and by 1915, there was significant damage to the grasses, shrubs, and trees that were being eaten by the deer.

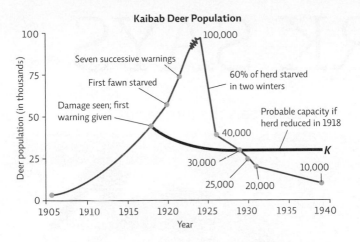

Kaibab Deer Population

Interpretation

1. With the initial level of predators, what was the probable carrying capacity (*K*) of the Kaibab Plateau for deer at the start of this story?

2. Once the population had passed its probable *K*, would it be able to sustain itself at the higher numbers? Explain your answer.

3. In what 2- to 3-year period did the deer population have the highest reproductive growth rate?

Advance Your Thinking

4. Explain the factors—growth, resistance, and density-dependent and density independent factors—that accounted for the changes in population numbers for the deer in 1905, 1915, and 1930.

5. Did Kaibab deer behave like an *r*-selected species or a *K*-selected species? Justify your answer.

6. Before the predators were exterminated, the habitat of the Kaibab Plateau consisted of meadows and forests with a wide diversity of grasses, shrubs, and trees. How do you think it probably looked in 1935, after 30 years of damage?

INFORMATION LITERACY EVALUATING INFORMATION

For over 100 years, up until the 1960s, wolves were extirpated in most of the United States. Over the past few decades, studies have indicated that wolves may have been more important to the functioning of an ecosystem than humans realized. Many groups have worked to reintroduce wolves in states such as Colorado, Montana, Idaho, North Carolina, Washington, and New Mexico. These groups sometimes encounter fierce resistance from other groups concerned with human safety and with possible financial impacts of wolves preying on livestock.

Read the article "Reintroducing the Gray Wolf in the U.S.," at www.actionbioscience.org/biodiversity/johnson.html. Then go to the main website for the group, at www.actionbioscience.org, and investigate the organization.

Evaluate the websites and work with the information to answer the following questions:

1. Is this a reliable information source?
   a. Does the organization give supporting evidence for its claims?
   b. Does it give sources for its evidence?
   c. What is the mission of this organization? Do you believe this mission is reasonable? Explain.

2. Now visit the Rocky Mountain Elk Foundation's website, www.rmef.org. In the site's search box, enter the word "wolves" and read at least two articles. Answer all the items under Question 1 for this website.

3. Finally, visit the Defenders of Wildlife website, www.defenders.org. In the site's search box, enter the word "wolves" and read at least two articles. Answer all the items under Question 1 for this website.

4. Do you believe that wolves should be reintroduced in a few isolated areas; in many areas, including those where human contact is frequent; or not at all? Justify your decision.

5. Do you believe that the hunting of wolves outside protected areas like Yellowstone National Park should be allowed? Justify your answer.

Find an additional case study online at http://www.macmillanhighered.com/launchpad/saes2e

WHAT THE STORK SAYS

A bird species in the Everglades reveals the intricacies of a threatened ecosystem

CORE MESSAGE

Ecological communities are complex assemblages of all the different species that can potentially interact in an area. All the pieces of the ecological community are connected; change one thing, and many others are affected. This means ecosystems are often negatively affected by human impact. Understanding the interconnections within the communities may allow us to better protect and even help restore damaged ecosystems.

AFTER READING THIS CHAPTER, YOU SHOULD BE ABLE TO ANSWER THE FOLLOWING **GUIDING QUESTIONS**

1

What is community ecology, and how is it studied? How do matter and energy move through ecological communities?

2

How do biotic and abiotic factors affect community composition, structure, and function?

Wood storks in Florida.
© Stephen Vincent/Alamy

3

How do species interactions contribute to the overall viability of the community?

4

In general, how do human actions affect the diversity of ecosystems? Why might this change in diversity make it difficult to restore damaged ecosystems?

5

How do ecosystems change over time through ecological succession? How can we use this knowledge to assist in ecosystem restoration?

James Rodgers steered his canoe toward a large cypress tree as sunlight trickled through the dizzy pattern of leaves overhead. The tree had several wood stork nests in it, and Rodgers and his assistant wanted to get a closer look at all of them. They were in the thick of a dense swamp near the northwestern edge of the Florida Everglades, and it was the height of breeding season for the storks—eggs had hatched, and nestlings everywhere were crying, loudly, for food. Rodgers was silent. He knew from experience that alligators patrolled the waters surrounding stork nests and that too much human disturbance could "flush" the adult storks—forcing them to flee in a hurry, which would leave their babies vulnerable to aerial predators.

The wood stork is an unassuming sort of bird: More than 1 meter (3 feet) tall, yes. But also covered with a mottled black-and-white coat of feathers—bland compared to some of its tropical neighbors. Despite the lack of majesty of the wood storks, however, Rodgers and others at the Florida Fish and Wildlife Service keep close tabs on their ranks.

Here's why: In the late 1970s, the number of nesting pairs of the bird plummeted to an all-time low of 4,500 or so. By the early 1980s, the bird had earned a spot on the endangered species list. It was then that Rodgers and his colleagues were first tasked with determining which of several factors (Reduced nesting habitat? Health of females? Damaged feeding grounds?) was most responsible for the decline of this particular bird. And it was through those research efforts—focused intently on the wood stork—that they found an entire ecosystem on the brink.

◉ **WHERE ARE THE FLORIDA EVERGLADES?**

The well-being of a species depends on the health of its ecosystem.

Community ecology is the study of how a given ecosystem functions—how space is structured, why certain species thrive in certain areas, and how individual species in the same community interact with one another and with their **habitat**.

This includes understanding how various species contribute to ecosystem services like pollination, water purification, and nutrient cycling (see Chapter 8). Wetlands like the Florida Everglades provide extremely important water management services, including the recharge of groundwater and the capture of contaminants and excess nutrients, enabling them to be stored or converted to safer forms. This process prevents those nutrients from reaching downstream fresh- and saltwater ecosystems. Most important to us humans, however, is the wetlands' contribution to flood control and provision of freshwater supplies. By capturing and storing large amounts of precipitation and runoff (overland flow of water) and then releasing it downstream slowly over time, wetlands significantly reduce peak flood levels during major rain events. This slow-moving water also has more time to soak into the ground, refilling groundwater supplies.

Community ecology also includes understanding the myriad ways in which we humans have altered

KEY CONCEPT 10.1

Community ecologists study the many populations that live and interact within a given area. Some species are particularly vulnerable to changes and are thus good indicators of an ecosystem's health.

various **ecosystems** and, in so doing, have changed the ways they function. In the Florida Everglades, which were heavily developed throughout the second half of the 20th century, infrastructure like roadways and canals have dramatically reordered the physical landscape. Meanwhile, all the pollutants that come with modern living—solid waste, agricultural chemicals, etc.—have upended the delicate balance of chemical and physical reactions that make this natural world function.

For the wood stork—a tall wading bird that weighs up to 3 kilograms (7 pounds)—it's all about food. During mating season alone, the birds consume an estimated 45 kilograms (100 pounds) of food—each. One captive bird ate more than 650 small fish in just 35 minutes. Indeed, their feeding habits alone make for great spectacle. They hunt almost exclusively in shallow, muddy, plant-filled water—just 15 to 50 centimeters (5 to 20 inches) deep but so cloudy that fish cannot be seen. They inch their way through these waters at a steady two-steps-per-second clip, sweeping their long,

narrow bills—which are kept precisely 8 centimeters (3 inches) agape, and submerged all the way up to the breathing passage—side to side in a relentless hunt for food. When a bill's methodical searching meets the sensation of a wriggling fish (or crayfish), it snaps shut with spectacular speed—in just 25 milliseconds, to be exact. It's the fastest reflex known in all vertebrates, and it enables the wood stork to capture prey that no other wading birds can access.

But for this tactile (or nonvisual) feeding method to work, the prey must be densely concentrated. This means wood storks need seasonally drying wetlands to forage—and lots of them. Even a small drop in feeding success can impact the ability of these colossal birds to successfully rear their young.

community ecology The study of all the populations (plants, animals, and other species) living and interacting in an area.

habitat The physical environment in which individuals of a particular species can be found.

ecosystems All the organisms in a given area plus the physical environment in which they interact.

↓ The inlets of Everglades National Park contain a unique mix of tropical and temperate plants and animals, including more than 700 plant and 300 bird species.

OTIS IMBODEN/National Geographic Creative

Joe Cavaretta/Sun Sentinel/MCT via Getty Images

↑ Canals that drained (and still drain) the Everglades allowed communities like Sunrise, Florida, to be built where Everglades once existed.

↑ Map of South Florida.

It's this sensitivity that makes the wood stork such a good **indicator species** for the Everglades. An indicator species is one that's particularly vulnerable to ecosystem perturbations. Because even minor environmental changes can affect them dramatically, they can warn ecologists of a problem before it grows. "It is much easier to follow one or two species than to try and monitor an entire ecosystem," says Rodgers, who is a wood stork specialist. "So if an indicator species can be identified, this makes it much easier to keep tabs on the health of the ecosystem."

Human alterations have changed the face of the Everglades.

indicator species A species that is particularly vulnerable to ecosystem perturbations, and that, when we monitor it, can give us advance warning of a problem.

food chain A simple, linear path starting with a plant (or other photosynthetic organism) that identifies what each organism in the path eats.

food web A linkage of all the food chains together that shows the many connections in the community.

producer An organism that captures solar energy directly and uses it to produce its own food (sugar) via photosynthesis.

consumer An organism that eats other organisms to gain energy and nutrients; includes animals, fungi, most bacteria.

South Florida once provided an ideal breeding ground for these amazing but picky birds. Before giving way to a hodgepodge of resorts, sugar plantations, and dense urban centers, the region was defined by an uninterrupted web of natural ecosystems, collectively known as the Everglades. Marshes, prairies, swamps, and forests stretched across some 10,000 square kilometers (4,000 square miles) of land. Each distinct community was connected by the same slow-moving water, which began at the southern edge of Lake Okeechobee and flowed south for 160 kilometers (100 miles) before emptying into the Florida Bay. The glacial pace of this water (it can take months or even years

for a given eddy to travel from lake to bay) across such a broad, shallow expanse (100 kilometers [60 miles] wide, and in some places just a few centimeters or inches deep) is known as *sheet flow*. The Everglades' nickname, "River of Grass," comes from the image of sheet flow through the region's iconic sawgrass marshes.

It was here that wood storks flourished. In the 1930s, an estimated 15,000 to 20,000 pairs nested throughout the southeastern United States—largely in South Florida, where foraging grounds were ideal. And because their plain black-and-white plumage was not lovely enough to attract bird hunters collecting feathers for fashionable ladies' hats, the storks thrived even as other wading bird stocks were decimated.

But it was not long after they first discovered the Everglades that American explorers began hatching plans to drain and then develop them. Swamps and muddy rivers choked with grass were seen as having no inherent value. "From the middle of the 19th century to the middle of the 20th, the United States went through a period in which wetland removal was not questioned," says University of Florida geographer and historian Christopher Meindl. "In fact, it was considered the proper thing to do." The Central and Southern Florida Project, authorized by Congress in 1948, set out to systematically drain the Everglades. And over time, a vicious new cycle was established: Humans would drain the swamps and replace them with towns and cities, and then a rash of floods would prompt more complete drainage (under the mandate of flood control and prevention), which would in turn be followed by even more development (on the newly drained land).

As the human population swelled in the region, water that once fed swamps and marshes was rerouted to the faucets of burgeoning developments. And as water levels changed, becoming deeper in some areas and completely disappearing in others, the total wading bird population plummeted—by 90% between the 1930s and 1990s.

Matter and energy move through a community via the food web.

As ecologists would soon discover, the loss of even one species can disrupt an entire ecosystem—from the health of giant wading birds right down to the movement of matter and energy.

Energy is the foundation of every ecosystem; it is captured by photosynthetic organisms and then passed from organism to organism via the **food chain**—a simple, linear path that shows what eats what. Any given ecosystem might have dozens of individual food chains. Linked together, they create a **food web**, which

shows all the many connections in the community. Both food chains and webs help ecologists track energy and matter through a given community. They can vary greatly in length and complexity between different types of ecosystems. But most share a few common features, and all are made up of the same basic building blocks—namely, producers and consumers.

Florida wood storks sit near the top of a food chain that begins with sawgrass, other plants like cypress and mangrove trees, and a mix of algae and bacteria known as periphyton. These photosynthetic organisms are all known as **producers**. Producers capture energy directly from the Sun and convert it to food (sugar) via photosynthesis (see Chapter 8). They are then eaten by a wide range of **consumers**—organisms that gain energy and nutrients by eating other organisms. Animals, fungi, and most bacteria and protozoa are consumers. **INFOGRAPHIC 10.1**

INFOGRAPHIC 10.1   EVERGLADES FOOD WEB

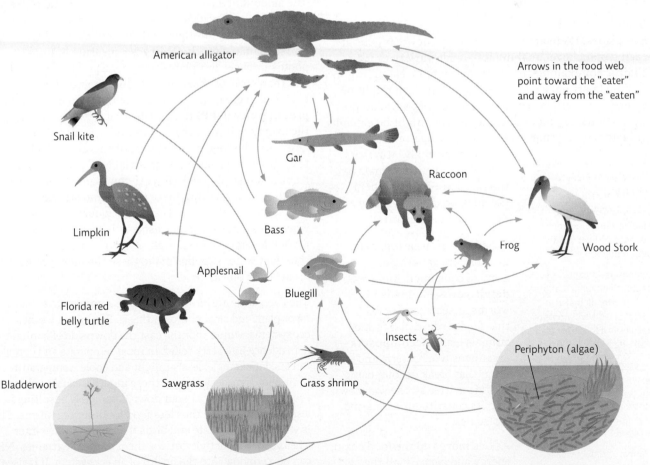

↑ The food web of the Everglades is very complex and varies among the different ecosystems found there. Periphyton algae mats form the base of the food web and may be the most important producers in the ecosystem. The American alligator is the main apex predator, though when young is also prey to various birds, fish, mammals, and even other alligators.

 Identify two or three food chains within this web that end with the alligator.

KEY CONCEPT 10.2

Producers capture energy, which is used or passed on to consumers via the food chain. The biomass of trophic levels decreases as one moves up the trophic pyramid because organisms use most of the energy taken in and pass only a small percentage on to the next level.

trophic level: Wood storks eating crayfish are feeding at trophic level 3, but when they eat small fish like bluegill, they are feeding at trophic level 4.

In an ecosystem as diverse and complex as the Everglades, there are dozens of different organisms at each trophic level, making for a wide variety of food chains. For example, periphyton might be eaten by grass shrimp (a primary consumer), which in turn might be eaten by bluegill (a secondary consumer), who fall prey to raccoons (tertiary consumers), who might be eaten by alligators (quaternary consumers). Alligators eat a wide variety of animals—turtles, fish, birds, even mammals like the raccoon—making them the apex predator in many Everglades food chains.

When any of these organisms die, they are eaten by an army of consumers known as **detritivores**—animals like worms, insects, and crabs that feed on dead plants and animals—and **decomposers**—organisms like bacteria and fungi that break decomposing organic matter all the way down into its constituent atoms and molecules.

As one moves up the food chain, energy and *biomass* (all the organisms at that level) decrease, creating a trophic pyramid. The reason is simple: Nearly every

trophic levels Feeding levels in a food chain.

detritivores Consumers (including worms, insects, crabs, etc.) that eat dead organic material.

decomposers Organisms such as bacteria and fungi that break organic matter all the way down to constituent atoms or molecules in a form that plants can take back up.

gross primary productivity A measure of the total amount of energy captured via photosynthesis and transferred to organic molecules in an ecosystem.

net primary productivity (NPP) A measure of the amount of energy captured via photosynthesis and stored in a photosynthetic organism.

These different feeding levels are known as **trophic levels**. Consumers are organized into trophic levels based on what they eat. *Primary consumers* eat producers; *secondary consumers* eat organisms on lower trophic levels, predominately feeding on primary consumers; *tertiary consumers* primarily (or exclusively) eat secondary consumers, and so on, ending with the apex predators in the last trophic level. Of course, many consumer species often feed at more than one

organism uses up the majority of its energy and matter in the complicated act of living. So when an organism is killed and consumed by a predator, it only passes on a small percentage of all the energy and matter it consumed during its lifetime.

INFOGRAPHIC 10.2

A pyramid's ultimate size is determined by its first trophic level—the one made up of photosynthesizing producers, namely plants. More plant growth means more food for primary consumers, which in turn means more food for those organisms above them, and so on up the food chain. The end result is a larger pyramid with more and larger trophic levels.

The amount of energy trapped by producers and converted into organic molecules like sugar is called *productivity* and is limited by sunlight and nutrient availability. **Gross primary productivity** is a measure of total photosynthesis. But plants use only a portion (actually less than 50%) of this energy to fuel their daily needs; the rest goes to growth. Scientists use the term **net primary productivity (NPP)** to describe that left-over energy—the energy that fuels the addition of new plant biomass (growth). The "net" is a measure of energy available to higher trophic levels. Terrestrial NPP is highest in tropical forests due to high year-round photosynthesis. Equaling or surpassing these tropical forests' NPP is that seen in wetlands and estuaries (aquatic habitats where rivers meet oceans), especially tropical ones. Favorable temperatures, high levels of nutrients, and abundant water supplies make these the most productive ecosystems on Earth.

The Everglades are blessed with long summer days that favor plant and algal photosynthesis. But in many other ecosystems, winter months cast the downside of sunlight dependence into stark relief. In most temperate and boreal forests, for example, less sunlight and cooler temperatures limit photosynthesis during these months, causing plant productivity to slow or shut down completely, resulting in less new growth and thus less food for other organisms. This is why bears hibernate and birds fly south for the winter: The lack of productivity drives them to these extremes. NPP can be a window into the health of an ecosystem. If it rises or falls unexpectedly, ecologists can look for the cause of the change, which might be an invasion by a non-native plant or a sudden drop in a producer population. Anything that alters

KEY CONCEPT 10.3

Matter also moves through the food chain and eventually makes its way back to producers, where it can be used again, thanks to the action of detritivores and decomposers.

INFOGRAPHIC 10.2   TROPHIC PYRAMID

↓ Energy enters at the base of the food chain in the first trophic level (TL) via photosynthesis and is passed on to higher levels as consumers feed on other organisms. This is shown as a pyramid (smaller on top) because only a small percentage of the energy is passed on to each higher level, with the majority being "lost" to the environment (usually as heat from the energy that the organism burns in day-to-day life before it is eaten). Most food chains have only four or five levels due to this progressive loss of energy. Though the actual amount varies from ecosystem to ecosystem, for illustration purposes, we show 10% passing on to the next higher level.

TROPHIC LEVELS (TL)

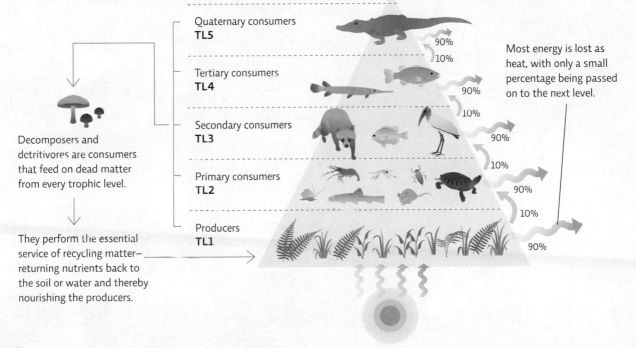

Quaternary consumers
TL5

90%
10%

Tertiary consumers
TL4

90%
10%

Most energy is lost as heat, with only a small percentage being passed on to the next level.

Secondary consumers
TL3

90%
10%

Decomposers and detritivores are consumers that feed on dead matter from every trophic level.

Primary consumers
TL2

90%
10%

They perform the essential service of recycling matter—returning nutrients back to the soil or water and thereby nourishing the producers.

Producers
TL1

90%

? Why are there seldom more than five trophic levels?

NPP can potentially affect organisms at every other level of the trophic pyramid.

As researchers discovered in the 1980s, it was a kink in the food chain that hurt the wood storks. While they feed on many things, they prefer fish—and not just any fish, but those between 2 and 15 centimeters (1 and 6 inches) long. Most fish need more than a single season to grow this big; in fact, they need wetlands that are flooded for longer than a year and only very rarely go completely dry. It turns out that as humans altered water cycles in South Florida, there were fewer and fewer such areas, and thus fewer fish for the storks to feed their young.

KEY CONCEPT 10.4

Ecosystems with a lot of habitat variety and niches can accommodate more species diversity, and this, in turn, increases the resilience of the community.

Communities, such as the ones found in the Everglades, are shaped by biotic and abiotic factors.

Much of Florida is made up of low, flat land that floods from June through September. During this period, Florida typically receives about 75% of its annual rainfall; water forms a vast sheet, covering thousands of acres, and small changes in depth can amount to large changes in surface area. Many wetland fish grow and reproduce in this expanding habitat. Then, as rains taper off, water begins to recede. And where they were once spread out, fish become concentrated in small ponds and *sloughs* (free-flowing channels of water that develop in between sawgrass prairies). Foraging storks follow these receding waters, which ecologists like to call "dry down," from upland ponds to lowland coastal areas, feeding on fish. They are so dependent on this water cycle, studies show, that their breeding cycle is regulated by water levels. Such profound connectedness—between landscape and life—is common in the Everglades.

Susan Cocking/Miami Herald/McClatchy-Tribune via Getty Images

↑ Volunteers and a park ranger canoe through mangroves in Biscayne National Park located just south of Miami, Florida. Originally slated to be included in the Everglades National Park, the Biscayne Bay area was cut from the proposed plan to make it easier to approve the formation of Everglades National Park. It achieved national park status itself in 1980.

niche The role a species plays in its community, including how it gets its energy and nutrients, what habitat requirements it has, and which other species and parts of the ecosystem it interacts with.

resilience The ability of an ecosystem to recover when it is damaged or perturbed.

species diversity The variety of species in an area; includes measures of species richness and evenness.

species richness The total number of different species in a community.

Such connections among species, and between species and their environment, give rise to ecosystem diversity, a measure of the number of species at each trophic level, as well as the total number of trophic levels and available niches. (See Chapter 12 for more on ecosystem diversity.) Each species occupies a unique **niche**—that is, a unique role and set of interactions in the community, how it gets its energy and nutrients, and its preferred habitat. If two different species tried to occupy exactly the same niche, one would outcompete the other. The less successful species has three choices: Leave the area, switch niches, or die out. Greater ecosystem diversity means more niches and thus more

ways for matter and energy to be accessed and exchanged. This generally increases a community's **resilience**—its ability to adjust to changes in the environment and return to its original state rather quickly.

Species diversity, which refers to the variety of species in an area, is measured in two different ways: species richness and species evenness. **Species richness** refers to the total number of different species in a community. **Species evenness** refers to the relative abundance of each individual species. In general, organisms at a higher trophic level will have fewer members than those at a lower trophic level, but organisms within

INFOGRAPHIC 10.3 SPECIES DIVERSITY INCLUDES RICHNESS AND EVENNESS

↓ The species diversity in an area is a measure of species richness (the total number of species) and species evenness (a comparison of the population size of each species). The Everglades contains forested areas known as hardwood hammocks. Each forest plot shown here contains 15 trees, but they differ in terms of species richness and evenness.

● Hackberry ● Red Maple ● Gumbo Limbo
● Mahogany ● Live Oak ● Cocopalm

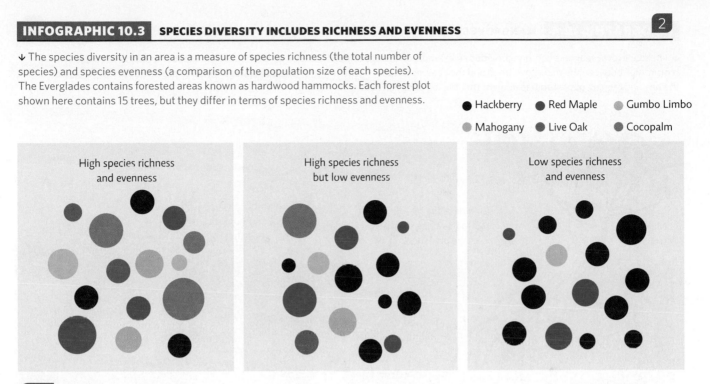

High species richness and evenness

High species richness but low evenness

Low species richness and evenness

? Which forest plot shown here would likely have the highest species diversity (richness and evenness) of birds? Explain.

the same trophic level should have relatively similar numbers. If they do, the community is said to have high species evenness. If, on the other hand, one or two species dominate any given trophic level, and there are few members of other species, then the community is said to have low species evenness. In such uneven communities, the less abundant species is at greater risk of dying out.

Both richness and evenness have an impact on diversity. In general, higher species richness and evenness makes for a more diverse community and a more intricate food web. Greater intricacy enables more matter and energy to be brought into the system and also makes the community less likely to collapse in the face of calamity. **INFOGRAPHIC 10.3**

A community's composition and diversity is also heavily influenced by its physical features. As physical features like temperature and moisture

change, so does community composition. This often happens in **ecotones**, places where two different ecosystems meet—like the edge between a forest and field or river and shore. The different physical makeup of these edges creates different conditions, known as **edge effects**, which either attract or repel certain species. For example, it is drier, warmer, and more open at the edge between a forest and field than it is further into the forest. This difference produces conditions favorable to some species but not others. Ecotones may also attract some species that use different aspects of the two adjacent communities; fish such as young snapper or grunts, for example, prefer to live in areas where seagrass beds are fairly close to a shoreline populated by mangrove trees. The mangrove "prop" roots, which anchor the trees into the wet, sandy ground below, offer the fish safety from predators during the day but are close enough to the seagrass beds where the snapper and grunts feed at night for easy "commuting." These fish are not found in coastal areas without the combination of protective coastal

KEY CONCEPT 10.5

Community composition is affected by the physical structure of the habitat, with some species preferring to inhabit ecotone regions where one habitat meets another (the edge) and others staying deep within one habitat (the core).

species evenness The relative abundance of each species in a community.

ecotones Regions of distinctly different physical areas that serve as boundaries between different communities.

edge effects The different physical makeup of an ecotone that creates different conditions that either attract or repel certain species (e.g., it is drier, warmer, and more open at the edge of a forest and field than it is further in the forest).

INFOGRAPHIC 10.4   MANGROVE EDGES   2

↓ The mangrove–seagrass ecotone provides an example of an edge effect. Fish such as immature gray snapper and bluestriped grunt "commute" between the mangrove trees and the seagrass beds. The proximity of these two areas is vital to provide both the protection during the day and feeding opportunities at night that these young fish need.

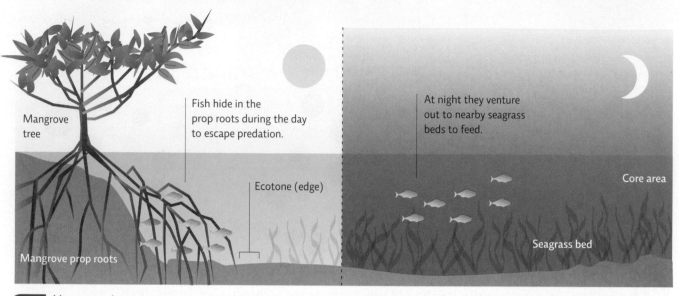

Mangrove tree

Fish hide in the prop roots during the day to escape predation.

At night they venture out to nearby seagrass beds to feed.

Core area

Ecotone (edge)

Seagrass bed

Mangrove prop roots

? Many coastal mangrove areas are being fragmented as stretches of mangrove are removed for residential or commercial development. What might happen to the gray snapper and bluestriped grunt populations if the mangrove trees are removed from part of the shoreline? What impact would this have on the seagrass beds?

mangrove trees and close-by, offshore seagrass beds. **INFOGRAPHIC 10.4**

Species that thrive in edge habitats like this are called **edge species**. Other species, those that can only be found deep within the core of a given habitat, are called **core species**. Some of the many species that find food and refuge in the seagrass, such as crustaceans, sea urchins, and worms, prefer to stay in core areas, where they are better hidden and protected from wave action or can make use of deeper sediment buildup in these inner areas. This fragmentation scenario plays out in ecosystems around the world, especially terrestrial ones such as forests where urban/suburban, agricultural, and industrial development creates patchworks of formerly expansive habitats. Edge species may thrive in these patchy habitats, but because core species will not venture out across the edge in search of new habitat, they are easily trapped by habitat fragmentation; we may eliminate these core species altogether if we don't leave enough core area behind. (See

---

edge species Species that prefer to live close to the edges of two different habitats (ecotone areas).

core species Species that prefer core areas of a habitat—areas deep within the habitat, away from the edge.

Chapter 12 for more on habitat fragmentation and core species.) **INFOGRAPHIC 10.5**

Changing community structure changes community composition.

Wood storks are spectacular fliers. From a perch, they spring their giant bodies into air in a single motion, then extend their necks and legs fully as they take flight. They can reach altitudes as high as 1,500 meters (5,000 feet) and can glide for miles without flapping their wings (a feat accomplished by riding vertical air currents—the currents support their weight and allow the storks to spiral upward). When foraging grounds dry up, or flood, or are converted into human developments, these aerial skills are pushed to the limit. Surveys found that some wood storks were flying farther and farther from their nesting habitat in search of foraging grounds—as much as 120 kilometers (75 miles) in some cases.

But they weren't the only ones to struggle in the newly developed region. As the natural landscape of the Florida Everglades was modified—as cities replaced swamps and roads replaced rivers—so too were the species

INFOGRAPHIC 10.5   EDGE EFFECTS

↘ Habitat structure influences where species live. Edge species, like deer, prefer habitats with forest and field edges, whereas core species, like some warblers (small birds), prefer the inner areas of forest and do not readily venture into edge regions.

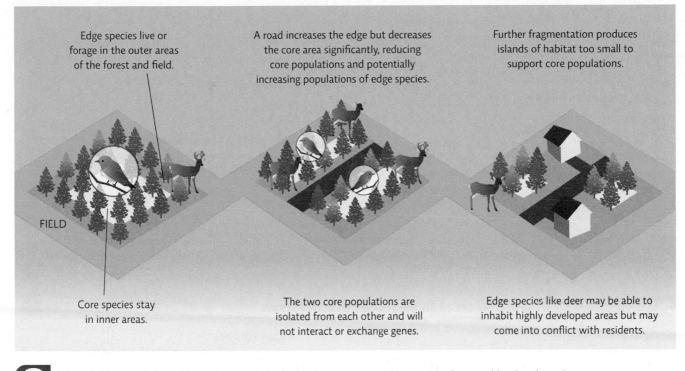

Edge species live or forage in the outer areas of the forest and field.

A road increases the edge but decreases the core area significantly, reducing core populations and potentially increasing populations of edge species.

Further fragmentation produces islands of habitat too small to support core populations.

FIELD

Core species stay in inner areas.

The two core populations are isolated from each other and will not interact or exchange genes.

Edge species like deer may be able to inhabit highly developed areas but may come into conflict with residents.

? Why might a wooded corridor that connects two habitat fragments need to be wider for a warbler than for a deer if we expect the bird to use it to travel between fragments?

interactions and thus the composition of the natural communities that remained.

Dams, dikes, and bridges installed to give humans total control over water levels—in any given portion of the Everglades at any given time—disrupted the flow of water like never before—causing some portions of the Everglades to stay too wet for too long and others to stay far too dry. As sloughs ran dry, key detritivores and decomposers like worms, grass shrimp, and microbial communities that had thrived there were decimated. This then led to the decline of the snakes, fish, alligators, turtles, and wading birds that fed on them.

Meanwhile, agricultural lands—sugar plantations, in particular—were doing as much damage as flood control efforts. The Everglades are a nutrient-poor ecosystem, especially low in phosphorus. This is fine for the plants and animals that live there; they have evolved and adapted to such conditions and are very effective at moving nutrients through the food chain. But, as early developers discovered, it's not so good for agriculture. To grow crops, farmers must add large quantities of synthetic nutrients to the mucky wetland soil. The runoff from

those nutrients has created vast algal blooms, from Lake Okeechobee and elsewhere, which have in turn choked off plant and animal life. As scientists recently discovered, phosphorus runoff is flushed into the canals, then pumped into the lake. When the lake drains, the phosphorus enters marshes and trickles through other ecosystems, changing nutrient levels and community composition of plant species along with it.

Sawgrass, typically the dominant species in the marsh, is well adapted to obtain phosphorus from the normally nutrient-poor waters. Cattail, another Everglades producer species with unique flowering spikes, normally prefers the marshes' edges, and its growth is limited by the lack of phosphorus. However, Danish ecologist Hans Brix and his colleagues have shown that when phosphorus levels increase, cattails can quickly outcompete sawgrass. In an experiment conducted by Brix, both plants took in more phosphorus from nutrient-enriched waters. But whereas cattails increased phosphorus uptake 10-fold, sawgrass increased only 5-fold. Therefore, the cattails can grow more quickly than sawgrass. In areas with nutrient enrichment, cattails have pushed beyond

KEY CONCEPT 10.6

Keystone species are particularly important to other members of their community, and if their numbers decline, many other species may be negatively affected.

also proven problematic. It turns out that mangrove trees are a **keystone species**—one that impacts its community more than its mere abundance would predict. It's a species that many other species depend on, and one whose loss creates a substantial ripple effect, disrupting interactions for many other species and, ultimately, altering food

their natural habitat, through ecotones and into neighboring communities, where they now grow in such dense mats that they're outcompeting sawgrass, choking off native invertebrates on the bottom of the food chain, and physically preventing birds and alligators from nesting.

Replacing mangrove forests with oceanfront resorts has

webs. From their natural habitat at the water's edge, mangrove "prop" roots stabilize the shoreline and provide shelter for a wide variety of fish. So when the mangrove forests are cleared, many other species suffer: the fish that hide among their roots, the fish that feed on those fish, and so on.

Alligators are also a keystone species in the Everglades, one that a great many species depend on during the dry season. As the waters recede, depressions made by alligators (gator holes) are some of the few places that still hold standing water. These holes become refuges for fish, invertebrates, and aquatic plants; they also become very attractive to the animals who feed on these aquatic creatures. Without gator holes, many species would not survive the dry season. **INFOGRAPHIC 10.6**

Wood storks also depend on the presence of alligators, but not just for the dry season gator holes. In the 1980s, Rodgers and his colleagues embarked on a comprehensive study of stork nests in an effort to see which types of trees the storks preferred to nest

INFOGRAPHIC 10.6   KEYSTONE SPECIES SUPPORT ENTIRE ECOSYSTEMS

3

→ Some species are especially important to their ecosystem. If a keystone species is lost or declines in number, the ecosystem could change drastically, and other species that depend on it may suffer or be lost.

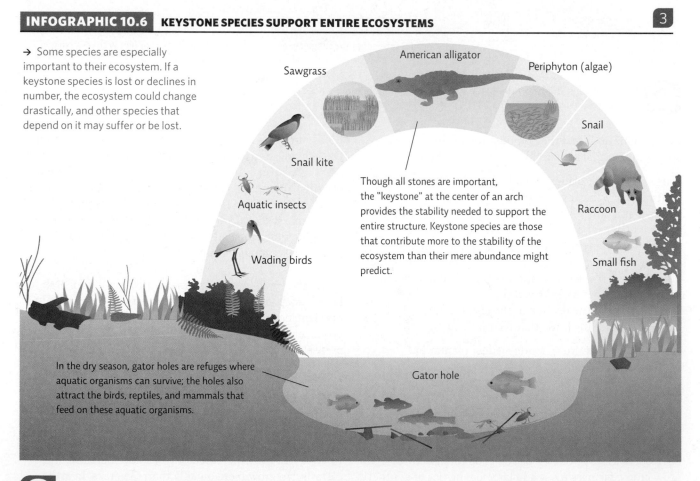

Though all stones are important, the "keystone" at the center of an arch provides the stability needed to support the entire structure. Keystone species are those that contribute more to the stability of the ecosystem than their mere abundance might predict.

In the dry season, gator holes are refuges where aquatic organisms can survive; the holes also attract the birds, reptiles, and mammals that feed on these aquatic organisms.

Sawgrass · American alligator · Periphyton (algae) · Snail kite · Aquatic insects · Wading birds · Snail · Raccoon · Small fish · Gator hole

Keystone species whose actions alter the habitat in a way that benefits other species are also called ecosystem engineers. Other than the alligator, identify a species that acts as an ecosystem engineer in its ecosystem and explain how its actions benefit other species.

in and whether the availability of those trees was impacting their ability to breed. "We went to 20 stork colonies," Rodgers remembers. "We measured every tree, recorded its species, size, cored it for age, noted its branching structure." The conclusion, reached after 5 years of painstaking work, can be summed up in a single sentence, Rodgers says: Wood storks will nest in just about anything, as long as it's surrounded by water that is patrolled by alligators. "Without the alligators, raccoons swim across, and climb up and destroy everything," Rodgers says. "Without the alligators, when predators get in, we've seen [the storks] abandon entire colonies."

KEY CONCEPT 10.7

The many interactions within and between species are critical to energy capture and flow and to matter cycling. Changes that interfere with these interactions can imperil many others and decrease overall the functioning of the ecosystem and the services it provides.

Species interactions are extremely important for community viability.

Communities are all about relationships. Successful communities are those where a certain balance has evolved between all the organisms living there. Species interactions serve many purposes; for example, they control populations and affect carrying capacity. Biodiversity (lots of species, lots of variety within a species) is important because more diversity means more ways to capture, store, and exchange energy and matter (see Chapter 12). But it is not sheer numbers that matter most; it is all the connections among species—how they help or hurt one another—that determine how and how well an ecosystem works. Each species is unique and thus interacts in its own unique ways with all the species around it.

Many species have adaptations that bind them to others, that allow them to coexist, or that facilitate **predation**. In the Everglades, for example, alligators have adaptations—like sharp teeth and powerful jaws—that allow them to stalk and capture prey while most of the fish they prey on have adaptations—like camouflage and a wary nature—that help them avoid capture.

Competition—the vying between organisms for limited resources—is another way that species interact. In general, it is subtle rather than outright fighting. *Intraspecific competition* (that which occurs between members of the same species) is generally stronger than *interspecific competition*

keystone species A species that impacts its community more than its mere abundance would predict, often altering ecosystem structure.

predation Species interaction in which one individual (the predator) feeds on another (the prey).

competition Species interaction in which individuals are vying for limited resources.

↓ Michael Korvela, of the South Florida Water Management District, with vegetation from an artificial marsh in Palm Beach, Florida, that filters pollutants from water bound for the Everglades.

↓ The heart of a functioning community is its species interactions. Some interactions are beneficial and others cause conflict, but all are important in keeping matter and energy flowing through an ecosystem.

MUTUALISM Both species benefit: The moth gains nutrition while the flower gets pollinated.

PARASITISM One species (the parasite) benefits and the other (the host) is harmed: An animal that has too many leeches will be weakened from the loss of blood.

COMMENSALISM One species benefits and the other is unaffected: The heron can catch twice as many fish when foraging alongside the ibis; this doesn't impact the ibis's ability to forage.

 Choose an ecosystem other than the Everglades and give examples of mutualism, commensalism, parasitism, predation, competition, and resource partitioning.

(that which occurs between members of different species). This is because members of the same species share exactly the same niche and thus compete for all resources in that niche, whereas members of different species may compete for only a single resource, like water.

Other neighbors—those that prey on the same food or inhabit similar niches—find a way to partition resources. That is, they divvy up the goods in a way that reduces competition and allows several species to coexist. For example, limpkins and snail kites (two Everglades birds that feed almost exclusively on apple snails) hunt in different regions of the Everglades. This strategy—known as **resource partitioning**—increases the ecosystem's overall capture of matter and energy and thus benefits the entire community.

There are other strategies, too, that keep an ecosystem functioning and strong.

Some of these interactions show a tremendous interdependency on the part of the participants. Known as **symbiosis**, these relationships can take one of three forms. The most commonly recognized form of symbiosis is **mutualism**, where both species benefit from the relationship. In the symbiotic relationship known as **commensalism**, one species benefits from the relationship, and the other is unaffected. Even **parasitism**, where one species benefits from the relationship and the other is negatively affected, is a form of symbiosis. **INFOGRAPHIC 10.7**

By ensuring that all populations persist, even as individuals die, these delicate checks and balances allow more energy to be captured and exchanged and thus increase the amount of biomass the ecosystem is able to produce.

As we've seen, the wood storks rely on several relationships to survive. Their ability to feed depends not only on very specific wetlands hydrology but also on the health and size of the fish they feed upon. And their ability to raise and fledge young depends not only on the availability of "moated" trees—those surrounded by water—but also on a healthy population of alligators that can patrol those moats for raccoons. When individual species are lost, or when a landscape is physically altered, the balance is tipped. And when that happens, things can fall apart. Fast.

resource partitioning
A strategy in which different species use different parts or aspects of a resource rather than compete directly for exactly the same resource.

symbiosis A close biological or ecological relationship between two species.

mutualism A symbiotic relationship between individuals of two species in which both parties benefit.

commensalism A symbiotic relationship between individuals of two species in which one benefits from the presence of the other, but the other is unaffected.

parasitism A symbiotic relationship between individuals of two species in which one benefits and the other is negatively affected.

PREDATION One species benefits (predator) and the other is harmed (prey): Alligators prey on a variety of animals and are prey themselves when young.

COMPETITION All participants are negatively affected: Apple snails, a variety of fish, and crustaceans all compete for periphyton algae as a food source.

RESOURCE PARTITIONING Both species benefit by partitioning a resource rather than competing for it: Though they both eat apple snails, snail kites and limpkins don't directly compete for them since each predator feeds in a different region of the Everglades.

Ecologists and engineers help repair ecosystems.

The federal 1992 Water Resources Development Act enlisted the U.S. Army Corps of Engineers to investigate the damage to the Everglades that resulted from nearly 50 years of unchecked expansion. The final report, published in 1999, acknowledged that the original Everglades (as we found them upon first exploration in the late 1800s) had been reduced by 50%. Human impact often simplifies ecosystems by decreasing the habitat variety normally seen, which leads to a decrease in species diversity. Constructed canals and levees had dramatically altered water levels, leaving some areas parched and others flooded. And poorly timed water releases were further starving ecosystems that had already been affected by hypersalinity, excessive nutrients (from agricultural runoff), and an ever-growing list of non-native species.

KEY CONCEPT 10.8

Human impact often reduces species diversity. It may be difficult to restore all the species and their connections when we try to repair ecosystem damage; therefore, our best course of action is to avoid the damage in the first place.

Ignoring these problems any longer could greatly imperil the 10 million people that had made their home in the region. "What folks finally realized when we reexamined the area was that the wetlands were this essential filter—they cleaned the water of pollutants," says Kim Taplin, a restoration ecologist who works for the U.S. Army Corps of Engineers, restoring the Florida wetlands. "So as the ecosystems have suffered, water quality has declined considerably. We're going to have millions of people with no clean water, unless we fix it." Avoiding the ecosystem damage in the first place is almost always our best option, but when damage is done, fixing it is the work of restoration ecologists. **Restoration ecology** is the science that deals with the repair of damaged or disturbed ecosystems. It requires a special blend of skills—not only biology and chemistry but also engineering and a heavy dose of politics.

In 2000, the U.S. Congress enacted the most comprehensive—and expensive—ecological repair project in history. The Comprehensive Everglades Restoration Plan, or CERP, included more than 60 construction projects to be completed over a 30-year period. The idea was to restore some of the natural flow of water through the Everglades and to capture a portion of the water that now flows to the ocean for South Florida cities and farms.

restoration ecology The science that deals with the repair of damaged or disturbed ecosystems.

One of the Army Corps' biggest challenges has been to take down at least part of the Tamiami Trail, a 240-kilometer (150-mile) stretch of U.S. highway that connects the South Florida cities of Tampa and Miami. The road, which was built in the 1920s, has proven to be one of the most serious barriers to freshwater flow in the region. It's also a heavily traveled, essential piece of human infrastructure that connects two major cities. That means the U.S. Army Corps must not only tear down the road but also build something in its place. "Most of our restoration projects involve building even more structures," says Tim Brown, project manager for the U.S. Army Corps of Engineers' Tamiami Trail project. "It's a delicate balance. We of course want to restore as much of the natural system as possible. But we are also charged with protecting lives and property, and in this case, that means building bridges."

> " *We're going to have millions of people with no clean water, unless we fix it.* " —Kim Taplin.

Dismantling the Tamiami Trail is not the only plan in the works. In 2008, the state of Florida agreed to buy U.S. Sugar Corporation and all of its manufacturing and production facilities in the Everglades Agricultural Area south of Lake Okeechobee for roughly $1.7 billion. State officials declared that they would allow U.S. Sugar to operate for 6 more years before shuttering facilities and beginning the work of restoration. After that, water flow from Lake Okeechobee would be funneled through a series of holding and treatment ponds that would release clean water into the Everglades, rehabilitating some 187,000 acres of land. But the agreement has been revised several times since then, as the economy has fluctuated and state officials and sugar executives have adjusted and readjusted exactly how many acres would be bought for exactly how many dollars. The proposed purchase was hotly debated, as many feared the cost would take money away from other Everglades restoration projects. In October 2010, just under 11,000 hectares (27,000 acres) were purchased for $197 million, with a 10-year option to acquire another 62,000 hectares (153,200 acres). However, U.S. Sugar is still farming part of the land, leasing it back from the state. The timetable for U.S. Sugar to stop farming the land has been pushed back three times since the 2010 deal was signed; farming is currently slated to end on the land in 2018, after which restoration can begin. In the meantime, the state of Florida failed to meet an October 2013 deadline to purchase the additional 153,200 acres from U.S. Sugar at only $7,400 per acre. The state still has the option to purchase the land up to the year 2020, but at market value.

Each facet of CERP has brought its own fresh round of debate over how best to balance the needs of a swelling human population against the importance of restoring and protecting a heavily degraded ecosystem. Of course, no one knows for certain what will work and what won't. The Everglades landscape has changed dramatically, in ways that not even the best scientists can reverse; decades of development will do that. To plan effective restoration efforts, then, scientists and engineers must be flexible; they must be willing to experiment and respond as conditions change—an *adaptive management* approach similar to that used to address stratospheric ozone depletion (see Chapter 2). **INFOGRAPHIC 10.8**

"The bottom line, though, is that there's only so much we can do," says Rodgers. "It's a lot just to figure out what the baseline was or should be. Some plant species have probably gone extinct, and some non-natives are virtually impossible to remove. What we can do is figure out what some of the big obstacles to recovery are, remove them, and after that, let nature take its course." For his part, Rodgers says he can't imagine that the great 1,000-breeding-pair wood stork colonies that early settlers described will ever return to South Florida. The landscape has been too dramatically altered, he says. Though, sometimes, of course, nature can surprise us.

Community composition changes over time as the physical features of the ecosystem itself change.

Though the changes the Everglades have experienced are extreme, changes to ecological communities are really the norm; nature is not static. Predictable transitions can sometimes be observed in which one community replaces another, a process known as **ecological succession**. **Primary succession** begins when **pioneer species** move into new areas that have not yet been colonized. In terrestrial ecosystems, these pioneer species are usually lichens—a symbiotic combination of algae and fungus. Lichens can tolerate the barren conditions. As time goes by and lichens live, die, and decompose, they produce soil. As soil accumulates, other small plants move in—typically sun-tolerant annual plants that live 1 year, produce seed, and then die—and the plant community grows. Gradually, the plant growth itself changes the physical conditions of the area—covering sun-drenched regions with broad,

ecological succession Progressive replacement of plant (and then animal) species in a community over time due to the changing conditions that the plants themselves create (more soil, shade, etc.).

primary succession Ecological succession that occurs in an area where no ecosystem existed before (e.g., on bare rock with no soil).

pioneer species Plant species that move into an area during early stages of succession; these are often *r* species and may be annuals—species that live 1 year, leave behind seeds, and then die.

INFOGRAPHIC 10.8 | **THE COMPREHENSIVE EVERGLADES RESTORATION PLAN**

4

↓ Darker green areas represent wetland areas or river floodplains; white arrows show overland water flow.

HISTORIC FLOW

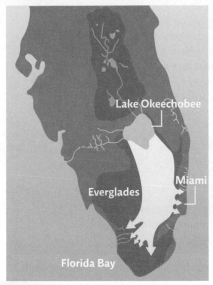

CURRENT FLOW

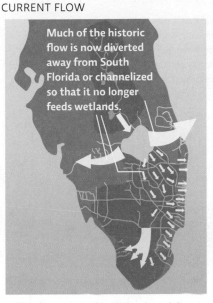

Much of the historic flow is now diverted away from South Florida or channelized so that it no longer feeds wetlands.

PROPOSED FLOW UNDER CERP

↑ Historically, the Everglades covered most of South Florida—more than 10,000 square kilometers (4,000 square miles).

↑ Projects to drain the wetlands and divert water to agricultural lands disrupted normal flow, drained about half of the wetlands, and resulted in water shortages for the downstream ecosystems and for people as well.

↑ The goal of the Comprehensive Everglades Restoration Plan (CERP) is to restore the flow of water back to some historic wetland areas through the removal of some canals and levees, as well as to capture some of the freshwater that drains into the ocean, benefiting both the ecosystems and the residents of South Florida.

? Explain how rerouting some of South Florida's water flow back through the center of the state, through restored wetlands, will help provide residents with more drinking water.

KEY CONCEPT 10.9

Over time, ecosystems naturally transition from one community to another in response to changing environmental conditions. An understanding of this ecological succession process can guide our restoration efforts.

shady leaves, for example. Since these conditions are no longer suitable for the plants that created them, new species move in, and those changes beget even more changes until the pioneers have been completely replaced by a succession of new species and communities.

Secondary succession describes a similar process that occurs in an area that once held life but has been damaged somehow; the level of damage the ecosystem has suffered determines what stage of plant community moves in. For example, a forest completely obliterated by fire may start close to the beginning with small herbs and grasses, whereas one that has suffered only moderate losses may start midway through the process

with shrubs or sun-tolerant trees moving in. The stages are roughly the same for any terrestrial area that can support a forest: first annual species, then shrubs, then sun-tolerant trees, then shade-tolerant trees. Grasslands follow a similar pattern, with different species of grasses and forbs (small leafy plants) moving in over time.

We tend to view this progression as a "repair" sequence. While we can certainly step in to assist in this natural progression to help a damaged ecosystem recover to a former state, the ecosystem is simply doing what comes naturally—responding to changing conditions.

Intact ecosystems have a better chance at recovering from, and thus surviving, perturbations. As mentioned earlier, ecosystems that recover quickly from minor perturbations are said to be resilient: They can bounce back. More species-diverse communities tend to be more resilient than simpler ones with fewer species because it is less

secondary succession Ecological succession that occurs in an ecosystem that has been disturbed; occurs more quickly than primary succession because soil is present.

INFOGRAPHIC 10.9   ECOLOGICAL SUCCESSION

FOREST ECOLOGICAL SUCCESSION DEPENDS ON SOIL AND LIGHT AVAILABILITY

Lichens can attach to bare, uncolonized rock surfaces; as they grow and die over the years, soil begins to form and accumulate, allowing mosses to move in.

Soil builds up over time, allowing larger plants to move in—first small herbs and grasses and, later, small shrubs.

Sun-tolerant trees such as pine and cedar move in when there is enough soil to support large root systems.

Taller plants produce shade, changing conditions and favoring different plant species.

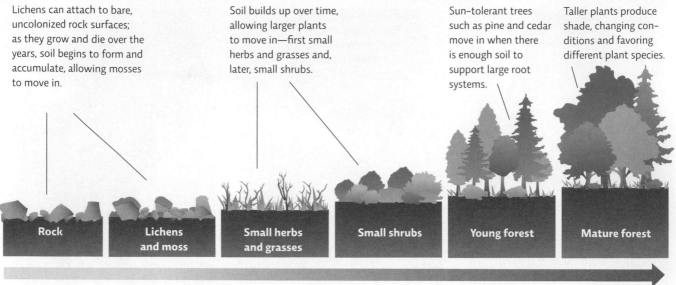

Rock | Lichens and moss | Small herbs and grasses | Small shrubs | Young forest | Mature forest

Time

↑ In terrestrial ecosystems, we see natural stages of succession occur whenever a new area is colonized or an established area is damaged. Sun-tolerant species give way to shade-tolerant ones as more soil is built up, supporting larger plant species.

EVERGLADES ECOLOGICAL SUCCESSION DEPENDS ON THE WATER LEVEL

When water is more than 50 centimeters (20 inches) deep, floating and submerged vegetation grow. As sediments collect, emergent grasses that are rooted underwater—but grow tall enough to emerge above the water surface—move in and establish the sawgrass marsh.

Water-tolerant cypress and willow move in and replace sawgrass as sediment becomes deeper and more stabilized.

Other species of trees, such as the pond apple, move in as sediment gets closer to the water surface.

If dry land emerges from the water, upland species such as oak move in and replace the water-tolerant species of the mixed swamp forest.

Water table (upper level of groundwater)

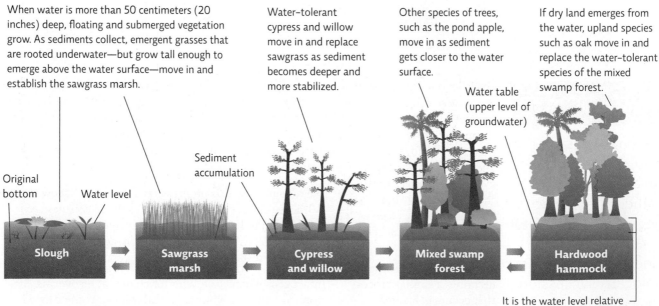

Original bottom | Water level | Sediment accumulation

Slough | Sawgrass marsh | Cypress and willow | Mixed swamp forest | Hardwood hammock

It is the water level relative to the land that determines which species move in.

↑ Ecological succession in the Everglades doesn't necessary follow the tidy predictable sequence seen in terrestrial ecosystems; in fact, periodic fires and the cyclic rainfall patterns may not support a predictable progression at all. Water levels are also important and influence which species move in. Succession in this area can actually go both ways: As ground level changes relative to the water level, an area might flood anew, or sediment buildup might continue to raise the land relative to the water level. In the absence of disturbance, succession will progress to the hardwood hammock forest when sediment builds up enough to expose dry land.

? Look at the forest successional stages shown in the top part of this diagram. Why can't small shrubs or young trees grow on the land shown supporting small herbs and grasses?

likely that the loss of one or two species will be felt by the community at large; even if some links in the food web are lost, other species are there to fill the void. Of course, if keystone species are lost, the community will feel the effect. The loss of the alligator from the complex Everglades community would impact many species and change the face of the ecosystem.

Some ecosystems remain in a constant cycle of succession; others eventually reach an end-stage equilibrium where the conditions are well suited for the plants that created them—for example, trees whose seedlings can grow in shady habitat. These species, which can persist if their environment remains unchanged, are called **climax species**. End-stage climax communities can stay in place until disturbance restarts the process of succession—although there is debate among scientists over whether any community ever reaches an end point of succession or continues to change and adapt.

Wetland areas also go through succession, responding to the presence of water and sediment depth. In the Everglades, each ecosystem is guided along this path by its own constellation of forces. Some, like the iconic sawgrass ecosystems, are fire adapted; fire returns them to early stages again and again, where the underwater roots of the emergent plants (those that are rooted underwater but grow above the waterline) such as sawgrass survive and quickly regrow. Others, if left undisturbed, would pass through successional stages of pioneers (grasses) to shrubs or small trees to larger species of trees, depending on the deposition of soil and proximity of the water table to the surface (the top of the groundwater in the area). Others still are guided by the engineering changes of

animals such as alligators, whose digging habits provide the foundation for an entire food chain. In each, though, the same general concept applies: As conditions change, other species better adapted to those conditions move in and displace previous residents. **INFOGRAPHIC 10.9**

However precarious their recovery might be, wood storks have indeed rebounded in recent years. Some say this rebound is the result of careful conservation efforts—including a restriction on development in certain areas—implemented under the U. S. Endangered Species Act. Others insist that it is merely the result of above-average rainfall in recent years. For his part, Rodgers sees another trend at work. Once again, he says, the storks are trying to tell us something. "They have shifted their center of distribution from South Florida to Central and North Florida," he says. "They're now spilling into Georgia and North Carolina—something we've never seen before." Rodgers suspects that the shift has something to do with the way climate is changing in the region, though he says much more research is needed before anyone can say for certain. "We're still trying to figure out what that means," he says. "But we know it's a clue to something."

climax species Species that move into an area at later stages of ecological succession.

Select References:

Brix, H., et al. (2010). Can differences in phosphorus uptake kinetics explain the distribution of cattail and sawgrass in the Florida Everglades? *BMC Plant Biology*, 10: 23.

Rodgers, J. A., et al. (1996). Nesting habitat of wood storks in North and Central Florida, USA. *Colonial Waterbirds*, 19(1): 1–21.

BRING IT HOME

PERSONAL CHOICES THAT HELP

The world is full of weird and wonderful species. Every year we discover new information about how intricate our biological communities are. By restoring habitats and increasing our understanding of the relationships between species, we can better ensure their long-term survival.

Individual Steps

• Visit a park or nature preserve and watch for signs of species interactions. Do you hear animals or birds? Can you see signs of predation or herbivory?

• Buy a Duck Stamp. Usually purchased by waterfowl hunters for license purposes, nonhunters can purchase a stamp, which supports wetland conservation in the National Wildlife Refuge System.

Group Action

• The Everglades case study is an example of a very extensive restoration project. Call your local park district or nature preserve to see what restoration work is happening in your area and how you can become involved.

Policy Change

• Follow the U.S. Fish and Wildlife Service Open Space blog to learn more about wildlife and issues facing conservation (http://www.fws.gov/news/blog).

Chris Matula/The Palm Beach Post/ZUMAPRESS.com/Newscom

ENVIRONMENTAL LITERACY UNDERSTANDING THE ISSUE

1 What is community ecology, and how is it studied? How do matter and energy move through ecological communities?

INFOGRAPHICS 10.1 AND 10.2

1. Community ecologists study:
   a. the relationships between species in a given area.
   b. the way species interact with their environment.
   c. the things that increase or decrease the number of species in an area.
   d. all of the above.

2. In ecological terms, a consumer is:
   a. any plant.
   b. any animal.
   c. any organism that eats other organisms.
   d. any animal that eats other animals.

3. Draw a simple food web for a natural area near you. Include producers and at least three levels of consumers, as well as detritivores and decomposers.

2 How do biotic and abiotic factors affect community composition, structure, and function?

INFOGRAPHICS 10.3, 10.4, AND 10.5

4. Edge effects:
   a. apply only to the largest and smallest members of a community.
   b. occur in the areas where two or more habitats meet.
   c. are beneficial for nearly all organisms.
   d. are harmful for nearly all organisms.

5. True or False: Ecosystem complexity increases as the variety of habitat and the number of species increases.

6. Cowbirds lay their eggs in the nests of smaller forest birds such as bluebirds. The bluebirds then spend the next several weeks caring for a huge baby cowbird, which quickly kills the bluebirds' own young. Cowbirds prefer open, disturbed areas near a forest and seldom venture far into the forest for any reason, even to lay eggs. Use the concept of edge effect to explain what happens to the populations of bluebirds when humans build roads, recreation areas, homes, and businesses in a large forest.

3 How do species interactions contribute to the overall viability of the community?

INFOGRAPHICS 10.6 AND 10.7

7. An example of mutualism is:
   a. a dog and a flea.
   b. an ant and a grasshopper.
   c. a butterfly and a flowering plant.
   d. a deer and a wolf.

8. How do mangrove forests fit the definition of a *keystone species*?

9. Explain why both species richness and species evenness are important for a healthy ecosystem.

4 In general, how do human actions affect the diversity of ecosystems? Why might this change in diversity make it difficult to restore damaged ecosystems?

INFOGRAPHIC 10.8

10. True or False: The construction of canals and levees to drain parts of the Everglades increased its ecosystem complexity.

11. It might be hard to restore damaged ecosystems like the Everglades because:
    a. there are no economic benefits to restoring the places like the Everglades.
    b. there is no early warning system in place to alert us that the area is being damaged.
    c. some important species might no longer be present.
    d. All of these are correct.

12. What happens to the net primary productivity and to the species diversity when humans disrupt wetlands by adding more nutrients, as in the example of agriculture in the Everglades?

5 How do ecosystems change over time through ecological succession? How can we use this knowledge to assist in ecosystem restoration?

INFOGRAPHIC 10.9

13. Ecological succession is important because it:
    a. allows ecosystems to respond to environmental changes.
    b. promotes competition and fair use of resources.
    c. eliminates K species.
    d. is a natural way to get rid of invasive species and replace them with native species.

14. An example of when secondary succession would occur in a particular area would be after:
    a. lichens have started to grow on a bare rock surface.
    b. a flood has removed much of the vegetation.
    c. hot ash from a volcano has completely burned and buried the area.
    d. a disease has reduced the top predator's population.

15. How might a restoration ecologist use an understanding of ecological succession to help repair a damaged area?

SCIENCE LITERACY WORKING WITH DATA

The Mississippi River lies on three sides of the city of New Orleans, curving around it. To the fourth side of the city lies Lake Pontchartrain, 65 kilometers (40 miles) long and over 50 kilometers (30 miles) wide; the lake connects directly with the Gulf of Mexico. Until 140 years ago, much of southern Louisiana was swampland, lower than sea level and much lower than the level of the Mississippi River. As the swamps were drained, levees were built, beginning in the 1700s, speeding up after 1880, and continuing today.

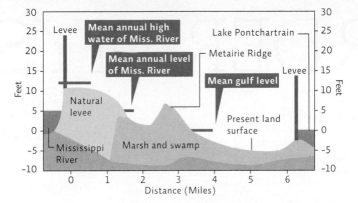

Interpretation

1. What is the difference in mean annual water level, in feet, between Lake Pontchartrain and the Mississippi River at New Orleans? Why is this?

2. New Orleans lies between the levees that hold back the Mississippi River and Lake Pontchartrain. What are the highest and lowest elevations of the city itself, in feet?

3. What type of former ecosystem lies underneath most of the city?

Advance Your Thinking

4. Without the levees throughout the southern half of the state, what would happen?

5. Why do southern Louisiana cemeteries feature above-ground tombs, crypts, and vaults rather than graves?

6. In 2005, Hurricane Katrina devastated New Orleans and much of southern Louisiana. In its aftermath, researchers studied the area, trying to understand what happened. What steps could Louisiana take to avoid a repeat of this flooding and damage in the future?

INFORMATION LITERACY EVALUATING INFORMATION

Woodlands are some of the most important communities in the United States. They are rich with wildlife, and we see them as places worth preserving. For over 70 years, Smokey the Bear has been telling us "Only YOU can prevent forest fires!" but the United States has had an active role in fighting and preventing forest fires for far longer. We began fighting wildfires in a systematic way around the turn of the 20th century; historic photographs show people building fire lines and shoveling dirt over smoldering spots in our woodlands. In the past 30 years, however, there have been increased discussions about the automatic response of immediately quelling all wildfires; some ecologists argue that some wildfires are helpful and should be allowed to burn. After all, change from storms, fires, and floods is part of the natural cycle of an ecosystem.

Your task is to learn about wildfires and decide for yourself what the best response should be.

1. Search the Internet for information about *fire ecology* to help develop an informed opinion about how we should respond to wildfires on wildlands. Topics to research include fire-adapted ecosystems, wildfire suppression, the cost of fighting wildfires, and prescribed fires. You may visit as many websites as you need, but you must visit at least four different websites for your research. Answer the following questions to evaluate whether each of the websites you visit is a reliable information source:
   a. Who are the authors of the information given in this article or on this webpage?
   b. Do the authors give supporting evidence for their claims?
   c. Do they give sources for their evidence? Are these reliable sources?
   d. Do you detect any strong biases for or against fighting wildfires? Explain.
   e. Summarize the position of this source regarding how we should respond to wildfires or what could be done to prevent wildfires.
   f. Based on your answers to the above questions, is this a reliable source of information? Explain.

2. Based on the information you obtained from the websites that you deemed to be reliable information sources, write an essay that addresses the following two questions. Provide evidence from your sources in support of your position.
   a. Should all wildfires on wildlands be fought immediately?
   b. What, if anything, should be done in an area to prevent or lessen a possible wildland wildfire?

Find an additional case study online at http://www.macmillanhighered.com/launchpad/saes2e

A TROPICAL MURDER MYSTERY

Finding the missing birds of Guam

CORE MESSAGE

The variety of life on Earth is a result of natural selection favoring the individuals within populations that are best able to survive in their particular environment. Given enough time, populations may be able to adapt to environmental changes, but extinction is also a natural part of this process, as less adapted populations are eliminated by better competitors or are lost due to natural disasters. Human activities can introduce changes so quickly that some populations of other species cannot adapt fast enough to survive and may go extinct.

AFTER READING THIS CHAPTER, YOU SHOULD BE ABLE TO ANSWER THE FOLLOWING **GUIDING QUESTIONS**

1

What is biological evolution, and how does natural selection allow populations to adapt to changes? Why is genetic diversity important to this process?

2

How did coevolution (or lack thereof) make most of Guam's bird species so vulnerable to the brown tree snake?

The brown tree snake
(*Boiga irregularis*). James Balog/
The Image Bank/Getty Images

3

How do random events influence the evolution of a population?

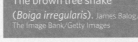

4

How do humans, intentionally or accidentally, affect the evolution of populations, and why do some scientists say that Earth is currently experiencing its "sixth mass extinction"?

5

What are some common misconceptions about evolution?

On a crisp December morning in 2013, representatives of several federal agencies met on Anderson Air Force Base in Guam—a South Pacific island and U.S. territory—to watch an experiment that sounded more like science fiction than science. As they looked on with binoculars, military personnel in a small fleet of helicopters dropped dead baby mice, one-by-one, into the surrounding jungle. The mice had been laced with acetaminophen and fitted with mini-parachutes. Some of them had also been implanted with tiny radio transmitters.

Though dead, the mice might still be thought of as paratroopers in a war that the island has been fighting for half a century, against a most elusive, yet devastating, enemy: brown tree snakes. The snakes are believed to have been accidently introduced from their native home on other Pacific islands to Guam through ships sometime back in the 1940s or 1950s; they have driven almost all of the island's native bird species to extinction, and it is feared that they will eventually sneak off the island through one of the base's aircraft, invade other islands, and devastate other bird populations in the region.

↓ U.S. Department of Agriculture wildlife specialist Tony Salas holds a brown tree snake at Anderson Air Force Base in Guam.

AP Photo/Eric Talmadge

As in any good military campaign, though, the scientist-generals in this war have studied their enemy well and are now making use of two of the tree snake's biggest weaknesses: their susceptibility to a common painkiller, acetaminophen, and their lack of pickiness at meal time. Acetaminophen is safe for human use (at the correct dose) but lethal to the snakes, who, unlike other species, are quite happy to eat prey that has already been killed.

The parachutes were employed to ensure that the mice would catch in the trees, where the snakes live and eat, rather than plummet to the ground, where they might be eaten by some other creature. The radio transmitters were meant to help the U.S. Department of Agriculture (USDA) scientists track the mice in order to determine whether they were eaten by the snakes and whether the snakes died as a result. It was a spectacular-looking feat to be sure. But desperate wars call for desperate measures, and no island has a tree-snake problem more desperate than Guam's.

"There really is no other place in the world with a snake problem like this," says Daniel Vice, assistant state director of the USDA's Wildlife Services in Hawaii, Guam, and the Pacific Islands. The December test drop was the fourth of its kind; mice were dropped over two 136-acre test plots. If it works—that is, if the snakes eat the mice, and the drug then kills the snakes—the project will be scaled up to cover more of the island. And if *that* works, operation mouse-drop will help address a problem that has plagued wildlife biologists for half a century and that for years was shrouded in mystery.

It was the late 1960s when the birds of Guam—some 18 native avian species that had long filled the forests with song—began dying off with disturbing speed. By the early 1980s, 4 species had gone extinct, and 10 others were in danger of joining them. Worst of all, researchers and wildlife experts had no clue why.

Back in the United States, in the winter of 1980, biologist Julie Savidge was gearing up to begin PhD work at the University of Illinois when she by chance attended a lecture in Redding, California, organized by the Wildlife

↑ Dr. Julie Savidge holding a Mariana Fruit-Dove. This species only occurs on certain islands within the Mariana Islands and the last sighting on Guam was in 1985. This bird was caught as part of an early blood sampling effort to see if exotic diseases might be causing the bird decline on Guam.

Society. A biologist from Guam had been invited to talk about the island's bird disappearances, and he noted—rather grimly—that no one had yet solved the devastating mystery. Savidge was immediately enthralled. "I went up to him afterwards and said, 'This sounds fascinating. Is there any way that I might be able to get out to Guam?'" Savidge recalls.

Two summers later, after much back-and-forth between Savidge and the Guam Division of Aquatic and Wildlife Resources, Savidge was hired to investigate the bird disappearances as part of her PhD dissertation.

At the time, Guam's scientists were convinced that diseases like blood parasites or pesticides were to blame for the avian losses. So Savidge began looking into those possibilities. But as soon as she started talking to Guam locals, she became skeptical of the idea. "When I would interview the natives, I'd be asking all these disease questions, and they'd say, 'Why are you asking about this? The real problem is the snakes,'" she explains.

The locals were certain that brown tree snakes (*Boiga irregularis*), 1- to 2-meter-long (3- to 6-foot-long), non-native snakes that had been accidentally introduced to the islands in military shipping cargo in the late 1940s and 1950s, were responsible for the birds' demise. Non-native species that cause ecological, economic, or human health problems and are hard to eradicate are considered **invasive species**, and they can cause significant damage in areas they invade. In fact, invasive species are the second leading cause of species endangerment, worldwide (see Chapter 13).

Oddly, Savidge's research was starting to show that, in fact, diseases weren't playing a factor in the bird deaths at all. When she and her colleagues sampled a variety of birds for bacteria, viruses, and parasites between 1982 and 1985, they found that the native birds were basically healthy. Curious about the tree snake hypothesis (an idea other scientists had dismissed), Savidge began to investigate, on her own time, whether these reptiles might be causing the **extinctions**. Surely it was something else; could a few snakes really obliterate a whole island's worth of birds?

Natural selection is the main mechanism by which populations adapt and evolve.

Before they started disappearing, the birds of Guam were a diverse and resplendent bunch. The island, about one-fifth the size of Rhode Island, was home to 18 native species of birds, each specially suited to life on the island.

Populations usually contain individuals that are genetically different from one another. According to the evolutionary theory first put forth by Charles Darwin and Alfred Russel Wallace, and subsequently supported by a

invasive species A non-native species (a species outside its range) whose introduction causes or is likely to cause economic or environmental harm or harm to human health.

extinction The complete loss of a species from an area; may be local (gone from an area) or global (gone for good).

◉ **WHERE IS GUAM?**

tremendous amount of evidence from a wide variety of scientific disciplines, **selective pressure** on a population—a nonrandom influence that affects who survives or reproduces—favors individuals with certain inherited traits over others (such as better camouflage, tolerance for drought, or enhanced sense of smell). These individuals have *differential reproductive success* compared to other individuals: They are best suited for their environment and leave more offspring than those who are less suited for their environment.

The traits that an environment favors are called **adaptations**, and the process by which organisms best adapted to the environment survive to pass on their traits is known as **natural selection**. Evolutionary biology helps us understand the diversity of life on Earth and how populations change over time. It is one of the pillars of biological science, and the vast amount of evidence in support of both the occurrence of evolution and the mechanisms by which it happens has elevated this explanation to the level of scientific theory (see Chapter 2).

For most populations, more offspring are born than can survive, since resources are limited and many species produce large numbers of young. Since only some individuals will survive, over time, the population will contain more and more of these better-adapted individuals and their offspring. Ultimately, this changes how common certain variants of **genes** (these variants are called *alleles*) are in the population: The frequency (percentage in the population) of some genes increases and that of others decreases. When this occurs, the population has experienced **evolution**, or changes in the **gene frequencies** within a population from one generation to the next. Natural selection may be *stabilizing*, *directional*, or *disruptive*, depending on which genetic traits are favored or selected against.
INFOGRAPHIC 11.1

It is important to note that *individuals* are selected for, but *populations* evolve; individuals do not change their own genetic makeup to produce new necessary adaptations, such as bigger size or pesticide resistance. If they get the opportunity to reproduce, they pass on their traits to the next generation. If they cannot tolerate environmental changes, as was the case with the first bird species to disappear from Guam (the bridled white-eye), they die or fail to reproduce and do not pass on their genes. Individuals may be able to adjust their behavior to accommodate environmental changes, but if a trait is not genetically controlled, and therefore is not heritable, it will not influence the composition of the next generation.

Populations need genetic diversity to evolve.

The ability of a population to adapt is a reflection of its tolerance limits, which largely depend on **genetic diversity**—different individuals having different alleles. A population that is highly diverse (has individuals with many different traits) is likely to have wider tolerance limits (see Chapter 8), which increases the population's potential to adapt to changes. This means it is more likely that some individuals will exist that can withstand (or even thrive in) the changes and that the population as a whole will survive. If a change occurs that produces a condition outside the range where individuals can survive and reproduce (for instance, the climate becomes too warm or too dry), the population will die out. Similarly, if a new challenge is presented, such as the introduction of a new predator or competitor, the survival of the population will depend on whether there are any individuals in the population who can effectively deal with the new species. If the snakes on Guam were indeed responsible for killing off the birds, any birds that happened to have effective snake-avoidance behaviors would have had a greater chance of survival.

Two main sources of variation can increase genetic diversity in a population. The ultimate

---

selective pressure A nonrandom influence that affects who survives or reproduces.

adaptation A trait that helps an individual survive or reproduce.

natural selection The process by which organisms best adapted to the environment (the fittest) survive to reproduce, leaving more offspring than less well-adapted individuals.

genes Stretches of DNA, the hereditary material of cells, that each direct the production of a particular protein and influence an individual's traits.

evolution Differences in the gene frequencies within a population from one generation to the next.

gene frequencies The assortment and abundance of particular variants of genes relative to each other within a population.

genetic diversity The heritable variation among individuals of a single population or within a species as a whole.

KEY CONCEPT 11.1

Natural selection is a mechanism by which a population can adapt to changes in its environment. Individuals that are best suited for their environment due to inherited traits leave more offspring (which also possess these traits) than other, less-suited individuals, producing subsequent generations that contain more individuals with the useful traits.

KEY CONCEPT 11.2

Individuals do not evolve; populations do. A population is said to have evolved if the assortment and frequency of genes in a descendant population are different from those in its ancestral population.

INFOGRAPHIC 11.1   NATURAL SELECTION AT WORK

↓ When the environment presents a selective force (e.g., a new predator, changing temperatures, change in food supply), natural selection is the primary force by which populations adapt. The survivors are those who were lucky enough to have genetic traits that allowed them to survive in their changing environment. (Others who did not possess the trait were not as likely to survive to reproduce.) Because survivors pass on those adaptations to their offspring, the gene frequencies of the population change in the next generation, which means some traits are more common and others are less common than they used to be. When this happens, the population is said to have evolved.

| Original population | | Next generation | Later generation |

  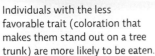 

Beetles resting on this tree vary in color.

Individuals with the less favorable trait (coloration that makes them stand out on a tree trunk) are more likely to be eaten.

Fewer dark individuals are born (though recombination might produce some from light or tan parents).

Over time, the population may be mostly or solely made up of tan individuals.

Genetic variation exists in the population: Individuals possess inherited differences.

Differential reproductive success: Not everyone will survive to reproduce.

Gene frequencies have changed: The population is evolving.

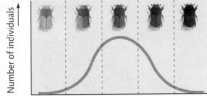

Number of individuals

Population of beetles of different colors

← Natural selection can have different outcomes, depending on what varieties in the population the environment favors or selects against.

—— Original population
—— Evolved population

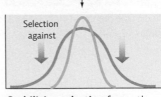 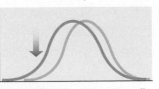 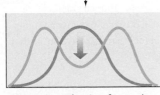

Selection against

Stabilizing selection favors the norm and selects against extremes.

Directional selection continually favors a particular extreme of the trait (bigger, darker, etc.).

Disruptive selection favors the extremes but selects against the intermediate forms.

 All trees are tan; tan beetles are favored.

 In areas with trees darkened from pollution, darker beetles are favored.

 In a forest with light and dark trees but no tan trees, tan beetles (the intermediate color) are not favored.

 Identify the gene frequencies of the "original population" for each color morph (dark gray, dark brown, light brown, dark tan, and light tan) by counting the number of each and expressing it as a percentage of the whole. Now do the same for the "later generation." Has evolution occurred? Explain.

KEY CONCEPT 11.3

Genetic diversity in a population is the raw material on which natural selection operates. The more diverse a population, the more likely there will be individuals present who can withstand or even thrive if environmental conditions change.

source of new variability is genetic *mutation*, a change in the DNA sequence in the sex cells that alters a gene, sometimes to the extent that it produces a new protein and, possibly, a new trait. Mutations are rare, but because DNA replication and repair occurs all the time, rare events do happen. When a mutation produces traits that are beneficial, they can quickly be passed on to the next generation, allowing the population to evolve to be better adapted to its environment. A second source of genetic variety occurs as eggs and sperm are made: *Genetic recombination* shuffles alleles around and sometimes produces individuals with new traits when a sperm fertilizes an egg.

The value of this genetic diversity is illustrated today in the example of the rock pocket mouse of the American

Southwest desert. Animals of this species have coats that are either light tan or a darker color. It turns out that coat color corresponds to a population's environment: Areas of light-colored rock contain populations with mostly tan mice, whereas darker mice inhabit black lava rock regions. Research by Hopi Hoekstra and colleagues at the University of Arizona has shown that coat color is determined by a single gene that comes in two different alleles. The dominant allele is designated by the uppercase letter D; the recessive allele is designated by the lowercase letter d. All individuals have two copies of the gene, and the color of their coat is determined by which two alleles they possess. Darker mice have at least one dominant allele (DD or Dd). Tan mice possess two recessive alleles (dd).

It is likely that coat color provides camouflage and protection from visual hunters, but only if the mouse is on a background of the same color. A study on deer mice (a similar species) showed that predatory owls are more successful at capturing mice on a contrasting background. This gives support to the conclusion that coat color is adaptive as camouflage and therefore is responsive to natural selection. **INFOGRAPHIC 11.2**

INFOGRAPHIC 11.2 EVOLUTION IN ACTION

1

↓ Natural selection produces populations with different gene frequencies (more or less of a particular gene variant or allele). For this to occur, there must be genetic variation (more than one allele for a given trait) and a selective pressure (a reason one variant is better than another in a given situation).

Different color morphs of the rock pocket mouse (*Chaetodipus intermedius*) are found on different-colored rocky outcroppings in the desert Southwest. An evaluation of the mice living on or near the Pinacate lava flow in southern Arizona represents the first documentation of the genetic basis (in this case, a single gene) for a naturally favored trait. The well-known peppered moth is another example in which different color variants are favored in different habitats but the genes responsible for that trait have not yet been identified.

Even though there is gene flow between dark and light populations that are close to one another, populations on tan rock have mostly tan individuals, and populations on dark rock have mostly dark individuals, suggesting a strong selective pressure that favors one color over the other.

Predatory owls are likely the selective pressure that favors different coat colors in different habitats.

Tan mice (the recessive trait, dd) predominate in light-colored rocky outcroppings.

Darker mice (the dominant trait, DD or Dd) predominate on darker lava rocks.

Gene flow

KEY CONCEPT 11.4

When two species each become the selective pressure that favors certain traits in each other, coevolution is occurring. Coevolution can produce species that are highly adapted to one another.

A special type of natural selection is known as **coevolution**. In coevolution, two species each provide the selective pressure that determines which of the other's traits is favored by natural selection. Predator and prey species usually evolve together, each exerting selective pressures that shape the other. As predators get better at catching prey, the only prey to survive are those a little better at escaping, and it is those individuals that reproduce and populate the next generation. This game of one-upmanship continues generation after generation, with each species affecting the differential survival and reproductive success of the other. The result can be a predator extremely well equipped to capture prey and prey extremely well equipped to escape. **INFOGRAPHIC 11.3**

If the birds on Guam were indeed eradicated by the invasive snake species, it was because the speed at which the eradication happened prevented the bird populations from potentially coevolving survival strategies to deal with the new snake population. The brown tree snake was already well adapted to preying on birds. But Guam's bird populations had never faced such a predator and had no natural defenses. It was an unfair fight. In fact, that the birds on Guam disappeared so quickly made it difficult to tease out the cause of their extinction at all. Savidge had to work backward to solve the mystery, putting the missing pieces together experiment by experiment. Her first step was to compare whether the distribution of the snakes on Guam matched up with the areas where birds had disappeared.

Based on what the residents told her, and what historical records like newspaper clippings reported, Savidge found that birds had begun to disappear first from southern Guam and that their disappearances matched up perfectly with when the brown tree snakes had begun populating

coevolution A special type of natural selection in which two species each provide the selective pressure that determines which traits are favored by natural selection in the other.

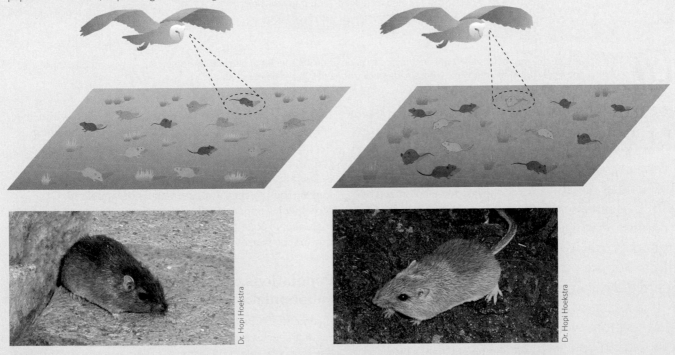

↓ A study done with dark and light varieties of deer mice revealed that owls caught twice as many opposite-colored mice (dark mice on a light background or tan mice on a dark background) as mice whose coloration matched their background, even in almost total darkness. Owl predation is therefore likely to be a strong selective pressure on coat color, driving directional selection that produces either light or dark populations of mice, depending on the background.

Dr. Hopi Hoekstra

Dr. Hopi Hoekstra

 If this mouse population migrated to an area with a red rock habitat with visual predators like hawks or owls, what could prevent the population from evolving into one with red coats?

INFOGRAPHIC 11.3   COEVOLUTION ALLOWS POPULATIONS TO ADAPT TO EACH OTHER   2

↓ As selection favored beetles closest to the tree color, only birds with the keenest eyesight feed well enough to survive and reproduce.

↓ Any beetle with an even better camouflage would escape predation and pass on its genes.

↓ This then favors birds with even keener eyesight who would feed well and pass on the sharp eyesight trait to their offspring.

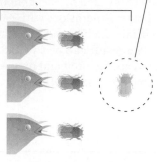

In this population, a few beetles have such good camouflage that they escape predation. These beetles are one step ahead of the birds. If a bird emerges that can see them, the birds will be one up on the beetles.

Individuals with better vision catch more beetles.

Most birds compete for the easy-to-see beetles, but any bird that can find the hidden ones will find more food—and thus selection will favor birds with even better eyesight.

No individual bird is present (yet) who can detect the new, camouflaged beetle variant.

? What do you predict would happen to the original beetle population if a species of bird with extremely keen eyesight (like those shown on the right side of the diagram) were accidently introduced into the beetle's habitat?

KEY CONCEPT 11.5

Because the brown tree snake coevolved with birds in its native habitat, it was very good at preying on birds. But the birds of Guam did not coevolve with a snake predator and therefore were not equipped to deal with a predatory snake.

the area. Ritidian, an area on the very northernmost tip of Guam, on the other hand, was the last area to have lost its birds, and it was also the last to be colonized by the snakes. "I found a close correlation between the bird decline and the expansion of brown tree snakes around the island," she says, which suggested to her that brown tree snakes really might be the culprit. **INFOGRAPHIC 11.4**

Some of Guam's bird species went extinct sooner

than others. For instance, the **endemic** bridled white-eye, the gregarious bird species that was extinguished first, happens to be very small, raising the possibility that the small size of these birds might have put them at a disadvantage. Larger species like flycatchers survived longer, though they, too, eventually disappeared. Other bird species experienced **extirpation**; the Guam rail, for instance, is gone from Guam, but other populations still live on the nearby island of Rota.

Populations can diverge into subpopulations or new species.

If some individuals of the bridled white-eye or other extinct species had been able to avoid the snake (perhaps due to a heritable trait that made them more wary of the predator), they might have given rise to new populations

INFOGRAPHIC 11.4   ENDANGERED AND EXTINCT BIRDS OF GUAM

2

↓ Of the 18 original native species on Guam, many are endangered or already extinct.

THE SPREAD OF THE SNAKE MAPPED WITH BIRD EXTINCTIONS OVER TIME

James Balog/The Image Bank/Getty Images

Brown tree snake
Boiga irregularis
The snake's range expanded about 1.6 km (1 mi) per year

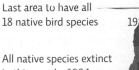

Last area to have all 18 native bird species — 1980s

All native species extinct in this area by 1984 — 1970s

GUAM

1960s

1950s

Bridled white-eyes and Guam flycatchers were last seen at this checkpoint in 1964.

Cocos Island —
Boiga is absent from Cocos Island; all bird species are present.

 ↑ The loss of birds followed the spread of *Boiga* as predicted. The last area to have all 10 species of native birds was the last area *Boiga* invaded.

Extinct
Bridled white-eye
The endemic Guam bridled white-eye was last seen in 1983 and is now extinct.

Extinct
Rufous fantail
Once common all over Guam, the endemic subspecies of the Rufous fantail is now extinct.

Extinct
Guam flycatcher
The endemic Guam flycatcher was last seen on Guam in 1985 and is now extinct.

Only Captive
Guam kingfisher
By 1986, only seven Guam kingfishers existed in the wild; a captive breeding program is under way.

Only Captive
Guam rail
The endemic Guam rail is extinct in the wild; a captive breeding program is under way.

In other islands
Mariana fruit dove
The Mariana fruit dove is extirpated on Guam but is still found on other islands.

H. Douglas Pratt, North Carolina State Museum of Natural Sciences

? Look at the progressive movement of the snake through Guam and the extinction of birds over the same time frame. Is this cause-and-effect evidence that the snake exterminated the birds, or is it correlational evidence? Explain.

that could cohabitate with the snake. Or, had they not been surrounded by sea, the Guam birds may have been able to relocate and find a safe haven from the brown tree snake.

When populations diverge because of isolation, food availability, new predators, or *habitat fragmentation* such that their members can no longer freely interbreed, new species may arise (*speciation*). This increases species diversity in a community and sometimes produces specialists that can exploit open niches. This separation may be physical (for

example, geographic boundaries the individuals won't cross) or may arise when something prevents some individuals from choosing others as mates, as may happen when individuals spend their time in different parts of their habitat.

However, not all evolution is driven in this manner. Random

endemic Describes a species that is native to a particular area and is not naturally found elsewhere.

extirpation Local extinction in one or more areas, though some individuals exist in other areas.

events play a role, too, typically by decreasing genetic diversity rather than increasing it.

In any population, some individuals will survive and mate, and others won't. For no other reason than pure chance, this random mating may increase or decrease the frequency of a particular trait, a process known as **genetic drift**. The loss of this trait doesn't have anything to do with any inherent advantage or disadvantage the trait offers; it is merely the chance loss of some gene variants because the individuals with these genes did not happen to mate. This can produce significant changes, especially in small populations where traits that may be present in just a few could be lost in a single generation if these individuals don't happen to reproduce.

Genetic drift can produce a new population that is different from the original population when only a subset of the original variants reproduce. The **bottleneck effect** occurs when a portion of the population dies, perhaps because of a natural disaster like a flood or because of a strong new selective pressure like the introduction of a new predator. The survivors then produce a new generation, and any alleles found only in the deceased individuals are lost from the population forever.

The **founder effect** is seen when a small group that contains only some of the gene variants found in the original population becomes physically isolated from the rest. Even if these individuals do not have the most adaptive traits, natural selection takes over, favoring the best adaptations in the founding group, and producing a subsequent population that is likely to be different from the original population. (See Chapter 12 for more on the effect of isolation.)

---

genetic drift The change in gene frequencies of a population over time due to random mating that results in the loss of some gene variants.

bottleneck effect The situation that occurs when population size is drastically reduced, leading to the loss of some genetic variants, and resulting in a less diverse population.

founder effect The situation that occurs when a small group with only a subset of the larger population's genetic diversity becomes isolated and evolves into a different population, missing some of the traits of the original.

endangered Describes a species that faces a very high risk of extinction in the immediate future.

Today, human impact increases instances of both the founder effect and the bottleneck effect. Much of what we do isolates populations into smaller groups, forcing them into these situations. **INFOGRAPHIC 11.5**

The pace of evolution is generally slow but is responsive to selective pressures.

In addition to genetic diversity, the size of the population also makes a difference in how quickly natural selection can produce a change in a population: Beneficial traits can spread more quickly in smaller populations simply because it is more likely that the individuals with the trait will find each other and mate (as long as it is not a population that is widely dispersed). Reproductive rate and generation time also influence how quickly a population can adapt to changes. Many problem species, like insect pests, are r-selected species and have fast generation times, which means they can often stay one step ahead of our efforts to control them. Many **endangered** species, on the other hand, are K-selected species, with longer generation times; therefore, they take longer to recover if population numbers fall. (For more on r- and K-selected species, see Chapter 9.)

The strength of the selective pressure also affects how quickly natural selection might produce a change in a population. One of the reasons the demise of birds in Guam was so stupefying was that it happened so quickly—particularly for the small birds, which were easiest for the snakes to eat. Larger birds disappeared later, when the snakes started eating their nestlings and eggs. While speciation can take thousands or millions of years, extinction can occur much more quickly if the rate of change exceeds the ability of the population to adapt.

To strengthen the case that the brown tree snake was indeed at the root of the bird extinction in Guam, Savidge had to perform a few more experiments. She needed to be sure, of course, that brown tree snakes actually *liked* eating birds. To do so, she set bird-baited traps all around the islands for the snakes and waited. "I put them in half a dozen locations throughout the island," she recalls. What she found shocked her: "In one area where the birds were extinct, 75% of my traps got hit within 4 nights," she recalls. On the other hand, in the baited traps she had set on Cocos Island, an island off the coast of Guam that is not populated by the snakes, all the birds survived.

Savidge knew, however, that birds weren't brown tree snakes' only prey. The reptiles also ate small mammals, so she checked to see whether these animals were also being adversely affected by the snakes' presence. In the 1960s, before the tree snakes had arrived, scientists on Guam had done a survey of small mammal density and had found, on average, 40 small mammals per hectare (2.5 acres) of land on the island. When Savidge did the same thing in the mid-1980s, she found only 2.8 animals per hectare. "My prediction was that I would see a decline in these

INFOGRAPHIC 11.5   **RANDOM EVENTS CAN ALTER POPULATIONS**

GENETIC DRIFT

↓ Random mating can eliminate some gene variants from the population not because these individuals were poorly adapted to their environment or less attractive to the opposite sex but because they were just unlucky and didn't mate.

 How is the bottleneck effect similar to and different from the founder effect?

A population contains a variety of individuals, but some gene variants are more common than others.

Random mating occurs, but some unlucky individuals don't find mates.

Subsequent generations may have different gene frequencies.

BOTTLENECK

↓ If something causes a large part of the population to die, leaving the survivors with only a portion of the original genetic diversity, the population may recover in size but will not be as genetically diverse as the original population.

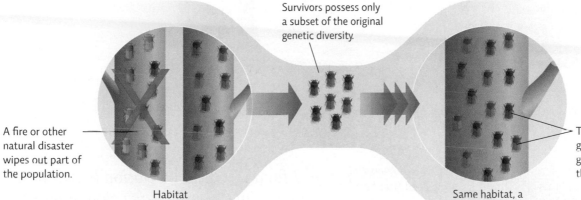

Survivors possess only a subset of the original genetic diversity.

A fire or other natural disaster wipes out part of the population.

The next generation is less genetically diverse than the original.

Habitat

Same habitat, a generation after the disaster

FOUNDER EFFECT

↓ If only a subset of a population colonizes a new area, and the subset becomes completely isolated from the original group, the new population will be less genetically diverse than the original.

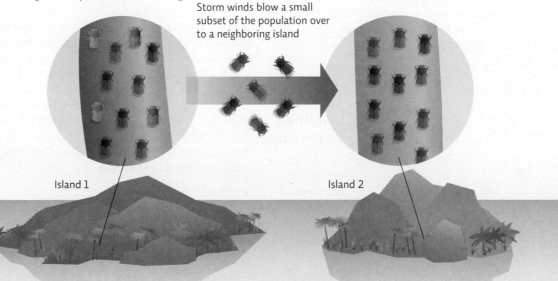

Storm winds blow a small subset of the population over to a neighboring island

Island 1

Island 2

JOEL SARTORE/National Geographic Creative

↑ A Guam rail (*Rallus owstoni*), a critically endangered species. The species' decline was caused by predation by the introduced brown tree snake. It is currently considered extinct in the wild, though reintroduction programs are under way.

rats and mice and shrews, and indeed, it was like a 94% decline," she explains. The findings suggested that brown tree snakes were devastating small mammal populations in addition to the birds. All in all, Savidge's three pieces of evidence—the fact that the geographic location of the snakes correlated strongly with the birds' disappearance, that brown tree snakes liked to eat birds, and that other small mammals also went missing after the snakes' arrival—convinced Savidge that she had finally solved the mystery of Guam's disappearing birds. Brown tree snakes, she concluded, were definitely the culprit.

background rate of extinction The average rate of extinction that occurred before the appearance of humans or that occurs between mass extinction events.

fossil record The total collection of fossils (remains, impressions, traces of ancient organisms) found on Earth.

threatened Describes a species that is at risk for extinction.

Extinction is normal, but the rate at which it is currently occurring appears to be increasing.

Extinction is nothing new on Earth. It is as constant and as common as evolution; by most estimates, more than 99% of all species that ever lived on the planet have gone extinct. Based on a critical analysis of

" *I found a close correlation between the bird decline and the expansion of brown tree snakes around the island.* "
—Julie Savidge.

the fossil record, scientists agree that there have been five *major extinction events*—when species have gone extinct at much greater rates than during intervening times, each event leading to the loss of 50% or more of the species present on Earth. The most infamous of these was the *K-T boundary mass extinction*, which occurred at the transition from the Cretaceous to the Tertiary period, 65 million years ago. Most scientists agree the K-T extinction event was set off by an asteroid impact in the Gulf of Mexico; 70% of all living species, including the dinosaurs, were wiped out. **INFOGRAPHIC 11.6**

But these kinds of events often lead to the emergence of new species, as other populations adapt to the open niches that are left. Cycles of extinction and evolution ultimately gave rise to the diversity of life we see on Earth today—estimates range from 3 to 100 million

INFOGRAPHIC 11.6   EARTH'S MASS EXTINCTIONS

4

↓ There have been five mass extinctions in Earth's history. Many scientists believe the high rate of extinction and endangerment seen today is leading to a sixth mass extinction.

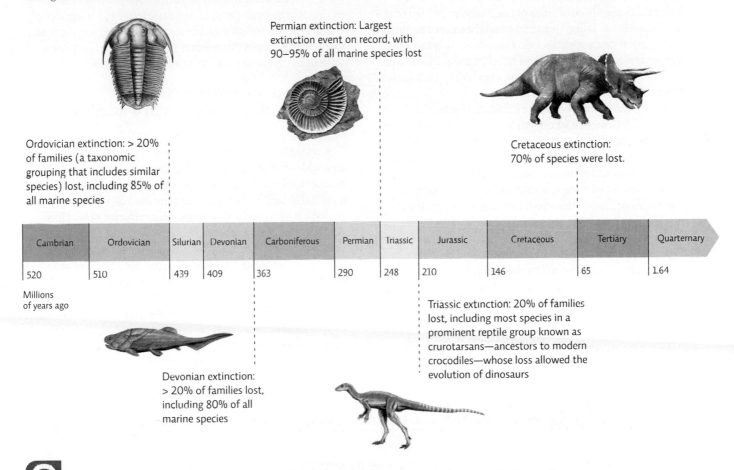

Permian extinction: Largest extinction event on record, with 90–95% of all marine species lost

Ordovician extinction: > 20% of families (a taxonomic grouping that includes similar species) lost, including 85% of all marine species

Cretaceous extinction: 70% of species were lost.

Cambrian	Ordovician	Silurian	Devonian	Carboniferous	Permian	Triassic	Jurassic	Cretaceous	Tertiary	Quarternary
520	510	439	409	363	290	248	210	146	65	1.64

Millions of years ago

Triassic extinction: 20% of families lost, including most species in a prominent reptile group known as crurotarsans—ancestors to modern crocodiles—whose loss allowed the evolution of dinosaurs

Devonian extinction: > 20% of families lost, including 80% of all marine species

? Why do we see mass extinction events occurring at transitions from one geologic time period to the next?

species. Throughout most of time, the **background rate of extinction**—the average rate of extinction that occurs between mass extinction events—has been slow. The **fossil record** tells us that, on average, 1 species out of every 1 million species goes extinct each year. In a world with 3 million species, this would be 3 species per year; if Earth is home to 100 million species, that would be 100 species per year.

Today, most scientists agree that swelling human populations have triggered a sixth major extinction event, one that we are witnessing right now. Plant and animal extinction rates are currently greater than the background rate of extinction. For example, British researchers Ian Owens of the Imperial College of London and Peter Bennett of the University of Kent analyzed the extinction risk for 1,012 **threatened** bird species and found that *habitat destruction* was cited as a risk factor in 70% of the cases, and other human interventions, such as the introduction of non-native

species or *overharvesting* were implicated in 35% of the cases. In some areas with high species diversity, such as a tropical rain forest, the extinction rate can be quite high: One estimate puts it as high as 27,000 extinctions per year in tropical rain forests that are being cut down. A look at the historic mammalian fossil record reveals that, on average, 1 mammal species has become extinct every 200 years, but in the past 400 years, we've documented 89 mammalian species extinctions—which is almost 45 times faster than the background rate.

KEY CONCEPT 11.7

Historically, species have gone extinct when they were replaced by better-adapted species; natural disasters have also caused mass extinctions. Today, human impact appears to be causing another mass extinction.

Estimates of just how rapidly we are losing species vary, in part because we don't know how many species exist and because it is hard to verify that a species actually is extinct. In an often-cited 1995 article published in the journal *Science*, Stuart Pimm of the Nicholas School of the Environment at Duke University and colleagues estimated that current rates of extinction range from 100 to 1,000 times greater than background rates. In 2014, Pimm and colleagues narrowed this window, concluding that current rates are 1,000 times greater than background. If all species currently threatened become extinct, this would raise the high-end estimate to 10,000 times greater. A 2011 report in *Nature* suggests that current methods overestimate extinction rates by as much as 160% but points out that even the low-end estimates are a cause for concern. And while scientists debate whether species extinctions are 100 times or 1,000 times faster than normal worldwide, it may be more useful to evaluate threats at a local level, like in Guam, and focus efforts on reducing species loss there. Especially on a small, isolated island, a rate that is even 10 times higher than normal is significant.

---

artificial selection A process in which humans decide which individuals breed and which do not in an attempt to produce a population with desired traits.

What we do know is that today's accelerated extinction is largely driven by human actions (see Chapters 12 and 13). As the human population increases, our impact is becoming much more devastating for other species. We remove the resources they need to survive, minimize their habitat ranges, introduce new predators or competitors, and strip them of their genetic diversity, all of which slowly eliminate them. In Guam, the near-total disappearance of birds between the 1960s and 1980s was a biological murder mystery of astounding proportions—one that illustrates just how vulnerable populations are to sudden changes.

Humans affect evolution in a number of ways.

The introduction of the brown tree snake to Guam was an accident: A snake hitchhiker crossed the ocean on a human tanker—unbeknownst to the crew—and landed in a veritable bird buffet. But humans also directly affect the evolution of a population through **artificial selection**. Artificial selection works the same way as natural selection but the difference is that the selective pressure is us. For many animal species, from pets to farm animals and plant species, humans choose who breeds with whom in an attempt to produce new individuals with the traits they desire. By doing this over many generations, people have accentuated certain plant and animal traits, sometimes to extremes. For instance, artificial selection created domestic dogs from their wolf ancestors. **INFOGRAPHIC 11.7**

INFOGRAPHIC 11.7   **HUMANS USE ARTIFICIAL SELECTION TO PRODUCE PLANTS OR ANIMALS WITH DESIRED TRAITS**   4

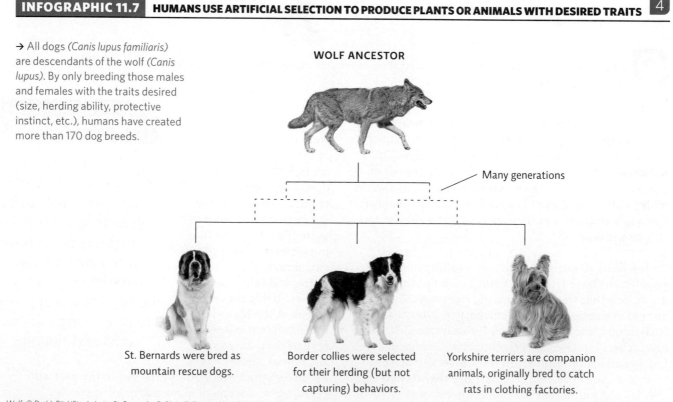

→ All dogs (*Canis lupus familiaris*) are descendants of the wolf (*Canis lupus*). By only breeding those males and females with the traits desired (size, herding ability, protective instinct, etc.), humans have created more than 170 dog breeds.

WOLF ANCESTOR

Many generations

St. Bernards were bred as mountain rescue dogs.

Border collies were selected for their herding (but not capturing) behaviors.

Yorkshire terriers are companion animals, originally bred to catch rats in clothing factories.

Wolf: © DaddyBit/ iStockphoto; St. Bernards: © GlobalP/Eric Isselée/ iStockphoto; Border collies: © GlobalP/Eric Isselée/ iStockphoto; Yorkshire terriers: © jimmyjamesbond/ iStockphoto

 Why can we say that artificial selection is goal directed but natural selection is not?

↑ Artificial selection has produced modern corn (maize) from teosinte, a grass native to Mexico. Early farmers began to domesticate corn about 10,000 years ago by planting only the teosinte plants that produced the largest and tastiest seed heads. These plants would cross pollinate to produce the next generation of plants. Artificial selection of the best plants continued and eventually produced the modern corn varieties with their large ears of corn that we are familiar with today.

KEY CONCEPT 11.8

Artificial selection works by the same mechanisms as natural selection except that humans choose which traits to keep and which to eliminate from a population through selective breeding. We can also inadvertently act as a selective pressure with any action that favors some traits over others, such as our use of pesticides or antibiotics that lead to resistant populations.

But evolution is ever at work. Pesticide- and antibiotic-resistant populations can emerge as an inadvertent human-influenced selection. When we apply a chemical that kills a pest or pathogen, some individuals survive because of their natural genetic resistance; that is, the individuals were already resistant even though they had never encountered the chemical. These survivors are then the only individuals who reproduce, producing the next generation that is also pesticide resistant, ultimately changing the frequency of resistant genes in the population (see Chapter 17).

Evolution can be a contentious subject for some people who feel it conflicts with other views they hold. Whether you are convinced by the physical, scientific evidence that evolution has occurred or that natural selection is a mechanism by which it occurs, it is important that

you base your criticisms of the theory on sound science and not on misconceptions of evolutionary theory. **TABLE 11.1**

Ultimately, by changing the environment rapidly, humans apply a number of new selective pressures on populations. Our changes have the capacity to be so great that natural selection simply cannot keep up: A new needed trait is not present in the population, or it cannot spread quickly enough to prevent a population collapse. The accidental introduction of the predatory brown tree snake is one such example; it introduced a major change, basically overnight, that was able to eat its way through the vertebrate populations before those populations could adapt.

Now, however, things are looking up for the birds of Guam. Thanks to efforts by the Guam Department of Agriculture and the USDA Wildlife Services, brown tree

TABLE 11.1   COMMON MISCONCEPTIONS ABOUT EVOLUTION   5

↓ The theory of evolution as a reasonable explanation for the diversity of life is contested by some people, but many of their arguments are based on misconceptions. Understanding the basic tenets of evolutionary theory is the first step in critically evaluating its validity.

Misconception	Evolutionary Explanation
The study of evolution seeks to explain the creation of life.	Evolution doesn't study creation or the origin of life. Evolution is the science that studies the "descent with modification" of life after it came into being.
Humans evolved from monkeys or apes.	Humans did not evolve from monkeys or apes. Humans, monkeys, and apes share a common ancestor, who lived 5 to 8 million years ago, but each group followed different evolutionary paths to their current forms.
Evolution is goal directed.	Evolution is not working toward a particular final version or trait. It is unpredictable because we cannot know which genetic variants will be available to natural selection, and we can't reliably predict selective pressures (environmental conditions and competition).
Evolution proceeds from simple to complex.	Simple is not always the ancestral condition. Parasites may be simple in design but many have evolved from nonparasitic, more complex forms.
Evolution produces perfect adaptations.	New traits are rarely "perfect." Species have a long history of ancestral adaptations, each giving way to the next. Natural selection works with whatever is available—it doesn't scrap the anatomy of an "arm" and reform a wing for a bat. It modifies existing structures for new functions. In addition, adaptations are often compromises between differing needs. For example, the flippers of a seal are better suited for swimming but make walking on land awkward.
Evolution is just a theory, and many scientists don't accept it.	The claim that evolution is "just a theory" reveals a misunderstanding of the concept of a scientific theory. "Theories" represent our most well-accepted explanations in science; to attain the status of theory, an explanation must have substantial evidence, ideally from multiple lines of evidence. Multiple polls show that well over 95% of scientists accept evolutionary theory as the best explanation for the diversity of life and how populations adapt to their environments.

? Some people reject evolution as a reasonable explanation for the diversity and ancestry of life on Earth, especially regarding humans. What reasons are most often offered for this rejection? Are these reasons scientific? Do they have to be?

KEY CONCEPT 11.9

There are many misconceptions about evolution among the general public, especially for those who feel it conflicts with personal beliefs. Understanding what science actually says evolution is (and is not) is an important first step to critically evaluate the soundness of evolutionary theory.

snakes are being controlled to allow the remaining bird populations to recover.

"We are using traps in northern Guam to control snakes in areas where we would like to reintroduce native species," explains Diane Vice, a wildlife biologist with the Guam Department of Agriculture. Her organization is also working to prevent the snakes from infiltrating Cocos Island, the atoll off of southern Guam, and an important haven for nesting sea birds, Micronesian starlings, and sea turtles. Scientists are also close to perfecting the mouse-drop strategy.

A follow-up study done by the Agriculture Department indicates that in earlier tests, the snakes did devour "dead neonatal mice adulterated with 80mg

acetaminophen" and that "baiting" with laced mice was effective in killing the snakes. The study also suggests that the impact on other species will be minimal; for one thing, the snakes have already wiped out the birds that might have also fallen prey to the dead mice. "One concern was that crows may eat mice with the toxicant," says William Pitt, of the U.S. Wildlife Research Center's Hawaii Field Station. "However there are no longer wild crows on Guam."

Vice says the goal is not to eradicate the snakes (which probably isn't possible) but to control and contain them so that they don't find their way to other islands, as they found their way to Guam. "The hope is to create safe habitat," Vice says, so that these beautiful native species can once again thrive.

Select References:

Clark, L., et al. (2012). Efficacy, effort, and cost comparisons of trapping and acetaminophen-baiting for control of brown treesnakes on Guam. *Human–Wildlife Interactions*, 6(2): 222–236. www.aphis.usda.gov/wildlife_damage/nwrc/publications/12pubs/clark122.pdf.

He, F., & S. P. Hubbell. (2011). Species-area relationships always overestimate extinction rates from habitat loss. *Nature*, 473(1747): 368–371.

Hockstra, H. E., J. G. Krenz, & M. W. Nachman. (2005). Local adaptation in the rock pocket mouse (*Chaetopidus intermedius*): Natural selection and phylogenetic history of populations. *Heredity*, 94(2): 217–228.

Owens, I. P. F, & P. Bennett. (2000). Ecological basis of extinction risk in birds: Habitat loss versus human persecution and introduced predators. *Proceedings of the National Academy of Sciences*, 97(22): 12144–12148.

Pimm, S., et al. (1995). The future of biodiversity. *Science*, 269(5222): 347–350.

Pimm, S., et al. (2014). The biodiversity of species and their rates of extinction, distribution, and protection. *Science*, 344(6187):1246752.

Savidge, J. A. (1987). Extinction of an island forest avifauna by an introduced snake. *Ecology*, 68(3): 660–668.

Vice, D. S., et al. (2005). A comparison of three trap designs for capturing brown treesnakes on Guam. *USDA National Wildlife Research Center*, Staff Publications, Paper 43, http://digitalcommons.unl.edu/cgi/viewcontent.cgi?article=1041&context=icwdm_usdanwrc.

BRING IT HOME

PERSONAL CHOICES THAT HELP

The astonishing variety of life found on Earth is a result of natural selection favoring those individuals within populations that are best able to survive in their particular environment. Given enough time, some populations may be able to adapt to environmental changes. However, human activities may disrupt natural ecosystems so that organisms cannot adapt fast enough to survive, and those organisms may go extinct. Conservation activities can help protect vulnerable organisms and ecosystems.

Individual Steps

• Live in an older, established area of your community. Suburban sprawl reduces habitat for wildlife, and reliance on cars causes greenhouse gas emissions that could result in species-threatening climate change.

• Save your pocket change and at the end of every year donate the money to a land, marine, or wildlife protection agency.

• Create a personal blog that includes photographs of wildlife, facts about current threats to plants and animals, and articles about conservation.

Group Action

• Throw a party in support of wildlife conservation. Take a collection at the door and donate the money to an organization that supports conservation.

Policy Change

• "Adopt an Organism." The U.S. Fish and Wildlife Service maintains a database of endangered plants and animals in every state. Research this database to find wildlife that interests you. Determine what agency, conservation group, or legislator you could contact and then start your own protection campaign. See what meetings, petitions, and legislation could impact your organism and get involved.

JOEL SARTORE/National Geographic Creative

ENVIRONMENTAL LITERACY UNDERSTANDING THE ISSUE

1 What is biological evolution, and how does natural selection allow populations to adapt to changes? Why is genetic diversity important to this process?

INFOGRAPHICS 11.1 AND 11.2

1. Evolution is defined as a change in _____ _____ in a population over time.

2. A population of butterflies used to have small, medium, and large individuals, but several years ago a non-native bird was introduced to the butterfly's habitat. It eats butterflies but only those that are medium size. Eventually, medium-sized butterflies became rare. This is an example of:
   a. disruptive selection.
   b. stabilizing selection.
   c. artificial selection.
   d. directional selection.

3. "When exposed to pesticides, some Japanese beetles become more pesticide resistant and harder to kill." There is something wrong with this statement. Reword it to more accurately represent what happens when pesticide application leads to a pesticide resistant population.

2 How did coevolution (or lack thereof) make most of Guam's bird species so vulnerable to the brown tree snake?

INFOGRAPHICS 11.3 AND 11.4

4. Which of the following is an example of coevolution?
   a. Polar bears and Arctic foxes both are white for camouflage on snow.
   b. Dolphins and whales have flippers that are similar in shape and function to the fins of a fish, allowing them both to swim efficiently.
   c. Moths that are preyed upon by bats can hear the ultrasonic sounds the bats use in hunting.
   d. Humans and chimpanzees share 98% of their DNA.

5. Suppose a native Guam bird species began to show evasive behaviors that allowed it to avoid the brown tree snake. Describe a potential coevolution scenario that might have allowed this adaptation to emerge in the population.

3 How do random events influence the evolution of a population?

INFOGRAPHIC 11.5

6. Ten thousand years ago, most members of Species A were killed by a series of volcanic eruptions. However, some members escaped; all modern members of Species A are descended from those 100 individuals. This is an example of:
   a. genetic drift.
   b. artificial selection.
   c. mass extinction.
   d. the bottleneck effect.

7. Why might a population, isolated by the founder effect, be more vulnerable to extinction than the original population from which they came?

4 How do humans, intentionally or accidentally, affect the evolution of populations, and why do some scientists say that Earth is currently experiencing its "sixth mass extinction"?

INFOGRAPHICS 11.6 AND 11.7

8. Artificial selection differs from natural selection in that:
   a. it cannot produce large changes over time, only minor ones.
   b. humans are the selective pressure, choosing which traits to favor.
   c. it does not depend on genetic variability in the population.
   d. it causes individuals to change their traits rather than causing a change in the population's gene frequencies.

9. Why do most scientists think that we are in the midst of a sixth mass extinction?
   a. The background extinction rate is approximately 20%.
   b. Most extinctions are occurring in the taiga and tundra.
   c. No new species are being discovered.
   d. Current extinction rates are greater than the background rate.

10. Why can't all species adjust to changes in their environment and avoid extinction?

11. Suppose you wanted to use artificial selection to produce a breed of hairless dogs for people who are allergic to dog hair. How would you go about doing this?

5 What are some common misconceptions about evolution?

TABLE 11.1

12. Which of the following is true of evolutionary processes?
    a. They are goal driven.
    b. New genes arise in response to environmental change.
    c. Evolution acts on existing genetic variation.
    d. Evolution proceeds from simple to complex.

13. Why do we not expect evolution by natural selection to produce perfect adaptations?

14. Why is it inappropriate to reject evolutionary explanations for the diversity of life because evolution is "just a theory?

SCIENCE LITERACY WORKING WITH DATA

The following graph depicts the relationship between numbers of extinctions and human population size since the 19th century.

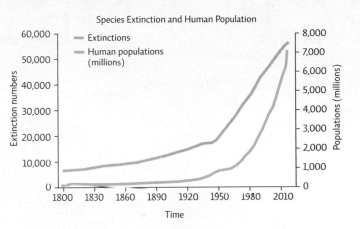

Species Extinction and Human Population

Interpretation

1. Describe what is happening to:
   a. the extinction rate over time.
   b. human population growth over time.

2. The two curves have been graphed together. What is the implication of presenting these data in this manner?

Advance Your Thinking

3. The y-axis is labeled "Extinction numbers." What taxonomic units are being measured? What if the taxonomic unit being evaluated here had been genus (a taxonomic group that can contain more than one species) or family (a taxonomic group that can contain more than one genus)? Would you be more concerned about the trend of extinctions or less? Explain.

4. What type of relationship is suggested by the figure: correlational or causal? What additional data would you like to see to support the graph's main point?

INFORMATION LITERACY EVALUATING INFORMATION

Life on Earth as we know it is a result of millions of years of evolutionary processes. These processes are not immune to changes in the environment; changes in evolutionary processes will ultimately result in changes in biodiversity, which will necessarily affect life on Earth, including humans.

Go to the website www.actionbioscience.org/evolution/myers_knoll.html and read the article "How Will the Sixth Extinction Affect Evolution of Species?" which discusses the effects of the sixth mass extinction on evolution. (If the URL above does not take you to the article, search for it by title at www.actionbioscience.org.)

Evaluate the website and work with the information to answer the following questions:

1. Is this a reliable information source? Does it have a clear and transparent agenda?
   a. Who runs this website? Does this person's/group's credentials make this source reliable/unreliable? Explain.
   b. Who are the authors? What are their credentials? Do they have the scientific background and expertise that lends credibility to the article?

2. In your own words, explain why the authors think that the current mass extinction will change evolutionary processes.
   a. What changes in particular do they think will impact future evolution?
   b. What types of evidence do they provide to support their arguments? Give specific examples.

3. The sixth mass extinction is attributed to human impact. Go to the International Union for Conservation of Nature (IUCN) "Species of the Day" website (www.iucnredlist.org/species-of-the-day/archives).
   a. Choose five species at random that are identified as vulnerable, endangered, or critically endangered. Read about each species and list the reasons for their endangerment.
   b. Create one master list of all the threats you encountered for your five species and categorize them as either *human caused* or *caused by natural events*. Which list is longer?
   c. Does human impact play a role in the endangerment of these species? What actions would be most useful to address these threats?

Find an additional case study online at http://www.macmillanhighered.com/launchpad/saes2e

PALM PLANET

Can we have tropical forests and our palm oil too?

CORE MESSAGE

The variety of life on Earth is tremendous. This biodiversity provides important ecological services to ecosystems; we depend on these services for things like food, medicine, and economic development. The decline of biodiversity has serious ramifications for other species as well as human well-being, so we should evaluate actions that threaten biodiversity and take steps to reduce the impact when possible.

AFTER READING THIS CHAPTER, YOU SHOULD BE ABLE TO ANSWER THE FOLLOWING **GUIDING QUESTIONS**

What is biodiversity, and why is it important? How many species are estimated to live on Earth, and which taxonomic groups have the most species?

How do genetic, species, and ecosystem diversity each contribute to ecosystem function and services?

In the past two decades, more than 20 million acres of rain forest have been cleared and planted as oil palm plantations JAMES P. BLAIR/ National Geographic Creative

3

What are biodiversity hotspots, and why are they important?

4

What role does isolation play in a species' vulnerability to extinction? How do habitat destruction and fragmentation threaten species?

5

How can we acquire the food, fiber, fuel, and pharmaceutical resources we need without damaging the ecosystems that provide those resources?

Even before his plane landed in Sumatra, Laurel Sutherlin could tell the scene awaiting him would be worse than expected. They were still 30,000 feet over the Java Sea, and already a sickly haze of yellow smoke seemed to be enveloping the aircraft. The 30-something naturalist and environmental crusader had traveled all the way from California to see firsthand the impact the burgeoning palm oil industry was having on this South Pacific island.

Palm oil comes from the waxy orange fruits that sprout from oil palm trees. Oil can be extracted from both the flesh and the seed of the fruits, and it makes up about 45% of the planet's edible oil production. In recent years, it has made its way into a mind-boggling list of consumer products—foods like chocolate, peanut butter, cereal, and biscuits, and also cosmetics, shampoos, and other household items such as mouthwash, diaper cream, and toilet cleaner. By some accounts, about half of all packaged food and household products in the United States contain palm oil. In the United States alone, palm oil consumption has increased six-fold since 2000, to the lofty sum of 1.25 million metric tons (1.4 U.S. tons) per year. About 95% of all that palm oil comes from Southeast Asia—in particular, from Malaysia and the Indonesian island of Sumatra, where Sutherlin's plane was now landing.

His eyes and nose were assaulted with smoke the instant he stepped off the plane. "I actually had to suppress an initial panic that I would suffocate from the smoke," he would later recall. In fact, he had been lucky to land at all; air traffic would be canceled later that day due to poor visibility. Sutherlin had heard stories about the devastation wrought by deforestation. But he was not prepared for this.

Since the early 1990s, more than 8 million hectares (20 million acres) of rain forest (about the size of Maine) have been cleared in Southeast Asia to make way for oil palm plantations. The environmental costs of this clearing are expected to be far-reaching. For one thing, such rapid and colossal loss of rain forest contributes to regional and global climate change and is almost certain to accelerate global warming. For another, forest loss contributes to soil erosion, water pollution, and flooding—not to mention a scarcity of resources on which intricate forest communities depend. Rampant air pollution from forest clearing poses a significant threat to human health in the region. But the most immediate consequences, by far, will be visited upon the region's wildlife. Fewer than 50% of the species normally found in the natural forest will be found in oil palm plantations. Mammals, in particular, tend to avoid the plantations; only about 10% of the original mammalian fauna are typically present. Scientists estimate that converting a forest into an industrial oil palm plantation results in the death or displacement of more than 95% of the orangutans living there. The habitat loss also impacts countless bird, insect, and plant species, many of which fail to thrive or even return at all because other species on which they depend are gone.

"When you look at a candy bar or package of crackers in a grocery store, or a jar of peanut butter in your kitchen, it's difficult to imagine that it has anything to do with orangutans going extinct," Sutherlin says. "But the link is actually quite direct."

Biodiversity provides a wide range of essential goods and services.

Rain forests, including those being cleared in Indonesia and Malaysia, contain the greatest concentration and variety of plant and animal terrestrial life-forms on Earth. This variety is called **biodiversity**, and it is the most unique and extraordinary feature of our planet. So great is the diversity of life on Earth that it is virtually impossible to know just how many species exist; in fact,

⊚ **WHERE IS SUMATRA?**

↑ A worker collects palm fruit clusters at an oil palm plantation in Sumatra. Oil can be extracted from both the flesh and the seed of the fruit.

KEY CONCEPT 12.1

The variety of life (biodiversity) on Earth is tremendous, but we have identified only a fraction of the species that exist. Insects make up the largest group, but we know much more about smaller groups such as plants and vertebrates.

many believe that the vast majority of all living species have yet to be discovered or identified by humans. So far science has identified about 1.9 million species, and common estimates for the total number of species on Earth range from 3 to 11 million (with one estimate as high as 100 million). Camillo Mora of Dalhousie University in Nova Scotia and colleagues estimate the total to be about 8.7 million. Most scientists agree that we have identified only a small fraction of the species that live on Earth. "The numbers are mind boggling," says Jim Miller, a scientist at the Missouri Botanical Garden in St. Louis. "It's almost unfathomable."

What we do know is that some life-forms are far more diverse than others; there are far more insects than there are vertebrates, for example, and there are relatively few mammals (fewer than 5,500 species) overall. **INFOGRAPHIC 12.1**

We also know that some regions of the world contain far greater concentrations of biodiversity—megadiversity—than other regions. Indonesia is one such region; it has more species of mammals, parrots, palms, swallowtail butterflies, and coral than any other country on the planet.

biodiversity The variety of life on Earth; it includes species, genetic, and ecological diversity.

INFOGRAPHIC 12.1   BIODIVERSITY ON EARTH

↓ We have identified about 1.8 million species so far, but our knowledge of Earth's total biodiversity is scant. Some researchers, especially those who work with tropical species and insects, believe the estimates of 3 to 11 million species are far too low; there may well be 5 million insect species alone. Of the species we have identified, insects far outnumber any other group of organisms. In fact, vertebrates (the group to which humans belong) likely make up only 1% of all creatures on Earth.

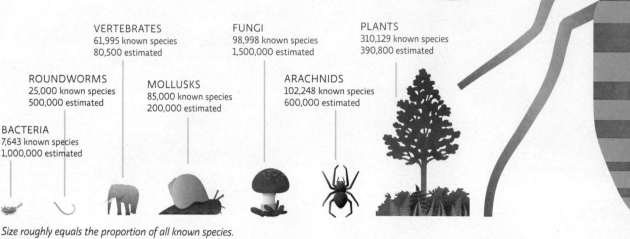

INSECTS
1,000,000 known species
5,000,000 estimated

VERTEBRATES
61,995 known species
80,500 estimated

FUNGI
98,998 known species
1,500,000 estimated

PLANTS
310,129 known species
390,800 estimated

ROUNDWORMS
25,000 known species
500,000 estimated

MOLLUSKS
85,000 known species
200,000 estimated

ARACHNIDS
102,248 known species
600,000 estimated

BACTERIA
7,643 known species
1,000,000 estimated

Size roughly equals the proportion of all known species.

How important do you think it is it to get an accurate accounting of the number of species on Earth?

Roughly 10% of all flowering plant species are found there. A recent census of a 3.4-square-kilometer (1.3-square-mile) plot on the Indonesian island of Siberut, for example, revealed 139 different species of trees. Indeed, tropical forests (and tropical coral reef ecosystems) are the most biodiverse in the world. John Terbough of the Duke University Center for Tropical Conservation reports that a 1-hectare (2.5-acre) research plot in an Amazonian rain forest contained close to 300 tree species. Compare this to the tree species diversity of the boreal forest that spans all of northern Canada—about 20 species.

Biodiversity is responsible for much more than the majesty of nature. For starters, it provides the key connections between individual species and between species and their environment. These connections help regulate the ecosystem as a whole: Insects pollinate flowers, for example. Photosynthetic organisms (plants on land, algae and phytoplankton in the sea) bring in energy, produce oxygen, and sequester carbon. Other organisms capture and pass along important nutrients like nitrogen and phosphorus. Others still help purify the air and water, and all species eventually become food for other creatures. Water also cycles through living things; in a forest, thousands of gallons a year are captured and passed along by each tree, releasing enough water vapor into the atmosphere to affect local rainfall. Life-giving soil forms as organisms decompose: Soil supports plant life whose roots, in turn, hold it in place, keeping it from being washed away in rains or floods. Meanwhile, predators and competitors keep each other in check, so that no single species grows too populous or gobbles up too many needed resources.

Biodiversity can also provide direct protection against disease in various ways. The incidence of Lyme disease, for example, is often lower in areas with higher biodiversity (especially areas with squirrels and possums,

ecosystem services Essential ecological processes that make life on Earth possible.

two species that effectively remove disease-carrying ticks and prevent the spread of the disease to other organisms). This is known as the *dilution effect*. Likewise, a study by John Swaddle of the University of California Santa Barbara and Stavros Calos of the College of William and Mary showed that fewer human cases of West Nile virus occur in eastern U.S. counties with higher bird biodiversity, possibly because some species are less effective at transmitting the virus than others—and the more species present, the less likely the virus will be transmitted to humans.

Biodiversity supplies cultural benefits as well—whether it is the enjoyment of a natural area for recreation or aesthetic appreciation, or a societal tradition rooted in nature.

Biodiverse ecosystems have economic value, too. Forests provide not only food, fuel, and building material but also pharmaceuticals. People use the chemicals they extract from plants and animals not only to attend to their individual human health but also as a source of income. Roughly half of all modern medicines—including medicine cabinet staples like aspirin and codeine, as well as most hypertension drugs and some cancer-fighting superstars—were originally derived from plants, many of

them originally traditional remedies used for centuries. Over generations, rural forest-based communities in Sumatra have perfected a sophisticated management of forest resources that produces hillsides of coffee, elegant groves of cinnamon trees, acres of terraced rice paddies, forests of rubber trees, and a healthy mixture of fruit trees, vegetables, tobacco, and other useful edible, commercial, and medicinal plants.

Ever since Robert Constanza's 1997 study that estimated the annual value of **ecosystem services** to be worth almost twice the annual gross domestic product of the entire world (close to $44 trillion in today's dollars), ecological economists have set out to quantify the monetary worth of ecosystem services to bring attention to this often underappreciated value. Even nature-based recreation is a multibillion-dollar business worldwide.

KEY CONCEPT 12.2

Biodiversity contributes to the health and well-being of ecosystems, which in turn benefits human populations by providing ecosystem goods and services as well as cultural and health benefits.

↓ Rescue workers from the animal charity Four Paws, found this orangutan mother holding her daughter tightly as the pair was being surrounded by a group of young men paid to hunt and kill orangutans in the area. The orangutans were rescued by the charity workers and relocated to a remote area in the rain forest.

Vier Ploten/Four Paws./RHOI/Rex Features via AP Images

INFOGRAPHIC 12.2   ECOSYSTEM SERVICES   1

↓ We depend on genetically diverse, species-rich communities to provide the goods and services we use every day. Impoverished ecosystems that have lost genetic, species, or ecological diversity cannot perform these tasks as well as highly diverse ecosystems can.

1. CULTURAL BENEFITS
Aesthetic
Spiritual
Educational
Recreational

2. HUMAN PROVISIONS
Food
Fiber products, such as cotton and wool
Fuel
Pharmaceuticals

3. ECOSYSTEM REGULATION AND SUPPORT
Nutrient cycling
Pollination and seed dispersal
Air and water purification
Flood control
Soil formation and erosion control
Climate regulation
Population control

2. Many forest products can be harvested.

1. Many people enjoy spending time in nature.

3. Shoreline vegetation protects the banks from erosion.

Danita Delimont/Gallo Images/Getty

?   In terms of reasons to value biodiversity, which of the beneficial ecosystem services listed here are the most important to you personally? Which ones do you think might be the most influential to society?

Whale watching, a pastime in more than 80 countries, brings in $1 billion annually. Visitors spent $12 billion at U.S. national parks in 2010. The U.S. Department of the Interior estimated the economic value of hunting, fishing, and wildlife viewing in 2001 to be $108 billion. (For more on valuing ecosystem services, see Chapter 6.)
INFOGRAPHIC 12.2

Of course, many people feel that the value of any given species goes beyond these **instrumental values** (i.e., the ecological, cultural, and economic benefits it brings). Many say that a species like the orangutan has **intrinsic value** (an inherent right to exist) and is therefore worth preserving, regardless of what benefits it might provide.

Biodiversity includes variety at the individual, species, and ecosystem levels.

Biodiversity can be broken down into three types of diversity that really represent three different levels of diversity, from the population up to the ecosystem. **Genetic diversity** is the heritable variation among individuals of a single population or within the species as a whole. This is diversity at the population level. Genetic diversity provides the raw material that allows populations to adapt to their environment (potentially leading to the emergence of new species). Without genetic diversity, even the most well-adapted species could be in danger of facing extinction if conditions change.

The importance of genetic diversity was tragically illustrated in 19th-century Ireland. At the time, potatoes were the main food source for one-third or more of the population. Potatoes grew well in the Irish climate, despite the plant not being native to Ireland—it hails from South America. On the slopes of the Andes mountains, thousands of different genetic varieties of potato plants grew, each adapted to the slightly different niche found at different latitudes and elevations, traits that were selected for by nature and later enhanced with artificial selection by hundreds of generations of Andean farmers. But only a few varieties of potato were imported to Ireland, and by the 1840s, only two types were planted, predominately one variety called the "lumper." These varieties grew well and fed millions until a fungus arrived (probably on ships from North America) in the mid-1800s and attacked potatoes. In the Andes, this fungus would damage some crops, but because all the potato varieties were so different, even if some plants died, others would survive. But in Ireland, without genetic diversity in the potato "population," most of the potato plants were identical, and so what killed one killed them all. The vulnerable lumper succumbed to the fungus, and the potato crop was wiped out for several years. Between 1845 and 1851, more than 1 million people died of starvation and 1 million more emigrated (many to the United States).

We can also find biodiversity at the community level— that is, **species diversity**. This is an accounting of the number of species living in an area (richness) and their distribution (evenness) (see Chapter 10). Some ecosystems naturally have higher species diversity than others. Areas with high species diversity often owe that variety to their **ecological diversity**—the variety of habitats, niches, and ecological communities in an ecosystem. Rain forests contain a wide variety of habitats and considerable physical complexity, creating lots of niches. Each of the many species living in a tropical rain forest occupies an individual niche, and they all contribute to the functioning of the ecosystem (including vital services like energy capture and nutrient cycling, decomposition, and biomass production). For example, a wide variety of plant species occupy every level of the rain forest, from the forest floor to the canopy, and they capture varying amounts of sunlight. This large producer base supports many trophic levels and many species within those trophic levels, increasing the efficient use and transfer of nutrients in the ecosystem. Few matter resources will go unused. The loss of species (extinction) in an ecosystem removes contributing members from that ecosystem, potentially impacting other species and the ability of that ecosystem to function and provide ecosystem services. **INFOGRAPHIC 12.3**

Tropical forests tend to be particularly flush with both ecological and species diversity, thanks largely to the abundant sunlight and climatic conditions conducive to growth. The forests that are currently being laid low in Sumatra are no exception. They have been left unhampered for so many millennia that these steamy amphibious ecosystems swarm with a cornucopia of life: elephants, orangutans, tapirs, tigers, and every manner of bird and beetle the human imagination can fathom. "The truth is, no one has any idea how many species used to live here," Sutherlin says. "Half the species in these forests have yet to be described to science."

In one stunning example of how ecological diversity yields biodiversity, researchers

instrumental value An object's or species' worth, based on its usefulness to humans.

intrinsic value An object's or species' worth, based on its mere existence; it has an inherent right to exist.

genetic diversity The heritable variation among individuals of a single population or within the species as a whole.

species diversity The variety of species, including how many are present (richness) and their abundance relative to each other (evenness).

ecological diversity The variety within an ecosystem's structure, including many communities, habitats, niches, and trophic levels.

INFOGRAPHIC 12.3 **BIODIVERSITY INCLUDES GENETIC, SPECIES, AND ECOSYSTEM DIVERSITY**

BIODIVERSITY

GENETIC DIVERSITY
Variations in the genes among individuals of the same species

SPECIES DIVERSITY
The variety of species present in an area; includes the number of different species that are present as well as their relative abundance

ECOLOGICAL DIVERSITY
The variety of habitats, niches, trophic levels, and community interactions

↑ Though these Delia butterflies are members of the same species, *Delias hyparete*, they show tremendous genetic diversity.

↑ Tropical forests have some of the highest plant and animal species diversity in the world.

↑ Indonesia's rain forest ecosystems contain complex, three-dimensional structures with varied habitats and multiple niches that support high species diversity and a complex community.

 Give an example of genetic diversity and species diversity from the grocery store produce section.

examined 14 different 1-hectare (2.5-acre) study plots in a cloud forest in Peru's Manu National Park, a biodiversity hotspot. The plots were spaced out across a ridge that climbed from 750 to 3,400 meters (2,500 to 11,000 feet) elevation, and the species diversity found in those altitude differences was stunning—more than 1,000 different tree species alone, in addition to other plants and animals.

endemic species A species that is native to a particular area and is not naturally found elsewhere.

biodiversity hotspot An area that contains a large number of endangered endemic species.

endangered species Species at high risk of becoming extinct.

Since areas with high ecological diversity offer so many unique habitats and niches, they often have a large number of **endemic species** that are specially adapted to that locale and naturally found nowhere else on Earth. They are most commonly found in small isolated ecosystems (islands, mountaintops) because the species' population members cannot easily disperse and share genes with other populations. **Biodiversity hotspots** are areas that have high endemism but also contain a large number of **endangered species** (those at risk of becoming extinct). Many are found in tropical regions.
INFOGRAPHIC 12.4

INFOGRAPHIC 12.4   **BIODIVERSITY HOTSPOTS**

↓ Biodiversity hotspots, areas with high numbers of endemic but endangered species, cover a small percentage of land and water areas but hold more than 40%–50% of all plant and vertebrate endemic animal species. Most hotspots are located in tropical biomes or in isolated terrestrial ecosystems, such as mountains or islands. Even small disturbances, such as a small farm plot or road that cuts through the area, can threaten endemic species that populate specialized niches in these hotspots.

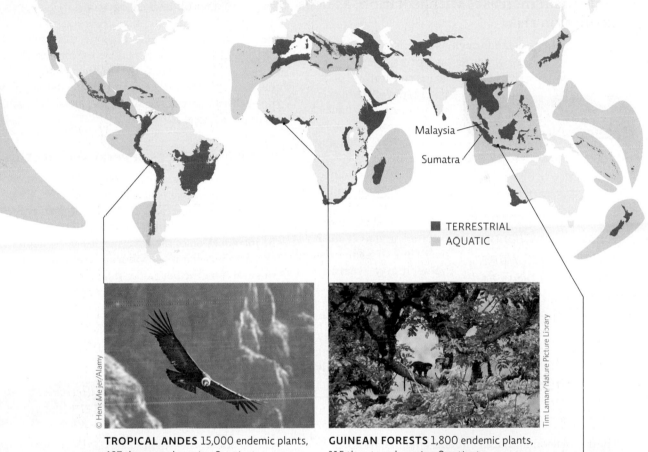

Malaysia

Sumatra

■ TERRESTRIAL
▨ AQUATIC

TROPICAL ANDES 15,000 endemic plants, 487 threatened species, 2 extinct

GUINEAN FORESTS 1,800 endemic plants, 115 threatened species, 0 extinct

POLYNESIA–MICRONESIA 3,074 endemic plants, 99 threatened species, 43 extinct

INDO-MALAYAN ARCHIPELAGO 15,000 endemic plants, 162 threatened species, 4 extinct

 Many hotspots are in tropical areas, but some are also found in areas further north and south of the equator. Why do you suppose so many non-tropical coastal areas are biodiversity hotspots?

The island of Sumatra is one such hotspot; it is home, for example, to the endemic Sumatran Tiger. The pygmy elephant is likewise endemic to the nearby island of Borneo. And the orangutan, one of humans' closest relatives on the evolutionary tree, is endemic to the region, with different types of orangutan found on the islands of Sumatra and Borneo. All three of these species are being driven to the edge of extinction by our pursuit of palm oil.

Endemism increases with isolation, as does extinction risk.

Species come to islands such as Sumatra in various ways: They may be blown in on storms, they may arrive as lost migrators, or the island may have broken off from a larger landmass at some point. Because these are rare events, once a species arrives, it is unlikely to be joined by other members of its species. The founding population is therefore isolated, and as it adapts over time to its new island home, it may eventually evolve into a new species (see Chapter 11 for more on the *founder effect*). For this reason, the number of unique species (the degree of endemism) generally increases with isolation. On the Hawaiian Islands, almost 3,800 km (2,400 miles) from the nearest mainland, 90% of native species are endemic.

> ## KEY CONCEPT 12.5
>
> Isolated populations are especially vulnerable to detrimental environmental changes because they cannot freely breed with other populations and thereby increase their genetic diversity and chances of survival.

Their isolation and high endemism make remote islands particularly vulnerable to species loss. The smaller the population (often a reflection of the habitat size), the more likely it is that random events such as fires or floods might exterminate the entire group. And isolation reduces the chances of re-colonization that might replace lost members or increase genetic diversity. Today, human impact contributes to isolation in the form of habitat fragmentation—producing habitat "islands" where before there were larger expanses of uninterrupted habitat. Small, completely isolated ecological communities are at the highest risk for species extinction. Research by Luke Gibson and colleagues showed that after dam construction flooded a Thailand forest, leaving behind some small islands covered in forest, almost all small mammal species were lost in as little as 5 years on the smallest islands (less than 5 hectares [12 acres] in size).

Isolation can be a recipe for disaster in the face of rapid environmental change. Stuart Pimm of the Nicholas School of the Environment at Duke University and his colleagues analyzed the fossil record and the living species on Pacific islands colonized by humans about 4,000 years ago and concluded that as many as half of the species might have been driven to extinction after humans arrived. In fact, even though they make up a tiny percentage of the land mass on Earth, islands have accounted for about half of all recorded extinctions in the past 400 years. In Hawaii alone, fully one-half of the indigenous flora faces immediate extinction. Such a high extinction rate threatens the fragile tapestry of life on these islands, from the soil and freshwater supplies to the health and economic future of the islands' residents. **INFOGRAPHIC 12.5**

Biodiversity faces several serious threats.

The world's biodiversity is threatened from many angles. The scientific consensus is that human actions are behind the high extinction rates of the recent past, especially since 1900. These threats include habitat destruction, the introduction of invasive species (see Chapter 11 for a detailed example), pollution, overharvesting (for consumption, the pet trade, and legal and illegal body parts, such as hides, tusks, and bones), and anthropogenic climate change. Many of these threats are related to overpopulation and affluence that leads to greater resource use. (See Chapter 13 for more on threats to biodiversity.)

At least part of the surge in palm oil use—and the habitat destruction it engenders—can be traced back to 2007, when then New York City Mayor Michael Bloomberg famously banned the use of hydrogenated oils, which contain high quantities of trans fats, in all New York City restaurants. The ban was well intentioned: Obesity rates were high in the city and in the United States as a whole, and trans fats are, well, fattening. They're also artery-clogging, heart attack–inducing, and

> ## KEY CONCEPT 12.6
>
> Human impact, especially habitat destruction and fragmentation, is a major threat to biodiversity, endangering ecosystems and the species (including humans) that depend on the ecosystem services these areas provide.

INFOGRAPHIC 12.5   ISOLATION CAN AFFECT POPULATIONS

↓ Isolation can increase the number of endemic species in an area because local populations do not "share" genes with other populations. Over time, an isolated population may diverge from its ancestral population as it becomes adapted to its immediate environment.

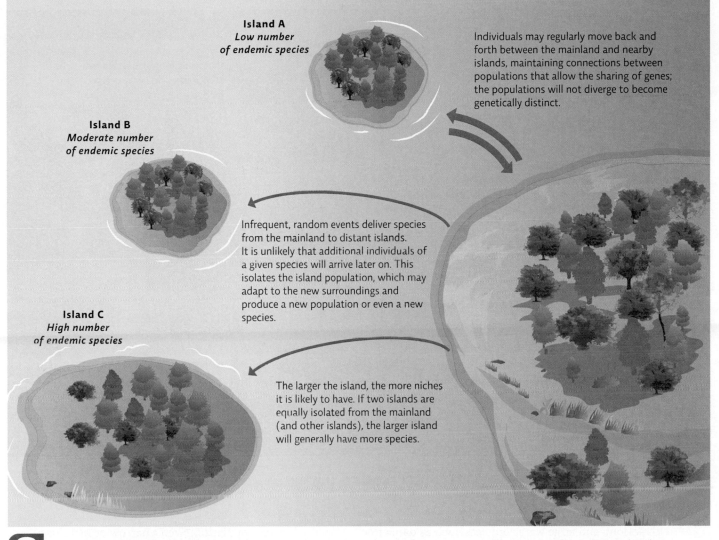

Island A
*Low number
of endemic species*

Individuals may regularly move back and forth between the mainland and nearby islands, maintaining connections between populations that allow the sharing of genes; the populations will not diverge to become genetically distinct.

Island B
*Moderate number
of endemic species*

Infrequent, random events deliver species from the mainland to distant islands. It is unlikely that additional individuals of a given species will arrive later on. This isolates the island population, which may adapt to the new surroundings and produce a new population or even a new species.

Island C
*High number
of endemic species*

The larger the island, the more niches it is likely to have. If two islands are equally isolated from the mainland (and other islands), the larger island will generally have more species.

? Why might populations of small animals, like salamanders that live only at high elevations on adjacent mountains, remain isolated from one another, even though there are no barriers between the two mountains?

all-around unhealthy to consume, and in 2007, they were pervasive. Seeing the writing on the wall, popular food brands, including Oreo and Dunkin' Donuts, followed Bloomberg's lead by voluntarily replacing hydrogenated oil with the next best alternative: palm oil.

Despite the sweet sugary tastes those brand names evoke, palm oil itself has no flavor. What it does have is a soft, thick texture that remains fantastically stable across many temperatures. This thickness and stability make palm oil so useful. It keeps donuts firm and Oreo cookie filling from melting into drizzle. It keeps cosmetics and detergents and peanut butter and Girl Scout cookies consistent and solid at room temperature.

Our palm oil obsession has contributed to the number-one cause of species endangerment: **habitat destruction**. Other drivers of habitat destruction include urban and suburban development, resource extraction (not just timber harvesting but also mineral mining and fossil fuel

habitat destruction The alteration of a natural area in a way that makes it unsuitable for the species living there.

INFOGRAPHIC 12.6   **OIL PALM PLANTATIONS ARE NOTHING LIKE NATURAL TROPICAL FORESTS**   4

↓ The tropical forests of Indonesia and Malaysia are among the most biodiverse in the world, home to thousands of endemic species. The lowland tropical forests of Sumatra alone have an estimated 30,000 different plant species and as many as 1,000 different tree species.

Indonesia's forests contain more than 15% of all the bird, reptile, and amphibian species on Earth and house rare mammals like the orangutan.

Trees contribute much of the moisture that helps form clouds and rain in the area. Areas with heavy deforestation receive less rainfall, altering local climates.

The forests provide important food, wood, medicinal, and other resources for tens of millions of local inhabitants.

Intact forests provide important water production, collection, and purification ecological services as well as erosion prevention.

The peat soils of lowland forests are important carbon sinks that help mitigate climate change.

Auscape/Universal Images Group/Getty Images

? Describe how tropical forests differ from oil palm plantations in terms of genetic, species, and ecosystem diversity.

extraction), and large water projects such as dams that destroy terrestrial and aquatic habitats. By obliterating certain parts of an ecosystem, habitat destruction makes that ecosystem physically less diverse, which then makes it less biodiverse. This loss of biodiversity, in turn, leads to a reduction of ecosystem services. **INFOGRAPHIC 12.6**

As mentioned earlier, the detrimental impacts of habitat destruction can also be touched off by the loss of even part of a larger habitat, known as **habitat fragmentation**. Deforestation that leaves patches of forest may not be suitable for species that need large expanses of forest. Some species will simply not cross over the deforested patch to disperse from one side of the patch to the other, so the patches effectively isolate them from other

parts of their population. Even a narrow or little-used road may isolate individuals on one side or the other. Wildlife corridors that connect fragmented habitats can be effective *conservation* tools. Specially constructed road overpasses or underpasses or ribbons of natural habitat that connect two habitat fragments can allow individuals to travel more freely and are vitally important in maintaining genetic diversity in the populations—especially for species that range widely and live at low population densities, such as large carnivores like grizzly bears and wolves.

Extinction doesn't have to be worldwide to have an impact. Species that become extinct locally but not everywhere are said to be **extirpated**. So, while a species may not technically be extinct (and perhaps not even

↘ Oil palm plantations have very low biodiversity, which means they are unable to provide the ecosystem services normally provided by the large, unfragmented forests they replace.

Oil palm plantations were a main contributor to Indonesia's high rate of deforestation—the highest in the world in the early 2000s. Indonesia lost 2.4 million ha/yr in both 2003 and 2004.

The plantations contain only one species of tree and a limited number of understory or canopy plants.

Oil palm plantations have only half as many bird species as the neighboring intact forest. Insect biodiversity is also lower there, but some non-forest species are present in higher numbers (i.e., bees).

Arthropod abundance is much lower in oil palm leaf litter (77.1% lower in one study) than in primary forest.

Open areas like plantations are prone to flooding and contribute to sediment pollution, damaging aquatic habitats and threatening water supplies.

© blickwinkel/Sailer/Alamy

identified as an endangered species), its extirpation can have serious impacts on its own ecosystem, potentially disrupting the connections that the ecological community depends on, triggering a cascade that threatens other species in the region. "Simple" ecological communities (ones with relatively few species) are much more affected by a single loss. But even "complex" ecosystems (those with many species and many filled niches) can be gravely imperiled if too many members are lost, especially if keystone species are lost. The loss of half of a native plant community is a loss that few, if any, ecosystems can endure without completely collapsing. Of course, once one ecosystem collapses, a new one may emerge. But such recovery takes a very long time.

In the tropics, deforestation for agricultural purposes is currently the leading cause of terrestrial habitat destruction. Fragmentation is also a problem for temperate forests due largely to timber harvesting and human development, but, interestingly, some temperate forests have increased in size in recent decades—in part because tropical forests have less political protection and are more easily exploited. Temperate forests in many areas have been restored and are managed for a variety of purposes, including

habitat fragmentation The destruction of part of an area that creates a patchwork of suitable and unsuitable habitat areas that may exclude some species altogether.

extirpated Describes a species that is locally extinct in one or more areas but still has some individual members in other areas.

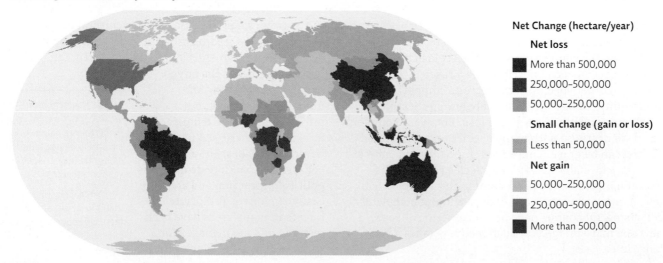

↑ Wildlife corridors like this one over the Trans-Canada Highway in Banff National Park in Alberta, Canada, allow animals safe passage over roads.

the planting of fast-growing trees that are harvested and then replanted. **INFOGRAPHIC 12.7**

To reach a patch of rain forest that had yet to be touched by destruction, Sutherlin and his colleagues traveled 10 hours through the night, from Riau to Jambi Province, then another 4 hours by car over horrendous dirt roads, to South Sumatra. From there, they took motorcycles, over winding-ribbon trails, through a desolate oil palm plantation to the edge of the peat lands, and then

continued by foot, "on a rough trail along a canal dug by loggers to remove logs from the forest." Sutherlin was thrilled when they arrived at the forest edge, sweaty and exhausted, to finally see some tall trees still standing. Monkeys howled in the distance. An electric blue butterfly swirled around his head. And spider-hunters, dollarbirds, and bulbuls circled overhead, flitting in and out of view. "Then, as if on cue," he says, "a chainsaw began to roar just out of sight, followed quickly by the terrible sound of trees crashing through trees to the ground."

INFOGRAPHIC 12.7 | GLOBAL FOREST CHANGE 4

↓ Deforestation has been highest in tropical areas in recent years. Afforestation projects, including forests planted and managed for timber harvesting, have actually led to an increase in forest cover in North America and China.

Net change in forest area by country, 2005–2010

Net Change (hectare/year)

Net loss
- More than 500,000
- 250,000–500,000
- 50,000–250,000

Small change (gain or loss)
- Less than 50,000

Net gain
- 50,000–250,000
- 250,000–500,000
- More than 500,000

 What areas show a net gain in forest cover on this map? Which country has experienced the highest net gain? What might be contributing to these gains?

> *Sustainable palm is the key to protecting our biodiversity, and our heritage.*
> —Raviga Sambanthamurthi.

Virtually all of the forest's inhabitants are facing annihilation. "The remaining populations of endemic Sumatran rhinos are widely considered to be the living dead," says Sutherlin. "Their habitat is too sparse, too fragmented and too disturbed, their numbers too few."

Sustainable palm oil may protect biodiversity.

The three most common cultivars (strains) of oil palm trees that are routinely planted are dura, pisifera, and a hybrid between the two called tenera. The tenera is considered the best of the three because it yields the most oil—30% more than either of its siblings, by most estimates. "They're known as farmers' gold," says Raviga Sambanthamurthi, a geneticist with the Malaysian Palm Oil Board.

Scientists knew for many years that these "golden" fruits were the result of variations of a single gene, Sambanthamurthi says. But they didn't know which gene or how to screen for it. That left farmers with some dismal options. The tenera variety is produced by manually cross-breeding the other two strains, which is easy enough. But farmers usually have to wait 5 or 6 years, until the oil palm plants bear fruiting bunches, before they can tell how successful they've been and how many tenera plants they've got.

So Sambanthamurthi was thrilled when she and her team of researchers discovered the SHELL gene, back in 2012, which is directly responsible for the tenera fruit, with its thin shell and high oil content. The discovery equips farmers with the ability to identify the tenera seeds before planting. "That puts years back on the clock," she says. "It enables farmers to produce more oil per hectare. And that means it reduces the pressure to clear virgin forest." It was the dream of making such a significant contribution as this that drove Sambanthamurthi to return to her native Malaysia and join the research division of the country's Palm Oil Board, after completing her PhD in genetics at the University College of London. "I knew if I was going to make a difference, it would be in palm oil," she said. "Sustainable palm is the key to protecting our biodiversity, and our heritage."

In Sumatra and elsewhere in Southeast Asia, the solution to protecting biodiversity will have to involve palm oil

KEY CONCEPT 12.7

Effective biodiversity protection programs must address the needs of humans as well as the ecosystems and species that are in danger.

↓ More than 90% of the native species on the Hawaiian Islands are found nowhere else, making these populations more vulnerable to extinction; if they are lost there, they are lost forever. The smaller the island, the smaller the population of a given species, another vunerability for extinction. Here, botanists work through the night setting up a protective fence around the last specimen of *Delissea undulata* found in the wild. This endangered plant was growing on the side of a collapsed lava tube, but had been knocked over by wind or animals and was dangling from its roots.

| **TABLE 12.1** | **PROTECTING BIODIVERSITY REQUIRES A CONSIDERATION OF ECONOMIC GOALS AND ENVIRONMENTAL NEEDS** | 5 |

↓ Deciding how to proceed with regard to palm oil requires that we evaluate the trade-offs of our choices and that we strike a balance between meeting our desire for palm oil and the survival of the ecosystems and local communities affected by its cultivation.

BENEFITS OF PALM OIL	DISADVANTAGES OF PALM OIL
Palm oil is the preferred dietary replacement for trans fats.	Though probably better than trans fats for one's health, it is still a fat whose consumption should be limited.
Palm oil is already widely used in many products.	The high demand for palm oil in products increases the need for more oil palm plantations.
Palm oil is inexpensive.	If palm oil is priced in a way that reflects its true costs, the prices of goods that contain it will increase.
Oil palm plantations can produce much more oil per acre than can other oil crops.	Oil palm plantations reduce biodiversity and decrease the ability of local ecosystems to provide ecosystem services.

ADDRESSING THE TRADE-OFFS

Cultivation of oil palms should be done in a way to minimize damage to local ecosystems and biodiversity, leaving some areas uncultivated as refuges for local biodiversity. The most productive palm varieties should be used to increase production per acre.

Manufacturers should only source palm oil that is certified as sustainably grown, and this information should be readily available to consumers.

Consumers should reduce their use of palm oil products in general and opt for products that contain certified sustainable palm oil to support companies that use it.

Consumers should be educated so that they understand that consumption of palm oil (in foods), like consumption of any other fat, must be moderate.

What role does the consumer play in the future course of the palm oil industry and what can manufacturers of products that contain palm oil do to help consumers make informed decisions? How can producing palm oil that is sustainably certified help address some of the disadvantages of using palm oil?

production. As Sambanthamurthi and others point out, it's the most productive oil-bearing crop at our disposal. It claims just 5% of the total vegetable oil acreage globally but accounts for the production of roughly one-third of all vegetable oil, making it the biggest single source of edible oil worldwide. Soybeans, by comparison, account for 41% of the total vegetable acreage and yet produce only 27% of the world's edible vegetable oil. "To get the same amount of soybean oil, you would have to use ten times as much land. So really, it's the best option we have environmentally as well."

There are ways to produce palm oil sustainably. To be certified sustainable by the nonprofit group Roundtable on Sustainable Palm Oil (RSPO), palm oil producers must adhere to certain practices that minimize the use of fire and pesticides, and they must take steps to prevent soil erosion and water pollution. The rights of workers (fair

wages and good working conditions) are a high priority, and the local community is given an opportunity to weigh in on proposed oil palm plantations. In addition, growers cannot clear old-growth forests or those with high biodiversity or cultural value to use for plantations. Sutherlin and others say the reasons for adopting those methods are clear: If we factor in the value of services provided by intact ecosystems, sustainably produced palm oil is significantly cheaper than non-sustainably produced product (*true cost accounting;* see Chapter 6). In fact, production methods that destroy ecosystems and drive species to extinction incur costs that we may never be able to repay. Only a careful and honest analysis of the trade-offs of using palm oil will help us decide how best to proceed. **TABLE 12.1**

But for oil producers to adopt better ways, demand will have to come from consumers. "Companies aren't hearing from their customers," Sutherlin says. "They tell us flat out 'We know it's bad. We know it's doing all these horrible things to the environment. But it's our cheapest option, and no one's threatening to boycott us over it.'" In 2013, Sutherlin's organization, the nonprofit Rainforest Action Network, launched the "Snack Food 20" campaign—a call on the top 20 global snack food companies, including Krispy Kreme, Kraft, Kellogg, Heinz, General Mills, and Nestle, to use only sustainably produced palm oil in its products (certified by the RSPO).

Others have taken up the cause as well. In 2010, New York Comptroller Thomas Di Napoli, who manages investments for the state's $153 billion pension fund, won pledges from companies like Dunkin' Donuts, Sara Lee, and Smucker's to only use sustainably produced palm oil. Unilever, maker of food and cleaning products, advertises that it is now sourcing only sustainable palm oil. And since at least 2007, two Michigan Girl Scouts have been lobbying the Girl Scouts U.S.A. to extract a similar pledge from the makers of Girl Scout cookies. Like so many other processed food manufacturers, the Girl Scouts switched to palm oil so the cookies would be free of trans fats. The organization says it would like to switch to sustainably produced palm oil, but so far there simply isn't enough of it out there. In any given year, Girl Scouts sell hundreds of millions of dollars' worth of cookies. That's a lot of palm oil. "Girls sell cookies from Texas to Hawaii and those cookies have to be sturdy," Amanda Hamaker, product sales manager for Girl Scouts USA, told the *Wall Street Journal* last year. Right now, only

about 6% of today's global supply is sustainably grown, but that is increasing; 2014 saw record increases in certified sustainable palm oil sales worldwide.

One thing is certain: There is still much value in saving these forests. Orangutans still swing freely through the canopies, and new species of lizards and birds continue to be discovered. Despite the destruction plaguing Indonesia's forests, all hope is not yet lost.

Select References:

Fitzherbert, E. B., et al. (2008). How will oil palm expansion affect biodiversity? *Trends in Ecology and Evolution*, 23(10): 538—545.

Gibson, L., et al. (2013). Near-complete extinction of native small mammal fauna 25 years after forest fragmentation. *Science*, 341(6153): 1508—1510.

Hadi, S. T., et al. (2009). Tree diversity and forest structure in northern Siberut, Mentawai islands, Indonesia. *Tropical Ecology*, 50(2): 315—327.

Mora, C., et al. (2011). How many species are there on Earth and in the ocean? *PLoS Biology* 9(8): e1001127. doi:10.1371/journal.pbio.1001127

Pimm, S., et al. (2006). Human impacts on the rates of recent, present, and future bird extinctions. *Proceedings of the National Academy of Sciences*, 103(29): 10941—10946.

Singh, R., et al. (2013). The oil palm SHELL gene controls oil yield and encodes a homologue of SEEDSTICK. *Nature*, 500(7462): 340-344.

Swaddle, J. P., & S. E. Calos. (2008). Increased avian diversity is associated with lower incidence of human West Nile infection: Observation of the dilution effect. *PLoS ONE*, 3(6): e2488. doi:10.1371/journal.pone.0002488

Teoh, C. H. (2010). *Key Sustainability Issues in the Palm Oil Sector*. Washington, DC: World Bank.

Turner, E. C., & W. A. Foster. (2009). The impact of forest conversion to oil palm on arthropod abundance and biomass in Sabah, Malaysia. *Journal of Tropical Ecology*, 25: 23—30.

BRING IT HOME

PERSONAL CHOICES THAT HELP

Species and habitats provide numerous benefits to people, including water and air purification, food sources, recreation, and medicine. Unfortunately, many species are facing threats at ever-increasing levels. The good news is that we as a society have a direct impact on these threats and can make changes to ensure the survival of many of our at-risk species.

Individual Steps

• Don't buy products made from wild animal parts such as horns, fur, shells, or bones. Only buy captive-bred tropical aquarium fish, not wild-caught fish.
• Visit the Rainforest Action Network website at http://ran.org/conflict-palm-oil and read about their Snack Food 20 campaign. Is this something you would support?
• Make your backyard friendly to wildlife, using suggestions from www.nwf.org.

• Install an Audubon Guide app on your smartphone, or buy a field guide to learn the plant and animal species in your area.

Group Action

• Work with faculty and other students to organize a bioblitz for a protected area in your region. A bioblitz, which is an intensive survey of all the biodiversity in the area, can generate a large amount of data to be used for habitat management and species protection.
• Join a citizen science program monitoring wildlife. Many regional conservation groups have monitoring opportunities and provide training. For national programs, see the Cornell Lab of Ornithology website at www.birds.cornell.edu and the Izaak Walton League website at www. iwla.org.

Policy Change

• The Endangered Species Act was the first U.S. legislation established to protect species diversity. To learn more about current challenges and updates to the program, visit www.epa.gov/espp.

Pete Mcbride/National Geographic Creative

ENVIRONMENTAL LITERACY UNDERSTANDING THE ISSUE

1 What is biodiversity, and why is it important? How many species are estimated to live on Earth, and which taxonomic groups have the most species?

INFOGRAPHICS 12.1 AND 12.2

1. Give a basic definition of *biodiversity*.

2. The total number of different species on Earth:
   a. is unknown, but insects are the most numerous species.
   b. is a few million, mostly bacteria and fungi.
   c. is more than 20 million, with half of them being plants.
   d. is less than 1 million, mostly vertebrates.

3. Why is biodiversity loss a concern?
   a. It primarily occurs in the developed world, where most of the world's population lives.
   b. It increases the degree of endemism in an area.
   c. It primarily affects well-known and charismatic species like elephants and orangutans.
   d. It disrupts ecological connections, potentially diminishing ecosystem services.

2 How do genetic, species, and ecosystem diversity each contribute to ecosystem function and services?

INFOGRAPHIC 12.3

4. True or False: As ecosystem diversity increases, so does species diversity.

5. All apples belong to the species *Malus domestica*. The wide variety of apples available in the produce department of your local grocery store is an example of _____ diversity.

6. An example of species diversity might be:
   a. the wide variety of coloration and tail size in guppies.
   b. the diverse habitat types and organisms inhabiting a deep lake, its edges, and the surrounding meadow and forest areas.
   c. the many different species inhabiting a swamp.
   d. None of the above.

7. Define genetic diversity and use the example of the potato blight in Ireland to explain the importance of genetic diversity to a population.

3 What are biodiversity hotspots, and why are they important?

INFOGRAPHIC 12.4

8. True or False: Biodiversity hotspots are areas with many endemic species that are well protected and not threatened with endangerment.

9. A species that naturally occurs in only one place is called:
   a. an endangered species.
   b. a hotspot species.
   c. an endemic species.
   d. a threatened species.

4 What role does isolation play in a species' vulnerability to extinction? How do habitat destruction and fragmentation threaten species?

INFOGRAPHICS 12.5, 12.6, AND 12.7

10. The leading human cause of species endangerment is _____ _____.

11. Why are many of the biodiversity hotspots around the world on islands?
    a. Islands accumulate species from many different areas.
    b. Populations of island species are isolated.
    c. Islands have more diverse habitats.
    d. There are more niches on islands.

12. Why are isolated populations more vulnerable to extinction than populations that are not isolated from each other?

5 How can we acquire the food, fiber, fuel, and pharmaceutical resources we need without damaging the ecosystems that provide those resources?

TABLE 12.1

13. True or False: Sustainable solutions to growing oil palms must consider the needs of consumers and nearby communities.

14. In terms of its impact on species and their environment, producing palm oil may be a better option than producing soybean oil because:
    a. palm oil is less expensive to grow.
    b. oil palm trees are not grown in areas with very many endangered species.
    c. palm oil is easier to extract from the seed than soybean oil.
    d. growers can produce more oil per hectare with palm plantations than soybean fields.

15. Evaluate the trade-offs of using palm oil as a replacement for trans fats and make a recommendation regarding our future path that considers economic, environmental, and societal needs.

16. How can you, as an individual, help maintain biodiversity worldwide? Justify your choices.

SCIENCE LITERACY WORKING WITH DATA

Habitat loss is currently the main driver of species endangerment and extinction, but habitat loss need not be complete to cause a problem; habitat fragmentation may also be an insurmountable problem for some species. Islands that are created when a river is dammed to form a reservoir provide instant habitat fragments. Luke Gibson and his team evaluated the number of small mammal species in large (10–56 hectares [25–140 acres]) and small (<10 hectares [<25 acres]) forested islands in Chiew Larn Reservoir of Thailand. Island sampling was done shortly after the reservoir was formed (about 6 years after isolation); the islands were sampled again about 26 years after isolation. Their results are below. (For comparison, on average, nine species were found on mainland (pre-reservoir) plots; the richness did not change in this mainland forest over the study period.)

Number of Small Mammal Species (Richness) Found on Islands 6 and 26 Years after Isolation

	Island (ID #)	Area of Island (hectares)	Species Richness (6 years)	Species Richness (26 years)
Large islands	6	56.3	12	5
	5	12.1	9	3
	9	10.4	7	1
Small islands	28	4.7	2	2
	7	1.9	3	2
	33	1.7	1	1
	3	1.4	2	1
	41	1.1	3	1
	39	1.0	3	1
	40	0.8	2	1
	2	0.4	2	1
	16	0.3	2	1

Interpretation

1. Before evaluating the data, draw a graph that compares species richness of large islands 6 years after isolation versus 26 years after isolation and that also shows the same for small islands. (Hint: Calculate the average species richness values for each group and draw a bar graph that allows you to directly compare the richness of large islands after both sampling periods to the richness of small islands after both sampling periods.)

2. Consider that species richness before isolation was nine. How does the species richness compare in large islands before isolation (use "mainland" data), 6 years after isolation, and 26 years after isolation? How does it compare for the small islands over those two sampling periods?

Advance Your Thinking

3. What might lead to the difference in species richness losses in large islands compared to small islands?

4. On all of the islands, the most common (and sometimes only) small mammal 26 years after isolation was a non-native rat. Could this have had an influence on the loss of the other, native species? Explain.

5. What conclusion can be drawn regarding the value of leaving behind fragmented forest landscapes for protecting species that live in the habitat fragments?

INFORMATION LITERACY EVALUATING INFORMATION

Visit the World Wildlife Fund (WWF) website (www.worldwildlife. org). Explore the website to learn about the organization and their work to protect endangered species.

1. What is the mission of the WWF? How do you know this?

2. Read about several of the species that WWF is protecting: Do they use a single species conservation approach, an ecosystem conservation approach, or both? Does their approach support their stated mission?

3. What scientific evidence does the organization provide to indicate that their work is effective at protecting species?

4. Read about the WWF's work in Borneo and Sumatra. What threats do they identify for this region and how are they helping species and the local communities there? Do you think their work is helping the region?

5. What is your overall opinion of the WWF: Would you do anything differently if you were in charge? Is this an organization you would support financially? Explain.

Find an additional case study online at http://www.macmillanhighered.com/launchpad/saes2e

A FOREST WITHOUT ELEPHANTS

Can we save one of Earth's iconic species?

CORE MESSAGE

Today, many species on Earth are in danger of becoming extinct, predominately due to human impact. Protection plans can focus on individual species or entire ecosystems; both methods have proven successful. A combination of national and international efforts is helping address biodiversity loss. Individuals can also help by making consumer choices that support sustainable use of biodiversity rather than actions that exploit and endanger species.

AFTER READING THIS CHAPTER, YOU SHOULD BE ABLE TO ANSWER THE FOLLOWING **GUIDING QUESTIONS**

1

What events or actions threaten species today, and what conservation status designations are used to identify the threat level? How does the loss or reduction of a species impact its ecosystem?

2

How do single-species conservation programs compare to ecosystem-based approaches?

Elephants in Garamba National Park in the Democratic Republic of Congo. Conservation groups say poachers are wiping out tens of thousands of elephants a year in Africa, with the underground ivory trade becoming increasingly militarized. © Tyler Hicks/The New York Times/Redux

3

How does conservation genetics contribute to the conservation of species?

4

What legal protections do threatened species have in the United States and internationally?

5

In what ways are local communities important to the success of conservation efforts? What role do consumers play in protecting biodiversity?

Deep in the African rain forest, an elephant has just been killed. Her corpse is found by park rangers; the body has been poked through with spears, the cheeks and trunk have been eaten away by scavengers. The rest was taken by the poachers who killed her: the tail sliced off to be made into bracelets of black elephant hair, which are popular with tourists, the eyes to be sold to traditional Chinese medicine shops, where their purported medicinal powers fetch a high price. The biggest prize, though—indeed, the whole reason for the killing—were the tusks, two gleaming protrusions of pure ivory that have been hacked off with an ax and are now making their way, by looping, circuitous routes, to jewelry shops in Manhattan and Beijing.

Elsewhere in the forest, the victim's herd is grieving. Elephants exhibit complex behaviors and are known for both their elaborate communication networks and their uncannily human mourning rituals. Even when they are miles apart, the members of a family maintain regular contact, communicating with sounds too low for human ears. When a family member dies, they process past the body—marching in slow, single file. Occasionally one elephant or another will place its foot over the heart of the deceased or prod the lifeless body with its trunk. They often remain with the deceased for days and may even cover the body with dirt and branches.

Elephant poaching in Africa is at its highest level in decades. A 2014 report on the illegal ivory trade revealed that in 2011 alone, authorities estimate that 25,000 African elephants were killed and 50 metric tons (55 U.S. tons) of ivory were seized—more than in any previous year. Half of that was seized in 14 large-scale (500-kilogram [1,100-pound] or larger) ivory shipments. Poaching numbers omit after 2011, but the amount of ivory confiscated increased; in 2013, more than 55 metric tons (62 U.S. tons) were seized. "Evidence is steadily mounting," says Tom Milliken, an elephant expert with Wildlife Conservation Society (WCS), that "African elephants are facing their most serious crisis since international commercial trade in ivory was first prohibited [in 1989]."

In some ways, the reasons for this uptick are obvious. Rising wealth and a growing market for elephant ivory in China (blocks made of ivory bearing a family's crest or signature are a popular status symbol throughout the country) and other nations are being abetted by a dramatic increase in road building in Africa. (Growing populations and rising consumption around the world have necessitated more and more resource extraction from previously untouched lands.)

> ❝ *African elephants are facing their most serious crisis since international commercial trade in ivory was first prohibited.* ❞
> —*Tom Milliken.*

But the plight of forest elephants, and the saga of those racing to save them, is also part of a much larger, more complicated story—laced with grave dangers, daunting odds, and heroic efforts. It's a story in which the world's biodiversity—and thus life as we know it—hangs in the balance.

◉ WHERE ARE THE ELEPHANT RANGES?

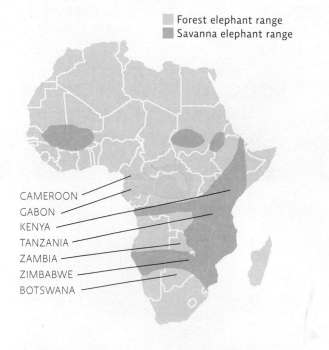

Forest elephant range
Savanna elephant range

CAMEROON
GABON
KENYA
TANZANIA
ZAMBIA
ZIMBABWE
BOTSWANA

Human impact is the main threat to species worldwide.

Conservation biology is the science of preserving biodiversity. Conservation biologists focus on protecting individual species and maintaining or restoring entire ecosystems. That means they also work intently to understand the threats facing species and ecosystems. Today those threats are legion. As has been detailed in numerous chapters throughout this book, we humans are changing the environment with unprecedented speed—so much so that many populations cannot keep up. We're clearing more land and building more roads to access more natural resources than ever before. And we're killing off an alarming number of plants and animals in the process. The details for any given species may differ, but the basic story is often the same: As more and more individuals die off, the genetic diversity of that species dwindles, and the population as a whole becomes increasingly vulnerable to extinction.

In general, conservation biologists agree that biodiversity the world over faces five main threats—threats tied to an expanding human population and increasing levels of affluence: overexploitation, pollution, climate change, the introduction of invasive species, and (the number-one cause) habitat destruction and fragmentation.

In recent decades, despite a rash of international attention to the problems of biodiversity loss, each of these factors has gotten worse. Overfishing has claimed some 85% of oyster reefs and as much as 90% of the world's populations of large predatory fish like tuna and cod (see LaunchPad Chapter 31); pesticides and mercury pollution have all but killed off the Mekong River dolphin; and climate change is threatening to bring the iconic polar bear in the Arctic Circle, not to mention many other species around the world, to disastrous ends. Meanwhile, invasive species are running amok from Alabama to Zimbabwe, thanks to intentional and accidental introductions and human travel around the globe. And habitat fragmentation is isolating populations and limiting their ability to migrate or disperse to new areas. It is also making commercially valuable wildlife— not only elephants with their ivory tusks but also tigers prized for their vibrant skins and rhinos whose horns are

conservation biology The science concerned with preserving biodiversity.

KEY CONCEPT 13.1

Habitat destruction is the biggest threat to species today. Other threats include invasive species, overexploitation, pollution, and climate change.

↓ Congolese soldiers and rangers patrolling Garamba National Park discover a poached elephant in a remote area of the park.

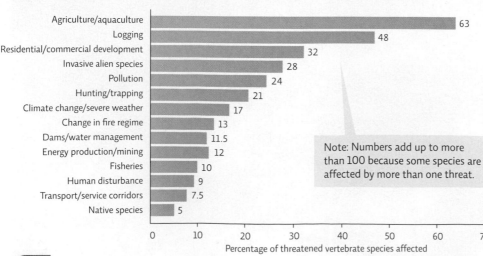

Habitat destruction and fragmentation

Humans change habitats on land and on river- and seabeds, both to harvest resources and to reclaim the area for agriculture or urban use. For example, in Southeast Asia, clearing forests for palm oil plantations has led to a loss of about 83% of bird and butterfly species. Habitat fragmentation splits up species' habitats, and many species have minimum space requirements. Forest elephants require large tracts of forest and are vulnerable to poachers in open, clear areas.

Climate change

A changing climate threatens species that cannot adapt or relocate to more suitable habitats. Species with specific habitat requirements (specialists) are particularly affected. For example, because ice cover on the Arctic Ocean has dropped as much as 34% in recent years, polar bears and ringed seals struggle to feed and reproduce.

Pollution

Pollution in the air and water is toxic to many species and damages habitats. For example, heavy pesticide use has brought the Mekong River dolphin to the brink of extinction.

Invasive species

Invasive species drive native species to extinction by outcompeting or preying on them. For example, the rapidly growing Japanese kudzu vine has spread over most of the American Southeast, blocking light and nutrients from native plants. Some estimate that it is spreading by 150,000 acres a year.

Overexploitation

Humans overharvest and deplete species populations (on land and sea). For example, about 85% of oyster reefs have disappeared since the late 1800s due predominantly to overharvesting.

Human Activities that Threaten Vertebrates

Category	Percentage
Agriculture/aquaculture	63
Logging	48
Residential/commercial development	32
Invasive alien species	28
Pollution	24
Hunting/trapping	21
Climate change/severe weather	17
Change in fire regime	13
Dams/water management	11.5
Energy production/mining	12
Fisheries	10
Human disturbance	9
Transport/service corridors	7.5
Native species	5

Percentage of threatened vertebrate species affected

Note: Numbers add up to more than 100 because some species are affected by more than one threat.

← Conservation biologists agree that there are five main threats to biodiversity. But each threat can be broken down into subparts. Here we show the subcategories of threats for vertebrates—the group that we know most about. This graph presents the percentage of threatened species according to specific threat.

Which categories on this graph represent habitat destruction/fragmentation? Does this data support the claim that habitat destruction/fragmentation is the leading cause of species endangerment?

KEY CONCEPT 13.2

Though most species have yet to be assessed, those that have been are assigned a conservation designation ranging from least concern to extinct.

According to the United Nations Environment Programme, one-third of all the plants and animals on Earth are now at risk of going extinct. And if we aren't worried about that, we should be. "Biodiversity . . . provide[s] a wide range of services to human societies," the WCS report's authors write. "Its continued loss, therefore, has major implications for current and future human well-being."

To avert the worst of those implications, we must figure out how best to preserve the biodiversity that's left. And to do that, we must first learn as much as possible about the ecosystems in question: What are the key requirements of its resident species? Are those requirements being met? If not, why not, and what other species are being threatened as a result? For conservation biologists working to save the forest elephants of Africa, answering those questions means, literally, following a trail of poop.

Human impact that threatens the forest elephant also puts its entire ecosystem at risk.

Steven Blake, an ecologist with WCS, followed several herds of forest elephants across several different wildlife refuges—in Congo, Cameroon, Gabon, and elsewhere— observing their behavior and collecting dung samples along the way for more than a decade. Here's what he learned: These elephants gobble up hundreds of pieces of fruit from a single tree. Then, as they walk on, they deposit the seeds of that fruit throughout the forest, with a generous helping of fertilizer. They can cross an entire national park in the space of 3 days, defecating roughly

used in traditional Chinese medicine—more vulnerable than ever to human predation. **INFOGRAPHIC 13.1**

To bring attention to the problem of **threatened species**—those that are at risk for extinction—the International Union for Conservation of Nature (IUCN) established the Red List of Threatened Species in 1963. The "Red List," as it is called, uses a series of designations to classify the seriousness of threat to a given species. Species are added to the list when they're at risk of becoming endangered, and they're removed when their status improves. The United States as a whole, its individual states, and other countries maintain their own lists of threatened species, using the same designations. **INFOGRAPHIC 13.2**

50 times along the way. Each pile of dung contains thousands of individual seeds from more than a dozen different plant species. It may be gross, but it's no trivial matter.

"Tropical forests are so diverse that a seed that lands near its parent plant has a suite of predators and pathogens waiting to nab it," Blake says. "So if you're a seed and you land under your parent, the probability of you surviving is almost zero." Without the elephants to disperse them, Blake says that plants with large fruits and seeds would eventually disappear.

INFOGRAPHIC 13.2   **CONSERVATION DESIGNATIONS**   1

↓ The International Union for Conservation of Nature (IUCN) maintains the Red List of Threatened Species, which identifies the conservation status of species worldwide, a process that helps conservationists focus efforts on the species most at risk. Status categories reflect extinction risk and range from *extinct* to *least concern*; numbers shown here indicate the number of species in that category as of October 2012.

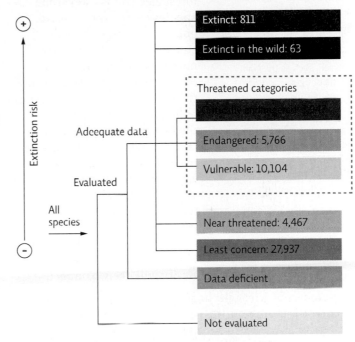

Look back at Infographic 12.1 to find the total number of named species to date. What percentage of named species have not received a designation (data deficient or not evaluated)?

KEY CONCEPT 13.3

When one species in an ecosystem is endangered, others that interact with or depend on it can be negatively impacted. This can cause a domino effect as more and more interacting species are impacted by the changes.

threatened species Species that are at risk for extinction; various threat levels have been identified, ranging from "least concern" to "extinct."

keystone species A species that impacts its community more than its mere abundance would predict.

single-species conservation A management strategy that focuses on protecting one particular species.

flagship species The focus of public awareness campaigns aimed at generating interest in conservation in general; usually an interesting or charismatic species, such as the giant panda or tiger.

ecosystem conservation A management strategy that focuses on protecting an ecosystem as a whole in an effort to protect the species that live there.

"The balance would be tipped in favor of those species that are dispersed by wind or other abiotic agents."

Of course, the impact elephants (or a lack of elephants) have on their ecosystem goes even further than that. In fact, the forest elephant is a **keystone species** (see Chapter 10): Not only does it disperse seeds and mobilize large amounts of nutrients with its feces, it also orders the physical structure of the forest—trampling vegetation, felling trees, and opening up the forest canopy in ways that facilitate plant growth and keep the ecosystem healthy. Savanna elephants also play a role as a keystone species. Without their grazing and trampling of young woody vegetation, the savanna grasslands that so many antelope and other species (and the predators who feed on them) depend on would quickly convert to shrublands. "Once they're gone," Blake says, "overall biodiversity will likely plummet."

Numbers of forest elephants are sharply declining (Blake and colleagues estimate that 65% of all forest elephants were killed between 2002 and 2013), making them more endangered than the savanna elephants that inhabit more open areas in Africa. But why exactly are elephants so imperiled in the first place? They have long been coveted for their ivory; why is poaching suddenly getting worse?

As it turns out, elephant feces provide researchers like Blake with some clues to that question as well. By diligently and systematically mapping the location of elephant dung across vast swaths of land, scientists have been able to detail the effects of newly created access points to various segments of forest. The elephants themselves proved elusive and difficult to count, but their dung provided a proxy census. The results of that census were a bit surprising.

"It's not the logging or mining that's doing the most damage," says Kate Evans, an animal behaviorist at the University of Bristol in the United Kingdom, and founder of the nonprofit Elephants for Africa. "It's the construction of roads." An endless network of new roads has cut its way into previously remote tracts of forest—slicing the landscape up into a patchwork of segregated fragments as they go. These roads have made elephant herds that were once protected by the dense maze of greenery easily accessible to poachers.

The elephants themselves are responding to the incursion. As Evans and others have documented, they travel 14 times faster than normal when crossing a road that is unprotected from poachers. Ultimately, as more and more roads crop up, their world shrinks. "If you put a 20-mile ring of death around your house, the chances are you won't want to go more than 15 miles from home," Blake explains. "And if that ring closes in, you're going to feel besieged. You won't be able to go to the places you need to, you won't be able to see your friends, you will become imprisoned, and most likely the food will start to run out. It's just like that with forest elephants."

There are multiple approaches to species conservation.

While the causes of forest elephant endangerment (road building and ivory poaching) may be clear, the question of what to do about it is anything but. Indeed, experts have long debated how to best protect any given species or ecosystem. Early programs often took the **single-species conservation** approach: They singled out well-known animals—known as **flagship species**—like pandas and condors, and focused on the specific threats those individual species faced, using a variety of methods, including captive breeding programs to increase population sizes and reintroducing the species to the wild. **INFOGRAPHIC 13.3**

In some ways, this approach has been a huge success: It has saved gray wolves, eagles, and even brown pelicans from complete decimation. And it's brought a great deal of attention and funding to conservation efforts. Indeed, the single-species approach is still employed today. Zoos around the world participate in Species *Survival Plans*, which use careful breeding programs to maximize genetic diversity. Among other things, this includes moving reproductive animals from zoo to zoo to introduce new genes into a breeding program or to minimize inbreeding.

The ultimate goal of these programs is to release animals back into the wild. But the tendency to focus mostly on those species (cute, furry, large) that are photogenic enough to capture the public's attention means that many, if not most, species in need of help fall through the

KEY CONCEPT 13.4

Single-species conservation programs focus on a specific species and have been very successful at protecting some interesting and charismatic species but have not been widely used for less visible or valued species.

INFOGRAPHIC 13.3 | SINGLE-SPECIES CONSERVATION APPROACH

↓ Species Survival Plans, administered by the Association of Zoos and Aquariums, focus on increasing the population size and genetic diversity of threatened populations. Work goes on in zoos and aquariums as well as in the wild.

CAPTIVE BREEDING PROGRAMS

© Corbis

← Individuals for captive breeding are carefully chosen to maximize genetic diversity in the managed population. In 1987, the last 22 California condors were captured and became part of a captive breeding program. Chicks are fed with condor puppets so that they do not associate humans with food, to better facilitate their eventual release. As of 2012, there were 450 individuals, 266 of them in the wild.

FIELD CONSERVATION WORK

© Marcus Bleasdale/VII/VII/Corbis

← Conservation groups work in the field to learn about and monitor species and to protect or restore habitat. They also work with local residents to prevent human—wildlife conflicts. Here, a small antelope known as a dik-dik is released from a poacher's snare in Kenya.

REINTRODUCTION PROGRAMS

CRISTINA QUICLER/AFP/Getty Images

← Individuals from captive breeding programs or other wild populations can be introduced to an area if suitable habitat and protection is in place. At the beginning of the 20th century an estimated 100,000 Iberian lynx roamed Portugal and Spain but by 2002 fewer than 150 individuals remained. Here an animal bred in captivity is released as part of a successful reintroduction program in Spain. Lynx are now breeding in the wild and the population in southern Spain is 300 and rising.

? Are you in favor of the reintroduction of predatory species (such as wolves) that could potentially harm humans or their pets or livestock? Explain.

cracks. (People are more likely to donate money to "Save the panda!" than to "Save the white warty-back pearly mussel!")

As long as the main problem facing a species is low population size, reintroduction campaigns can be very effective, but this is insufficient if other threats—such as a degraded habitat or heavy poaching pressure—still exist. Indeed, such campaigns can give a false impression that we are saving many species, when in fact habitat destruction continues to decimate the vital habitat that supports innumerable species. For these reasons, many conservationists advocate an **ecosystem conservation** approach. This means identifying entire ecosystems—often *biodiversity hotspots*—that are at risk and taking

Ecosystem conservation programs focus on protecting the habitat within an ecosystem, which helps protect all the species that live there.

steps to restore or rehabilitate them. The goal of **ecosystem restoration** is to return the ecosystems to their original states (or as close to their original states as possible). It may involve reforestation projects, removal of non-native species, restoration of a natural river's flow patterns, or **remediation** (the cleanup of pollution); the restoration goal depends on the ecosystem in question.

In the Costa Rican rain forest, conservation biologist Daniel Janzen has worked for nearly 40 years on restoring the area in and around Guanacaste National Park, forest land that had been converted to pastures and agricultural land. The largest forest restoration project of its kind, Janzen's plan involved protecting the area from invasive, non-native grasses and fire; replanting trees; and allowing pioneer species to move back into the cleared lands from neighboring forests. Through this process of secondary succession (see Chapter 10), forest plant communities moved back into the area. All told, the project has restored more than 1,000 square kilometers (400 square miles) of pasture land to forest and is now home to an estimated 235,000 species. While restoration programs like this are needed, efforts that prevent destruction in the first place are more likely to protect a greater number of species. The tropical dry forests of Guanacaste will take several hundred years to return to their original mature forest state, and we will never know how many species were lost on their way back to recovery.

Proponents point out that by focusing on the ecosystem as a whole, the entire community benefits—not just the species we knew were endangered but also those that we didn't know were there. In order to implement an ecosystem approach to conservation, we must know how to identify when an ecosystem is in danger. Ecologists are developing metrics (factors that are measured) to assess ecosystem quality, such as *species richness* and *evenness* (see Chapter 10), soil health, water quality, plant community composition, and the abundance of non-native species.

Ecologists often monitor an **indicator species**—a species that is particularly vulnerable to ecosystem perturbations—to keep track of an ecosystem's health. A type of ecosystem conservation known as **landscape conservation** draws on this idea, but instead of following one species, it examines several indicator species in what is known as a *landscape species suite*. The species are specifically chosen to include a group that, together, uses all the vital areas within the ecosystem. The idea behind this approach is that if you monitor these species together and work to protect them as a group, you will simultaneously be protecting the entire ecosystem in which they reside, including all the species that live around them. **INFOGRAPHIC 13.4**

Either way, says Evans, the forest elephant is a good bet for conservationists. As a keystone species, it protects and facilitates an endless array of other species. A single-species approach that focuses on preserving the forest elephant will, by necessity, work to restore and protect its habitat and in doing so confer protection on the other species that reside there. A similar strategy proved successful in the Pacific Ocean at the beginning of the 20th century. By the end of the 1800s, sea otter populations there had fallen drastically because of overhunting (otter pelts were highly valued); this led to an overpopulation of their prey, sea urchins, which feed on kelp. As sea urchin populations grew, the kelp forests were decimated, removing habitat and food for myriad species. After the International Fur Seal Treaty banned the hunting of sea otters in 1911, the otter populations recovered, which brought the sea urchin populations back in check, allowing the kelp to recover and benefitting all the species of the kelp forest.

As for the forest elephant—a highly intelligent mammal with a complex social system—its plight has captured human hearts and thus shone a light on the oft-forgotten African rain forest. Still, to stop the ivory trade from obliterating this particular species, scientists will have to employ a range of tools, some of them much more complicated than dung.

To Sam Wasser, a conservation biologist at the University of Washington, tracking the ivory trade is a bit like playing a shell game where one must guess which shell the bean is under. By sequencing ivory DNA from elephants all across Africa, Wasser's Seattle laboratory has mapped the ivory trade—from the slaughtered elephants in Africa to the curio shops in New York and Beijing. What Wasser's lab has found has astounded nearly everyone involved in elephant conservation. It turns out that

---

ecosystem restoration The repair of natural habitats back to (or close to) their original state.

remediation Restoration that focuses on the cleanup of pollution in a natural area.

indicator species A species that is particularly vulnerable to ecosystem perturbations, and that, when we monitor it, can give us advance warning of a problem.

landscape conservation An ecosystem conservation strategy that specifically identifies a suite of species, chosen because they use all the vital areas within an ecosystem; meeting the needs of these species will keep the ecosystem fully functional, thus meeting the needs of all species that live there.

conservation genetics The scientific field that relies on species' genetics to inform conservation efforts.

↓ Protecting the entire ecosystem where an endangered species lives helps protect all the species that live there—even those that conservationists didn't know were endangered or threatened. One way to do this is to identify and meet the needs of a landscape species suite, a group of species that, between them, use most of the vital resources needed by other species in their ecosystem.

 Forest elephants, chimpanzees, and mountain gorillas have been proposed as a landscape species suite for the northeastern forests of Congo. Most of the other species in the ecosystem will benefit when elephants, chimpanzees, and gorillas—and their habitat—are protected.

Elephants and gorillas prefer slightly open forests with a wide variety of understory plants. Many other species also use these plants as food or habitat.

All three species avoid roads or rivers with heavy human use.

Elephants need the food, mineral, and water resources found near watering holes and wetlands. These areas also attract gorillas and other wildlife but not chimps.

Chimps prefer older, closed-canopy stands with many mature fruiting trees. Protecting this habitat will benefit the many other species that also feed and live in this part of the forest.

Elephants need corridors wide enough to allow safe passage to other protected areas. This benefits gorillas, chimps, and other core species that will not use a narrow corridor.

Elephants, gorillas, and other species that feed on fast-growing plants are attracted to logged areas that are regrowing. Chimps avoid these areas.

? Why were these three species—the elephant, chimpanzee, and gorilla—chosen to be the landscape species suite for the conservation of the forests of northeastern Congo?

countries regarded as veritable success stories—Gabon and Tanzania, for example—where poaching was thought to be all but eradicated, are actually poaching hotspots. "By shipping through all these completely not-intuitive routes," Wasser says, "the poachers evade detection, because you can't tell where it comes from." Moreover, virtually all of the ivory making its way to Wasser's lab originated in just a handful of places. "We used to think that so much ivory—even before the recent peak, authorities were seizing tons and tons of the stuff every year—we used to think it must be coming from all over the place, because how else could you possibly get that much ivory?" Wasser says. "But the DNA evidence showed that we were way wrong."

Wasser's work falls under the umbrella of **conservation genetics**—the scientific field that relies on species' genetics to inform conservation efforts. Through DNA analysis, conservation biologists can determine the amount of genetic diversity within a population, or

KEY CONCEPT 13.6

Scientists can use conservation genetics to help identify endangered populations and to track illegal sale or trade of endangered species.

the kinship between separate groups—that is, whether they are part of one extended population or represent distinct populations that don't interbreed—or even whether a given population is part of an endangered species. **INFOGRAPHIC 13.5**

For example, IUCN scientists are analyzing anatomical and genetic data in order to determine whether the forest elephant and the savanna elephant of Africa are two different species. (Currently they are considered two distinct subspecies: *Loxodonta africana cyclotis* [forest elephant] and *Loxodonta africana africana* [savanna elephant].) The forest elephant is several feet shorter than the latter, on average (2.5 meters [8 feet] versus 4 meters [12 feet] for full-grown males); eats much more fruit; and has straighter, slimmer tusks and smoother skin that allow it to move easily through dense forests. The savanna elephant, on the other hand, being much larger and literally rougher around the edges, prefers wide-open grasslands. The two had been grouped together—by poachers, laypeople, and conservationists alike—for centuries. But it turns out they have no more genetic overlap than the Indian elephant and the now-extinct woolly mammoth from which it evolved.

INFOGRAPHIC 13.5   TRACKING POACHERS BY USING CONSERVATION GENETICS

↓ Scientists have identified five genetically distinct elephant populations in Africa by analyzing samples of their dung. Law enforcement officials can use this information to identify the population that a confiscated tusk came from by matching its DNA to the dung reference map, helping them track down poachers. This also helps identify regions of high poaching activity.

1 DNA microsatellites
Researchers track DNA microsatellites—regions on a gene where small sequences of DNA subunits called nucleotides (shown here by letter designations) repeat over and over. Microsatellites are not a functional part of a gene and are not used produce a protein.

Microsatellite

2 DNA fingerprinting
Because they are not functional, microsatellites can accumulate changes without harming the individual. They are passed on to offspring and the microsatellites in different breeding populations of African elephants are distinct enough to serve as a fingerprint of that population.

3 Tracking ivory
A reference map of elephant populations in Africa has been created based on microsatellite analysis of dung from each population. The micro satellite pattern in ivory can be compared to this map, identifying poaching "hotspots."

Dung sampling locations of reference populations

The DNA fingerprint of a confiscated tusk can be compared to the fingerprint of known populations to identify where in Africa the tusk (and elephant) came from.

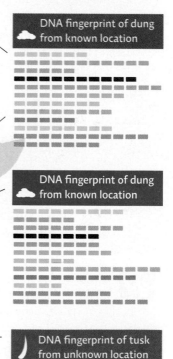

Each colored band represents a different microsatellite region—banding patterns differ for different populations.

DNA fingerprint of dung from known location

DNA fingerprint of dung from known location

DNA fingerprint of tusk from unknown location

Compare the banding pattern for the unknown sample (the tusk) to the DNA fingerprint of the two dung samples. Which population did the tusk come from?

TABLE 13.1 LEGAL PROTECTION FOR SPECIES

↓ A variety of national laws and international treaties (conventions) offer protection for species both inside and outside "protected areas." In the United States, there are laws that specifically protect particular groups (e.g., wild horses or eagles) or specific habitats (e.g., wetlands), but the two broadest laws, both passed in the 1970s, are the main federal statutes that protect species. Internationally, there are many treaties, such as the ones mentioned here, that protect habitats or place restrictions on the use or harvesting of species.

U.S. NATIONAL LAWS	
Marine Mammals Protection Act (1972)	Protects all marine mammals (no killing, capture, or harassment without authorization)
Endangered Species Act (1973)	Mandates protection for "listed" species. Listing is a cumbersome process, and many species don't make it to the list due to budgetary concerns rather than need.

INTERNATIONAL TREATIES	
Convention on International Trade in Endangered Species of Wild Fauna and Flora (1973)	Regulates the sale and trade of endangered or threatened species or products (175 signatory nations)
Convention on Biological Diversity (1992)	The 193 signatory nations agree to pursue goals of biodiversity conservation, sustainable use of biodiversity, and equitable sharing of genetic resources (crops and livestock).

? U.S. environmental laws like the ESA are considered "citizen laws"—citizens or citizen groups (like nonprofit organizations) can sue the federal government to enforce the law if they feel it is not doing so. The Canadian version of this law (Species at Risk Act) does not have this provision. Which law do you think is stronger (i.e., more successful at protecting species)?

One new technology just getting off the ground is the use of *conservation drones* (unmanned aerial vehicles). Drones are being used for research (to census difficult-to-track populations) and for conservation efforts (to identify poachers or even scare elephants away from villages). Simply the presence of drones has been known to scare poachers away.

Legally mandated protection can aid in species conservation

Whether African elephants represent one species or two has huge implications for conservation biology. The **Convention on International Trade in Endangered Species of Wild Fauna and Flora (CITES)**—an international treaty that regulates global trade of selected species—banned the trade of African elephant ivory in 1990. Since then, various governing bodies across Africa and Asia have been lobbying to ease those restrictions. So far, they have had little success, but with two distinct species, Blake says, they may finally have their way. "The forest elephant has been gravely imperiled by poaching," he says. "But the savanna elephant, by many accounts, is doing just fine. So, if they are two different species, CITES could now open the trade in savanna elephants." Demand would then rise, he says. And ivory prices would, too. "Rising demand and rising prices will make it even more profitable for black marketers to operate, and ultimately, the forest elephants will suffer even greater decimation."

There are also international treaties that protect species and ecosystems around the world. The 1992 **Convention on Biological Diversity (CBD)** attempts to reach even further than CITES, supporting conservation and sustainable use of all biological diversity, not just endangered species. But because CBD is less concrete than CITES—it does not outline any specific targets or provide a mechanism to reach its goals, leaving it up to each signatory to determine how best to proceed—it has been less successful.

National laws also protect endangered species. The two main laws in the United States are the Marine Mammal Protection Act and the better-known **Endangered Species Act (ESA)**. Passed in 1973, the ESA mandates that listed species—those that have been officially declared threatened or endangered—be protected through a range of federally funded and scientifically proven strategies. Those strategies include the conservation of natural habitats as well as the breeding, relocation, and/or reintroduction of captive animals into the wild. The law has been controversial since its passage. It has certainly succeeded in bringing back several iconic species from the brink of extinction—the bald eagle and the American alligator, to name just two. But its effectiveness has also been perpetually hampered by a variety of constraints—including landowner disputes and what critics describe as colossal funding shortfalls. **TABLE 13.1**

Of course, CITES has troubles of its own—especially where the

Convention on International Trade in Endangered Species of Wild Fauna and Flora (CITES) An international treaty that regulates the global trade of selected species.

Convention on Biological Diversity (CBD) An international treaty that promotes sustainable use of ecosystems and biodiversity.

Endangered Species Act (ESA) The primary federal law that protects biodiversity in the United States.

KEY CONCEPT 13.7

Legal protection for threatened species includes national laws and international treaties as well as the establishment of protected areas.

Zambia and Tanzania petitioned CITES for permission to sell their stockpiled ivory—that which had come from elephants that died naturally or that were killed before the ban was implemented. The ivory was just sitting there, the countries' representatives argued. The elephants were already dead, and the countries themselves could really use the money.

In the past, several African nations had, with permission from CITES, auctioned off their confiscated and stockpiled

African forest elephant is concerned. Member nations meet every few years to review the status of various species and to modify their regulation, if necessary—a process that critics say is as fraught as the ESA's with political maneuvering, budget woes, and other practical difficulties.

At a 2010 conference, ivory (some of which had come from culled animals—those killed off deliberately when herds exceed park capacity). Selling this ivory was supposed to help elephant conservation efforts by reducing the market for illegally killed animals (and by increasing funds for conservation initiatives). But in the end, poaching only increased.

Opponents to the Zambian and Tanzanian request argued that such allowances would only bolster the illegal trade, which they said was already rampant in both of the petitioning nations. Zambia and Tanzania denied the charge, insisting that poaching had been all but eliminated within their borders. The conference grew heated, and the two sides quickly reached an impasse.

And then Wasser presented his data: DNA analysis showing that much of the ivory seized that year—more than 60% of that which had been smuggled to ports in China and Thailand and New York—did indeed come from Tanzania and Zambia.

The petition was quashed. Since that time several countries, including China and the United States, have destroyed many

---

INFOGRAPHIC 13.6   GLOBAL PROTECTED AREAS

4

↓ Protected areas come in all shapes and sizes and vary according to what level of protection the area receives (e.g., a wildlife refuge that allows hunting of some species versus a nature preserve that does not allow any hunting). About 13% of land on Earth has some protected status (only 1.6% of the oceans are protected; see LaunchPad Chapters 29 and 31 for more on marine protected areas).

PERCENTAGE OF TERRESTRIAL PROTECTED AREAS BY EACH REGION (2011)

Less than 10%
10%–30%
30%–50%
More than 50%

 Which area of the United States has the highest percentage of protected area? Why do you suppose this area has the most protection?

tons of ivory stockpiles (by crushing and/or burning), hoping to send a clear message to potential poachers or consumers that ivory trafficking will not be tolerated.

Protected areas provide another legal avenue for conservation. **Protected areas** are clearly defined geographic spaces on land or at sea that are recognized, dedicated, and managed to achieve long-term conservation of nature. Every nation in the world has protected areas. They include a range of designations. *National parks* are set aside primarily for human recreation. *Wildlife refuges* and *wilderness areas* are generally open to visitors, as well as to hunting and fishing, but are not commercially developed (i.e., they have no restaurants, hotels, or other human accommodations). *Nature preserves* (also called *nature* or *game reserves*) are closed to hunting and fishing; their main goal is to protect wildlife.

Evidence is mounting that protected areas are helping. For example, after a serious coral bleaching event, Kanton Island reef in the South Pacific was designated a marine protected area. Its remarkable recovery (in less than a decade) is attributed to the protection of the entire ecological community (see LaunchPad Chapter 29). Likewise, the African white rhinoceros, a species once believed to be extinct, is thriving in protected areas; its conservation status has been upgraded to vulnerable. **INFOGRAPHIC 13.6**

As with legal safeguards like the ESA and CITES, protected areas have limits. According to the WCS, elephant poaching is lower—and overall elephant populations are dramatically higher—in national parks and game reserves, especially in the eastern and southern regions of Africa, where the savanna elephant is thriving. But most experts agree that protected areas are not enough to save much of anything in the long run. The physical condition of such reserves, and the type and amount of species that can survive inside them, are changing, says Evans. "Whether it's because of global warming, or road building, or other factors, we've already got huge percentages of elephants living outside these protected

protected areas Geographic spaces on land or at sea that are recognized, dedicated, and managed to achieve long-term conservation of nature.

↓ While the number of protected areas has increased dramatically in recent decades, from about 5 million square kilometers (2 million square miles) in 1975 to almost 25 million square kilometers (10 million square miles) in 2011, species extinction rates have not dropped; they are actually increasing as pressure outside of protected areas increases. Protection of the habitats of the most vulnerable species is a must if we want to effectively address species extinctions.

GLOBAL PROTECTED AREAS, 1911–2011

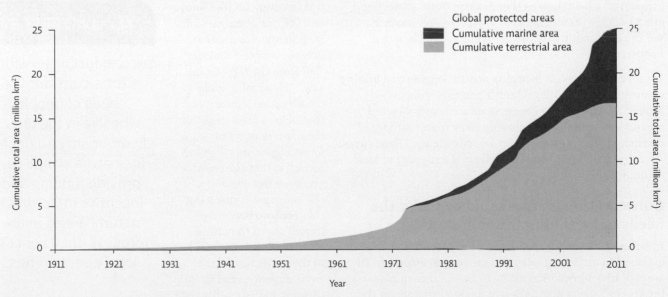

 Starting with 1961–1971, estimate how many million square kilometers were added in each 10-year period. In which decade was the most land area designated as protected?

Bullit Marquez, File/AP Photo

↓ A steamroller crushes 5 tons of confiscated elephant tusks in the Philippines in a public ceremony to bring attention to the problem of elephant poaching and to ensure the tusks did not re-enter the ivory black market.

areas. And what happens then? When those protected areas become places that they can't live? We haven't yet thought about what we are going to do when that happens."

For its part, the WCS is urging private logging and mining companies to curb illegal hunting in their concessions, especially when those concessions are near protected areas. "We need to start moving away from that idea of distinct areas, towards a place of coexistence," Evans says. And that will never happen without the support of local communities.

Conservation plans should consider the needs of local human communities.

In Botswana's forests, peace seems to reign. Elephants, gathered in a clearing, play and groom and engage in all sorts of histrionics. According to Evans, the country boasts the largest population of forest elephants on the continent. There are several reasons for this success: For one thing, the human population is much lower here than it is in countries like Kenya and Gabon, where poaching is epidemic. For another, Botswana's president has "led from the top," Evans says. "He's really made conservation a priority and that helps a great deal." But the main reason Botswana's forest elephants are doing so well is that the country's economy has come to rely on them. It turns out that **ecotourism**—low-impact travel to natural areas that contributes to the protection of the environment—is a big and thriving industry: Tourists will pay good money to see wildlife and well-preserved wild areas, especially if they know their money is helping conservation efforts. "The people of Botswana have made a very proactive decision," Evans says. "For their own long-term survival, they see wildlife as a crucial sustainable resource."

Therein lies the key: For conservation programs to work, they must protect not only the species in question but also the humans who share its habitat.

A 2009 drought in Kenya, for example—one of the worst droughts in living memory—caused the region's tribes to lose many of their cows and crops; meanwhile, the price of ivory soared. Brokers just across the Tanzania border were paying around $20 a pound at the time for raw ivory—a deal too good for any starving family to ignore. The bushmeat trade in central Africa offers another example. The loss of agricultural land to environmental degradation and armed conflict has driven more area residents into natural areas to kill wild animals for meat, which in turn diminishes wildlife, increases access to the area, and spreads zoonotic diseases (see Chapter 5) as more people come into contact with wild animals. Steps to improve agriculture, increase access to nutritious food, and resolve political disputes would do more to mitigate these problems than any law prohibiting bushmeat trade.

Ecotourism is not the only means to such ends. **Debt-for-nature swaps**—in which a wealthy nation forgives part of a developing nation's debt, in return for which the developing nation pledges to protect certain ecosystems—are another way to appeal to the needs of those living nearest the species and ecosystems that are so embattled. (So far, almost $3 billion in debt has been forgiven, and tens, if not hundreds, of millions of acres of wilderness have been preserved through such programs.) Nonprofit

ecotourism Low-impact travel to natural areas that contributes to the protection of the environment and respects the local people.

debt-for-nature-swaps Arrangements in which a wealthy nation forgives the debt of a developing nation in return for a pledge to protect natural areas in that developing nation.

← Officials from the Wildlife Conservation Society and the Manhattan District Attorney's office showcase illegal elephant ivory products seized from two New York jewelry store owners who pled guilty for selling and offering for sale illegal elephant ivory valued at more than $2 million.

Bebeto Matthews/AP Photo

TABLE 13.2   MANY ROUTES TO CONSERVATION

↓ Conservation of biodiversity does not have to depend solely on regulations, laws, and treaties. There are other effective approaches to protecting species and their ecosystems. Some of these are market driven (there is a financial incentive to pursuing them), while others rely on individuals to voluntarily give of their time or money.

METHOD	EXAMPLE
Ecotourism: Low-impact travel to natural areas that contributes to the protection of the environment and respects the local people.	A community, region, or nation may find that tourists who come to see intact ecosystems bring in more money than would be gained by harvesting resources. Ecotourism in Costa Rica brings in more than $3 billion annually.
Valuing ecosystem services: Assessing the monetary value of an intact ecosystem to inform how best to use an area.	In Thailand, the value that a mangrove swamp provides in terms of local harvestable goods and value to offshore fisheries is about six times more than the economic value of shrimp farms that displace the mangroves. If the cost of replacing the mangroves is factored in, shrimp farming would cost money, not generate it.
Debt-for-nature swaps: The preservation of natural areas is funded by an agreement in which a nation forgives part of the debt of a developing nation, or a nonprofit agency pays off the debt, in return for the developing nation's pledge to put a percentage of that saving toward conservation projects.	So far, more than $3 billion in debt has been forgiven in these programs, with almost $1 billion funding conservation in debtor countries. Between 1991 and 1998, the United States forgave $1.4 billion in debt for seven South American countries, generating around $170 million in conservation funding.
Nonprofit organizations: Voluntary membership supports a large number of conservation-oriented organizations, which employ thousands of scientists, educators, legal advisors, and policy experts around the world.	The World Wildlife Fund (WWF) was the first wildlife-focused organization. In 2011, 85% of its donations went to conservation projects. The Sierra Club was founded in 1892 and has many college chapters; it seeks to protect American wilderness and promote environmental protection.
Land trusts and conservation easements: Land trusts are nonprofit organizations set up to protect private nonprofit land under their care. Conservation easements represent a legal agreement between a landowner and land trust or government that permanently limits land development and restricts use of that land to help preserve the conservation value of the land.	Using the money from public donations, the nonprofit organization Nature Conservancy has purchased about 6 million hectares (15 million acres) in the United States, permanently protecting that land from future development. It also funds programs worldwide that protect another 40 million hectares (100 million acres).
Consumer choices: Consumers can influence actions taken by businesses and industries by purchasing products obtained in a way that does not harm species or, better yet, that helps conservation efforts. They can also vote with their dollars by choosing not to buy products whose acquisition harms biodiversity.	Avoid purchasing exotic pets (they may be wild-caught) or any items made from endangered species, such as ivory products, animal skins, and collectable specimens (e.g., butterflies under glass). Buying fair trade, shade-grown coffee and chocolate is another way to promote habitat protection and protect biodiversity in the regions where the crops are grown.
Opportunities for citizen scientists: Volunteer work is an important part of keeping natural areas in good shape. There are a wide variety of educational, research, and wildlife/habitat management opportunities that allow individuals to contribute to conservation work.	The All Taxa Biodiversity Inventory of the Great Smoky Mountains National Park and the Chesapeake Bay Program, aimed at restoring the bay, are two examples of robust programs that allow citizens to pitch in and help in conservation efforts.

? Would you be willing to participate in citizen science projects? If so, what kind of activities would appeal to you?

organizations have also been major players in conserving species and their habitats. **TABLE 13.2**

Consumers also play a major role. The first step is to become informed about the threats to species and the consumer choices that help protect biodiversity. For example, some actions harm endangered species (such as buying wild-caught tropical birds or fish, or ivory products), whereas other choices support the sustainable use of natural areas, like buying sustainably grown coffee and palm oil products (see Chapter 12) or sustainably harvested fish (see LaunchPad Chapter 31). Buying or using

fewer resources (from gasoline to sunglasses to sneakers) means less impact on the natural areas (and their resident species) from which the resources are extracted or made. Finally, reducing one's contribution to climate change will reduce this threat to Earth's biodiversity (see Chapter 21.)

For his money, Blake says better enforcement of existing laws and a marketplace that reflects the true costs of natural resources would go a long way toward curbing poaching. "Bad environmental companies need to be penalized and good ones rewarded," he says. Yes, this

KEY CONCEPT 13.9

The average person can help protect biodiversity by making wise consumer choices that support the sustainable use of natural areas and that avoid contributing to any of the main causes of species endangerment.

will drive up the cost of certain goods—low prices cannot be maintained if companies have to invest in good road planning, stop poaching, and engage in other environmentally and socially friendly practices. But, as with any other natural resource—from water to timber to coal and oil—higher prices might be the key to better conservation and lower overall costs in the long run.

However, Evans says, education may be the most crucial element of all. "We want the communities living in close proximity to elephants to understand that they provide crucial ecosystem services," she says. "We also want people in developed countries to understand that even if they never set foot on the African continent, or see an elephant up close, their decisions have a direct impact on these ecosystems."

Because in the end, it's demand from these countries not only for ivory products but for timber and other resources—that's driving and facilitating the trade.

Ultimately, then, the story of the forest elephant ends not in a shrinking African forest, or even in a Chinese port city, but much closer to home: in a jewelry shop in a bustling Manhattan neighborhood. That's where federal agents recently discovered $2 million worth of illegal ivory. In 2012, the shop's owners pled guilty to selling ivory without permits; under plea bargain, they agreed to forfeit the products and to pay $55,000 in donations to the WCS for use in elephant conservation and anti-poaching efforts. It was a rare victory for conservationists, one that underscored a crucial point for Wasser, Evans, Blake, and others. "It's a global problem that requires very local solutions," Evans says. In the end, preserving biodiversity—even in the jungles of Africa—begins right here at home.

Select References:
Baillie, J. E. M, et al. (2010). *Evolution Lost: Status and Trends of the World's Vertebrates.* London: Zoological Society of London.
Blake, S., et al. (2007). Forest elephant crisis in the Congo Basin. *PLoS Biology,* 5: e111. doi:10.1371/journal.pbio.0050111
CITES. (2014). *Elephant Conservation, Illegal Killing and Ivory Trade: Report to the Standing Committee of CITES.* Sixty-fifth meeting of the Standing Committee. Geneva, Switzerland, July 7–11, 2014.
Stokes, E. J., et al. (2010). Monitoring great ape and elephant abundance at large spatial scales: Measuring effectiveness of a conservation landscape. *PLoS ONE,* 5: e10294. doi:10.1371/journal.pone.0010294
United Nations Environment Programme. (2012). *Global Environmental Outlook—5 Report,* http://unep.org/geo/pdfs/geo5/GEO5_report_full_en.pdf.
Wasser, S. K., et al. (2008). Combating the illegal trade in African elephant ivory with DNA forensics. *Conservation Biology,* 22: 1065–1071.

BRING IT HOME

PERSONAL CHOICES THAT HELP

A recent study estimates that nearly 9 million different species are found on Earth. With such a stunning amount of biodiversity, it is hard to imagine that we could possibly be in danger of losing a significant number of these species to extinction. Yet the extinction rates will continue to grow if we do not consider serious changes.

Individual Steps
• Gain an appreciation for the tremendous diversity we have by watching a documentary of an ecosystem far away from you. Make a list of all the species featured in the documentary and then check their status at www.iucnredlist.org.

• Avoid purchasing any pets that are listed as endangered species, even those bred in captivity, as this can increase the threat to the species in the wild. If you do choose to buy an exotic pet that is not an endangered species, request certification that it is captive bred.
• Consider contributing to an organization that works to protect species or their habitat; check out options at www.fws.gov/endangered/what-we-do/ngo-programs.html.

Group Action
• Identify county and state parks and preserves around you. Find out what specific actions they take to maintain the

biodiversity of the ecosystems there, such as removing invasive species and cultivating native ones. Form a volunteer group to assist in these efforts.
• Discover which endangered species live in your state at www.fws.gov/endangered/. Make flyers and posters with images and information about these species and their habitats. Share them around your school or community.

Policy Change
• The goal of the U.S. Endangered Species Act is the protection of species diversity. To learn more about current challenges and updates to the program, visit www.epa.gov/espp.

ENVIRONMENTAL LITERACY **UNDERSTANDING THE ISSUE**

 1 What events or actions threaten species today, and what conservation status designations are used to identify the threat level? How does the loss or reduction of a species impact its ecosystem?

INFOGRAPHICS 13.1 AND 13.2

1. True or False: The most serious conservation designation for a threatened species is "endangered."

2. The leading cause of species endangerment and extinction is:
   a. invasive species.
   b. pollution.
   c. climate change.
   d. habitat destruction.

3. Explain how the decline of the forest elephant affects other species that share its ecosystem.

 2 How do single-species conservation programs compare to ecosystem-based approaches?

INFOGRAPHICS 13.3 AND 13.4

4. True or False: In a landscape conservation approach, the single most important indicator species in a habitat is targeted for protection. By protecting or restoring the habitat it needs, other species in the area will also be protected.

5. How does ecosystem conservation differ from single-species conservation?
   a. The ecosystem approach has been more successful at attracting the public's attention to conservation needs.
   b. The ecosystem approach focuses on a single "charismatic" species—large, furry, and photogenic.
   c. The ecosystem approach involves restoring and protecting an entire habitat and all the species within it.
   d. All of the above.

6. Which approach to protecting forest elephants do you advocate, and why?

 3 How does conservation genetics contribute to the conservation of species?

INFOGRAPHIC 13.5

7. Conservation genetics:
   a. relies on analysis of species' DNA to make conservation decisions.
   b. is used only for captive breeding of endangered species.
   c. is relevant for single-species approaches to conservation but not ecosystem approaches.
   d. is used by poachers to identify species.

8. Local community members are protesting a proposed development because it would destroy a population of sunflowers they believe is a listed endangered species. Developers claim that this flower is found throughout the area and is not endangered. How would conservation genetics help settle this dispute?

 4 What legal protections do threatened species have in the United States and internationally?

TABLE 13.1

9. True or False: Human recreation, hunting, and fishing are prohibited in wildlife refuges and wilderness areas in order to protect the threatened species and ecosystems within.

10. Which of the following is true of the Endangered Species Act?
    a. It is very well funded by the government, due to the public's strong support for it.
    b. It has not been effective in saving any species from extinction.
    c. It is controversial due to its restrictions on landowners.
    d. It mainly protects endangered species outside the United States.

11. How can CITES protect endangered species if it only regulates international trade? Do you think this is sufficient to protect endangered species? Why or why not?

12. Distinguish between these types of protected areas: national parks, wildlife refuges and wilderness areas, and nature preserves. Why might it be useful to have these different types of protected areas?

 5 In what ways are local communities important to the success of conservation efforts? What role do consumers play in protecting biodiversity?

INFOGRAPHIC 13.6 AND TABLE 13.2

13. True or False: Land trusts and conservation easements protect private land from future development.

14. Wildlife camera safaris have replaced hunting safaris in popularity in Africa. (Photos are taken instead of animal lives.) Some of the money from these safaris is going to conservation and to local communities. This is an example of:
    a. ecotourism.
    b. debt-for-nature programs.
    c. a land trust.
    d. valuing ecosystem services.

15. Identify some actions you could personally take that would contribute to conservation of biodiversity.

SCIENCE LITERACY **WORKING WITH DATA**

The International Union for Conservation of Nature (IUCN) has maintained an inventory, called the "Red List," of the extinction risk for the world's species since 1963. Look at the following IUCN graph and answer the next five questions.

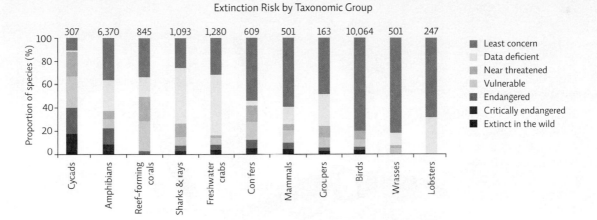

Extinction Risk by Taxonomic Group

Interpretation

1. Describe what the graph shows about the proportion of cycad species in each category.

2. Compare the proportion of threatened (critically endangered, endangered, vulnerable) amphibians to birds.

3. Compare the number of threatened amphibians to birds (Hint: The total number of species per group is at the end of each bar in the graph).

Advance Your Thinking

4. Identify the three groups that contain the highest proportion of species that are "data deficient." Why do you think less is known about species in these particular groups? How would you predict that extinction risk estimates would change if more were known about these species—that is, would a higher proportion be threatened, or would more be of "least concern"?

5. Many North American species of birds are migratory, spending winters in tropical regions (such as Central or South America) and summers in North America. Habitat destruction due to deforestation is one of the most common threats many of these species face. Suppose you spearheaded a program that returned adequate habitat to the migratory birds in your area. Why might this not be enough to allow the species to recover?

INFORMATION LITERACY **EVALUATING INFORMATION**

Everyone has limited time and money, and choosing how to invest that time and money to support biodiversity issues can be complex. Should you fund an organization that protects individual species or ecosystems? How much money is reasonable to spend on administration and fundraising? What conservation focus do you prefer?

Evaluate the following websites and work with the information:

1. Visit the World Wildlife Fund (WWF) website (www.worldwildlife. org) and the Nature Conservancy website (www.nature.org). For each, answer the following questions:
   a. What is the mission of the organization?
   b. Is the website up-to-date? Does it appear to be accurate? Reliable? Explain.
   c. Does the organization use a single-species conservation approach, an ecosystem approach, or both? Support your answer.
   d. What does the organization claim about the percentage of its operating budget it spends on programs versus on other costs? What evidence does it provide to support this claim? Is the evidence sufficient?

   e. How long has the organization been operating? What are some of the successes it claims?

2. Go to Charity Navigator (www.charitynavigator.org).
   a. Search for World Wildlife Fund. Does the information on the percentage of the operating budget spent on programs, fundraising, and administration match the claims of the WWF website? If there is a discrepancy, why might that be? Which source would you trust, and why?
   b. Search for Nature Conservancy. How does the budget information on the Charity Navigator site compare to that on the Nature Conservancy site?
   c. How does the Nature Conservancy compare to WWF in terms of money spent on programs versus on fundraising and administration? Be specific.
   d. Would these differences lead you to choose to donate to one organization over the other? Why or why not? What other information would you use to make a decision about which organization to support?

Find an additional case study online at http://www.macmillanhighered.com/launchpad/saes2e

CORE MESSAGE
Freshwater is a precious but limited resource, and it is essential to life. Some regions consume water faster than it is replenished. And, unfortunately, water is not evenly distributed across the globe; many people worldwide lack access to enough clean water. Methods are available to recover and purify otherwise dirty water, but we also need to use water more wisely.

AFTER READING THIS CHAPTER, YOU SHOULD BE ABLE TO ANSWER THE FOLLOWING **GUIDING QUESTIONS**

1

How is water distributed on Earth, and what are the sources of freshwater? How does water cycle through the environment?

2

What are the causes and consequences of water scarcity?

TOILET TO TAP

A California county is tapping controversial sources for drinking water

Clean drinking water is a rare thing in many places on Earth.
ConstantinosZ/iStock/360/ Getty Images

3

What is an aquifer, how does it receive water, and what problems emerge when too much water is removed?

4

What are some of the ways that our wastewater is treated to make it potable or safe to release into the environment?

5

What can be done to address water scarcity issues?

One of the most exciting moments in Shivaji Deshmukh's career as a water engineer came one bright, sunny day in January 2008. He had gathered with staff from the Orange County Water District (OCWD) in Anaheim, California, to watch for the first time as former sewage water, cleaned using state-of-the-art techniques, was pumped into underground drinking water sources. It was the beginning of a groundbreaking project designed to help save the region from ongoing, and frightening, water shortages.

"It's basically this drought-proof supply of water," says Deshmukh. "Nobody else has done it. Nobody thought a community could support it, because they would be too grossed out by it."

The water that Deshmukh and other engineers watched seep into the region's underground water stores that day in 2008 was purified **wastewater**—including sewage and used water from homes and industrial sites. Understandably, when many residents first heard about the project, they were concerned.

But that same month, Deshmukh and other OCWD staff attended a dedication ceremony for the Orange County Groundwater Replenishment System (GWRS), at the water treatment plant in Fountain Valley, California, along with hundreds of other people, including various community groups, to honor the massive project. Having that support from the community was key to the project's success, says Deshmukh—but getting it hadn't been easy.

◉ **WHERE IS ANAHEIM, CALIFORNIA?**

Water is one of the most ubiquitous, yet scarce, resources on Earth.

Even though Earth is covered in more than 1.4 billion cubic kilometers (370,000,000 trillion gallons) of water—about 75% of its surface—only about 1/100 of 1% of that water is usable by humans.

Water provides many important ecosystem services that animals and plants require to live. Up to 75% of the human body, for instance, consists of water. But humans need liquid **freshwater** (which has few dissolved ions such as salt); ocean water is too salty for human consumption and is toxic in large doses.

Complicating things, nearly 80% of the freshwater on the planet is trapped in ice caps at the poles and glaciers around the world, which contain more than 35 million cubic kilometers (9,000,000 trillion gallons)—enough to fill 140 billion Olympic-sized swimming pools. **INFOGRAPHIC 14.1**

Wherever there is water, it is constantly moving through the environment via the **water cycle** (hydrologic cycle). Heat from the Sun causes water to evaporate from **surface waters** (rivers, lakes, oceans) and land surfaces. At the same time, plant roots pull up water from the soil and then release some into the atmosphere in a process called **transpiration**. Plants with deep roots, like trees, may bring up thousands of gallons of water a year, releasing much of this to the atmosphere. Altogether, the combination of **evaporation** and transpiration—*evapotranspiration*—sends more than 66,000 cubic kilometers of water vapor into the atmosphere every year, equivalent to 17,000 trillion gallons. Once aloft, that water condenses into

wastewater Used and contaminated water that is released after use by households, industry, or agriculture.

freshwater Water that has few dissolved ions such as salt.

water cycle The movement of water through various water compartments such as surface waters, atmosphere, soil, and living organisms.

surface water Any body of water found above ground, such as oceans, rivers, and lakes.

transpiration The loss of water vapor from plants.

evaporation The conversion of water from a liquid state to a gaseous state.

KEY CONCEPT 14.1

Only 3% of water on Earth is freshwater, and very little of that is accessible to humans. Fortunately, even this small percentage represents a large amount of water.

Courtesy of Mark Greening/Orange County Water District

↑ Crystal-clear purified water from the Groundwater Replenishment System is piped to the Orange County Water District's percolation ponds in Anaheim, California.

INFOGRAPHIC 14.1 **DISTRIBUTION OF WATER ON EARTH**

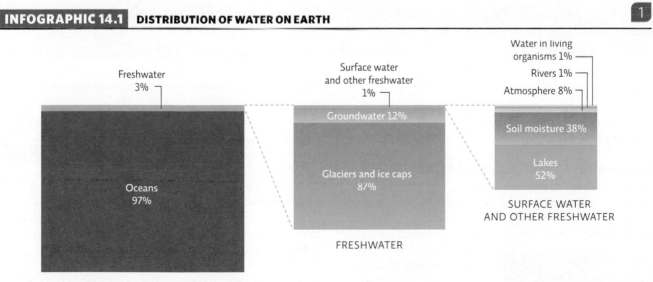

Freshwater 3%

Surface water and other freshwater 1%

Water in living organisms 1%
Rivers 1%
Atmosphere 8%

Groundwater 12%

Soil moisture 38%

Oceans 97%

Glaciers and ice caps 87%

Lakes 52%

TOTAL GLOBAL WATER

FRESHWATER

SURFACE WATER AND OTHER FRESHWATER

↑ Most of the water on Earth is found in the oceans, and most of the freshwater is tied up in ice and snow. Only about 0.001% of all of Earth's water is available for us to use, but with more than 1,300 trillion liters (350 trillion gallons) of water on the planet, that is still a lot of water.

 Based on this diagram, what percentage of the total water supply on Earth is found in groundwater? What percentage is found in rivers?

INFOGRAPHIC 14.2   **THE WATER CYCLE**   1

↓ Water cycles between liquid and gaseous forms as it moves through space and time. Ocean water (which we cannot use) is converted to freshwater when it evaporates and falls back to Earth as precipitation, refilling freshwater surface and underground water supplies. Liquid freshwater is a renewable resource as long as we don't use it faster than it is naturally replenished.

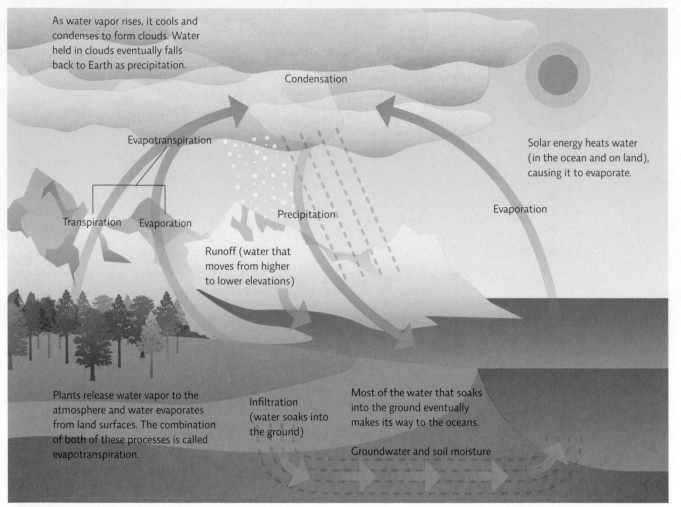

As water vapor rises, it cools and condenses to form clouds. Water held in clouds eventually falls back to Earth as precipitation.

Condensation

Solar energy heats water (in the ocean and on land), causing it to evaporate.

Evapotranspiration

Evaporation

Transpiration   Evaporation

Precipitation

Runoff (water that moves from higher to lower elevations)

Plants release water vapor to the atmosphere and water evaporates from land surfaces. The combination of both of these processes is called evapotranspiration.

Infiltration (water soaks into the ground)

Most of the water that soaks into the ground eventually makes its way to the oceans.

Groundwater and soil moisture

The trees of tropical rain forests are said to contribute as much as half of the rain that falls back on the forest. Explain how this occurs.

clouds (**condensation**) and may fall back to Earth as **precipitation** (rain, snow, sleet, etc.). **INFOGRAPHIC 14.2**

Almost all precipitation ends up falling on the oceans, and a tiny remainder falls on land. This latter portion is the part humans can harvest for their own use. We access freshwater from lakes and rivers (surface water) and from **groundwater**. Worldwide, the biggest drain on freshwater supplies isn't the water used to shower, flush the toilet, and wash dishes. Globally, about 70% of all freshwater withdrawals go to agriculture. Industry is the next biggest consumer of water. Domestic use accounts for only 10% of freshwater usage.

People don't always live near abundant sources of freshwater, making access a vital issue. Around the world, many areas suffer from **water scarcity**—not having sufficient access to clean water supplies. In some dry regions, there is simply not enough to meet needs; many arid nations like those of the Middle East, parts of Africa, and much of Australia face water shortages as a way of life. The Middle Eastern countries of Bahrain, Qatar, Kuwait, and Saudi Arabia have the lowest per capita water availability in the world, but these oil-rich nations can

KEY CONCEPT 14.2

Water cycles through the environment via the water cycle, a process that constantly recycles water on Earth.

KEY CONCEPT 14.3

Water scarcity can be physical or economic. Poor sanitation causes health problems and can contribute to scarcity when it contaminates local water sources.

afford to invest in costly technology to access water (like facilities to remove salt from seawater). In other areas, particularly in Sub-Saharan Africa, there may be enough water, but people do not have the money to purchase it or dig wells to access it. People in these areas may be getting by on just a few gallons of water a day—and that water may not even be safe to use. By far, per capita domestic (household) use of water in developed nations is much higher than that in developing nations; in the United States the average person uses more than 300 liters (about 80 gallons) of water a day.

The United Nations estimates that as many as 3.5 billion people—half of the world's population—lack access to enough clean water; 2.5 billion lack access to sufficient sanitation facilities (safe disposal of human waste). In developing nations where water and funding for basic sanitation are scarce, people use nearby surface waters to meet their basic cooking, drinking, and washing needs. These waters can be contaminated with raw sewage, which increases the chance for disease transmission. According to the World Health Organization (WHO), more than 1.1 trillion liters (300 billion gallons) of raw sewage enters the Ganges River of India every minute. In Africa, almost 3,000 people die each day from waterborne diseases like cholera and typhoid fever as a result of poor sanitation and contaminated water.

As human populations increase, so will scarcity and sanitation issues; according to the United Nations, two of every three people will face water shortages by 2025. **INFOGRAPHIC 14.3**

condensation The conversion of water from a gaseous state (water vapor) to a liquid state.

precipitation Rain, snow, sleet, or any other form of water falling from the atmosphere.

groundwater Water found underground in aquifers.

water scarcity Not having access to enough clean water.

INFOGRAPHIC 14.3   GLOBAL WATER USE AND ACCESS

↓ Globally, agriculture is our biggest user of water and also the sector with the largest amount of waste. There is much room for improvement in how we use water in all sectors.

↓ Individuals in more developed countries use far more water per person than those in less developed countries. In some areas, individuals must make do with only a few gallons a day.

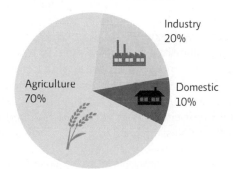

GLOBAL WATER USE BY SECTOR

Industry 20%
Agriculture 70%
Domestic 10%

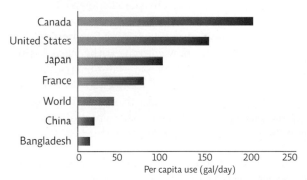

DOMESTIC WATER USE

Canada
United States
Japan
France
World
China
Bangladesh

Per capita use (gal/day)
0   50   100   150   200   250

↓ Around the world, more than 750 million people lack access to clean water, and as many as 3.5 billion have access to some, but not enough clean water to meet their needs. An additional 2.6 billion have no access to sanitation.

■ **Physical water scarcity**
Use of water is exceeding sustainable limits.

■ **Approaching physical water scarcity**
With more than 60% of river flow withdrawn, these areas will face water shortage in the near future.

■ **Economic water scarcity**
Access to water is limited by the ability to pay for it, not by its physical scarcity.

■ **Little or no water scarcity**
Water supplies are abundant and there are no economic constraints to access it.

■ **Not estimated**

How might steps to address physical water scarcity be different from steps to address economic water scarcity?

The United Nations estimates that as many as 3.5 billion people—half of the world's population—lack access to enough clean water.

Like other communities around the world, California depends on many sources of water.

Around the world, each region faces unique water challenges. In California, freshwater flows into the northern part of the state when the Sierra Nevada Mountain snowpack melts in the spring. This snowmelt provides as much as one-third of California's water. But as Earth's climate changes, the state could lose much of its snowpack. Indeed, in the 2009 "water year" (California tracks its yearly totals from July of one year to June of the next), California's precipitation was 20% below average, and the snowpack was 40% below its average size. The 2014 water year was the driest on record in California; precipitation was 55% below normal, and the spring snowpack was 97% below normal.

---

aquifer An underground, permeable region of soil or rock that is saturated with water.

infiltration The process of water soaking into the ground.

water table The uppermost water level of the saturated zone of an aquifer.

Even in a good snowpack year, the state faces major water issues, explains Deshmukh, now working at the West Basin Municipal Water District in Carson, California. Two-thirds of California's water is located in the northern part of the state, but two-thirds of the state's residents live in the south, he explains. At the moment, the state ships some of the northern water to the south via pipes and canals, which costs money and uses a great deal of electricity. And if there is an earthquake or another natural disaster, that transport system could be cut off.

Many people (not just in California) draw their water from an underground region of permeable soil or porous rock saturated with water, called an **aquifer**. These stores receive water from rainfall and snowmelt that soaks into the ground through **infiltration**. Plant roots take up some of the water along the way, but much of the water continues to move downward, filling every available space in the aquifer. As the water trickles down, it becomes naturally filtered by rocks and soil, which trap bacteria and other contaminants as the water passes by. The top of this water-saturated region, referred to as the **water table**, rises and falls due to seasonal weather changes.

The depth of Orange County's groundwater varies, says Deshmukh. At the coast, the aquifer is 6 to 90 meters (20 to 300 feet) deep, but further inland, at its deepest, the groundwater extends about 900 meters (3,000 feet) deep. Water quantity is often measured in terms of acre-feet—the amount needed to cover an acre in water to a depth of 1 foot (30.5 centimeters), which is equivalent to more than 1.1 million liters (300,000 gallons). One acre-foot of water is enough for two American families for 1 year. Deshmukh estimates that nearly 5,000 acre-feet of water

KEY CONCEPT 14.4

Aquifers are refilled when water soaks into the ground, but hard land surfaces in urban and suburban areas limit infiltration.

↓ Shivaji Deshmukh of the Orange County Water District spearheaded the groundwater replenishment system project that purifies wastewater and injects it back into the local aquifer.

Ric Francis/AP Photo

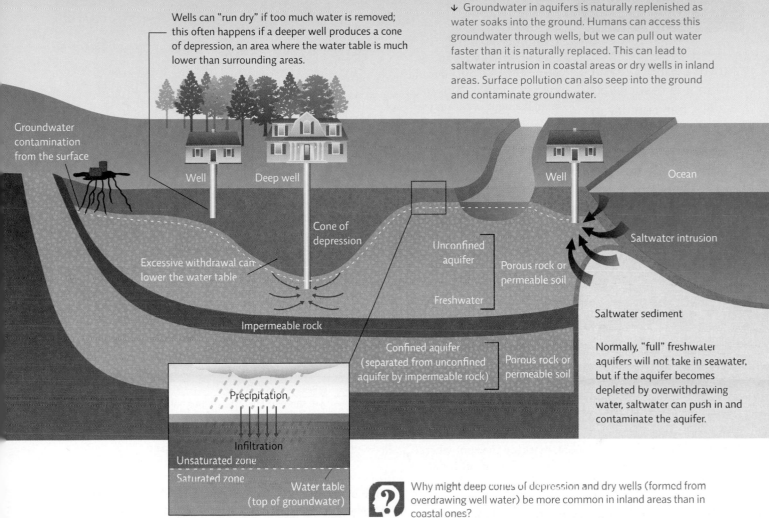

Wells can "run dry" if too much water is removed; this often happens if a deeper well produces a cone of depression, an area where the water table is much lower than surrounding areas.

↓ Groundwater in aquifers is naturally replenished as water soaks into the ground. Humans can access this groundwater through wells, but we can pull out water faster than it is naturally replaced. This can lead to saltwater intrusion in coastal areas or dry wells in inland areas. Surface pollution can also seep into the ground and contaminate groundwater.

Groundwater contamination from the surface

Well

Deep well

Well

Ocean

Cone of depression

Saltwater intrusion

Unconfined aquifer

Porous rock or permeable soil

Excessive withdrawal can lower the water table

Freshwater

Saltwater sediment

Impermeable rock

Confined aquifer (separated from unconfined aquifer by impermeable rock)

Porous rock or permeable soil

Normally, "full" freshwater aquifers will not take in seawater, but if the aquifer becomes depleted by overwithdrawing water, saltwater can push in and contaminate the aquifer.

Precipitation

Infiltration

Unsaturated zone

Saturated zone

Water table (top of groundwater)

Why might deep cones of depression and dry wells (formed from overdrawing well water) be more common in inland areas than in coastal ones?

is accessible from the deepest part of the aquifer. But that deep subterranean water is harder to get, and it costs more to pump it out of the ground than it costs to remove the groundwater closer to the surface.

In addition to withdrawals for agriculture, industry, and personal use, anything that reduces infiltration will reduce the rate at which the aquifer refills and thus decreases the amount of water we can sustainably remove. Infiltration is hampered in urban and suburban settings because of all the hard surfaces, such as roads and buildings; even a typical suburban lawn is so compacted from the home construction process that very little water infiltrates the ground. Urban and suburban designs that provide ways for water to soak into the ground—such as permeable pavement and *rain gardens*—can help refill aquifers as well as help prevent flooding events after heavy rainfalls. (For more on rain gardens, see Chapter 15.)

But in California, excess withdrawals from the aquifer were not leading to depleted aquifers and dry wells. Instead, the proximity to saltwater was actually threatening the freshwater supply.

Decades ago, Orange County officials discovered to their dismay that saltwater was seeping into some of the region's aquifers, putting those precious freshwater stores in jeopardy. Groundwater levels are typically higher than sea level, so saltwater doesn't infiltrate aquifers. But as freshwater was pumped out of the county's aquifer inland, the groundwater level dropped, so salty ocean water had started to enter the coastal edge of the aquifer to the west, where it bordered the Pacific Ocean. In Orange County, some aquifers are confined by geologic faults, which prevent ocean water from entering at some points—but not everywhere. It was in these unconfined coastal aquifers that **saltwater intrusion** was becoming a problem. **INFOGRAPHIC 14.4**

To stem the influx of saltwater, in 1975 the Orange County Water District started pumping highly treated (purified) sewage wastewater into injection wells. At about 19 million liters (5 million gallons) a day, the

saltwater intrusion The inflow of ocean (salt) water into a freshwater aquifer that happens when an aquifer has lost some of its freshwater stores.

If water is removed from aquifers faster than it is resupplied, wells can run dry in inland areas or become contaminated with saltwater in coastal areas.

underground injection created a curtain of freshwater along the California coast that prevented salty ocean water from seeping into the county's aquifer. This water management program was the first to pump treated wastewater into the ground, says Deshmukh.

Most residents—even those who thought about water every day—had no idea that water that had recently been flushed down someone's toilet was being cleaned and then pumped into the area's groundwater supply.

Untreated wastewater can contaminate freshwater sources and is a serious health risk worldwide.

Jack Skinner is a medical doctor and lifelong surfer based in Orange County, whose house in Newport Beach is supplied with water from the county's wells. His second home is in Laguna Beach, surrounded by water; in the morning, he and his wife would watch from their oceanview apartment as the Sun would light up a pole stationed about a mile offshore, marking the spot where a nearby sewer treatment plant discharged its **effluent**. "Every morning when we got up, we would see that marker." In the early 1980s, he began experiencing eye infections while distance swimming along the coast. He decided to learn more about the impact of releasing wastewater into the ocean and how the wastewater was being treated.

Sewage can carry pathogens like viruses and disease-causing bacteria, and surfers get exposed to these when they surf in sewage outflow that is untreated. Public health officials monitor drinking and recreational waters for the presence of **coliform bacteria**. Since many coliforms are intestinal bacteria, their presence could indicate fecal contamination of the water.

Over the years, Skinner had become a self-proclaimed "troublemaker," making sure that wastewater was highly treated before being discharged into the Pacific Ocean, and speaking publicly about water pollution.

One day, the Orange County Sanitation District (OCSD) called Skinner and asked him to

effluent Wastewater discharged into the environment.

coliform bacteria Bacteria often found in the intestinal tract of animals; monitored to look for fecal contamination of water.

↑ An aquifer core sample shows the porous nature of a limestone rock formation. Groundwater can accumulate within and move through the limestone pores and cracks of the aquifer. Wells tap into these formations and pump out the water held there.

Courtesy of the St. John's River Water Management District, Palatka, Fla.

come in for a meeting. The OCSD wanted to consult with him about a massive new project that that would need the support of the community in order to succeed. What people at the meeting said made Skinner very concerned.

In the mid-1990s, Orange County water sanitation engineers and hydrologists found that they faced a problem of water scarcity and, at times, the opposite problem—too much water.

As an increasing number of people moved to the region, they used more water, creating excess wastewater. The OCSD already operated an 8-kilometer (5-mile) "outfall," or underground pipeline that takes treated sewage water past California's beaches and out to the Pacific Ocean. The OCSD also had another 1.6-kilometer (1-mile) long outfall to the Santa Ana River, but it was not enough to handle heavy rainfall, which could wash sewage through and out of a facility before it could be adequately treated. About 375 million liters (100 million gallons) a day of partially treated sewage water was already flooding the river. Nearly five times that amount would reach the river during a major storm. The county was considering building another pipeline to carry excess water to the Pacific Ocean.

But such an endeavor would be incredibly expensive. So the OCSD considered a project that would solve both the problem of too much wastewater and too little freshwater. It decided to expand the existing system that used treated wastewater to protect groundwater from infiltration by the ocean. But folks at the agency knew they would face community opposition.

Since the 1970s, the county had been pumping only small amounts of treated wastewater into the ground—about 19 million liters (5 million gallons) per day, and only at the coasts. This new project would expand that amount to 260 million liters (70 million gallons) per day, and it would take place not just at the seawater barrier. The actual amount of treated wastewater that would make its way into people's homes would vary but would ultimately average 15% in north and central Orange County, says Deshmukh. "All of a sudden, it became a significant component of the water supply."

Once the project was conceived, OCSD began intensive community outreach, such as giving presentations in front of community groups about water scarcity and the need for new sources and placing representatives at big events. "Any chance we got we would talk about it," says Deshmukh. Part of the multiyear initiative included outreach to community leaders, such as Skinner.

When Skinner initially learned about the project, and the fact that OCSD had been doing a smaller version at the coast for decades, he was uneasy. "Since the late 1970s, my family had unknowingly been drinking that water,"

↓ A sinkhole swallows three cars during a heavy rainstorm in Chicago, Illinois, in 2013. Aquifers can be weakened when they lose water (due to drought or when well water is removed faster than it is replaced) that supports the open spaces in an aquifer. A heavy rain event may add enough pressure to the overlying ground to cause the aquifer to collapse and open up a sinkhole.

Zbigniew Bzdak/Chicago Tribune/McClatchy-Tribune via Getty Images

Wastewater can be decontaminated using high-tech methods that use advanced filtration and harsh chemicals or low-tech methods that mimic the way wetlands purify water.

Skinner says. "I thought of my daughter, who lives in the area. I thought about my grandson, Robbie. He would be drinking the water, and is probably right now. I wanted to be sure it was safe."

So Skinner set about learning how the wastewater would be treated. The state health department appointed him to serve on a state committee that would review the treatment process for the recycled wastewater, and Skinner began speaking with experts about the techniques.

Wastewater is simply water that has been used and disposed of (by individuals, businesses, or industry), and it can contain a wide variety of contaminants, from food waste to sewage to chemical pollutants. Prior to the 20th century, most wastewater in the United States (and elsewhere) was simply released to the environment, untreated. As populations swelled and wastewater amounts increased, this untreated sewage became a major source of contamination, harming the environment and human populations. Today, before it can be safely released into the environment, it must be cleaned, or "treated."

Typically, **wastewater treatment** includes initial steps that filter the water then send it to settling tanks where much of the remaining suspended solids sink to the bottom. The water continues on to other tanks where bacteria digest much of the remaining organic matter. At this point it may be clean enough to release into the environment.

wastewater treatment The process of removing contaminants from wastewater to make it safe enough to release into the environment.

potable Clean enough for consumption.

wetland An ecosystem that is permanently or seasonally flooded.

dam A structure that blocks the flow of water in a river or stream.

reservoir An artificial lake formed when a river is impounded by a dam.

Skinner learned that the first step of the cleaning process used by Orange County consisted of microfiltration, in which microscopic, strawlike fibers, 1/300 the diameter of a human hair, filter out many suspended solids, bacteria, and other viruses because only water can pass through the center of the fibers. Then, to render wastewater **potable**—safe for humans to drink—engineers would perform a crucial step known as reverse osmosis. During this process, they use

pressure to force water through a plastic membrane. The pores of this membrane are so tight, explains Deshmukh, that salt and other contaminants (such as pharmaceutical drugs and toxic chemicals) do not pass through, but water does. After reverse osmosis, the water is exposed to ultraviolet (UV) light, which kills any remaining viruses and bacteria. At completion, the final product is cleaner than state and federal regulations require.

Another California community, Arcata, tackled its water purification problems in a different way. Rather than construct a typical wastewater treatment facility to handle sewage that had been contaminating nearby Humboldt Bay, they repurposed a retired landfill near the coast by converting it to a **wetland**—an ecosystem that is permanently or seasonally flooded. A slow river meanders through the wetland where organisms there purify it; to them, it's not "sewage," it's food. The Arcata facility depends on nature to perform the job of water purification. The water discharged into the ocean is very clean, and the health of the bay ecosystem has improved. But while the Arcata facility addresses the need to decontaminate sewage, it does not take the extra steps to produce potable water like Orange County's GWRS.
INFOGRAPHIC 14.5

The GWRS went online in January 2008, and now Orange County's water district taps its artificially replenished groundwater to deliver clean freshwater to people for drinking, irrigation, and other uses. As of spring 2014, it had recycled more than 5 trillion liters (136 billion gallons) of water. Every day, approximately 260 million liters (70 million gallons) of recycled water are pumped into wells or percolation basins in Anaheim, where sand and gravel naturally purify the water further as it trickles down into the region's aquifers—enough to meet the needs of nearly 600,000 residents.

Solving water shortages is not easy.

Though only about 1% of Earth's freshwater is surface water, these lakes and rivers are an important source of water for many communities. To ensure an ongoing freshwater resource, communities often invest in **dams**. These barriers slow the flow of rivers and create **reservoirs**, large bodies of water that hold freshwater for a variety of uses (freshwater source, flood control, electricity production). In the United States, these often become recreation sites and fishing

Water scarcity can be addressed by storing water (dams) and desalinating seawater, and with new solutions like using purified wastewater. All these solutions come with trade-offs.

INFOGRAPHIC 14.5   **HOW IT WORKS: WASTEWATER TREATMENT**

4

↳ Sewage must be treated before it can be safely released to the environment. Most communities use chemical- and energy-intensive high-tech methods, but systems that mimic nature can also effectively purify water.

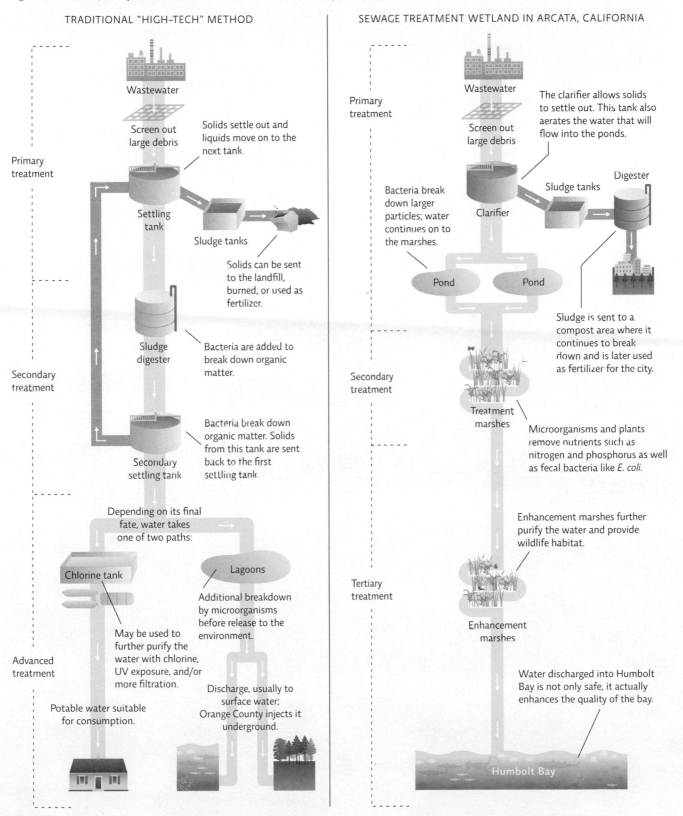

TRADITIONAL "HIGH-TECH" METHOD

Primary treatment

Wastewater

Screen out large debris

Settling tank — Solids settle out and liquids move on to the next tank.

Sludge tanks

Solids can be sent to the landfill, burned, or used as fertilizer.

Sludge digester — Bacteria are added to break down organic matter.

Secondary treatment

Secondary settling tank — Bacteria break down organic matter. Solids from this tank are sent back to the first settling tank.

Depending on its final fate, water takes one of two paths:

Chlorine tank — May be used to further purify the water with chlorine, UV exposure, and/or more filtration.

Lagoons — Additional breakdown by microorganisms before release to the environment.

Advanced treatment

Potable water suitable for consumption.

Discharge, usually to surface water; Orange County injects it underground.

SEWAGE TREATMENT WETLAND IN ARCATA, CALIFORNIA

Primary treatment

Wastewater

Screen out large debris

The clarifier allows solids to settle out. This tank also aerates the water that will flow into the ponds.

Clarifier

Bacteria break down larger particles; water continues on to the marshes.

Sludge tanks

Digester

Pond   Pond

Sludge is sent to a compost area where it continues to break down and is later used as fertilizer for the city.

Secondary treatment

Treatment marshes — Microorganisms and plants remove nutrients such as nitrogen and phosphorus as well as fecal bacteria like *E. coli.*

Tertiary treatment

Enhancement marshes further purify the water and provide wildlife habitat.

Enhancement marshes

Water discharged into Humbolt Bay is not only safe, it actually enhances the quality of the bay.

Humbolt Bay

 Compare the traditional high-tech method of wastewater purification to the wetland system for wastewater purification. What do they have in common? How are they different?

resources. California depends on reservoirs (both in and outside the state) for much of its water supply. But while reservoirs are a valuable resource, they lose an enormous amount of water every day through evaporation. In fact, thanks to lower-than-normal rainfall and snowfall and high water demand, water levels in just about every reservoir in northern California were well below average in 2014.

Depending on temperatures, atmospheric conditions, and the surface area of a reservoir or lake, wide-open bodies of water can evaporate thousands of gallons of water a day in desert settings where it's hot and dry. The California Department of Water Resources calculated that freshwater reservoirs in the South Coast region lost more than 200 billion liters (53 billion gallons) to evaporation in 2000 alone. Worldwide, reservoirs lose more water to evaporation than is used for industry and domestic purposes combined.

The construction of dams can also spark political conflicts. In the Middle East, Turkey's plans to build 22 dams that pull water from the Tigris and Euphrates rivers for agriculture and electric power will impact its neighbors Syria and Jordan downstream. With too little water available for too many people, this hotspot may be the site of future conflict.

---

desalination The removal of salt and minerals from seawater to make it suitable for consumption.

↓ This wetland marsh in Arcata, California, is actually part of a wastewater treatment system that uses nature to help purify sewage. The wetland is now an Audubon birding sanctuary.

David Howell

↑ Revelstoke Dam is one of four dams on the Columbia River in British Columbia. A fifth penstock turbine was recently added adjacent to the original four shown here (the long tubes on the front face of the dam), giving the dam a generating capacity of around 2500 megawatts. The sixth penstock may be installed and operational by 2019, making this the most powerful dam in British Columbia.

Of course, Californians are lucky enough to have plenty of water all along the coast. But removing salt and other minerals from seawater—a process known as **desalination**—is expensive and uses a large amount of energy. Still, thousands of desalination plants worldwide operate today to meet some of the water needs of their regions. The largest such facilities in the world are in the Middle East, some of which are processing around 750 million liters (200 million gallons) of water per day—about 10 times the volume of the largest U.S. plants (which are in Tampa Bay, Florida, and El Paso, Texas). In late 2012, California began construction on a $1 billion desalination plant that would become the largest facility in the United States, expected to produce 190 million liters (50 million gallons) per day; other California facilities are in the planning stages.

TABLE 14.1   WATER-SAVING TECHNOLOGIES AND ACTIONS   5

↓ Our water usage can be reduced by using new water-efficient technologies and by making behavioral changes that don't waste water. This can be as easy as simply turning off a faucet when not using the water, taking shorter showers, and using appliances like dishwashers only when they are full.

Old Technology	New Technology	Behavioral Change
Toilet 6 gallons/flush	**Low-Flow Toilet** 1.3 gallons/flush	Don't flush tissues—use the trash. Flush liquid waste less frequently.
Shower 3.8 gallons/minute **Bath** 35 gallons	**Low-Flow Shower Head** 2.3 gallons/minute	Take a "Navy" shower: Turn off the shower head except to rinse (some shower heads come with a convenient valve that allows you to switch off the water without turning it off at the source).
Faucet 5 gallons/minute	**Low-Flow Faucet** 1.5 gallons/minute	Don't leave the faucet running while brushing your teeth, shaving, or washing your face.
Washing Machine 40 gallons/load	**Washing Machine (Energy Star)** 22 gallons/load	Don't wash a clothing item unless it needs it (those jeans can probably be worn several times before washing) and run the washer only when it is full.
Dishwasher 9 gallons/load	**Dishwasher (Energy Star)** 4 gallons/load	Run the dishwasher only when it is full and limit the amount of rinsing you do before loading dishes into the dishwasher; if you have a new dishwasher, rinsing isn't needed.

Estimate how long your typical shower lasts and then calculate how much water you would use over the course of a year if you used a traditional, 3.8-gallon/minute shower head. Do the same calculation for the low-flow shower head. How many gallons per year would you save by switching to a low-flow shower head? Compared to the water used in a typical shower using a traditional shower head, how much would you save in a year if you reduced the duration of your shower by half and used a low-flow shower head?

Conservation is an important "source" of water.

The GWRS project was expensive: The total price tag to build the system came to about $481 million from federal, state, and local funding. It is currently being expanded to create another 114 million liters (30 million gallons) per day.

An easier and cheaper way to maintain water supplies is simply not to waste so much. For example, water-saving irrigation methods limit loss to evaporation and runoff, thus significantly reducing the water that is used—by as much as half, according to Tess Russo of the Columbia Water Center at Columbia University. This has the added advantages of protecting surface waters and of preventing soil salinization (the buildup of salt as water evaporates), a common problem in dry climates. Choosing to plant crops more suited to the environment and water availability will also decrease agricultural water use. Many industrial processes are now designed to reuse water rather than discharge it into the environment.

The average U.S. citizen uses about 300 liters (80 gallons) of water per day in the home. Small individual changes in the household can save a lot of that water. But it is important to remember that our personal use of water is not limited to direct use of water from the tap. Water is also used on our behalf by industry to produce the products and energy that we consume (indirect use). This brings the average daily use of water up to 7,500 liters (2,000 gallons) per person in the United States. Reducing our use of resources and making wiser consumer choices with the purchases we make can reduce our **water footprint**. **TABLE 14.1** and **INFOGRAPHIC 14.6**

water footprint The water appropriated by industry to produce products or energy; this includes the water actually used and water that is polluted in the production process.

KEY CONCEPT 14.8

Conservation can effectively address scarcity and includes the use of water-saving technologies, behavioral changes that decrease water use, and consumer choices that minimize our water footprint.

INFOGRAPHIC 14.6 REDUCING OUR WATER FOOTPRINT

↓ Understanding how we use water on a daily basis helps us make decisions about ways to save water. Much of our water usage can be reduced by buying less stuff and by choosing products with lower water footprints for those things we do choose to purchase. Using less energy through conservation and using more energy-efficient products will also reduce our water footprint.

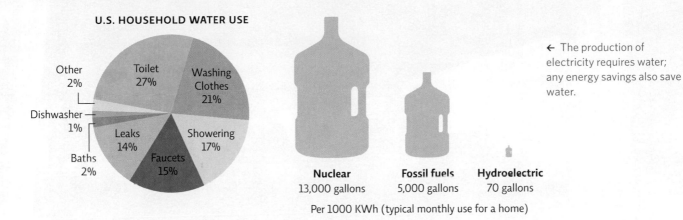

U.S. HOUSEHOLD WATER USE

Other 2%
Toilet 27%
Washing Clothes 21%
Dishwasher 1%
Leaks 14%
Baths 2%
Faucets 15%
Showering 17%

Nuclear 13,000 gallons
Fossil fuels 5,000 gallons
Hydroelectric 70 gallons

Per 1000 KWh (typical monthly use for a home)

← The production of electricity requires water; any energy savings also save water.

GALLONS OF WATER NEEDED TO PRODUCE 1 POUND OF FOOD

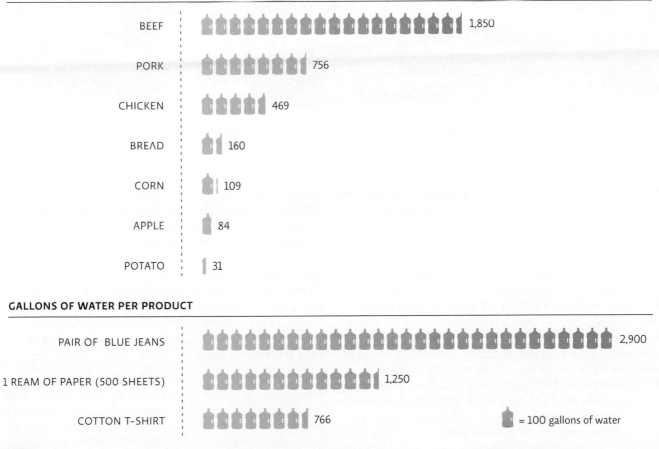

BEEF — 1,850
PORK — 756
CHICKEN — 469
BREAD — 160
CORN — 109
APPLE — 84
POTATO — 31

GALLONS OF WATER PER PRODUCT

PAIR OF BLUE JEANS — 2,900
1 REAM OF PAPER (500 SHEETS) — 1,250
COTTON T-SHIRT — 766

= 100 gallons of water

How much water could be saved in 1 week if a person who normally eats ¼ pound of beef daily made a change to eat ¼ pound of beef twice a week and ¼ pound of chicken the other 5 days of the week? How much water would be saved in a year?

↑ The Groundwater Replenishment System in Fountain Valley, California, solves two problems at once: water scarcity and dealing with wastewater. This $480 million water treatment system converts the sewage water of Orange County into drinking water, producing more than 200 million litres of drinking water every day.

In the meantime, supplementing potable water supplies with recycled water is an innovative way to help ameliorate ongoing water issues, says Channah Rock, a water-quality specialist and assistant professor at the University of Arizona. It's rare to find initiatives like Orange County's, she says, but several communities—in Arizona, California, Nevada, and Florida, for instance—are reusing recycled water for nonpotable use, such as for irrigating landscapes and crops, filling fountains and fire hydrants, and flushing toilets. Communities are trying to "match the quality of water with the right use of water," she says. Since people are prohibited from drinking the water that is used for irrigation, for example, says Rock, it's not necessary to subject that water to the same advanced treatment processes as are used for potable water.

Public perception of recycled water projects actually varies, says Rock, depending on how familiar or confident people are with the treatment process and how plagued their community is by water-scarcity

issues. In Arizona, which has regions affected by drought, a recent statewide survey found that Arizona residents generally supported most potential uses of recycled water and felt that it was very important that their community use recycled water to help meet its water needs. To many people, she says, what matters is that the water ultimately meets regulatory standards designed to protect public health. In fact, Rock says, the "toilet to tap" phrase is incredibly misleading because it leaves out the testing, treatment, and scrutiny that take place in between.

One California community may be taking the toilet to tap approach even further. An ambitious project in San Diego would send highly purified wastewater directly to the tap rather than injecting it into a larger water source such as an aquifer or reservoir; this project is not yet approved (by residents or the state of California).

The GWRS has received many accolades, including the prestigious 2008 Stockholm Industry Water Award. In

2010, just 10 years after Deshmukh finished graduate school, his work was profiled in *National Geographic*. It's an achievement that fills Deshmukh with pride. "This is water that's normally just wasted in the ocean. For the first time, it was being added to the water basin, cleaner and at a higher quantity than we'd done before."

Once Skinner learned how the water would be treated, spoke to scientists about the effectiveness of the treatment process, and learned that the project would have continuous oversight, he was reassured. "I feel they have successfully addressed my concerns," he

says. Skinner and Deshmukh even worked together to create videos describing the project. Today, Skinner, his wife, daughter, and grandson consume the recycled water. "I think it's safe for my family to drink."

Select References:

Heffernan, O. (2014). Bottoms up. *Scientific American*, 311(1): 69-75.

Hoekstra, A., & A. Chapagain. (2007). Water footprints of nations: Water use by people as a function of their consumption pattern. *Water Resources Management*, 21(1): 35–48.

United Nations World Water Assessment Programme. (2014). *The United Nations World Water Development Report 2014: Water and Energy.* Paris: UNESCO.

BRING IT HOME

PERSONAL CHOICES THAT HELP

Regardless of whether our water comes from an aquifer or a local reservoir, we can make those water sources last longer by taking steps to use our water as efficiently as possible.

Individual Steps

- If you have a smartphone, download a water usage tracking app. Once you have a baseline, try to reduce it by 10%.
- Time your shower and try to reduce it by 1 to 2 minutes.
- Have a container by the sink or shower to catch water while it warms up; make sure not to get soap in it. Use this water for watering plants both inside and out.

Group Action

- Install a rain barrel at home. Rain barrels allow people to use the rain that falls on the roof of a building to water plants as opposed to letting it run off into the storm drain. If you live in a dorm or an apartment, see if you can get permission to have a rain barrel installed.

Policy Change

- Do you know where your water comes from? Talk to a city representative to find out where your water comes from and what steps are being taken to make sure it lasts as long as possible.
- Encourage local policy makers to ban the watering of lawns or restrict the use of water for landscaping to certain days of the week.

igor kisselev/iStockphoto/360/Thinkstock/Getty Images

ENVIRONMENTAL LITERACY UNDERSTANDING THE ISSUE

 1 How is water distributed on Earth, and what are the sources of freshwater? How does water cycle through the environment?
INFOGRAPHICS 14.1 AND 14.2

1. Approximately 75% of Earth's surface is covered with water. The amount of that water that is drinkable by humans is:
   a. nearly all of it.
   b. about half of it.
   c. perhaps 1/10 of it.
   d. much less than 1%.

2. Draw a flowchart of the water cycle. (Don't copy from the book; create your own small drawing.) Follow a single water molecule from a cloud through some portion of the cycle, including a living organism, and back to a cloud.

 2 What are the causes and consequences of water scarcity?
INFOGRAPHIC 14.3

3. True or False: The sector that uses the highest percentage of water globally is industry.

4. When access to water is limited only by one's ability to pay for it, this is known as:
   a. physical water scarcity.
   b. virtual water scarcity.
   c. economic water scarcity.
   d. unpredictable water scarcity.

5. Why do the problems of water scarcity and unsanitary water conditions often occur together?

 3 What is an aquifer, how does it receive water, and what problems emerge when too much water is removed?
INFOGRAPHIC 14.4

6. The uppermost water level of the saturated zone of an aquifer is found is known as the _____ _____.

7. True or False: Most aquifers are not underground lakes of water but rather are regions of porous rock saturated with water.

8. Infiltration is:
   a. made easy in urban and suburban areas by all of the lawns.
   b. made harder in urban and suburban areas by roads, buildings, and lawns.
   c. the process of removing particulate matter from sewage.
   d. what happens when seawater enters a freshwater system.

9. If too much water is removed by a well in coastal areas:
   a. the water table will rise.
   b. a cone of depression will form.
   c. the aquifer might collapse.
   d. saltwater can seep into the aquifer.

 4 What are some of the ways that our wastewater is treated to make it potable or safe to release into the environment?
INFOGRAPHIC 14.5

10. True or False: Potable water is water that is safe to consume.

11. The main difference between high-tech and low-tech methods of wastewater treatment is that high-tech methods:
    a. use toxic chemicals to purify water.
    b. use bacteria to break down solids.
    c. filter water at the start of the process.
    d. are less expensive.

12. What are some of the things that communities in the United States do to deal with their wastewater?

 5 What can be done to address water scarcity issues?
INFOGRAPHIC 14.6 AND TABLE 14.1

13. Many people say that desalination (removing the salt from ocean water) is the obvious way around our water shortages. This is:
    a. becoming relatively easy and inexpensive and will be common soon.
    b. inexpensive now, but it is difficult to route the water from the coasts to inland areas.
    c. still very expensive and uses a great deal of energy.
    d. not being done at this time.

14. Dams can be used to store water and increase water supplies, but they come with trade-offs such as:
    a. habitat destruction.
    b. loss of water from evaporation.
    c. regional conflicts downstream from the dam.
    d. A and B.
    e. A, B, and C.

15. One creative way that some communities deal with wastewater is:
    a. diverting it, unfiltered, to agricultural fields that need irrigation
    b. using microfiltration, then bottling and selling it as mineral water.
    c. pumping it through a separate water system for people to use for laundry and for yard irrigation.
    d. evaporating it in gigantic reservoirs, creating additional clouds and rain.

16. How did Orange County address the problems of water scarcity, aquifer saltwater intrusion, and high volumes of wastewater? Would you support this solution in your own community? Explain.

SCIENCE LITERACY WORKING WITH DATA

The World Health Organization's (WHO's) *Global Burden of Disease* analysis provides a comprehensive and comparable assessment of mortality and loss of health due to diseases, injuries, and risk factors for all regions of the world. The overall burden of disease is assessed using the disability-adjusted life year (DALY), a time-based measure that combines years of life lost due to premature mortality and time lived in states of less than full health.

Interpretation

1. According to the map, which continent appears to have most problems with the lack of sanitation?

2. What is the DALY status attributable to water and hygiene in most of the world's developed countries?

3. In your own words, describe the DALY status in Eurasia, one of the most populous areas on Earth.

Advance Your Thinking

4. Explain why people in many places in the United States complain of water shortages, while this map indicates that they have access to clean water and sanitation.

5. If much of Africa is grasslands and jungle, why do the inhabitants have so little access to clean water and sanitation?

6. Some countries that border oceans are able to afford desalination plants to improve their access to clean drinking water. If all countries could afford it, would this solve the worldwide problem? Explain.

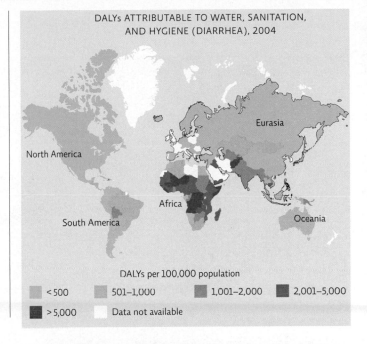

DALYs ATTRIBUTABLE TO WATER, SANITATION, AND HYGIENE (DIARRHEA), 2004

INFORMATION LITERACY EVALUATING INFORMATION

In the summer of 1989, Dr. Noah Boaz and his archaeological Earth-watch crews were excavating a site of ancient human habitation along the Semliki River, which runs by the border between Zaire (now the Democratic Republic of Congo) and Uganda. They could not, however, just drink the water from the river, or even swim in it. They had to filter the water a gallon or two at a time and then add chemicals to it in order to remove the waterborne parasites and pathogens. Bathing required wearing shoes and keeping their eyes, nose, and mouth out of the water. The nearby villagers did drink the water, and they had endemic health problems.

More than 750 million people do not have access to clean drinking water; as many as 3.5 billion do not have access to enough clean water. The results affect all aspects of life in developing countries: According to Water.org, a child dies every 20 seconds from a water-related illness, and women in some water-stressed areas spend several hours every day collecting water for their families' basic needs.

There have been many suggestions about ways to improve access to clean water. One of the problems is that in many areas, the lack of access is coupled with a lack of the electricity, developed roads, machines, and equipment necessary to be able to support digging municipal wells and providing pumping stations, reservoirs, and pipelines.

Go to the Global Water website (globalwater.org) and explore the links under "Projects." Then go to the Water.org website (water.org) and look at some of the featured projects.

Evaluate the websites and work with the information to answer the following questions:

1. Are these authors/sponsoring groups reliable information sources?
   a. Do they give supporting evidence for their claims?
   b. Do they give sources for their evidence, as well as clear explanations?
   c. What is the mission of the organization? How do you know this?
   d. Does the organization appear to have a workable solution or solutions?

Now search the Internet for information about two low-tech filtration devices: the LifeStraw and the PlayPump. Potentially useful sites include the bottlelessvancouver website (bottlelessvancouver.wordpress.com) and the HowStuffWorks website (science.howstuffworks.com).

2. Evaluate the proposed solutions for:
   a. price.
   b. ease of use.
   c. whether they would be portable or stationary.
   d. whether they include pumps for underground water or can clean only surface water.

3. Does either of the proposed solutions stand out as a good option for remote or undeveloped areas, such as the ones featured in the Global Water website or the Water.org website? Explain your answer.

Find an additional case study online at http://www.macmillanhighered.com/launchpad/saes2e

INTO THE GULF

Researchers try to pin down what's choking the Gulf of Mexico

CORE MESSAGE

Water pollution decreases our usable water supplies, harms wildlife and human life, and is largely caused by human actions. Some types of pollution may be easier to address than others, but in general, we can decrease water pollution by controlling what we discharge into water bodies, restoring forested areas, and limiting the use of potential pollutants.

AFTER READING THIS CHAPTER, YOU SHOULD BE ABLE TO ANSWER THE FOLLOWING **GUIDING QUESTIONS**

1

In general, what is water pollution? What is the difference between point source and nonpoint source pollution, and what are some common sources of each?

2

What is eutrophication? How can water pollution from fertilizer or animal waste cause eutrophication and ultimately kill aquatic life?

The Mississippi River is fed by thousands of smaller streams and rivers as well as runoff from the surrounding landscape. When the river eventually empties into the Gulf of Mexico, it deposits sediments and fertilizers it picked up along the way which fuels greenish phytoplankton blooms in the Gulf, so large they are visible on this satellite image.

© spacephotos.com/Age Fotostock

3

What is a watershed, and how does it affect the *quality* of surface water as well as the *quantity* of groundwater?

4

Why is nutrient pollution from agricultural sites a bigger problem today than in years past?

5

What actions can be taken to address water quality and quantity problems in the Gulf of Mexico watershed and in other areas?

Back in 1974, when he had just begun his career at Louisiana State University, biologist Eugene Turner took a 15-foot skiff out along the Gulf Coast to survey the water. He brought a handheld oxygen meter along with him. Other researchers had measured oxygen levels in the same waters and come up with some disturbingly low numbers—levels low enough to essentially "suffocate" any aquatic organism that couldn't relocate to more oxygen-rich waters. But none of them had followed up, and Turner was curious. Were those earlier measurements wrong? Flukes? And if they weren't, what did it mean for the Gulf's ecosystem?

VICTORIA LOE/KRT/Newscom

↑ Louisiana State University biologist Nancy Rabalais taking water samples on a research vessel in the Gulf of Mexico.

Sure enough, Turner's own readings came up low as well—much lower than expected. His curiosity deepened: What would cause low oxygen levels in these waters? He suspected the myriad oil rigs in nearby waters might have something to do with it. But due to the complicated nature of water pollution, he also knew that the true culprit could be hiding hundreds, or even thousands, of miles away.

Different types of water pollution degrade water quality.

Water pollution is the addition of any substance to a body of water that might degrade its quality. The list of such substances, or pollutants, is depressingly long: Industrial chemicals like polychlorinated biphenyls (PCBs) and raw sewage get dumped directly into a body of water or wash into it from the land. Meanwhile, contaminants like mercury and acid-forming precursors, along with other air pollutants from fossil fuel combustion or industry, fall back to Earth with the rain and flow as **stormwater runoff** into rivers, streams, lakes, and seas. Nutrients (from fertilizers and animal waste) and pesticides also enter from farm and lawn runoff. Sediments from soil erosion can flow into surface waters from farms, construction sites, or heavily eroded stream banks that are no longer shored up by a well-rooted plant community. Water pollution is not limited to the introduction of chemicals or sediment; municipal trash often finds its way to rivers, streams, and oceans (see Chapter 7). The heated water released into surface waters near power plants causes *thermal pollution*, raising the temperature of the water enough to impact many of the organisms that live in that stretch of river, lake, or ocean. Even groundwater sources can become polluted from underground chemical storage tanks or from the movement of surface pollutants down through the soil.

There are two classes of pollution, defined by how they are delivered to the water. **Point source pollution** is water pollution whose discharge source can be clearly identified (that is, one can "point" to the "source"). This includes pollution from large discharge pipes of

Lynn Berts/NRSC/USDA

↑ Unprotected farm fields lose topsoil as well as farm fertilizers and other potential pollutants when heavy rains occur.

KEY CONCEPT 15.1

Water pollution may come from readily identifiable sources such as discharge pipes (point sources) or from more dispersed sources such as stormwater runoff or atmospheric fallout (nonpoint sources).

wastewater treatment plants or industrial sites. The discharge itself is known as *effluent*.

Some of this effluent has the capacity to be quite dangerous. In most cities, sewer pipes run alongside streams because engineers assume that pipe sewage, if placed alongside the stream, will flow in the same direction and therefore reach treatment plants via gravity. But in many urban areas, those pipes are old and beginning to leak. With such a leak, raw sewage flows directly into the stream. The problem is particularly

bad in cities of the developing world, where waterborne *pathogens* (disease-causing organisms) in raw or partially treated human and animal waste represent a leading cause of sickness and death.

But the good news is that because point sources can be easily identified, they can also—at least hypothetically—be remedied.

As researchers would soon discover, the low oxygen levels in the Gulf of Mexico were due to a different and an even more challenging problem: **nonpoint source pollution**. The origin of nonpoint source pollutants are not easily identifiable. Though some arrive by air, most enter

water pollution The addition of any substance to a body of water that might degrade its quality.

stormwater runoff Water from precipitation that flows over the surface of the land.

point source pollution Pollution from discharge pipes (or smoke stacks) such as that from wastewater treatment plants or industrial sites.

nonpoint source pollution Runoff that enters the water from overland flow.

KEY CONCEPT 15.2

The influx of excess nutrients into a body of water may spur algae growth and bacterial population explosions, which ultimately result in hypoxia severe enough to harm aquatic life.

the water from overland flow (stormwater runoff); this means they can come from any part of the land that drains into a given body of water.

INFOGRAPHIC 15.1

Nonpoint source pollutants can include everything from humanmade toxicants to natural substances such as silt, sand, and clay, which can enter the water as *sediment pollution* (eroded soil that is washed into the water through runoff). To be sure, these substances deliver valuable nutrients to aquatic ecosystems. But excessive amounts of them can cloud the water, making it hard for sunlight to penetrate and thus disrupting photosynthesis. Sediment pollution can also harm organisms directly by clogging gills. And, when it covers the sea or river bottom, sediment can smother the nooks and crannies that serve as habitat or spawning areas.

Something in the Gulf waters—some pollutant—was causing the levels of **dissolved oxygen (DO)** to plummet—a condition known as **hypoxia**. Turner and his colleague Nancy Rabalais (the two would later marry) knew that hypoxia was a serious problem. Water can hold only a limited amount of oxygen, much less than found in air. So even a small decrease can have immediate effects on aquatic life. Even underwater organisms need oxygen to survive. Like terrestrial beings, they use it in the process of cellular respiration.

Month after month, year after year, in bigger and better-equipped boats, Turner and Rabalais surveyed the water, mapping out a hypoxic zone that grew from 40 km² (15 square miles) in 1988 to more than 15,000 km² (5,800 square miles) in 2013 (that's about the size of Connecticut). As they quickly learned, the oil rigs were not the main problem. An excess of the nutrients nitrogen and phosphorus were triggering a process known as **eutrophication** (or, more precisely, *cultural eutrophication*, since human activities were the source of these nutrients).

dissolved oxygen (DO)   The amount of oxygen in the water.

hypoxia   A situation in which a body of water contains inadequate levels of oxygen, compromising the health of many aquatic organisms.

eutrophication   A process in which excess nutrients in aquatic ecosystems feed biological productivity, ultimately lowering the oxygen content in the water.

"Imagine stretching a giant sheet of plastic wrap from the Mississippi River's mouth, straight across to Galveston [Texas]," Turner says. "Now imagine sucking all the air out and leaving the whole ecosystem there to suffocate."

Here's how eutrophication works: Because nitrogen and phosphorus fuel plant growth, extra amounts trigger explosions of algae. Though algae produce oxygen through photosynthesis, they also block sunlight from reaching underwater plants, ultimately blocking much more photosynthesis than they conduct. This imbalance causes the plants at the bottom of the water to die en masse. When that happens, the turbidity (cloudiness) of the water increases: Dead and dying plant roots can no longer secure the river or seabed in shallow areas, and bottom sediments can easily enter the water column if disturbed by waves or fast-moving water. This disturbance reduces photosynthesis even more.

From there, it gets even worse. As the plants die, they are consumed by bacterial decomposers, triggering yet another bloom—a bacterial one. The bacterial populations

> ❝ *Imagine stretching a giant sheet of plastic wrap from the Mississippi River's mouth, straight across to Galveston [Texas]. Now imagine sucking all the air out and leaving the whole ecosystem there to suffocate .* ❞ —Eugene Turner.

◉ **WHERE IS THE GULF OF MEXICO WATERSHED?**

GULF OF MEXICO WATERSHED

GULF OF MEXICO

INFOGRAPHIC 15.1   **MAJOR CAUSES OF WATER POLLUTION**

↓ Water pollution comes from a variety of point and nonpoint sources. Although the main threat to the Chesapeake Bay is excess nutrient and sediment runoff, the EPA identifies pathogens as the leading cause of impaired waters in the United States as a whole. Found in sewage or animal waste, pathogens can come from both point and nonpoint sources.

POINT SOURCES

Some industrial and agricultural sources discharge pollutants directly into a body of water.

NONPOINT SOURCES

A variety of sources contribute pollutants that can run off the surface of the land during rainfall and enter the water; air pollutants can fall directly with the rain.

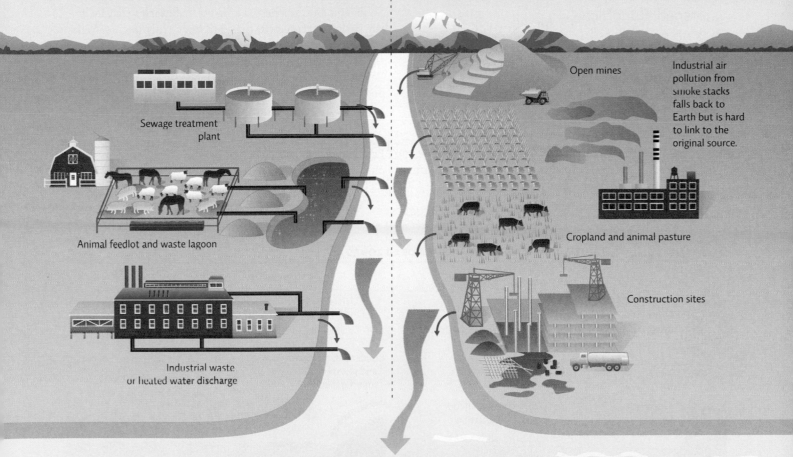

Open mines

Industrial air pollution from smoke stacks falls back to Earth but is hard to link to the original source.

Sewage treatment plant

Animal feedlot and waste lagoon

Cropland and animal pasture

Construction sites

Industrial waste or heated water discharge

↓ Various pollutants contribute to the many miles of U.S. surface waters that are "impaired"—waters that are too polluted to meet state or local water quality standards.

LEADING CAUSES OF IMPAIRED SURFACE WATERS IN THE UNITED STATES (2012)

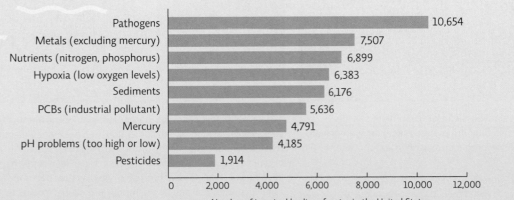

Cause	Number
Pathogens	10,654
Metals (excluding mercury)	7,507
Nutrients (nitrogen, phosphorus)	6,899
Hypoxia (low oxygen levels)	6,383
Sediments	6,176
PCBs (industrial pollutant)	5,636
Mercury	4,791
pH problems (too high or low)	4,185
Pesticides	1,914

Number of impaired bodies of water in the United States

What type of pollution predominantly contributes to the hypoxic zone in the Gulf of Mexico: point source or non-point source?

KEY CONCEPT 15.3

All the land area over which water could potentially flow and empty into a body of water is that water body's watershed. Runoff can pick up pollutants in the watershed and deliver them to the water body.

A watershed is simply an area of land over which rain and other sources of water flow to drain into a given river, lake, or stream. It includes all the smaller streams that empty into the river and their drainage areas. **INFOGRAPHIC 15.3**

increase rapidly, consume oxygen as they digest the dead plants, and quickly deplete any remaining oxygen in the water. **INFOGRAPHIC 15.2**

But where was all that extra nitrogen and phosphorus coming from in the first place?

The source of pollution can be hard to pinpoint.

The entire Mississippi River **watershed** drains into the Gulf of Mexico.

The watershed feeding the Gulf of Mexico stretches well past the Louisiana basin and the Mississippi delta. It stretches all the way up to the northernmost reaches of the continental United States. Indeed, to glimpse the causes of—and possible solutions to—the hypoxia uncovered by Rabalais and Turner, we must travel to Minnesota, where the Mississippi River begins.

It's tough to imagine that the homesteads of the far north—a place of corn fields and dairy farms—have anything to do with the rollicking backwater swamps of Louisiana. But it is here, along the tributaries that feed the nation's farmlands, that our story truly begins. "The nitrogen that's killing the Gulf starts here, gets added to the ecosystem here, in these farms and fields," says Alex Echols, former head of the U.S. Fish and Wildlife Service and a consultant to the Minnesota-based Sand County Foundation (an environmental group). "This is where it's all coming from."

Echols likes to think of nitrogen and phosphorus as characters in a story. "Nitrogen is like a big healthy [college] boy," he says. "He's smart and he's got potential. If we set him on the right path, he can grow the world

---

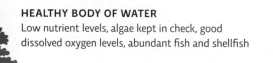

INFOGRAPHIC 15.2   EUTROPHICATION CAN CREATE DEAD ZONES   2

HEALTHY BODY OF WATER
Low nutrient levels, algae kept in check, good dissolved oxygen levels, abundant fish and shellfish

Clear water with good sunlight penetration

UNHEALTHY BODY OF WATER
High algae and bacterial growth, low dissolved oxygen, and loss of some aquatic life

Sediment, nitrogen, and phosphorus enter as runoff pollution.

❷ Less sunlight can penetrate the algal blooms and sediments.

❶ Nutrients cause algal blooms.

Dissolved oxygen levels drop.

❸ Underwater photosynthesis decreases, plants die, and oxygen levels drop.

Sediments cloud the water and coat surfaces.

❹ As algae die, decomposers (bacteria) increase in number and use more of the available oxygen, causing dissolved oxygen levels to drop even more.

↑ Healthy bay water is relatively clear, with abundant sea life. Oysters and menhaden fish filter out particles, keeping water clean, while submerged vegetation produces oxygen to support healthy fish and crab populations.

 Why is it that in eutrophic waters hypoxia can develop from both a decline in oxygen production and an increase in oxygen consumption?

↑ Sediments that cloud the water or an influx of nutrients that causes an algal bloom (eutrophication) can prevent sunlight from reaching underwater plants, producing hypoxic regions. An influx of organic material (waste, dead organisms) can also cause hypoxia; growing bacterial populations use oxygen as they decompose the organic matter. Fish and larger, mobile organisms may be able to leave the area; others cannot and will suffer in the hypoxic waters.

↓ Anything that happens in the watershed can potentially affect the quality of a body of water as well as the quantity of groundwater. This is especially important in terms of nonpoint source pollution that originates on land. Mapping the watershed is an important tool in watershed management. Here the watershed of this river and coastal area is outlined as a black dashed line.

If you could draw a line from hilltop to hilltop around a river and its tributaries (the streams that feed into it), you would be outlining its watershed. Water on the other side of the dotted line flows away from the watershed.

Land uses that prevent infiltration reduce the rate at which aquifers are refilled.

If you find pollution in the water at this point and you have mapped the watershed, you know where to look for the source of the pollution (upstream, in the watershed) and where not to look (downstream, or outside the watershed).

Any rain that falls on the river side of the edge of the watershed could flow downhill into the river or one of its tributary streams.

? Suppose you discover that the stream by your house appears to be polluted (i.e., unusual smell, lots of dead fish, noticeable pollution, etc.). Why might it be helpful to have mapped the watershed of this stream?

green and make it better. But it's easier and more fun for him to slip on down to Louisiana and party." Phosphorus is a bit of a follower: It generally stays bound up in the soil but can also be carried away when that soil starts to move.

The route that this "college boy" travels is well mapped. In fact, a variety of experiments have shown how water flows through a watershed's plants, soils, and streams. It turns out that while runoff pollution can directly affect surface water *quality* by delivering pollutants, it also impacts the *quantity* of groundwater. When it

KEY CONCEPT 15.4

Land uses that decrease runoff and increase infiltration help to recharge aquifers because the water soaks into the ground rather than reaching the nearest surface water as runoff.

rains, some rainfall soaks into the soil, infiltrating the ground below; some eventually reaches groundwater in the **aquifer**, which people can tap into to create a well. If the groundwater is deep enough, infiltration can act as a filtering system that purifies the water. However, if the water table is close to the surface, pollutants can still make it all the way to groundwater; for example, nitrate pollution from fertilizer runoff can contaminate well water enough to be life threatening, especially to young children. (For more on aquifers, freshwater, and the water cycle, see Chapter 14.)

But human land uses have altered the way water drains through watersheds and waterways. Pavement and even suburban lawns enhance runoff by preventing water from soaking into the ground, reducing the rate at which aquifers are refilled; they can

watershed The land area surrounding a body of water over which water such as rain can flow and potentially enter that body of water.

aquifer An underground, permeable region of soil or rock that is saturated with water.

↑ Algal blooms along the shore of Simon Lake in Naughton, Ontario, are still occurring despite steps to reduce the amount of phosphorus entering the lake from various sources. Spring runoff still delivers unacceptable levels of phosphorus to the lake but lake turnover may also be releasing stored phosphorus from sediments, triggering the algal blooms.

be rapidly depleted if we remove well water faster than it is replaced. Many meandering waterways have been channelized (made straight) to divert water toward or away from land areas as needed. Wetlands, too, have been completely drained to allow urban or suburban development or for agriculture. These activities also speed the flow of water across and out of a watershed, decreasing infiltration and bringing with the water whatever industrial, agricultural, or municipal pollutants it encounters on its trip downhill. In the Mississippi River watershed, this includes runoff pollution from roadways, urban areas, and industries throughout the region, as well as animal waste and excess fertilizer from hundreds of thousands of acres of farmland.

In all, some 1.5 million metric tons of nitrogen and phosphorus now flow through the Mississippi River watershed and into the Gulf of Mexico every year. And the Gulf is not the only body of water so beset. According to the United Nations, nutrient pollution is the leading type of water pollution worldwide. And according to a 2008 study, hypoxic zones now plague more than 400 systems and affect more than 245,000 square kilometers (95,000 square miles) of water around the world. "They are for sure a leading source of trouble for marine ecosystems globally," says Robert Diaz, a researcher at the Virginia Institute of Marine Sciences and the study's lead author. Of those 400 hypoxic zones, 166 are located in the United States.

"We have far more nitrogen moving off the landscape than we did historically," Echols says. "And there are three main reasons for that." The first is that we've changed what we grow—from foods like alfalfa and oats that don't leak much nitrogen, to foods like soybeans and corn that do release lots of nitrogen back into the soil. The second is the rampant increase in the use of synthetic nitrogen for fertilizer throughout the modern agricultural era. "That's not necessarily a bad thing," Echols says. "There are certainly benefits to the way we've increased productivity; but those pluses come with minuses and this is a huge, huge minus." Synthetic fertilizers, combined with atmospheric byproducts of fossil fuel burning, have essentially doubled the amount of usable nitrogen in the environment.

KEY CONCEPT 15.5

Nutrient pollution is a serious problem for surface waters in some areas. Reducing overall fertilizer use and taking steps to reduce runoff and erosion can reduce this pollution.

But the most interesting cause of nitrogen increases, Echols says, is this: We have massively replumbed the natural hydrology of our watersheds. To be sure, there were good reasons for this. A lot of our prime agricultural land is not exactly prime; it's too wet in the spring, which means that seeds would rot rather than grow were it not for some feats of engineering that drain the subsurface water out of the soil. Drainage systems enable farmers to get the crop in earlier and to guard against crop losses during flooding. Saturated ground does not soak up any rainfall; rather, the rain runs off the surface. Drainage systems reduce surface runoff, which in turn prevents the loss of sediment and thus of phosphorus (because, remember, phosphorus is bound up in the soil).

"By installing a drainage system, you can increase the land's productivity by 20–30 percent," Echols says. "That means more food for people to eat, and more income for farmers. But at the same time you've just established a very efficient mechanism for moving water off the field, and keep in mind, nitrogen is water soluble. Water is its ticket down to the party in Louisiana." And those drainage pipes (commonly called "tile lines" because they were originally made of short, perforated clay pipes known as tiles) end up being its passageway.

Echols and others say that tile lines may also be *our* passageway—to addressing the problem of nutrient runoff. "It's going to take huge, tectonic shifts—in federal policies, and in the global food economy, and even in the culture—to move farmers off corn and soybean; to move them off synthetic fertilizers; to move them off fossil fuels," Echols says. "But managing the tile lines? We can do that. We can do that farm by farm."

Addressing eutrophication begins in the farm field.

Jeffrey Strock, a soil scientist at the University of Minnesota, is used to farmers insisting that their land would be perfect for one study or another. They want to increase crop production, or try to understand why a particular field is struggling. They come to Strock and ask for research. So at first Strock was a bit dismissive of Brian Hicks, the proprietor of Nettiewyynt Farm, who approached him at a drainage workshop in 2003. "Brian and his father wanted to try controlled drainage on their land," Strock recalls. "They said, 'Hey why don't you come do some studies back on our farm? We've got the perfect land for it.' And I said, 'If I had a dime for every farmer that said that, I'd be rich.'" But a year later, Hicks approached him again. And so Strock drove by the farm and took a look. And sure enough, the land was perfect: a snowbound expanse of 1,500 acres that straddles the Cottonwood River, which feeds the Minnesota, which in turn feeds the Mississippi. The farm had been in the Hicks family since about 1886. That meant the farm's history was well documented. And because Hicks and

↑ Jeffrey Strock, a soil scientist at the University of Minnesota, points out the satellite uplink that delivers data from sensors in the soil (like temperature, moisture, and greenhouse gas emissions) to his research center.

↑ Strock and Hicks working in the fields to collect data on soil drainage.

his father used to run cattle and had yet to convert some of their pastureland to corn and soybean, there was still virgin prairie left. That meant Strock and his team could use untouched land to install and compare two different drainage treatments.

"I had two goals," says Hicks. "On the environmental side, I wanted to reduce the amount of nitrogen we were sending down into the river. And on the business side, I wanted to improve my crop yields."

Strock and Hicks started by dividing one field into two zones. On the east side: conventional drainage with 4-foot-deep pipes that removed excess water from the soil at a steady rate. On the west side: controlled drainage, where the depth of the water would be made higher or lower (via a small wooden box that would serve as a dam), depending on the time of year and the needs of the crop. The hope was that the controlled drainage site would store more water—and nutrients—in the top layer of soil, which would then be available during crop production.

"As far as I know," says Strock, "this was the very first controlled drainage project in the state of Minnesota." By 2008, the Hicks farm had been fully instrumented, and the farmer–scientist team was ready to start taking measurements.

Here's what they've found so far: during normal non-drought years, a 50% reduction in nutrient loss and a slight increase in yield (between two and five bushels—around 10%) for the controlled drainage zone compared with the conventional drainage zone. "There have been a couple occasions where conditions have been perfect and we've gotten a 23 bushel yield increase," Hicks says. "Just one season with a yield like that can pay for the whole system."

"We don't always get the agronomic benefit [of higher crop yields]," Strock says. "But we have consistently reduced the amount of nutrients running off the field. So that environmental benefit is definitely there."

And this environmental benefit is needed. The use of synthetic fertilizers has increased dramatically since their introduction in the mid-20th century, coinciding with the occurrence of more and bigger hypoxic regions. While direct measurements of the size of the hypoxic zone in the Gulf of Mexico have only been made since 1985, Turner and others have reported hypoxia in various regions of the Gulf since the 1970s. But researchers can also look back in time for evidence of links between hypoxia and fertilizer use; sediment core analysis shows that while some late summer hypoxia existed in the Gulf before the middle of the 20th century, it has increased in area and duration since the 1950s. (The layers of sediment cores reveal the types of organisms alive in years past—those that could tolerate hypoxic conditions and those that could not.)

Strock's research has also shown that *how* one farms also makes a difference. Alternative farming techniques that focus on preventing soil erosion and increasing nutrient uptake by plants leaves less nitrogen behind in the soil to wash away. For example, a 3-year direct comparison between a conventional test plot and one farmed using organic and soil conservation techniques significantly reduced both the flow of water out of drain pipes and the amount of nitrogen in that water. **INFOGRAPHIC 15.4**

Watershed management is the key to reducing hypoxic zones.

In 2008, the newly formed Gulf of Mexico Watershed Nutrient Task Force set a goal: to reduce the hypoxic zone down to 5,000 km² (1,900 square miles). That same year, Rabalais and Turner concluded that the Gulf ecosystem was becoming more sensitive to nutrient loads. "It's not the same system as the 1960s," Rabalais says. "It's taking less nutrients now to fuel the hypoxia."

Scientists are still assessing the impact of nutrient runoff on the Mississippi River watershed's various ecosystems. One way to do this is a process known as **biological assessment**, the simple act of looking at what lives in a given ecosystem. In some cases this means simply netting, identifying, and counting **benthic macroinvertebrates**, such as insects and crayfish that inhabit the bottoms of the smaller streams that feed into the Mississippi River. If a stream is unhealthy, there won't be many organisms present that are sensitive to pollutants. The abundance and diversity of pollution-tolerant and pollution-sensitive species in the sample can be used to "rate" the stream quality. Poor stream quality can sometimes indicate that nutrient runoff or other toxins are polluting the water or that sediments are smothering needed stream-bottom habitat.

According to a 2008 report from the Gulf Task Force, hypoxia is causing "long term ecological changes in species diversity and a large scale, often rapid change in the ecosystem's food web that will be difficult to impossible to reverse." This is not surprising, given the condition of other hypoxic zones around the world. For example, after centuries of nutrient runoff from the Danube River, rich mussel beds that had populated the

biological assessment The process of sampling an area to see what lives there as a tool to determine how healthy the area is.

benthic macroinvertebrates Easy-to-see (not microscopic) arthropods such as insects that live on the stream bottom.

INFOGRAPHIC 15.4   ADDRESSING THE PROBLEM OF FERTILIZER RUNOFF

↓ The use of synthetic fertilizers (like nitrogen) has increased since they were introduced in the 1940s. The size of the summer Gulf hypoxic zone (defined as water with less than 2 ppm O_2) has also increased along with it. Various lines of evidence suggest that while some hypoxia naturally existed in the Gulf, it has increased in area and duration since the 1950s, especially in recent years. The zone is largest in years with abundant rainfall due to a decrease in runoff but in most years exceeds the Task Force goal of no more than 5,000 km² (1900 square miles).

↓ Strock compared the loss of nitrogen draining out of a conventionally farmed test plot (corn and soybeans using recommended amounts of synthetic fertilizer) to an adjacent alternative test plot that was farmed using a variety of organic management techniques including planting a wider variety of crops (increases the amount of nitrogen removed from the soil), using only animal manure as fertilizer (releases nitrogen more slowly than synthetic fertilizers), and planting a crop during the winter months (holds the soil in place, reducing erosion). In the 3 years of the study, the alternative methods significantly reduced the movement of nitrogen out of the soil into the drainage water that would eventually reach surface waters as a pollutant.

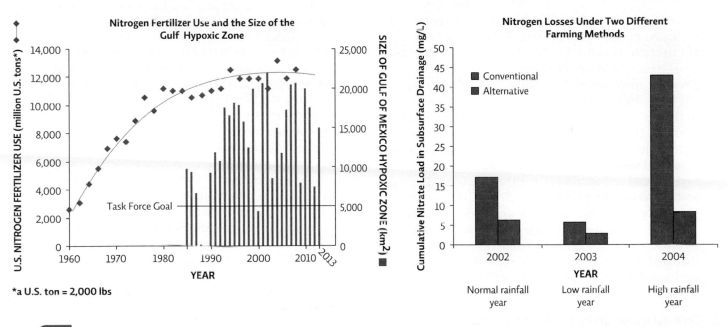

*a U.S. ton = 2,000 lbs

According to the first graph shown here, what kind of relationship is shown between nitrogen fertilizer use and the size of the Gulf of Mexico hypoxic zone?

Black Sea for centuries disappeared. Food chains broke, big fish died off, and a seafood economy that had thrived since the Byzantine era evaporated. The sea recovered a bit when the Iron Curtain countries collapsed, fertilizer subsidies fell, and the steady flow of nutrients from the Danube ceased, like a faucet being turned off. Today, the sea is home to a thriving anchovy industry. But the big fish and rich oyster beds have never recovered. In the United States, Chesapeake Bay, home to another hypoxic zone, is struggling mightily against the same fate.

Still, Strock and others point out that the term *dead zone* is a misnomer. "It's a low-oxygen zone," Strock says. "There are still things living there. And we still have time to save that ecosystem."

Water quality has always been a concern but perhaps never more so than in 1969. At the time, there were no restrictions on the release of industrial chemical pollutants into bodies of water. That year, the Cuyahoga River in Cleveland, Ohio, made headlines when it caught fire (and not for the first time) because so much oil and other flammable industrial pollutants floated on the surface of the water. This dramatic event helped spur the passage of the 1972 **Clean Water Act (CWA)**, which regulates industrial and municipal (such as sewage treatment plants) point source pollution, with the goal of making all environmental waters "fishable and swimmable." It does this by setting **pollution standards**—allowable levels of

Clean Water Act (CWA) U.S. federal legislation that regulates the release of point source pollution into surface waters and sets water quality standards for those waters. It also supports best management practices to reduce nonpoint source pollution.

pollution standards Allowable levels of a pollutant that can be present in environmental waters or released over a certain time period.

KEY CONCEPT 15.6

Good watershed management can reduce nonpoint source pollution. For example, well vegetated riparian areas reduce runoff and act as nutrient sinks.

a pollutant that can be present in environmental waters or released over a certain time period.

For example, the U.S. Environmental Protection Agency (EPA), which administers the CWA, protects the public against pathogens in *recreational* waters (rivers, lakes, coastal areas)—the leading cause of impaired waters in the United States and a major problem worldwide. Globally, exposure to pathogen-polluted water is one of the leading causes of infection; 2 million people die each year of diarrheal diseases linked to unsafe water. The United Nations cites the provision of clean water and sanitation as one of its major development goals (see Chapter 5). In the United States, the EPA allows only very small amounts of fecal bacteria such as *E. coli* in recreational waters; the presence of fecal bacteria indicates sewage or animal waste contamination. Violations are investigated and addressed at either the federal or state level. (The Safe Drinking Water Act mandates that absolutely no pathogens are allowed in drinking water.) Today, about 65% of U.S. waters meet the fishable/swimmable goal, more than double the number that met that goal prior to the passage of the CWA. (However, we are still far from the goal of 100% compliance with CWA guidelines.)

"We did a good job on point source [pollution]," Echols says. "Back in the late 1960s, we said 'thou shall not pollute,' and we made rules and we made people follow them."

Nonpoint source pollution is a bit trickier. The CWA does not specify, for example, how much nitrogen fertilizer a farmer can apply. And this is not feasible because recommended amounts vary from farm to farm, and runoff potential varies according to rainfall, terrain, and even the crop that is planted. But Echols, Strock, Hicks, and others are working to fix it, just the same. Although they differ in their preferred approaches, most scientists agree that **watershed management**—management of what goes on in an area around streams and rivers—will be the key to saving the Gulf. They've begun to create *best management*

practices (agreed-upon actions that minimize pollution problems caused by human actions) to reduce the amount of pollution being delivered to the Mississippi River. In agriculture, many of these practices focus on ways to decrease the amount of chemicals applied to land areas in the first place and to reduce the potential for soil erosion and runoff.

Recent CWA amendments advocate the use of best management practices and provide funding for their implementation. However, watershed management is not easy to do; after all, many individuals and groups are engaged in many different land uses within any watershed. Identifying how each should address its contribution to nonpoint source water pollution—much less enforcing compliance—is no easy task.

One of the key steps is to restore the watershed's **riparian areas**, the land areas close to the water, by maintaining or planting vegetated buffer zones that slow runoff and give the rainwater time to soak into the ground. "Most of our riparian areas—in the Mississippi River watershed, anyway—are not functioning as riparian areas anymore," Echols says. "We can't really fix hypoxia until we fix that." Healthy riparian areas are widely recognized as critical to maintaining good water quality, and projects are under way across the United States to restore and revegetate the riparian areas and watersheds of rivers and streams. In the late 1990s, New York City invested billions of dollars to restore and protect areas in the Catskills and the Delaware watershed supplying its water; thanks to this investment, the city avoided the need to construct high-tech and costly filtration systems. **INFOGRAPHIC 15.5**

Another agricultural best management practice is frequently testing the soil to ensure that only the correct type and amount of fertilizer is added as needed; tilling fertilizer directly into the soil (rather than spreading it on the surface) also reduces the potential for loss due to runoff or wind. Farmers can also use a variety of erosion prevention methods, such as planting winter crops in the off season to hold soil in place or planting trees as windbreaks to reduce soil erosion from wind. (For more on sustainable farming techniques, see Chapter 17.) Farmers are also going high-tech, pursuing *precision agriculture*. They are using GPS technology to guide farm equipment and reduce overlap when working a field, thus reducing disturbance of soil that might lead to erosion. Farmers are also using precision agriculture to guide site-specific applications of chemicals within a single field to minimize the amount used.

Of course, not all of hypoxia's solutions will come from the farm belt. In suburban areas, lawns can be major nonpoint sources of nitrogen pollution if

watershed management Management of what goes on in an area around streams and rivers.

riparian areas The land areas close enough to a body of water to be affected by the water's presence (for example, areas where water-tolerant plants grow) and that affect the water itself (for example, provide shade).

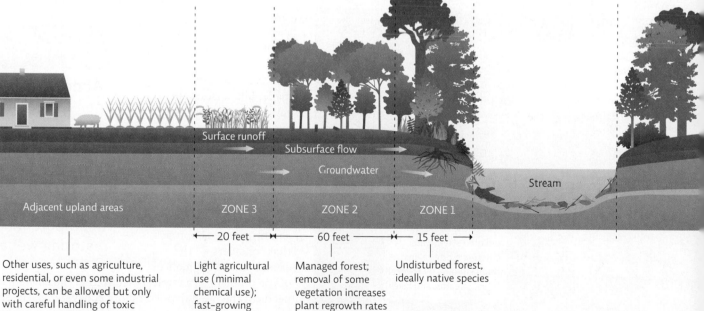

Plants are nutrient sinks—they store nutrients in their tissues, reducing the amount free to enter the water.

Ground vegetation slows runoff and allows water to soak into the ground. This keeps the water and any pollutants it may carry out of the stream and increases infiltration, which replenishes groundwater.

Shade provided by overhanging trees cools the water; important because cooler water can hold more oxygen than warmer water.

Plant roots stabilize banks and prevent erosion and the deepening of the channel, which can speed water flow and lead to further erosion.

Trees provide food; leaf litter that falls in the water is a main nutrient source for aquatic organisms, including nitrifying bacteria, which break down nitrates, reducing the amount sent downstream to the bay.

Jochen Schlenker/Robert Harding World Imagery/Getty Images

↑ The area next to a body of water that impacts that water (provides shade and nutrients) and is itself impacted (water-tolerant species live here) is the riparian area. A well-vegetated riparian area reduces the runoff that reaches a body of water by slowing the water's movement across the land so that it soaks into the ground rather than flow into the stream.

↓ Shown here are the U.S. Department of Agriculture's recommendations for land use in the riparian area. This includes setting aside at least 75 feet of land in managed and undisturbed forest. More may be required for suitable protection in areas with steeper terrain—that is, where runoff flow would be faster.

Surface runoff

Subsurface flow

Groundwater

Stream

Adjacent upland areas

ZONE 3

ZONE 2

ZONE 1

← 20 feet → ← 60 feet → ← 15 feet →

Other uses, such as agriculture, residential, or even some industrial projects, can be allowed but only with careful handling of toxic chemicals and fertilizer.

Light agricultural use (minimal chemical use); fast-growing grasses trap nutrients that might run off from farms or lawns.

Managed forest; removal of some vegetation increases plant regrowth rates and maximizes nutrient uptake from the soil and subsurface flow.

Undisturbed forest, ideally native species

 In a managed riparian area, why not allow zone 1 to have managed forest like zone 2?

INFOGRAPHIC 15.6 INCREASING INFILTRATION OF STORMWATER

↓ Stormwater that doesn't soak into the ground can enter storm drains that flow directly to rivers and streams, or cause floods, especially in heavily built-up urban settings. Anything that increases infiltration can help avoid these stormwater problems.

Trees slow and allow infiltration of runoff.

A green roof has vegetation over a waterproof layer, which can trap some water and reduce roof runoff.

Redirected downspout

Rain barrel captures runoff from roof.

Rain gardens capture gutter and lawn runoff.

Storm drain

Curb cutouts reduce street runoff by diverting it to the ground, where it can infiltrate.

All these efforts help reduce the amount of polluted stormwater entering storm drains that lead directly to local rivers, streams, and lakes.

 What are some other building or infrastructure methods that could reduce the flow of stormwater? (Consider things like roof size, road width, other pavement options, etc.)

homeowners apply too much fertilizer. However, the grass also sequesters nitrogen because of its long growing season, preventing it from flowing to the nearest stream. Echols encourages homeowners to limit fertilizer use on lawns and to plant native plants and grasses that do not need fertilizer. Planting a rain garden of water-tolerant plants in low-lying areas that tend to flood in rain events can also help capture water and reduce runoff.

In urban areas, replacing some hard surfaces with porous surfaces such as green space or permeable pavers reduces runoff by allowing water to seep through. For this reason, Chicago has changed a large part of its alley pavement to porous concrete. Green roofs can also help capture rainwater and slow or prevent its release to the environment. Installing curb cutouts in roadways can direct stormwater flow onto natural areas where the water has a chance to soak into the ground rather than flowing directly into a storm drain, which often takes it to a nearby water body. **INFOGRAPHIC 15.6**

Water pollution is a wicked problem with many causes and consequences, as well as multiple stakeholders who often have different opinions about what actions to take. But the success of the CWA in controlling point source pollution in the United States shows that progress can be made. Likewise, steps to address watershed health and reduce the potential for runoff are reducing nonpoint source pollution.

To address hypoxia in the Gulf of Mexico, the Gulf Task Force has identified a multipronged approach that considers all the causes and consequences of the problem as a triple-bottom line: the *environmental* needs of the coastal and offshore ecosystems, the *social* impacts on human communities affected throughout the watershed, and the *economic* realities that require us to prioritize our actions. **INFOGRAPHIC 15.7**

KEY CONCEPT 15.7

Urban and suburban areas are significant contributors of stormwater runoff, but many steps can be taken to reduce runoff and increase infiltration.

KEY CONCEPT 15.8

Addressing water pollution includes identifying where problems exist, reducing pollution at the source, watershed management to reduce runoff potential, and restoring wetlands.

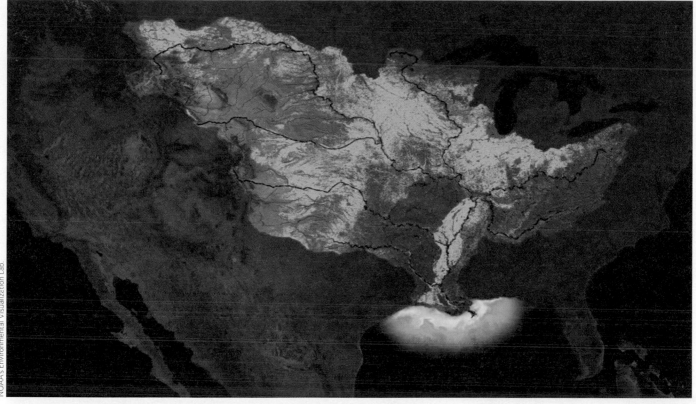

↑ Cities (shown in red) and farmland (shown in green) within the watershed of the Mississippi River fuel the Gulf of Mexico's hypoxic zone in late summer.

INFOGRAPHIC 15.7 GULF OF MEXICO REGIONAL ECOSYSTEM RESTORATION 5

GOAL	STRATEGY	GOAL	STRATEGY
Restore and conserve habitat.	• Restore natural flow of rivers and streams to wetland and delta regions. • Restore riparian areas and wetland habitats. • Implement measures to restore seagrass beds and protect seagrass from boat propeller damage.	Replenish and protect marine resources.	• Restore and manage coral reefs and oyster beds; sustainably harvest fish and shellfish. • Track sentinel species such as sea oats, pelicans, bluefin tuna, and oysters to monitor progress or problems. • Minimize or eliminate invasive species such as lionfish and nutria.
Restore water quality.	• Implement better regulation of industrial point source pollution. • Reduce runoff from animal operations by fencing off streams and construct wetlands to capture runoff. • Implement precision fertilizer application to reduce overuse.	Enhance community resilience.	• Improve coastal protections against storm damage. • Promote low-impact community growth plans. • Enhance education and outreach.

 How is the task force considering the triple-bottom line with their recommendations to reduce the size of the gulf hypoxic zone?

© Bettmann/CORBIS

↑ Firefighters stand on a bridge over the Cuyahoga River in Cleveland, Ohio, to spray water on a tugboat as a fire—started in an oil slick on the river—moves toward the docks at the Great Lakes Towing Company site. This 1952 blaze, one of 13 fires on the river since the late 1800s, was the most costly, destroying three tugboats, three buildings, and the ship-repair yards.

Peter Wynn Thompson/The New York Times/Redux

← "Green alleys" like this one in Chicago are those that have been resurfaced with permeable pavement to allow rainwater or snowmelt to pass through the pavement and soak into the ground rather than run off into storm drains which would eventually empty into nearby Lake Michigan.

A 2013 report from the Gulf of Mexico Nutrient Task Force suggested that while it is unlikely that the goal of reducing the hypoxic zone to 5,000 km² will be met by 2015 (the hypoxic zone averaged about 14,000 km² in recent years), it remains a reasonable target for the near future if there is an acceleration in programs that address the problem.

These types of control efforts are gaining increasing significance in a world where human population is still growing and the need for higher agricultural productivity grows with it. Meeting these needs and protecting our water at the same time may get even more difficult due to climate change that is reducing agricultural productivity in many areas worldwide, a trend that is expected to get worse in the future.

INTO THE GULF 295

Select References:

Diaz, R. J., & R. Rosenberg. (2008). Spreading dead zones and consequences for marine ecosystems. *Science*, 321(5891): 926–929.

Environmental Protection Agency. (2011). "Gulf of Mexico Regional Ecosystem Restoration Strategy," www.epa.gov/gcertf/pdfs/GulfCoastReport_Full_12-04_508-1.pdf.

Oquist, K. A., et al. (2007). Influence of alternative and conventional farming practices on subsurface drainage and water quality. *Journal of Environmental Quality*, 36(4): 1194–1204.

Rabalais, N., et al. (2002). Beyond science into policy: Gulf of Mexico hypoxia and the Mississippi River. *BioScience*, 52(2): 129-142.

BRING IT HOME

PERSONAL CHOICES THAT HELP

We are facing not only shrinking supplies of easily accessible water but also the potential degradation of this resource due to pollution. By changing products and modifying common practices, we can improve our water quality for years to come.

Individual Steps

• Read your city's water quality report to see which pollutants are prevalent in your area.

• Decrease your use of chemicals (fertilizers, pesticides, harsh cleaners, etc.) that will end up in the water supply. For alternatives to traditional yard chemicals, see www.safelawns.org.

• Always dispose of pet waste properly. In high quantities, it acts as an oxygen-demanding waste and can also spread disease.

Group Action

• Marking storm drains with "Don't Dump" symbols can remind people not to dump waste liquids down sewer drains. If the drains in your area are not marked, talk to city officials to see if you and other volunteers can mark them.

Policy Change

• August is National Water Quality Month. Take steps every day to reduce your water pollution and help raise awareness in August by writing a letter to your local newspaper outlining simple steps people can take to improve water quality.

Robert Brook/Science Source

ENVIRONMENTAL LITERACY UNDERSTANDING THE ISSUE

 In general, what is water pollution? What is the difference between point source and nonpoint source pollution, and what are some common sources of each?
INFOGRAPHIC 15.1

1. Water pollution is:
   a. found only in surface water near cities.
   b. primarily excess nutrients from lawns, farms, and animal feedlots.
   c. usually from excess carbon being added to the system.
   d. contaminants or excess nutrients in surface water and in groundwater.

2. Fertilizer from your lawn and motor oil from the leaky oil pan on your car are examples of:
   a. nonpoint source pollution.
   b. point source pollution.
   c. eutrophication.
   d. pathogenesis.

3. Compare and contrast three typical point source pollutants and three nonpoint source pollutants from the area where you live.

 What is eutrophication? How can water pollution from fertilizer or animal waste cause eutrophication and ultimately kill aquatic life?
INFOGRAPHIC 15.2

4. True or False: Hypoxic waters are waters with extremely low levels of oxygen.

5. The following items are parts of the process of eutrophication.
   1. Algae quickly reproduce, blocking sunlight to underwater plants.
   2. Bacteria decompose organic matter (dead algae).
   3. Underwater plants die.
   4. Excess nutrients enter a body of water.
   What is the correct order of these parts in the process?

6. Many pastures have ponds to provide water to the livestock. By summer's end these ponds are often covered in a thick green scum. If the ponds are stocked with fish, the fish can all die. What causes the green scum to form, and what might cause the fish to die?

 What is a watershed, and how does it affect the *quality* of surface water as well as the *quantity* of groundwater?
INFOGRAPHIC 15.3

7. A watershed includes:
   a. only the land that would be underwater in a normal rainfall year.
   b. the surface water and the underground aquifer.
   c. all the uphill land surrounding a river and the streams that can feed water into that river.
   d. all the land downhill from a river that could potentially be flooded.

8. You are investigating a die-off of fish in the local mountain lake, Lake Pleasant. It is fed by Hilltop Stream, and the Happy Valley River comes out of the lake and heads down toward Happy Valley. Do any of the following have something to do with the die-off? Justify your answers.
   - The new manufacturing plant just over the hill in the next valley
   - The cattle feedlot on the edge of the Happy Valley River
   - The recent construction of a new subdivision on the eastern slopes above Lake Pleasant
   - The lovely older neighborhood with wide lawns, a golf course, and a large grassy park on the western slope above the lake

 Why is nutrient pollution from agricultural sites a bigger problem today than in years past?
INFOGRAPHIC 15.4

9. Nitrogen and phosphorus pollution in the Gulf of Mexico could originate from:
   a. anywhere in the Mississippi River watershed.
   b. areas 100 miles or less upstream from the Gulf.
   c. only farms that lie in the floodplain or delta of the Mississippi River.
   d. agricultural areas but not suburban or urban areas.

10. According to Alex Echols, what three changes to agriculture contribute to the increase of nutrient pollution to the Mississippi River? What could be done to reduce each?

 What actions can be taken to address water quality and quantity problems in the Gulf of Mexico watershed and in other areas?
INFOGRAPHICS 15.5, 15.6, AND 15.7

11. True or False: The Clean Water Act focuses primarily on reducing nonpoint source pollution.

12. In a biological assessment of stream health, scientists collect bottom-dwelling insects and assess water quality based on:
   a. physical parameters important to life, such as temperature and pH.
   b. the chemical composition of insects captured.
   c. the diversity and abundance of the species present.
   d. the reproductive health of the insects, as determined by the number of eggs and larvae found.

13. Which of the following would be the best land use for a riparian area in terms of reducing runoff pollution?
   a. A paved surface that is smooth and unobstructed
   b. A well-manicured lawn
   c. A forested area with lots of native trees and shrubs
   d. An agricultural field such as a corn field

14. Rain gardens can be planted in areas to:
   a. reduce the amount of stormwater runoff reaching streams.
   b. prevent infiltration.
   c. capture rainfall so it can be used by the homeowner.
   d. direct runoff to storm drains.

SCIENCE LITERACY WORKING WITH DATA

The application of road salt in cold climates can result in runoff pollution that can damage aquatic ecosystems. Researchers at Saint Mary's University in Halifax surveyed local roadside ponds (some of which were occupied by amphibians and some of which were not) to determine chloride concentration and species richness of amphibians. They also investigated the vulnerability of five amphibian species to sodium chloride (NaCl) by determining the LD_{50}, the dose or concentration of chloride that kills 50% of the population.

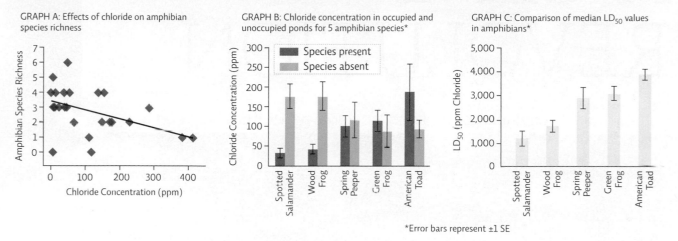

GRAPH A: Effects of chloride on amphibian species richness

GRAPH B: Chloride concentration in occupied and unoccupied ponds for 5 amphibian species*

GRAPH C: Comparison of median LD_{50} values in amphibians*

*Error bars represent ±1 SE

Interpretation

1. Look at Graph A above. In a single sentence, explain the relationship between chloride concentration and species richness.

2. Look at Graph B. Which species of amphibian is least affected by the presence of chloride in the water? Explain your reasoning.

3. Look at Graph C. Which species is the most sensitive to chloride? Which is the least sensitive?

Advance Your Thinking

4. Look at Graph B and Graph C. Does the LD_{50} for each species shown in Graph C correlate with the data (presence or absence of each species according to chloride concentration) shown in Graph B? Explain.

5. Suggest two actions that could be taken to reduce the chance that road salt would enter ponds in runoff.

INFORMATION LITERACY EVALUATING INFORMATION

Settling ponds, also known as holding ponds, are locations where contaminated water is allowed to stand so that particulates suspended in the water settle to the bottom of the pond. The water then evaporates or is drawn off, leaving behind the unwanted sediment, which can be from a mine, a quarry, a manure pit at an animal facility, industrial wastewater, stormwater, or other sources.

Go to www.technology.infomine.com/sedimentponds to see an overview of some of the concerns about building and maintaining a settling pond for surface mines.

Evaluate the authors of the article and work with the information to answer the following questions:

1. Do an Internet search and read two or three news articles about the toxic sludge spill near Kolontár, Hungary.
   a. What is the most recent article you can find about this, and what does it say about the incident?
   b. How similar is the information in the articles?
   c. Do the authors give any sources cited for the "facts" presented?

2. Now do an Internet search using Google Scholar or another scholarly database. Select a peer-reviewed article that you can access in its entirety (not just the abstract; look for a PDF link). Read the article and summarize its purpose and main points.
   a. How does this article compare to the news articles you read? Compare and contrast the types of information regarding scope of coverage, intended audience, clarity of information, and reliability of information (whether reliable sources are cited).
   b. In general, when would you go to a news source for information on a topic like this, and when would you go to a scholarly source?

3. What could be done to prevent another incident like what happened at Kolontár from happening again?

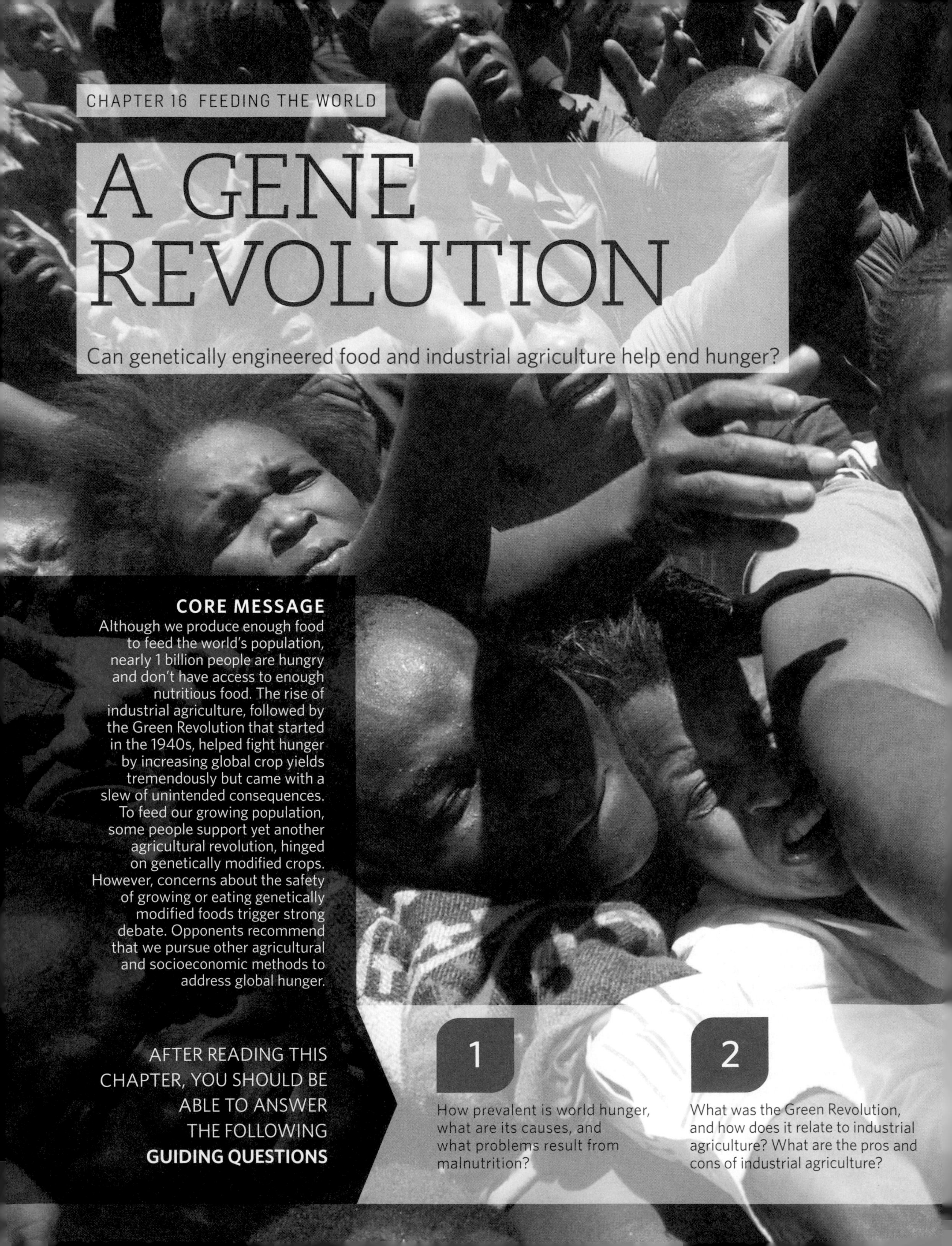

A GENE REVOLUTION

Can genetically engineered food and industrial agriculture help end hunger?

CORE MESSAGE

Although we produce enough food to feed the world's population, nearly 1 billion people are hungry and don't have access to enough nutritious food. The rise of industrial agriculture, followed by the Green Revolution that started in the 1940s, helped fight hunger by increasing global crop yields tremendously but came with a slew of unintended consequences. To feed our growing population, some people support yet another agricultural revolution, hinged on genetically modified crops. However, concerns about the safety of growing or eating genetically modified foods trigger strong debate. Opponents recommend that we pursue other agricultural and socioeconomic methods to address global hunger.

AFTER READING THIS CHAPTER, YOU SHOULD BE ABLE TO ANSWER THE FOLLOWING **GUIDING QUESTIONS**

1

How prevalent is world hunger, what are its causes, and what problems result from malnutrition?

2

What was the Green Revolution, and how does it relate to industrial agriculture? What are the pros and cons of industrial agriculture?

Desperate residents plead to workers from the Kenyan Red Cross for food during distribution in Nairobi in 2008. The crowd eventually broke through the gate but was chased back by police with whips. YASUYOSHI CHIBA/ AFP/Getty Images

3

What is the gene revolution, and how do researchers get desired traits into crops using this new technology?

4

What are the trade-offs of using genetically modified organisms in agriculture?

5

What are some low-tech (non-GMO, non-industrial) options for increasing food supplies? What role should industrial agriculture and GMOs play in solving world hunger?

It was in Burkina Faso—a tiny, landlocked West African country, virtually unknown to the developed world—that a brewing global crisis finally came to a head. On February 22, 2008, riots broke out in the country's two major cities. Angry protesters clogged the streets—shouting, throwing rocks, and flipping cars. Soldiers mobilized to restore order, but the chaos only spread from there. In neighboring Côte D'Ivoire, tear gas was employed, and dozens were injured; in Cameroon, some two dozen people were killed; in Egypt, a single boy was shot in the head. Before long, the violence spilled across Africa's borders. Protesters in Yemen torched police stations and blocked roads. In Bangladesh, they smashed cars and buses, vandalized factories, and ultimately injured dozens of bystanders.

What were these people rioting over? Food. It had become too expensive. In the 2 preceding years, the cost of rice had risen by 217%; wheat by 136%; and corn and soybeans by 125% and 107%, respectively. In the 7 months leading up to the riots, those prices had then doubled. Such increases were not as much of a problem in the United States or Europe, where even the poorest one-fifth of households spends just 16% of their budget on food. But in the cities that had descended into violence, families were already spending between 50% and 75% of their income on basic staples—rice and beans and bread and milk. Such dramatic and rapid cost increases had pushed those goods out of their reach and, in so doing, had nudged too many people toward the brink of starvation. "For countries where food comprises from half to three quarters of [income] consumption," explained World Bank President Robert Zoellick, "there is no margin for survival." In all, he said, some 22 countries were at risk of violent revolts if prices did not soon stabilize.

But what had destabilized them in the first place? Newspaper columnists, politicians, and cable news anchors trotted out the usual suspects. Some blamed extreme weather events: A drought in Australia, a heat wave in California, and flooding in India had decimated crops of wheat and maize and supplies of beef. Some speculated that diverting corn to the production of biofuels might have led to the supply shortages. Others pointed to the global financial crisis and to rising fuel prices: Because large, industrial farms rely on fossil fuel–powered machines and chemical inputs derived from fossil fuels, the cost of grain is tethered to the costs of oil and natural gas. Yet others focused on the growing demand for meat from Asia's expanding middle class: The more grain we devote to cattle rearing, the less we have to sell as food, and the more expensive it thus becomes (see LaunchPad Chapter 30).

But experts said that the trouble with the global food supply goes much further than that—beyond isolated weather events and the slumping global economy, to the very heart of global food policy.

World hunger and malnutrition are decreasing but are still unacceptably high.

While global food prices have stabilized since the 2008 riots, and grain reserves are currently well stocked, much of the world is still struggling to feed itself. The world currently produces enough food to feed the human population—enough to provide each person with more

◉ **WHERE IS BURKINA FASO?**

BURKINA FASO

CÔTE
D'IVOIRE   BENIN

CAMEROON

KENYA

UGANDA

INFOGRAPHIC 16.1   **WORLD HUNGER**   1

↓ Though we currently produce enough food to feed the world's population, almost a billion people are undernourished. Of those who are underfed, 98% live in developing nations. Even in wealthy nations there are those who do not have access to enough food or to a balanced diet.

PERCENTAGE OF POPULATION UNDERNOURISHED IN THE DEVELOPING WORLD (2011–2013)

Pakistan: Rising food prices and unemployment linked to policies of the new coalition government; conflict in western regions

Mongolia: Political changes due to the breakup of the USSR and extreme weather (harsh winters, dry summers)

Haiti: Natural disaster (2010 earthquake) and deforestation

Kenya: Ongoing drought and poor government response

Burundi: Civil war ended in 2005 but recovery is slow; between 2011–2013 it had the highest percentage of undernourished people in the world: 67%

<5%

5–14.9%

15–24.9%

25–34.9%

35% or over

No data

→ Of developing nations, those in protracted crisis (a prolonged ecological or political crisis that seriously affects the well-being of a large percentage of the population) have a chronic inability to acquire sufficient food supplies, leading to undernourishment, elevated infant mortality rates, and stunted growth in children.

Percentage undernourished

37%

15%

13%

40

30

20

10

0

Countries in protracted crisis

China and India

Other developing countries

Developing countries (2005–2007)

? Why do you think so many countries in Sub-Saharan Africa have a large percentage of their population that is undernourished, whereas the countries in northern Africa have less than 5% of their population undernourished?

than 2,800 calories per day, according to a 2011 analysis by the United Nations Food and Agriculture Organization (FAO). Yet in 2013, the United Nations estimated that 842 million people (about 12% of the global population) suffered from *undernutrition*—meaning that they did not consume enough calories to meet daily needs. According to the World Food Programme of the United Nations, hunger and poor nutrition is the leading health risk worldwide and causes almost half of all deaths in children under 5 years of age annually. **INFOGRAPHIC 16.1**

The FAO has set itself the task of eradicating hunger (and extreme poverty along with it). Its goal is to cut the percentage of underfed people around the world in half from the 1990–1992 level by 2015 and to one day achieve total food security. **Food security** is defined as all people at all times having physical, social, and economic access to sufficient safe and nutritious food. The FAO's mission is not just a matter of food quantity but

food security Having physical, social, and economic access to sufficient safe and nutritious food.

INFOGRAPHIC 16.2 **MALNUTRITION**

↳ Malnutrition is defined as a state of poor health that results from inadequate or unbalanced food intake. This includes diets that don't provide enough calories (and thus are nutrient deficient) as well as those that may provide enough (or even too many) calories but are deficient in one or more nutrients.

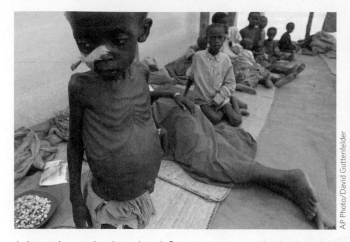

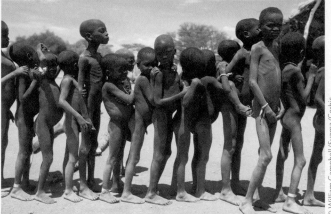

A diet with enough calories but deficient in protein can lead to kwashiorkor (usually in children)—a disorder that produces symptoms such as a bloated belly, loss of muscle mass, and lethargy. The symptoms are a result of the body breaking down its own protein in an attempt to keep vital processes functioning.

Caloric and protein deficiency can result in the condition known as wasting disease, or marasmus, most commonly seen in children. Victims are very skinny and frail, very susceptible to infection and disease, and experience stunted growth and developmental problems (including mental retardation).

 Compare and contrast kwashiorkor and marasmus.

KEY CONCEPT 16.1

Malnutrition is a worldwide problem: Almost 1 billion people don't take in enough calories and/or nutrients; 1.5 billion consume too many calories.

---

malnutrition A state of poor health that results from inadequate caloric intake (too many or too few calories) or is deficient in one or more nutrients.

---

also of improving food quality. A healthy diet contains a variety of foods that provide the proteins, carbohydrates, and fats needed for good health, as well as enough calories to meet daily energy needs. The food that supplies these calories must also supply micronutrients such as vitamins and minerals.

When a person's diet falls short of these basics— when he or she does not consume enough protein or vitamins—that person is said to be malnourished. **Malnutrition** can serve as a prelude to a whole host of diseases, from blindness (a result of vitamin A deficiency) and anemia (iron or vitamin B12 deficiency) to wasting (or marasmus) and kwashiorkor. **INFOGRAPHIC 16.2**

Overnutrition, or the consumption of too many calories, is also considered malnutrition. By some estimates, about 1.5 billion people around the world are overnourished and are thus vulnerable to another set of nutrition-related conditions—namely obesity and type 2 diabetes. Contrary to intuition, this is also a problem of the poor. Cheaper foods tend to make us unhealthy. And because those foods are often low in essential nutrients, people who consume too many calories are often just as malnourished as those who consume too few.

Of course, to conquer global hunger, we must grapple with more than just nutrients and calories. Indeed, we must tackle a vast array of problems that lie at the root of the brewing food crisis. Political instability and ecological degradation play significant roles, to be sure. As shown in Infographic 16.1, undernourishment is three times greater in developing nations experiencing a prolonged armed conflict, drought, or natural disaster than it is in countries not in such crises. But there are other contributors, too— namely, social disempowerment and poverty. Even in the poorest countries (or especially in the poorest countries), there are groups of people who have access to food and groups of people who do not. To reach the FAO's lofty goals, we must first understand why and how that came to be.

And to do that, we must travel back through recent history, to the last time global hunger was the stuff of headlines.

Agricultural advances significantly increased food production in the 20th century.

It was the late 1960s. Global population was soaring; food crops were flatlining in some parts of the developing world and plummeting in others. India especially seemed to be hovering on the brink of a massive famine.

To stave off such catastrophe, the international community had launched the **Green Revolution**—a coordinated global effort to eliminate hunger by bringing modern agricultural technology to developing countries in Asia. Working across the globe, scientists, farmers, and world leaders introduced India and China to chemical pesticides, sophisticated irrigation systems, synthetic nitrogen fertilizer, and modern farming equipment—technologies that most industrialized nations had already been using for decades. They also introduced some novel technology—namely, new **high-yield varieties (HYVs)** of staple crops like maize (corn), wheat, and rice. HYVs have been selectively bred to produce more grain than their natural counterparts, usually because they grow faster or larger or are more resistant to crop diseases.

If all that were not enough, developed countries like the United States also began implementing a litany of agricultural policies—tax breaks, government subsidies, and insurance plans—that encouraged their own farmers to plant crops "fencerow to fencerow" and thus added substantially to the world's food supply.

In the short term, the effort was a huge success. The combined force of HYVs, existing technology, and new food policies resulted in a 1,000% increase in global food production and a 20% reduction in **famine** between 1960 and 1990. Today, most experts credit the initiative with the fact that we are now producing enough food to feed every one of the planet's 7 billion or so mouths.

So why are so many people still going hungry?

As it turns out, the Green Revolution owes its success to modern **industrial agriculture**. Though industrial agriculture can be very productive, it has had some unintended consequences. Industrial agriculture is dependent on machinery, as well as synthetic fertilizers (chemicals that boost plant growth beyond what the soil could naturally support) and pesticides (chemicals that protect crops from pests that might harm or consume the crop before harvest). In many cases, mechanized irrigation, drawn from surface water or groundwater supplies, is also used and allows farmers to grow crops that could not be sustained by local rainfall. This mechanization and these chemical farming "inputs" increase efficiency and produce tremendous amounts of food on a large scale. Typically, a single, genetically uniform crop (monoculture) is planted, making it possible for fewer farmers to farm larger tracts of land using heavy equipment like tractors and combine harvesters.

But farming in this fashion produces environmental problems that just get worse over time. For one thing, excessive use of chemical inputs and large-scale monoculture operations have degraded the soil, making it less fertile (and in need of even more fertilizers). It has also contributed to the twin problems of water pollution and water scarcity. Large-scale irrigation can also deplete water supplies; in hot, arid areas, high rates of evaporation can leave a salt residue on soils which impedes plant growth and even renders some soils unusable. Widespread pesticide use is leading to pesticide-resistant pests that are harder to kill, in addition to exposing farm workers and consumers to toxic chemicals. Farm equipment runs on fossil fuels, and even the fertilizers and pesticides are derived from fossil fuel, made from natural gas or petroleum. This requires more fossil fuel extraction, with all its accompanying environmental issues (see Chapters 18 and 19). There are also local social and economic impacts: Fewer farmers on larger farms threaten to make the family farm a thing of the past, impacting entire communities. (For more on industrial farming, see Chapter 17 and LaunchPad Chapter 30).

The Green Revolution also led to a heavy reliance on just a few high-yield varieties of each crop, which has led to a dramatic reduction in agricultural biodiversity. This is no small matter. Without a diverse gene pool, the global food supply is much more vulnerable to pests that can't be controlled by agrochemicals: When all plants in the field are genetically identical, what kills or damages one will probably kill or damage them all. Moreover, as the old varieties fall out of use—as farmers stop producing and conserving the seeds that have been bred through traditional agriculture

---

KEY CONCEPT 16.2

Modern industrial agriculture produces huge amounts of food but leads to ecological and social problems. The Green Revolution exacerbated these problems.

Green Revolution A plant-breeding program in the mid-1900s that dramatically increased crop yields and paved the way for mechanized, large-scale agriculture.

high-yield varieties (HYVs) Strains of staple crops selectively bred to be more productive than their natural counterparts.

famine A severe shortage of food that leads to widespread hunger.

industrial agriculture Farming methods that rely on technology, synthetic chemical inputs, and economies of scale to increase productivity and profits.

INFOGRAPHIC 16.3   **THE PERKS AND PROBLEMS OF INDUSTRIAL AGRICULTURE AND THE GREEN REVOLUTION**   2

↓ The Green Revolution transformed agriculture into industrial model we see today. Through the selective breeding of the most productive plants, high-yield varieties of crops were devloped that more than doubled production per acre (when grown with inputs like fertilizer and pesticides). This increased food supplies worldwide but brought with it a new set of problems.

AREA HARVESTED OVER TIME, IN WORLDWIDE PRODUCTION OF GRAIN

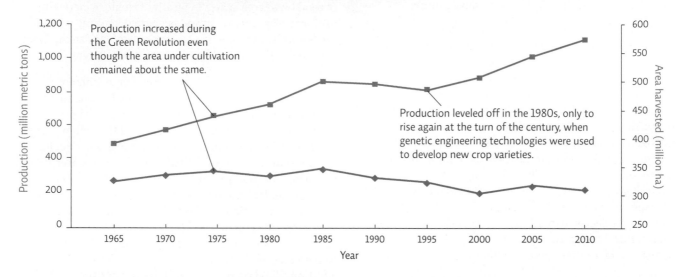

Production increased during the Green Revolution even though the area under cultivation remained about the same.

Production leveled off in the 1980s, only to rise again at the turn of the century, when genetic engineering technologies were used to develop new crop varieties.

THE TRADE-OFFS OF INDUSTRIAL AGRICULTURE

Neil W. Zoglauer/Moment/Getty Images

Mmphotos/Photodisc/Getty Images

Noel Hendrickson/Digital Vision/Getty Images

ADVANTAGES	DISADVANTAGES
• Large-scale farming • Higher yields/acre • Crops can be grown on nutrient-poor soil (with fertilizers) • Crops have fewer blemishes (pesticides) • Crops can be grown in water-poor areas (with irrigation) • Monocultures are more efficient to grow and harvest • Less labor intensive	• Dependence on mechanization (expensive) • Large quantities of fossil fuels are used (as fuel and as raw material to make fertilizers and pesticides) • Soil quality is degraded • Water pollution and scarcity • Monocultures are more vulnerable to pests and disease • Toxic residues on food • Fewer jobs • Decreases the number of family farms, which affects local communities

 Which of the advantages of industrial agriculture do you feel is its greatest strength? Why? Which of the disadvantages of industrial agriculture worries you the most? Why?

over thousands of years—many valuable genetic traits face the possibility of being lost forever. **INFOGRAPHIC 16.3**

No one can dispute the fact that the Green Revolution fed, and continues to feed, a lot of people. Unfortunately,

not everyone has benefited from this agricultural transformation. Africa, a continent plagued by hunger and largely bypassed by the Green Revolution, has nonetheless fallen prey to a different set of the Green Revolution's unintended consequences: a lack of *food self-sufficiency*

↑ Runoff from farmland near the Santa Maria River in California.

↑ A toxic crust of salt and other minerals builds up in heavily irrigated fields in Colorado.

(the ability of an individual nation to grow enough food to feed its people), a lack of *food sovereignty* (the ability of an individual nation to control its own food system), and ultimately, a lack of food security.

Here's why: As industrialization and farm subsidies enabled (mostly American) farmers to produce vast surpluses of wheat, corn, and soybeans, the global marketplace was flooded with cheap food from the developed world. Farmers in countries like Burkina Faso could not compete with such cheap and plentiful food imports—plagued as they were by land degradation (drought, soil erosion, water shortages) and armed conflict (violent clashes destroy existing crops and prevent new ones from being planted).

So they converted much of their farmable land to **cash crops** like coffee and cocoa. Rather than feed local populations, these commodities are exported for profit (which is usually held by those in power). Thus a system where grain was locally produced and supplied gave way to a system where it was imported from thousands of miles away. And as they became dependent on food imports, developing countries found themselves at the mercy of forces far beyond their own borders.

And so it was that by 2008, Australian droughts and American agricultural policies had left the people of West Africa to riot in the streets for want of bread. We are indeed making enough food to feed the world. We just aren't getting it to the people who need it most.

Into this morass an additional 3 billion people will soon be born. Global population is expected to reach 10 billion by 2050; experts say that to feed that many mouths, we will need to produce twice as much food as we are now producing. That will mean either farming more land or devising more production-boosting technologies. To farm more would mean clearing more forests, and thus destroying more natural habitats, species, and ecosystem services—a dire prescription at a time when the planet is already flirting with ecological disaster. How then, will we make enough food to feed the future?

The next Green Revolution may be a "gene" revolution.

As Robert Paarlberg and his driver made their way up one of Uganda's countless narrow dirt roads, a collection of mud huts encircled by a patchwork of small and midsize fields came into view. The driver stopped, and the pair made their way through dry bush. A young boy tending goats spotted them from the distance and ran toward the huts to inform his kin of impending visitors. When they reached the village, Paarlberg and his driver-translator were greeted by a group of women and children: curious, welcoming, and eager to chat with outsiders.

Paarlberg, a Harvard-based political scientist, had been traveling throughout a northern swath of Sub-Saharan Africa—visiting Kenya, Benin, Cameroon, and now Uganda—in an effort to understand the countries' opinions and attitudes about genetically modified crops. The women's hospitality reminded him of the farmers he'd known as a boy, growing up in Indiana, but the rest of the scene contrasted sharply with such memories. In Indiana, crops were invariably bountiful: seas of grain billowing beneath an endless sky. In the evening,

KEY CONCEPT 16.3

In less developed countries, converting farmland that once grew food for local communities to cash crop agriculture reduces the food self-sufficiency and the food sovereignty of the country.

cash crops Food and fiber crops grown to sell for profit rather than for use by local families or communities.

↑ Food insecurity plagues millions of people, especially in drought and war-torn regions of the world. In rural Africa, women do much of the farming with rudimentary tools. Here, a young mother with her baby on her back works with other women of her Rwandan village to prepare their field for planting.

Per-Anders Pettersson/Photonica World/Getty Images

farmers—most of them hardy, cigar-smoking men—leaned against colossal tractors and other such equipment and bragged about the profits their crops were reaping.

> *Global population is expected to reach 10 billion by 2050; experts say that to feed that many mouths, we will need to produce twice as much food as we are now producing.*

In West Africa, most of the farmers were women and school-aged children (conspicuously not in school). On this particular farm, crops of maize struggled to pop up through parched soil—starved for both water and nutrients. Cassava withered under the strain of a viral infection, and a single cow suffered from untreated sores. Still, the women were warm and chatty. They did not have chemical fertilizers, they told him. Or water pumps. Or any kind of machinery. In fact, there was no electricity. And with the nearest road several kilometers away, everything they bought or sold had to be transported in or out of the tiny farming village on foot. Through the translator and the children who spoke some English, the women fretted about the weather, and crop prices, and how much grain they had in storage. These women knew absolutely nothing of the emerging technology Paarlberg had come to the region to discuss.

As their name suggests, **genetically modified organisms (GMOs)** are organisms that have had their genetic information modified in a way that does not occur naturally. Scientists have been using such organisms for decades to produce insulin, antibiotics, and other important medicines. By inserting select genes into certain kinds of bacteria, they can coax those microbes into churning out large quantities of medically important compounds—like tiny living drug factories. In the 1990s researchers began applying the same technology to food crops such as corn, rice, wheat, and soybeans. By using bacteria as "vehicles" to transfer genes for desirable traits (like pest resistance, drought tolerance, and increased nutrient production) from one species to another, they have created a new suite of genetically modified food crops—plants that can grow more plentifully and thrive in a wider range of habitats than they ordinarily would. Organisms that have been altered in this way are known as **transgenic organisms. INFOGRAPHIC 16.4**

genetically modified organisms (GMOs) Organisms that have had their genetic information modified to give them desirable characteristics such as pest or drought resistance.

transgenic organism An organism that contains genes from another species.

KEY CONCEPT 16.4

Genetically modifying crops (or animals) to contain useful traits may boost growth and expand the growing range of some crops but comes with environmental and ethical concerns.

INFOGRAPHIC 16.4 | MAKING A GENETICALLY MODIFIED ORGANISM

↓ Genetic material can be transferred from one organism to another. This can give organisms new traits that they haven't naturally evolved but that may prove desirable. For example, the bacterium *Bacillus thuringiensis* (Bt) produces a toxin that kills many types of insect pests. If the gene responsible for that toxin is inserted into a plant's genetic code, that plant will gain the ability to produce the toxin and will be able to ward off pests itself.

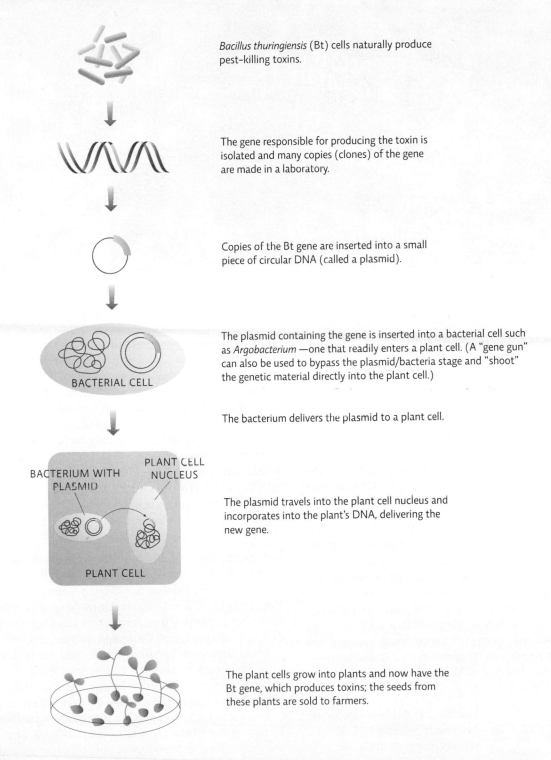

Bacillus thuringiensis (Bt) cells naturally produce pest-killing toxins.

The gene responsible for producing the toxin is isolated and many copies (clones) of the gene are made in a laboratory.

Copies of the Bt gene are inserted into a small piece of circular DNA (called a plasmid).

The plasmid containing the gene is inserted into a bacterial cell such as *Argobacterium* —one that readily enters a plant cell. (A "gene gun" can also be used to bypass the plasmid/bacteria stage and "shoot" the genetic material directly into the plant cell.)

BACTERIAL CELL

The bacterium delivers the plasmid to a plant cell.

BACTERIUM WITH PLASMID   PLANT CELL NUCLEUS

The plasmid travels into the plant cell nucleus and incorporates into the plant's DNA, delivering the new gene.

PLANT CELL

The plant cells grow into plants and now have the Bt gene, which produces toxins; the seeds from these plants are sold to farmers.

 Two different bacteria species are used to produce the Bt plants shown here. What is the role of each in the process?

↑ More than 75% of processed food in the United States contain GMOs. Only certified organic food is guaranteed to be GMO free.

ROBYN BECK/AFP/Getty Images

In the United States, more than 75% of processed food contains GMOs, including 85%–90% of our corn, soybeans, and cotton. Most of these crops have an herbicide-tolerance (HT) gene added that enables them to withstand huge doses of herbicides (chemicals that kill weeds). This trait enables farmers to douse the field with herbicide and kill the weeds without threatening their harvest. Other crops, known as Bt crops, have been engineered to better resist pests; they contain a gene from *Bacillus thuringiensis*, a naturally occurring bacterium that produces a toxin that kills some insect pests. Some crop GMOs even contain both an HT gene and a Bt gene.

Scientists are also working to add genetically modified animals to our food supply. The biotechnology company AquaBounty has developed *AquAdvantage* salmon—a genetically modified Atlantic salmon that grows much faster than normal (thanks to genes from the larger Chinook salmon). The Food and Drug Administration determined that the fish is safe to eat in a 2012 assessment but has yet to give the company approval to bring the fish to market. If approved, this would be the first genetically modified (GM) animal food product sold in the United States.

Green Revolution 2.0 Programs that focus on the production of genetically modified organisms (GMOs) to increase crop productivity.

Paarlberg and others (including Microsoft founder and famed philanthropist Bill Gates) believe that GM crops will go a long way toward helping African farmers achieve food self-sufficiency. In fact, GMOs form the basis of what some farmers and scientists like to think of as **Green Revolution 2.0**, or the *Gene* Revolution—the next battle between human hunger and innovation, this time waged in Africa as well as Asia, and against not only hunger but poverty and sickness as well. "When you put the right tools in farmers' hands, the results can be magical," Gates said during a speech in Rome, Italy, in February 2012. "A submergence-tolerant seed capable of surviving in flood waters can grow into a rice surplus that [allows parents to afford to take] a sick child to the doctor. . . . And when that child grows into a healthy and educated young mother, the future is bright."

But are expensive GMO seeds and the chemical inputs they need a reasonable solution in impoverished areas, on small farms? Researchers Matin Qaim and Shahzad Kouser of the Georg-August-University of Goettingen looked at farmers with small farms in an area of India facing food shortages. Qaim and Kouser studied farmers that switched from traditional methods of growing cotton to growing Bt cotton. The farmers in the 6-year study that grew Bt cotton increased their income, and this correlated with a diet higher in calories and nutritional value for the farm families. In fact, fewer of the farm families that grew Bt cotton were classified as food insecure (7.93%) compared to families that grew traditional cotton (19.94%). The researchers acknowledged that GMOs are not the sole solution to world hunger but believe they can be part

TABLE 16.1   TYPES AND EXAMPLES OF GMOs

TYPE OF GMO	EXAMPLES (ON THE MARKET OR UNDER DEVELOPMENT)
HT (herbicide-tolerant) crops are not killed by the herbicide, so the herbicide can be sprayed on the crop/soil directly, where it will kill weeds but not the crop.	Bromoxynil-tolerant canola and cotton Glyphosate- (Roundup-) tolerant corn, cotton, soybeans Imidazolinone-tolerant wheat
Bt crops contain a gene from *Bacillus thuringiensis*, a naturally occurring bacterium that produces a toxin that kills some pests.	Bt corn, potatoes, and cotton
Nutritionally enhanced food: Genes inserted can increase the amount of a particular nutrient or allow the crop to produce a nutrient it would not normally produce.	Golden rice—produces more vitamin A than average rice
Genetically modified animals are being developed for the food supply.	AquAdvantage Salmon—faster growth Pigs that produce more omega-3 fatty acids (a healthy fat)

of the solution, even in less-developed countries. It remains to be seen if replacing traditional farming methods with industrial ones will be sustainable in these areas in the long run.

Proponents of the technology point to the possibility of not only increasing our food supply with GMOs but also improving it. GMO foods could be nutritionally enhanced with the addition of genes to produce needed vitamins. The most notable example is "golden rice," which is genetically modified to produce extra beta carotene (a precursor to vitamin A), to help address vitamin A deficiency, a leading cause of blindness in children.

TABLE 16.1

Of course, for the people of Africa to benefit from this new technology, they would have to first be made aware of its existence and then be persuaded to use it. And as Paarlberg was discovering in his travels, neither of those would happen easily. Not only were most local farmers completely unaware of GM technology, but government officials, who did know about it, were exceedingly wary.

Concerns about GMOs trigger strong debate.

To be sure, genetic engineering is a decades-old technology. But genetically modified food is unlike any of the other products that have been produced with it. With the production of genetically engineered medications, the transgenic organism remains confined to a flask in a lab; GM crops are out in the real world, growing and sharing genes with other organisms, and being eaten by both animals we are raising for food and directly, by humans—many of whom are very uneasy about the prospect.

Since the first genetically modified foods appeared in the early 1990s, a vocal contingent of activists—mostly in the United States and Europe—have rallied against such products—dubbing them "Franken Foods" and strongly contesting their safety. What if humans proved allergic to some of the proteins whose genes were being spliced in, they asked? What if negative health effects took years to manifest? Today, even as GM crops permeate the U.S. food supply, other countries (including Japan and much of the European Union [EU]) continue to fiercely resist the technology. As of 2014, only two genetically modified crops were authorized for planting in EU member states, and only one of them is currently planted—a Bt corn variety. Individual countries have the option of banning cultivation of any EU-authorized GMO within their borders.

One enduring concern is that the genes introduced into genetically modified crops could be passed to wild plants through cross-pollination—that is, they could escape into the natural world and be incorporated into other plants for which they were not intended. For example, if an HT gene were accidently transferred to a weed species, it could enable the weed to grow more aggressively and outcompete other plants, including our crops. This has already happened in the United States. So far, 16 weed species have acquired a gene for herbicide tolerance. These so-called superweeds, including giant ragweed and pigweed, can be found in 22 states and can tolerate all the herbicide a farmer can spray. They take over entire fields, stop combines, and are tough to clear by hand.

Another fear is that crops engineered to resist pests could inadvertently repel beneficial insects, or that they might give rise to a new population of Bt-resistant pests much more quickly than would traditional pesticide spraying.

KEY CONCEPT 16.5

Concerns about GMOs include health and environmental effects, and the worry that the use of patented GMOs could give a few large corporations unprecedented control over global agriculture.

The rise of secondary pests (those not affected by Bt) is also a concern in some cases. For example, several studies have shown that the amount of pesticides applied to Bt cotton was as great as the amount of pesticide used on traditional cotton because the elimination of Bt-susceptible pests enabled other Bt-resistant pests to thrive.

Raising up only a few plant varieties (as we did in the Green Revolution) could further degrade crop biodiversity precisely at a time when genetic and species diversity could not be more important. Only a deep genetic storehouse would allow us to develop crops that can cope with global climate change and other environmental challenges in the future.

For consumers, the data are less clear. Some critics express deep concern over possible allergens or toxic substances in GM foods, though such claims have not yet been well studied. Purported advantages of GMOs—like better-tasting or more nutritious foods—while feasible with current technology, have yet to hit the market. There are even some concerns that the nutritional content of some GM foods might decline; increasing the production of one nutrient in a GMO might affect the ability of the food to produce other nutrients. Currently in the United States there are no federal labeling requirements for informing consumers that foods contain GMOs (though certified organic food must be GMO free). In June 2014, Vermont became the first state to pass a law requiring GMO labels; the Grocery Manufacturers Association called the bill "critically flawed" and promised to sue the state in federal court in an effort to overturn the law.

Finally, opponents also worry about the prospect of putting even more of our food supply under the control of a few multinational corporations. In the United States, companies like Monsanto have been known to tightly guard their GM seeds. Because they are patented, farmers are not permitted to save seeds from one year to the next but instead must repurchase them year after year. Worse yet, when GM crops grown by one farmer have accidentally (that is, naturally) cross-pollinated another farmer's field, the second farmer has been held liable and forced to pay for the seeds, even if she didn't plant them— and even if she doesn't want to grow GM crops. "Food security in private hands is no food security at all," U.S. Senator Tom McGovern said, "because corporations are in the business of making money, not feeding people."

Proponents argue that in Africa especially, where individual countries lack the capacity for large-scale research and development, private investment is essential to agricultural development. "The country programs, agencies, and research centers don't have expertise," Gates said. "And they don't have time to build it from scratch." If we are ever to develop technological solutions to the challenges of developing world agriculture, he said, we will have to rely on the expertise of the private sector.

INFOGRAPHIC 16.5

Gates and others point out that technological innovation is exactly how developed countries like the United States turned Dust Bowl—era food shortages into vast surpluses in less than a generation. It is unfair, they say, to tell impoverished countries to forgo the technologies that we ourselves have benefitted from, especially while the people in those other countries are still starving. "This postmodern resistance to agricultural science makes considerable sense in rich countries, where science has already brought so much productivity to farming that little more seems needed," Paarlberg said. "It becomes dangerous, however, when exported to countries in Africa where farmers remain trapped in poverty."

Meanwhile, in the shadows of this global debate, many African farmers have been devising their own solutions to food insecurity.

It will take a combination of strategies to achieve global food security.

The central plateau of Burkina Faso is marked by poor soils, low crop yields, and very little rain. In many places, the land is so parched that it has crusted over into what local farmers call zippelle: hard, dry cake. To grow anything here takes a special kind of ingenuity. In the 1980s, with the help of the FAO, farmers revived an ancient strategy known as the Zaï pit. During the dry season (between November and May), farmers dig thousands of small pits—5,000 to 10,000 of them per hectare (12,000 to 25,000 per acre), each about 30 centimeters (12 inches) wide and 15 centimeters (6 inches) deep. Once the pits are dug, each one is filled with roughly half a kilogram (about a pound) of organic matter. After the first rainfall, the organic layer is covered with a thin layer of soil, and seeds are placed in the middle of the pit. The top of the pit is then ridged, so that when the rain does fall, it accumulates in the pit and hydrates the seed.

Zaï pits are not the only technology being employed in this landlocked, desert country. Researchers and farmers have also worked together to adapt cutting-edge micro-dose fertilization technology to suit local needs. By applying small and precise quantities of fertilizers close to each seed at planting, they only apply what the plants can actually use: Wheat crops grown in Burkina Faso now require one-tenth of the fertilizer used on their U.S.-grown counterparts; corn crops require one-twentieth.

INFOGRAPHIC 16.5   THE TRADE-OFFS OF GMOS

↓ As with other environmental issues, the advantages of using GMOs must be weighed against the disadvantages when deciding whether to use GMO crops (or animals). The conclusion may differ in different areas, depending on available resources and the needs of the people in that region.

GMO TRAIT	ADVANTAGES	DISADVANTAGES

HERBICIDE TOLERANCE

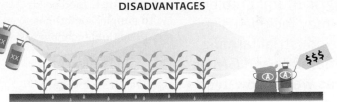

- Weeds are controlled by simply spraying herbicide.
- Higher yields/acre due to less competition from weeds
- Less weed competition can reduce the need to till the soil, reducing soil erosion and fossil fuel use.

- More herbicides are used since they do not harm the crop (potential water pollutant).
- The farmer is locked into using the herbicide the crop can tolerate.
- The company producing the seed and herbicide has an unchallenged market (no competitors), so prices can be high.
- "Superweeds" can develop when weeds acquire herbicide resistance genes.

PEST RESISTANCE

- Less pesticide may be used because crops can defend themselves from pests.
- Higher yields/acre due to less damage from pests.

- Weeds can grow faster and taller if they acquire the trait for pest resistance.
- The crop may repel beneficial insects like pollinators.
- Pesticide-resistant pest populations may increase.
- Secondary pests (those not affected by the GMO trait) may increase.
- Because the crop can tolerate it, more pesticide can be applied to better kill pests or weeds.

OTHER ISSUES

CORPORATE FUNDING

- Corporations have absorbed the tremendous cost and time of developing useful GMO products, saving taxpayers money.
- Corporations take on the costs and logistics of distribution.

- The high seed prices reflect the corporation's time and financial investment.
- The seeds are patented: Farmers must purchase new seeds every year (as opposed to harvesting them for future use).
- Fields of organic produce accidently cross-pollinated by GMOs no longer qualify as organic, and farmers can be sued for having GMO plants in their fields.

FOOD SECURITY AND SAFETY

- Less potential for food shortages
- Potential for more nutritious food (i.e., golden rice).

- Avoiding GMO foods is difficult; they are prevalent in processed food and there are no label requirements in the United States.
- Research on the health safety of GMOs is lacking but there are concerns about:
    - the potential for allergic reactions to new components
    - the presence of toxic substances in GMO foods
    - the decline of the nutritional quality of the food

Do you think that corporations with the patents for GMO crops should be able to sue farmers whose fields contain GMO plants from cross-pollination in the field? Explain.

This significantly reduces fertilizer waste and brings the cost of fertilization down to an affordable level.

In addition to using Zaï pits and microfertilization, farmers here have planted trees to retain soil and stave off desertification, and they are conserving livestock manure and applying it to fields. And cooperative groups have been established throughout the region to manage village cereal banks and community wells. The result has been an increase in grain production that matches some of the first Green Revolution successes in Asia and belies the despair of the 2008 food riots. "So much has been written about the disappointments in African agriculture that it is easy to overlook the successes," said Steve Wiggins of the Overseas Development Institute in London. "In some parts of the continent and for particular crops and activities, there have been veritable booms in farming."

Indeed, a growing litany of similar successes have been reported from across the continent. The details of each

KEY CONCEPT 16.6

Solutions to agricultural problems may include using industrial farming methods and GMOs as well as low-tech traditional methods and crop varieties.

differ, but the central point is often the same: Across the vast African continent, at least some individual communities are meeting their food needs with simple, sometimes centuries-old technologies.

To those opposed to GMOs, these low-tech successes underscore a key argument—namely, that there are other solutions that are more cost-effective and that can address not only the need to produce food but also the need to reduce farming's impact on the natural environment and to rehabilitate damaged areas. Instead of giving industrial agriculture another tool to use—GM seeds that will need lots of fertilizer and cost lots of money—proponents of a more sustainable approach to farming say we should teach farmers to plant nitrogen-fixing trees and crops that would furnish the soil with nutrients and help improve crop yields. (See Chapter 8 for more on nutrient cycling.) Farmers should be encouraged to plant many different crops instead of just one so that the soil is replenished naturally. And communities should create seed banks

and seed-sharing programs that enable farmers to both use and preserve the wild relatives of our crops (known as the *wild type*) and the traditional plant varieties—those that over generations have become adapted to each individual area. (See Chapter 17 for more on sustainable agriculture and seed banks.)

Critics also contend that government policies should be examined as well. For example, rather than subsidize cash crops for export, policies could provide support for farms that grow some crops for local consumption. And instead of offering tax incentives to agribusiness giants like Monsanto, investments in infrastructure could be made: Creating and repairing roads would dramatically enhance farmers' ability to get supplies to the farm and to get fresh produce to market—while it's still fresh.

Low-tech solutions have a key advantage over higher-tech ones: They are more accessible to women. Women make up a large percentage of the agricultural workforce—43% worldwide—and possess considerable knowledge about successful farming techniques in their home regions. But in developing countries especially, they are much less likely than men to have government support or the financial means to purchase the inputs required for industrial agriculture. Because of this gender bias, programs that focus on high-tech solutions tend to bypass women and local farmers like the ones that Paarlberg surveyed in his travels. **INFOGRAPHIC 16.6**

INFOGRAPHIC 16.6   LOW-TECH FARMING METHODS CAN ALSO HELP ADDRESS FOOD INSECURITY   5

↑ Seed exchanges, like this one sponsored by SEED, a non-profit community group in South Africa, helps farmers use and preserve a wide variety of locally adapted seeds.

↑ Zaï pits help seedlings grow in dry areas such as the Sahel region of Africa. The pits, dug by hand and mostly by women, collect the sparse rainfall and help seeds stay hydrated.

 Explain the importance of preserving the wild relatives and traditional plant varieties (i.e., the seeds from as many corn varieties as possible) rather than focusing on just a few crop varieties and letting the others die out.

Ultimately, of course, the countries of Africa will need much more than Zaï pits or GMOs to attain food security: They will need financial support, infrastructure, equipment, and training. And even with those essentials in place, the solutions still won't be straightforward. Different regions—and different farmers within the same region—face different agricultural challenges. In some cases, GMOs may represent the best option. In others, HYVs may work better, and in others still, native varieties may be the right choice. Whatever methods are pursued, they must also protect the natural environment so as not to jeopardize future productivity. And they must be made available to all people, not just an empowered few. That means issues of environmental sustainability and social justice must be considered alongside issues of productivity per hectare.

One thing nobody disputes is this: Helping countries like Uganda and Burkina Faso achieve food self-sufficiency will be the key to lifting them out of poverty. "Most extremely poor people in the developing world get their food and income from farming small plots of land," Gates explained. "Many others live in big cities and need access to inexpensive food to be healthy and productive. So helping small farmers grow more food sustainably is the best way to fight hunger and poverty over the long term."

Select References:

Azadi, H., & P. Ho. (2010). Genetically modified and organic crops in developing countries: A review of options for food security. *Biotechnology Advances,* 28(1): 160–168.

Indian Council of Agricultural Research. (2011). *CICR Vision 2030.* Nagpur, India: Central Institute for Cotton Research.

Paarlberg, R. (2008). *Starved for Science: How Biotechnology Is Being Kept Out of Africa.* Cambridge, MA: Harvard University Press.

Paarlberg, R. (2010). GMO foods and crops: Africa's choice. *New Biotechnology,* 27(5): 609–613.

Patel, R. C. (2012). Food sovereignty: Power, gender, and the right to food. *PLoS Med,* 96: e1001223.

Qaim, M., & S. Kouser. (2013). Genetically modified crops and food security. *PLoS ONE,* 8(6): e64879.

BRING IT HOME

PERSONAL CHOICES THAT HELP

The challenge of meeting the nutritional needs of the entire population seems overwhelming, but there are plenty of opportunities to make an impact within your local community. Two of the best ways to have an impact on those facing food insecurity are through education and financial support.

Individual Steps
• There are thousands of food banks all over the country that collect and distribute food to those who need it the most. Donate your time or extra nonperishable food items. Find the closest food bank at www.feedingamerica.org.
• Plant your own vegetable garden and donate part of the harvest to your local food bank. Some areas even have community gardens that rent space during the growing season.
• Make a monetary donation to an international nonprofit organization that provides food aid or agricultural assistance such as UNICEF (www.unicef.org) or Heifer International (www.heifer.org).

Group Action
• Work with parents, community leaders, and education leaders to develop a summer food service program for school-age children on free or reduced-price lunch programs. Find contacts and resources at www.fns.usda.gov/cnd/summer/.

Policy Change
• GMOs may be part of the solution because they can increase crop yields around the world, but they may cause some unintended consequences. Contact your federal representative: Express your concerns and discuss your rep's position on the regulation, testing, and labeling of GMOs.

Todd Anderson/The New York Times/Redux

ENVIRONMENTAL LITERACY **UNDERSTANDING THE ISSUE**

1

How prevalent is world hunger, what are its causes, and what problems result from malnutrition?

INFOGRAPHICS 16.1 AND 16.2

1. True or False: The main cause of world hunger today is underproduction of food.

2. Food security is:
   a. having sufficient safe and nutritious food freely available to everyone in the population.
   b. having enough money to buy sufficient safe and nutritious food.
   c. all people having access to enough calories to survive.
   d. all people, at all times, having physical, social, and economic access to sufficient safe and nutritious food.

3. Malnutrition is a leading cause of illness and death around the world, and it can be caused by:
   a. too few calories.
   b. too many calories.
   c. insufficient micronutrients.
   d. all of the above.

2

What was the Green Revolution, and how does it relate to industrial agriculture? What are the pros and cons of industrial agriculture?

INFOGRAPHIC 16.3

4. The Green Revolution:
   a. increased world food supplies but introduced new problems.
   b. was based on traditional, locally adapted crop species.
   c. used crop plants that required less fertilizer and pesticides.
   d. created food security for the global population.

5. What did the Green Revolution accomplish, and how did it do so? Discuss some of the unintended consequences of the methods used in the Green Revolution.

3

What is the gene revolution, and how do researchers get desired traits into crops using this new technology?

INFOGRAPHIC 16.4 AND TABLE 16.1

6. Which of the following is *not* an example of a genetically modified organism (GMO)?
   a. Golden rice
   b. High-yield variety (HYV) corn
   c. Bt cotton
   d. AquAdvantage salmon

7. Proponents of the "Gene Revolution" believe that:
   a. humans should be genetically modified to require less food.
   b. genetically modified (GM) crop plants are useful, but GM animals are not.
   c. GMOs are needed to achieve global food security.
   d. GMOs will be useful in developed countries but not in developing countries.

8. Explain how genetic engineering is used to create plants or animals with more desirable traits.

4

What are the trade-offs of using genetically modified organisms in agriculture?

INFOGRAPHIC 16.5

9. True or False: The problems created by GMOs are much different than those created by industrial agriculture.

10. Critics say that GMOs are dangerous because:
    a. herbicide-tolerant genes may migrate to other species and create "superweeds."
    b. beneficial insects as well as crop pests are killed.
    c. they place too much power in the hands of large corporations.
    d. all of the above are true.

11. Explain how planting a genetically modified crop with a trait for pest resistance could lead to the use of less pesticide in some cases but could lead to the use of more in others.

5

What are some low-tech (non-GMO, non-industrial) options for increasing food supplies? What role should industrial agriculture and GMOs play in solving world hunger?

INFOGRAPHIC 16.6

12. An advantage that low-tech farming methods have over high-tech farming methods in developing countries is that low-tech methods are
    a. more accessible to women.
    b. more likely to be subsidized by the government.
    c. more useful for growing cash crops.
    d. preferred by male farmers in these areas.

13. What is the value of keeping biodiversity in our crops high?

14. Compare the high-tech agriculture methods (HYVs and GMOs) with the low-tech suggestions given in the chapter. What path do you think a nation like Burkina Faso should pursue to achieve food security? Support your answer.

SCIENCE LITERACY WORKING WITH DATA

The Food and Agriculture Organization (FAO) of the United Nations monitors world hunger and has set a goal of decreasing the percentage of underfed people to 10% of the world's population by 2015.

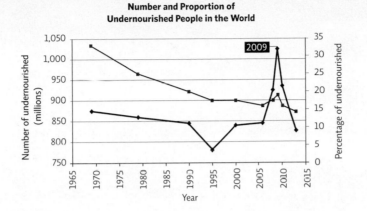

Number and Proportion of Undernourished People in the World

Interpretation

1. Describe in one sentence what the graph shows about change in undernourishment over time. Why are both lines shown here?

2. What accounted for the decline in the number and proportion of undernourished people from 1970 to 2000?

3. What happened to the percentage of underfed people between 2008 and 2009? What do you think led to this change?

Advance Your Thinking

4. Do you think the FAO will meet its goal of halving the percentage of underfed people from 1992 to 2015, from 20% to 10%? Why or why not?

5. In looking at both the number and percentage of underfed people, do you think there has been a complete recovery from the 2008 world food crisis? Why or why not?

6. Extrapolate these lines into the future, based on whether the Gene Revolution is successful and whether it is not. Which scenario do you think is more likely? Explain your reasoning.

INFORMATION LITERACY EVALUATING INFORMATION

Oxfam is an international not-for-profit organization dedicated to addressing issues of poverty and injustice around the world. Visit the Oxfam website (www.oxfam.org).

Evaluate the website and work with the information to answer the following questions:

1. Is the website a reliable information source?
   a. Does Oxfam give supporting evidence for its claims?
   b. Does the organization give sources for its evidence?
   c. Why was Oxfam originally founded in 1942, and what is its mission today?

2. Watch the videos about aid on Oxfam's website—"Does Aid Work" at www.oxfam.org/en/video/2010/does-aid-work and "Good Aid: The Video" at www.oxfam.org/en/multimedia/video/2010-good-aid-video (Search for the videos by title on the Oxfam website if these urls are not active.) Summarize the arguments made about the success of aid. Which of the reasons do you agree with most strongly, and why? Which do you find less compelling?

3. Review Oxfam's campaign on agriculture. Select two articles on food prices and read them. For each article answer the following questions:
   a. Do you agree with Oxfam's position presented in this article? Explain.
   b. Identify a claim Oxfam makes and the evidence the organization gives in support of this claim. Is it sufficient? Where would you attempt to find more evidence to support or refute the claim?

4. Review Oxfam's campaign to "transform a broken food system"—the GROW campaign. Read the information found in the links found on the GROW webpage such as "What is GROW?", "The Issues" and "FAQs" and answer the following questions:
   a. Summarize the solutions put forth by the campaign.
   b. Do you agree that this is a reasonable approach? What would you do differently?
   c. What is Oxfam's position on GMOs? Is it consistent with the organization's overall goals?

Find an additional case study online at http://www.macmillanhighered.com/launchpad/saes2e

CORE MESSAGE
Achieving food security for all people requires that we build a sustainable food system. Sustainable agriculture methods can help us grow crops within the means of the local ecosystem and without degrading the soil, water, and biodiversity that support that growth. Changes in consumer purchasing practices to support more sustainably produced food will give these foods the economic backing their producers need to succeed, while providing consumers with healthier food.

AFTER READING THIS CHAPTER, YOU SHOULD BE ABLE TO ANSWER THE FOLLOWING **GUIDING QUESTIONS**

1
What is sustainable agriculture, and how does it differ from industrial agriculture?

2
What environmental problems result from the use of industrial agriculture methods such as the application of synthetic fertilizers and pesticides?

FARMING LIKE AN ECOSYSTEM

Creative solutions to feeding the world

An ancient Japanese rice farming practice offers a more sustainable approach to growing the crop. Rice fields like this one across the globe produce millions of tons of rice each year.

Moment/Getty Images

3

What are some examples of methods employed by those pursing sustainable agriculture, and how do these methods decrease the impact of farming?

4

What role does the consumer play in helping build a sustainable food system?

5

What are the advantages and disadvantages of sustainable farming? Can it feed the world?

If there's one thing Greg Massa and his wife Raquel Krach hate, it's weeds—all varieties, but especially the azolla—an insidious, fernlike plant that grows on the surface of water. Each spring, azolla plants invade the couple's rice farm, snaking their way through the dense, muddy paddies that stretch for miles along the Sacramento River near Chico, California. They strangle young rice plants and force Greg into an endless and tedious battle.

The rivalry—Massa versus azolla—has spanned three generations. Greg's great-grandfather, Manuel Fonesca, planted the family's (and some of California's) first rice crops in 1916, on the same land that Greg and Raquel now manage. Back then, rice farming was a hard and uncertain life; Manuel was largely powerless against the azolla, which in some seasons claimed his entire crop.

By 1962, when Greg's father, Manuel, took over, human ingenuity and modern science had completely changed the nature of the fight. Heavy doses of chemical herbicides enabled him to obliterate the weed. And specially bred higher-yield rice varieties developed during the Green Revolution (see Chapter 16), along with modern farming equipment and a heavy dose of chemical **fertilizers** and **pesticides**, made the family farm both efficient and profitable. Of course, that modern approach, known as **industrial agriculture**, has its own problems. It relies on cheap fossil fuel energy and huge amounts of water. It also leads to a progressive degradation of the environment.

---

fertilizer A natural or synthetic mixture that contains nutrients that is added to soil to boost plant growth.

pesticide A natural or synthetic chemical that kills or repels plant or animal pests.

industrial agriculture Farming methods that rely on technology, synthetic chemical inputs, and economies of scale to increase productivity and profits.

sustainable agriculture Farming methods that can be used indefinitely because they do not deplete resources, such as soil and water, faster than they are replaced.

organic agriculture Farming that does not use synthetic fertilizer, pesticides, GMOs, or other chemical additives like hormones (for animal rearing).

Greg and Raquel wanted to find a better, more sustainable way. **Sustainable agriculture** is farming that meets the needs of the farmer and society as a whole without compromising the environment or future productivity. The techniques used will maintain or even enhance the environment. They often do this by mimicking the traits of a sustainable ecosystem: They rely on renewable energy and local resources for inputs, and they depend on biodiversity to trap energy, deal with waste, and control pest populations. **INFOGRAPHIC 17.1**

So when Greg and Raquel took over in 1997, they converted a portion of their farm to a sustainable farming method known as **organic agriculture**. Instead of using synthetic fertilizers and pesticides, organic agriculture employs more natural, or "organic," techniques in the growing of crops—such as using manure as fertilizer and luring in natural predators to control pests. In addition, genetically modified organisms (GMOs) cannot be grown on organic farms (see Chapter 16). This type of farming uses fewer or no chemicals and in some cases may even produce food that is more nutritious. A 2014 meta-analysis of 343 studies showed that organic food had levels of antioxidants as much as 50% higher than that in conventionally grown crops. For example, research by Washington State University soil scientist John Reganold showed that organically grown strawberries had a longer shelf life and a higher level of antioxidants than conventionally grown

KEY CONCEPT 17.1

The goal of sustainable agriculture is to raise food without damaging the environment or future productivity while operating ethically with regard to animals and local communities.

◉ **WHERE IS CHICO, CALIFORNIA?**

INFOGRAPHIC 17.1 **SUSTAINABLE AGRICULTURE**

↓ According to the U.S. Department of Agriculture, sustainable agriculture is farming that uses only limited amounts of nonrenewable resources (like fossil fuels) and does not degrade the environment or the well-being of people or society as a whole.

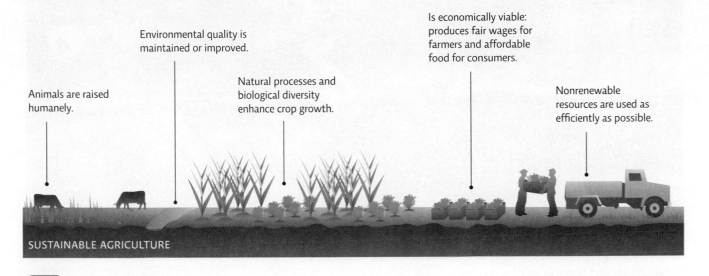

Environmental quality is maintained or improved.

Is economically viable: produces fair wages for farmers and affordable food for consumers.

Animals are raised humanely.

Natural processes and biological diversity enhance crop growth.

Nonrenewable resources are used as efficiently as possible.

SUSTAINABLE AGRICULTURE

Organic farming is a subset of sustainable farming. What do these farming methods have in common, and how do they differ?

KEY CONCEPT 17.2

Industrial agriculture is very productive but relies on large amounts of fossil fuels and water, degrades the soil and local environment (reducing future productivity), and is criticized for treating animals unethically.

berries. The meta-analysis also showed that, on average, conventionally grown food had four times more pesticide residue than organic food. But organic farming forced Greg and Raquel to battle weeds much as the first Manuel had: with great difficulty.

The trick was to lower water levels enough to kill water-loving weeds but not so much that the rice crop also died. Each day, Greg would wade into the paddies to see how the rice plants were faring against the azolla. Some weeks, he worried the entire crop would die. After a few seasons, the Massas started to despair: How could they make their farm environmentally friendly without losing their livelihood to an army of mangy weeds?

The Massas' story is the story of modern farming; it's the story of how we feed ourselves. And on a planet where population is exploding, climate is shifting, and energy and water resources are running low, it's also a story of constant change.

Modern industrial farming has advantages and disadvantages.

The changes that helped Greg's father thrive also ushered in a whole new way of growing crops, known as **monoculture** farming. Instead of growing a mix of plants, or growing different crops each season, farmers began growing the same single crop year after year. Before long, farms that had been populated by a variety of crops morphed into industrial operations, focused on just one crop. Thus, biodiversity was replaced by specialization, and farm ecosystems that more closely resembled nature were replaced by operations completely dependent on technology.

"In the 1920s, half of Iowa's farms produced 20 commodities each," says Fred Kirschenmann, a Distinguished Fellow at Iowa State University's Leopold Center for Sustainable Agriculture. "Today 80% of the state's cultivated land is exclusively corn or soybean. Farming systems that were once supported by complexity and diversity of species have now been replaced by reliance on inputs." Worldwide, 90% of our food comes from just 15 crop species and 8 species of livestock.

The same monoculture approach has also been applied to rearing

monoculture A farming method in which a single variety of one crop is planted, typically in rows over huge swaths of land, with large inputs of fertilizer, pesticides, and water.

↑ Cattle in a feedlot pen.

livestock. In a **concentrated animal feeding operation (CAFO)**, livestock are raised in confined spaces, with a focus on raising as many animals in a given area as possible. In the case of cattle, they are fed a nutrient-rich diet of grain and soybean, supplemented with some hay; the animals are generally not permitted to graze or roam free. CAFOs are highly productive; they minimize the amount of land that is used. Since about the mid-1900s, high-density feedlot operations like this have been the most common method of raising cattle in the United States. Today, poultry (chickens and ducks) and pigs are also predominantly raised in CAFOs. (See LaunchPad Chapter 30 for more on CAFOs.)

The advantages of this approach are obvious: A single crop (or type of animal in a confined space) is much easier to manage and to mass produce, and with greater ease comes not only greater efficiency and greater profits but also drastically greater amounts of food for a planet that has never seemed to have enough.

In recent years, however, the disadvantages of industrial agriculture have become equally apparent. For starters, both monocrop agriculture and CAFOs contribute heavily to global warming. Clearing so much land to grow food reduces the amount of carbon that can be sequestered by natural vegetation through photosynthesis. Also, fertilizer and livestock emit greenhouse gases. In fact, according to the UN Food and Agriculture Organization, livestock is responsible for some 18% of anthropogenic greenhouse gases. (See Chapter 21 for more on global warming and climate change.)

Worldwide, 90% of our food comes from just 15 crop species and 8 species of livestock.

In monoculture farming, the crop that is chosen is not necessarily locally adapted. Instead of choosing the crop best suited to the existing ecosystem, farmers focus on the crops with the highest market demand and thus the highest dollar value. Because these crops are not locally adapted, and because the volume of plants grown increased exponentially, the average farm becomes heavily dependent on external inputs—water, pesticides, and fertilizer—added to the farm from outside its own ecosystem.

Fertilizers boost growth because they provide nutrients that plants need, such as nitrogen and phosphorus. (See Chapter 8 for more on nutrient cycles.) And while heavy doses of fertilizer can indeed boost crop production, the excess nutrients (whatever the plants don't use) can soak into the ground and contaminate the groundwater or be easily washed from fields by rain and modern irrigation

concentrated animal feeding operation (CAFO) A method in which large numbers of meat or dairy animals are reared at high densities in confined spaces and fed a calorie rich diet to maximize growth.

cultural eutrophication Nutrient enrichment of an aquatic ecosystem that stimulates excess plant growth and disrupts normal energy uptake and matter cycles.

KEY CONCEPT 17.3

Fertilizer use can increase productivity but can degrade soil and contaminate nearby bodies of water.

INFOGRAPHIC 17.2   THE USE OF FERTILIZER COMES WITH TRADE-OFFS

ADVANTAGES

DISADVANTAGES

INCREASE PRODUCTIVITY

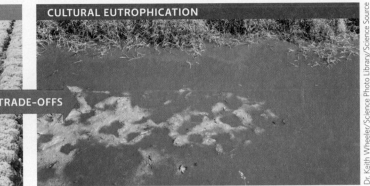

CULTURAL EUTROPHICATION

TRADE-OFFS

↑ Fertilizer can greatly increase soil productivity and is required to support the growth of the high-yield varieties grown today. Soil nutrients like nitrogen and phosphorus are often in limited supply and plant growth slows if one nutrient starts to run out. The addition of extra nutrients overcomes this deficiency and can boost growth.

↑ Runoff pollution that contains fertilizer can also cause algal blooms and result in the build up of dead organic material, both of which reduce the amount of oxygen in the water. This can kill many more aquatic organisms , even fish.

GROW CROPS IN MARGINAL SOILS

DEVELOP A DEPENDENCE ON FERTILIZER

TRADE-OFFS

↑ Fertilizers help crops grow in areas that may not otherwise be able to support agriculture. This may be the only way to farm in many areas of the world and would help increase local food supplies.

↑ The extra plant growth that fertilizers support can pull other nutrients out of the soil, depleting soils further and requiring even more fertilizer in the future.

? How could the disadvantages of fertilizer use be managed to reduce their impact?

practices. The excess, or *runoff*, enters waterways and can ultimately create hypoxic, or oxygen-poor, conditions that threaten aquatic life. This process, known as **cultural eutrophication**, has been a significant problem in the United States; nutrient runoff from farms in states within the Mississippi River watershed has created a summertime dead zone in the Gulf of Mexico (see Chapter 15). In addition, synthetic fertilizers are made from fossil fuels (natural gas and, to a lesser extent, petroleum), both of which are in limited supply and are environmentally damaging to extract, process, and ship.

INFOGRAPHIC 17.2

Monoculture crops are also especially vulnerable to pests or disease: Pest populations can explode when they encounter acres and acres of a suitable crop on which to feed, and a single infestation can wipe out the entire crop because what kills one plant will likely kill them all. To

deal with this, farmers have turned to pesticides (many of which are made from petroleum), but this also has proven problematic. To be sure, pesticides kill plant and animal pests and thus dramatically reduce the number of crops lost each year to infestation. But because they are toxic, pesticides also pose a threat to human and ecosystem health. And as scientists have discovered, pest populations can develop **pesticide resistance**, an unintentional example of artificial selection (see Chapter 11). Herbicide-resistant weeds and insecticide-resistant insects are cropping up all around the world. Such resistance encourages us to employ more drastic measures, in the form of higher doses or more toxic chemicals; as resistance to that next pesticide develops, the cycle repeats itself. It's like an arms

pesticide resistance   The ability of a pest to withstand exposure to a given pesticide; the result of natural selection favoring the survivors of an original population that was exposed to the pesticide.

INFOGRAPHIC 17.3   EMERGENCE OF PESTICIDE-RESISTANT PESTS

↓ Exposure to a pesticide will not make an individual pest resistant; it will likely kill it. However, if a few pests survive because they happen to be naturally resistant, they will breed and their offspring (most of which are also pesticide resistant) will make up the next generation. Over time, the original pesticide will no longer be effective and will have to be applied at a higher dose or a different pesticide will have to be used.

Pests attack a crop and multiply quickly. There is usually genetic diversity in the population, with some pests being killed more easily than others.

Pesticide is applied to kill the pests.

Only a few survive, including those that are pesticide resistant.

The survivors reproduce.

THE NEXT GENERATION
Survivors reproduce and pass on their traits to offspring.

Pesticide is applied again.

Every time pesticide is applied, it kills vunerable individuals and leaves resistant ones behind to reproduce.

A large, pesticide–resistant population now infests the crop.

How could application of a pesticide actually lead to an increase the size of the pest population? (Hint: Think about the predators of the pests.)

race between humans and pests, with the deck stacked in favor of the pests. **INFOGRAPHIC 17.3**

Even water inputs have created problems. It turns out that the use of irrigation can result in soil salinization; as water evaporates from the soil, salts are left behind and eventually become so concentrated that they impede crop growth. And if all that weren't enough, the colossal machines used to plow endless fields of corn and soybean have compacted the soil, making it harder for plants to take root and grow.

On top of that, the quality of the food itself can also suffer. Several recent studies have shown a decline in the nutrient content of food grown on soils subjected to years of industrial farming. In one of the first such studies, University of Texas researcher Donald Davis compared the nutritional content of 43 fruits and vegetables grown in 1950 to those grown in 1999; he noted a decline in six nutrients (protein, calcium, phosphorus, iron, and vitamins B2 and C). Davis suspects that this is due to the impoverished soil in which the crops are grown.

Another underappreciated consequence of modern

KEY CONCEPT 17.4

Pesticide use can help fight pests but is toxic to other species, including humans, and can lead to pesticide resistant pest populations.

↑ Seed bank in Svalbard, Norway.

monoculture farming is that fewer and fewer varieties of crops are planted. In the past, each region used locally adapted crop varieties developed over hundreds, even thousands, of years by area farmers—crops that were adapted to the local climate, soil conditions, pests, etc. But as independent farmers joined the industrialized food production system, their tendency was to plant the single variety of corn or wheat or rice that was currently the highest producer. This has led to an erosion of genetic diversity, which translates into a loss of the genetic raw material that allows crops to respond to changes such as the arrival of a new fungal pest or a drier climate. (See Chapters 12 and 16 for more on the importance of genetic diversity to agriculture.)

Without a wealth of crop varieties to choose from, plant breeders will have a harder time developing new strains to meet specific needs. The fear is that, if not planted, many plant varieties could be lost forever. **Seed banks**, such as one located in the high Arctic in Svalbard, Norway, are one solution to this dilemma. There seeds are kept in cold storage, housed in a tunnel carved out of a mountain; even if the power goes out, these seeds will stay frozen. The Svalbard

bank holds more than 700,000 varieties of crops from all over the world. Community seed banks and exchanges may be of even more immediate use to local farmers, and many of them are springing up in Asia and Africa.

CAFOs come with their own set of problems: They are environmentally taxing and ethically questionable. While the high-grain diets fed to confined animals certainly maximize growth, they are not necessarily good for the animals' health. Such diets may cause liver and stomach disorders, increased susceptibility to infections, and even death. In addition, crowded living conditions (which some view as inhumane) make CAFO animals more vulnerable to disease, forcing farmers to use heavy doses of antibiotics. Such heavy use contributes to antibiotic resistance, which threatens the health of the animals and humans.

As modern farmers face these enormous challenges (and global warming and energy and water shortages along with them), the story of how we feed ourselves

seed banks Places where seeds are stored in order to protect the genetic diversity of the world's crops.

↑ Greg Massa in his rice fields.

is changing yet again. This time, we may have to consider not just high-tech solutions but look back to the natural world for answers. In searching for a new weapon against the azolla, Raquel found a Japanese rice farmer who was doing just that.

Mimicking natural ecosystems can make farms more productive and help address some environmental problems.

Takao Furuno was, by most standards, a very successful industrial rice farmer, with annual yields among the highest in southern Japan. But it was a tough grind. Each year he was forced to put all his earnings back into the next year's crop—insecticides, herbicides, irrigation, and fertilizer—so that despite his success, he and his family were left with very little for themselves at the end of each season.

In searching for a better way, he turned, as he often did, to his forebears to see what he could learn from their knowledge. He was surprised to discover that they used to keep ducks in their rice paddies. Like most other rice farmers, Furuno considered ducks a pest, albeit a slightly cuter one than the azolla. Adult ducks eat rice seeds before they have a chance to grow, and as they forage, they trample young seedlings into the mud. This disturbance creates open patches of water, which in turn invites more ducks. "If you're not careful, you end up with a big problem pretty quickly," says Raquel.

But ducklings, Furuno soon realized, were too small to do such damage; for one thing, their bills were not big or strong enough to extract seeds from mud. Instead, they ate bugs and weeds. Azolla was one of their favorites.

Furuno's forebears also grew loaches—a type of fish—in their paddies. The loaches would also eat azolla and could be harvested and sold as food.

Together the ducklings and loaches would keep the weed from strangling the rice crop; but they would not completely eliminate the azolla the way a heavy dose of pesticides would. Furuno quickly discovered that, when kept at this benign concentration, the azolla (which contains symbiotic bacteria that produce a usable form of nitrogen) actually fertilized the rice. In fact, between the nitrogen from the azolla and the duck and fish droppings, he soon found that he no longer needed to spend money on synthetic fertilizer.

When raising ducks, fish, and rice crops together, Furuno discovered that the root crowns (where the root meets the stem) of rice plants increased to about twice the size that they had been in his old industrial system. A larger root crown meant more rice. "We're not exactly sure why the crowns grew," Furuno told an audience

KEY CONCEPT 17.5

Modeling a farm after an ecosystem (agroecology) to include a variety of plants and animals can boost productivity and protect or even enhance the local environment.

INFOGRAPHIC 17.4   AGROECOLOGY: THE DUCK/RICE FARM

↓ Takao Furuno's farm is a self-regulating, multiple-species system that naturally meets the needs of the farm ecosystem. All of the species play a role in the system, helping each other and boosting overall production.

THE METHOD

| Rice seedlings are planted in flooded rice paddies. | Ducklings are introduced to eat weeds and provide "fertilizer." | Fish are introduced to eat weeds and provide "fertilizer." | Azolla is introduced to add nitrogen; ducklings and fish keep the azolla from growing too much. |

THE FINAL PRODUCT: AN INTEGRATED SYSTEM

THE HARVEST

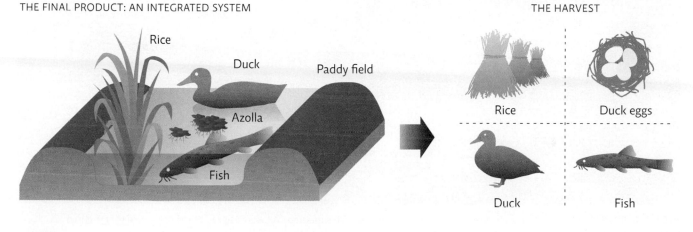

Rice
Duck
Paddy field
Azolla
Fish

Rice   Duck eggs

Duck   Fish

? What normal industrial inputs can be averted by growing rice using the duck/rice farm model?

of American farmers at a recent convention in Iowa. "But the ducks seem to actually change the way the rice grows. It's got something to do with the synergy of the whole system." Furuno's operation is an elegant example of biomimicry—a farm operating like a natural ecosystem. This type of farming, known as **agroecology**, considers both ecological concepts (modeling a farm after an ecosystem) and the value of traditional farming. **INFOGRAPHIC 17.4**

There were other financial gains, too. Duck eggs, duck meat, and fish all fetched a good price in the market. And because he was no longer using pesticides, Furuno could also grow fruit on the edges of his rice field. (He opted for fig trees, as he could harvest the figs annually without having to replant.)

Furuno's farm is an example of **polyculture**— intentionally raising more than one species on a

given plot of land. In the decade and a half since Furuno began duck/rice farming, his rice yields have increased by 20–50%, making his among the most productive farms in the world, nearly twice as productive as conventional farms. This kind of success is especially important for farmers in many developing countries, who struggle to produce enough food for current populations. Increasing production with lesser dependence on expensive inputs and using methods that enhance rather than diminish environmental quality can help communities become more self-sufficient and help them achieve *food security* (see Chapter 16).

agroecology A scientific field that considers the area's ecology and indigenous knowledge and favors agricultural methods that protect the environment and meet the needs of local people.

polyculture A farming method in which a mix of different species are grown together in one area.

↑ Cows at this farm are grass fed and graze using the rotation method. This method is closer to a traditional ecological system than conventional farming.

The Bangladesh Rice Research Institute, which has evaluated Furuno's method and independently verified his success, recommends the technique to Bangladeshi farmers. And by now, some 10,000 Japanese farmers have followed Furuno's lead; his method is also catching on in China, the Philippines, and California.

Furuno's method of duck/rice farming looked promising to the Massas. Not only would the rice plants rely on natural fertilizers and natural pest control, but the duck eggs and meat produced would be more humanely grown than those produced by factory farms. The ducks would grow up in ponds, not crammed together on slats in a barn without access to swimming water. They would get to splash around and express their "duckiness," as Raquel puts it.

When the ducks first arrived on the Massa farm, they were just 24 hours old, cotton-ball-sized tufts of yellow feathers. The Massa children cared for them in wooden crates in their barn. But as Raquel soon learned, small ducklings grow mighty quickly. In just 2 weeks, they were large enough to turn loose. Furuno had advised stocking about 100 ducklings per acre, but Greg and Raquel did not want to sacrifice that much land for this first attempt, so they fenced off just 1,000 square meters (a quarter acre) instead. This amount of space provided plenty of room for their 120 ducklings to swim and forage but, as it turned out, not enough food to support

KEY CONCEPT 17.6

Integrated pest management techniques can often effectively control pests while minimizing or eliminating the use of chemical pesticides.

them. "They quickly ate all the weeds in the field," says Greg. They also trampled some of the rice plants in their pond because their section of the field was too small. But even as they ran out of weeds to eat, the baby ducks stayed away from the rice plants, just as Furuno had insisted they would.

Sustainable techniques can control pests, protect soil, and keep farm productivity high.

Both of Greg and Raquel's azolla-control methods—reducing water levels and employing ducks as natural predators—are examples of **integrated pest management (IPM)**, or the use of a variety of methods to help reduce a pest population. The goal of IPM is to successfully control pests while minimizing or eliminating the use of chemical toxins. First, the farmer must examine the life cycle of the pest and the pest's interactions with the environment to identify the best way to deal with the pest. In general, IPM techniques fall into four categories: cultural control, biological control, mechanical control, and chemical control.

In sustainable agriculture, farmers use a combination of cultural, mechanical, and biological controls to deal with pest problems, and they resort to using chemicals only if these methods don't adequately deter the pests. If a pesticide is going to be used, the preference is for natural, biodegradable chemicals that are toxic only to a limited group of organisms. For example, pyrethrum, a compound naturally produced by a flower in the chrysanthemum family, is directly toxic to insects but not to mammals; it is certified for organic agriculture because it is naturally produced (not a synthetic human creation) and because it breaks down quickly and does not linger in the environment. However, while it does kill

> ### KEY CONCEPT 17.7
>
> Sustainable agriculture draws on a variety of traditional farming methods that can protect or improve soil and reduce pest problems. Many of these methods can be used on large and small scales.

integrated pest management (IPM)   The use of a variety of methods to control a pest population, with the goal of minimizing or eliminating the use of chemical toxins.

↑ Ducklings in the Massa Organics rice fields.

Courtesy Massa Organics

terracing The process of leveling land into steps on steep slopes; reduces soil erosion and runoff down the hillside.

contour farming Farming on hilly land in rows that are planted along the slope, following the lay of the land, rather than oriented downhill.

reduced-tillage cultivation Planting crops in soil that is minimally disturbed and that retains some plant residue from the previous planting.

cover crop A crop planted in the off-season to help prevent soil erosion and to return nutrients to the soil.

pest insects, it is also toxic to *good* insects like honeybees, so its use is avoided during times of pollination. Synthetic pesticides, like those commonly used in industrial agriculture, are not acceptable for use on certified organic crops but may be part of an IPM plan for conventionally raised crops. **INFOGRAPHIC 17.5**

Greg and Raquel were well versed in the problems of modern agriculture. Before settling in California to take over the family farm, they had worked as tropical ecologists in Costa Rica, where they learned about traditional, non-industrial farming methods that help protect the soil and keep productivity high without the use of synthetic fertilizers or pesticides.

Soil is vital to life on Earth; it supports the growth of plants and the animals who feed on those plants. Its formation is very slow (it can take more than 1,000 years to form 2.5 centimeters [1 inch] of topsoil) and depends on myriad soil organisms (see LaunchPad Chapter 27). Modern farming can contribute to a decline in soil fertility and to the direct loss of soil though erosion. However, some traditional (preindustrial) methods can actually help restore or protect soil. For example, methods such as **terracing** and **contour farming** can decrease soil erosion on sloped land. Other techniques that protect the soil include **reduced-tillage cultivation** methods, as well as planting a **cover crop** in the off-season to prevent exposed soil from washing away. This last approach has the

↓ Harvesting wheat with four John Deere combines.

↓ Controlling pests is important for our agricultural yields as well as for our health and for the health of our pets. Rather than using harsh methods in an attempt to completely eliminate the pest (which rarely works anyway), a combination of less hazardous methods can often reduce pest numbers to manageable levels.

INTEGRATED PEST MANAGEMENT HAS SEVERAL STEPS

1. IDENTIFY TRUE PESTS
Not all "bugs" and "weeds" are pests—some plants and animals are innocuous or actually beneficial. By working to control actual pests only, we save time and money, and we help the environment by maintaining diversity and preventing toxic pollution.

2. SET AN ACTION THRESHOLD AND MONITOR PESTS
The pest population size that is unacceptable must be identified. We may have zero tolerance for some pests (e.g., fleas and ticks in our homes) but be able to tolerate small populations of crop pests; we will only act if the action threshold is reached.

3. DEVELOP AN ACTION PLAN
This may include a variety of methods, each of which aids in pest control by excluding, discouraging, or killing pests. The goal is to control the pest while avoiding or minimizing the use of chemical control agents which may be toxic and have unwanted health and environmental effects.

PREVENTION AND CONTROL METHODS

CULTURAL Usually the first method chosen as part of the action plan; involves cultivation techniques that minimize the habitat or food source for the pest so that other control methods can then be used to adequately control the pests.

BanksPhotos/E+/Getty Images

↑ Strip cropping minimizes potential food for pests that don't disperse over great distances, lessening the chance of an outbreak of pests.

MECHANICAL Relies on methods that physically exclude, trap, repel or remove pests or weeds; can be labor intensive but often inexpensive and may be particularly useful in developing countries with plentiful labor but little cash.

Cosmo Condina/Getty Images

↑ Netting can keep out birds and rabbits that would eat the crops. Reducing water levels in rice paddies to kill azolla is an example of mechanical control.

BIOLOGICAL Introducing predators, sterile males, or plants that repel the pest; technique works best if it follows cultural and mechanical steps.

Henrik_L/iStock / 360/Getty Images

↑ Ladybugs, predatory beetles that eat pests such as this aphid, can actually be purchased and released, or steps can be taken to attract them naturally to the area. The ideal control agent is a specialist whose preferred food is the pest in question, and who does not attack nontarget species.

CHEMICAL Applying chemicals that kill or repel pests; a last resort that is used only if the other three methods cannot control the pests.

Abid Katib/Getty Images

↑ To minimize health and environmental concerns, the preferred chemical is the one that can do the job while being the least toxic and the most degradable.

 What pest control methods did the Massas use in their rice fields? Identify each as cultural, biological, mechanical, or chemical.

© ZUMA Press, Inc/Alamy

↑ Biochar is added to landscapes and agricultural fields to improve soil. It provides nutrients to soils but its biggest advantages may come from the fact that it stays in the soil a long time. This creates long-lasting habitat for useful soil organisms and helps keep carbon sequestered in the soil (useful in combating climate change).

crop rotation Planting different crops on a given plot of land every few years to maintain soil fertility and reduce pest outbreaks.

strip cropping Alternating different crops in adjacent strips, several rows wide; helps keep pest populations low.

biochar A form of charcoal that is produced when organic matter is partially burned and that can be used to improve soil quality.

added advantage of helping to restore fertility. Soil fertility can also be enhanced and pests controlled using **crop rotation** and **strip cropping**. In addition, systems like Furuno's that combine animal and plant rearing, return animal waste—a natural fertilizer—to the soil. **INFOGRAPHIC 17.6**

For millennia, humans have employed methods to *amend* soil to improve its fertility. The addition of fertilizers (natural or synthetic), organic matter (to loosen compacted soil), and substances such as lime (to alter the pH of the soil) are all soil amendments that can address soil

problems. But recently a traditional soil amendment from the tropics is gaining interest—biochar. **Biochar** is a form of charcoal produced when organic matter (usually agricultural waste) is partially burned. The use of biochar over thousands of years helped create the fertile terra preta soils in the Amazon basin, where thin, poor, acidic soils predominate. Not only does biochar provide nutrients to poor soils, it also helps the soil store carbon (it does not decompose readily so "holds onto" much of its carbon), impeding its release into the atmosphere and reducing its contribution to climate change.

INFOGRAPHIC 17.6 **SUSTAINABLE SOIL MANAGEMENT PRACTICES**

↓ Many traditional methods are useful for sustainable agriculture because they focus on protecting the soil, the heart of successful farming.

J. IRWIN/Robertstock/Aurora Photos

CONTOUR FARMING When farming on hilly land, rows are planted along the slope, following the lay of the land, rather than oriented downhill to reduce the loss of water and soil after a rainfall.

Bill Barksdale/AGStockUSA

REDUCED TILLAGE Planting crops into soil that is minimally tilled reduces soil erosion and water needs (it reduces water evaporation). It also requires less fuel because of less tractor use.

voraorn/iStock/360/Getty Images

TERRACE FARMING On steep slopes, the land can be leveled into steps. This reduces soil erosion and allows a crop like rice to stay flooded when needed.

Michael Melford/Getty Images

CROP ROTATION Planting different crops on a given plot of land every few years helps maintain soil fertility and reduces pest outbreaks since pests (or their offspring) from the year before will not find a suitable food when they emerge in the new season.

Andrew Holt/Photographer's choice/Getty Images

STRIP CROPPING Alternating different crops in strips that are several rows wide keeps pest populations low; it is less likely the pests will travel beyond the edge of a strip and they may not find another row of this crop.

voraorn/iStock/360/Getty Images

COVER CROPS During the off-season, rather than letting a field stand bare, a crop can be planted that will hold the soil in place. Nitrogen-fixing crops like alfalfa that improve the soil are often chosen.

 Which of these sustainable soil practices help reduce soil erosion, which help improve soil fertility, and which help reduce pest outbreaks?

Consumers also have a role to play in helping to bring about a sustainable food system.

In rice farming, the Massas saw a chance to implement, on their own land, all the concepts and theories they had learned as ecology students and refined and discussed as scientists. "We wanted a farm where success was measured not just in crop yield, but in the overall health of the land," Greg says. "A place where we would count profits, but we would also count the number of sandhill cranes and California quail we saw populating the area."

To achieve their goals, the Massas installed a recirculation system to reclaim and reuse irrigation water. They planted native oak trees along field borders to serve as a natural windbreak (windbreaks prevent soil from being carried away by wind erosion), and they installed nest boxes for wood ducks, barn owls, and bats so that those wild animals would keep area pests in check. "The idea was to restore as much of the natural biodiversity as possible," says Greg, "so that we would not need artificial inputs to run the system." They also took their sustainable ideals to the next level and built a straw house to live in—made of 2-foot-thick (0.5 meters) walls of rice bale, coated on both sides with plastic or stucco—that can withstand the unforgiving heat of a Chico summer and maintain a steady temperature almost entirely by itself. The house is fireproof, rodent proof, and, as Greg likes to joke, bulletproof, too.

The Massas also wanted to create a farm that contributed to the formation of a local community food system; they planned to sell a portion of their rice at local outlets like farmers' markets and food co-ops. More and more consumers are buying food from local farmers; local agriculture supports local economies and provides fresher and thus healthier food to consumers. Because transportation depends on fossil fuels, the more **food miles** a product travels before reaching the consumer, the greater the **carbon footprint** of that food. And much of our food has traveled quite far—about 2,400 kilometers (1,500 miles), on average.

food miles The distance a food travels from its site of production to the consumer.

carbon footprint The amount of carbon released to the atmosphere by a person, a company, a nation, or an activity.

greenwashing Claiming environmental benefits about a product when the benefits are actually minor or nonexistent.

However, transportation is not the main use of fossil fuels when crops are raised industrially. Research by Christopher Weber and Scott Matthews of Carnegie Mellon University determined that about 90% of the carbon footprint for food grown using conventional industrial methods is from the production of the crop (fuel for equipment, raw materials for pesticides and fertilizer production), not its transport. Therefore, buying organically grown produce—even from far away—may reduce the carbon footprint more so than buying locally grown industrial crops. Of course, the best option is to choose organic foods that are locally grown. Likewise, because beef raised in industrial CAFOs has one of the highest carbon footprints of all agricultural products (see LaunchPad Chapter 30), one of the best things you can do is to replace at least some of the beef you eat with chicken, pork, fish, or meatless dishes.

Consumers are becoming more aware that things like food miles and the way food is raised matter for the environment, their communities, and their own health. Organic foods and ethically raised animal products are claiming a bigger share of consumer dollars annually. But this has opened the door to **greenwashing**—making claims about the environmental benefits of sustainably raised or organic foods that are misleading. (For example, organic cookies are probably not really healthier than those made with conventional ingredients, and "cage-free" eggs may still be from chickens living in overcrowded conditions.) Consumers need to be diligent about evaluating claims and make informed decisions about what to purchase. **INFOGRAPHIC 17.7**

> ### KEY CONCEPT 17.8
>
> Consumers can support sustainable agriculture with their purchases. Buying locally grown, organic food is the best option for reducing the carbon footprint of that food.

A sustainable food future will depend on a variety of methods.

Furuno and the Massas are not the only ones experimenting with polyculture. Indeed, many other farmers and scientists across the country are turning to agroecology—working to develop mixed agriculture systems, where instead of planting a single crop, they grow a mix of different species that better replicates the normal ecological community makeup of a given region. Evidence is mounting that such systems can increase a farm's productivity.

A 2010 report by the International Livestock Research Institute concluded that mixed polyculture farms—ones that, like Furuno's and the Massas', grow both plants and livestock—hold the most promise for intensifying food production worldwide. "It is not big efficient farms on high potential lands but rather one billion small, mixed, family farmers tending rice paddies or cultivating maize

INFOGRAPHIC 17.7 **CONSUMER CHOICES MATTER**

4

↓ Because the growing and transport of our food impacts the environment and our own health so much, choosing foods produced in a way that has a lower impact makes a difference. This also supports sustainable agriculture as an economic endeavor, helping the farmers and communities pursuing these methods.

CONSIDER HOW YOUR FOOD IS RAISED...

© Angel Franco/The New York Times/Redux

Industrially grown food is usually cheaper but has a higher environmental impact and a high carbon footprint (more fossil fuels are used to produce it) than sustainably grown food. The best way to reduce the carbon footprint of the food you buy is to opt for organically grown food when possible (no fossil fuel–derived fertilizers or pesticides).

...AND HOW FAR IT IS SHIPPED

Mindy Schauer/The Orange County Register/ZUMA Press

Even though more fossil fuels go into the *production* of industrially grown foods than in shipping it to market, buying food produced closer to home decreases the transportation part of the carbon footprint. For example, in Iowa, the average grocery store apple has traveled 1,726 miles, and the average head of broccoli, 1,846 miles, while locally grown produce has traveled 56 miles on average. So while transportation does not represent the main way fossil fuels are used in agriculture, it still has a significant impact.

PERSONAL FOOD CHOICES

↓ Buying organic food not only reduces the carbon footprint of the food, it is also healthier for you. But it can get expensive, so if your buying dollars are limited, consider steering your purchases away, when you are able, from the "dirty dozen"—the 12 fruits and vegetables most likely to be contaminated with pesticide residue. The "clean 15" are the products least likely to have pesticide residue, so if you can't afford to buy all organic produce, buy these from the regular produce shelf—but always wash all produce well before eating or cooking!

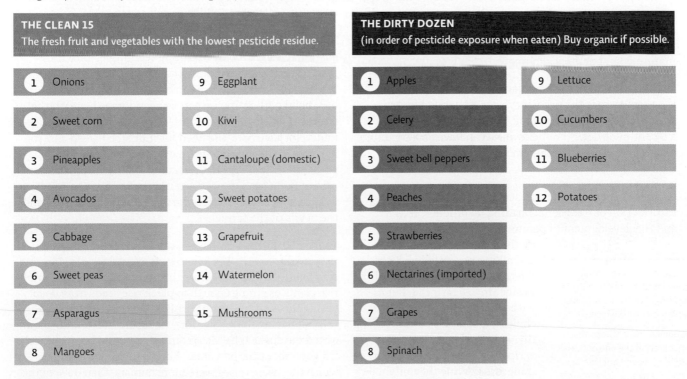

THE CLEAN 15
The fresh fruit and vegetables with the lowest pesticide residue.

1 Onions	9 Eggplant		
2 Sweet corn	10 Kiwi		
3 Pineapples	11 Cantaloupe (domestic)		
4 Avocados	12 Sweet potatoes		
5 Cabbage	13 Grapefruit		
6 Sweet peas	14 Watermelon		
7 Asparagus	15 Mushrooms		
8 Mangoes			

THE DIRTY DOZEN
(in order of pesticide exposure when eaten) Buy organic if possible.

1 Apples	9 Lettuce
2 Celery	10 Cucumbers
3 Sweet bell peppers	11 Blueberries
4 Peaches	12 Potatoes
5 Strawberries	
6 Nectarines (imported)	
7 Grapes	
8 Spinach	

? What information would you like to see on food labels that would allow you to make wise consumer choices when buying food? Does food labeling need to be regulated so that certain terms (e.g., organic, free range) is specifically defined?

and beans while raising a few chickens and pigs, a herd of goats or a cow or two . . . [who are] likely to play the biggest role in global food security over the next several decades," the institute's executive director, Knut Hove, wrote in the report. "These 'mixed extensive' farms make up the biggest . . . and most environmentally sustainable agricultural system in the world."

Meanwhile, at the Land Institute in Kansas, Wes Jackson is working on a more ambitious plan that he says will correct a mistake made well before the Green Revolution or the rise of factory farms: He wants to replace virtually all of our existing grain crops—which are **annual crops**—with **perennial crop** varieties. Early farmers domesticated annual plants—which grow, produce seeds, and die in a single year—because they could manipulate them to produce higher yields from one year to the next (by selecting the best producers and junking the rest or by crossbreeding plants with good traits). But annuals have to be replanted every season, which requires labor and fossil fuel energy and necessitates tearing up the ground and disrupting the delicate balance of soil ecosystems.

Perennials, on the other hand, can be harvested year after year without disturbing the soil to replant—which means heavy equipment is used less often to manage perennial crops. And the deep roots that perennials develop not only hold soil in place, but they tap much farther down into the soil than their annual counterparts, allowing them to access more of the soil's water, thus dramatically reducing the amount of irrigation needed. Less herbicide is needed for perennial plots as well; weeds do not readily sprout and grow among the established plants. This makes perennials especially attractive for regions with marginal land or an arid climate.

"We hope to advance and enlarge upon the idea that the ecosystem is the necessary conceptual tool for truly sustainable grain agriculture," Jackson told the *Atlantic* in a recent interview. "We believe we can have an agriculture where management by human intervention is greatly reduced."

As these many examples illustrate (agroecology, the resurrection of traditional farming techniques, IPM, and the development of perennial crops), a multitude of methods qualify as sustainable, and all can potentially help in the transition from industrial to sustainable farming. But as with all other environmental choices, it shouldn't be surprising that sustainable and organic agriculture have their own set of trade-offs: While they might be more environmentally friendly than high-input industrial

annual crops   Crops that grow, produce seeds, and die in a single year and must be replanted each season.

perennial crops   Crops that do not die at the end of the growing season but live for several years, which means they can be harvested annually without replanting.

methods, they may also be more expensive and in some cases produce less food per acre of land. But Greg and Raquel were determined to at least try. **TABLE 17.1**

Can sustainable farming methods feed the world?

For the Massas, ducklings were part of an ongoing search for solutions to the challenges of modern rice farming. They ended up with duck meat to sell, and though they didn't take any precise measurements of yields during their first trial run, their rice crops did not appear to suffer at all. "We learned a couple of things," Raquel says. "The ducks trampled some of the rice in their pond, which would not have been a problem if we had used a larger section of the field. I also think we used the wrong breed of duck. They were a little too large to move effectively between the dense plants, and they were not active enough in their foraging activities. These ducks were bred to sit around all day and eat and gain weight quickly for industrial meat production. We are currently researching which breeds to try next."

> **KEY CONCEPT 17.9**
>
> Sustainable agriculture comes with trade-offs, but many feel that the disadvantages are less problematic than are those of modern industrial agriculture, especially in the long term.

The Massas have continued on to incorporate more sustainable agroecology farming techniques into their farm. Organic almond orchards are managed like natural woodlands with a diverse understory (that attracts beneficial pollinating insects); sheep forage beneath the trees, providing fertilizer and helping to maintain the understory without the need for industrial weed control methods. (They used to burn out weeds by hand with a propane flamer.) Pigs have become another permanent part of their farm, moved around the farm via temporary fencing to feed on whatever area best fits their needs and the needs of the farm.

At the end of that first season, the Massas harvested both rice and duck meat. The key to successful duck/rice farming is to remove the ducks before they get big enough to trample rice plants or strong enough to pluck rice seeds from deep within the mud. The hard part isn't knowing when to do this but having the resolve for what comes next: killing and eating the ducks. The Massa family struggled but ultimately felt good about the outcome. "I know the conditions in which they were raised were more humane than 99% of the meat ducks in this country," says Raquel. "They had it good and you can taste that in the finished product."

TABLE 17.1   **ADVANTAGES AND DISADVANTAGES OF SUSTAINABLE AGRICULTURE**   5

For	Advantages	Disadvantages
The Consumer	• Food is fresher, tastier, and healthier (better nutrient profile, no pesticide residue if organic, etc.). • Gains satisfaction in making a more ethical and environmentally sound choice.	• Sustainably grown crops may be more expensive. • Greenwashing can mislead consumers. • Organic produce may have more blemishes. • Shelf life of organic produce is shorter (not waxed, not picked before ripe).
Farmer and Environment	• Using fewer inputs of water and fossil fuels saves money and causes less environmental damage. • Soil is not degraded and may be enhanced. • Less use of toxic chemicals benefits the environment and local communities. • More genetic diversity and species diversity makes it less likely that a pest outbreak or other problem will decimate the entire crop.	• May be more labor intensive. • Crops grown sustainably may not be as productive per acre as industrially farmed crops (in the short term). • Fewer government subsidies are available for sustainable agriculture compared to those for industrially grown crops. • The certification process for getting crops to be labeled as organic takes time and is costly to farmers.
Society	• Many sustainable methods are less expensive so are suitable for developing nations. • Methods are available that minimize water need—useful in arid areas. • Local production of food can increase food security (see Chapter 16).	• Research is needed to identify best methods and crops for a given area. • Farmers need training to implement these systems (though if indigenous methods are used, it may be the locals who educate the researchers).

? In your opinion, which advantages listed above are the most important? Which disadvantages are the most troublesome? Explain.

Select References:
Baranski, M., et al. (2014). Higher antioxidant and lower cadmium concentrations and lower incidence of pesticide residues in organically grown crops: A systematic literature review and meta-analyses. *The British Journal of Nutrition*: doi:10.1017/S0007114514001366
Davis, D. R., et al. (2004). Changes in USDA food composition data for 43 garden crops, 1950 to 1999. *Journal of the American College of Nutrition*, 23(6): 669–682.

Hossain, S. T., et al. (2005). Effect of integrated rice-duck farming on rice yield, farm productivity, and rice-provisioning ability of farmers. *Asian Journal of Agriculture and Development*, 2(1): 79–86.
Reganold, J. P., et al. (2010). Fruit and soil quality of organic and conventional strawberry agroecosystems. *PLoS ONE*, 5(9):e12346.doi:10.1371
Weber, C. L., & H. S. Matthews. (2008). Food-miles and the relative climate impacts of food choices in the United States. *Environmental Science and Technology*, 42(10): 3508–3513.

BRING IT HOME

PERSONAL CHOICES THAT HELP

While a typical supermarket may seem to present a dizzying array of food choices to the consumer, a look at the ingredient labels betrays our increasing reliance on growing monocultures of common strains of corn, soy, and wheat. These choices will change only if you, the consumer, demand it.

Individual Steps

• Carefully examine the labels on the food you buy. As your food budget allows, opt for food products that are organically grown and, if available, locally produced.

Group Action
• Organize a community garden that specializes in heirloom varieties of vegetables that might not be found in the local grocery stores. Start by requesting a seed catalog from www.seedsavers.org or www.rareseeds.com.
• Research specific farming practices that more closely mimic those found in natural ecosystems, such as Joel Salatin's Polyface Farm (www.polyfacefarms.com).
• Subscribe to a community-supported agriculture (CSA) farm and receive a weekly supply of sustainably grown

produce. For a list of CSAs in your area, see www.localharvest.org.

Policy Change
• Identify bodies of water in your area that may be impacted by cultural eutrophication from fertilizer runoff. Meet with local officials and propose ordinances limiting fertilizer use by homeowners, golf courses, or other possible sources of the pollution.

ENVIRONMENTAL LITERACY UNDERSTANDING THE ISSUE

1 What is sustainable agriculture, and how does it differ from industrial agriculture?
INFOGRAPHIC 17.1

1. True or False: To qualify as sustainable, agriculture practices must maintain or improve the quality of the environment.

2. Identify at least three ways that the duck/rice farm mimics a sustainable ecosystem.

2 What environmental problems result from the use of industrial agriculture methods such as the application of synthetic fertilizers and pesticides?
INFOGRAPHICS 17.2 AND 17.3

3. Which of the following is *not* an advantage of farming with monocultures?
   a. Higher yields
   b. Ease of planting and harvest
   c. Fewer pest outbreaks
   d. Works well on a large scale

4. The use of synthetic fertilizers on fields where crops are grown can:
   a. increase crop yields.
   b. contribute to water pollution.
   c. deplete the soil of other nutrients.
   d. A and B.
   e. A, B, and C.

5. Explain how the use of chemical pesticides can lead to the emergence of a pesticide-resistant population of pests. How could it also lead to the emergence of a population even bigger than the original pest population?

3 What are some examples of methods employed by those pursing sustainable agriculture, and how do these methods decrease the impact of farming?
INFOGRAPHICS 17.4, 17.5, AND 17.6

6. Techniques used in sustainable agriculture practices include:
   a. crop rotation to minimize crop pests.
   b. frequent use of herbicides to keep weeds down.
   c. flooding crops frequently with abundant water to reduce pests and increase runoff of excess nutrients.
   d. abundant use of fertilizers to allow plants to reach their full capacity.

7. *Biological control* in an integrated pest management (IPM) system might include:
   a. the application of organic herbicides and pesticides.
   b. using species such as fish to eat weeds or pests.
   c. removing habitat that might harbor pests.
   d. all of the above.

8. The presence of ducks and azolla in Takao Furuno's rice cultivation technique has shown that:
   a. ducks must be eliminated from rice fields because they eat the rice seed before it can grow.
   b. invasive species like ducks and azolla diminish rice yields.
   c. crops such as rice cannot succeed without azolla to provide shade and ducks to provide pollination.
   d. restoring some biodiversity to rice fields reduces pest damage and increases rice yields.

4 What role does the consumer play in helping build a sustainable food system?
INFOGRAPHIC 17.7

9. One major way to reduce the carbon footprint of the foods you eat is to:
   a. eat local, industrially produced food.
   b. eat produce shipped in from countries where it can be grown "in season."
   c. eat foods that are grown using organic methods, even if not locally grown.
   d. only eat foods that are grown with conventional industrial methods.

10. In regard to agriculture, the "dirty dozen" refers to:
    a. the 12 pesticides organic farmers are banned from using.
    b. the 12 conventionally grown fruits or vegetables highest in pesticide residue.
    c. the 12 most destructive crop pests.
    d. the 12 countries that use the highest amounts of pesticides on crops.

5 What are the advantages and disadvantages of sustainable farming? Can it feed the world?
TABLE 17.1

11. True or False: All sustainably grown food can be certified organic.

12. Which of these is *not* an advantage of sustainable agriculture?
    a. Fresher, better-tasting produce that may be healthier to eat
    b. Maintained or enhanced soil quality
    c. More genetic diversity in crops
    d. Higher productivity per acre than with conventional methods

13. What advantage do perennial crops have over annual crops?
    a. Perennial crops don't need fertilizers.
    b. The soil is not disturbed to replant perennial crops each year.
    c. Perennial crops have more shallow roots and can access irrigation water better than annual crops.
    d. Perennial crops are more tolerant of herbicides than annual crops.

14. Look over the disadvantages of sustainable agriculture in Table 17.1. Propose actions that could address each.

SCIENCE LITERACY **WORKING WITH DATA**

The data in the following graph come from the Farming Systems Trial® (FST) research study conducted by the Rodale Institute. This side-by-side comparison of corn and soybean crops grown under organic and industrial agricultural systems was started in 1981 and is one of the longest-running studies of its kind.

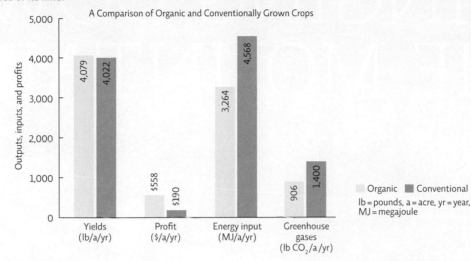

A Comparison of Organic and Conventionally Grown Crops

Interpretation

1. What does this graph show?

2. Calculate the following for conventional and organic systems: profit per unit of yield, energy input per unit of yield, and greenhouse gas emissions per unit of yield. How do the two systems compare on these three parameters?

Advance Your Thinking

Hint: To answer the following questions, it might be helpful to access the actual FST report, at http://rodaleinstitute.org/our-work/farming-systems-trial/farming-systems-trial-30-year-report/.

3. According to the Food and Agricultural Organization of the United Nations, "Organic agriculture has the potential to secure a global food supply, just as conventional agriculture does today, but with reduced environmental impact." How do the data from the FST study support this statement?

4. According to the FST report, even in drought years, the yields for organic corn were approximately 31% greater than those for conventional (non-drought-resistant) varieties. At the same time, genetically engineered drought-tolerant varieties had yields that were no more than approximately 13% greater than conventional (non-drought-resistant) varieties. Why might this be the case? Why is this an important finding?

5. The FST report indicates that crops grown using organic methods produced yields equivalent to those of conventional crops, even though they had more weed competition in their fields. Why might this be the case? What makes this an important finding?

INFORMATION LITERACY **EVALUATING INFORMATION**

Which seed varieties are best? There are groups, such as Renewing America's Food Traditions (RAFT) (www.albc-usa.org/RAFT/), that say we should preserve the original biodiversity and the heirloom varieties that our ancestors grew. Other groups believe that selective breeding and hybridization create superior crops, the downside being that farmers must purchase new seeds every year. Bayer is a huge conglomerate that sells a variety of hybrid seeds and chemicals to enhance growth (www.bayercropscience.us). Another group, AgBioWorld (www.agbioworld.org), says that it is neutral and reports the news on agricultural practices. Look at each website to help analyze the organizations.

Evaluate the websites and work with the information to answer the following questions:

1. Evaluate the agendas of the three organizations as well as the accuracy of the science behind their positions on heirloom varieties versus hybrid crops.

a. Who runs each website? Do the person's/organization's credentials make the information presented on food and agriculture issues reliable or unreliable? Explain.

b. What is the mission of each website? What are the underlying values? How do you know this?

c. What claims does each website make about the current problems in food production and what the future of agriculture should be? Are their claims reasonable? Explain.

d. How do the websites compare in providing scientific evidence in support of their assessment of agriculture and their position on the role of heirloom varieties versus hybrid crops? Is the evidence accurate and reliable? Explain.

2. How do the three organizations compare in engaging you, as a citizen, in agricultural policy? Do you think that citizen involvement in policy issues is necessary and effective? Explain your responses.

Find an additional case study online at http://www.macmillanhighered.com/launchpad/saes2e

BRINGING DOWN THE MOUNTAIN

In the rubble, the true costs of coal

CORE MESSAGE

Human society runs on energy, and coal continues to be a major reliable energy resource. However, coal mining causes irreversible environmental degradation, and the by-products of mining and burning coal pose significant environmental and health risks. Despite these drawbacks and the availability of renewable, cleaner energy sources, coal's availability and industrial presence keep it a major energy player. Researchers are developing ways to lessen the impact of burning coal, but as long as we continue to use it, much of the impact of mining will remain.

AFTER READING THIS CHAPTER, YOU SHOULD BE ABLE TO ANSWER THE FOLLOWING **GUIDING QUESTIONS**

1

How important is coal as an energy source, and how is it used to generate electricity?

2

What is coal, how is it formed, and what regions of the world have coal deposits that are accessible?

Leveling a mountain for coal in Appalachia. The mountaintop is blown off and dumped into valleys to leave sprawling, terraced, barren lands where there were once diverse temperate forests. © George Steinmetz/Corbis

3

What methods are used to mine coal, and what are the advantages and disadvantages of each?

4

What are the advantages and disadvantages of burning coal?

5

What new technologies allow us to burn coal with fewer environmental and health problems? How can mining damage be repaired, and how effective is this restoration?

A thousand feet above the foothills of central Appalachia, near the Kentucky–West Virginia border, a four-seater plane ducks and sways like a tiny boat on an anxious sea. It's windier than expected, and Chuck Nelson, a retired coal miner seated next to the pilot, grips the door to steady his nerves. The passengers have come to survey the devastation wrought by mountaintop removal—a type of mining that involves blasting off several hundred feet of mountaintop, dumping the rubble into adjacent valleys, and harvesting the thin ribbons of **coal** beneath.

At first, the landscape looks mostly unbroken; mountains made soft and round by eons of erosion roll and dip and rise in every direction, carrying a dense hardwood forest with them to the horizon. But before long, a series of **mountaintop removal** sites come into view. Trucks and heavy equipment crawl like insects across what looks like an apocalyptic moonscape: decapitated peaks and acres of barren sandstone and shale. Smoke curls up from a brush fire as the side of an existing mountain is cleared for demolition. Orange and turquoise sediment ponds—designed to filter out heavy metal contaminants before they permeate the water downstream—dot the perimeter.

Here and there, tiny patches of forest cling to some improbably preserved ridge line. "That's where I live," Nelson says, forgetting his air sickness long enough to point out one such patch. "My God, you would never know it was this bad from the ground." The aerial tour has reached Hobet 21, which, at more than 52 square

↓ The controversial Spruce No. 1 coal mine in West Virginia was originally given a permit authorizing it to dump strip mining waste into 11 kilometers of creeks and onto 800 hectares of land. In 2011, the Environmental Protection Agency rejected the permit on the basis of the "irreversible damage" that would be inflicted to the streams, groundwater, and land, including: "the elimination of all fish, killing of birdlife, reduction of habitat value, and risk of human illness."

Antrim Caskey

↑ Seams of coal exposed at a mountaintop removal mining site in Welch, West Virginia.

◉ WHERE IS APPALACHIA?

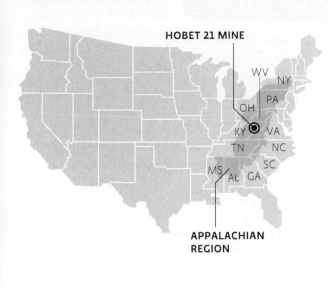

HOBET 21 MINE

WV
NY
OH
PA
KY
VA
TN
NC
MS
SC
AL   GA

APPALACHIAN
REGION

kilometers (20 square miles), is the region's largest mining operation. So far, sites like this one have claimed roughly 400,000 hectares (1 million acres) of forested mountain, across just four states: Kentucky, West Virginia, Virginia, and Tennessee. But there is still more coal to mine. It could mean the obliteration of hundreds of thousands of more acres in the coming years.

To stem this tide of destruction, environmental activists have sued the coal industry, the state of West Virginia, and the federal government. They argue that mountaintop removal mining destroys biodiversity, pollutes the water beyond recompense, and threatens the health and safety of area residents. And by obliterating the mountains, they say, it also obliterates the culture of Appalachia.

Coal industry reps have countered by decrying the loss of jobs, tax revenue, and business the already impoverished region would suffer if the mines were to close under the weight of too much regulation. They also point out that the culture of Appalachia is as bound to coal mining as it is to the mountains. Both sides count area residents, including miners, among their ranks.

At the heart of the issue is coal itself—our country's dirtiest, and most abundant, energy source—the one most responsible for rising CO_2 levels from electricity production and also the one we rely on most heavily and the one we are consuming most rapidly. As the Appalachian reserves dwindle, debates raging throughout these decimated foothills are reverberating across our energy-addicted nation.

The world depends on coal for most of its electricity production.

Simply put, coal equals energy. **Energy** is defined as the capacity to do work; like all other living things, we humans need it, in a biological sense, to survive.

But we also need it to run our societies: to heat and cool our homes; operate our cell phones, lamps, and laptops; fuel our cars; and power our industries. Most of our energy comes from **fossil fuels**—nonrenewable carbon-based resources, namely coal, oil, and natural gas—that were formed over millions of years from the remains of dead organisms.

Worldwide, we used more than 7.5 billion metric tons (8.3 billion U.S. tons) of coal in 2012, the vast majority of it to generate electricity. **Electricity** is a natural form of energy (lightning and nerve impulses are electrical) that we have learned to create on demand; we produce it in a central location and send it out via transmission lines to where we want it to go. More

coal A fossil fuel that is formed when plant material is buried in oxygen-poor conditions and subjected to high heat and pressure over a long time.

mountaintop removal A surface mining technique that involves using explosives to blast away the top of a mountain to expose the coal seam underneath; the waste rock and rubble is deposited in a nearby valley.

energy The capacity to do work.

fossil fuels Nonrenewable resources like coal, oil, and natural gas that were formed over millions of years from the remains of dead organisms.

electricity The flow of electrons (negatively charged subatomic particles) through a conductive material (such as wire).

KEY CONCEPT 18.1

Coal is the leading fuel used for electricity production. It is burned to heat water to produce steam; the steam turns a turbine connected to a generator, producing electricity.

than 40% of electricity generation worldwide, and 39% in the United States, involves burning coal. (Americans especially love their electricity: In 2013, we used 19.5% of all the kilowatt-hours generated in the world; only China surpassed the United States, using 22%.)

Coal-fired power plants work by feeding pulverized coal into a furnace to generate heat, which then powers a system that produces electricity. It takes roughly 0.5 kilogram (1 pound) of coal to generate 1 kilowatt-hour (kWh) of electricity; that's enough to run ten 100-watt

incandescent light bulbs for an hour, an energy-efficient refrigerator for 20 hours, or an older, less efficient refrigerator for 7 hours. The average U.S. family of four uses about 11,000 kWh of electricity per year. That comes out to around 1,375 kilograms (3,000 pounds) per person per year.

So, how does coal stack up against other fossil fuel energy sources? On one hand, it produces more air pollution than any other fossil fuel. On the other, it is safer to ship, cheaper to extract, and in the United States at least, more abundant by far. In fact, the United States has 10 times more coal than it does oil and natural gas combined; in 2012 alone, we mined more than 930 million metric tons (1.0 billion U.S. tons). Most of it came from Wyoming, which leads the nation as a coal producer, and the Appalachian mountains, which follow behind as a close second. While we exported some of that yield, the vast majority was used to power American households and businesses. **INFOGRAPHIC 18.1**

INFOGRAPHIC 18.1   **HOW IT WORKS: ELECTRICITY PRODUCTION FROM COAL**   1

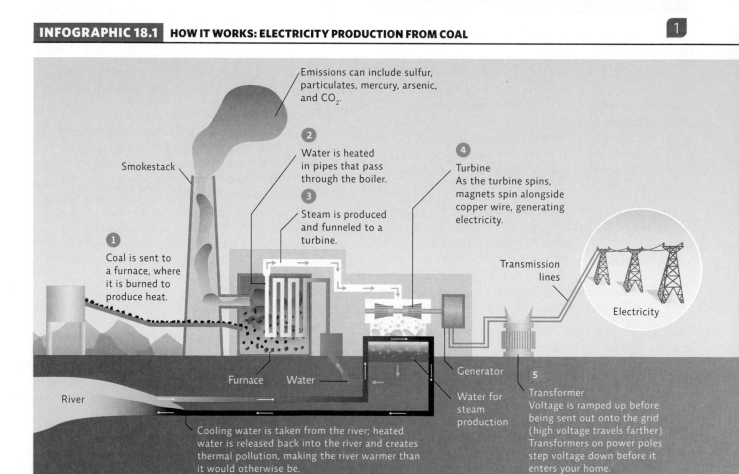

↑ The most common way to generate electricity is to heat water to produce steam; the flow of steam turns a turbine inside a generator to produce electricity. This schematic shows TVA's coal-fired Kingston plant in Tennessee, which generates 10 billion kilowatt-hours a year by burning 13,000 metric tons of coal a day, supplying electricity to almost 700,000 homes.

 What other methods could be used to spin a turbine and generate electricity?

↑ A train carrying coal leaves a mountaintop removal mining site and travels through the backyards of homes in Welch, West Virginia.

©Les Stone/The Image Works

In terms of net energy, or **energy return on energy investment (EROEI)**—a metric that allows us to compare the amount of energy we get from any individual source to the amount we must expend to obtain, process, and ship it—coal is neither the best nor the worst. In terms of electricity production, it has an EROEI of about 18:1 (18 units of energy produced for every 1 unit consumed), compared to 7:1 for natural gas and 5:1 for nuclear power. Wind has a better EROEI, at 20:1, and hydroelectric has the best EROEI, at 40:1.

There can be no denying the blessings of coal: This sticky black rock has powered several waves of industrialization—first in Great Britain and the United States, now in China—and in so doing has shaped and reshaped the world as we know it. But as time marches on, the costs of those blessings have become all too apparent. They include an ever-growing list of health impacts—from birth defects to black lung disease—and an equally lengthy roster of environmental costs—not only the destruction of Appalachia, but also the pollution of Earth's atmosphere with CO_2 and other greenhouse gases. (See Chapter 21 for more on greenhouse gases and climate change.)

This litany of paradoxes has given rise to a deep national ambivalence. While we are consuming more electricity, and burning more coal, than at any other time in our history, applications for new coal-fired power plants have been rejected left and right in recent years by determined citizens and local governments from Maryland to Minnesota. "We're caught in a catch-22," says Scott Eggerud, a forest manager with West Virginia's Department of Environmental Protection. "On one hand, it's like we need the stuff to live; on the other hand, we see that it's kind of killing us." Nowhere is this catch-22 more pronounced than in the foothills of central and southern Appalachia.

Coal forms over millions of years.

The Appalachian Mountains were born a few million years before the rise of the dinosaurs, when Greenland, Europe, and North America hovered near the equator as a single giant landmass bathed in a dense tropical swamp. The northwestern bulge of what is now Africa pressed into the easternmost edge of North America, pulverizing the colliding continental margins and forcing a colossal mass of land upward, into a mountain range as high as the Himalayas. It was the last and greatest of three violent clashes that joined all the world's land into a single supercontinent, upon which stegosauruses and velociraptors would eventually roam.

energy return on energy investment (EROEI) A measure of the net energy from an energy source (the energy in the source minus the energy required to get it, process it, ship it, and then use it).

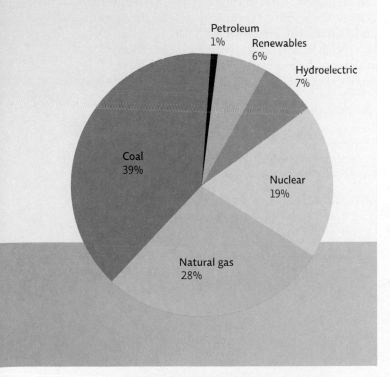

U.S. ELECTRICITY GENERATION BY FUEL, 2013

Petroleum 1%
Renewables 6%
Hydroelectric 7%
Nuclear 19%
Natural gas 28%
Coal 39%

↑ Coal is the main fossil fuel used to produce electricity in the United States, though its use decreased 6% between 2010 and 2013 due mainly to an increase in nuclear power. Increasing the use of renewable fuels and taking steps to improve energy efficiency and conservation could decrease the role coal plays in energy production.

KEY CONCEPT 18.2

Coal is a fossil fuel, formed over long periods of time when dead plant material was buried and subjected to high heat and pressure.

produced *peat*—a soft mash of partially decayed vegetation. As time passed and more and more layers of sediment were laid down over the peat, pressure and heat compressed it into the denser rocklike material that we know as coal. **INFOGRAPHIC 18.2**

This same story—tectonic upheaval followed by deep and rapid burial of organic material, followed by the material's slow compaction into coal—has played out in numerous places around the globe. As a result, coal is found everywhere, though of course some places have more than others. Europe and Eurasia hold about 35% of the world's reserves, while the Asian Pacific has about 32%, and North America holds 27.5%. The United States has more coal than any other country with almost 27% of the world's total reserves, followed by Russia (18%) and China (13%).

overburden The rock and soil removed to uncover a mineral deposit during surface mining.

The fallout from this cataclysm—the gradual accretion of decomposing swamp vegetation compressed and baked by heat and time—established the Appalachian coal beds. As the swamp plants died out, they were buried under a mud so thick it kept oxygen out. Instead of being fully decomposed by bacteria, their remains

INFOGRAPHIC 18.3

The Appalachian beds were once among our most bountiful

reserves; seams as tall as a man wound for miles through the mountainside and made for easy harvesting. But after 150 or so years of mining, those reserves have dwindled noticeably. At current rates of usage, proven coal reserves (those we know are economically feasible to extract) should last about 120 years—longer if deeper reserves can be accessed. And as the layers of minable coal have grown thinner and harder to reach, the coal industry has become both more sophisticated and more destructive in its approach to extracting the coal.

KEY CONCEPT 18.3

Coal deposits are located throughout the world, though some areas have a larger supply than others.

Mining comes with a set of serious trade-offs.

Hobet 21, which has claimed at least 5,000 hectares (12,000 acres) of land, was once the site of several adjacent peaks. To get at the coal beneath those peaks, miners began by clear-cutting the forest above. Next, they drilled holes deep into the side of the mountain (some holes up to 120 meters [400 feet] deep), set dynamite in those holes, and blasted as much as 300 meters (1,000 feet) of mountain into a mass of rubble known as **overburden**. The miners repeated this process several times, until the layers of coal were exposed. Then, using buckets big enough to hold 20 midsized cars, they scooped that rubble into staircase-shaped mounds that filled in entire neighboring valleys from the ground up—all told some

---

INFOGRAPHIC 18.2   COAL FORMATION   2

↓ Coal is formed over long periods of time as plant matter is buried in an oxygen-poor environment and subjected to high heat and pressure. Places with substantial coal deposits that are retrievable with current technology are called coal reserves.

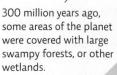
300 million years ago, some areas of the planet were covered with large swampy forests, or other wetlands.

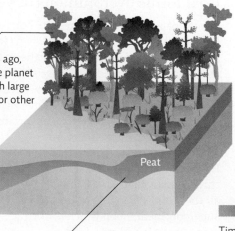

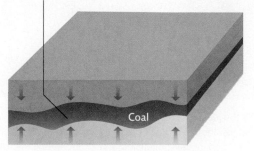
Over time, pressure built up as more sediment was laid down; increased pressure and heat converted the peat to a soft coal (lignite); in areas with enough pressure and heat, lignite was converted to harder varieties of coal (bituminous and anthracite).

Time

Some vegetation died and was submerged in oxygen-poor sediments of the swamp. With limited oxygen, decomposition slowed tremendously, and peat formed.

 Peat is found in many areas of the world today, such as the peat forests of Indonesia and the peat bogs of Ireland. Will this peat eventually turn into coal?

↑ Seven metric tons of overburden (the rock, soil, and ecosystem waste) are moved for every metric ton of coal extracted.

INFOGRAPHIC 18.3   MAJOR COAL DEPOSITS OF THE WORLD

2

→ Coal reserves are not evenly distributed around the world. The United States has 28% of the world's coal reserves, with much of its best, low-sulfur bituminous, coal found in Appalachia.

 Anthracite and bituminous coal   ⬤ Lignite

 Why aren't coal reserves found everywhere?

KEY CONCEPT 18.4

Mountaintop removal is a cost-effective surface mining method used to obtain coal, but it creates tremendous environmental damage and employs fewer miners than other methods.

surface mining A form of mining that involves removing soil and rock that overlays a mineral deposit close to the surface in order to access that deposit.

subsurface mines Sites where tunnels are dug underground to access mineral resources.

acid mine drainage Water that flows past exposed rock in mines and leaches out sulfates. These sulfates react with the water and oxygen to form acids (low-pH solutions).

72 million metric tons (80 million U.S. tons) of overburden each year. The process obliterated the forest habitat, buried countless streams, and permanently reordered the land's natural contours. **INFOGRAPHIC 18.4**

This mountaintop removal mining is just one form of **surface mining**. The other type, known as *strip mining*, employs a similar process: Workers use heavy equipment to remove and set aside overburden so that they can harvest the coal beneath. When they finish mining one strip of land, they return the overburden to the open pit and move on to a new strip. Strip mines are used in areas like Wyoming, where the coal is close to the surface and the ground above is fairly level.

With their reliance on explosives and heavy equipment, surface mines are a far cry from the underground mines, also called **subsurface mines**, that sustained the Nelson family for so many generations. "When our daddies were mining, back in the '40s and '50s, the seams were as tall as full-grown men," says Nelson, who since retiring has become a spokesperson for the anti-mining Ohio Valley Environmental Coalition. "So you could get at 'em the old-fashioned way, with pickaxes and sledgehammers." Those days of plenty are gone, he says. Many of the coal seams that remain are too thin to be culled by human hands.

To be sure, subsurface mines (which still make up about 60% of all coal mines worldwide, and about 50% of those in the United States) come with their own challenges. Water seeps easily into tunnels, and as it does, toxins leach from the surrounding rocks into the gathering pools. Sulfate in particular produces **acid mine drainage**, which goes on to contaminate soil and streams and has become a major problem with both active and closed mines. In 2011, Duke University ecologist Emily Bernhardt and her colleagues reported that not only is acidic water directly toxic to many aquatic plants and animals, it also alters the nutrient cycle of streams in ways that reverberate all the way up the food chain. Subsurface mines also account for 10% of all methane release in the United States; methane is a potent greenhouse gas. **INFOGRAPHIC 18.5**

But subsurface mines also come with some advantages. Unlike surface mines,

---

INFOGRAPHIC 18.4   MOUNTAINTOP REMOVAL

3

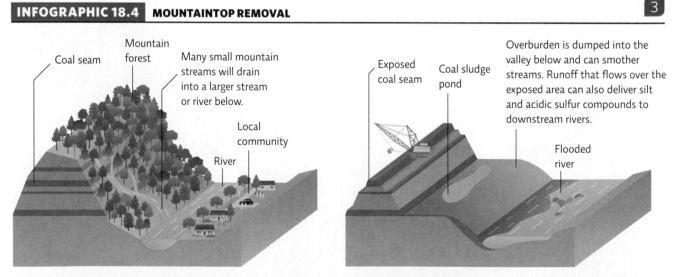

↑ Surface mining techniques are used when coal seams are close to the surface, or in the case of Appalachia, when the coal seams are thin and can't be efficiently accessed via subsurface mining. In Appalachia, the forests are first clear-cut and then explosives are used to blast away part of the mountain. Heavy equipment then digs through debris, dumping the overburden (soil and rock) into the nearby valley, burying streams as the valley is filled in. The exposed coal is dug out and some processing is done on site. Coal sludge left over from processing is stored in ponds on the mining site.

 Why is the overburden dumped into the valley below when it is known that it damages streams and valley habitat?

INFOGRAPHIC 18.5   SUBSURFACE MINING

3

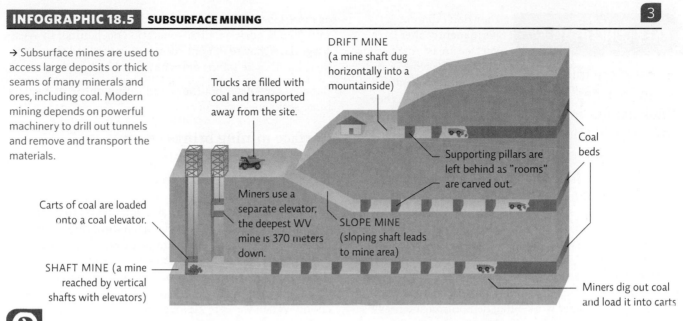

→ Subsurface mines are used to access large deposits or thick seams of many minerals and ores, including coal. Modern mining depends on powerful machinery to drill out tunnels and remove and transport the materials.

Trucks are filled with coal and transported away from the site.

DRIFT MINE (a mine shaft dug horizontally into a mountainside)

Supporting pillars are left behind as "rooms" are carved out.

Coal beds

Carts of coal are loaded onto a coal elevator.

Miners use a separate elevator; the deepest WV mine is 370 meters down.

SLOPE MINE (sloping shaft leads to mine area)

SHAFT MINE (a mine reached by vertical shafts with elevators)

Miners dig out coal and load it into carts

? Why isn't the coal found in the regions where mountaintop removal is used removed with subsurface mining instead?

MINING HAZARDS

COAL DUST
Far more miners die from pneumoconiosis (black lung disease) caused by breathing coal dust than from mining accidents. In 2007, the Centers for Disease Control and Prevention reported 525 coal miner deaths from pneumoconiosis, compared to 18 accidental deaths.

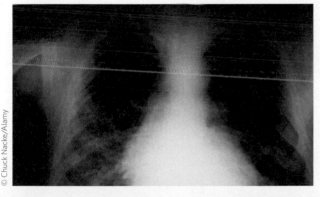

EXPLOSIONS AND MINE COLLAPSE
Methane gas fumes and coal dust are the most common causes of mine explosions; 362 miners died in the West Virginia Monongah coal mine explosion in 1907—the worst mining accident in U.S. history; 29 died after an explosion in the Upper Big Branch mine in 2010.

FIRE
Some underground coal mine fires have been burning for hundreds of years; a fire in an abandoned mine in Centralia, Pennsylvania, burning since 1962, has forced the abandonment of the town.

TOXIC FUMES
A vigil being held for 12 miners who died from carbon monoxide poisoning after an explosion trapped them in the Sago Mine in West Virginia in 2006.

KEY CONCEPT 18.5

Subsurface mining is used to access deep, thick coal seams. It is less environmentally damaging and employs more workers than surface mining but is a hazardous job.

they don't disrupt or permanently alter large surface areas. And because much of the work is still done with workers rather than by machines, they employ more people. In Appalachia, 100,000 mining jobs were lost between 1980 and 1993 as underground mining gave way to mountaintop removal. In Kentucky alone, mining jobs are down 60% in recent years, even though coal production is on an upswing. The loss of traditional mining jobs has bred yet another controversy in the region. Coal industry reps say that the riskier deep-mining jobs are being replaced with higher-paying, safer work—like demolition and heavy equipment operation. But that has not alleviated tension

as more jobs are lost than replaced. "It's a double insult," says Tim Landry, a fourth-generation deep miner in West Virginia. "They're not only destroying the land that we love, but they're taking our jobs away, too."

But a loss of jobs is not the community's only—or even its most serious—concern.

Surface mining brings severe environmental impacts.

Bob White is a tiny unincorporated village on the edge of Charleston, West Virginia—just one segment of an endless trickle of trailers and shotgun houses that hug the mountain on each side of long winding valley after long winding valley. Maria Gunnoe, a 40-something waitress, and the daughter, sister, wife, niece, and aunt of coal miners, has lived there, on the same property, all her life. She remembers having free run of the mountain as a child. "When we were kids we used to roam deep in the holler," she says, referring to the mini-valleys that snake

↓ Maria Gunnoe became an environmentalist after the removal of over 800 hectares as the result of mountaintop mining in her area caused flooding to her home and property, poisoned her well water, and made her daughter sick. Gunnoe received the Goldman Environmental Prize for her organizing efforts in her southern West Virginia community.

through the region's foothills. "We had access to all the resources—food, medicine, water—that these mountains provided."

Things are different now. Gunnoe's children run into big yellow gates and No Trespassing signs wherever they go. The mountains, she says, have been closed off for blasting. And in the past decade, several million tons of overburden—an unimaginable mass—have been dumped into the valleys around Bob White.

The upheaval has had a noticeable impact on area residents. For one thing, the loss of forest and the compaction of so much soil has increased both the frequency and severity of flooding; without trees, and with the soil so compressed, the ground can't absorb water. "They've filled in like 15 of the valleys around me, and floods are about three times more serious than they ever were before," she says One 2003 flood nearly swallowed her entire valley—house, barn, family, and all.

Floods aren't the only problem. In fact, it's the blasting that most scares Gunnoe. It fills the air with tiny particles of coal dust—easily inhaled and full of toxic substances like mercury and arsenic. Studies show a higher-than-typical incidence of respiratory illnesses in mining communities. "When they were clearing the ridgeline right behind us, we'd get blasted as much as three times a day," Gunnoe says "There were days when we'd have to just stay inside because you couldn't breathe out there. And now, my daughter and I both get nosebleeds all the time."

The health impacts are far reaching. According to a 2011 study by Melissa Ahern of Washington State University, children in those communities are more likely to suffer a range of serious birth defects, including heart, lung, and central nervous system disorders. And children aren't the only ones affected; a 2012 study by Ahern and her colleague Michael Hendryx of West Virginia University found increased rates of leukemia and lung, colon, kidney, and bladder cancer in Appalachian counties with mountaintop removal sites compared to non-mining counties.

In addition to filling the air, these toxic substances also permeate the region's rivers, streams, and groundwater. In a 2005 **environmental impact statement** on mountaintop removal mining, the U.S. Environmental Protection Agency (EPA) reported that selenium levels exceed the allowable limits in 87% of streams located downhill from mining operations. Toxic substances were as much as eight times higher in streams near mined areas than in streams near unmined areas. When tests revealed dangerous levels of selenium

in the stream behind Gunnoe's house, she and her neighbors started getting their water from town. "We can't trust the water in our own streams anymore," she says "It's sad, but this is just not the same place that I grew up in."

On top of all these immediate threats to human health and safety, area residents like Gunnoe worry about long-term damage to the region's natural environment. They are not alone. Around the world, in fact, mining operations are a major cause of environmental degradation. To be sure, geological resources—those like copper, gold, iron ore, and coal, that are buried deep underground and so must be dug up—are essential to the functioning of modern society. But harvesting them unleashes a dangerous mix of toxic substances that are normally sequestered away from functioning ecosystems, and that living things—including humans—are not generally well adapted to handle.

> *" We can't trust the water in our own streams anymore, It's sad, but this is just not the same place that I grew up in."* —Maria Gunnoe

At surface mines throughout Appalachia, dangerous quantities of iron, aluminum, selenium, and other metals have leached out from the blasted overburden and may be starting to bioaccumulate and biomagnify up the food chain. Scientists and fishers alike have noted an alarming decline in certain fish that populate area streams and rivers. And recent surveys indicate that biodiversity has decreased in direct proportion to the concentration of such metals in the water. This loss of aquatic life also affects forest life since many terrestrial animals feed on insects like mayflies and dragonflies that begin their life in the water.

The nutrient cycles in ecosystems surrounding mined areas have also been altered—in some cases dramatically. Extra sulfates, released from blasted rock, have increased nitrogen

environmental impact statement A document outlining the positive and negative impacts of any federal action that has the potential to cause environmental damage.

and phosphorus availability, and in so doing have led to eutrophication. And as sulfate levels rise, so do populations of sulfate-feeding bacteria. These microbes transform sulfate into hydrogen sulfide, which is toxic to many aquatic plant species.

Meanwhile, overburden has completely buried more than 3,200 kilometers (2,000 miles) of streams and destroyed an untold range of natural habitats in the process. And destroying habitats invariably threatens local species diversity. Biodiversity in the streams of Appalachia is rivaled only by that in the streams found in rain forests; the diversity of trees is also second only to the tropics. With these ecosystems destroyed on more than 500 mountains in Appalachia, researchers predict the permanent loss of many local plant and animal species.

Slurry impoundments—reservoirs of thick black sludge that accompany each mining operation—are among the most controversial features of the landscape created by mountaintop removal. They too are a consequence of thinner coal seams. "The thinner seams are messier," says Randal Maggard, a mining supervisor at Argus Energy, a company that has several mining operations throughout Appalachia. "It's 6 inches of coal, 2 inches of rock, 8 inches of coal, 3 inches of rock, and so on. It takes a lot more work to process coal like that." To separate the coal from the rock, miners use a mix of water and magnetite powder known as *slurry* that coal can float in. Once the sulfur and other impurities have been washed out, the coal is sent for further processing, and the slurry by-product is pumped into artificial holding ponds.

Maggard insists that the impoundments are safe. "Before we fill it, we have to do all kinds of drilling to test the bedrock around it," he says "And then a whole slew of chemical tests on top of that—all to make sure the barrier is impermeable." Still, area residents worry about a breach. With good reason.

In October 2000, the dike holding back the slurry at one of these ponds failed, pouring more than 1.1 billion liters (300 million gallons—30 times the Exxon Valdez spill) of toxic sludge into the Big Sandy River of Martin County, Kentucky. The contamination killed all life in some streams and eventually reached the Ohio River, more than 32 kilometers (20 miles) away. The sludge was 1.5 meters (5 feet) deep in some places. The EPA would register it as one of the worst environmental disasters in the history of the eastern United States. More than a decade later, sludge could still be found a few inches under the river sediments.

More recently, in January 2014, a chemical spill into the Elk River near Charleston, West Virginia, contaminated local water supplies with thousands of liters of methylcyclohexane methanol (MCHM), a chemical used to wash coal to remove some of its impurities before burning. The health effects of MCHC are poorly understood; only a few studies have been done on the chemical, and those were done by the chemical industry that produces it; none were peer reviewed. Based on two small studies that examined toxicity in laboratory rats, a safe level of MCHC in drinking water was set at 1 part per million—the Centers for Disease Control and Prevention's (CDC's) "best guess" at a safe level. (However, the CDC recommended that pregnant women avoid using the water until no MCHM was detected in water samples "due to limited availability of data, and out of an abundance of caution.") Following the spill, roughly 300,000 residents were told not to drink or bathe in their tap water; emergency room visits for rashes, diarrhea, nausea, and vomiting increased following the spill. Water bans lasted up to 10 days in some communities, but many residents continued to use bottled water for months, not trusting the water coming from their taps.

Can coal's emissions be cleaned up?

Of course it's not just the mining and processing of coal that pollute the environment.

When coal is burned to produce heat energy, it releases a range of toxic substances that damage the environment and threaten human health: gases (sulfur dioxide, carbon monoxide, nitrogen oxide, and planet-warming carbon dioxide), heavy metals (such as mercury and arsenic), radioactive material (uranium and thorium), and particulate matter (soot) with particles small enough to irritate lung tissue or even enter the bloodstream if inhaled. Until December 2011, there were no regulations to limit how much of these toxic air emissions power plants could release.

According to the UN's 2013 Global Mercury Assessment, worldwide, coal burning accounts for the release of an estimated 474 metric tons (520 U.S. tons) of mercury into the atmosphere (24% of the total anthropogenic mercury release globally); the EPA estimates that 50% of mercury emissions in the United States are from power plants. Mercury released into the atmosphere can be transported around the world; between 20% and 30% of the mercury found in North America originates in areas as far away as East Asia. Much of this mercury finds its way into the world's oceans. In fact, the mercury content of the top 100 meters (330 feet) of oceans has doubled over the past

↑ In 2008, almost 4 billion liters of coal fly ash slurry surged into homes and waterways near Harriman, Tennessee, after breaching a coal ash containment pond. It was the largest fly ash release in U.S. history. Many residents have been permanently displaced.

100 years due to anthropogenic emissions. And this extra mercury is finding its way into the organisms (including humans) that consume seafood. The mercury content of ancient and present-day samples of teeth, bone, and hair were analyzed and revealed a dramatic and rapid increase of mercury between preindustrial and modern times; levels in the late 1900s were roughly 10 times greater than levels before the Industrial Revolution. This has ecological consequences but also has direct impacts on human health: The main way humans are exposed to mercury is by eating contaminated fish, especially top predators like tuna that biomagnify mercury (see Chapter 3). The EPA's new Mercury and Air Toxic Standards will reduce these emissions by 90% and are predicted to save up to $90 billion in human health costs by 2016 (while only costing $9.6 billion to implement).

Coal-fired power plants also generate tons of toxic fly ash—fine ashen particles made up mostly of silica. Some of that ash is diverted to industry, where it's used in concrete production. But the majority of it is buried in hazardous waste landfills or stored in open ponds like the ones used to contain the slurry waste from mining. In 2008, following a heavy rain event, a 15-meter-tall (50-foot-tall) dike from a pond holding coal fly ash from the TVA Kingston Fossil Plant in Tennessee failed, releasing 4.2 billion liters (1.1 billion gallons) of fly ash into nearby rivers and coating vast expanses of riverside land. The EPA estimates that cleanup would cost $1.2 billion—and more when property damage and lawsuit settlements were factored in. Federal regulators have since identified more than 100 other ash ponds at risk for breach.

INFOGRAPHIC 18.6

INFOGRAPHIC 18.6 ENVIRONMENTAL AND HEALTH PROBLEMS OF MINING AND BURNING COAL

4
Blasting can damage home foundations and wells and, in rare cases, trigger rockslides. In 2005, a 3-year-old child was killed when a boulder crashed through his house and landed on his bed.
FORMER MOUNTAIN CONTOUR
Typical U.S. Coal Power Plant (500 megawatts)
ANNUAL INPUTS
2.2 billion gallons of water
1.4 million metric tons of coal (140,000 railroad cars)
ANNUAL POLLUTION PRODUCED INCLUDES ROUGHLY
130,000 metric tons of toxic ash
450 metric tons of small particles
9,000 metric tons of nitrogen oxides (NO_x)
9,000 metric tons of sulfates (SO_2)
80 kilograms of mercury
100 kilograms of arsenic
3.5 million metric tons of CO_2
Thermal pollution to nearby waterways
AFTER MOUNTAINTOP REMOVAL
Mountaintop removal permanently damages habitat and streams. In Kentucky alone, more than 2,200 kilometers (1,400 miles) of streams have been damaged or destroyed and 4,000 kilometers (2,500 miles) of streams polluted.
COAL POWER STATION
Home water wells close to reclaimed mines (150 meters [500 feet] or closer) receive significantly higher mine drainage than wells farther away, exceeding EPA allowances for arsenic, lead, iron, and sulfate.
FLY ASH POND

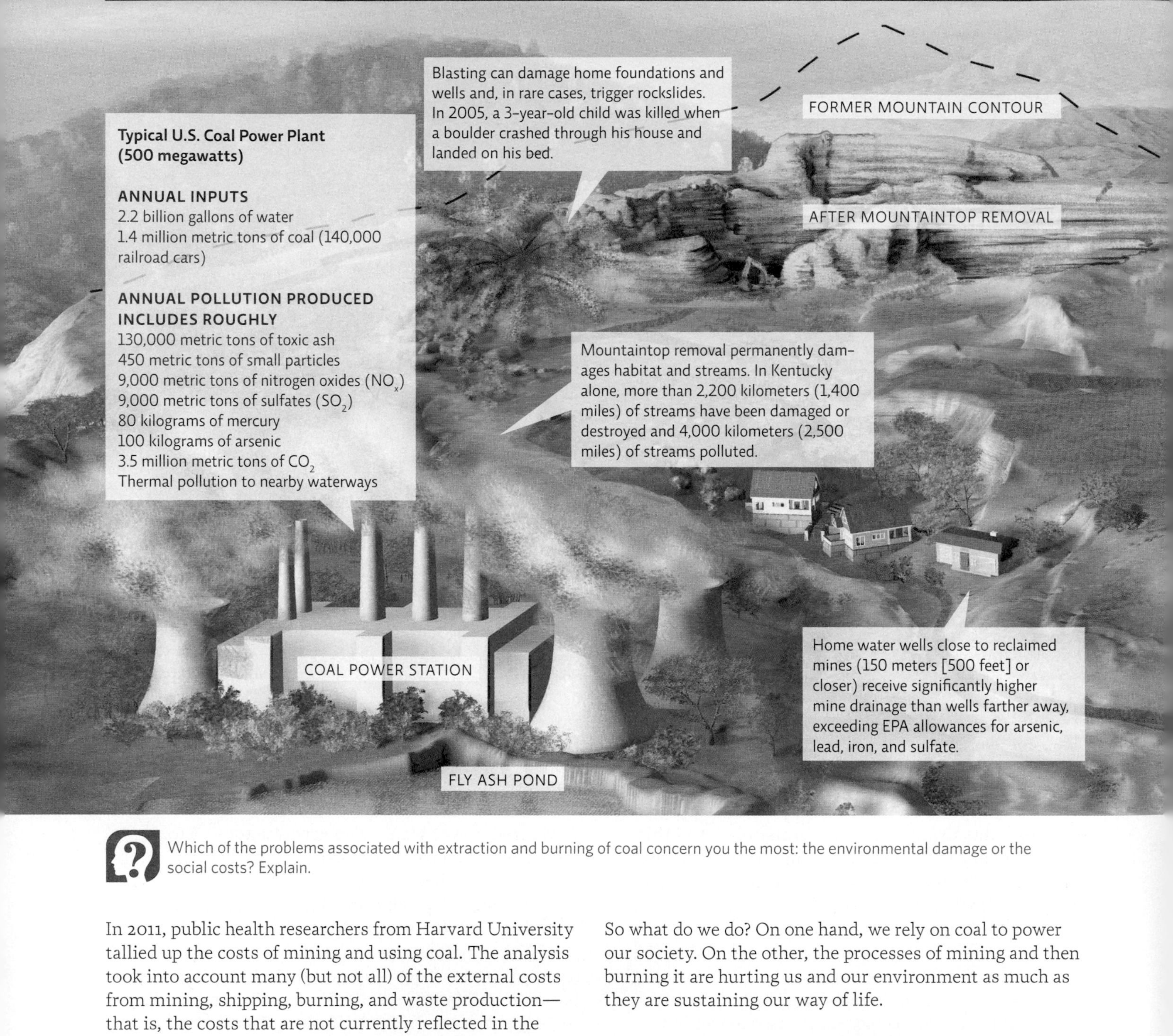

Which of the problems associated with extraction and burning of coal concern you the most: the environmental damage or the social costs? Explain.

In 2011, public health researchers from Harvard University tallied up the costs of mining and using coal. The analysis took into account many (but not all) of the external costs from mining, shipping, burning, and waste production—that is, the costs that are not currently reflected in the market cost of coal. They estimated that coal costs the American public between \$300 and \$500 billion a year in externalized costs (health, environmental, and property costs). This amounts to an extra \$0.10 to \$0.26 per kWh in external costs—in some cases more than twice what consumers actually pay.

So what do we do? On one hand, we rely on coal to power our society. On the other, the processes of mining and then burning it are hurting us and our environment as much as they are sustaining our way of life.

One potential solution is clean coal technology—technology that minimizes the amount of pollution produced by coal. For example, scientists and engineers around the world are working on ways to capture the gases emitted from burning coal. We already have the capacity to capture some emissions like

↓ The use of coal contributes to environmental and health problems when it is mined or burned. Surface mining, especially mountaintop removal, produces the most environmental damage. Miners and people living in coal mining communities are most at risk for health problems, but air pollution from burning coal affects individuals and ecosystems far removed from the actual power plant.

More than 500 mountains and 570,000 hectares (1.4 million acres) of forest have been damaged or destroyed in Appalachia since mountaintop removal began in the 1970s. More than 200 species are impacted by the practice; 40 or more are already rare or endangered. Habitat loss from mountaintop removal is believed to be the main reason that Cerulean warblers (a songbird) have decreased as much as 80% in some areas of West Virginia.

SUBSURFACE COAL MINE

VALLEY FILL

Piles of mine waste fill valuable plant and animal habitats.

Loss of vegetation and the compaction of soil increases stormwater runoff; stormwater flows over mined or reclaimed areas have been measured at two to three times higher than nearby forested areas.

Water and soil contamination from acid mine drainage; 87% of streams tested downstream from mines contained toxic selenium levels that exceeded EPA allowances.

+HOSPITAL

There is a greater incidence of birth defects and lung, heart, and kidney disease in West Virginia counties with coal mining than in those without; the incidence of disease was also correlated with the volume of coal mined in the county.

mercury, particulate matter, and sulfur (see Chapter 20).

The next big challenge is **carbon capture and sequestration (CCS)**—the capture and storage of CO_2 in a way that prevents it from reentering the atmosphere. There are many approaches to CCS; some of them promise to capture more than 90% of CO_2 emissions from coal-fired power plants. But most are still in the research and development stage, and so far, progress has been hindered by the cost of capturing the carbon and by a dearth of good storage options. The extra energy needed to implement CCS must also be considered: Estimates for the additional energy needed range from 25% to 40%—this means 25–40% more coal that must be mined, processed, transported, and burned. **INFOGRAPHIC 18.7**

Another emerging clean coal technology involves chemically

carbon capture and sequestration (CCS) The process of removing carbon from fuel combustion emissions or other sources and storing it to prevent its release into the atmosphere.

INFOGRAPHIC 18.7   HOW IT WORKS: CARBON CAPTURE AND SEQUESTRATION (CCS)

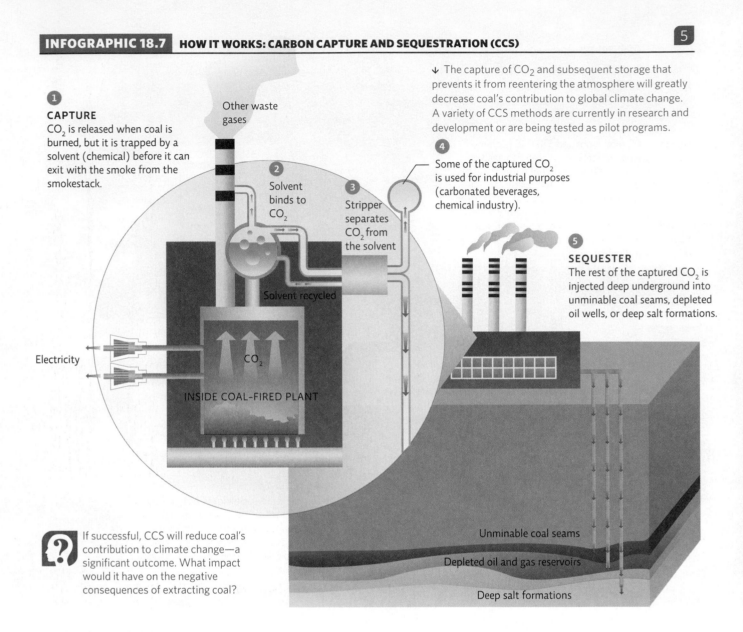

① **CAPTURE**
CO_2 is released when coal is burned, but it is trapped by a solvent (chemical) before it can exit with the smoke from the smokestack.

Other waste gases

↓ The capture of CO_2 and subsequent storage that prevents it from reentering the atmosphere will greatly decrease coal's contribution to global climate change. A variety of CCS methods are currently in research and development or are being tested as pilot programs.

② Solvent binds to CO_2

③ Stripper separates CO_2 from the solvent

④ Some of the captured CO_2 is used for industrial purposes (carbonated beverages, chemical industry).

Solvent recycled

Electricity

CO_2

INSIDE COAL-FIRED PLANT

⑤ **SEQUESTER**
The rest of the captured CO_2 is injected deep underground into unminable coal seams, depleted oil wells, or deep salt formations.

If successful, CCS will reduce coal's contribution to climate change—a significant outcome. What impact would it have on the negative consequences of extracting coal?

Unminable coal seams

Depleted oil and gas reservoirs

Deep salt formations

KEY CONCEPT 18.7

The negative impact of burning coal can be addressed by capturing pollutants before they are released or converting coal to a liquid or gaseous fuel that burns cleaner.

removing some of coal's contaminants before burning it. While it still contains some toxic substances, the final product (a liquid or gaseous fuel, depending on the process used), is cleaner than the original coal. Of course, this process requires energy (thus lowering coal's EROEI) and generates hazardous waste (from the toxic substances found in the coal) that still has to be dealt with.

In fact, as critics are quick to point out, coal can never be truly clean. Each of these "clean coal" technologies produces its own toxic by-products, and none of

them eliminate the need to mine coal from deep within Earth's surface.

Even if we can't easily avoid using coal or make it totally clean, we can certainly try to use less of it. One way to accomplish this is to design and use more energy-efficient appliances. Less than 40% of the energy from coal is converted to electricity; in addition, the amount of that energy that goes toward powering appliances depends on how efficient those appliances are. We can also reduce our coal consumption through simple conservation efforts like turning off lights and electronics when we're not using them. Every kilowatt-hour of electricity conserved means roughly 0.5 kilograms (1 pound) less of coal being burned. And a growing number of public utilities now offer "green power" programs in which users can choose to buy electricity that has been generated by solar or wind power instead of coal. (For more on energy efficiency and conservation, see Chapter 23.)

Reclaiming a closed mining site helps repair the area but can never re-create the original ecosystem.

In some ways, closed mines—those where all the coal has been harvested—look even more alien than do active sites like Hobet 21. Instead of the natural sweep of rolling hills, staircase-shaped mounds covered with what looks like lime green spray paint join one mountain to the next. Atop some of them, gangly young conifers, evenly spaced, strain toward the Sun. The spray paint is really hydroseed, and most of the gangly trees are loblolly pine, a non-native hybrid that foresters are trying to grow in the region.

Such efforts represent the coal industry's attempt to honor the U.S. Surface Mining Control and Reclamation Act, which in 1977 mandated that areas that have been surface mined for coal be "reclaimed" once the mine closes. **Reclamation** requires that the area be returned to a state close to its pre-mining condition.

At some surface mines, at least, reclamation is straightforward (albeit labor-intensive) work: If the mined area was originally fairly flat, the reclamation process includes filling the site with the stockpiled overburden and contouring the site to match the surrounding land. This relaid rock is then covered with topsoil saved from the original dig. Sometimes, alkaline material such as limestone powder is sprinkled overtop to neutralize acids that have leached into the soil. Vegetation, usually grass, is then planted, leaving other local vegetation to move in on its own. According

reclamation The process of restoring a damaged natural area to a less damaged state.

KEY CONCEPT 18.8

Surface coal mines must undergo reclamation once they are closed; this reduces the environmental damage of mining but does not eliminate it.

↓ A valley in West Virginia after mining shows none of the original forest, ridges, or streams that were once found there. The grass is grown from "hydroseed," a seed-chemical mixture sprayed onto surfaces to bind the soil to prevent erosion. When hydroseed is applied to rock (often done in highway construction), it grows into a plush lawn of grass that dies after a month or two.

Antrim Caskey

INFOGRAPHIC 18.8   MINE SITE RECLAMATION

↓ U.S. federal law requires that after a surface coal mine ceases operations, the land must be restored to close to its original state. However, mining sites never truly or fully recover. After the coal is removed, the area is recontoured to produce a slope, and grasses are planted. To replace the mountain streams that were buried by the mining removal process, new channels are constructed to accommodate water flow—but these channels do not support the diverse biological communities that once existed. No matter how much care goes into reclamation, the area will not support a mountain forest community like it once did. However, if done well, a new ecological community may develop.

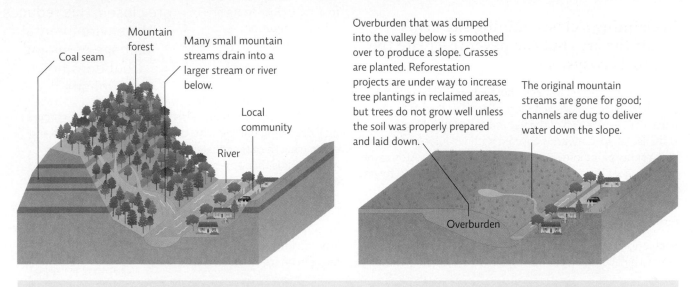

Coal seam

Mountain forest

Many small mountain streams drain into a larger stream or river below.

Local community

River

Overburden that was dumped into the valley below is smoothed over to produce a slope. Grasses are planted. Reforestation projects are under way to increase tree plantings in reclaimed areas, but trees do not grow well unless the soil was properly prepared and laid down.

The original mountain streams are gone for good; channels are dug to deliver water down the slope.

Overburden

Reclaimed land may be suitable for other land uses such as agricultural, residential, commercial, or industrial, or can be set aside as a natural area. While there have been some successes on smaller-level reclamation sites, more complex reclamation procedures are required to address mountain sites. Projects, such as the Kempton Project in Tucker County, West Virginia, that reclaimed 60 acres at a cost of $2.3 million, are rare but show that progress is being made. At Kempton, toxic deposits were removed, native trees and grasses were replanted, streams were relocated, and natural wetlands were installed to neutralize acids from two seeps connected to the former mine site. Though not the mountain it once was, the area is considered a success for mountaintop removal reclamation.

? If you lived near an areas where mountaintop removal mining was being done, would you be satisified with the reclamation process and final product? What are the advantages and disadvantages of this type of reclamation?

to the Mineral Information Institute, more than 1 million hectares (2.5 million acres) of coal-mined area have been successfully reclaimed using this strategy.

But reclamation has always been a controversial idea. For one thing, the Appalachian forests were created over eons. In fact, in some ways, they were born of the same calamity that laid the coal beneath the mountains. As the supercontinent drifted, carrying the Appalachian range well north of the equator, a temperate hardwood forest replaced the swamps and grew over time into one of the most biologically diverse ecosystems on the planet. In some areas, a single mountainside may host more tree species than can be found in all of Europe, not to mention songbirds, snails, and salamanders that exist nowhere else on Earth. No amount of human restoration can re-create the mountain habitat or bring back the species diversity and nuanced ecosystem relationships born of millions of years of evolution.

From a practical point of view, rebuilding a mountain is considerably more difficult than filling in a strip mine where the original land was relatively flat. Critics say that so far, there's no evidence that a site as expansive and

as thoroughly destroyed as Hobet 21 could ever be truly reclaimed. In one survey, Rutgers ecologist Steven Handel found that trees from neighboring remnant forests did not readily move in to recolonize mountaintop removal sites, largely because of problems with the soil: At some mines, there was simply not enough of it; at others, it has not been packed densely enough for trees to take root. Those problems have technical solutions, Handel says but so far, the seemingly simple act of changing reclamation protocols has been stymied by politics.

And trees are just one facet of reclamation. What about all those streams that were buried? The 1973 Clean Water Act prohibits the discharge of materials that bury a stream or, if unavoidable, requires mitigation efforts that return the stream close enough to its original state such that the overall impact on the stream ecosystem is "non-significant." Federal regulations also require that damaged streams be restored. Industry reps argue that they are indeed working to rebuild streams: Once the overburden has been reshaped and smoothed over, they dig drainage ditches and line them with stones in a way that resembles a stream or river. But so far, research shows—and most ecologists agree—that such channels don't perform the ecological functions of a stream. "They may look like streams," says ecologist Margaret Palmer from the University of Maryland. "But form is not function. The channels don't hold water on the same seasonal cycle, or support the same aquatic life, or process contaminants out of the water—all things a natural stream does."

INFOGRAPHIC 18.8

In Appalachia, the arguments over when and where and how to mine for coal are quickly boiling down to a single intractable question: Once it's all gone, how will we clean up the mess we've made? For a story that has played out over geologic time, the question is more immediate than one might think. In West Virginia, coal reserves are expected to last another 50 years, at best. That means no matter what regulations the government imposes, or what methods the coal companies resort to, the day of reckoning will soon be upon us.

Select References:
Ahern, M., et al. (2011). The association between mountaintop mining and birth defects among live births in central Appalachia, 1996–2003. *Environmental Research*, 111(6): 838–846.

Ahern, M., & M. Hendryx. (2012). Cancer mortality rates in Appalachian mountaintop coal mining areas. *Journal of Environmental and Occupational Science*, 1(2): 63–70.

Bernhardt, E. S., & M. A. Palmer. (2011). The environmental costs of mountaintop mining valley fill operations for aquatic ecosystems of the central Appalachians. *Annals of the New York Academy of Sciences*, 1223(1): 39–57.

Epstein, P.R., et al. (2011). Full cost accounting for the life cycle of coal. *Annals of the New York Academy of Sciences: Ecological Economics Review*, 1219(1): 73–98.

Hendryx, M., & Ahern, M. (2008). Relations between health indicators and residential proximity to coal mining in West Virginia. *American Journal of Public Health*, 98(4): 669–671.

Inman, M. (2013). The true cost of fossil fuels. *Scientific American*, 308(4): 58–61.

Pond, G., et al. (2008). Downstream effects of mountaintop coal mining: Comparing biological conditions using family- and genus-level macroinvertebrate bioassessment tools. *Journal of the North American Benthological Society*, 27: 717–737.

United Nations Environment Programme (UNEP). (2013). Global Mercury Assessment 2013: *Sources, Emissions, Releases and Environmental Transport*. Geneva, Switzerland: UNEP Chemicals Branch.

U.S. Environmental Protection Agency. (2005). *Final Programmatic Environmental Impact Statement on Mountaintop Mining/Valley Fills in Appalachia*. Philadelphia, PA: EPA.

Weakland, C. A., & P. B. Wood. (2005). Cerulean warbler (Dendroica cerulea) microhabitat and landscape-level habitat characteristics in southern West Virginia. *The Auk*, 122(2): 497–508.

BRING IT HOME

PERSONAL CHOICES THAT HELP

Although coal is one of our most abundant fossil fuels, its drawbacks are significant. They include CO_2 emissions, the release of air pollutants that cause environmental problems such as acid rain, health problems such as asthma and bronchitis, and massive environmental damage from the mining process. One way to minimize the impact of coal is to reduce consumption of electricity.

Individual Steps
- Always conserve energy at home and in your workplace.

- Turn off or unplug electronics when not in use.
- Put outside lights on timers or motion detectors so that they come on only when needed.
- Dry clothes outdoors in the sunshine.
- Turn the thermostat up or down a couple of degrees in summer and winter to save energy and money.

Group Action
- Organize a movie screening of *Coal Country* or *Kilowatt Ours*, which present issues related to coal mining and mountaintop removal from many perspectives.

Policy Change
- The Appalachian Regional Reforestation Initiative is a great example of how groups, sometimes with very different objectives, can work toward a common goal. Go to http://arri.osmre.gov to see how this coalition of the coal industry, citizens, and government agencies are working to restore forest habitat on lands used for coal mining. Visit http://beyondcoal.org to find out about events in your state and for the opportunity to weigh in on the decommissioning of outdated coal power plants and the building of new ones in the United States.

ENVIRONMENTAL LITERACY UNDERSTANDING THE ISSUE

 How important is coal as an energy source, and how is it used to generate electricity?
INFOGRAPHIC 18.1

1. True or False: Worldwide and in the United States, coal is used to produce more than half of all electricity.

2. Which of the following is *not* true about coal?
   a. It produces more air pollution than other fossil fuels.
   b. It is difficult to ship.
   c. Extraction is relatively inexpensive.
   d. The United States has an abundant supply of it.

3. Coal can be used to generate electricity by:
   a. sending a current though the pulverized coal.
   b. melting the coal and sending its electrons across conductive wires.
   c. pouring liquefied coal over a turbine, which triggers the release of electrons from copper wire.
   d. burning the coal to make steam, which turns a generator that makes electricity.

 What is coal, how is it formed, and what regions of the world have coal deposits that are accessible?
INFOGRAPHICS 18.2 AND 18.3

4. How long are coal reserves expected to last in West Virginia?
   a. 50 years
   b. 100 years
   c. 500 years
   d. Forever

5. Which country has the largest coal reserves?
   a. China
   b. The United States
   c. Canada
   d. Russia

6. In your own words, describe the process of coal formation. Why is coal a finite resource?

 What methods are used to mine coal, and what are the advantages and disadvantages of each?
INFOGRAPHICS 18.4 AND 18.5

7. Surface strip mining techniques are used when:
   a. mines are located close to human settlements.
   b. coal is located in thin seams in mountainous areas.
   c. thick coal seams are close to the surface in relatively flat areas.
   d. it is too dangerous to use explosive charges to access the coal.

8. Compare and contrast the two coal mining methods used in Appalachia: mountaintop removal and subsurface mining.

 What are the advantages and disadvantages of burning coal?
INFOGRAPHIC 18.6

9. True or False: If we paid the true costs of using fossil fuels like coal, the price of our electricity would likely go up.

10. Emissions from a coal-fired power plant include all of the following *except*:
    a. oxygen.
    b. carbon dioxide.
    c. mercury.
    d. arsenic.

11. An advantage of subsurface (underground) coal mining is that it:
    a. employs more people than surface mining techniques and thus creates more jobs.
    b. is safer for the miners than surface mining techniques.
    c. is cheaper than surface mining techniques.
    d. All of these are advantages.

12. What health problems are associated with coal burning?

 What new technologies allow us to burn coal with fewer environmental and health problems? How can mining damage be repaired, and how effective is this restoration?
INFOGRAPHICS 18.7 AND 18.8

13. True or False: Mine reclamation projects at mountaintop removal sites focus on re-creating the mountain habitat and then reintroducing native plants and animals.

14. Which of the following is an example of reclamation?
    a. Reworking coal mines to extract more coal
    b. Replanting the site of mountaintop removal with grass and pine trees
    c. Returning land to the people who originally owned it
    d. Filling in a subsurface mine

15. Define and describe the process of carbon capture and sequestration. Why is this process necessary? What are the costs and benefits of the process?

SCIENCE LITERACY WORKING WITH DATA

The following graphs depict fatalities associated with coal mining and metal/nonmetal mining (M/NM). Note: In Graph B, the number of fatalities is shown at the top of each bar; the y-axis depicts the rate of fatalities.

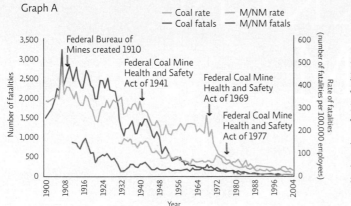

Graph A

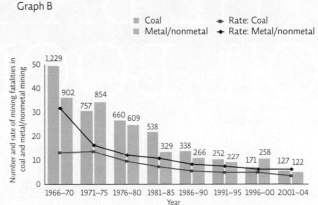

Graph B

Interpretation

1. In Graphs A and B, why are both the absolute number of fatalities and the rate of fatalities presented? Which is the fairer measure to use?

2. Based on Graphs A and B, would you rather work in the coal industry or the metal/nonmetal industry? Why?

3. Why do the numbers in Graph B appear to be larger than those in Graph A? For example, compare the numbers for 1968 in both graphs. What conclusion can you make about graphing data in general?

Advance Your Thinking

4. The graphs represent fatality data. Do you know what caused the fatalities, based on the graph? What do you assume these fatalities are related to? In your opinion, based solely on these numbers, would you risk being a miner, provided that the pay and benefits were good?

5. Based on the information in Graph A, what can you say about fatalities and federal regulations? Do you think it would be safe to dispense with federal regulations and allow mine owner/operators to determine safety standards and practices, now that they know the benefits of safety regulations?

INFORMATION LITERACY EVALUATING INFORMATION

The EPA is responsible, in part, for issuing mining permits, based on the predicted environmental damage from the mines. As with many other issues in environmental science, there is often disagreement between government regulators and companies about the extent of environmental damage. Federal regulators argue that some practices are so harmful to the environment that they should be halted; companies argue that halting certain practices will result in severe economic loss.

Go to www.npr.org/templates/story/story.php?storyId=122297492. Click on the "Listen to the Story" link, review the photo gallery, and read the story about mountaintop removal.

Evaluate the website and work with the information to answer the following questions:

1. Is this a reliable information source? Does it have a clear and transparent agenda?
   a. Who runs this website? Do this group's credentials make it reliable or unreliable? Explain.
   b. Is the information on the website up to date? Explain.

2. When was this story aired, and on what program? What type of information is reported in this story?

3. Now read an ABC News story about mountaintop removal at http://abcnews.go.com/US/community-rallies-mountaintop-removal-mining/story?id=13897050#.IxWpurKacZJ.
   a. Who runs this website? Is it comparable to the other website you looked at? Why or why not?
   b. Compare the content of the two stories. Which do you think provides more in-depth coverage of the issue?
   c. Now look for links to other stories about mountaintop removal. Do both sites have links to other websites that you could use to learn more if you wished? Are the links more useful or relevant on one site than the other?

4. One of the links provided on the ABC site is to Massey Energy, which declined to be interviewed for the story.
   a. Why do you think Massey Energy would decline to be interviewed?
   b. Do a brief Internet search for Massey Energy. What types of stories appear? Do a brief review of the stories. Do these stories give you any more insight into Massey Energy's motives for declining an interview?

Find an additional case study online at http://www.macmillanhighered.com/launchpad/saes2e

THE BAKKEN OIL BOOM

Is fracking the path to energy independence?

CORE MESSAGE

Oil and natural gas are vital fuels for modern society and important raw materials for a wide range of products. However, environmental damage is caused at every stage of their acquisition and use. In addition, conventional supplies of oil and natural gas are becoming depleted. Untapped sources of these fossil fuels do exist but are more difficult and costly (economically and environmentally) to extract, especially some unconventional sources. We should weigh the pros and cons of pursuing these unconventional sources as we make decisions about how best to meet our energy needs.

AFTER READING THIS CHAPTER, YOU SHOULD BE ABLE TO ANSWER THE FOLLOWING **GUIDING QUESTIONS**

1
How are fossil fuels formed, and why are they considered nonrenewable resources? Where are current proven oil and natural gas reserves found, and what is meant by "peak oil"?

2
How are conventional oil and natural gas extracted? How is crude oil refined and used?

Natural gas is flared off as oil is pumped from a fracking well in the Bakken formation located close to a private home near Watford City, North Dakota. © Jim West / Alamy

3

What are some of the environmental costs of finding, extracting, and using oil and natural gas from conventional reserves?

4

What unconventional sources of oil and natural gas exist, and how are they extracted?

5

What are the trade-offs of pursuing unconventional oil and natural gas sources? What role do you think these resources should play in meeting future energy needs?

"It is as you imagine it: Vast. Open. Windy. Stark. Mostly flat. All but treeless. Above all, profoundly under-populated, so much so that you might, at times, suspect it is actually unpopulated. It is not. But it is heading there."

So went the opening paragraph of a 2006 feature article in the *New York Times Sunday Magazine* about the demise of the prairie states in general and of North Dakota in particular. The entire region, it seemed, was suffering a mass exodus: houses and churches and schools and stores, all completely abandoned.

That was 2006. Less than a decade later, the same sleepy patch of country is in the middle of a modern-day gold rush—an oil boom, to be precise. Journalists today describe local grocery stores barely keeping shelves stocked, town movie theaters being so crowded that people are sitting in the aisles, and housing costs approaching those of the most expensive cities. Towns once home to 1,200 residents now claim 10 times as many. "With the way it is now," Jeff Keller, a natural resource manager with the Army Corps of Engineers, told Pro Publica in 2012, "you're getting to the crazy point."

Fossil fuels are a valuable, but nonrenewable, resource.

Fossil fuels form over millions of years when organisms die and are buried in sediment under conditions that slow down decomposition tremendously. In time, as pressure and heat increase, the buried material goes through a chemical transformation to form **oil**, **natural gas**, or coal. (See Chapter 18 for more on coal.) Even though they are produced by natural processes, fossil fuels are considered a **nonrenewable resource** because we are using them up in a fraction of the time it takes for them to form. In other words, natural processes cannot possibly keep pace with human consumption. **INFOGRAPHIC 19.1**

Everywhere, countries are looking for new sources of oil.

fossil fuel A variety of hydrocarbons formed from the remains of dead organisms.

oil A liquid fossil fuel useful as a fuel or as a raw material for industrial products.

natural gas A gaseous fossil fuel composed mainly of simpler hydrocarbons, mostly methane.

nonrenewable resource A resource that is formed more slowly than it is used or that is present in a finite supply.

The heart of this U.S. oil boom is the Bakken formation, a roughly 500,000-square-kilometer (200,000-square-mile) shale formation deep underground that resides mostly in North Dakota but also stretches westward into Montana and north into the Canadian provinces of Saskatchewan and Manitoba. (Shale is a type of sedimentary rock that is very dense and does not easily let the oil and natural gas trapped within escape.) It's the largest

⊚ WHERE IS THE BAKKEN FORMATION?

continuous oil accumulation the U.S. Geological Survey (USGS) has ever assessed. And according to the agency, it contains as much as 167 billion barrels of crude oil in the North Dakota portion alone; the entire formation contains perhaps 400 billion barrels. (A barrel of oil is equivalent to 159 liters, or 42 gallons.) The United States has other

↓ Rows of mobile homes house oil workers in Williston, North Dakota, where housing shortages drive up the cost of rent. In 2014 Williston was the most expensive place to rent a home in the United States.

REUTERS/Ben Garvin/Newscom

INFOGRAPHIC 19.1   HOW OIL AND NATURAL GAS FOSSIL FUELS FORM   1

 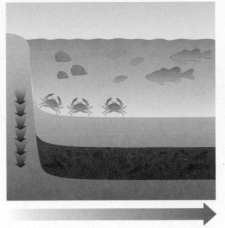 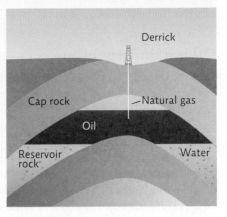

↑ Long ago, some marine organisms died and were buried in sediment. This burial excluded oxygen, and decomposition was greatly slowed down.

↑ As sediments accumulated, the partially decomposed buried biomass was subjected to high heat and pressure. Over the course of millions of years, it was chemically converted to oil or natural gas.

↑ Since oil and natural gas are lighter than water, they flow upward in porous reservoir rock until stopped by a layer of dense cap rock. We tap these deposits by drilling into the porous rock reservoirs.

 How is the formation of oil and natural gas similar to and different from the formation of coal? (See Infographic 18.2.)

oil-rich shale formations, including the highly productive Eagle Ford formation in Texas. Together, Bakken and Eagle Ford produced 96% of all U.S. shale oil in 2012.

How many of the barrels at Bakken can actually be harvested has been a matter of intense speculation. Bakken oil is made up almost exclusively of **tight oil**—shale oil that's trapped inside impermeable rock deposits. Tight oil is difficult to access not only because the oil is trapped inside rock but because the deposits themselves tend to be buried very deep—thousands of meters below ground, in many cases. It also tends to be widely dispersed throughout the rock. In 2013, the USGS estimated that 7.4 billion barrels could be recovered from the Bakken oil fields, largely thanks to **fracking (hydraulic fracturing)**—a drilling technique that involves detonating explosive charges to create fractures in the rock and then injecting mass quantities of a highly pressurized water and chemical mixture into the factures to liberate the oil (and/or natural gas) trapped inside.

In 2014, roughly 1 million barrels of crude oil were being recovered each day from Bakken via fracking. That's not much in the grand scheme of things: In 2013 the United States consumed almost 19 million barrels of oil daily. But in the short term, experts say, fracking has transformed the domestic energy debate. As of 2013, the Bakken oil field was producing more than 10% of all oil produced in the United States, and North Dakota had gone from being the nation's ninth-largest oil producer to being its second, right after Texas. As a result, its unemployment rate had dropped precipitously, to around 1%.

But booms tend to be mixed blessings at best, and the fracking boom is no exception. In and around the cities close to the Bakken oil fields, such as Williston, North Dakota, economic gains on the one hand have come at the expense of rampant housing shortages, rising crime, and environmental degradation on the other. Fracking is water intensive and requires a blend of toxic chemicals to work properly. Area residents can't help but worry about their drinking water, not to mention the air and soil quality around drilling sites. They also worry about their way of life: Can a regional culture built on solitude and serenity survive the boom, not to mention the bust that will inevitably follow?

And those concerns now echo across a nation, as oil and natural gas reserves from North Dakota to Texas to Pennsylvania are harvested using fracking technology. "The implications are already reverberating far beyond North Dakota," *National Geographic* wrote in 2013. "Bakken-like shale formations occur across the United States,

KEY CONCEPT 19.1

Fossil fuels form when organic matter is buried and subjected to high temperature and pressure. Because we use them much faster than they form, they are considered nonrenewable resources.

tight oil Light (low-density) oil in shale rock deposits of very low permeability; extracted by fracking.

fracking (hydraulic fracturing) The extraction of oil or natural gas from dense rock formations by creating factures in the rock and then flushing out the oil/gas with pressurized fluid.

indeed, across the world. The extraction technology refined here is in effect a skeleton key that can be used to open other fossil fuel treasure chests."

Oil is a limited resource.

Oil is a liquid fossil fuel made up of hundreds of types of *hydrocarbons*, organic compounds of hydrogen and carbon. Hydrocarbons take many forms—solid, liquid, or gas— and we have developed methods for extracting them from the depths of Earth. Oil is found some 0.3 to 9 kilometers (0.2 to 5.5 miles) below the surface, where rigs have to dig deep in order to reach it. At that depth, it does not exist in thick black pools. Instead, if you could look down into an oil reservoir, you'd see rock. **Crude oil** is found as tiny droplets wedged within the microscopic open spaces, or pores, inside rocks.

Oil deposits are found in various regions around the world. **Proven reserves** are the amount of a fuel that is economically feasible to extract from a deposit using current technology. **Conventional oil reserves** hold crude oil that can be extracted with traditional oil wells. Such reserves are not evenly distributed around the planet, leading to political problems among countries that have the oil, like those in the Middle East, and those that do not have enough to meet their own needs, like the United States. At current rates of extraction and use, known reserves are expected to last another 40 to 50 years, though this number is uncertain because reports of oil reserves tend to be questionable. (Many nations keep their reserves estimates secret.)

Natural gas reserves are also finite. The Middle East has the most conventional reserves of natural gas, with about 41% of the world's reserves; the United States has only 6%. Total world conventional reserves of natural gas are expected to last 60 to 100 years at current rates of use. But as with oil, natural gas use is increasing at about 2% per year. At this rate, natural gas could run out much sooner if new supplies, or methods to access currently inaccessible supplies, are not found. **INFOGRAPHIC 19.2**

INFOGRAPHIC 19.2 **PROVEN OIL AND NATURAL GAS RESERVES** 1

↓ Different regions of the world have different amounts of oil and natural gas reserves. Here we show the proven reserves by region as a percentage of the world total. (Geographic areas are shown in different colors to more easily distinguish each region.)

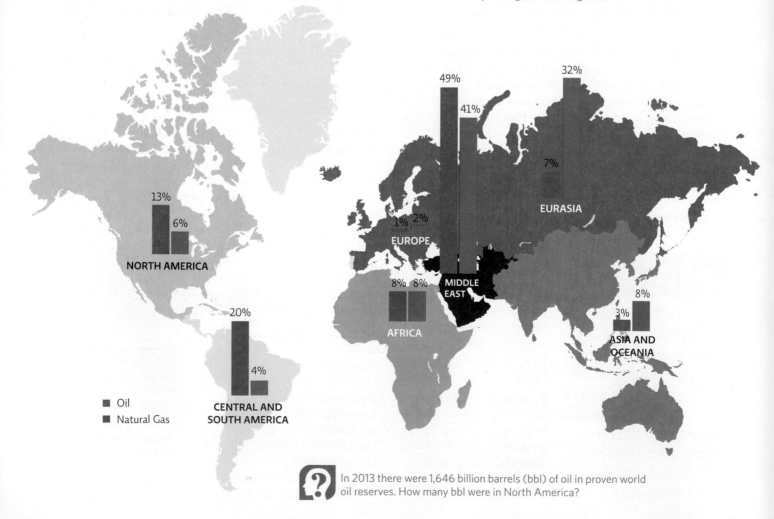

In 2013 there were 1,646 billion barrels (bbl) of oil in proven world oil reserves. How many bbl were in North America?

INFOGRAPHIC 19.3 | PEAK OIL

OIL PRODUCTION FROM CONVENTIONAL SOURCES BY REGION

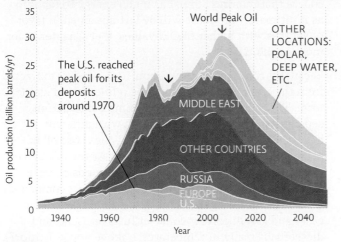

← Peak oil represents the point in time when oil reaches its highest production levels and then steadily and terminally declines. It is likely that the world has reached, or will soon reach, peak oil for conventional reserves (red arrow). The dip (black arrow) seen in the early 1980s represents an oil glut; conservation efforts during the oil crisis of the 1970s led to reduced demand, but we have since resumed an upward course of oil consumption.

How much oil did Russia produce in 1990? (Hint: It was *not* 7.5 billion barrels; see Appendix 2 for pointers on reading area graphs.)

KEY CONCEPT 19.2

The largest conventional oil reserves are in the Middle East. We have likely passed peak oil for conventional reserves; production has peaked and will decline over time.

World demand for oil rises more than 2% each year, and because conventional oil reserves are finite, it is possible that we have already passed **peak oil**, the moment in time when oil reaches its highest production levels before it steadily and terminally declines. Some say peak oil occurred in 2002, others in 2006, and others still that it won't be reached until 2020. In any case, once we're past peak oil, the demand will outstrip the supply, and this will have international economic repercussions. Most economists agree that the days of cheap oil are gone. The sooner we prepare for a world with less oil, the easier the transition to a different energy economy will be. **INFOGRAPHIC 19.3**

Weaning ourselves off fossil fuels will not be easy. Populations in developed and developing countries alike are utterly dependent on fossil fuels. And the problems that oil, natural gas, and coal bring are significant and increasing. Unfortunately, there are no easy answers to the question of how best to power our societies. Every energy option at our disposal (see Chapters 18–23 and LaunchPad Chapter 32) comes with advantages and disadvantages, and no single energy source can easily replace fossil fuels. This is what makes meeting our energy needs a *wicked problem*.

New **unconventional reserves** of oil and natural gas, such as those that can be extracted using fracking technology, may increase supplies in the short term but will not extend them indefinitely. They are, however, changing the global oil and natural gas market as more and more is being produced in the United States.

Conventional oil and natural gas reserves are tapped by drilling wells.

Initially, after a new oil well is drilled, nature does most of the work. There is significant pressure on oil located deep underground from millions of tons of rocks pressing down and from Earth's heat, which causes gases around the oil and rock to expand. So when a well is first drilled, oil naturally flows upward, escaping like air gushing out of a balloon. This is known as *primary production*, and about 5% to 15% of the oil can be recovered in this first phase.

Eventually, the pressure on the trapped oil will decrease, so injection wells are drilled near the oil well, and huge hoses pump water through the injection wells into the ground. This increases the pressure on the remaining oil

crude oil A mix of hydrocarbons that exists as a liquid underground; can be refined to produce fuels or other products.

proven reserves A measure of the amount of a fossil fuel that is economically feasible to extract from a known deposit using current technology.

conventional oil reserves Light- or medium-density crude oil deposits; extracted by pumping.

peak oil The moment in time when oil reaches its highest production levels and then steadily and terminally declines.

unconventional reserves Deposits of oil or natural gas that cannot be recovered with traditional oil/gas wells but may be recoverable using alternative techniques.

KEY CONCEPT 19.3

Oil and natural gas often occur together in rock formations and are extracted from conventional reserves using oil and gas wells.

and forces more oil to the surface. During this phase of *secondary production*, an additional 20% and 40% of a reserve's oil can be recovered.

But even after maximal primary and secondary production (up to 55% of the total oil in the reserve), there is a lot of oil left in the ground. *Tertiary production* methods like injecting steam or carbon dioxide into the reservoir allow the additional recovery of up to 15% of the reserve's oil, for a total production of up to 70% of the oil found in the reserve. **INFOGRAPHIC 19.4**

With each more intensive production method, the *energy return on energy invested* (EROEI) goes down, and the cost goes up. A comparison of the EROEI for the extraction of different energy sources gives producers and investors useful information that helps determine which energy sources are worth pursuing. An energy source that requires almost as much energy to tap into as it ultimately produces will be left uptapped in favor of sources with higher EROEI values. (See Chapter 18 for more on EROEI.)

Because it contains so many impurities once it's extracted from the ground, crude oil must be refined into a usable form. The refining process, called simple distillation, separates the oil into fractions, each with different hydrocarbon components. To do this, the crude oil is heated and put into a column or tower. As the oils boil off and rise in the column, the larger compounds condense back to liquid form and are collected. The first compounds to boil off include lighter liquids, like liquid petroleum gases and gasoline. Next come jet fuel, kerosene, and diesel fuel. Finally, at temperatures above 600°C (1,100°F), the heaviest products, including grease, wax, asphalt, and residual fuel oil, can be recovered.

INFOGRAPHIC 19.4   HOW IT WORKS: CONVENTIONAL OIL AND NATURAL GAS WELLS   2

↓ Oil is obtained by drilling through layers of dense rock to reach the reservoir below. At first, oil may easily flow due to the relief of pressure caused by the drill hole. Pumpjacks are used to mechanically pump out more oil when the oil stops flowing freely. Natural gas is also harvested from conventional deposits in this way.

Derrick

Engine

Drill

Drill bit

Groundwater

Deep groundwater

Oil

Pumpjack

Storage and processing

High pressure boiler

Water or steam injection

 Why aren't pumpjacks needed during the initial retrieval of oil from a newly drilled well?

INFOGRAPHIC 19.5   **PROCESSING CRUDE OIL**

↓ Crude oil is refined into a variety of chemical products by heating it in a tower. Different-sized carbon compounds "boil off" (vaporize) at different temperatures and heights within the tower and are collected as separate products.

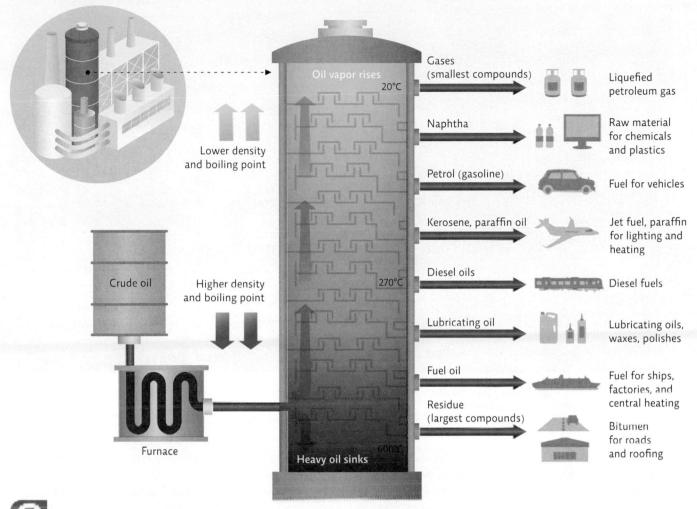

Which of these is the longer hydrocarbon (longest molecule): naphtha or lubricating oil? How do you know?

Many of the products of distillation, called **petrochemicals**, are used not just as fuel but as raw materials in the production of industrial organic chemicals, pesticides, plastics, soaps, synthetic fibers, explosives, paints, and medicines. The production of a desktop computer, for example, consumes 10 times the computer's weight in fossil fuels, mostly oil. **INFOGRAPHIC 19.5**

The importance of these fuels and petrochemical products to modern society cannot be overstated; they permeate our everyday lives and improve our quality of life. A shortage like the one that occurred during the oil crisis of the 1970s can send a nation into a panic and change global politics. It is no wonder we are constantly in search of new supplies, but as they become scarcer, the extent to which we will go to extract them has some people worried.

Fossil fuel extraction and use comes at a high environmental cost.

Oil extraction is not just financially expensive; it also has environmental consequences at almost every stage. In order to find oil, companies send seismic

> **KEY CONCEPT 19.4**
>
> Crude oil is refined by distillation to produce a variety of petrochemicals; about half of all oil goes to the chemical industry to produce plastics and other products.

petrochemicals Distillation products from the processing of crude oil such as fuels or industrial raw materials.

↑ The 1,300-kilometer Trans-Alaska Pipeline delivers oil across Alaska to the southern port of Valdez. An even longer pipeline has been proposed to take oil from Canada's oil sands to Texas.

waves into the ground that bounce back to reveal the location of possible reserves. Doing this in ocean areas disorients marine wildlife; in 2009, ExxonMobile had to abandon its oil exploration in Madagascar because more than 100 whales had beached themselves, presumably as a result of these seismic exploratory methods.

KEY CONCEPT 19.5

Fossil fuels are incredibly useful and versatile chemicals, but their extraction, processing, transport, and use damage the environment and affect human health.

Drilling can affect wildlife, too. Politicians have long debated the merits of drilling in the Arctic National Wildlife Refuge (ANWR), 8 million hectares (20 million acres) of protected wilderness that was created by Congress under the Alaska National Interest Lands Conservation Act of 1980. The USGS estimates that parts of ANWR located on the northern Alaskan coast could harbor between 5 and 16 billion barrels of technically retrievable crude oil and natural gas reserves. But drilling there would have serious environmental consequences. The coastal area where drilling would take place is home to Arctic foxes, caribou, polar bears, and migratory birds, all of whose habitats could be disturbed by drilling.

Of course oil spills—an unintended but frequent consequence of oil drilling and transport—can have immediate and long-term effects. Thankfully, large-scale accidents like the 2010 *Deepwater Horizon* oil platform explosion and oil spill in the Gulf of Mexico are rare, but unfortunately, they can be devastating. The *Deepwater Horizon* spill eventually covered 930 square kilometers (360 square miles) of ocean, impacted roughly 1,800 kilometers (1,100 miles) of coastlines, and infiltrated and damaged coastal wetlands. The oil threatened the lives of many species, including sea and shore birds; thousands of dead or oiled birds were recovered. In addition, it threatened several species of endangered sea turtles, which feed and live in the waters affected by the spill; hundreds of dead turtles were collected in the area. Exposure to spilled oil causes deformities in young fish, including important commercial species such as bluefin tuna that spawn in the area; numbers of bluefin larvae fell by 20% after the spill. Deep-water coral communities as far as 22 kilometers (14 miles) from the spill site show dead and injured coral, with damage indicative of exposure to oil. And while bacteria did help break down and digest much of the oil, a 2014 study conducted by Florida State University researchers found that the bacteria did not consume the more highly toxic chemicals in the oil (a class of chemicals known as polycyclic aromatic hydrocarbons).

The *Deepwater Horizon* disaster also disrupted the livelihoods of residents and those in the fishing and tourism industries. The closure of commercial fisheries resulted in an estimated loss of more than $4 billon, about 40% of an average year's revenue. Some fisheries, notably the oyster fishery, have yet to recover their pre-disaster productivity. Tourism was also affected: Although it's difficult to really put a price on the financial losses to all the large and small business affected, losses were estimated at $3.8 billion over a 3-year period. Litigation

↑ Boats hose down a massive fire on the oil rig *Deepwater Horizon*, 50 miles southeast of the tip of Louisiana in the Gulf of Mexico. The explosion killed 11 workers and released more than 200 million gallons of oil into the Gulf.

© John Mosier/ZUMApress.com

and criminal trials are still under way regarding the *Deepwater Horizon* disaster.

The hazards of oil are not limited to exploration, extraction, and transportation of oil. Burning fossil fuels releases a variety of air pollutants that are linked to negative ecosystem and health impacts. It is also the number-one anthropogenic contributor to climate change. (See Chapter 20 for more on air pollution and Chapter 21 for more on climate change.) In addition, working with oil is a hazardous undertaking, even when the required precautions are taken. Workers in the petroleum chemical industry have much higher rates of cancer, rashes, heart disease, and various other health problems than the general population. This hazard extends to people living near oil refineries and petrochemical plants; rates of cancer, birth defects, headaches, and asthma in such areas exceed national averages. The reality is that working with fossil fuels is a hazardous undertaking, even when state and federal regulations are met, and they often are: Most workers do not intentionally or carelessly expose communities or the nearby environment to hazards.

INFOGRAPHIC 19.6

Natural gas (predominantly methane) is an important alternative to oil for some purposes, like generating electricity and heat. But according to the U.S. Energy Information Administration (EIA), which tracks and analyzes energy data for the Department of Energy, natural gas consumption is expected to increase more than 18% by 2030, an increase with which our current conventional reserves just can't keep up.

The extracting and processing methods for natural gas and oil are very similar. Natural gas sometimes flows freely up wells due to underground pressure, but most natural gas contains impurities and must be refined before it can be used. It is often "flared off" at the oil well in areas where capturing the gas is uneconomical (usually because the wells are not close to pipelines or processing facilities that could transport and process the captured gas), a waste of a potentially useful resource. Natural gas is considered the cleanest fossil fuel because it releases much less carbon dioxide (a greenhouse gas linked to climate change) and other air pollutants than oil or coal do when burned. Ironically, natural gas is itself a potent greenhouse gas. Capturing it or flaring it off is preferable to releasing it into the atmosphere.

Energy producers are turning to unconventional reserves of oil and natural gas.

This is not North Dakota's first oil boom. Its first oil boom was actually back in the 1950s, when there were still plenty of conventional oil reserves to be tapped. Of

INFOGRAPHIC 19.6   ENVIRONMENTAL COSTS OF OIL

3

↓ Oil has environmental costs at every stage—from exploration for reserves, to extraction, processing, transportation, and burning. If these, often externalized, costs were added to the price of our fuel and petroleum-based goods, the prices of these products would be much higher.

WILDLIFE DISRUPTION

Sonic exploration for oil caused whales to beach near Busselton, Australia.

HABITAT LOSS

An open-pit tar sands mine in Alberta where boreal forest once stood.

EXPLOSIONS

The *Deepwater Horizon* burns.

AIR POLLUTION

Burning oil is a major source of air pollution in Mexico City, causing health problems, environmental damage, and global warming.

OIL SPILLS

A severely oiled pelican being rescued in Louisiana. The bird lived.

OCCUPATIONAL AND COMMUNITY HAZARD

Working in or living near a petrochemical facility is correlated with higher rates of cancer and birth defects, likely due to hazardous chemicals released into the air.

? What could be done to decrease each of the impacts shown here?

INFOGRAPHIC 19.7 — FRACKING FOR NATURAL GAS OR OIL **4**

↓ Unconventional deposits of natural gas and oil can be accessed via fracking. Concerns about surface and groundwater contamination make this a contentious resource in populated areas.

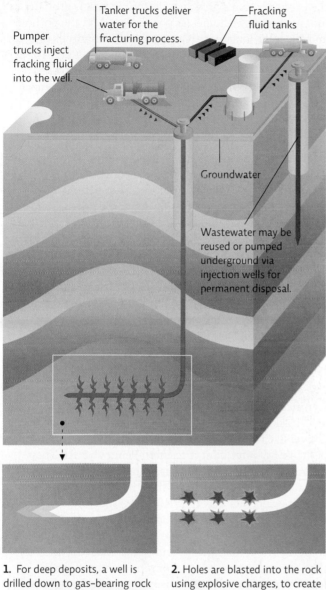

Tanker trucks deliver water for the fracturing process.

Fracking fluid tanks

Pumper trucks inject fracking fluid into the well.

Groundwater

Wastewater may be reused or pumped underground via injection wells for permanent disposal.

1. For deep deposits, a well is drilled down to gas-bearing rock and then extended horizontally.

2. Holes are blasted into the rock using explosive charges, to create large fractures.

3. A slurry of sand, water, and chemicals is pumped into the fractures, to further crack open the rock, hold the fractures open, and make the gas more readily extractable.

4. Natural gas can now flow through the more permeable, fractured rock and be extracted.

course, that boom was much slower and less dramatic. By the time the oil dried up, there were fewer than 1,000 wells in the state. This time around, oil companies are drilling some 25,000 wells *per year*.

The tight oil they're harvesting is considered an unconventional source of oil. It was not previously accessible because the oil, though abundant, is widely dispersed within the rock formations where it is found and cannot be retrieved using conventional oil wells. But those reserves can be tapped with fracking.

Fracking involves drilling thousands of feet straight down into ancient rock beds and then turning the drilling apparatus to drill horizontally into the gas- or oil-bearing rock formation. Several horizontal wells are drilled, in all directions, and each may reach 1.5 kilometers (1 mile) in length. As drilling progresses, steel pipes are inserted into the wellbores and encased in several inches of concrete. Once all the pipes and casings are in place, explosive charges are detonated to perforate the pipe and casing of each horizontal well and to create openings, or "fractures," in the rock through which oil can flow. Next, between 15 and 20 million of liters of water containing a mix of sand and chemicals (some of them toxic) are injected into the well under high pressure. The chemical additives serve various purposes: Disinfectants kill bacteria (there are bacteria deep within rock!) that might foul fracking equipment. Hydrochloric acid helps dissolve minerals and forms initial cracks in the rock. Formic acid and acetaldehyde prevent corrosion of the pipes, and petroleum distillates help reduce friction. Detergents and salts are added as lubricants and stabilizers, and alcohols or ethylene glycol (antifreeze) serve as winterizing agents. The sand enters the small fractures, holding them open, while the water helps flush out the oil or natural gas in the deposit. The water mixture returns to the surface, followed by the oil and/or gas. The concrete casing is thickest close to the surface and in the groundwater region to prevent the chemicals from leaking into the surrounding soil and water. **INFOGRAPHIC 19.7**

Fracking technology is used to access unconventional reserves of natural gas that occur in a wide variety of places throughout the United States; the most productive unconventional reserve is the Marcellus Shale that underlies much of Pennsylvania and New

KEY CONCEPT 19.6

Some oil and natural gas deposits that cannot be accessed with conventional wells can be retrieved via fracking.

 Suppose you are a homeowner. An oil company approaches you, offering to lease your land in order to install a fracking well. What questions would you have for the company as you consider what to do?

York. Currently, the United States leads the world in fracking, recovering enough natural gas and oil to change the global market for these fossil fuels. In 2001, fracking for natural gas only accounted for 2% of total U.S. natural gas production; in 2013, the United States was the leading exporter of natural gas, and fracking gas made up 40% of its total production. This increased production caused natural gas prices to fall, which made it a more attractive fuel for U.S. electricity production—even replacing some coal. Global oil markets, too, have responded to the large quantities emerging from the fracking oil fields of North Dakota and Texas with slightly lower and more stable prices.

While productive, the oil and gas shale deposits aren't the only types of unconventional reserves; there's also crude bitumen—a heavy black oil that is often trapped in sticky, dense conglomerations of sand or clay known as **tar sands**, or **oil sands**. Alberta, Canada, is home to the largest tar sand reserve; with 170 billion barrels of oil, it is the third-largest proven oil reserve in the world. In 2012, Alberta produced 1.9 million barrels of synthetic crude oil per day from the tar sands—much of it shipped to the United States.

There is another unconventional type of oil known as **kerogen shale**, or **oil shale** (not to be confused with shale oil, referred to as *tight oil* in this chapter to help avoid confusion). Kerogen shale deposits are compressed sedimentary rocks that contain kerogen, an organic compound that is undergoing the chemical change into hydrocarbons; it exists as dry matter but can be converted to an oil-like liquid when the rock is heated. Primarily on federal lands in the United States, kerogen shale deposits are huge, with an estimated 1.23 trillion barrels of oil, almost 40 times the nation's proven oil reserves. Currently, retrieving kerogen shale is not economically feasible. **INFOGRAPHIC 19.8**

Pursuing unconventional reserves comes with a high environmental cost.

The upsides of the Bakken boom are not difficult to see or imagine. For one thing, such a mammoth supply of domestic oil means, for the time being at least, less reliance on foreign oil. For another, the boom has stimulated a once-declining regional economy. "It's that long term investment that's important," Ron Ness, president of the North Dakota Petroleum Council, told Fox News in April 2014. "The Bakken has been a huge asset to the state, the region and the nation." The

tar sands/oil sands Sand or clay formations that contain a heavy-density crude oil (crude bitumen); extracted by surface mining.

kerogen shale/oil shale Rocks that contain kerogen, a mix of solid organic material that can be converted to a liquid fossil fuel for extraction.

prominent business magazine *Forbes* agreed: "If 2013 taught us anything going forward, it's that fracking oil is a big deal for growing state economies."

But in North Dakota, the downsides are equally apparent. In fact, long-time North Dakota residents say it's not just the economy that's growing. The crime rate is growing, too. And so is the homeless population. And the potholes and air pollution are increasing as well. Hundreds of trucks per day carry sand, water, chemicals, and drilling equipment down roads that once saw barely 100 cars in a month. In 2011 Pro Publica reported that McKenzie County alone—an area of just 6,000 or so people—required nearly $200 million in road repairs.

Another problem is that the amount of oil being fracked has outpaced the existing capacity to transport it. Without a pipeline to rely on, oil companies have resorted to using railways. But the sheer volume of oil to be transported is bogging down the rail system nationwide and shipping tight oil this way is a hazardous undertaking. It turns out that tight oil is "light oil"—it consists of more shorter-chain hydrocarbons than typical crude oil and contains methane (natural gas) dispersed throughout. This means that tight oil is more volatile than crude oil and is similar to jet fuel in its combustibility. In the summer of 2013, the brakes on a train carrying Bakken oil failed just outside a tiny lakeside village in Quebec; the resulting crash leveled 30-some buildings and claimed 47 lives.

The environmental costs of fracking are equally significant. The presence of methane in local wells near fracking sites is a concern. A 2013 study by Duke University researchers looked at methane contamination of drinking water in northeastern Pennsylvania. They found that contamination was highest in areas closest to fracking wells. While it is true that regions with methane deposits may normally have some methane in the groundwater, the researchers' chemical analysis of the contamination linked it to fracking, not to natural sources. The researchers acknowledge that it is unlikely that contamination comes from the fractured shale formations themselves because they are so far removed from groundwater sources. Contamination more likely comes from leaking pipes or improperly handled fracking water. Improved drilling methods and stricter fracking water storage requirements are decreasing that risk, industry experts say.

The water footprint and production of toxic wastewater is also a concern. The volume of water required to frack a well is huge—around 11 million liters for a Bakken well. The chemicals added to the water make up only about 0.5% to 1% of the mixture by volume (about 90% is water and another 9% to 9.5% is sand or other material to hold open the cracks), but many of the chemical additives are

INFOGRAPHIC 19.8   UNCONVENTIONAL OIL AND NATURAL GAS RESERVES IN THE UNITED STATES AND CANADA   4

↓ There are vast reserves of nonconventional petroleum and natural gas deposits like tar sands, tight oil, and shale gas, but they are more costly and environmentally damaging to extract than conventional deposits and have all the pollution problems associated with burning petroleum.

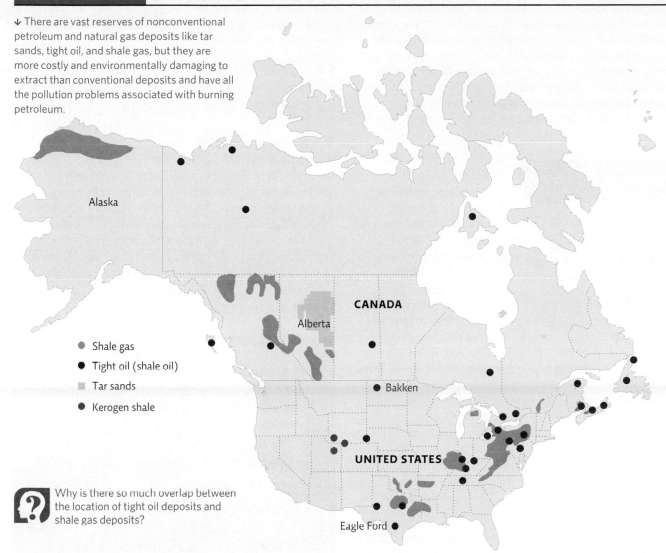

Alaska

CANADA

Alberta

- Shale gas
- Tight oil (shale oil)
- Tar sands
- Kerogen shale

Bakken

UNITED STATES

❓ Why is there so much overlap between the location of tight oil deposits and shale gas deposits?

Eagle Ford

toxic. In addition, the total volume of toxic material is not insignificant: 1% of 11 million liters is still a lot of toxic material.

During fracking, between 20% and 40% of the water injected into the ground resurfaces with the oil or natural gas, bringing the chemicals along with it. This wastewater also contains contaminants picked up from the rock such as salt and small amounts of naturally occurring radioactive materials. Bakken wastewater is 10 times saltier than seawater. Most wastewater is injected into deep wells, but some is still stored in open pits; in the past, it was sprayed on roadways. To determine whether spills or direct disposal of fracking wastewater on land surfaces was potentially damaging to the land, Mary Beth Adams of the U.S. Department of Agriculture Forest Service applied fracking wastewater to a small patch of forest in a 2008 experiment. Within a few days of application, ground

vegetation died, and after about a week, many of the trees began to lose their leaves; 2 years after the application of wastewater, more than half of the trees had died. Though land application is not allowed in North Dakota, spills of wastewater and oil are significant: More than 6.4 million liters of wastewater and 2.6 million liters of oil were spilled between February 2010 and July 2011. In North Dakota, new state laws prohibit storing wastewater in open pits; some wastewater is recycled and used to frack other wells, but most is stored in injection wells.

Wastewater injection wells have their own environmental issues. Concerns that link fracking to an uptick in earthquakes are not centered around detonating explosive charges deep underground or pumping in the pressurized water mixture but focus on injection wells used to dispose of fracking wastewater. Oklahoma, for example, is experiencing an unprecedented increase in earthquake frequency. An area within a 320 kilometer (200-mile) radius of Oklahoma City experienced 31 earthquakes (2.0 magnitude or greater) in a 1-week period in July 2014. To put this into perspective, whereas the state experienced only 6 earthquakes between 2000 and 2008, it experienced 6,345 earthquakes between 2010 and 2013. Research published by Katie Keranen of Cornell University linked this earthquake swarm to nearby wastewater injection wells.

KEY CONCEPT 19.7

Tar sands mining is the most energy- and water-intensive way to extract oil, as well as the most environmentally damaging.

Other unconventional sources of energy also have unique drawbacks. Tar sands deposits can be mined and the oil recovered, but the process is energy and water intensive and produces three or four times the amount of greenhouse gases as conventional oil and gas production. David Schindler, an ecologist at the University of Alberta in Canada, has spent years studying the effects of tar sands extraction on the water quality of the Athabasca River, which cuts through the heart of one of Alberta's biggest tar deposits and mining sites. Research by Schindler and his graduate student Erin Kelly showed that concentrations of 13 toxic compounds were higher in the river's tributaries located downstream of tar sands mining than they were upstream, which suggests that the extraction is at least partly responsible for them.

Tar sands oil also requires more processing steps than conventional crude oil, and the resultant thick crude bitumen is difficult to ship; it will flow only in heated pipelines. The proposed Keystone XL pipeline that would transport tar sands oil south to the Gulf Coast refineries of Texas is supported by industry advocates but strongly opposed by many environmental and citizen groups. Other pipeline routes to the west and east coast of Canada have also been proposed; these too are opposed by those who live in the areas through which the pipelines would pass. Also of concern are the huge lakes of acidic and toxic wastewater that oil companies store at the mining sites as a by-product of the refining process. The pools are so large that they can be seen on satellite images from space.

Is fracking the answer to our energy needs?

It's tempting to think of the oil made available by fracking technology as infinite, or at least plentiful enough that we can stop worrying about peak oil and **energy independence** for

energy independence Meeting all of one's energy needs without importing any energy.

the next several lifetimes. The numbers, after all, are mind-boggling: We have more than 700 billion barrels of tight oil in the Bakken, Eagle Ford, and other known reserves—and that's just in the United States.

But, of course, just a tiny fraction of those totals is actually recoverable—somewhere between 1% and 2% in the Bakken and Eagle Ford shales, or between 7 and 14 billion barrels. "At the high end of the estimates, predicted production from Bakken and Eagle Ford together amounts to perhaps a two year oil supply for the United States at 2011 consumption rates," says Raymond T. Pierrehumbert, a professor and geophysicist at the University of Chicago. "That's significant, but not a game changer."

KEY CONCEPT 19.8

Fracking has positive economic impacts, but negative environmental and social impacts make it contentious.

↑ Fred Mayer of Candor, New York, uses a charcoal grill lighter to ignite water running from his kitchen faucet. His water has been contaminated with methane ever since a fracking operation began nearby.

AP Photo/The Post Standard, Mike Greenlar

INFOGRAPHIC 19.9   THE TRADE-OFFS OF FRACKING

↘ Fracking for oil and gas in shale deposits can allow us to extract large amounts of oil and natural gas, extending supplies and reducing our dependence on imports. But several trade-offs that must be considered when determining whether or how best to proceed with fracking.

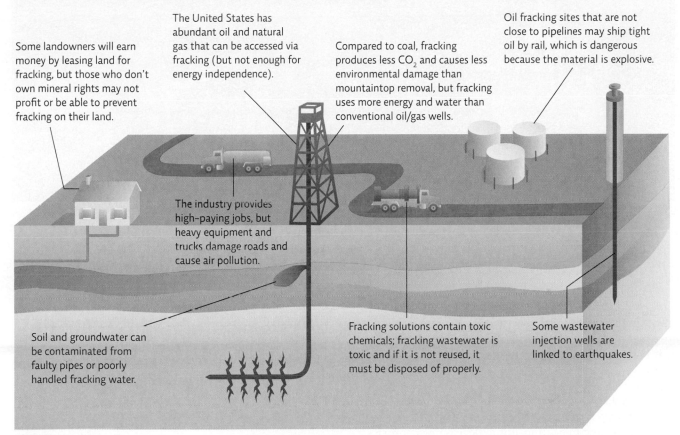

Some landowners will earn money by leasing land for fracking, but those who don't own mineral rights may not profit or be able to prevent fracking on their land.

The United States has abundant oil and natural gas that can be accessed via fracking (but not enough for energy independence).

Compared to coal, fracking produces less CO_2 and causes less environmental damage than mountaintop removal, but fracking uses more energy and water than conventional oil/gas wells.

Oil fracking sites that are not close to pipelines may ship tight oil by rail, which is dangerous because the material is explosive.

The industry provides high-paying jobs, but heavy equipment and trucks damage roads and cause air pollution.

Soil and groundwater can be contaminated from faulty pipes or poorly handled fracking water.

Fracking solutions contain toxic chemicals; fracking wastewater is toxic and if it is not reused, it must be disposed of properly.

Some wastewater injection wells are linked to earthquakes.

? The negative impacts of fracking are listed here. What are the advantages to the United States of fracking for unconventional sources of oil and natural gas? Do the advantages outweigh the disadvantages?

Because fracking wells can access only a small part of the oil deposit that surrounds them, many, many wells are needed to tap an entire deposit. This diminishes the EROEI because creating each new well requires an expenditure of energy. In addition, the productivity of any given well is likely to drop off rapidly after the first couple years and then trail off into nothing, slowly, over decades.

"Technological developments have made it possible to tap into tight oil," says Pierrehumbert. "But these are not the same kinds of technological developments that have given us ever more powerful computers and cellphones at ever declining prices. Oil production technology is giving us ever more expensive oil with ever diminishing returns for the ever increasing effort that needs to be invested."

Indeed, experts have begun to suggest that the Bakken oil fields may already be close to, if not slightly past, their peak. More and more wells are being drilled each year, and they are yielding less and less oil overall. "Tight oil is headed for a Red Queen's Race," says Pierrehumbert. "You have to keep drilling and drilling and drilling just to keep your production in the same place." **INFOGRAPHIC 19.9**

Oil consumption drives extraction.

Not all countries use energy equally. Until recently, the United States, for instance, used more energy than any other country. Its overall primary energy use tripled from 1949 to 2009, according to the EIA. In 2009, China surpassed the United States as the biggest consumer of energy, though on a per capita basis, the United States still leads the way, with each person using roughly three times as much energy as the average person in China. Most of this energy comes from fossil fuels—coal, oil, and natural gas, with crude oil as the dominant source. The Houston oil field services company Baker Hughes Inc. estimates that as of February 2011, the United States had 1,713

" *You have to keep drilling and drilling and drilling just to keep your production in the same place.* " *—Raymond T. Pierrehumbert*

↑ A fracking operation along U.S. Hwy 85 just south of Watford City, North Dakota.

conventional rigs searching the country for oil or gas. Nevertheless, we import most of what we use: In 2012, the United States produced only 12.4% of the world's oil but consumed 20.6% of it, importing much of our oil from Canada, Venezuela, and Saudi Arabia.

The good news is that in the United States, oil consumption is decreasing, according to a 2009 report from IHS Cambridge Energy Research Associates. The report predicts that because of changing demographics and improved vehicle efficiency, U.S. oil use peaked in 2005 and is now in decline. But in developing countries such as China and India—the world's second- and fourth-largest consumers of oil, respectively (with Japan coming in third)—oil consumption is still rapidly increasing, forcing the United States and other countries to extract as much oil as possible, regardless of the consequences.

Our economy and lifestyle are dependent on access to affordable energy, or **energy security**. Yet there are many reasons that our energy supplies might become unreliable or unaffordable. These include dwindling supplies, increasing demand, dependence on energy imports from politically unstable countries or those that might stop exporting energy resources to us for political reasons, competition from other countries, and a cartel or monopoly increasing prices or decreasing oil supplies.

In the United States, many of these problems already exist. As supplies dwindle and global demand continues to increase, easily extracted oil becomes less available, forcing companies to use more powerful and costly—and potentially more dangerous—technologies.

To be sure, how to best meet our future energy needs is a wicked

energy security Having access to enough reliable and affordable energy sources to meet one's needs.

problem—we need an abundant, reliable source of energy to power our societies. (See Chapter 1 for more on wicked problems.) But each choice comes with its own trade-offs. Phasing out the use of fossil fuels would certainly cut down on air pollution, climate change, and myriad other environmental problems. But the costs and availabilities of various non–fossil fuel sources, in some cases, make them less than ideal alternatives. Even by choosing the best of our admittedly problematic fossil fuel choices (and which

KEY CONCEPT 19.9

Accessing unconventional oil and natural gas reserves will not address long-term energy needs, but it may buy time to transition to more sustainable sources of energy.

↓ The oil boom in North Dakota has brought jobs and money to the area, but the high volume of truck traffic causes almost constant traffic congestion and dangerous driving conditions on two-lane roads that were not constructed to handle two-lane traffic heavy truck traffic.

Eugene Richards

is best is debatable), are we just delaying our transition to a carbon-free energy society? A lot is at stake in this wicked problem—huge investments and even bigger profits, the health and well-being of human communities and ecosystems, and the need to keep society running smoothly without facing an energy shortfall.

So how do we increase our energy security and reduce our dependence on fossil fuels so that we can avoid using dangerous extraction methods? There are many possibilities, but two important strategies are diversification and conservation. We can focus efforts on acquiring a variety of domestic sources of energy (including alternative fuels such as biofuels) and importing energy from multiple suppliers. At the same time, we can reduce energy needs through increased conservation and energy efficiency. (See Chapter 23 and Launchpad Chapter 32 for more on alternative energy sources and energy efficiency.) Iceland, for instance, is increasing its energy security by focusing on energy independence, such that it will meet all of its energy needs without imports by the year 2050.

In April 2014 the Bakken oil fields of western North Dakota reached a milestone: They yielded up their billionth barrel of crude oil. Only time will tell how many more barrels remain to be harvested and over how many more years. In the meantime, the greatest gift of Bakken may be the time it buys us to develop other domestic energy sources. "It will only do us good if we use this transitional period wisely," says Pierrehumbert. "We're in for a hard landing if we don't use our current prosperity to pave the way for a secure energy and climate future."

Select References:

Dobb, E. (2013). America strikes new oil. *National Geographic*, 223(3): 28–59.

Hughes, J. D. (2013). A reality check on the shale revolution. *Nature*, 494(7437): 307–308.

Jackson, R. B., et al. (2013). Increased stray gas abundance in a subset of drinking water wells near Marcellus shale gas extraction. *Proceedings of the National Academy of Sciences*, 110(28): 11250–11255.

Kelly, E. N., et al. (2010). Oil sands development contributes elements toxic at low concentrations to the Athabasca River and its tributaries. *Proceedings of the National Academy of Sciences*, 107(37): 16178–16183.

Keranen, K. M., et al. (2014). Sharp increase in central Oklahoma seismicity since 2008 induced by massive wastewater injection. *Science*, DOI:10.1126/science.1255802.

Malakoff, D. (2014). The gas surge. *Science*, 344 (6191): 1464–1467.

Stolstad, E. (2014). Will fracking put too much fizz in your water? *Science*, 344 (6191): 1468–1471.

BRING IT HOME

PERSONAL CHOICES THAT HELP

Environmental and health concerns that surround our acquisition and use of fossil fuels continue to grow as we dig deeper and use more extreme methods to obtain oil and natural gas, highlighting our dependence on nonrenewable energy resources. By decreasing our oil and natural gas use, we can reduce the pressure on oil companies to pursue sources of oil that have a greater potential for environmental damage such as deep-water drilling or oil and natural gas deposits like those in the tar sands and shales.

Individual Steps

• Minimize your fuel use when driving by planning ahead to condense shopping trips and errands and reduce total miles driven and by parking as far as you can from your entrance and getting some exercise instead of wasting gas. Try not driving: See where you can safely walk, bike, or use public transport.

• Reduce your use of disposable plastics like water bottles and single-serve food containers and always recycle plastics when possible.

• Turn down your thermostat in the winter to reduce your energy use.

Group Action

• Organize a carpool system in your community to reduce the number of single-passenger car trips.

• Organize a screening of a documentary on energy such as *Tipping Point: The Age of the Oil and Sands, Gasland* or *Gasland 2*, or *FrackNation*. Discuss the positions presented and evaluate the evidence given in support of claims that are made. Do you detect any bias?

Policy Change

• Work with your school's administration to encourage public transportation and increased bike usage on your campus.

• If you live in a state where fracking occurs, research fracking regulations in your state and voice your opinion about existing or proposed regulations.

© Jan Sonnenmair/Aurora Photos/Alamy

ENVIRONMENTAL LITERACY UNDERSTANDING THE ISSUE

1 How are fossil fuels formed, and why are they considered nonrenewable resources? Where are current proven oil and natural gas reserves found, and what is meant by "peak oil"?

INFOGRAPHICS 19.1, 19.2, AND 19.3

1. What is the correct definition of fossil fuel *proven reserves*?
   a. Humanmade storage areas for emergency supplies of fossil fuels
   b. Areas where there are known stores of untapped fossil fuels held in reserve for the future
   c. Known sources of fossil fuels that are both economically and technologically recoverable
   d. Sources of fossil fuels that have not yet been discovered but that geologists have predicted to exist

2. The world will reach "peak oil" when:
   a. we have managed to find and tap all oil reserves.
   b. oil production reaches its highest levels and then starts to decline.
   c. we develop techniques to extract oil without environmental damage.
   d. oil can be extracted from all hard-to-reach sources.

3. If oil and natural gas are formed from once-living organisms, why are they considered to be nonrenewable resources?

2 How are conventional oil and natural gas extracted? How is crude oil refined and used?

INFOGRAPHICS 19.4 AND 19.5

4. True or False: On average, 95% to 100% of the oil in a conventional oil reserve can be extracted by using primary, secondary, and tertiary techniques.

5. Examine Infographic 19.5. According to the data, which of the following has the highest boiling point?
   a. Jet fuel
   b. Gasoline
   c. Naphtha
   d. Fuel oil

6. Describe the process of conventional oil extraction, including primary production, secondary production, and tertiary production.

3 What are some of the environmental costs of finding, extracting, and using oil and natural gas from conventional reserves?

INFOGRAPHIC 19.6

7. True or False: If we internalized all the costs of finding, extracting, and using oil, the prices of fuel and petroleum-based products would go up.

8. Burning any fossil fuel generates air pollution, but the fossil fuel that releases the least amount of air pollution when burned is:
   a. coal.
   b. oil.
   c. natural gas.
   d. tar sands oil.

9. Compare the environmental impacts of finding, extracting, and using oil and rank these steps in terms of environmental damage and threats to human well-being. Which step would you address first in an effort to reduce the impact of finding, extracting, and using oil? Why?

4 What unconventional sources of oil and natural gas exist, and how are they extracted?

INFOGRAPHICS 19.7 AND 19.8

10. Why is the United States so interested in extracting tight oil from shale deposits?
    a. It has abundant supplies.
    b. These deposits are easier to extract than conventional oil supplies.
    c. Shipping tight oil is easier and safer than shipping conventional crude oil.
    d. Tight oil deposits are not contaminated with methane.

11. Which unconventional oil currently being commercially produced has the lowest EROEI and the highest carbon footprint?
    a. Tight oil
    b. Tar sands oil
    c. Crude oil
    d. Kerogen shale

12. Explain the process of fracking. Why is fracking used for some oil and natural gas reserves and not others?

5 What are the trade-offs of pursuing unconventional oil and natural gas sources? What role do you think these resources should play in meeting future energy needs?

INFOGRAPHIC 19.9

13. True or False: Fracking uses a lot of water, but the wastewater produced is either recycled and used again or purified and pumped back into aquifers.

14. Some homeowners suspect that methane-contaminated water is linked to nearby fracking wells. If this is true, what is the likely source of this contamination?
    a. Leaky pipes near groundwater supplies or improperly handled fracking fluid
    b. The initial drilling of the well
    c. Installation of the well pipes and concrete casings
    d. Detonation of explosive charges in the horizontal sections of the well

15. Distinguish between the actions a country might take when pursuing energy independence versus energy security.

16. Since conventional sources of fossil fuels are running out, do you think accessing our abundant unconventional fossil fuels is the best path the United States or the world can take for energy security? Explain.

SCIENCE LITERACY **WORKING WITH DATA**

Methane (CH_4) can normally occur in upper levels of the soil because it is produced by bacteria and can accumulate to high levels over time in some areas. Ethane (C_2H_6), however, is only created under high temperature and pressure, such as might occur deep underground at the site of shale gas deposits. Robert Jackson and colleagues suspected that methane contamination in water from wells in northeastern Pennsylvania might be linked to nearby fracking gas wells. For their study, they asked the question "Does being close to fracking gas wells increase the likelihood that the water will contain methane or ethane contamination?" They measured the amounts of methane and ethane in 141 water wells close to and far away from fracking wells. Look at Graph A and Graph B and answer the following questions.

Graph A: Methane Concentrations in Water Wells

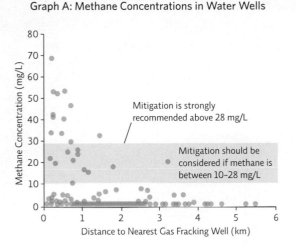

Graph B: Ethane Concentrations in Water Wells

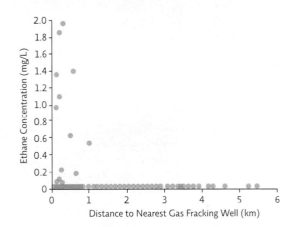

Interpretation

1. What was the hypothesis that these researchers tested in this experiment?

2. For each chemical (methane and ethane), identify the relationship between its level in well water and the water well's proximity to fracking wells. Do these relationships represent causation or correlation? Explain.

3. How many water wells had levels of methane that fell within the range where mitigation should be considered? How many fell within the range where mitigation is strongly recommended?

Advance Your Thinking

4. Do all wells that are very close to a fracking well (within 0.5 kilometer) show evidence of methane or ethane contamination? (Are there any close-by wells with no contamination?) Does this negate the conclusion that methane and ethane contamination are due to fracking in the area?

5. Why might methane be high in water even in areas that are not close to fracking wells?

6. Why did researchers need to measure ethane in water wells to strengthen their conclusion that the water contamination is linked to fracking activity?

INFORMATION LITERACY **EVALUATING INFORMATION**

The United States produces a lot of oil but still imports a lot from Canada, our biggest source of imported oil. For the most part, this is tar sands oil from Alberta. Transport of the mining products is accomplished through the transcontinental Keystone pipeline that currently runs from Alberta into the United States, but it does not continue on all the way to refineries on the Gulf Coast. Oil companies would like to expand the pipeline and have proposed the Keystone XL, which would transport oil from Canada to Texas.

Search for information about the Keystone XL pipeline and other proposed ways to ship tar sands crude oil. Evaluate the articles you read and answer the following questions:

1. Visit at least three websites that *advocate* for the construction of the Keystone XL pipeline and three websites that *oppose* it. For each website, answer the following questions about the organization or authors responsible for the information you read:
   a. Who runs the website or wrote the article? Do the credentials of this individual or group make the person or group reliable or unreliable? Explain.
   b. Does this individual or group have a clear and transparent agenda?
   c. Do you detect any bias or logical fallacies? If so, identify them.
   d. Is this a reliable information source? Explain.

2. Briefly summarize what you believe are the most important pros and cons of the Keystone XL pipeline. Based on your reading, what decision would you suggest that the president of the Unites States make regarding this project? Justify your conclusions.

Find an additional case study online at http://www.macmillanhighered.com/launchpad/saes2e

CORE MESSAGE

Air quality issues span the globe and have serious health effects for humans and other organisms; they also take an economic toll on individuals, industry, and society. Though there are natural sources of air pollution, most of it is caused by human actions. Pollutants generated in one area can travel great distances to affect other places. Policies that restrict air pollution and new technology to reduce it at the source can help us address air pollution problems.

AFTER READING THIS
CHAPTER, YOU SHOULD
BE ABLE TO ANSWER
THE FOLLOWING
GUIDING QUESTIONS

1

What is air pollution, and what is its global impact? What are the main types and sources of outdoor air pollution?

2

What are the health, economic, social, and ecological consequences of air pollution?

THE YOUNGEST SCIENTISTS

Kids on the frontlines of asthma research

With almost 7 million children affected, asthma is the most common chronic illness among children in the United States.

© 13/ballyscanlon/Ocean/Corbis

3

What are the causes and consequences of acid deposition, and how is it an example of transboundary pollution?

4

What are the main sources of indoor air pollution, and what can be done to reduce these pollutants?

5

How can air pollution be reduced, and what are the trade-offs of reducing it?

For 10-day stretches during 2003 and 2004, 45 asthmatic kids from smoggy Los Angeles County, some as young as 9, carried more than books in their backpacks. The kids, from two regions of the county, wore backpacks containing small monitors that sampled the air around them continuously as they went about their daily lives—going to class, playing with friends, having dinner with their families.

Children have a lot to teach us about **asthma**. It's not just that the respiratory disorder more commonly afflicts kids than adults—which it does. (1 in 10 American children has asthma, compared to about 1 in 13 adults.) Or that research on children with asthma is on the rise. No, in fact, children are actually conducting some of the seminal research on asthma themselves.

And those personal air monitors are far more accurate than measurements taken at local monitoring stations, explains Ralph Delfino, an epidemiologist at the University of California, Irvine, who with his colleagues recruited the students to help collect the data. "A monitoring station can be many miles from where the subject lives, where they go to school, etc.—so that measurement may not represent their actual exposure very well," he says. The air monitors used in Defino's study detected levels of harmful pollutants in the air surrounding each individual child.

The children selected for the experiment were all currently being treated for mild to moderate asthma, and the area they all lived in had significant vehicle air pollution. Each child wore a backpack containing a monitor that would continuously sample the air around the child for 10 consecutive days; the monitor measured the amount of small particles and nitrogen dioxide (NO_2), both of which are commonly present in vehicle emissions and are known to irritate lung tissue. Exposure to these pollutants can trigger asthma symptoms such as wheezing, coughing, or shortness of breath in sensitive individuals.

Ten times a day, the children exhaled into a special bag that assessed their breath for nitric oxide (NO), a chemical marker of airway inflammation—a telltale symptom of asthma.

When the 10 days were over, Delfino compared the types of pollutants to which the kids were exposed with their nitric oxide levels at similar points in time. The goal was to paint a picture of the types and amounts of pollutants that exacerbate asthma.

asthma A chronic inflammatory respiratory disorder characterized by "attacks" during which the airways narrow, making it hard to breathe; can be fatal.

air pollution Any material added to the atmosphere (naturally or by humans) that harms living organisms, affects the climate, or impacts structures.

◉ **WHERE IS LOS ANGELES, CALIFORNIA?**

Asthma is a respiratory ailment marked by inflammation and constriction of the narrow airways of the lungs. Delfino's work is important in part because asthma is one of the most common chronic childhood diseases in the United States and other developed nations, and it's a major cause of childhood disability. The United States and the United Kingdom have the highest incidence of asthma, with more than 8% of each population diagnosed as asthmatic in 2011. Developing nations are also seeing a rise in asthma, especially in urban centers. In all areas, asthma rates are likely underdiagnosed.

In the United States, asthma is the leading cause of school absences and, hence, lost revenue for public schools, whose federal funding is based on attendance. The prevalence of childhood asthma more than doubled from 1980 to the mid-1990s, and though it has leveled off in recent years, with 6.8 million children and 18.7 million adults diagnosed, it still remains at historically high levels. (Because there are more adults than children, overall more adults have asthma, but a larger percentage of children have the disease.)

Researchers believe that **air pollution**—contaminants, either from natural sources or human activities, that cause health or environmental problems—may play a key role in the recent asthma spike. In addition, air pollution has been linked to cancer, infection, and other respiratory diseases—and it not only harms humans but

The Los Angeles skyline is
obscured by smog.
Justin Lambert/ Getty Images

also plants, other animals, and even buildings, bridges, and statues.

As far back as the 1930s, scientists recognized a link between outdoor air pollution and human illness. In 1930, for instance, 63 people died and 1,000 were sickened in Belgium when a temperature inversion—a situation that occurs when the temperature is higher in upper regions of the atmosphere than in the lower, causing pollutants to become trapped near Earth's surface—led to a sudden spike in lower atmospheric sulfur levels. And 1952 was the year of the famous Great Smog in London, England, when pollutants trapped in the lower atmosphere killed 4,000 people. **INFOGRAPHIC 20.1**

There are many different types of outdoor air pollution.

In many urban areas and in developed nations today, much of the air pollution comes from vehicle exhaust

and industry emissions, including emissions from coal-fired power plants.

While less-developed nations do have outdoor pollution, their biggest problem is often indoor air pollution from small particles released through burning solid fuels such as charcoal, wood, or animal waste. However, only in the past 25 years have scientists really been able to tease out the link between air pollution and asthma.

Outdoor air pollution includes chemicals and small particles in the atmosphere that can be either natural in origin—arising from natural events like sandstorms, volcanic eruptions, or wildfires—or come from humans—such as pollution released from factories

> ### KEY CONCEPT 20.1
>
> Air pollution is a serious problem that causes millions of deaths worldwide each year.

INFOGRAPHIC 20.1   AIR POLLUTION IS A WORLDWIDE PROBLEM   1

↓ The World Health Organization (WHO) recognizes air pollution as a major threat to human health, especially cardiovascular disease (such as coronary heart disease and stroke) and respiratory disease (including chronic obstructive pulmonary disease [COPD], lung cancer, asthma, and respiratory infections). Outdoor air pollution caused and estimated 3.7 million premature deaths in 2012; indoor air pollution caused 4.3 million premature deaths.

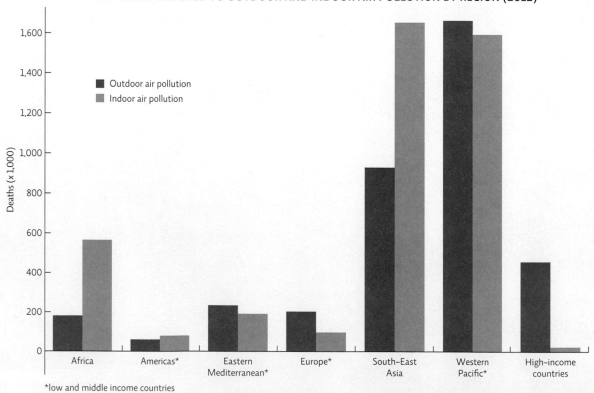

DEATHS ATTRIBUTED TO OUTDOOR AND INDOOR AIR POLLUTION BY REGION (2012)

 It is easy to see how breathing unhealthy air affects the respiratory system, but why does air pollution affect the cardiovascular system?

↑ Controlled burns of agriculture fields, like this one of an asparagus field in California, help clear land for more planting but release particulate matter (small particles) into the air, contributing to respiratory distress in sensitive individuals.

and vehicles during the combustion of fossil fuels or from burning any biomass such as wood, crop waste, or garbage. Of these anthropogenic sources, **primary air pollutants** are pollutants released directly from both mobile sources (such as cars) and stationary sources (such as industrial plants). In addition, some primary air pollutants react with one another or with other chemicals in the air to form **secondary air pollutants.** For example, **ground-level ozone** forms when nitrogen oxides (NO and NO₂—together expressed as NOx) released during fossil fuel combustion react with atmospheric oxygen in the presence of sunlight. **Smog**—a term that's a combination of the words *smoke* and *fog*—is hazy air pollution that contains a variety of primary and secondary pollutants.

Pollution can also move from the troposphere (closest to Earth) up into the stratosphere, a much thinner layer of the atmosphere that extends from 11 to 50 kilometers (7 to 31 miles) above Earth. This region contains the "ozone layer," an area with high ozone (O_3) concentrations. As we saw in Chapter 2, ozone in the stratosphere is significant because it serves as Earth's sunscreen, blocking some of the dangerous ultraviolet (UV) radiation from the Sun. Air pollution can have grave impacts on this layer; for instance, chlorofluorocarbons (CFCs)—compounds that contain carbon, chlorine, and fluorine—can travel up into the stratosphere and destroy ozone.

Don't confuse ground-level ozone pollution with stratospheric ozone depletion. These are two very different problems, though they deal with the same molecule—O_3. Ozone in the stratosphere is a good thing, but ozone at ground level is a problem—you don't want to breathe it in as it can directly damage the sensitive tissue of the lungs. Even plants are damaged by the corrosive action of ozone.

KEY CONCEPT 20.2

Air pollution is caused by natural and anthropogenic sources and includes emissions that are directly harmful (primary pollutants) and those that are converted to harmful forms (secondary pollutants).

primary air pollutants Air pollutants released directly from both mobile sources (such as cars) and stationary sources (such as industrial and power plants).

secondary air pollutants Air pollutants formed when primary air pollutants react with one another or with other chemicals in the air.

ground-level ozone A secondary pollutant that forms when some of the pollutants released during fossil fuel combustion react with atmospheric oxygen in the presence of sunlight.

smog Hazy air pollution that contains a variety of pollutants, including sulfur dioxide, nitrogen oxides, tropospheric ozone, and particulates.

↓ There are many sources of outdoor air pollution, both natural and anthropogenic. These sources release primary pollutants, some of which may be converted to different chemicals (secondary pollutants). Prevailing winds transport pollution that reaches the upper troposphere or stratosphere around the globe; no area is immune to air pollution. Agricultural and industrial pollutants have been found in Arctic and Antarctic air, delivered by these prevailing winds.

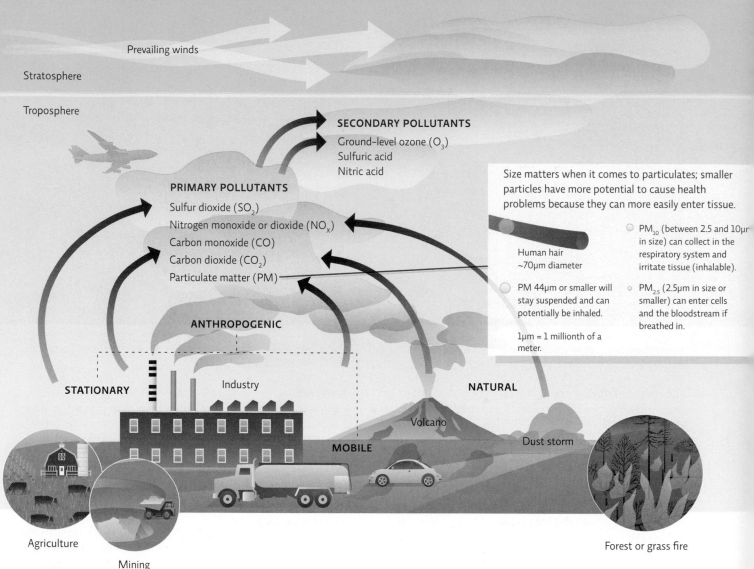

Prevailing winds

Stratosphere

Troposphere

SECONDARY POLLUTANTS

Ground-level ozone (O_3)
Sulfuric acid
Nitric acid

PRIMARY POLLUTANTS

Sulfur dioxide (SO_2)
Nitrogen monoxide or dioxide (NO_x)
Carbon monoxide (CO)
Carbon dioxide (CO_2)
Particulate matter (PM)

ANTHROPOGENIC

STATIONARY

Industry

NATURAL

Volcano

MOBILE

Dust storm

Agriculture

Mining

Forest or grass fire

Size matters when it comes to particulates; smaller particles have more potential to cause health problems because they can more easily enter tissue.

Human hair ~70μm diameter

PM 44μm or smaller will stay suspended and can potentially be inhaled.

1μm = 1 millionth of a meter.

PM_{10} (between 2.5 and 10μr in size) can collect in the respiratory system and irritate tissue (inhalable).

$PM_{2.5}$ (2.5μm in size or smaller) can enter cells and the bloodstream if breathed in.

 Identify the sources of pollution shown in this infographic as either natural or anthropogenic.

Air pollution is responsible for myriad health and environmental problems.

Outdoor air pollution, in all forms, is one of the most dramatic contributors to asthma. When the Environmental Protection Agency (EPA) started regulating air pollution, the agency did not know the extent to which air pollution could affect health; in fact, the EPA administrator noted at the time that the agency's clean air regulations were "based on investigations conducted at the outer limits of our capability to measure connections between levels of pollution and effects on man." Nevertheless, in 1971, the EPA set standards for the most common but problematic pollutants—known as *criteria air pollutants*. Five of these were chemical air pollutants: carbon monoxide, sulfur dioxide (SO_2), nitrogen oxides (NO_x), lead, and the secondary pollutant ground-level ozone; standards were also set for **particulate matter (PM)**—particles or droplets small enough to remain aloft in the air for long periods of time. Although all particulates reduce visibility, it is the smallest particles—those with a diameter less than 2.5 micrometers (μm), about 1/40 the diameter of a human hair—that aggravate asthma and other chronic lung diseases and increase the risk for death. **INFOGRAPHIC 20.2**

With the exception of lead (whose levels have dropped significantly since the passage of the Clean Air Act), these criteria pollutants are still considered the most problematic and health-threatening pollutants in the United States. A 1993 study published by Harvard University researchers helped establish the link between air pollution and impaired health. A follow-up study in 2006 estimated that particulate pollution—which includes soot, ash, dust, smoke, pollen, and small, suspended droplets (aerosols)—accounts for 75,000 premature deaths per year and showed a clear dose–response effect: The higher the air pollution, the higher the risk for death. **INFOGRAPHIC 20.3**

The World Health Organization (WHO) estimates that 8 million people die prematurely each year as a result of exposure to air pollution. Respiratory ailments are common because particulates from soot and smog damage respiratory tissue and increase susceptibility to infection. In addition, research has shown that asthma rates are higher in people who breathe polluted air—and children are particularly sensitive because they breathe in more air for their size than adults do and because developing tissue is more vulnerable.

Delfino's team of researchers followed up their first study with a second in 2007 that included 53 students with asthma (ages 9–18) with air monitors strapped to their backs. The students also had to breathe into detectors that recorded how much air they were able to blow out from their lungs at once—a measure of lung function. The results suggested that high levels of particulate pollution actually decrease lung function. More recent studies have shown a positive correlation between the amount of air pollution exposure and lung inflammation as well as the number of hospital visits.

Lungs are particularly vulnerable to these small particulates because they get so much exposure (we breathe all the time) and the tissue itself is delicate. Irritants like particles, dust, and pollen can cause the lungs to produce excess mucus in an attempt to trap and expel the irritant. The lining of the airways can become inflamed and swell; in people with asthma, the irritation may trigger muscle contractions that close off the airway completely. Particles smaller than 2.5 μm can actually penetrate cells of the lungs or enter the

KEY CONCEPT 20.3

Air pollution causes health problems, damages structures, reduces visibility, and contributes to stratospheric ozone depletion and climate change.

particulate matter (PM) Particles or droplets small enough to remain aloft in the air for long periods of time.

INFOGRAPHIC 20.3   THE HARVARD SIX CITIES STUDY LINKED AIR POLLUTION TO HEALTH PROBLEMS 2

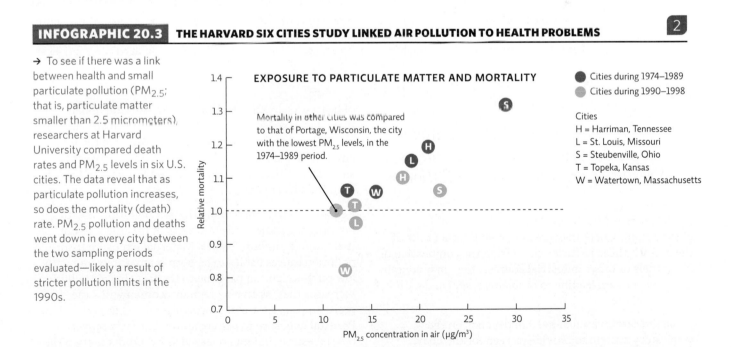

→ To see if there was a link between health and small particulate pollution (PM$_{2.5}$; that is, particulate matter smaller than 2.5 micrometers), researchers at Harvard University compared death rates and PM$_{2.5}$ levels in six U.S. cities. The data reveal that as particulate pollution increases, so does the mortality (death) rate. PM$_{2.5}$ pollution and deaths went down in every city between the two sampling periods evaluated—likely a result of stricter pollution limits in the 1990s.

EXPOSURE TO PARTICULATE MATTER AND MORTALITY

Mortality in other cities was compared to that of Portage, Wisconsin, the city with the lowest PM$_{2.5}$ levels, in the 1974–1989 period.

- Cities during 1974–1989
- Cities during 1990–1998

Cities
H = Harriman, Tennessee
L = St. Louis, Missouri
S = Steubenville, Ohio
T = Topeka, Kansas
W = Watertown, Massachusetts

Relative mortality (y-axis)
PM$_{2.5}$ concentration in air (μg/m³) (x-axis)

In this graph, what does it mean to say that the relative mortality rate of Steubenville, Ohio, was 1.3? Identify the concentration of particulate matter in Steubenville, Ohio, during the first time period (1974-1989) and the morality relative to the comparison city, Portage, Wisconsin. Now do the same for Steubenville in the second time period (1990-1998). How much did each parameter change over the course of the study? For every 100 people who died in Portage between 1990-1998, how many died in Steubenville?

Chang W. Lee/The New York Times/Redux

↑ Derrick Reliford, 14, in his Bronx, New York, home. At 10 years old, he participated in the New York University study to measure how much pollution Bronx children were exposed to. The researchers found that students in the South Bronx were twice as likely to attend a school near a highway as were children in other parts of the city. The South Bronx is home to some of the highest asthma hospitalization rates for children in New York City.

The World Health Organization (WHO) estimates that 8 million people die prematurely each year as a result of exposure to air pollution.

bloodstream, where they are delivered to other cells of the body. If these particles come from the combustion of fossil fuels or other industrial sources, they may contain toxic substances, leading to problems associated with toxic exposure.

Since the cardiovascular system depends on the respiratory system to provide oxygen for the body, anything that impairs the lungs also harms the cardiovascular system, which might explain the fact that people living in polluted areas also have higher rates

of heart attacks and strokes. Cancer rates are higher in people exposed to air pollution, too; exposure to secondhand smoke and exposure to radon are the leading environmental causes of lung cancer, and exposure to smog and vehicle emissions is linked to increased risk for lung and breast cancer.

Researchers have also found that air pollution affects not only the adults and children who breathe it in but it also affects unborn babies. Maternal exposure to air pollution causes babies to be born prematurely and with low birth weight, in part due to the fact that air pollution increases the risk that a pregnant woman will suffer from preeclampsia, a form of pregnancy-related hypertension. Prenatal exposure to *polycyclic aromatic hydrocarbons* (PAHs)—air pollutants released as by-products from the combustion of wood, tobacco, coal, or diesel—is linked to birth defects. "We have learned that PAHs end up in the placenta and fetus," explains Beate Ritz, an epidemiologist at the University of California, Los Angeles, School of Public Health.

But humans aren't the only creatures suffering the ill effects of air pollution. Many animals suffer the same respiratory distress as humans: All lung (and gill) tissue is very vulnerable to air pollution. Invertebrates as a group, especially aquatic ones, seem to be more directly impacted by air pollution (toxic effects or reproductive declines) than are vertebrates. But aquatic vertebrates like fish and amphibians are certainly feeling severe impacts. Declines in North Atlantic salmon populations have been linked to air pollution–induced water acidity. The higher acidity causes aluminum to build up in water; aluminum is toxic to fish, especially juvenile salmon.

Plant tissues are also vulnerable to pollutants like smog and ozone, which cause direct damage to sensitive cell membranes. Exposure can damage a leaf's ability to photosynthesize, preventing healthy growth and compromising its survival. Lichens are particularly vulnerable to air pollution such as nitrogen emissions (NO_x), and their decline is seen as a warning of the potential for damage to forests and crops. Together with changes in soil chemistry— which can hinder plant growth—pollution damage to plant crops will ultimately cause global crop yields to fall. Estimates of crop yields predict decreases for soybean, wheat, and maize ranging from 4% to 26% by 2030, amounting to dollar losses in the billions.

Finally, pollution damages buildings and monuments. Acid in polluted rain literally eats away at limestone and marble structures; it can etch glass and damage steel and concrete, causing billions of dollars of damage per year. Smog, SO_2, and ground-level ozone pollution also lower visibility by creating haze, a concern for areas that depend on tourism. On hazy days, for instance, it can be impossible to see across the Grand Canyon.

Outdoor air pollution has many sources.

Where does outdoor air pollution come from? The burning of fossil fuels commercially, industrially, and residentially contributes heavily to outdoor air pollution. Coal and oil burning releases emissions that can produce smog that contains a variety of pollutants, including SO_2, NO_x, particulates, and ground-level ozone.

Factories, incinerators, and mining operations also release pollutants. Industrial pollution releases **point source pollution**, so named because it is possible to identify the pollution's exact point of entry into the environment. Hypothetically, point source pollution is easier to monitor and regulate than is **nonpoint source pollution**—pollution from dispersed or mobile sources like vehicles and lawn mowers.

Agriculture is yet another source of nonpoint source outdoor pollution. Toxic pesticides sprayed on crops can become airborne and drift as far as 30 kilometers (20 miles); confined animal feeding operations produce significant odor problems and particulate pollution; and animal waste contributes to global warming by releasing the greenhouse gas methane.

In his study in Los Angeles County, Delfino found that particles contained in diesel fuel exhaust were among the worst asthma culprits. In addition, particulate levels were higher in Riverside, one of two regions he tested; the researchers concluded that Riverside had more pollution because it was downwind of the main urban areas in Los Angeles.

In addition to the 6 criteria pollutants, the EPA also recognizes 187 hazardous air pollutants that can have adverse effects on human health, even in small doses. These toxic substances may cause cancer or developmental defects, or they may damage the central nervous system or other body tissues. They include *volatile organic compounds* (VOCs), a variety of chemicals that readily evaporate but don't dissolve in water. VOCs are released by natural sources such as wetlands, and household products including paint, carpets, and cleaners; the main outdoor source is fossil fuel combustion. And in 2007, the EPA ruled that greenhouse gases are air pollutants, giving the agency the authority to regulate carbon dioxide (CO_2) emissions. **TABLE 20.1**

KEY CONCEPT 20.4

In addition to the criteria pollutants, the EPA regulates VOCs and mercury, and it is developing a program to regulate CO_2, an air pollutant linked to climate change.

The air we breathe affects our lungs, especially those of children.

If anyone was born to study air pollution, it was Kari Nadeau. Growing up near smoggy Newark, New Jersey, she suffered as a child from terrible asthma and allergies, which she always suspected were related to the

point source pollution Pollution that enters the air from a readily identifiable source such as a smokestack.

nonpoint source pollution Pollution that enters the air from dispersed or mobile sources.

TABLE 20.1 SOURCES AND EFFECTS OF AIR POLLUTANTS

Criteria Air Pollutant	Source	Health/Environmental Effects
Carbon monoxide (CO) From incomplete combustion of any carbon-based fuel (and most combustion is incomplete)	Vehicles, forest fires, volcanoes	Interferes with red blood cells' ability to carry oxygen; causes headaches and can lead to asphyxia (death)
Sulfur dioxide (SO_2) From natural sources and fossil fuel combustion.	Industry, volcanoes, dust	Respiratory irritant; harms plant tissue and can be converted to sulfuric acid, which damages plants, aquatic creatures, and concrete structures
Nitrogen oxides (NO_x = NO and NO_2) From the reaction of nitrogen in fuel or air with oxygen at high temperatures (usually during combustion of a fuel)	Vehicles, industry, nitrification by soil and aquatic bacteria	Respiratory irritant; can increase susceptibility to infection; can be converted to nitric acid, which damages plants, aquatic organisms, and even concrete structures; overfertilizes ecosystems; can cause eutrophication (see Chapter 15)
Ground-level ozone (O_3) Formed from reactions between NOx and VOC with oxygen in the presence of sunlight	Vehicles are the main source of NOx; VOCs can be released from manufactured products or can be released directly by industry. Some are also released by trees and other plants.	Respiratory irritant that reduces overall lung function; can reduce photosynthesis in plants
Particulate matter (PM) Tiny airborne particles or droplets, smaller than 44 micrometers. The smaller the particle, the more dangerous it is for tissue.	Released during the combustion of any fuel or activity that produces dust; also produced by forest fires, dust storms, and even sea spray	Respiratory irritant; can reduce respiratory and cardiovascular function; reduces visibility; particles can end up in aquatic or terrestrial ecosystem supplying nutrients or acids that can harm organisms that live there
Lead (Pb) Additive to gasoline, paint, and other solvents; phased out of the U.S. gas supply in the 1970s and officially banned in 1996	Lead-based paint in older homes and from other countries; leaded gasoline; soil erosion and volcanoes	Damages nervous, excretory, immune, reproductive, and cardiovascular systems; can accumulate in soils and in the tissues of organisms and can biomagnify up a food chain (see Chapter 3)
Other EPA-Regulated Air Pollutants	**Source**	**Health/Environmental Effects**
Volatile organic compounds (VOCs) Organic molecules (hydrocarbons) that easily evaporate	Solvents, paints, glues, and other organic chemicals; plants naturally release VOCs	Those from human sources can be directly toxic or disruptive to living organisms, including humans; contribute to ground-level ozone formation
Mercury (Hg) Naturally occurring element	Burning coal; mining and smelting operations; forest fires and volcanoes	A major neurotoxin that can disrupt development in embryos and young children; can bioaccumulate in individuals and biomagnify up the food chain
Carbon dioxide (CO_2)	Burning carbon-based fuels such as fossil fuels; forest fires and normal decomposition	Nontoxic so no health effects at normal levels of exposure; greenhouse gas that contributes to climate change, affecting ecosystems worldwide

 What types of pollution listed here are you exposed to on a daily basis, and what are their sources?

pollution surrounding her. Her mom was a public health school nurse, and her dad worked for the EPA. "I always had these questions lingering about how much the environment affects people with asthma or allergies," she recalls. Was air pollution a culprit in her respiratory woes? Her instincts told her yes.

And now her research does, too. An associate professor of pediatric immunology and allergy at Stanford University, Nadeau recently uncovered something surprising: When she looked at the blood collected from kids who came to her clinic from Fresno, California—kids who frequently had terrible asthma—she saw that they had different-looking immune systems than other California kids. Specifically, their immune systems' "peacekeeping" functions didn't work as well in keeping asthma-producing inflammation at bay. This observation piqued her curiosity: "I thought, what's different in Fresno? So I went on the Internet and searched and saw that Fresno is the second-most polluted city in the country," she says.

Nadeau is now collaborating with scientists in Fresno to study the link between immune system regulatory cells, asthma, and air pollution. To understand the differences among kids from Fresno and other areas in California, she collected blood from 71 asthmatic children who had spent their entire lives in Fresno, 30 healthy children from Fresno, 40 asthmatic kids from less-polluted Palo Alto, and 40 healthy kids from Palo Alto. She found that all the kids who grew up in Fresno had far higher levels of common pollutants in their blood than did the Palo Alto kids—and the higher their pollutant levels were, the more likely they were to have asthma.

Based on her research, Nadeau thinks that air pollutants stifle the activity of genes responsible for maintaining normal immune system function—and that by doing so, they increase the risk for asthma and allergies. Indeed, in a 2012 study, Nadeau and colleagues looked at pairs of identical (monozygotic) twins—one with and one without asthma. Because identical twins share the same genes, differences (such as the immune system function or occurrence of asthma) can be attributed to differences in their environment. The researchers found suppression of key immune system factors in each twin with asthma compared to his or her non-asthmatic sibling. A similar result was found for twins exposed to secondhand smoke (another significant air pollutant) compared to those not exposed. Twin studies such as these are invaluable in clearly showing a link between environmental exposure to air pollution and health problems.

Other studies also support the link between asthma symptoms and air pollution. More children in the South Bronx are hospitalized for asthma than anywhere else in New York State, and since many Bronx children live or attend schools adjacent to congested highways, Bronx Congressman José Serrano wondered if the two factors might be related. In 2002, he asked New York University environmental scientist George Thurston if he would be willing to conduct a study to find out. "We thought about it for a nanosecond, and then said, 'sure,'" Thurston recalls. In a study reminiscent of Delfino's, Thurston recruited 40 South Bronx fifth graders to tote wheeled backpacks containing personal air monitors for a month while rating their respiratory symptoms three times a day. "You rolled it, so it wasn't really that heavy," Derrick Reliford, one of the students in the study told the *New York Times*. "They were the rock stars of the class—everybody wanted to help them with the backpacks," Thurston recalls.

The children came from four different schools, two of which were close to a highway and two of which were not. Thurston found that, sure enough, the children who went to schools or lived closer to highways were exposed to more air pollution—in particular, diesel fuel exhaust—and they also had more severe respiratory symptoms.

Low-income or minority areas often have some of the worst air. This raises questions of **environmental justice**—the concept that access to a clean, healthy environment is a basic human right. Sources of major pollution like power plants or waste incinerators are often placed in areas where residents have less ability to fight for their rights—less money, less education, little or no voice in local government. In some cases, even when socioeconomic status is accounted for, minority communities still face more exposure to pollution than average, an example of **environmental racism**. A 2002 study conducted in southern California by Brown University researcher Rachel Morello-Frosch found that a person's risk for developing cancer from exposure to polluted air increased as income decreased. And, in general, cancer risk was higher for minorities (Asian, African American, Latino) than for the majority (Caucasians), no matter what the income level.

Children of low-income families are at particular risk: As in the Bronx, their homes and schools are near major roads or factories, and they often come and go to school during rush hour,

KEY CONCEPT 20.5

Air pollution is often especially bad in minority and low-income areas, raising concerns that it is an environmental justice issue.

environmental justice The concept that access to a clean, healthy environment is a basic human right.

environmental racism A form of racism that occurs when minority communities face more exposure to pollution than average for the region.

KEY CONCEPT 20.6

Acid deposition is a secondary pollutant that results from fossil fuel burning. This pollution can travel long distances and can harm plants and animals that are exposed.

---

acid deposition Precipitation that contains sulfuric or nitric acid; dry particles may also fall and become acidified once they mix with water.

transboundary pollution Pollution that is produced in one area but falls in a different state or nation.

when traffic is heaviest and smog forms.

Traveling pollution has far-reaching impacts.

One of the most problematic characteristics of air pollution is that it moves. Air pollution produced in one city can end up harming humans and other species halfway around the globe. For example, as much as half of the air pollution that falls on the Great Smoky Mountains of Tennessee and North Carolina originates in the Ohio Valley, where it is released by tall smokestacks of coal-burning electrical power plants. Prevailing winds bring the pollution southeast, and the tall mountains in the southern Appalachians eventually stop it. There, it not only pollutes the air but also produces **acid deposition**—sulfur and nitrogen emissions that react with oxygen and water to form acids that can fall back to ground as acid rain or snow. Acidification of soil due to acid deposition can change the soil chemistry and mobilize toxic metals such as aluminum, hindering plants' ability to take up water.

Acids leach nutrients from the soil, too, reducing the amount of calcium, magnesium, and potassium available to plants in topsoil. Taken together, these impacts can decrease plant growth, weaken plants so they are more vulnerable to disease or pests, and even kill them. Many aquatic organisms are also vulnerable to the acidification of their water habitat, especially the eggs and young of many fish and amphibians. This acid deposition is also a problem throughout the northeastern United States. (See LaunchPad Chapter 29 for more on the pH scale and ocean acidification.) **INFOGRAPHIC 20.4**

Evidence that air pollution can travel long distances can be found in the far north. Prevailing air currents pick up pollutants from the western United States, conveying them all the way to Lake Laberge in Canada's Yukon Territory, where the moisture condenses, forms clouds, and falls on the lake as rain or snow. Thus, even the most isolated regions on Earth are vulnerable to the effects of air pollution because atmospheric and hydrologic circulation moves chemical and particulate pollutants around the globe.

Appalachian acid rain, pollution in Lake Laberge, and stratospheric ozone depletion are **transboundary pollution** problems because regions that suffer from the pollution are not necessarily the ones that released the pollutants. This means that even if an area does not produce pollution itself, its air may still be toxic. With the EPA's Air Quality Index available online, people can search for up-to-date air quality reports about any U.S. region. The index also alerts local communities about air quality problems from ground-level ozone and particulate pollution.

↑ These trees in an experimental test plot show damage from the ozone pumped out of the tall vertical pipes (white) that ring the area.

↑ Exposure to acid rain has resulted in yellowing and loss of needles, decreasing overall photosynthesis and stunting the growth of these conifers at high elevations in the Austrian Alps.

INFOGRAPHIC 20.4 ACID DEPOSITION

↓ Burning fossil fuels releases sulfur and nitrogen oxides. These compounds react in the atmosphere to form acids. Acid rain, snow, fog, and even dry particles can fall to Earth as acid deposition, with the potential to alter the pH of lakes and soil, damaging plant and animal life.

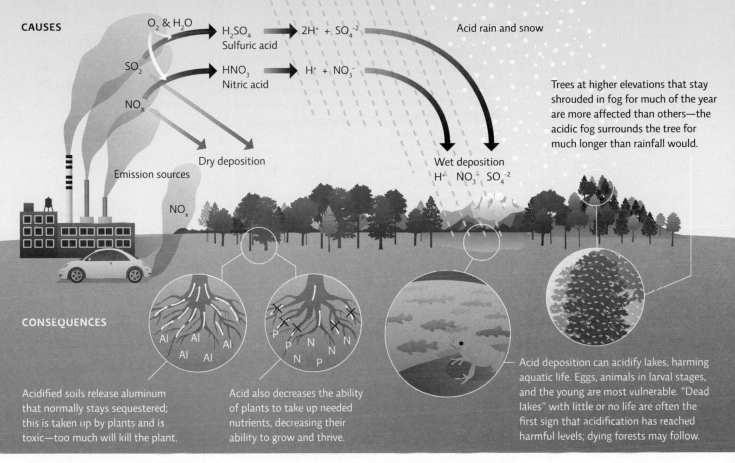

CAUSES

O_2 & H_2O

H_2SO_4
Sulfuric acid

$2H^+ + SO_4^{-2}$

Acid rain and snow

SO_2

HNO_3
Nitric acid

$H^+ + NO_3^-$

NO_x

Dry deposition

Emission sources

NO_x

Wet deposition
H^+ NO_3^- SO_4^{-2}

Trees at higher elevations that stay shrouded in fog for much of the year are more affected than others—the acidic fog surrounds the tree for much longer than rainfall would.

CONSEQUENCES

Al Al Al
Al Al

P N N
P N
N P

Acidified soils release aluminum that normally stays sequestered; this is taken up by plants and is toxic—too much will kill the plant.

Acid also decreases the ability of plants to take up needed nutrients, decreasing their ability to grow and thrive.

Acid deposition can acidify lakes, harming aquatic life. Eggs, animals in larval stages, and the young are most vulnerable. "Dead lakes" with little or no life are often the first sign that acidification has reached harmful levels; dying forests may follow.

ACID DEPOSITION HAS DECREASED

↓ Restrictions imposed by the Clean Air Act have helped decrease acid deposition in the United States. Smokestack scrubbers remove sulfur from coal burning, reducing the SO_2 released. Emission-control technologies on vehicles, such as the catalytic converter, convert dangerous combustion by-products to safer emissions (such as converting NO_x to N_2). This reduces, but doesn't eliminate, these dangerous emissions.

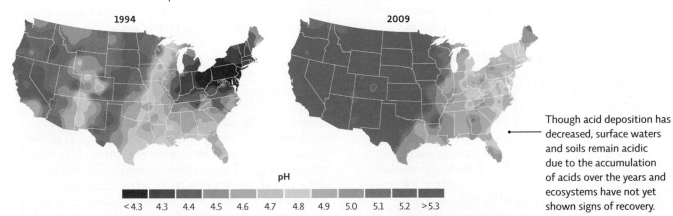

pH OF PRECIPITATION IN THE UNITED STATES

1994

2009

pH

| < 4.3 | 4.3 | 4.4 | 4.5 | 4.6 | 4.7 | 4.8 | 4.9 | 5.0 | 5.1 | 5.2 | > 5.3 |

Though acid deposition has decreased, surface waters and soils remain acidic due to the accumulation of acids over the years and ecosystems have not yet shown signs of recovery.

? Why do you think that acid deposition has been a bigger problem in the eastern United States than the western part of the country? Why might acid rain be increasing in some western areas?

Indoor air pollution is also a significant health threat.

As Nadeau discovered, outdoor air quality substantially impacts human health. However, we breathe air indoors as well as outdoors, and indoor air quality is a growing concern among public health scientists. In fact, people living in affluent, developed nations may find that their greatest exposure to unhealthy air comes from indoors. This is because so much time is spent indoors in homes, schools, or the workplace; these areas contain many potential air pollution sources. For instance, cigarette smoke causes significant health problems, including eye, nose, and mucous membrane irritation; lung damage, which can exacerbate or cause asthma; and lung cancer. Items in our home, like paint, cleaners, and furniture, release VOCs, which can also cause health problems.

KEY CONCEPT 20.7

Homes trap or are the source of many indoor air pollutants. Better ventilation and alternative building or household materials can significantly reduce this pollution.

Outdoor pollutants can also find their way into our buildings. Radon is a naturally occurring radioactive gas produced from the decay of uranium in rock. It can seep through the foundations of homes and accumulate in basements; exposure to radon can cause lung cancer. Not every area has the type of rock that produces radon, but buildings constructed over areas where soil or groundwater is contaminated with VOCs, such as areas with underground chemical storage tanks, also present an infiltration risk. **INFOGRAPHIC 20.5**

In developing countries, where many people cook and heat with open fires, smoke and soot from burning wood, charcoal, dung, or crop waste are major sources of indoor pollution. Kirk Smith, a professor of environmental health at the University of California, Berkeley, has found that indoor fires increase the risk of pneumonia, tuberculosis, chronic bronchitis, lung cancer, cataracts, and low birth weight in babies born of women who are exposed during pregnancy. "Considering that half the world's households are cooking with solid fuels, this is a big problem," Smith says.

command-and-control A type of regulation that involves setting an upper allowable limit of pollution release that is enforced with fines and/or incarceration.

Clean Air Act (CAA) First passed in 1963 and amended in 1990, a U.S. law that authorizes the EPA to set standards for dangerous air pollutants and enforce those standards.

"Current estimates are that indoor fires cause the premature death of 1.5 to 2 million women and children per year." Many nonprofit organizations are stepping up to meet this problem with a simple $50 solar cooker that allows people to cook food without building a fire. This technology has the added advantage of not depleting local biomass resources for fuel.

KEY CONCEPT 20.8

In developing countries, air pollution mainly comes from indoor cooking fires. It can be reduced by using cleaner fuels and solar ovens.

We have several options for addressing air pollution.

Solutions such as installing better ventilation systems in our homes and providing solar ovens to individuals in developing countries will help address indoor air pollution, but outdoor air pollution requires a more regional, national, and even international approach. Since air pollution often travels to areas that do not produce significant amounts of pollution themselves, regulating air pollution is a particular challenge. How does one country regulate pollution that travels through the atmosphere from another country?

In developed countries, the original approach to dealing with air pollution from human activities was to spread it out; the slogan was "The solution to pollution is dilution." Factories, power plants, and other point sources built tall smokestacks to send emissions high into the atmosphere so that they wouldn't pool at the site of production. The idea was that if dispersed, the amount of pollution in any one area would be too low to cause a problem. But this approach simply doesn't work: Industry releases too much pollution, and air circulation patterns cause some areas to get more than their share of pollution.

Eventually it became clear that regulation would be necessary. The typical approach in the 1970s was **command-and-control** regulation, a type of regulation that involves setting national limits on how much pollution can be released into the environment and imposes fines or even brings criminal charges against violators who release more than is allowed. An example of command-and-control regulation in the United States is the **Clean Air Act (CAA)**. It sets a maximum amount, or *air quality standard*, for emissions of pollutants or for the presence of pollutants in ambient air. States are responsible for monitoring air quality as well as for

INFOGRAPHIC 20.5   SOURCES OF INDOOR AIR POLLUTION

↓ For most people, the greatest exposure to air pollution comes from being indoors. There are many sources of air pollution in a home or other building, as these structures tend to trap pollutants, keeping concentrations high. One can reduce exposure by avoiding or limiting the use of carpets, upholstered items, and furniture made with toxic glue and formaldehyde. Safer cleaners and low-VOC paints are readily available. Simple behaviors like taking off your shoes before entering the house and using a vacuum equipped with a HEPA filter will also help—which is especially important if you have indoor pets. Good ventilation and properly working heating and air conditioning units help keep indoor air pollutants at bay.

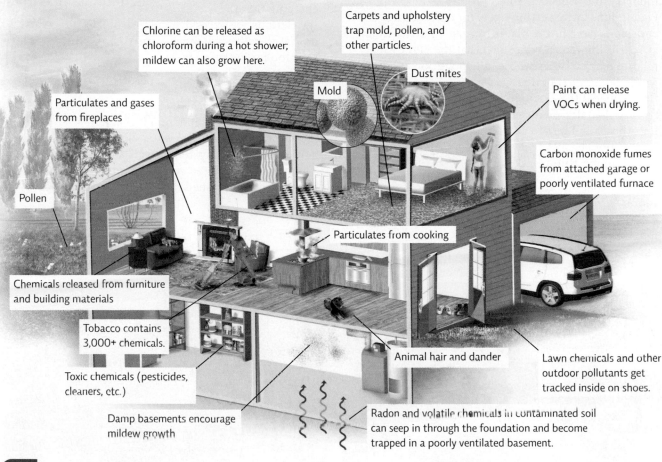

Chlorine can be released as chloroform during a hot shower; mildew can also grow here.

Carpets and upholstery trap mold, pollen, and other particles.

Dust mites

Mold

Particulates and gases from fireplaces

Paint can release VOCs when drying.

Pollen

Carbon monoxide fumes from attached garage or poorly ventilated furnace

Particulates from cooking

Chemicals released from furniture and building materials

Tobacco contains 3,000+ chemicals.

Animal hair and dander

Lawn chemicals and other outdoor pollutants get tracked inside on shoes.

Toxic chemicals (pesticides, cleaners, etc.)

Damp basements encourage mildew growth

Radon and volatile chemicals in contaminated soil can seep in through the foundation and become trapped in a poorly ventilated basement.

What are your main sources of indoor air pollution, and what could you do to reduce them?

developing, implementing, and enforcing compliance plans approved by the EPA. As a result of the CAA, the United States has seen major reductions in common air pollutants. Removing lead from gasoline, for instance, reduced lead air pollution by 98% from 1970 levels. Sulfur pollution has also been significantly reduced. And the Mercury and Air Toxic Standards approved in 2011, which limit the release of mercury, acid gases, and other pollutants from power plants, are expected to prevent 130,000 cases of serious asthma and as many as 11,000 premature deaths annually.

The CAA is, however, now under attack. Because regulation is based on legislation, it is subject to political wrangling; several congressional bills have been introduced that would limit the EPA's ability to regulate air quality. Of particular concern right now is the regulation of carbon dioxide (CO_2). Although it is naturally occurring, and previously considered harmless, CO_2 has now been strongly linked to climate change (see Chapter 21). In a landmark case, the Supreme Court gave the EPA the authority to regulate CO_2 as a pollutant in 2007, but the EPA immediately faced political opposition. In 2010, the EPA approved greenhouse gas emission standards (including CO_2, methane, and N_2O) for light-duty vehicles (cars and trucks) that require new vehicles to produce less greenhouse gas emissions; the government started phasing in these new regulations in 2012, and will continue to do so until 2016.

Setting new U.S. standards for coal-fired power plants or other large CO_2 emitters is proving more difficult as these regulations are strongly opposed by the coal industry and others. The EPA proposed the Clean Power Plan in June 2014, setting a goal for cutting carbon emissions from power plants by 30% by 2030. In July 2014, the Supreme Court ruled that the EPA cannot require new facilities whose only pollution emissions would be CO_2 to obtain CO_2 permits; however, the Court upheld the EPA's ability to require any existing facility that already must acquire permits for emissions for other pollutants to also obtain permits for CO_2 emissions. The EPA hopes to have a regulatory plan in place by June 2016.

The fact that our air is cleaner today than it was in the 1960s—even with a larger U.S. population and more industry—is evidence that regulations can be effective. Still, these improvements are costly to industry, and such costs are usually passed on to consumers. For this reason, many individuals and groups oppose such policies, charging that the restrictions are excessive or that the government goes too far in trying to regulate emissions. Some environmentalists worry that if we weaken or dismantle the environmental legislation that protects our air and water (and, by extension, our health and ecosystems), we face the return of a highly contaminated environment, compromised health, and diminished ecosystem function and services.

KEY CONCEPT 20.9

Cleaner fuels and smokestack scrubbers can reduce industrial air pollution. Regulations and economic incentives can spur innovation but may raise the cost of providing energy or doing business.

↓ Statue in Trafalgar Square, in London, England, shows erosion that exceeds normal weathering and is likely due to acid rain.

© Ernie Janes/Balance/Photoshot/ZUMAPRESS.com

↑ Mass transit options that decrease the number of cars on the road will reduce air pollution. Buses that run on compressed natural gas emit fewer emissions overall but the particulates they release are very small so, while better than a traditional diesel bus, they are still not pollution free.

In addition to command-and-control regulation, there are other ways to curb pollution. One example is **green taxes** on environmentally undesirable actions, such as an extra tax on low-mile-per-gallon vehicles. **Tax credits**, reductions in the amount of tax one pays in exchange for environmentally beneficial actions, fall on the other side of the spectrum. Tax credits encourage consumers to pursue options that might be more expensive than conventional options (such as the purchase of a hybrid automobile). As more people buy the products, the industries that make them can scale up and bring down prices.

Governments also offer **subsidies**, free money or resources intended to promote environmentally friendly activities. And with **cap-and-trade**, also called *permit trading*, a government or regulatory agency sets an upper limit on emissions for a pollutant on a nationwide or regional level and then gives or sells permits to polluting industries. Users that reduce their pollution emissions below what their permit allows can sell their remaining credits to other users that exceed their allotments. Over time, pollution levels can be reduced as the cap—or limit—is lowered. A cap-and-trade program successfully reduced sulfur pollution from coal-fired power plants in the United States in the 1990s. A downside to cap-and-trade programs is that pollution can become concentrated in areas where industries choose to buy additional permits rather than reduce emissions.

green tax A tax (fee paid to government) assessed on environmentally undesirable activities.

tax credit A reduction in the tax one must pay in exchange for some desirable action.

subsidies Financial assistance given by the government to promote desired activities.

cap-and-trade Regulations that set an upper limit for pollution emissions, issue permits to producers for a portion of that amount, and allow producers that release less than their allotment to sell permits to those who exceeded their allotment.

INFOGRAPHIC 20.6　APPROACHES TO REDUCING AIR POLLUTION　

↓ Many approaches can be used to lessen air pollution, including technology to reduce emissions before a fuel is burned (see Infographic 18.7) and technology to capture emissions after a fuel is burned, as shown below. Some policy tools, such as cap-and-trade, encourage the use of "best available control technologies," that is, the current technology that releases the lowest amount of pollution. There are economic costs to implementing these changes, but some benefits include new jobs, a competitive advantage for industries that can successfully reduce emissions, and a healthier society and ecosystem.

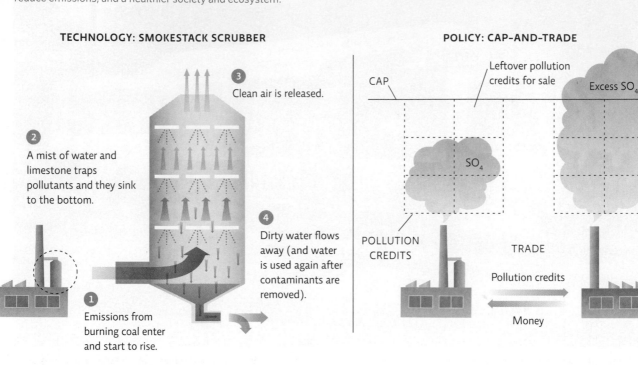

TECHNOLOGY: SMOKESTACK SCRUBBER

③ Clean air is released.

② A mist of water and limestone traps pollutants and they sink to the bottom.

④ Dirty water flows away (and water is used again after contaminants are removed).

① Emissions from burning coal enter and start to rise.

POLICY: CAP-AND-TRADE

CAP

Leftover pollution credits for sale

Excess SO₄

SO₄

POLLUTION CREDITS

TRADE

Pollution credits

Money

↑ A variety of methods are available to capture harmful pollutants before they leave the smokestack. Smokestack scrubbers send the emissions though a mist of water and limestone to trap contaminants and prevent their release.

↑ In cap-and-trade programs, a cap (upper limit) is set for the total pollution that can be released in an area, and pollution credits are issued to each producer that identify how much pollution each can release. If a producer implements changes that reduce pollution emissions below their total credit, it can sell its leftover credits to another producer that exceeds its allotment.

[?] How could a cap-and-trade program for sulfur pollution lead to lower pollution in one area and higher pollution in others?

Technology can also play a big role in improving air quality. To improve indoor air quality, we can install air filters and better ventilation systems. To curb pollution emissions from manufacturing, we can use end-of-pipe solutions like scrubbers, filters, electrostatic precipitators, and catalytic converters to trap pollutants before they are released. In addition, technology can inspire cleaner methods for extracting energy out of fossil fuels, as is the case with "clean coal" technologies described in Chapter 18. **INFOGRAPHIC 20.6**

Mitigating or preventing air pollution costs money, but many feel it is money well spent because it prevents far

greater losses down the line—especially in terms of human health. The Centers for Disease Control and Prevention estimates that, in 2007, the cost of asthma to the United States was $56 billion. According to a 2003 study published in the *Journal of Allergy and Clinical Immunology*, the average annual cost of care for an asthma patient is $4,912, with 65% of that going to medications, hospital admissions, and nonemergency doctor visits. The remaining 35% goes to indirect costs like lost time at work. Nadeau, Delfino, and others hope their research helps policy makers realize just how useful curbing air pollution can be. "We're talking about the air we breathe," Thurston says. "There's nothing more communal than that."

Select References:

Cisternas, M. et al., (2003). A comprehensive study of the direct and indirect costs of adult asthma. *The Journal of Allergy and Clinical Immunology*, 111(6): 1212–1218.

Delfino, R. J., et al. (2008). Traffic-related air pollution and asthma onset in children: A prospective cohort study with individual exposure measurement. *Environmental Health Perspectives*, 116(10): 550–558.

Dockery, D., et al. (1993). An association between air pollution and mortality in six US cities. *New England Journal of Medicine*, 329(24): 1753–1759.

Laden, F., et al. (2006). Reduction in fine particulate air pollution and mortality, extended follow-up of the Harvard Six Cities Study. *American Journal of Respiratory and Critical Care Medicine*, 173(6): 667–672.

Morello-Frosch, R., et al. (2002). Environmental justice and regional inequality in southern California: Implications for future research. *Environmental Health Perspectives*, 110(2): 149–154.

Nadeau, K., et al. (2010). Ambient air pollution impairs regulatory T-cell function in asthma. *Journal of Allergy and Clinical Immunology*, 126(4): 845–852.

Runyon, R. S., et al. (2012). Asthma discordance in twins is linked to epigenetic modifications of T cells. *PloS ONE*, 7(11): e48796.

Spira-Cohen, A., et al. (2011). Personal exposures to traffic-related air pollution and acute respiratory health among Bronx schoolchildren with asthma. *Environmental Health Perspectives*, 119(4): 559–565.

BRING IT HOME

PERSONAL CHOICES THAT HELP

Individuals can have an effect on air quality by researching the threats to their area, making appropriate behavior changes, and supporting legislation that limits the production of air pollutants.

Individual Steps

• Reduce your exposure to indoor air pollution by reducing your use of harsh cleaning products, synthetic air fresheners, vinyl products, and oil-based candles.

• Avoid outdoor exercise during poor air quality days. Go to http:// airnow.gov to find the local air quality forecast.

• Buy a radon detector and carbon monoxide detector for your house to keep your family safe.

Group Action

• Organize a "car-free day" at your school, community, or workplace to reduce emissions from vehicles.

• Work with community leaders and businesses to sponsor a "free public transit" day.

Policy Change

• If your community does not have public transit, ask community leaders to investigate bringing it to your area.

• Many groups are working to improve our air quality. Find one in your region and see what issues it is addressing. For a list of national and regional organizations, go to www.inspirationgreen .com/air.

Goodshoot/Thinkstock

ENVIRONMENTAL LITERACY UNDERSTANDING THE ISSUE

1

What is air pollution, and what is its global impact? What are the main types and sources of outdoor air pollution?

INFOGRAPHICS 20.1 AND 20.2

1. The WHO estimates that about _____ people die each year due to air pollution.

2. True or False: Ozone is useful in the stratosphere but is a dangerous pollutant when located in the troposphere.

3. Air pollution that results when chemicals in the atmosphere react to form a new pollutant is called:
   a. primary pollution.
   b. secondary pollution.
   c. point source pollution.
   d. particulate pollution.

2

What are the health, economic, social, and ecological consequences of air pollution?

INFOGRAPHIC 20.3 AND TABLE 20.1

4. Given the relationship between asthma and air pollution, where would you raise a family to decrease the risk of asthma?
   a. In an area with low VOCs but moderate to high particulates
   b. In an urban area
   c. Away from major highway systems
   d. In a valley where most people use wood to heat their homes

5. The placement of polluting industries close to minority or low-income areas is an example of a violation of:
   a. environmental justice.
   b. federal law.
   c. EPA clean air standards.
   d. transboundary pollution.

6. Describe the types of problems that air pollution causes in ecosystems and human health.

3

What are the causes and consequences of acid deposition, and how is it an example of transboundary pollution?

INFOGRAPHIC 20.4

7. True or False: Acid deposition has decreased across much of the eastern United States, and ecosystems are showing strong signs of recovery.

8. High-elevation trees are often more affected by acid deposition than trees lower on a mountain because the high-elevation trees:
   a. have more shallow root systems that are easily harmed by acids.
   b. grow in areas where acids bind aluminum in the soil so that the trees can't access it.
   c. live in colder areas and are more easily stressed than other trees.
   d. are often shrouded in acidic fog, which means they are exposed to acids longer than trees that only receive acid in rain or snow.

9. Why are industrial pollutants found in even the most remote places on Earth?

4

What are the main sources of indoor air pollution, and what can be done to reduce these pollutants?

INFOGRAPHIC 20.5

10. True or False: In most areas of developing countries, indoor pollution is more of a problem than outdoor pollution.

11. Which of these actions would best address the main cause of indoor pollution in developing countries?
    a. Using more wood and less charcoal in cooking fires
    b. Using emission control devices on vehicles to reduce air pollution on nearby roads
    c. Distributing solar ovens
    d. Making homes more airtight to keep out pollution from outside

12. Using the information presented in Infographic 20.5, explain why indoor air pollution is a cause of growing concern, especially with regard to health problems.

5

How can air pollution be reduced, and what are the trade-offs of reducing it?

INFOGRAPHIC 20.6

13. Which of the following is considered a penalty for not reducing pollution rather than an incentive or a reward for acting in a way that reduces pollution?
    a. Tax credit
    b. Green tax
    c. Subsidy
    d. All of the above

14. Describe the policy of cap-and-trade. What are the advantages and disadvantages of this policy option?

SCIENCE LITERACY WORKING WITH DATA

The graph shown here indicates levels of ground-level ozone and particulate matter that exceeded national reference levels (the levels above which health or ecosystem problems occur) in areas and cities in Canada.

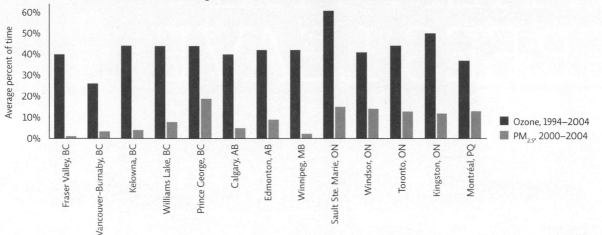

Percent of Time Smog Pollutant Reference Levels Were Exceeded in Select Canadian Areas

Interpretation

1. What does the y-axis represent? Choose a city and describe the data for that city.

2. The legend states that PM$_{2.5}$ data is given for 4 years. PM stands for *particulate matter*. The number 2.5 represents the size of the particulate. From a health standpoint, why are the PM$_{2.5}$ values reported?

3. How many years of data are graphed for ozone? For particulate matter? Does the difference in the amount of time over which the data have been collected make a difference in your interpretation of these data?

Advance Your Thinking

4. Which city would be the worst for your health, based on its levels of pollution? Why?

5. Based on the type of pollution present, what can you predict about the causes of pollution in Sault Ste. Marie versus Montreal? (Hint: Find the two cities on a map and compare the sizes of the cities, their weather, and their primary industries.)

INFORMATION LITERACY EVALUATING INFORMATION

The EPA is tasked with regulating pollutants in the United States. As part of this process, the agency collects and records data for many pollutants, but not all of them. The federal EPA is assisted in this endeavor by state EPAs. However, it is impossible to collect air quality data about every locality in the United States, so most data is collected in and around cities.

Go to www.stateoftheair.org. Enter your zip code in the "Report Card" box to get a report about air quality in your area. If no air quality monitoring stations exist in your area, choose your state and look at the data for the county closest to you. Record these data. Then, across the top bar, click on "Key Findings" and read about how the grades were calculated for each county. Finally, click on "Health Risks" and read about the specific health risks associated with both ozone and particulate matter.

Evaluate the website and work with the information to answer the following questions:

1. Is this a reliable information source?
   a. Does the organization have a clear and transparent agenda?
   b. Who runs the website? Do this organization's credentials make it reliable or unreliable? Explain.

2. What grade did your area receive for both ozone and particulates?
   a. Based on what you read about how the grade was determined, do you feel the grading system is too lax or too strict?
   b. Why does the American Lung Association advocate for a stricter system?

3. Based on what you know about the levels of pollution in your area and the effects of ozone and particulates on human health, do you believe that the regulations of the Clean Air Act should be loosened, tightened, or remain the same? Should more areas be monitored, or is it sufficient to monitor only large cities? Why?

Find an additional case study online at http://www.macmillanhighered.com/launchpad/saes2e

WHEN THE TREES LEAVE

Scientists grapple with a shifting climate

CORE MESSAGE

One of the biggest environmental problems facing humanity today is climate change. Evidence overwhelmingly points to the fact that climate is rapidly changing and that human activity is responsible for the changes. Climate change is impacting species, ecosystems, and the health and well-being of people around the globe, with more changes to come. Science can help us evaluate the changes that are happening, investigate causes, and provide information to help make sound policy for dealing with changing climate.

AFTER READING THIS CHAPTER, YOU SHOULD BE ABLE TO ANSWER THE FOLLOWING **GUIDING QUESTIONS**

What is the difference between climate and weather? Why is a change of a few degrees in average global temperatures more concerning than day-to-day weather changes of a few degrees?

What is the physical and biological evidence that climate change is currently occurring?

Flames engulf the sky and the ancient forest of the North Woods of Minnesota in a forest fire that occurred unusually early in the season. Layne Kennedy

3

What natural and anthropogenic factors affect climate, and which are implicated in the climate change we are experiencing now? How might positive feedback loops affect climate?

4

How do scientists determine past and present temperatures and CO_2 concentrations? What evidence suggests that climate change is due to human impact?

5

What are the current and potential future impacts of climate change? What actions can we take to respond to a world with a changing climate?

Lee Frelich was examining a 700-year-old cedar tree when he first noticed the smoke curling up into the sky over Minnesota's great North Woods. Within an hour, his entire view was filled, so Frelich, an ecologist with the University of Minnesota, and his companions—a photographer and a journalist who had cajoled him into taking them on a tour of the iconic boreal forest—trekked up to the north end of Ham Lake, away from the calamity. They would remain trapped there, amid dense, ancient stands of spruce and fir, for 3 full days, as fire claimed some 30,000 hectares (75,000 acres) around them.

The fire's magnitude was not surprising. Neither, really, was the fact that the flames laid bare a patch of forest that had not burned since 1801.

What was surprising was the timing. It was the first weekend in May—unusually early in the year for such a tremendous fire, especially given that the foot-thick ice had just broken apart on the lake a few days before. Frelich's hiking companions, who knew that such fires tended to come in late summer, were surprised. Frelich wasn't. To him it was just one more not-so-subtle reminder that **climate change** was rapidly throwing this ancient landscape into flux.

Other reminders have become commonplace: earlier springs, later winters, some tree species dying off, others popping up in unexpected places. Frelich and his colleagues worry that if current projections hold true, the forests themselves could vanish—converted by stress and time into scrubland or savanna.

Minnesota is not alone. In fact, the great North Woods are but one example of a whole planet in distress. In Africa and the American West, prairies are giving way to deserts (see LaunchPad Chapter 27). Ocean ecosystems are being affected by the twin problems of warmer temperatures and acidic waters. (See LaunchPad Chapter 29 for more on ocean acidification.) At the poles and higher altitudes, ice is melting at unprecedented rates, causing flooding in some areas and diminishing freshwater supplies in others. Precipitation patterns are shifting, with some areas getting drier and others wetter. In many places around the world, heat waves are becoming more frequent and more extreme. Wildfires like the one that scorched the North Woods are increasing in many areas around the globe due to hot, dry conditions turning forests into tinderboxes. Declining crop yields linked to temperatures or precipitation changes are evidence that agriculture is also being negatively affected. And all over the world, biodiversity is being threatened on a scale not seen since the last mass extinction 65 million years ago (see Chapter 13). Scientists say that all these changes are occurring as global climate warms with unprecedented speed. What this might mean for the future of our planet is something they are still trying to figure out.

Climate is not the same thing as weather.

Weather refers to the meteorological conditions in a given place on a given day, whereas **climate** refers to long-term patterns or trends. Put another way, the actual temperature on any given day is the weather, while the range of expected values, based on location and time of year, is the climate. We use what we know about a region's climate to predict the weather: Seasonal shifts come at about the same time each year, and winter lows and summer highs generally hover close to expected norms.

Weather can and does vary—sometimes considerably—from one day to the next. But no single weather event—no individual storm, flood, drought, or wildfire (not

◉ **WHERE ARE THE NORTH WOODS OF MINNESOTA?**

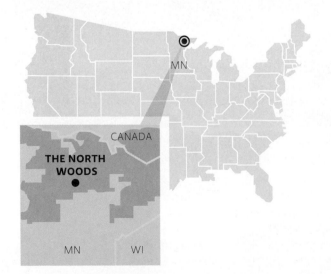

THE NORTH WOODS

CANADA

MN

MN   WI

↑ University of Minnesota ecologist Lee Frelich surveys the damage to the burnt forest immediately following the Ham Lake fire.

Layne Kennedy

KEY CONCEPT 21.1

Weather refers to the day-to-day conditions outside your window; *climate* describes the long-term weather patterns expected in a given area.

there but higher high temperatures, more and longer heat waves, earlier springs, and later winters.

To be sure, climate change is not a new phenomenon. In fact, Minnesota's climate has been in flux for thousands of years. Fossil plant and pollen records show that, after the North American ice cap retreated—some 10,000 years ago—the climate warmed so dramatically that tree species' ranges shifted northward at a rate of 50 kilometers (30 miles) per century. Pines, oaks, and other deciduous species replaced the spruce trees that had covered most of the region. As summers became warmer, water levels fell, prairie plants took root, and birches

even one as colossal and ill-timed as the one that trapped Frelich and his friends)—can be attributed to **global warming**. Global warming is the province of climate. It refers to a rapidly shifting range of average temperatures that scientists have measured in myriad locations around the world—not just a few warmer days here and

and pine moved north. Then, about 6,000 years ago, the climate cooled a bit, and trees began migrating south and west once again. (Such shifts are sometimes called *tree migration*, but a more accurate term is *tree range migration*, since these shifts are really changes in the range of a species.)

Scientists say organisms must respond much more quickly to current climate change. Instead of having thousands of years, organisms that inhabit a given region—not only the boreal forest around Ham Lake but ecosystems everywhere—might have just a few decades to adapt or migrate as climate change makes their current homes uninhabitable.

Evidence of global climate change abounds.

By all accounts, the forests of northern Minnesota are places of uncommon majesty. Moose and deer meander through an ocean of trees—some of them close to 1,000 years old—that grow out

climate change Alteration in the long-term patterns and statistical averages of meteorological events.

weather The meteorological conditions in a given place on a given day.

climate Long-term patterns or trends of meteorological conditions.

global warming The observed and ongoing rise in the Earth's average temperature that is contributing to climate change.

of colossal, hundreds-of-feet-high granite hills whose outcrops reflect the colors of day: pale, soft pink when the Sun rises and deep, somber rouge when it sets. The serenity belies an unsettling truth: This forest, which has existed for more than 3,000 years—since the early days of the Roman Empire—and even inspired the Wilderness Act of 1964, could vanish within the next century.

To Frelich's well-trained eye, the signs are obvious. He has spent his entire adult life trekking through this ancient landscape, observing it and cataloguing the changes—inch by inch, leaf by leaf. So it's no surprise that when he looks on the placid landscape, he sees a catastrophe unfolding. In one patch of forest, scrawny young birch trees bud several weeks ahead of schedule. In another, adult birches are dying off rapidly, leaving a graveyard of bony white trunks. "A long growing season is not good for this tree," Frelich explained, "because it goes hand in hand with warmer soil, which the paper birch doesn't tolerate well." Meanwhile, red maple, a temperate species that grows as far south as Louisiana but is far less common up north, is thriving in northeast Minnesota's Sea Gull Lake area. "When you have red maples growing as much as 4 feet in a single year," said Frelich, "you are not talking about a boreal climate anymore."

Regional climate data correlates well with the changes Frelich sees. In the past two decades, springtime has come ever earlier to the region—a week or two sooner than the historical average, according to the climatology office of the Minnesota Department of Natural Resources. Nine of the state's 20 warmest years have been recorded since 1981. And for the first time in recorded history, Minnesota logged three mild winters in a row, each with record highs: 1997, 1998, and 1999.

Those data correspond to larger global trends. Each of the past three decades has been warmer than any decade since the late 1800s, when reliable temperature measurements began. Overall, 2000 to 2010 is the warmest decade on record since climatologists started keeping records back in 1850. The 10 warmest years have all occurred since 1998, with 2005 and 2010 tying for warmest individual year. The global land average temperature increased by 0.85°C (1.53°F) from 1880 to 2012, with most of that warming occurring since 1950.

A degree or two might not seem like much. But even such seemingly small changes in climate can have tremendous impacts on weather and, thus, on natural ecosystems and human societies. **INFOGRAPHIC 21.1**

INFOGRAPHIC 21.1   **A CHANGE IN AVERAGE TEMPERATURE: WHY DO ONLY A FEW DEGREES MATTER?**   `1`

↓ A shift of only a few degrees in the average global temperature will likely result in more frequent and extreme heat waves. Compared to the mid-20th-century average, we have increased about 1.7°F (~1.0°C). To put this in perspective, at the end of the last ice age 10,000 years ago, Earth's average temperature was only about 5.5°F (3.0°C) colder. At that time there was an ice sheet 1.6 kilometers (1 mile) thick as far south in North America as Chicago.

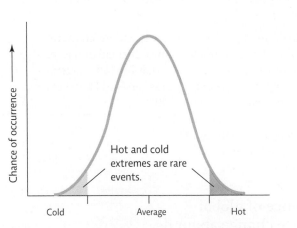

In any stable climate time period (decade, century, millennium), there is some climate variability. In some years, the average temperature is hotter or colder than in others but, on average, extremely cold and hot years occur infrequently. Most often, temperatures fall somewhere in the middle.

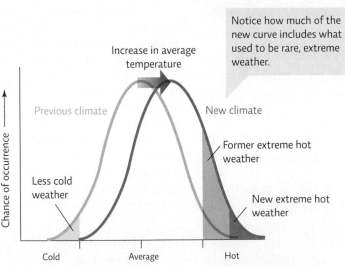

An increase of a few degrees in the "average temperature" shifts the entire curve to the right. This means more years that have the former "extreme" hot weather and new records for extreme heat. In some areas, climate change could also produce colder–than–average temperatures (widen the entire bell curve) by causing shifts in the jet stream or ocean currents that bring cold air to a region.

Climate change could also produce a new temperature curve that is broader and flatter. What would this mean for cold and hot temperature extremes?

In fact, we are already seeing some big effects, especially in the Arctic, which is particularly vulnerable to climate change because warming that occurs there causes ice to melt, which triggers additional warming. Warming in the Arctic is also affecting weather around the world. Recent studies suggest that climate change is altering the Northern Hemisphere's polar jet stream, slowing it down and making it "wavier," meaning the peaks and troughs of the jet stream are extending farther north and south as the jet stream moves from west to east. This brings colder-than-normal air farther south and can slow down the movement of weather systems, sometimes causing them to stall and dump excessive amounts of snow or rain in one area. Essentially, cold arctic air is reaching farther south than normal, more often than normal, and for longer periods of time, contributing to extreme winter weather events such as those seen in North America and Europe in recent years.

So how quickly is the Arctic warming? In 2011, researchers at Cambridge University evaluated the weather patterns of Ellesmere Island, Nunavut, in Canada's Arctic, and found that spring and summer temperatures were 11° to 16°C (20° to 29°F) higher than in previous years—making the climate similar to a boreal climate 1,600 to 2,200 kilometers (1,000 to 1,400 miles) farther south. But as large as that increase was, it was not entirely surprising.

In 2007 and some subsequent years, researchers have measured record losses of sea ice (ice that floats on top of the ocean) in the region, including the record low sea ice extent that occurred in September 2012. Based on the current rate of melting, the entire Arctic could be completely ice free in the summers come 2035.

And it's not just arctic ice that's melting. Overall, sea ice is decreasing about 3% to 4% per decade. In Antarctica, the Larsen B Ice Shelf—a colossal sheet of sea ice, more than 210 meters (700 feet) thick and as big as Rhode Island—collapsed in 2002 in just a couple weeks. This stunning spectacle, which took climatologists by surprise, was repeated in 2008, when the Wilkins Ice Shelf (also in Antarctica) collapsed, again in the space of 2 weeks.

Land-based ice (glaciers) in high-altitude and high-latitude areas is also melting at unprecedented rates. And as ice on land melts, sea levels are rising—by an average of 10 to 20 centimeters (4 to 8 inches) during the 20th century. About half of this rise was due to land-based ice melt and the other half to thermal expansion— the expansion of water molecules as they heat up. Rising sea levels have already displaced hundreds of thousands of coastal-dwelling people around the world. A 2014 NASA report concluded that glaciers in West Antarctica are melting faster than previously realized and appear to have passed a "point of no return" in which complete glacial melt may be unstoppable. Total melting of these glaciers would raise sea level by as much as 1.2 meters (4 feet) over the next few centuries.

When evaluating the evidence for climate change (e.g., its occurrence, its causes, its effects), it is important to do so scientifically and place a level of certainty on conclusions. The United Nations (UN) and the World Meteorological Organization established the **Intergovernmental Panel on Climate Change (IPCC)** to do just this in 1988. The IPCC is made up of thousands of scientists from around the world. They evaluate all the climate science that is published through peer review and compile it into a cohesive series of publications that explain what is understood about the current state of the climate; they also make recommendations that may inform government action. The IPCC recognizes five levels of confidence that its conclusions are correct: very low, low, medium, high, and very high. **INFOGRAPHIC 21.2**

KEY CONCEPT 21.2

A change of just a few degrees in average temperature can result in a climate with a larger number of extreme weather events than normal and new record temperatures.

KEY CONCEPT 21.3

A warming planet should see warmer average temperatures, melting land and sea ice, rising sea levels, and precipitation changes; all of these are currently being observed.

A variety of factors affect climate.

So why is all this climate change happening?

In the past few decades, scientists have discovered that the levels of certain gases in Earth's atmosphere are on the rise. These gases, called **greenhouse gases**—which include carbon dioxide, methane, and nitrous oxide— trap heat and help warm Earth, in a process known as the **greenhouse effect**. To be sure,

Intergovernmental Panel on Climate Change (IPCC) An international group of scientists that evaluates scientific studies related to climate change to thoroughly and objectively assess the data.

greenhouse gases Molecules in the atmosphere that absorb heat and reradiate it back to Earth.

greenhouse effect The warming of the planet that results when heat is trapped by Earth's atmosphere.

↳ We know that a variety of factors can alter global temperature, but what is the physical evidence that temperatures have actually increased and that the climate is changing? In other words, what do we predict we would see if warming were occurring, and what do we actually see when we test those predictions?

WARMER TEMPERATURES
If climate is indeed warming, we expect to see warmer global temperatures, on average, than in the recent past (more temperature anomalies in the direction of warming).

OBSERVED CHANGES IN SURFACE TEMPERATURE, 1901–2012

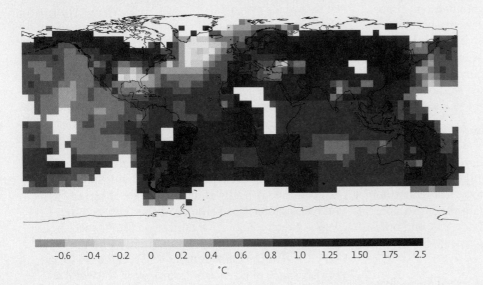

-0.6 -0.4 -0.2 0 0.2 0.4 0.6 0.8 1.0 1.25 1.50 1.75 2.5
°C

Since 1901, temperatures have increased over most of Earth's surface *(virtually certain)*. Higher-latitude areas generally show more warming than midaltitude areas. (Areas shown in white are those for which there is not enough data to calculate "robust estimates" of changes over this time period.)

OBSERVED GLOBALLY AVERAGED COMBINED LAND AND OCEAN SURFACE TEMPERATURE ANOMALY, 1850-2012

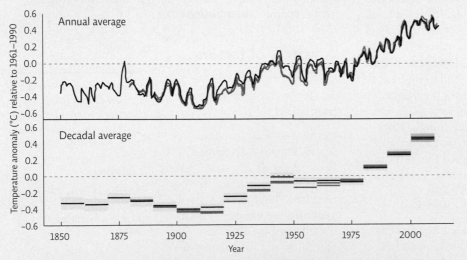

The top graph shows the mean value for annual anomalies compared to the mean temperature of 1961–1990 (shown as the 0.0 dashed line); three data sets were evaluated and shown here in different colors. The bottom graph shows the decadal averages of up to three data sets. Climatologists generally look for directional changes over at least 30 years before concluding that the changes represent a trend.

WARMER TEMPERATURES SHOULD LEAD TO

MELTING ICE
If temperatures are warming, we would expect to see more ice melt.

SEA LEVEL RISE
We would also expect to see an increase in sea level as land-based ice melts and as warmer seawater expands.

CUMULATIVE LOSS OF GLACIER MASS

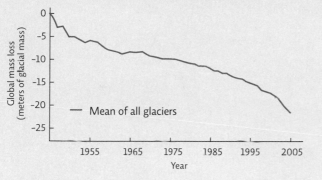

Since the middle of the 20th century, glaciers around the world have a net loss of ice *(high confidence)*. Permafrost is also thawing in high latitude and high altitude regions *(high confidence)*.

SEA LEVEL CHANGE OVER TIME
(RELATIVE TO THE 1900–1905 AVERAGE)

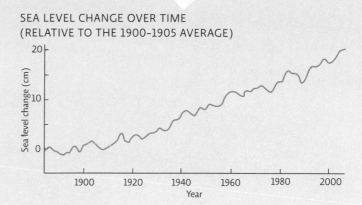

Sea level has indeed risen 10–20 cm over the last century and the rate of increase has accelerated since 1993 *(high confidence)*.

WEATHER EXTREMES
A warmer climate causes more water to evaporate from Earth's surfaces and is expected to produce more extreme weather (e.g. heat waves and tropical storms) but also should produce unusual weather as the atmosphere redistributes this heat and moisture around the planet. Areas close to large bodies of water are expected to be wetter (more rain or snow), whereas inner continental areas are expected to be drier as warmer temperatures increase the loss of water from soil, but not enough to fall back down down as rain.

PRECIPITATION CHANGES Since 1951 some areas have received more precipitation than normal while others have received less. The data set is incomplete in many areas (shown as white here). Precipitation has increased in mid-latitude areas *(high confidence)* but the data are weaker for other areas *(low confidence* that precipitation has changed in those areas).

INCREASE IN STORMS Evidence for stronger storms is less robust than other lines of evidence or climate change. Globally, there is not enough data to conclude that the number of tropical cyclones has increased *(low confidence)*, but in the Atlantic Ocean the IPCC concludes with *very high confidence* that there has been an increase in the number of strong cyclones (category 3 or higher).

OBSERVED CHANGE IN ANNUAL PRECIPITATION OVER LAND
(1951–2010)

MAJOR TROPICAL CYCLONES (HURRICANES)
IN THE ATLANTIC OCEAN

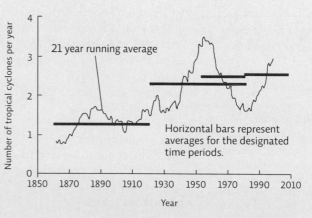

INFOGRAPHIC 21.3   THE GREENHOUSE EFFECT

↓ Life on Earth depends on the ability of greenhouse gases in the atmosphere to trap heat from the Sun and warm the planet. More greenhouse gases, however, mean more trapped heat and a warmer planet (an enhanced greenhouse effect).

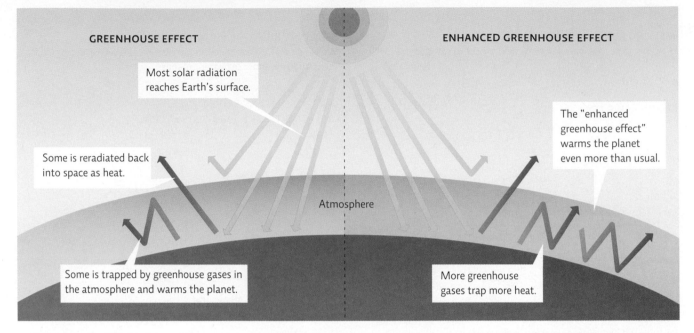

GREENHOUSE EFFECT

Most solar radiation reaches Earth's surface.

Some is reradiated back into space as heat.

Atmosphere

Some is trapped by greenhouse gases in the atmosphere and warms the planet.

ENHANCED GREENHOUSE EFFECT

The "enhanced greenhouse effect" warms the planet even more than usual.

More greenhouse gases trap more heat.

↓ Carbon dioxide (CO_2) is the greenhouse gas released by human actions that currently has the biggest impact on climate. Because different greenhouse gases have different abilities to trap heat, their heat-trapping capacity is expressed as CO_2 equivalents—the amount of CO_2 that would produce the same warming. For example, since a molecule of methane (CH_4) traps 25 times as much heat as CO_2, 1 methane molecule is equivalent to 25 CO_2 molecules. Much less methane is released in tonnage compared to CO_2, but because each molecule of methane is so much more potent a greenhouse gas than CO_2, it too has a big impact on climate.

GREENHOUSE GASES: RELATIVE CONTRIBUTIONS TO GLOBAL WARMING

64%	17%	13%	6%
Carbon dioxide (CO_2)	Methane (CH_4)	Halocarbons (CFCs and HCFCs)	Nitrous oxide (N_2O)

Use the concept of the greenhouse effect to explain how Earth's surface could warm up even if the Sun's output does not change.

the greenhouse effect is a good thing: Without it, the average temperature on Earth would be around −18°C (0°F)—that's about 34°C (60°F) colder than the planet's current average temperature! But starting in the 1980s, scientists began to see evidence of an enhanced greenhouse effect, which they have linked to the release of greenhouse gases from burning fossil fuels and other industrial and agricultural practices—in other words, from human activities.
INFOGRAPHIC 21.3

Greenhouse gases are just one type of **radiative forcer**, or factor that can affect global climate. Another kind of forcer that plays a role in present warming trends is **albedo**, the ability of a surface to reflect away solar radiation. Light-colored surfaces, like glaciers and meadows, have a high albedo: They reflect sunlight, and thus heat, away from the

radiative forcer Anything that alters the balance of incoming solar radiation relative to the amount of heat that escapes out into space.

albedo The ability of a surface to reflect away solar radiation.

positive feedback loop Changes caused by an initial event that then accentuate that original event (e.g., a warming trend gets even warmer).

KEY CONCEPT 21.4

Greenhouse gases trap incoming solar radiation and warm the atmosphere. Adding more greenhouse gases enhances this greenhouse effect and warms the planet.

INFOGRAPHIC 21.4   ALBEDO CHANGES CAN INCREASE WARMING VIA POSITIVE FEEDBACK   3

↓ Albedo is a measure of the reflectivity of a surface. The lighter colored the surface, the higher the albedo. Unreflected (absorbed) light is reradiated as heat, so surfaces with a low albedo release more heat to the atmosphere than do high-albedo surfaces.

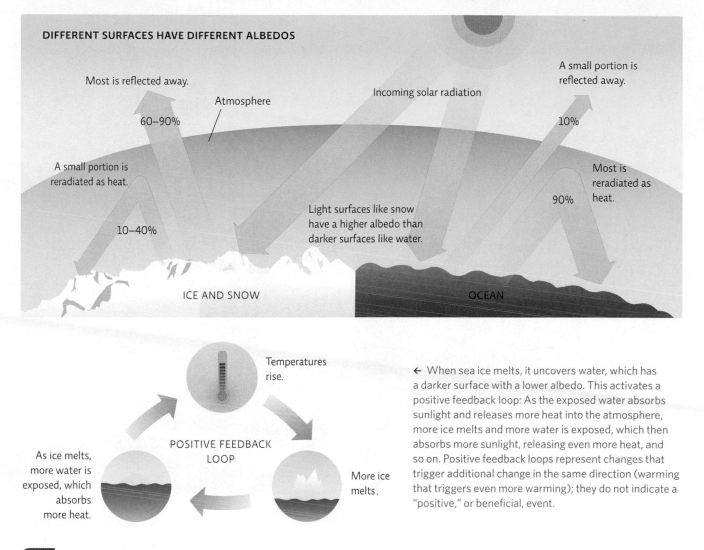

DIFFERENT SURFACES HAVE DIFFERENT ALBEDOS

Most is reflected away.

Atmosphere

Incoming solar radiation

A small portion is reflected away.

60–90%

10%

A small portion is reradiated as heat.

Most is reradiated as heat.

90%

10–40%

Light surfaces like snow have a higher albedo than darker surfaces like water.

ICE AND SNOW

OCEAN

Temperatures rise.

POSITIVE FEEDBACK LOOP

As ice melts, more water is exposed, which absorbs more heat.

More ice melts.

← When sea ice melts, it uncovers water, which has a darker surface with a lower albedo. This activates a positive feedback loop: As the exposed water absorbs sunlight and releases more heat into the atmosphere, more ice melts and more water is exposed, which then absorbs more sunlight, releasing even more heat, and so on. Positive feedback loops represent changes that trigger additional change in the same direction (warming that triggers even more warming); they do not indicate a "positive," or beneficial, event.

One suggestion to combat global warming is to replace dark rooftops with light-colored ones. How would this help reduce warming?

planet's surface. Darker surfaces, like water and dark asphalt, have low albedo: They absorb sunlight, and heat along with it, and then reradiate that heat back to the atmosphere. As surfaces with high albedo are replaced by those with low albedo, not only does the planet warm, but a **positive feedback loop** can be triggered.

Glaciers provide a good example of positive feedback: As temperatures rise, glaciers melt, and ice (with a high albedo) gives way to water (with a

low albedo). Because this new watery surface absorbs more heat than the former icy surface, the region warms even faster—replacing even more ice with water. And the cycle continues. **INFOGRAPHIC 21.4**

There are other positive feedback loops, too. In the Arctic, for example, the upper levels of permafrost (land that normally remains frozen year-round) are melting during the summer months. When these areas thaw, they release stored carbon, adding more greenhouse gases to the

KEY CONCEPT 21.5

The albedo, or reflectivity, of a surface affects surface temperatures and climate. Decreased albedo can increase warming via positive feedback.

↑ Icebergs 60 meters (200 feet) tall, formerly part of the Greenland Ice Sheet, float into the North Atlantic Ocean. Because this ice used to be on land, it contributed to sea level rise when it fell into the ocean.

atmosphere. These gases warm the area further, causing even more permafrost melt.

There are also other natural forcers, including clouds. Some have a high albedo and thus work to cool the planet; others trap reradiated heat from the planet's surface and have a warming effect. If warming temperatures cause the formation of more of the high-albedo clouds, this could trigger cooling—a **negative feedback loop**. Right now, clouds have a net cooling effect; whether this trend will continue in the future remains to be seen. **INFOGRAPHIC 21.5**

negative feedback loop Changes caused by an initial event that trigger events that then reverse the response (e.g., warming leads to events that eventually result in cooling).

The oceans play a major role in moderating climate change. While average global atmospheric and surface temperatures climbed steadily beginning in the mid-1970s, around 2003, temperatures began to stabilize, even while atmospheric greenhouse gas concentrations continued to rise. Scientists are finding evidence that much of

KEY CONCEPT 21.6

A wide variety of climate forcers can warm or cool the planet. Positive forcers currently outweigh negative forcers, which results in a net warming effect.

INFOGRAPHIC 21.5 CLIMATE FORCERS

→ A variety of factors can warm or cool the planet. Positive forcers have a warming effect; negative forcers cool the climate. Greenhouse gases trap heat in the atmosphere and warm it, while aerosols like sulfate emissions cool it. While there are differences in the confidence that values given for forcers are accurate, it is virtually certain that the anthropogenic forcing is positive rather than negative.

Aerosols like sulfur released from coal power plants are negative forcers. What impact will the efforts to remove sulfur from power plant emissions have on potential warming? Is this a reason to stop our efforts to prevent the release of sulfur? Explain.

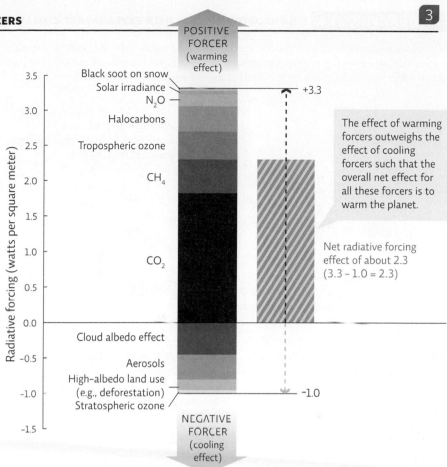

POSITIVE FORCER (warming effect)

Black soot on snow
Solar irradiance
N_2O
Halocarbons
Tropospheric ozone
CH_4
CO_2

+3.3

The effect of warming forcers outweighs the effect of cooling forcers such that the overall net effect for all these forcers is to warm the planet.

Net radiative forcing effect of about 2.3 (3.3 – 1.0 = 2.3)

Radiative forcing (watts per square meter)

Cloud albedo effect
Aerosols
High-albedo land use (e.g., deforestation)
Stratospheric ozone

–1.0

NEGATIVE FORCER (cooling effect)

→ Climate change research in Alaska monitors CO_2 release from thawing permafrost, along with tundra growth. The project, led by Ted Schuur of University of Florida, has found that in the short term, CO_2 released from melting permafrost leads to increased growth of local tundra vegetation. But in the longer term, the thawing leads to increased atmospheric loading of CO_2 as more is released than can be taken up by the vegetation.

Martin Shields/Science Photo Library/Science Source

INFOGRAPHIC 21.6 | MILANKOVITCH CYCLES HELP EXPLAIN PAST CLIMATE T CHANGE

↓ Warm periods and ice ages of the past can be attributed in part to Earth's position in space relative to the Sun. Earth has three different cycles that can each have an impact on climate. The current warming we are experiencing cannot be explained by any of these cycles—Earth is currently not in a part of any cycle in which it would have greater warming.

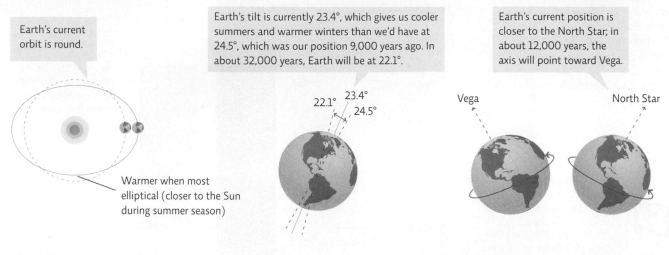

Earth's current orbit is round.

Earth's tilt is currently 23.4°, which gives us cooler summers and warmer winters than we'd have at 24.5°, which was our position 9,000 years ago. In about 32,000 years, Earth will be at 22.1°.

Earth's current position is closer to the North Star; in about 12,000 years, the axis will point toward Vega.

Warmer when most elliptical (closer to the Sun during summer season)

ORBITAL ECCENTRICITY The shape of Earth's orbit around the Sun varies over a 100,000-year cycle from mostly round to more elliptical.

AXIAL TILT The angle of Earth's tilt as it spins on its axis changes in a 41,000-year cycle. The greater the angle, the greater the extremes between seasons (hotter summers and colder winters).

AXIAL PRECESSION Earth "wobbles" on its axis, changing not the angle but the direction the axis points in a 20,000-year cycle. This changes the orientation of Earth to the Sun and affects the severity of the seasons. When Earth is tilted toward Vega, it is also tilted toward the Sun during summer, making summers hotter in the Northern Hemisphere.

 Why do scientists conclude that Earth's axial tilt is not responsible for our current warming?

that "missing heat" may be in the world's oceans, but there is a limit to how much more heat the oceans can hold, and they could release much of the heat they have sequestered. If it is released, we could be in for a jump in temperature that would continue the previous upward trend.

Volcanic eruptions and changes in solar irradiance (e.g., sunspot cycles) are also considered natural forcers, as both have been known to impact climate in the past, though only over short time frames and not as severely as greenhouse gases. Scientists also believe that **Milankovitch cycles** (predictable long-term cycles of Earth's position relative to the Sun) played an important role in earlier climate change events such as the Pleistocene ice ages. **INFOGRAPHIC 21.6**

To figure out how much any given forcer or feedback loop is contributing to current warming, or to predict what future climate might look like based on what we are seeing now, climate scientists must do more than monitor current atmospheric conditions; they must also gather data (on temperature, CO_2 levels,

Milankovitch cycles Predictable variations in Earth's position in space relative to the Sun that affect climate.

etc.) from the distant past. They do this by studying a wide variety of clues that have been left behind—ice and sediment cores, tree rings, coral reef growth layers, even fossilized mud at the bottom of lakes and rivers (see Chapter 1). These sources tell us that temperatures and CO_2 levels have varied over time and that these two parameters are positively correlated: As one increases or decreases, so does the other. These various sources have consistently corroborated one another: Ice cores paint the same picture as tree rings, and coral reef layers confirm what pollen sediments tell us. This consistency enables us to trust their overall story.
INFOGRAPHIC 21.7

Climate scientists have used this wealth of current and historical data to develop *climate models*—computer programs that allow them to make future climate projections by plugging in all the current values (for temperature, CO_2, global air circulation

KEY CONCEPT 21.7

Past climate changes are correlated with natural forcers such as the Milankovitch cycles, but these cycles do not account for current warming.

↓ We use a variety of methods to measure temperature and CO_2 levels. Current temperatures are monitored directly using instruments such as thermometers and satellites; a variety of analytical equipment is used to measure CO_2 levels in the air. Charles Keeling began taking precise CO_2 measurements at the Mauna Loa Observatory, Hawaii, in 1958, a location chosen because the air there received no local pollution to compromise the data. We can also estimate values from the past indirectly by evaluating physical evidence such as sediment and ice cores, and tree-ring data. Ice cores are particularly helpful; scientists can date the sections by counting the annual layers. They then measure atmospheric gases like CO_2 from the bubbles in the ice core sample to determine levels at the time the ice was laid down; measuring the ratio of 2 isotopes of oxygen, O-16 and O-18, gives a very accurate estimate of temperature at the time.

? Explain how increasing atmospheric levels of CO_2 are both a cause and effect of warming.

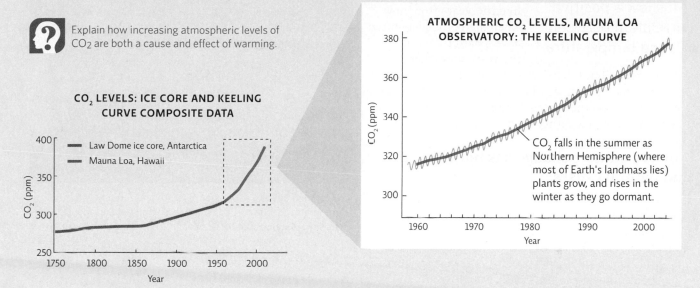

CO₂ LEVELS: ICE CORE AND KEELING CURVE COMPOSITE DATA

— Law Dome ice core, Antarctica
— Mauna Loa, Hawaii

ATMOSPHERIC CO₂ LEVELS, MAUNA LOA OBSERVATORY: THE KEELING CURVE

CO_2 falls in the summer as Northern Hemisphere (where most of Earth's landmass lies) plants grow, and rises in the winter as they go dormant.

↓ A comparison of historic CO_2 levels and temperatures, as determined from the Antarctic Vostok ice core, shows that the two parameters have been closely aligned over the past 400,000 years. It turns out that the relationship between CO_2 and temperature is one of cause and of effect. In the far past, natural events such as differences in Earth's orbit triggered warming, resulting in the release of more CO_2 (an effect), which then caused even more warming (a cause). Today, humans are the source of much of the extra CO_2 (and other greenhouse gases) being released. In this case, the CO_2 release is preceding the warming—it is the initial cause. The bottom line is that no matter the reason for the release of extra greenhouse gases such as CO_2, temperatures change as CO_2 increases and decreases.

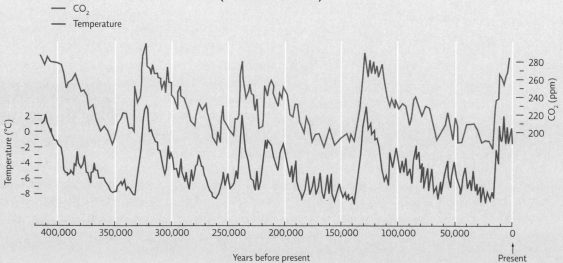

TEMPERATURE AND ATMOSPHERIC CO₂ CONCENTRATION OVER THE PAST 400,000 YEARS (VOSTOK ICE CORE)

— CO₂
— Temperature

416 CHAPTER 21: CLIMATE CHANGE

A variety of methods are used to measure past and present atmospheric CO_2 levels and temperatures. All show a positive correlation between CO_2 and temperature.

patterns, etc.). These models are used to see how altering the value of certain parameters (say, increasing the amount of atmospheric CO_2) might impact future climate. It may seem ironic that climatologists can predict what the climate will be like 100 years from now, when meteorologists often have a hard time getting the weekly weather forecast right. But scientists say that climate is actually easier to predict than weather. Climate refers to general trends, while weather is much more specific; it is more like a close-up view rather than a view from afar.

Current climate change has both human and natural causes.

Evaluating the wide range of scientific studies that relate to climate change is a monumental task. Here's what everyone agrees on so far: Earth's atmosphere is changing dramatically and with alarming speed. Analysis of CO_2 trapped in ice cores reveals that current levels are higher than at any other time in the past 800,000 years, and the levels are rising at an increasing rate. In 2013, the average atmospheric concentration of CO_2 was 396 ppm (parts per million); in May 2014 it topped 400 ppm. This is considerably higher than the preindustrial level of about 280 ppm—a level that had been maintained for millennia.

Based on all the clues they have gathered and analyzed, scientists agree that all the natural forcers combined are not enough to account for the rapid climate change that is currently under way. Only when we consider both natural and **anthropogenic** (related to human actions) forcers together do current trends make sense. In fact, the vast majority of this change is due to human activities, especially the burning of fossil fuels, and the consequent release of greenhouse gases like CO_2 into the atmosphere.
INFOGRAPHIC 21.8

For most of modern history, the United States has been the biggest emitter of CO_2. In 2007, however, China took the lead: The proliferation of coal-fired power plants that has fueled the country's recent economic and industrial growth has also released copious amounts of CO_2 into the atmosphere. China now releases nearly 30% more CO_2 than the United States (though the United States has one of the highest per capita carbon footprints in the world, and much of the CO_2 released in the 20th century came from U.S. sources).

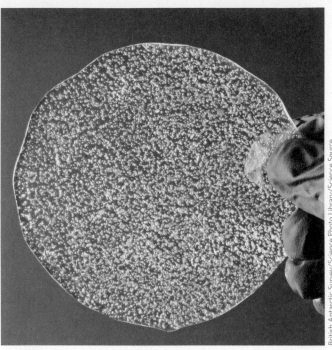

↑ A researcher holds a thin slice of ice from an ice core extracted from Antarctica. The core contains trapped air bubbles that can reveal information about the atmosphere and temperatures of the past.

British Antarctic Survey/Science Photo Library/Science Source

Many climate scientists set the upper limit of CO_2 that we should not cross (to avoid substantial negative effects, like the melting of the Greenland Ice Sheet and other events that we cannot reverse) at 450 ppm. At our current pace—featuring rapid fossil fuel consumption combined with sluggish efforts to curb our emissions—we will easily surpass that 450 ppm threshold before the end of the 21st century. But some scientists place the upper limit much lower, at 350 ppm—meaning we must not just reduce the amount of CO_2 we release, we must bring down current atmospheric levels.

The reality is that we have already set into motion a chain of events that is dramatically changing the face of the planet: Glacial and permafrost melt, ocean acidification, and loss of carbon sinks and vital habitats due to deforestation are positive feedback cycles kicking CO_2 accumulation into high gear, accelerating the pace of global warming.

But while the causes of climate change are scientifically well established, the future is still riddled with uncertainty. Forest ecologists like Frelich point

Current warming cannot be explained without accounting for both natural and anthropogenic climate forcers.

---
anthropogenic Caused by or related to human action.

INFOGRAPHIC 21.8 | WHAT'S CAUSING THE WARMING?

4

↓ Climate scientists use computer models (multiple mathematical equations) that take into account the major factors that are known to have affected past climates in order to see what might be responsible for recent warming. Data about natural and anthropogenic factors can be fed into a computer model separately and then together to see which circumstances match up with the warming that has been observed.

COMPUTER MODELS' RECONSTRUCTION OF PAST TEMPERATURES

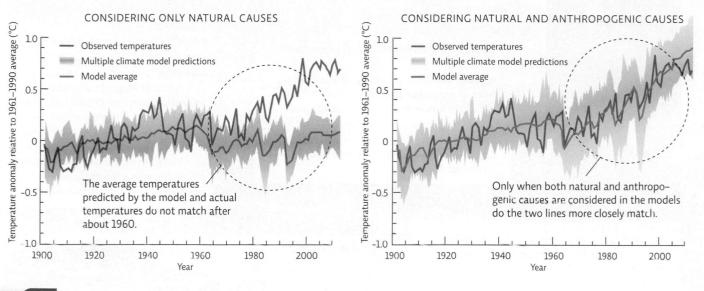

CONSIDERING ONLY NATURAL CAUSES

— Observed temperatures
▨ Multiple climate model predictions
— Model average

The average temperatures predicted by the model and actual temperatures do not match after about 1960.

CONSIDERING NATURAL AND ANTHROPOGENIC CAUSES

— Observed temperatures
▨ Multiple climate model predictions
— Model average

Only when both natural and anthropogenic causes are considered in the models do the two lines more closely match.

What would you expect the purple line and shaded area generated by the model in the left-hand graph in this infographic to look like, relative to the blue line (observed temperatures) if current warming could be explained by natural causes?

out that one of the biggest uncertainties is the effect that changes in mean annual temperature and precipitation will have on the world's forests. Some species may be able to migrate as temperatures rise and their optimum temperature range shifts northward—but which ones? Will spruce and fir species, which thrive in colder environments, disappear from the landscape? Will iconic and economically important trees like sugar maple and jack pine "move" to Canada? Will important ecological relationships be fractured as some species move or adapt while others (perhaps prey species or pollinators) do not?

Somewhere, buried in the reams of data that scientists like Frelich have spent decades accumulating, lie clues to answering these questions.

Some tree species are already migrating north, but it doesn't mean they will survive.

Like Frelich, Chris Woodall has spent his entire adult life studying the great forests—both temperate and boreal—that stretch from the northeastern United States well into Canada. Unlike Frelich, Woodall spends most of his time in front of a computer screen, crunching numbers. He works for the U.S. Department of Agriculture's (USDA's) *Forest Inventory and Analysis Program*, which maintains roughly 100,000 permanent plots throughout

the region—segments of forest where the USDA monitors a host of variables, from temperature and precipitation to tree growth and sapling density. It's a tremendous database, and for the past decade, scientists have used it to develop computerized models of how a warmer climate might change forests in the future. But until recently, no one had looked at whether these forests are already changing.

Woodall's logic is simple: Mature trees tell you where the current range is. Seedlings tell you where that range will be in the future. By comparing the ratio of seedlings to mature trees, one should be able to say whether any given species is on the move. Using the most recent data collected from 30 states and some 66,000 inventory plots, Woodall compared tree-seedling densities to forest biomass for more than two dozen tree species.

He was astounded by what he found. For most of those species—spruce, jack pine, sugar maple, and several others—the mean location for seedlings was significantly further north than their associated mature trees. "It's like if they had a gravity center, it would be pulling them northward," he said.

Whether these trees will actually survive in new locations is unknown. While studies of the average

↓ Species have evolved to live and thrive in certain habitats. If the climate is changing enough to alter ecosystems, we expect to see species responding by changing where they live or the timing of important temperature-dependent biological events. Species' responses such as shifting ranges or earlier blooming and hatching may be the best evidence that climate is actually changing; it is unlikely that these temperature-dependent events would change in this way if the planet were not getting warmer.

SOME SPECIES ARE MOVING TO HIGHER ALTITUDES

A 2008 study of the elevation distribution of 171 forest plants looked at where plants were found as well as the optimum elevation for growth. Data were compared for plants from two different time periods: 1905–1985 and 1986–2005. On average, plants shifted their range to higher elevations 29 meters per decade.

SOME IMPORTANT COMMUNITY CONNECTIONS ARE UNCOUPLED

Caterpillars, an important food source that birds feed their young, hatch based on temperature cues; the caterpillars are hatching 15 days sooner than they did in 1985. Pied flycatchers, birds that migrate north to their breeding grounds based on day–length cues (and thus have not changed their migration timing), are arriving too late to take advantage of peak caterpillar hatching.

FLOWERING PLANTS ARE BLOOMING EARLIER

Since many biological events are linked to the arrival of spring, we would expect to see earlier blooming, leaf out, and reproduction in species sensitive to temperature cues. One study showed that in Alberta, Canada, aspen, an early spring bloomer, blooms 2 weeks earlier now than normal. Other plants that normally bloom later in spring bloomed 0–6 days earlier.

? Why are species responses such as these considered good evidence that climate is changing?

locations for trees and saplings show northward migration, subsequent studies show contraction, or a loss of trees, at the northern edges. "The obvious question is, 'Well, how does that reconcile with the findings on mean, that they are moving north?'" Woodall said. "The answer is that it's like buffalo rushing off a cliff. As climate warms, you've got all these species gravitating to the northernmost edges of their traditional ranges. But that doesn't mean they will survive there long term."

Back in the North Woods, Frelich is working to understand why. So far, he's identified several forces that seem to be working in concert. "Warming triggers a whole cascade of events," he said. "Factors that make

tree ranges shift northward, and factors that prevent those trees from thriving in their new, more northerly habitats."

One of the biggest factors, he said, is deer, which have proliferated like mad in recent years. "In a warmer climate," Frelich said, "you'd expect the maple to advance in the understory, so that as the spruce die off, the maple are ready to take over. Likewise in the south: As maple move northward in response to warming, oak should move in to fill the void." But it turns out that deer like maple much more than they like spruce, and they like oak even more than maple. "So in places where the deer population is very high, the trees are having a hard time adapting to climate change because the deer are eating up all the early migrators."

And as climate warms, other stresses abound: Snowpack melts earlier, causing more severe water deficits in summer, right when trees need extra water to survive. The whole landscape dries out, creating conditions that favor intense fires and stressing trees so much that they

KEY CONCEPT 21.10

Species' responses to climate change such as range shifts provide strong evidence that climate is actually changing in a way that affects ecosystems.

SOME SPECIES ARE MOVING TO HIGHER LATITUDES

Several studies report a northward shift in the range of several, but not all, species of birds that have been studied. For example, the blue-gray gnatcatcher has extended its breeding range more than 300 kilometers northward since the 1970s.

The sugar maple was one of 11 out of 15 species that showed an average northern shift in range of 21 kilometers. Most seedlings are found in the northern-most regions, a sign that this species is migrating north.

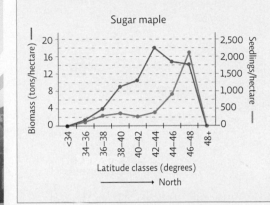

SOME SPECIES' POPULATIONS ARE EXPANDING

Many species of insect pests are on the increase, resulting in the decimation of forests around the world. Forests across the Northern Hemisphere are affected by several species of bark beetles. The pest populations normally die back in the winter, helping to keep them in check and allowing the trees to recover. Winter temperatures no longer get cold enough to kill the beetles in some areas, allowing the beetles to thrive year round. The U.S. Forest Service estimates that more than 4 million acres have been affected in the western United States.

> " *It's like buffalo rushing off a cliff. As climate warms, you've got all these species gravitating to the northernmost edges of their traditional ranges.* "
> —Chris Woodall.

become easy prey for beetle infestation. Pine beetles are a natural part of the life cycle in western forests, but the current outbreak, under way for more than a decade, is unlike anything seen before. "It used to get down to 40 below, every couple years, and that would keep things in check by killing the beetles off," said Frelich. "But that isn't happening anymore."

There are many other indicators that species are responding to climate change. In fact, that they are responding is evidence itself that climate is changing.

Because communities are complex assemblages of many species, there are concerns that important community connections will become uncoupled as species respond in different ways to climate change. **INFOGRAPHIC 21.9**

Climate change has environmental, economic, and health consequences.

The Boundary Waters Canoe Area surrounding Ham Lake—where Frelich and his colleagues were trapped—is the most heavily used chunk of the National Wilderness Preservation System. The boreal forest, its lakes, hiking trails, and breathtaking wildlife entice some 200,000 visitors every year, providing roughly 18,000 tourism jobs that pay a total of $240 million in wages. Global warming and shifting tree ranges threaten all that.

"Everyone's worried about losing the forests," Frelich said. "Resort owners, people who own cabins up there, local outfitters that rent camping gear—and the tourists themselves. If the forests burn too much, or if they

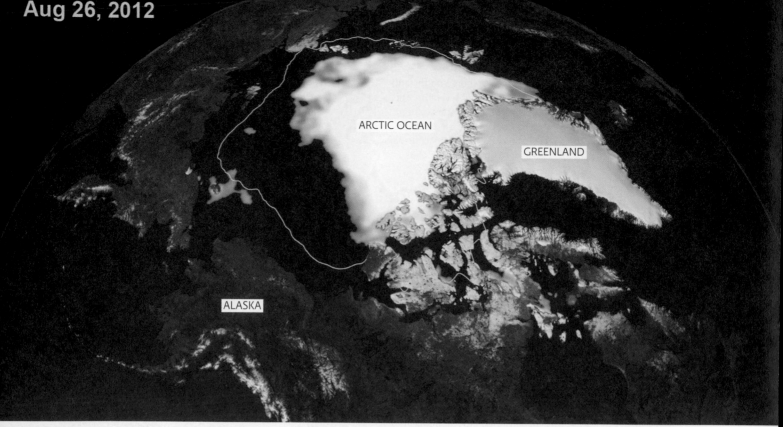

ARCTIC OCEAN

GREENLAND

ALASKA

NASA/Goddard Space Flight Center Scientific Visualization Studio The Blue Marble data is courtesy of Reto Stockli (NASA/GSFC)

↑ NASA satellite data reveals that 2012's minimum sea ice extent in the Arctic Ocean, reached on August 26, 2012 as depicted here, was far smaller than the 30-year average (in yellow) and is part of an emerging trend. The lowest ice extents were recorded between 2007 and 2012, with 2012 being the lowest in recorded history (very high confidence.)

descend into savanna, or can no longer support the iconic wildlife—like moose, lynx, and boreal owls—that people have come to expect, tourism will dry up. Because no one will want to vacation there."

And it's not just the tourism industry that will suffer in a warmer North Woods. The region as a whole supports about 1,000 forestry and logging jobs, which pay about $50 million in wages. On top of that, some individual species have become industries unto themselves—for example, the sugar maple.

Maple syrup production is heavily dependent on climate. The flow of sweet, sticky tree sap that eventually covers your pancakes is governed by changes in air pressure, which are in turn governed by changes in air temperature. When the temperature drops below freezing, the tree acts as a giant suction system, pulling the sap out of its branches, down into its roots. When the temperature rises above freezing, this action is reversed: The pressure gradient forces sap up from the roots, through the

> ### KEY CONCEPT 21.11
>
> The effects of climate change are varied and are already being felt. Though there are some positive impacts, there will likely be more losers than winners.

branches and out any holes—including ones that syrup makers have drilled for taps.

Traditionally, climate in the northern United States—from Minnesota to Maine—has provided the optimal freeze–thaw patterns for this process, which syrup makers call "sugaring." But in recent years, the transition from winter to spring has accelerated, leaving fewer freeze–thaw cycles and less sap overall. Meanwhile, across the border in Canada, warmer daytime temperatures have increased the number of freeze–thaw cycles, and some experts say that Canada is already in the middle of a syrup boom. "If current trends continue," Woodall said, "it's not impossible that the entire industry could one day be lost to Canada."

For Woodall, the stakes are both more basic and more terrifying than the loss of any given industry: Societies that don't protect their forests fail, he said. And it's easy to see why. "Forests stabilize soil and clear water of pollutants," he said. "In fact, the vast majority of Americans drink water that comes from a forested watershed. That means trees are as crucial to our survival as the water we drink." The loss of forests is also another positive feedback loop that threatens to exacerbate climate change: Forests are an important carbon sink: They store trillions of tons of CO_2 in their plants and soils. When they are burned, or die and decompose, much of that carbon is released into the atmosphere. (See

LaunchPad Chapter 28 for more on the ecosystem services of forests.)

To be sure, some people and places will benefit from climate change. In Greenland, for example, warmer temperatures have enabled farmers to grow a wider variety of crops than they have been able to grow in the past. Warmer weather during the summer months has also opened the Northwest Passage in some recent years—a long-sought-after shipping route through the Arctic Ocean—which would significantly reduce the transport time for ships that otherwise have to take the southern route through the Panama Canal. High-latitude land in Canada and Siberia will likely become warmer and more habitable—lessening the incidences of cold-related health problems and deaths.

But other areas will suffer more harm than good, and many climate-related impacts are being felt now. Extreme weather has begun to claim both human lives and valuable crops. Coastal flooding is already affecting low-lying areas from Bangladesh to New Orleans. Wildfires are breaking records for size and destruction across arid regions of the United States, southern Europe, and Australia. On the whole, global crop productivity is decreasing, especially for maize, rice, and wheat. Though impacts such as crop declines, wildfires, and floods most certainly are affected by a number of factors, many have been linked directly to the changing climate.

Desertification and drought are accelerating in many regions where precipitation levels are declining, including the southern and Sahel regions of Africa, much of southern Asia, the Mediterranean region, and the western United States. At the same time, some regions, such as northern Scandinavia and parts of North America, are experiencing more precipitation, and much of that precipitation is coming in heavy rain and snow events. For example, the northeastern United States has experienced a 58% increase in the number of days with very heavy precipitation in the past 50 years. The type of precipitation matters as well; the Pacific Northwest is receiving more rain and less snow, which is resulting in less snowpack to feed rivers during spring thaws. This is already impacting local water supplies and salmon populations that depend on mountain streams for spawning runs.

With rising global temperatures, infectious tropical diseases have begun to migrate north of their traditional ranges. Dengue fever, for example, has made its way from regions with more tropical climates into Texas and Florida. A recent study by University of Michigan ecologists also showed that in warmer years, malaria spread (as predicted) out of lowland areas to higher latitudes in Ethiopia and Colombia.

Species worldwide are also being impacted by climate change. Though other factors contribute to species declines, in many cases climate change is a leading threat, often because it is occurring too rapidly for some populations to successfully adapt. The golden toad of Costa Rica's cloud forest may have been the first species to go extinct due to climate change. Climate change has also been implicated in the decline or local extinction of many other species, such as various tropical coral species, Adélie penguins, the emblematic quiver tree of southern Africa, and the orange-spotted filefish, which has gone locally extinct in several coral reef communities in the Pacific Ocean. In fact, a 2004 evaluation made by a team led by University of York scientist Chris Thomas estimated that 15% to 37% of all species on Earth are "committed to extinction," based on a midrange climate-warming scenario.

Confronting climate change is challenging.

Jack Rajala's timber company owns 14,000 hectares (35,000 acres) of commercial forest land in Itasca County, Minnesota. In recent years, he's started doing things differently—namely, deliberately thinning out his paper birches in an effort to cultivate more oaks and white pines. This is not to say that the paper birch isn't valuable. But with massive die-offs under way throughout the region, Rajala needs to hedge his bets. "We think we can still facilitate birch," he said. "But it may be an understory tree, not a canopy tree anymore."

Rajala's strategy is called *resistance forestry*; it includes a handful of techniques aimed at maintaining existing species in their current locations, even as the climate shifts. For example, prescribed burns that mimic historic fire patterns might bolster the ranks of fire-dependent species like paper birch, black spruce, and jack pine, allowing them to spread over a wider area and enhancing their genetic diversity. Planting seeds instead of saplings also helps species hold their ground; natural selection favors the hardiest field-grown seedlings and so may yield a population better able to survive environmental stresses.

Scientists refer to such efforts, which are intended to minimize the extent or impact of climate change, as **mitigation**. Mitigation includes any attempt to seriously curb the amount of CO_2 we are releasing into the atmosphere—either by using carbon capture techniques to remove the greenhouse gas from our air and sequester it underground or by consuming fewer fossil fuels to begin with. In 2004, Princeton University researchers Stephen Pacala and Robert Socolow proposed a "stabilization wedge" strategy—a step-by-step implementation of currently available technology; each step could prevent the release of 1 billion tons of carbon. At the time of their paper's publication, Pacala and Socolow estimated that any 8 of the 15 steps, or "wedges," would stabilize CO_2 in the atmosphere

mitigation Efforts intended to minimize the extent or impact of a problem such as climate change.

KEY CONCEPT 21.12

Responding to climate change will require both steps that try to reduce future warming (mitigation) and steps to deal with inevitable warming (adaptation).

at close to 525 ppm in the next 50 years. Today, more wedges would be needed (since more CO_2 is in the air), especially if we aim for a target of 450 ppm CO_2. But the good news is that we already have at our disposal the means to seriously reduce CO_2 emissions. **INFOGRAPHIC 21.10**

On a national or global scale, mitigation efforts can be facilitated in a variety of ways, such as command-and-control regulations that limit greenhouse gas release; tax breaks; green taxes (in this case, **carbon taxes**); and market-driven programs such as carbon cap-and-trade (see Chapter 20). Financial incentives that encourage the development and use of non-carbon fuels and more energy-efficient technology will be critical (see Chapters 22, 23, and 24).

No matter which strategies we employ, curbing greenhouse gas emissions will take a coordinated global effort, meaning that world superpowers like the European Union, the United States, and China will have to cooperate with the developing nations of the world. So far, efforts have been fraught with obstacles and lack of cooperation. In 1997, an international treaty called the Kyoto Protocol was ratified by every UN nation except the United States. The treaty set different but specific targets for the reduction of CO_2 emissions for various countries; the United States objected because the protocol set much higher reduction requirements for developed countries (which were responsible for most of the historic emissions) than it did for developing countries.

Additional criticisms of Kyoto underscore the trouble with confronting such a global problem. Some of those who opposed Kyoto said that it went way too far in curbing greenhouse gas emissions; they argued that placing any kind of limit on CO_2 would hurt the economy because it would force industries to spend money updating their infrastructure, limit the amount of work they could do, and place them at a disadvantage compared with countries that had lower reduction targets under the treaty.

carbon taxes Governmental fees imposed on activities (such as fossil fuel use) that release CO_2 into the atmosphere.

precautionary principle Acting in a way that leaves a safety margin when the data is uncertain or severe consequences are possible.

Other critics said that Kyoto did not go far enough. Given the overwhelming evidence, these critics felt we needed to set much higher reduction targets to make any dent in the problem. They also felt that setting concrete, legally binding reduction targets in developed countries like the United States

would actually stimulate the economy because it would force companies in those countries to develop new, cleaner, more efficient technologies that other countries would then buy.

Taking steps to address greenhouse gas emissions will certainly cost money, but a 2014 study by researchers at the Massachusetts Institute of Technology showed that these steps will actually save money when the health benefits of controlling air pollution are factored in (since steps that reduce greenhouse gas emissions also reduce our exposure to particulate matter and ground-level ozone). For example, the study reported that while a carbon cap-and-trade system might cost the United States $14 billion to implement, it would save 10 times that amount thanks to decreased health care costs and fewer employee sick days. In a press release, lead author Tammy Thompson said, "If cost–benefit analyses of climate policies don't include the significant health benefits from healthier air, they dramatically underestimate the benefits of these policies."

Combatting climate change challenges the bedrock of modern civilization: energy use. Some argue that applying the **precautionary principle** now could help avoid, or at least lessen, some of the most serious consequences of a changing global climate.

The conflicting views on Kyoto also illustrate why climate change is considered a *wicked problem*—one that is resistant to resolution because it is fraught with complexity, change, incomplete information, and lack of sociopolitical acceptance (see Infographic 1.3). Indeed, the contention over climate change resides in the industrial and political sphere. Energy corporations and their backers have spent millions of dollars on campaigns designed to sow seeds of doubt. Naomi Oreskes and Erik Conway, science historians at Harvard University and California Technical Institute, respectively, reported in their 2010 book, *Merchants of Doubt*, that these campaigns are spearheaded by some of the same individuals who, at the behest of the tobacco industry, mounted campaigns that effectively raised doubt that smoking was a health hazard. Among other things, they claimed that science had insufficient evidence to conclude that smoking was bad for one's health.

These same tactics give fodder to climate skeptics. While an estimated 97% of scientists agree with the conclusion that climate is changing and that this change is due to human impact, a small but vocal percentage of scientists, along with individuals from industry and some conservative "think-tanks," are managing to stymie effective or far-reaching U.S. responses to climate change. They do so by raising doubt, dragging out long-discounted arguments, demanding more certainty before acting, and spending heavily on political campaigns and lobbying efforts.

INFOGRAPHIC 21.10 **FUTURE CLIMATE CHANGE DEPENDS ON OUR CURRENT AND FUTURE ACTIONS** 5

↓ In 2013, the IPCC produced its 5th Assessment Report based on an evaluation of 9,200 peer-reviewed scientific studies on climate change. That report devised four scenarios to predict future global temperatures depending on how our atmosphere might change during this century. Each scenario, called a Representative Concentration Pathway (RCP), represents the potential impact of a different level of radiative climate forcing in the year 2100 (e.g. RCP 2.6 = 2.6 W/m^2 of radiative forcing in 2100). Lower RCP values mean less radiative forcing and less climate warming. The level of radiative forcing (and amount of warming) we actually see will depend on societal factors such as how quickly countries reduce fossil fuel use and share emission reduction technologies with other nations, and on natural factors such as feedback loops in the Arctic and how much CO_2 the ocean can absorb.

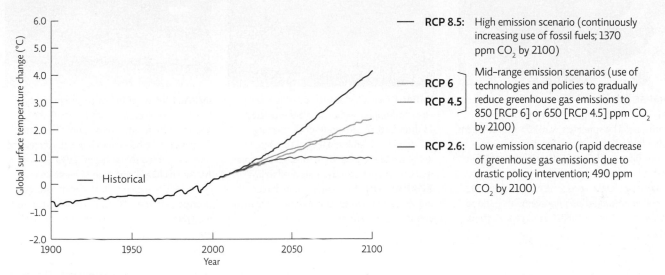

RCP 8.5: High emission scenario (continuously increasing use of fossil fuels; 1370 ppm CO_2 by 2100)

RCP 6
RCP 4.5 Mid-range emission scenarios (use of technologies and policies to gradually reduce greenhouse gas emissions to 850 [RCP 6] or 650 [RCP 4.5] ppm CO_2 by 2100)

RCP 2.6: Low emission scenario (rapid decrease of greenhouse gas emissions due to drastic policy intervention; 490 ppm CO_2 by 2100)

MITIGATION STRATEGIES HELP REDUCE FACTORS THAT LEAD TO CLIMATE CHANGE

↓ We can take steps to curb climate change by reducing emissions of greenhouse gases and by making better resource and land-use decisions. This will lessen the eventual peak warming we might experience. In 2004, Pacala and Socolow estimated that employing any 8 of 15 potential stabilization wedges, which each reduce CO_2 emissions by 1 gigaton (1 billion tons) per year over the next 50 years, would allow the atmosphere to stabilize close to 500 ppm. However, the longer we wait, the more "wedges" we will need.

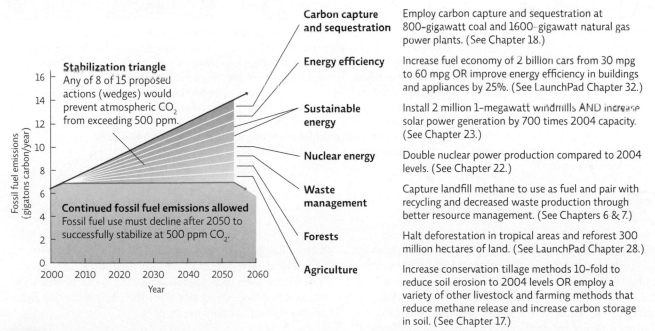

Carbon capture and sequestration — Employ carbon capture and sequestration at 800-gigawatt coal and 1600-gigawatt natural gas power plants. (See Chapter 18.)

Energy efficiency — Increase fuel economy of 2 billion cars from 30 mpg to 60 mpg OR improve energy efficiency in buildings and appliances by 25%. (See LaunchPad Chapter 32.)

Sustainable energy — Install 2 million 1-megawatt windmills AND increase solar power generation by 700 times 2004 capacity. (See Chapter 23.)

Nuclear energy — Double nuclear power production compared to 2004 levels. (See Chapter 22.)

Waste management — Capture landfill methane to use as fuel and pair with recycling and decreased waste production through better resource management. (See Chapters 6 & 7.)

Forests — Halt deforestation in tropical areas and reforest 300 million hectares of land. (See LaunchPad Chapter 28.)

Agriculture — Increase conservation tillage methods 10-fold to reduce soil erosion to 2004 levels OR employ a variety of other livestock and farming methods that reduce methane release and increase carbon storage in soil. (See Chapter 17.)

 Which stabilization wedge do you think would be the easiest to accomplish? Which would be the hardest?

↓ No matter how hard we try, we will not be able to avoid some future warming, as greenhouse gases emitted in the 20th century will continue to impact climate into the future. Adaptation strategies equip us to adjust to the inevitable warming that will occur.

HEALTH IMPACTS

John-Michael Maas/AFP/Getty Images/Newscom

IMPACT Tropical diseases like dengue fever and malaria are already moving to higher latitudes. The spread of waterborne pathogens should lead to increased incidence of infectious diseases; fewer cold-related deaths but more heat-related deaths.
ADAPTATION Improve disease surveillance, implement sanitation improvements in flood-prone areas, and establish emergency action plans.

CROP PRODUCTIVITY

Jim Richardson/National Geographic/Getty Images

IMPACT Overall, global crop yields have already dropped slightly and are projected to drop further with additional warming. Worldwide, an additional 6 million people are projected to be undernourished due to a reduction in per capita caloric availability.
ADAPTATION Use erosion-control techniques to improve agricultural productivity; choose crops to fit new conditions.

COASTAL EROSION AND FLOODING

© US Air Force Photo / Alamy

IMPACT Increases in health and property losses are projected due to sea level rise and more frequent, intense rainfall events. Damage due to hurricanes is already increasing and is expected to double by 2100.
ADAPTATION Relocation of some coastal communities may be necessary; construct protective barriers like seawalls and restore wetlands in coastal areas to protect inland areas.

BIODIVERSITY LOSSES

© Paul Souders/Corbis

IMPACT Some species may benefit and expand their ranges, but local or global extinctions are occurring and expected due to the inability to migrate, the lack of suitable habitat to migrate to, or the loss of other species on which they depend.
ADAPTATION Wildlife and habitat management to provide migration corridors or relocation assistance; protect vulnerable habitats from further human impact.

DROUGHT

Svan Torfinn/Panos

IMPACT The proportion of land area in severe drought is increasing due to increases in the evaporative loss of soil moisture and changes in precipitation patters. The western United States may be in the grips of its worst drought ever.
ADAPTATION Focus on methods to capture and conserve water, including desalinization in coastal areas; practice pollution prevention to increase and protect water supplies.

FIRE RISK

Layne Kennedy

IMPACT Fire has already increased in some areas; it has more than doubled in boreal North America and is projected to increase even more in the future.
ADAPTATION Pursue better fire-prevention management, including prescribed burns, thinning of forests to reduce combustible material, and improved fire-response plans.

 Which impacts concern you the most? Explain.

The Kyoto Protocol expired on December 31, 2012; so far, despite annual meetings, the international community has had no success in drafting a replacement treaty. The 2011 UN Climate Change Conference in Durban, South Africa, resulted in a legally binding agreement to adopt a future international climate treaty by 2015; subsequent annual conferences are laying the groundwork for a new treaty. Some nations (including Germany and the United Kingdom) have made their own progress toward CO_2 reductions, but other nations (including China and India) have seen increases in annual CO_2 emissions. U.S. emissions dropped slightly after the 2007 recession but are beginning to creep back up.

Meanwhile, change is coming to the North Woods. And as birches die off and maples try to expand their territory, as moose falter and pine beetles thrive, those who know the woods best say that mitigation will not be enough.

adaptation Efforts intended to help deal with a problem that exists, such as climate change.

"Resisting climate change at this point is like paddling upstream," said Frelich. "It might buy us some time, but it's not going to save the day." So, he said, we need to start thinking about **adaptation**: responding to the climate change that has already occurred or will inevitably occur. For human societies at large, this means taking steps to ensure a sufficient water supply in areas where freshwater supplies may dry up; it means planting different crops or shoring up coastlines against rising sea levels; it means preparing for heat waves and cold spells and outbreaks of infectious diseases. **INFOGRAPHIC 21.11**

In the North Woods, it might mean facilitation—moving tree species to entirely new ranges where they don't currently grow, based on the notion that the speed of climate change will make it impossible for natural tree migratory processes, such as seed dispersal, to occur. "The idea is that if we want the forest to adapt, we will have to help it along," said Frelich.

Facilitation has no shortage of critics, many of whom say such tinkering is both dangerous and unnecessary. "Facilitation is my nightmare," said John Almindinger, a forest ecologist with Minnesota's Department of Natural Resources. "That we'll start to believe we're smart enough to figure out how to move things. It's sheer hubris." Besides, he said, many if not most tree species seem to be moving just fine on their own, along traditional forest migration routes. So far, the U.S. Forest Service agrees; the agency does not allow such bold interventions as planting pines inside the wilderness.

Still, some skeptics are coming around to the idea. "We've changed the landscape through development and agriculture," said Peter Reich, a colleague of Frelich at the University of Minnesota. "And we've changed the climate, too, with fossil fuel consumption. So we might now need to change the way we manage wild lands to compensate."

On this much, everyone seems to agree: If northern Minnesota is to remain fully forested in the coming century, something will have to be done. "We see it already," said Rajala. "The impact of climate change will be too big to just let nature take its course."

Select References:
Beaubien, E., & A. Hamann. (2011). Spring flowering response to climate change between 1936 and 2006 in Alberta, Canada. *BioScience,* 61(7): 514–524.
Frelich, L., & P. Reich. (2010). Will environmental changes reinforce the impact of global warming on the prairie–forest border of central North America? *Frontiers in Ecology and the Environment,* 8(7): 371–378.
Hitch, A., & P. Leberg. (2007). Breeding distributions of North American bird species moving north as a result of climate change. *Conservation Biology,* 21(2): 534–539.
IPCC. (2013). *Climate Change 2013: The Physical Science Basis. Contribution of Working Group I to the Fifth Assessment Report of the Intergovernmental Panel on Climate Change* [Stocker, T. F., et al (eds.)]. Cambridge, UK: Cambridge University Press.
NOAA National Climatic Data Center. (2012). *State of the Climate: Global Analysis for Annual 2012.* www.ncdc.noaa.gov/sotc/global/2012/13.
Oreskes, N., & E. Conway. (2010). *Merchants of Doubt.* New York: Bloomsbury Press.
Pacala, S., & R. Socolow. (2004). Stabilization wedges: Solving the climate problem for the next 50 years with current technologies. *Science,* 305(5686): 968–972.
Thomas, C. D., et al. (2004). Extinction risk from climate change. *Nature,* 427(6970): 145–148.
Woodall, C. W., et al. (2009). An indicator of tree migration in forests of the eastern United States. *Forest Ecology and Management,* 257(5): 1434–1444.

BRING IT HOME

PERSONAL CHOICES THAT HELP

The effects of climate change are already being felt by humans, other species, and ecosystems around the globe. Individuals and community groups can make choices that decrease the production of greenhouse gases and increase the removal of CO_2 from the atmosphere. This will show policy makers that citizens are interested in preventing global climate change.

Individual Steps

• Do your part to reduce carbon emissions by conserving energy. Walk or ride a bike instead of driving a car. Share a ride with a coworker rather than driving alone. Negotiate with your employer to telecommute. Live close to where you work or go to school. Reduce your heating and cooling energy use and always turn off electronics and lights when not in use.
• If your utility company offers renewable energy, buy it.
• Reduce the carbon footprint of your food by decreasing the amount of feedlot-produced meat you eat. Buy your food as locally as possible to reduce energy used in transportation.
• Go to www.terrapass.com to see how you can offset your CO_2 production from your car, your house, and your airplane travel.

Group Action

• Volunteer to help build a zero-energy Habitat for Humanity home.
• Organize a community lecture on climate change with a local university expert or meteorologist as the speaker.
• Organize an event at your school or community to raise awareness about global climate change and ways to prevent it. Go to http://350.org to join a current campaign and get other program ideas.

Policy Change

• Consider writing, calling, or visiting the offices of your legislators and sharing your views about supporting funding for research and development of clean and renewable sources of energy. In addition, ask that they support the funding of science, especially efforts to understand and confront climate change.

ENVIRONMENTAL LITERACY UNDERSTANDING THE ISSUE

1 What is the difference between climate and weather? Why is a change of a few degrees in average global temperatures more concerning than day-to-day weather changes of a few degrees?

INFOGRAPHIC 21.1

1. Day-to-day changes in meteorological conditions are known as _____, whereas long-term patterns of meteorological conditions are known as _____.

2. True or False: A warmer climate should result in more heat extremes and in new record high temperatures.

3. In the winter of 2010, the northeastern part of the United States had several large snowstorms that resulted in record high snowfall amounts. How does this weather fit in with the notion of global climate change?

2 What is the physical and biological evidence that climate change is currently occurring?

INFOGRAPHICS 21.2 AND 21.9

4. Recent sea level rise is attributed to:
   a. melting glaciers.
   b. thermal expansion of water.
   c. melting icebergs.
   d. a and b.
   e. a, b, and c.

5. Which of the following is least compelling line of evidence in that climate is indeed changing?
   a. An increase in the number of hurricanes
   b. An increase in global average temperature
   c. Temperature-dependent shifts in the ranges of species
   d. Sea level rise

6. Outline the evidence for climate change. Do you feel that this evidence supports the conclusion that climate is changing? Explain.

3 What natural and anthropogenic factors affect climate, and which are implicated in the climate change we are experiencing now? How might positive feedback loops affect climate?

INFOGRAPHICS 21.3, 21.4, 21.5, AND 21.6

7. True or False: Current warming can be explained by the Milankovitch cycles.

8. Which of the following has the greatest albedo?
   a. A forest
   b. A light-colored roof
   c. A dark asphalt road
   d. The surface of the ocean

9. Define *greenhouse gases*. What human actions have led to an increase in the amount of greenhouse gases in the atmosphere? What has been the result?

10. What is the difference between a positive feedback loop and a negative feedback loop? Give a climate-related example of each.

4 How do scientists determine past and present temperatures and CO_2 concentrations? What evidence suggests that climate change is due to human impact?

INFOGRAPHICS 21.7 AND 21.8

11. True or False: Current CO_2 levels in the atmosphere can be measured, but there are no good methods for determining CO_2 levels in the distant past.

12. What relationship is seen between temperature and atmospheric CO_2 levels?
    a. They are positively correlated.
    b. They are negatively correlated.
    c. They are not correlated in any meaningful way.

13. Compare and contrast the major radiative forcers, including both those that are natural and those that are humanmade. Overall, which forcers are currently having the greatest effect on global climate? Are they natural or produced by human actions?

5 What are the current and potential future impacts of climate change? What actions can we take to respond to a world with a changing climate?

INFOGRAPHICS 21.10 AND 21.11

14. True or False: The amount of future warming that Earth will experience in the next 100 years depends on the choices we make now.

15. Trying to decrease the extent or impact of future climate change is known as _____, whereas taking steps to adjust to current or inevitable climate change is known as _____.

16. According to the stabilization wedge strategy proposed by Pacala and Socolow we:
    a. must reduce atmospheric CO_2 concentrations below the 350ppm to avoid disastrous consequences.
    b. can continue to burn fossil fuels for the next 50 years if we reduce greenhouse gas emissions in other ways.
    c. can use a variety of currently available strategies to prevent atmospheric CO_2 from exceeding 500 ppm.
    d. should focus our efforts on adapting to inevitable climate change rather than mitigation

17. Describe the types of problems that global climate change causes for human health. Which do you feel is likely to cause the biggest problem? Why?

SCIENCE LITERACY WORKING WITH DATA

The following graphs show 2 of the 15 northern species evaluated in Chris Woodall's study of tree-range migration mentioned in this chapter. The total standing biomass and the total number of seedlings of each species are shown at different latitudes within the study area.

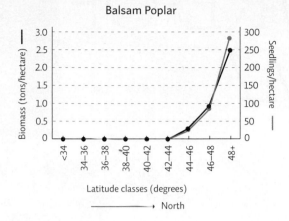

Balsam Poplar

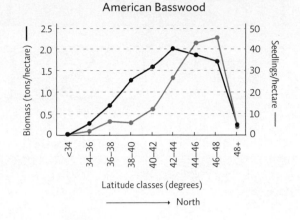

American Basswood

Interpretation

1. What does the purple line on each graph represent? What does the blue line represent?

2. There are 111 kilometers between two adjacent latitude lines. How far apart are the latitude classes shown here? How many kilometers wide is the study site (from latitude 34° to 48°)?

3. Look at each graph to determine at which latitude class each species shows the most biomass per hectare. Which tree species has more standing biomass at its peak: balsam poplar or American basswood? How can you tell?

Advance Your Thinking

4. Is either of these tree populations exhibiting a range migration shift? Present the evidence for your conclusions.

5. Of the two species shown here, which would you expect to see at higher altitudes on a mountainside and which at lower altitudes? Assuming that both species can migrate, what do you predict will happen to the populations of these two species if the climate warms a little? If it warms a lot? Explain your answers.

INFORMATION LITERACY EVALUATING INFORMATION

Among scientists, there is broad consensus (97%) that climate change is significantly caused by human activity. Yet in a 2014 Gallup poll, 61% of the public are either not convinced or deny that humans are causing global warming. Members of the public get their information from a variety of media sources and information posted on the Internet. How can there be such a large disconnect between scientists and the public?

Go to the Global Warming Hoax page (www.globalwarminghoax .com/news.php?) and read the entry "Antarctic Sea Ice for March 2010 Significantly Greater Than 1980."

Evaluate the website and work with the information to answer the following questions:

1. Is this a reliable information source? Does it have a clear and transparent agenda?
   a. Who runs the website? Do this person's/group's credentials make the site reliable or unreliable? Explain.
   b. What is the primary message of the website? What evidence is it providing in the short article on Antarctic sea ice?
   c. Do you have any questions about the data presented? If so, what are they?

Now go to the Skeptical Science website (www.skepticalscience .com). Click on the link "Most Used Climate Myths."

2. Is this a reliable information source? Does it have a clear and transparent agenda?
   a. Who runs the website? Do this person's/group's credentials make the site reliable or unreliable? Explain.
   b. What is the primary message of this website? What types of evidence does it provide to support its message?
   c. Click on the "Antarctica is gaining ice" link. Read the article and compare the main point of the article to the article on the Global Warming Hoax site.
   d. Which explanation and website do you find more credible? Why?

Find an additional case study online at http://www.macmillanhighered.com/launchpad/saes2e

THE FUTURE OF FUKUSHIMA

Can nuclear energy overcome its bad rep?

CORE MESSAGE

Nuclear energy can be harnessed to create tremendous amounts of power, and with concerns over fossil fuel supplies and climate change, nuclear energy has the potential to be an increasingly important part of the world's energy future. However, there are serious safety concerns with nuclear power, including vulnerability to natural disasters, radioactive waste disposal, and potential for weapons production.

AFTER READING THIS CHAPTER, YOU SHOULD BE ABLE TO ANSWER THE FOLLOWING **GUIDING QUESTIONS**

1

What are radioactive isotopes, and why are they important for nuclear power?

2

What types of radiation are produced when an isotope decays? How is the rate of decay of a radioactive atom measured?

The Japanese authorities originally declared a 20-kilometer (12.5 mile) evacuation area around Fukushima, an exclusion zone which may only be entered under government supervision. Four months after the explosion, residents in protective suits are briefed before being escorted to their homes to retrieve a few small items.
AP Photo/David Guttenfelder

3

How is nuclear energy harnessed to generate electricity in a fission reactor? How safe are nuclear reactors?

4

How dangerous is radiation that is released from radioactive material? What problems are associated with nuclear waste?

5

In a trade-off analysis of nuclear power, what factors must be considered, and what is your own conclusion regarding the future role of nuclear power?

The Fukushima Daiichi Nuclear Power Station is a maze of steel and concrete, perched right on Japan's Pacific coast, just 240 kilometers (150 miles) north of Tokyo. Its six nuclear reactors supplied some 4.7 GW (1 gigawatt = 1 billion watts) of electric power to the country, making it one of the largest nuclear power plants in the world. On March 11, 2011, when a magnitude 9.0 earthquake struck 130 kilometers (80 miles) north of the plant, there were more than 6,000 workers inside. The quake caused a power outage, and in the darkness, chaos ensued: Men and women groped desperately for ground that would not stabilize beneath their hands and feet, and they shouted in panic as steel and concrete collided around them. When the shaking stopped, emergency lights came on, revealing a cloud of dust. But that was only the beginning of the disaster.

The earthquake had erupted beneath the ocean floor, triggering a tsunami that would arrive at the plant in two distinct waves. The first wave was not big enough to breach the 10-meter-high (33-foot-high) concrete wall that had been built between the plant and the sea. But the second wave, a fearsome mass of water that came 8 minutes later, was. At four stories tall, the water wall bulldozed a string of protective barriers, sent buses and cars and trucks careening into pipes and levers and control panels, and eventually settled, in deep black pools, around the reactors themselves.

These were the strongest earthquake and largest tsunami in the country's long memory. Together, they would claim some 20,000 lives along a 400-kilometer (250-mile) stretch of coast (roughly equal to the distance between Maine and Manhattan). But as the ground steadied and the water subsided, the world's attention would quickly turn to a third disaster, even more precarious and potentially deadly than the first two: the risk of nuclear meltdown at Daiichi.

It's no surprise that the story of nuclear power pivots on calamity. Ever since its potential was first demonstrated, humankind has scurried relentlessly between two competing goals—the desire to harness nuclear energy for our own ends and the impulse to protect ourselves from its destructive capacity. When the ground trembled beneath Fukushima, concerns over greenhouse gases and global warming had been pushing much of the world—including the United States—toward the former. As the people of Japan scrambled to respond, the world watched closely.

nuclear energy Energy released when an atom is split (fission) or combines with another to form a new atom (fusion).

nuclear fission A nuclear reaction that occurs when a neutron strikes the nucleus of an atom and breaks it into two or more parts.

The heat of nuclear reactions can be harnessed to produce electricity.

In some ways, electricity generated using **nuclear energy** is no different than other forms of thermoelectric power (those that use heat to produce electricity). Just like power plants that run on oil or coal, nuclear plants use heat to boil water and produce steam, which is then used to generate electricity. The difference, really, is in where that heat comes from. Coal and oil plants create it by burning fossil fuels. In the thermonuclear production of electricity, heat is produced through a controlled nuclear reaction—usually a **nuclear fission** reaction.

◉ **WHERE IS FUKUSHIMA, JAPAN?**

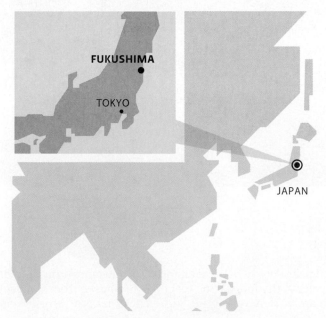

AP Photo/David Guttenfelder

↑ The crippled Fukushima Daiichi Nuclear Power Station 10 months after the disaster.

→ Satellite image of the damaged Fukushima Daiichi Nuclear Power Station on March 14, 2011.

Photo by Digital Globe via ABACAPRESS.COM/Newscom

INFOGRAPHIC 22.1 ATOMS AND ISOTOPES   1

→ All matter is made up of atoms. Each atom is made up of subatomic particles: *protons* and *neutrons* in the nucleus (center) of the atom, make up the mass of the atom. Orbiting around the nucleus are much smaller particles called *electrons*. Different combinations of these subatomic particles produce specific *elements*—a chemical substance made up of only one kind of atom. The number of protons, the *atomic number*, is unique to each element. For example, any atom with only 2 protons is an atom of the element helium. The sum of the number of protons and neutrons gives an element its *mass number*. Helium, shown here, is an element that has 2 protons, 2 neutrons, and 2 electrons.

THE ATOM

HELIUM ATOM

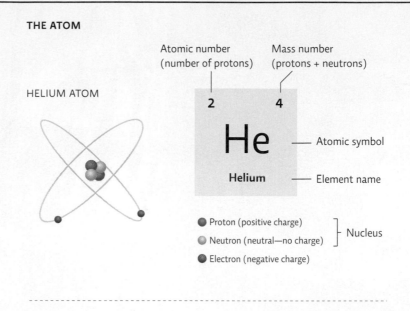

Atomic number (number of protons)

Mass number (protons + neutrons)

2   4

He — Atomic symbol

Helium — Element name

● Proton (positive charge)
◐ Neutron (neutral—no charge) } Nucleus
● Electron (negative charge)

→ *Isotopes* are atoms that have the same atomic number (number of protons) but a different number of neutrons, and thus a different mass number. An atom with 92 protons is uranium; it can have 146 neutrons, and is called U-238 (92 protons + 146 neutrons = 238). Another uranium isotope, U-235, has 143 neutrons (92 protons + 143 neutrons = 235).

 Uranium also exists as uranium-233. How many protons and neutrons does it have? Do you think it is more or less stable than U-235? Explain.

ISOTOPES

URANIUM-238

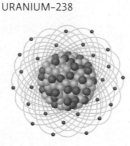

● 92 protons ◐ 146 neutrons
More neutrons—heavier

URANIUM-235

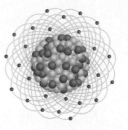

● 92 protons ◐ 143 neutrons
Fewer neutrons—lighter and less stable

Nuclear fission reactions are those that result in the splitting of an atom. Specifically, these reactions involve a special type of atom known as a *radioactive isotope*. Some elements can exist in two or more forms; each form has the same number of protons and electrons but a different number of neutrons and, hence, a different atomic mass; these different versions of the atom are called **isotopes**. **INFOGRAPHIC 22.1**

Most isotopes are stable, meaning they do not spontaneously lose protons or neutrons. But some are **radioactive**: They emit subatomic particles and heat energy (radiation) in a process known as *radioactive decay*.

isotopes Atoms that have different numbers of neutrons in their nucleus but the same number of protons.

radioactive Atoms that spontaneously emit subatomic particles and/or energy.

radioactive half-life The time it takes for half of the radioactive isotopes in a sample to decay to a new form.

Radioactive decay is measured in half-lives. An isotope's **radioactive half-life** is the amount of time it takes for half of the radioactive material in question to decay to a new form. So after one half-life, 50% of the material will decay; in the next half-life, 50% of what's left (or 25% of the original amount) will then decay. After 10 half-lives, just 0.1% of the original radioactive material is left. **INFOGRAPHIC 22.2**

Most nuclear reactors use uranium, which has several isotopes. Uranium-238 (U-238) is the most stable and is the most abundant form of uranium; it makes up roughly

KEY CONCEPT 22.1

Radioactive isotopes are the starting material for the thermonuclear production of electricity.

↓ The rate of decay for a given radioactive isotope is predictable and expressed as a *half-life*—the amount of time it takes for half of the original radioactive material (*parent*) to decay to the new *daughter* material (a new isotope, or even a new atom if protons are lost). Radioactive isotopes and their daughter radioactive isotopes continue to decay until they form a stable isotope that no longer loses particles. For instance, U-238 decays initially to thorium-234, which itself will decay over time. The entire decay sequence of U-238 includes progression through at least 13 isotopes (each step with its own half-life that ranges from milliseconds to thousands of years) until the final isotope decays into lead-206, a stable atom.

RADIOACTIVE DECAY

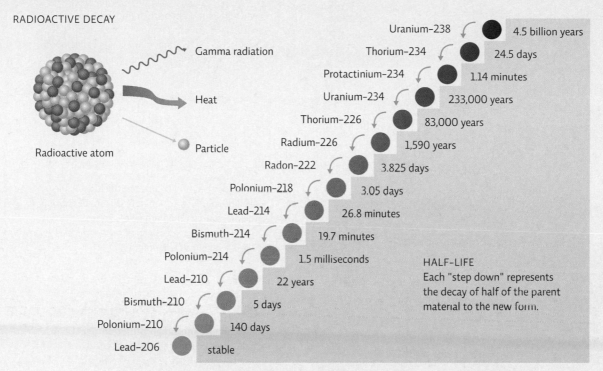

Uranium-238	4.5 billion years
Thorium-234	24.5 days
Protactinium-234	1.14 minutes
Uranium-234	233,000 years
Thorium-226	83,000 years
Radium-226	1,590 years
Radon-222	3.825 days
Polonium-218	3.05 days
Lead-214	26.8 minutes
Bismuth-214	19.7 minutes
Polonium-214	1.5 milliseconds
Lead-210	22 years
Bismuth-210	5 days
Polonium-210	140 days
Lead-206	stable

Gamma radiation

Heat

Particle

Radioactive atom

HALF-LIFE
Each "step down" represents the decay of half of the parent material to the new form.

RADIOACTIVE HALF-LIFE

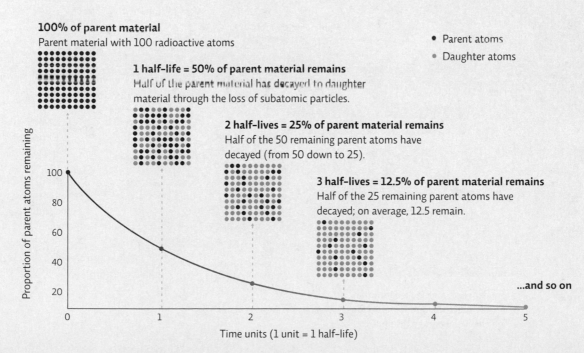

100% of parent material
Parent material with 100 radioactive atoms

● Parent atoms
● Daughter atoms

1 half-life = 50% of parent material remains
Half of the parent material has decayed to daughter material through the loss of subatomic particles.

2 half-lives = 25% of parent material remains
Half of the 50 remaining parent atoms have decayed (from 50 down to 25).

3 half-lives = 12.5% of parent material remains
Half of the 25 remaining parent atoms have decayed; on average, 12.5 remain.

Proportion of parent atoms remaining

100
80
60
40
20

0   1   2   3   4   5

...and so on

Time units (1 unit = 1 half-life)

What percentage of the parent material will be left after 5 half-lives?

INFOGRAPHIC 22.3   NUCLEAR FUEL PRODUCTION

↓ Uranium ore (rock that contains uranium) is mined and goes through many stages of processing to produce fuel suitable for a nuclear reactor. The process creates hazardous waste at every step.

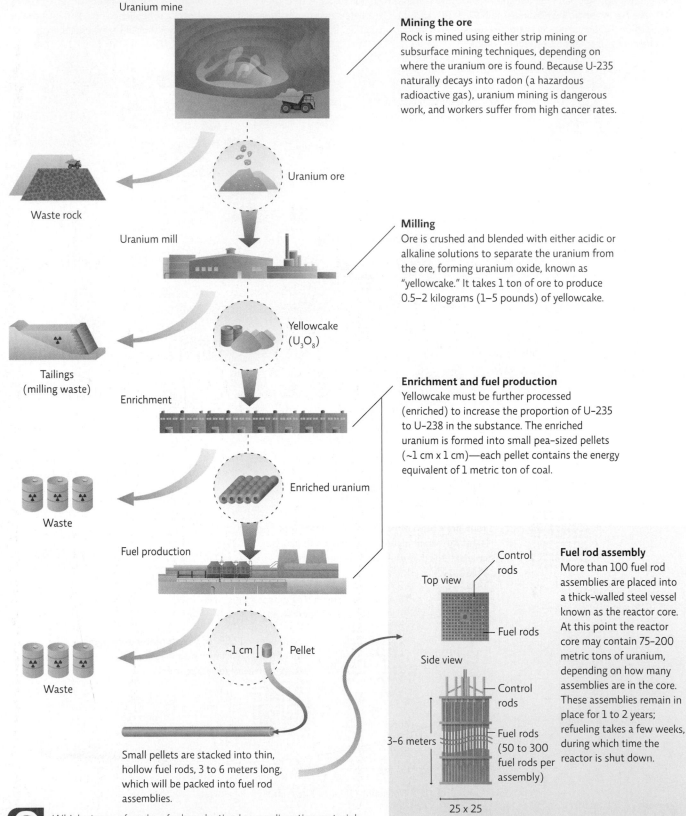

Mining the ore
Rock is mined using either strip mining or subsurface mining techniques, depending on where the uranium ore is found. Because U-235 naturally decays into radon (a hazardous radioactive gas), uranium mining is dangerous work, and workers suffer from high cancer rates.

Milling
Ore is crushed and blended with either acidic or alkaline solutions to separate the uranium from the ore, forming uranium oxide, known as "yellowcake." It takes 1 ton of ore to produce 0.5–2 kilograms (1–5 pounds) of yellowcake.

Enrichment and fuel production
Yellowcake must be further processed (enriched) to increase the proportion of U-235 to U-238 in the substance. The enriched uranium is formed into small pea-sized pellets (~1 cm x 1 cm)—each pellet contains the energy equivalent of 1 metric ton of coal.

Fuel rod assembly
More than 100 fuel rod assemblies are placed into a thick-walled steel vessel known as the reactor core. At this point the reactor core may contain 75–200 metric tons of uranium, depending on how many assemblies are in the core. These assemblies remain in place for 1 to 2 years; refueling takes a few weeks, during which time the reactor is shut down.

Small pellets are stacked into thin, hollow fuel rods, 3 to 6 meters long, which will be packed into fuel rod assemblies.

Which stages of nuclear fuel production have radioactive material present?

INFOGRAPHIC 22.4   NUCLEAR FISSION REACTION

↓ Fission, or the breaking apart of atoms, begins when an atom like U-235 is bombarded with a neutron. This breaks the atom into other smaller atoms and releases free neutrons, which in turn hit other U-235 atoms, causing them to split and release neutrons, and so on. The reaction in the fuel assembly is controlled by the insertion of control rods of nonfissionable material which absorb some of the free neutrons.

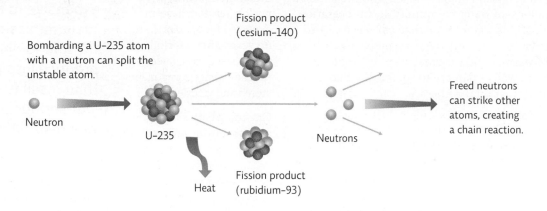

Bombarding a U-235 atom with a neutron can split the unstable atom.

Neutron

U-235

Heat

Fission product (cesium-140)

Fission product (rubidium-93)

Neutrons

Freed neutrons can strike other atoms, creating a chain reaction.

How do control rods help control the chain reaction in a nuclear fuel assembly?

KEY CONCEPT 22.2

Radioactive decay is measured in terms of a half-life and can vary in duration for different isotopes, from milliseconds to billions of years.

200 years, but the worry that we would run out of nuclear fuel is less of an issue than it is with fossil fuels; other isotopes can also be used. **INFOGRAPHIC 22.3**

A nuclear fission chain reaction begins when U-235 in the fuel rods is deliberately bombarded with neutrons. This bombardment makes the uranium nucleus unstable, causing it to split into a variety of two or more smaller atoms and releasing two or three additional neutrons in the process. These newly released neutrons then hit other U-235 atoms, causing them to split and release even more neutrons, and so on.

KEY CONCEPT 22.3

The production of nuclear fuel involves mining and several processing steps, all of which generate hazardous waste.

Unlike the type of nuclear reaction at work in a nuclear bomb, which uses much more radioactive

99% of Earth's total supply. But it's U-235, the most reactive form, that is mined from Earth, processed into nuclear fuel (which, like all other mining work, involves both safety and environmental hazards), and packed into the **fuel rods** that are used in facilities like Fukushima. Estimates for supplies of uranium range from 80 to

material, setting off a massive chain reaction that is almost instantaneous, the reactions at nuclear power plants are highly controlled. **Control rods**—made of materials such as boron or graphite that absorb neutrons—are placed in the fuel rod assembly between the fuel rods to control the speed of the reaction. They can be added (to slow down or stop the reaction) or removed (to make it go faster). **INFOGRAPHIC 22.4**

Even controlled, this chain reaction releases a tremendous amount of heat—10 million times more than would be released by burning a comparable amount of coal or oil. The heat is used to boil water, which produces steam, which turns turbines that create electricity.

All types of thermoelectric power take a lot of water; that's why power plants are sited near rivers and oceans. But at the moment, nuclear power requires the most—on average, around 2,500 liters per megawatt hour (MWh), compared with 1,900 liters per MWh for coal and 600 liters per MWh for natural gas. (A MWh is the production of 1 megawatt—1 million watts—over an hour's time.) That means a typical

KEY CONCEPT 22.4

In a nuclear fission reaction, U-235 is bombarded with neutrons to split the atoms; this releases more neutrons, which leads to a self-perpetuating chain reaction.

---

fuel rods Hollow metal cylinders filled with uranium fuel pellets for use in fission reactors.

control rods Rods that absorb neutrons and slow the fission chain reaction.

1,000-MW nuclear reactor requires 2,500,000 liters of water per minute to flow through the cooling system. Some reactors use much more. For example, each of the two 845-MW reactors at the Calvert Cliffs nuclear power plant in Maryland requires 4,500,000 liters (that's 1,200,000 gallons) of water per minute during operation. Though most of this water (more than 95%) is returned to the source (river or ocean), there are problems with the release of warmer-than-normal water back into the environment, as well as damage to aquatic life that gets trapped in or against intake filters.

The reason for all that water is simple: With nuclear energy, water is needed not only to produce steam but also to keep spent fuel rods cool and to prevent the reactor from overheating. (Remember that heat is produced from radioactive decay of fission by-products, and it continues even after the reactor is shut down and fission stops; therefore, spent fuel needs constant cooling.) Without water to cool them, fuel rods can melt, releasing large amounts of radioactivity; the fuel rod metal casing can also get hot enough to react with steam in a way that produces highly explosive hydrogen gas.

Nuclear energy has a troubled history.

Nuclear energy is the most concentrated source of energy on Earth. Its fearsome power was first demonstrated in 1945, when the U.S. military dropped atomic bombs over the Japanese cities of Hiroshima and Nagasaki. The bombs brought an end to World War II, but they also wreaked havoc on an entire nation of civilians: radiation sickness, cancers that killed slowly and to which young children were especially vulnerable, infertility in some, and birth defects in others.

In 1953, President Eisenhower made his famous "Atoms for Peace" speech, laying out a plan by which this destructive force could be harnessed for good: Instead of building bombs, we would produce cheap, reliable energy. In the years that followed, nuclear physics indeed gave rise to a litany of technologies that have benefitted humankind, from radiocarbon dating to X-rays to radiation therapy for cancer.

But while there are more than 400 nuclear power plants around the world, nuclear energy itself remains mired in controversy.

Proponents argue that uranium ore (uranium-containing rock) is both more abundant and produces a more efficient fuel than any fossil fuel. For example, 1 kilogram of uranium produces the same amount of energy as about 100,000 kilograms of coal. And the operating costs, per kilowatt-hour, for nuclear power are comparable to those of coal.

Of course, as critics are quick to point out, that estimate does not factor in the great expense of building, maintaining, and then decommissioning nuclear plants. In the United States, it costs about $4 billion to build a nuclear reactor (about twice the cost of building a coal power plant), and anywhere from $200 million to $1 trillion to decommission one. (Most nuclear reactors have an expected life span of 40 to 60 years, after which they need to be disassembled and the radioactive components stored and guarded.)

One thing both sides agree on is that using nuclear energy is a cleaner way to produce electricity. The processes of generating it emit much less CO_2 than the analogous processes for any fossil fuel, and nuclear power creates virtually none of the other problematic combustion by-products, like sulfur dioxide, nitrogen oxides, and particulate matter. According to the Department of Energy, switching from fossil fuels to nuclear energy would be the single most effective way to reduce greenhouse gas emissions in the United States. In the United States alone, existing nuclear power plants already prevent roughly 650 million metric tons of CO_2 per year from being released. Worldwide, they prevent close to 2.5 billion metric tons of CO_2 from entering the atmosphere.

And despite some persistent fears, research suggests that living near a nuclear power plant is actually safer than living near a coal-fired one. A 2011 study found no increased risk of birth defects for those living within 10 kilometers (6 miles) of a nuclear facility compared to the risk for those living farther away. Meanwhile, epidemiologist Javier García-Pérez has found that in Spain, the number of cancer-related deaths does increase as one moves closer to coal-fired power plants. "You still have some environmental hazards, from mining uranium and from radioactive waste and water," says Charles Powers, a professor and nuclear energy scientist at Vanderbilt University. "But on balance, nuclear is far cleaner than any fossil fuel."

Still, the debates over safety remain unresolved. Proponents point out that, considering the number of existing plants and the length of time they have been operating, accidents have been exceedingly few and far between. But opponents say that such safety claims ignore two key points: the vulnerability of nuclear power plants to natural disasters (which at Fukushima led to nuclear meltdown and the release of radioactive material into the environment) and the potential for nuclear fuel to be stolen and weaponized. The radioactive waste produced by nuclear power plants is another huge safety issue: It's extremely dangerous; there's a lot of it, and we have yet to come up with a plan for disposing of it safely.

KEY CONCEPT 22.5

Nuclear reactors are expensive to build and decommission. They are safe to operate and produce less pollution than fossil fuel plants, but serious problems can result if things go wrong.

France now generates more than 75% of its electricity from nuclear power. But while the United States has the most nuclear reactors of any nation, Americans themselves have been divided over nuclear energy since the 1980s, after two infamous nuclear accidents made global headlines. The first was a partial meltdown at the Three Mile Island plant near Middletown, Pennsylvania, in 1979 (due to an electrical failure followed by a flurry of operator errors). The second was a full nuclear meltdown at the Chernobyl reactor in what is now Ukraine, in the spring of 1986.

The Three Mile Island incident did not result in any deaths or major public health problems. The Chernobyl meltdown was considerably more severe. A steam explosion triggered by a test that went awry sent tremendous amounts of radiation wafting over much of western Russia and Europe. More than one-fifth of the surrounding farmland remains unusable to this day, and the World Health Organization estimates that the radiation will ultimately be responsible for some 4,000 deaths when all is said and done. That figure does not include cancer-related deaths, which range from 60,000 (according to one European report) to nearly 1 million (according to a Russian report).

The radioactive waste produced by nuclear power plants is another huge safety issue: It's extremely dangerous; there's a lot of it, and we have yet to come up with a plan for disposing of it safely.

For their part, the Japanese were terrified of nuclear power after the bombings of Hiroshima and Nagasaki. (The popular Godzilla movies were actually based on a fictional reptile that had been mutated by a nuclear reaction and was coming to exact his revenge on humankind!) But as their country entered its own era of industrialization and economic growth, they were forced to overcome those fears. "Japan had no other natural energy source," says Frank N. von Hippel, nuclear physicist and arms control expert at Princeton University. "Nuclear was their only ticket to becoming a world superpower."

Nuclear accidents can be devastating.

On the day of the quake, each of the three operating reactors at the Fukushima plant held about 25,000 fuel rods, each rod about 4 meters (12 feet) long and filled with pellets of enriched uranium. The power outage had not only plunged the plant into darkness but also stopped the normal delivery of water to those reactors.

As the world watched, workers at the plant tried everything they could think of to get cold water on the hot fuel. They tried to bring in fire trucks and emergency power vehicles, but the quake and tsunami had rendered the roads impassable. In one desperate attempt, some workers even searched the parking lot for vehicles that might have survived the tsunami with their batteries intact. But it was all to no avail.

Not only were the rods in danger of melting and of producing hydrogen gas, but without a steady supply of coolant, they were rapidly boiling away

↓ Police guard a checkpoint at the edge of the exclusion zone leading to the town of Minami Soma, just north of Daiichi. The sign reads "Keep Out."

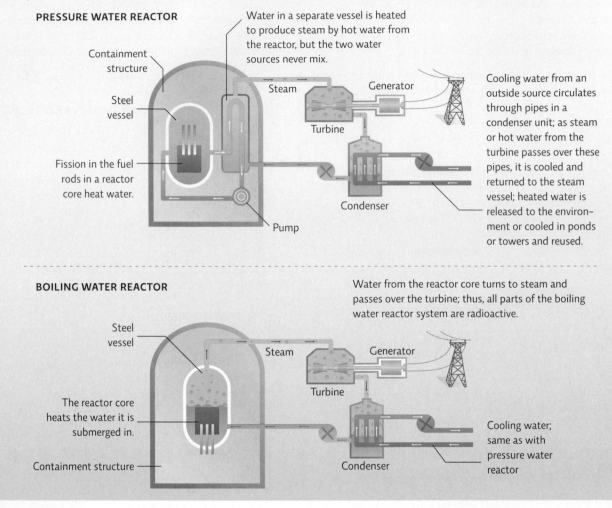

INFOGRAPHIC 22.5   **HOW IT WORKS: NUCLEAR REACTORS**

↓ The most common type of nuclear power plant is a pressurized water reactor (PWR); 265 of them are in operation around the world. The reactor at Fukushima is a boiling water reactor (BWR), one of 94 in the world. Both designs use fuel assemblies with control rods and use water as a cooling and steam source. Temperatures are kept "down" to about 1,400°C, well below the melting temperature of the uranium fuel (2,800°C) and the metal casing of the fuel rods (2,200°C). If the reaction is not kept cool with circulating water, a meltdown can occur. Even if "shut down" by inserting all the control rods to absorb the neutrons and stop the chain reaction, heat will still be produced by the natural decay of the isotopes (they are not being split by bombardment; they are simply spontaneously losing particles).

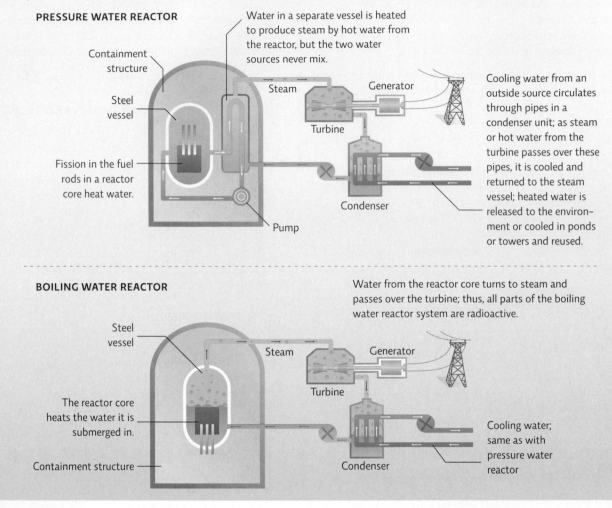

PRESSURE WATER REACTOR

Containment structure

Steel vessel

Fission in the fuel rods in a reactor core heat water.

Water in a separate vessel is heated to produce steam by hot water from the reactor, but the two water sources never mix.

Steam

Generator

Turbine

Pump

Condenser

Cooling water from an outside source circulates through pipes in a condenser unit; as steam or hot water from the turbine passes over these pipes, it is cooled and returned to the steam vessel; heated water is released to the environment or cooled in ponds or towers and reused.

BOILING WATER REACTOR

Water from the reactor core turns to steam and passes over the turbine; thus, all parts of the boiling water reactor system are radioactive.

Steel vessel

Steam

Generator

Turbine

The reactor core heats the water it is submerged in.

Containment structure

Condenser

Cooling water; same as with pressure water reactor

? How is a nuclear reactor similar to a coal-fired power plant? How are they different?

all the existing water, thereby creating huge steam pressure in the reactor. To prevent an explosion, the steam would have to be released through a vent. But the design of the Fukushima reactors made this an especially tricky feat.

There are several different types of fission reactors. The most common type worldwide is a *pressurized water reactor* (PWR), where the steam that turns the turbine is not exposed to radiation: Nuclear fission in the core heats water under pressure (like a pressure cooker); that radioactive water enters a pipe that passes through, and thus heats, a separate container of water (which is not exposed to radiation); it is the steam from this

radiation-free water that turns the turbine. The reactors at Fukushima, however, were *boiling water reactors* (BWRs); BWRs produce steam in the reactor core itself. This means that both the steam and the turbine become radioactive in the process. **INFOGRAPHIC 22.5**

It also meant that opening the vent at the Fukushima reactor would be akin to pumping radiation straight into the air. Still, not opening it would almost certainly be worse: If any one of the reactors exploded, it would release much, much more radiation than the steam from the vent. "It was this horrible double-edged sword," says von Hippel. "Exactly the kind of situation that we always feared with the BWRs—a lose–lose."

INFOGRAPHIC 22.6   **RADIOACTIVE ISOTOPES CAN RELEASE ONE OR MORE OF THREE DIFFERENT KINDS OF RADIATION**   4

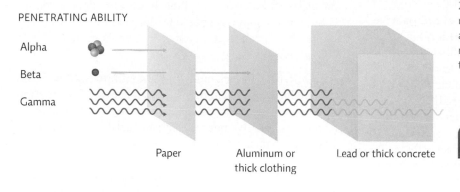

Alpha radiation
2 protons and 2 neutrons are lost (helium nucleus).

Beta radiation
An electron is lost.

Gamma radiation
In the process of decaying, additional energy may be released in the form of high-energy electromagnetic radiation (shorter wavelengths than visible light; similar to X-rays).

PENETRATING ABILITY

Alpha

Beta

Gamma

Paper   Aluminum or   Lead or thick concrete
               thick clothing

← The different types of radiation have differing abilities to penetrate surfaces. Alpha particles don't travel far and can't even penetrate paper; they do not penetrate skin but can be harmful if inhaled or ingested in food or water. The much smaller beta particles can penetrate the upper layers of the skin but are easily stopped by a thin sheet of aluminum or very heavy clothing. High-energy gamma rays can penetrate the deepest but can be stopped by thick or very dense material such as 20 centimeters (8 inches) of concrete or 2.5 centimeters (1 inch) of lead. Exposure to radiation, especially gamma rays, can damage a wide variety of body organs, leading to radiation sickness; it can also cause mutations that lead to cancer or cause birth defects.

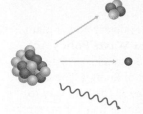

Why were the people of Fukushima Prefecture concerned about the alpha radiation released with the reactor meltdown? After all, it cannot even penetrate paper!

KEY CONCEPT 22.6

Alpha, beta, and gamma radiation vary in their ability to penetrate substances; while all are dangerous, gamma radiation is the most hazardous to health.

heavy firefighter uniforms, along with oxygen tanks and masks, in an effort to shield themselves from alpha and beta radiation. Both alpha and beta radiation come from particles. (Alpha particles are basically a helium nucleus; beta particles are essentially electrons.) Neither type can penetrate the skin very well. But if they enter the body through ingestion, both kinds can linger for years, causing organ damage and cancer.

None of the equipment would protect the workers from gamma radiation, which is much more energetic than alpha or beta particles and can easily penetrate walls, skin, and other surfaces. The only way for workers to avoid overexposure to gamma rays would be to get in and

On the morning of March 12 (day 2 of the disaster), six workers gathered in the plant's main control room, where each of them swallowed several iodine pills; by loading their thyroid glands with iodine beforehand, they hoped to prevent their bodies from absorbing radioactive iodine from the steam that would pour out of the vent. They donned thick,

out as quickly as possible. Dosimeters—mini radiation detectors attached to each fire suit—would let them know, with loud, alarmlike beeps, when they had exceeded the legal limits of exposure. **INFOGRAPHIC 22.6**

As the Sun climbed high over the ravaged coastline, the team of six headed into the first reactor to search in darkness for the vent. The plan was to work in teams of two and to relay their efforts so that no one worker was left hovering over the vent for too long. It took the first team 11 minutes to find the manual gate valve, crank it open a quarter of the way, and retreat. With the vent open partway, the radiation levels climbed so fast that the second team could not even reach it; in just 6 minutes, one of them absorbed a dose of radiation that surpassed the maximum limit allowed for 5 years. The third team did not even try.

That afternoon, the building housing reactor number 1 exploded, sending concrete and steel debris flying through the air and setting off a chaotic, and seemingly irreversible, chain of events: Radiation levels around the plant climbed exponentially, deterring efforts to vent the other two reactors. In the days that followed, both exploded.

The generation of nuclear waste is a particularly difficult problem to address.

The reactors weren't the only problem. Experts around the world, especially those observing from the United

States, were particularly concerned about Fukushima's radioactive waste.

To understand why, it helps to know a little about radioactive waste. In general, there are two kinds: low-level radioactive waste and high-level radioactive waste. **Low-level radioactive waste (LLRW)** is material that has low amounts of radiation relative to its volume and can usually be safely buried. This includes clothing, gloves, tools, etc., that have been exposed to radioactive material. The United States produces about 60,000 cubic meters (2 million cubic feet) of LLRW per year, all of which is stored at just four sites—in South Carolina, Texas, Utah, and Washington.

KEY CONCEPT 22.7

Radioactive waste is actually more radioactive than the original fuel. We currently have no long-term storage plan for high-level radioactive waste.

High-level radioactive waste (HLRW) is another story. As its name suggests, it's more reactive than LLRW; in fact, because so many radioactive by-products are created in fission reactions, HLRW is actually much more radioactive than the original fuel rods. The United States produces about 2,000 metric tons (2,200 U.S. tons) of HLRW per year and has some 65,000 metric tons (72,000 U.S. tons) of the stuff in storage right now; this estimate includes both spent fuel rods and waste from the nuclear weapons we've produced.

So far, we have no safe, reliable, long-term storage plan for this waste. Spent fuel rods are stored onsite in steel-lined pools, where at least some isotopes—those with short half-lives—can decay to safe levels. Isotopes with longer half-lives require more time to reach this point. Because the isotopes produced in the fission reaction or by the decay of these isotopes are all mixed together, the waste has to be stored for as long as it takes the longest half-life material to decay to safe levels—in some cases more than 2 million years.

Almost all of the short-term storage pools in the United States are full; 55 sites are currently storing HLRW in large steel casks that are surrounded by concrete, waiting for longer-term storage options to become available. So far, we don't have

low-level radioactive waste (LLRW) Material that has a low level of radiation for its volume.

high-level radioactive waste (HLRW) Spent nuclear reactor fuel or waste from the production of nuclear weapons that is still highly radioactive.

U.S. Nuclear Waste Policy Act (1982) The federal law which mandated that the federal government build and operate a long-term repository for the disposal of high-level radioactive waste.

any. "Waste has been one of the biggest sore points in the debate over nuclear energy," says Powers. "It's the one thing we really don't have even a good theoretical solution to."

In 1982, 25 years after the first U.S. nuclear power plant started operation, the **U.S. Nuclear Waste Policy Act** was passed, assigning responsibly for nuclear waste disposal to the federal Department of Energy. The law identified deep underground storage as the best way to dispose of HLRW and established guidelines for research and development into permanent repositories. Preferred sites would be geologically stable and able to hold the waste for tens of thousands of years to allow time for the waste to decay to acceptable levels of radioactivity. Groundwater supplies should be deep below any storage tunnels to prevent water contamination from the radioactive material.

The United States selected Yucca Mountain in Nevada as the site for a long-term repository and began construction in 1994, but the project has been repeatedly stymied by opponents who live near the site and activists everywhere who oppose nuclear power. After 15 years and billions of dollars, President Obama halted construction in 2009, a move that fueled even more bickering. Opponents of the Yucca Mountain repository heralded the move as long overdue. But supporters argued that because the Yucca repository was authorized by an act of Congress, only Congress had the power to cancel the project. The debate promises to continue: Even if a long-term storage option is approved and constructed, we still have the problem of how to safely transport HLRW across the country—a dilemma not likely to be resolved in the near future.

INFOGRAPHIC 22.7

Meanwhile, back at Fukushima, radioactive waste posed yet another deadly threat. Each of the six reactors had its own swimming pool–like container where used radioactive fuel was stored. The pools were full of years' worth of waste, and they relied exclusively on water to prevent overheating. If just one of those pools ran dry, the consequences would be catastrophic: By some estimates, a worker standing next to a dry pool could receive a fatal dose of radiation in just 16 seconds.

On March 13, just as experts around the world were contemplating such an event, a fire broke out around the spent-fuel pool near reactor Number 4.

Responding to a nuclear accident is difficult and dangerous work.

By March 14, three days after the earthquake and tsunami, all but a handful of workers had evacuated. The world media would dub the cohort that stayed behind "The Fukushima 50" (though in reality hundreds of

INFOGRAPHIC 22.7 | **RADIOACTIVE WASTE**

4

↓ Radioactive waste does not just come from nuclear power plants but is also generated by industry, the medical field, research laboratories, and weapons production. All of these users produce low-level radioactive waste (LLRW); nuclear power and weapons production is responsible for almost all of the high-level radioactive waste (HLRW). Though much of this material is highly dangerous and remains so for centuries, we currently have no safe way to dispose of it.

Justin Jin/Panos Pictures

HLRW is a by-product of nuclear fission. The United States lacks a long-term storage facility for HLRW; most spent fuel rods are currently stored onsite at nuclear power plants in steel-lined pools. Some have been moved out of the pools and into dry casks and stored above ground. European and Asian nations are looking into the possibility of joint disposal facilities, though little progress has been made.

← Technicians load highly enriched uranium fuel assemblies into casks at the Institute of Nuclear Physics in Kazakhstan.

© SuperStock/Alamy

LLRW includes contaminated items such as clothing, filters, gloves, and other items exposed to radiation. Short-half-life LLRW can be stored until it is no longer radioactive and then disposed of as regular trash. LLRW with a longer half-life is stored in casks; in the United States, LLRW is sent to one of four U.S. storage facilities (four other sites are closed) where it is buried underground; some must be shielded with concrete or lead containment before burial.

← Above-ground casks hold longer-half-life LLRW (also known as intermediate-level radioactive waste) as Chalk River Nuclear Laboratory in Ontario. These casks can be easily monitored and the material retrieved for disposal once it has decayed to safe levels.

©David Hancock-Skyscans/Auscape/The Image Works

MILL TAILINGS Mining and crushing uranium ore produces small-particle (sandlike consistency) waste. It contains low levels of radioactive isotopes with long half-lives, such as radium, thorium, and uranium. In the United States, piles of mill tailings are stored near milling facilities and covered with clay and rock to prevent the release of radioactive material into the atmosphere—by law, they must remain covered for at least 200 years.

← Mine tailings are stored in a large open pit at the Ranger Uranium Mine in Australia; rainwater fills the pit during the rainy season, and there are concerns that the water could overflow the dam in heavy rain years.

 Do you think we should continue to use nuclear power, or increase its usage, if we do not have a plan for long-term storage of HLRW?

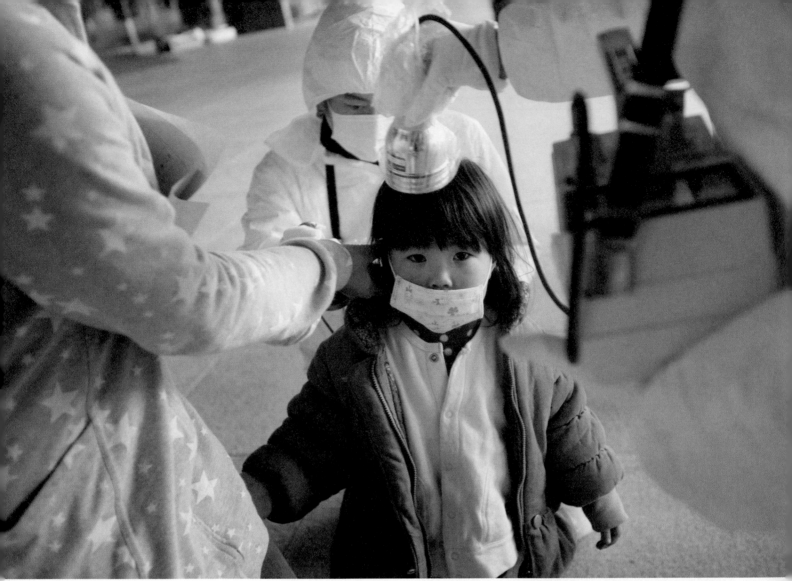

↑ Displaced people who were evacuated from Minamisoma, Futaba, and other towns located near the Fukushima Daiichi Nuclear Power Plant are checked for traces of radiation before they are permitted to enter a sports facility in Fukushima City, where about 1,200 evacuees found temporary shelter.

workers stayed behind to help). The radiation levels in the main control room of reactor Number 2 had climbed so high that those remaining workers had to rotate out at regular intervals to avoid being poisoned with radiation.

Around the world, nuclear experts worried aloud over what might happen next: Would there be another explosion? Would the spent-fuel pools run completely dry? Would fuel that had already melted into a heap at the bottom of some reactors now melt through the steel vessels and react with the concrete below? And how far would the wind carry all the radioactive vapors that were escaping into the atmosphere? Experts in the United States said it was at least possible that the vapors could reach the edges of Tokyo—the world's largest city, with a population of 35 million.

Meanwhile, helicopters were scooping up buckets of water from the sea and trying to dump them on the reactors. First, they were deterred by high radiation levels. They bolted lead plates to the choppers' bottoms and tried again, but strong winds blew most of the water askew of the Daiichi plant. Eventually, fire engines made their way through. Using hoses designed for jet-fuel fires, they were able to get some water to the places where it was most needed.

The saving move, however, was already in the works by then. "Even as buildings exploded," the *New Yorker* reported, "some of the workers had hooked up a train of fire trucks capable of generating enough pressure to inject water directly into the fuel cores. The drenching continued around the clock, and it went on for months. The process was ungainly, and it produced millions of gallons of radioactive waste that is dangerous and proving difficult to store, but it probably did more than any other measure to avert a far worse disaster."

The impacts of nuclear accidents can be far reaching.

On March 16, members of the U.S. Nuclear Regulatory Commission (NRC) determined that anyone more than 80 kilometers (50 miles) from the plant was probably safe from atmospheric radiation emitted by the disaster, based on computer models they had developed.

The finding did not quell public fears. In fact, long after the likely dispersal of radiation had been mapped and made widely known, people in many communities were fiercely divided about whether to evacuate. Local newspapers published daily radiation levels alongside weather reports, and average people sorted as best they could through reams of dense technical information—not all of it reliable—as they tried to make decisions about the health and safety of their families.

For months, it seemed, one new shocking discovery or scandalous revelation followed the next: The Japanese government discovered stores selling beef, spinach, and other food containing small amounts of radiation and scrambled to recall those products. Authorities warned parents not to give local milk to their children because children's cells grow faster than those of adults, and that makes them more vulnerable to the effects of radiation.

↓ The Gösgen Nuclear Plant is located within the town of Däniken, Switzerland. Steam rising from a cooling tower greets children as they walk home from school.

Mark Henley/Panos Pictures

Distrustful of government reports and frustrated by the lack of consistent information, average citizens began collecting and testing their own soil and water—sometimes getting disturbing results. On October 14, for example, the *New York Times* reported that citizen groups had found 20-odd spots in and around Tokyo that were contaminated with potentially dangerous levels of radioactive cesium.

The impact on the natural environment has been more difficult to determine. "It depends so much on luck," says Powers. "Even if everyone involved in the response does everything perfectly, we're still at the mercy of the wind." At first, the wind at Fukushima Daiichi blew steadily out to sea. But eventually it turned inland, to the northwest, carrying all those radioactive vapors with it. Rain and snow captured some particles, laying them deep in the mountains, forests, and streams around Fukushima. It will take decades before we know the full effect they will have on the ecosystems they have become a part of. But 5 months after the quake, Japan's central government acknowledged that the area within 3 kilometers (1.8 miles) of the plant will likely be uninhabitable for decades.

Though radiation levels in surrounding areas are falling, parts of the reactor buildings were still not safe to enter some 3 years after the event. Tokyo Electric Power Company (Tepco), the public utility that operated Fukushima, has been criticized for acting slowly and withholding information about local contamination and the clean-up operation. An estimated 300 million liters of radioactive water pumped out from the reactor buildings have been stored in hastily erected above-ground storage tanks. In 2014 a storage tank spill released 90,000 liters of highly radioactive water. The year before, more than 260,000 liters of radioactive water leaked out of the reactors and reached the Pacific Ocean. The construction of an "ice wall" began in 2014 in an attempt to prevent more groundwater from seeping into the plant and becoming contaminated. Pipes containing super cooling fluid are being sunk into the ground to freeze the groundwater around the entire complex.

Will there be any long-term effects on residents and workers as a result of the accident? A 2014 study published in the *Proceedings of the National Academy of Sciences* estimates only a very slight increase in the lifetime risk of cancer for people living in Fukushima Prefecture at the time of the disaster, a level that the authors say is "unlikely to be epidemiologically detectable." This may be good news, but as of February 2014, some 136,000 evacuees still cannot (or will not) return home. Residents are being allowed to reoccupy some areas (areas with acceptable levels of radiation); other areas with slightly higher radiation levels have been opened for day visits (to clean, rebuild, and restore). Remediation of the areas continues, with the hope of allowing more residents to return. Still, officials expect at least half of the area within 20 kilometers (12.5 miles) of the nuclear power plant to remain closed for the foreseeable future.

Will nuclear power play a role in future energy?

The March 2011 tsunami would become the most expensive natural disaster in human history, with losses estimated at $300 billion. In its wake, countries around the world began rethinking their plans to expand their own nuclear energy programs. Germany resolved to move up the date for its planned phase-out of all nuclear power plants by 10 years, to 2022. Austria, Italy, and Switzerland also reconsidered. In the United States, however, experts fretted in a more undecided way. Critics pointed out that if the Indian Point Power Plant, 56 kilometers (35 miles) north of Times Square in New York City, necessitated the evacuation of an 80-kilometer radius (as the Daiichi plant had), it would include some 20 million people. Proponents, meanwhile, held fast to their contention that nuclear power would be an essential, unavoidable part of weaning the planet from its fossil fuel dependence. Under proper conditions, they insist, it can be a completely safe resource; the U.S. Navy, for example, has been safely using nuclear energy to power vessels since 1954, without incident.

Meanwhile, back in Japan, just 2 months after the quake and tsunami, the prime minister responded to public pressure and global criticism by calling for a temporary shutdown of one nuclear plant in the country's center, where scientists have estimated an 87% chance of a big quake sometime in the next three decades. Other plants that were closed for routine maintenance were told to hold off on reopening so that, by summer's end, less than one-third of Japan's 54 reactors were still running.

KEY CONCEPT 22.8

The advantages (low air pollution, ample supplies) and disadvantages (radioactive hazards) of nuclear power must be weighed to determine whether or how much to pursue it.

In the wake of the nuclear power plant closures, Japan responded with an unprecedented campaign to reduce overall energy usage in the country. These efforts were able to replace 15% of Japan's normal electricity use—about half of what the nuclear power plants had provided—allowing the county to avoid electricity shortfalls and blackouts.

At first the shutdowns were short-lived. Once factory owners began warning that such drastic power cuts would quickly lead to a recession, many of the deactivated nuclear power plants sprung back to life by order of the same prime minister who had closed them, only to close again after safety concerns overrode the need for power. By May 2013, no reactors were in use. To replace the lost nuclear production, Japan ramped up its fossil fuel power plants (its main source of electricity even before the disaster) to supply more than 90% of its electricity. Driven, in part, by its lack of domestic sources of fossil fuels, Japan's government intends to resume nuclear power production after stricter safety measures are put into place. In July 2014, Sendai Nuclear Power Station became the first Japanese nuclear power plant to receive preliminary approval to start up under the country's new, more stringent safety requirements.

Less than a year after the fires had cooled, experts would agree that the Daiichi disaster was something of a draw. "People who support nuclear energy can't argue anymore that the risk is nonexistent," says Powers. "But at the same time, neither can opponents say that a meltdown would automatically result in a zillion immediate casualties."

In a paper published before the disaster at Fukushima, British nuclear engineers Robin Grimes and William Nuttall argued that nuclear power could enjoy a "renaissance" and increase its contribution to electricity production in the future, if only we improved our nuclear technology. They pointed to new "third-generation" BWR and PWR reactor designs (and others) available now, and other technologies ("fourth generation") in development that should be safer and more efficient, producing more electricity with less fuel. Some designs also produce less of the waste material that could be used to make nuclear weapons, reducing concerns about weapons proliferation.

The third-generation designs in use today are safer, according to safety assessments, and include more fail-safe responses (technical responses that automatically kick in if a problem occurs). But, of course, the 2011 earthquake and tsunami have illustrated the need to reevaluate the potential damage of natural disasters,

especially with regard to siting future nuclear power plants away from potentially dangerous areas such as close to active fault lines or flood zones. The renaissance would also require better waste-handling options, including the reprocessing of spent fuel (which would decrease the waste produced and increase overall efficiency).

Given the problems associated with fossil fuels, it is likely that nuclear power will continue to have a place in our energy mix, especially with global electricity use projected to double by 2030. But in making choices about how to pursue our energy future, we must consider the costs and benefits of all our energy options. A benefit analysis must evaluate how well a particular energy source meets our energy needs. A cost analysis must consider not just the monetary cost of getting kilowatts delivered to our homes but also the environmental and social costs associated with every step of the energy source's life, from acquisition to production to delivery to waste disposal. In addition, as mountaintop removal, oil spills, climate change, and nuclear meltdowns demonstrate, a risk assessment must also take place. We must answer two very crucial questions: How risky is the venture (an assessment we can do with at least some degree of accuracy)? and How much risk are we willing to take? **TABLE 22.1**

"There is definitely a certain weighing that has to take place," says Powers. "Global warming on one hand, nuclear accidents like Fukushima on the other."

The future of nuclear energy is uncertain.

On August 23, 2011, a magnitude 5.8 earthquake erupted beneath the state of Virginia, causing the North Anna Nuclear Power Plant there to tremble with much greater force than its reactors were designed to withstand: Dry casks, each weighing more than 100 metric tons and filled with spent fuel rods, shifted inches. It was the region's largest quake in more than a century, and the first time such a calamity had struck an American nuclear power station. Just 5 days later, when a category 5 hurricane by the name of Irene struck the East Coast, workers at three other nuclear power plants noticed that emergency sirens had failed to function properly. And at one plant—the Indian Point Power Plant, the one closest to Manhattan—a discharge canal carrying (nonradioactive) water from the cooling system overflowed due to the high river levels.

On September 9 that year NRC staff suggested ordering power plants to review their ability to survive quakes and floods "without unnecessary delay." Safety procedures that have been implemented include increasing the

TABLE 22.1   NUCLEAR POWER: TRADE-OFFS

↓ What role will nuclear power play in the future? Like all of our other energy options, nuclear power has advantages and disadvantages that must be weighed when making this decision.

ADVANTAGES	DISADVANTAGES
Costs • Operating costs are comparable to those of a fossil fuel power plant. • The technology is available now.	• Nuclear power plants are much more expensive to build ($4 billion to build a fission reactor), maintain, and decommission ($200 million to $1 trillion to decommission after a 40- to 60-year life span) than fossil fuel power plants.
Electricity Production • Power production can be increased or decreased to meet demand (up to the capacity of the facility). • Dependable amounts of electricity can be produced—no worries about nighttime or cloudy days (solar power) or windless days (wind power).	• Though power production can be altered to meet demand, it cannot be done as quickly as it can be done with a fossil fuel facility. • Large amounts of water are needed in the process for cooling.
Pollution and Safety • No CO_2 is released during operation, so nuclear power does not contribute to climate change (though some CO_2 is released during mining and processing). • No particulates, sulfur, or nitrogen pollution are released during operation, so it is much cleaner than burning fossil fuels to generate electricity.	• Radioactive waste is very hazardous, and we still have no long-term plan for dealing with the waste that is produced from the fission reaction. • Shipping waste (by truck or rail) is also a safety concern and vehemently opposed by those who live on the transport route. • Though accidents are rare in the history of nuclear power, like those at Chernobyl and Fukushima, they can have extremely serious consequences and long-term impacts when they occur.
Supplies • Uranium supplies are good, enough for 80–200 years, and other isotopes can also be used; some reactor designs actually *produce* fuel during the reaction process. • Fuels used for nuclear reactors are energy rich: One small uranium pellet contains the same amount of energy as 1 metric ton of coal.	• Mining and processing ore for nuclear fuel produces hazardous waste; surface mining damages the environment and can pollute air and water. • Some methods of nuclear power production produce radioisotopes that could be used in nuclear weapons production; facilities for the processing of fuel could hide weapons programs.

[?] Which advantage of nuclear power do you consider to be its strongest advantage? Which of its disadvantages is its biggest detractor? Explain.

availability of back-up power sources and ensuring that these power sources are not located in areas that might flood (such as basements). "Expedited" seismic and flooding evaluations were also ordered; modifications will be made if deemed necessary.

Because of the Fukushima disaster, the focus has shifted from not merely equipping nuclear facilities to withstand expected natural disasters (storms, earthquakes, floods) but to bolstering the power plants to withstand *worst-case scenario* events.

With some modifications already in place, an NRC accident analysis released in 2013 concluded that all U.S. reactors are designed to withstand complete loss of power. Indeed, in 2011 and 2012, U.S. reactors withstood power losses and damage inflicted by a record 226 tornadoes that hit the Southeast in a 24-hour period, hurricanes

(including Hurricane Irene and Hurricane Sandy), and a magnitude 5.8 earthquake in Virginia. In all cases, the facilities shut down according to safety protocols, without serious incident.

Select References:

García-Pérez, J., et al. (2009). Mortality due to lung, laryngeal and bladder cancer in towns lying in the vicinity of combustion installations. *Science of the Total Environment,* 407(8): 2593–2602.

Grimes, R. W., & W. J. Nuttall. (2010). Generating the option of a two-stage nuclear renaissance. *Science,* 329(5993): 799–803.

Harada, K. H., et al. (2014). Radiation dose rates now and in the future for residents neighboring restricted areas of the Fukushima Daiichi Nuclear Power Plant. *Proceedings of the National Academy of Sciences,* 111(10): E914–E923.

Nuclear Energy Institute. (2013). "U.S. Government and Nuclear Energy Industry Response to the Fukushima Accident," www.nei.org/Knowledge-Center/Backgrounders/White-Papers.

Queißer-Luft, A., et al. (2011). Birth defects in the vicinity of nuclear power plants in Germany. *Radiation and Environmental Biophysics,* 50(2): 313–323.

BRING IT HOME

PERSONAL CHOICES THAT HELP

Nuclear energy has been rebranded as "green energy" because it does not emit greenhouse gases. Technology has improved the safety of nuclear facilities; however, there are still safety issues and valid concerns over the long-term storage of nuclear waste. In addition, cost, national security, and uranium supplies make nuclear power a complicated energy solution.

Individual Steps

• Use the facility locator on the U.S. Nuclear Regulatory Commission website (www.nrc.gov) to find out if you have nuclear reactors where you live.

Group Action

As you would expect, the policies endorsed by a particular group depend on the group's overall view of nuclear energy. To see two examples, check out the following sites and see how they compare.

• Visit the Nuclear Energy Institute (www. nei.org), which is a pronuclear organization, to see proposed legislation regarding increasing our current nuclear energy production.

• Visit the Nuclear Energy Information Service (http://neis.org), which is a non-profit antinuclear organization committed to a nuclear-free future.

Policy Change

• What is your opinion about the necessity of nuclear energy in the United States? The U.S. Department of Energy website (www. ne.doe.gov) maintains articles, updates, and the latest policy initiatives regarding nuclear energy in America. This resource can help you understand current policy, funding issues, and upcoming legislation.

AP Photo/Toby Talbot

ENVIRONMENTAL LITERACY UNDERSTANDING THE ISSUE

1

What are radioactive isotopes, and why are they important for nuclear power?

INFOGRAPHIC 22.1

1. The number of _____ is unique to a given element and gives an element its atomic number.

2. Atoms that have the same number of protons and electrons but different numbers of neutrons are called:
    a. radioactive.
    b. ions.
    c. isotopes.
    d. subatomic.

2

What types of radiation are produced when an isotope decays? How is the rate of decay of a radioactive atom measured?

INFOGRAPHIC 22.2

3. After four half-lives, about how much radioactive parent material is left?
    a. 25%
    b. 40%
    c. 0.5%
    d. 6.25%

4. In a nuclear fuel assembly, control rods are used to regulate the nuclear reaction. But even if all the control rods are inserted into the fuel assembly, heat is still produced. Explain why.

3

How is nuclear energy harnessed to generate electricity in a fission reactor? How safe are nuclear reactors?

INFOGRAPHICS 22.3, 22.4, AND 22.5

5. True or False: Barring a major accident, it is less hazardous to your health to live near a nuclear reactor than a coal-fired power plant.

6. Nuclear fission is a reaction that:
    a. splits an atom, releasing energy.
    b. combines two or more atoms, producing energy.
    c. results in large explosions.
    d. is required to make atoms radioactive.

7. Using Infographic 22.3, describe the steps used in mining and processing uranium to the point where it is packed into fuel rods.

4

How dangerous is radiation that is released from radioactive material? What problems are associated with nuclear waste?

INFOGRAPHICS 22.6 AND 22.7

8. There are different types of radiation. Which one is most energetic and can therefore penetrate many surfaces, including skin?
    a. Alpha radiation
    b. Beta radiation
    c. Gamma radiation
    d. Particle radiation

9. Which of the following is an example of low-level radioactive waste?
    a. Worker clothing and gloves
    b. Tools used in the nuclear facility
    c. Spent fuel rods
    d. a and b are both examples of LLRW
    e. a, b, and c are all examples of LLRW

10. How are spent fuel rods currently disposed of in the United States?

5

In a trade-off analysis of nuclear power, what factors must be considered, and what is your own conclusion regarding the future role of nuclear power?

TABLE 22.1

11. What problem related to nuclear power do we not yet have a workable solution for?
    a. Finding a substitute for uranium once it is used up
    b. Safe disposal of HLRW
    c. Producing stream to turn the turbine that is not radioactive
    d. Controlling the fission reaction

12. What are the pros and cons tof generating energy from a nuclear source versus a fossil fuel source (coal)?

13. From an economic standpoint, which type of electricity production—nuclear or fossil fuel—is less expensive? Explain. Remember to include both internal and external costs in your answer.

SCIENCE LITERACY WORKING WITH DATA

The following graph depicts the number of accidents that occur in a variety of U.S. industries.

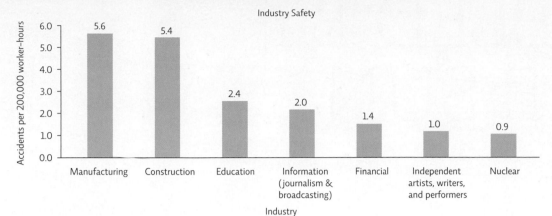

Industry Safety

Interpretation

1. What does the height of each bar represent?

2. Based on the graph, which two industries have the highest accident rates?

3. Based on the graph, which two industries have the lowest accident rates?

Advance Your Thinking

4. To determine whether there is bias in this graph, answer the following questions:
   a. Which industries are represented? Are they comparable?

b. Do the data for nuclear workers include accidents that occur as a result of mining, processing, and production?

c. Which of these industries do you think has the strictest OSHA (Occupational Safety and Health Administration) regulations? Might that account for differences in the number of accidents?

5. If you had access to safety data from all industries, which industries would you show in the graph to obtain a fair comparison to the nuclear industry?

INFORMATION LITERACY EVALUATING INFORMATION

The Nuclear Regulatory Commission (NRC) is tasked with regulating nuclear power in the United States and "protecting people and the environment." The members of the commission "formulate policies, develop regulations governing nuclear reactor and nuclear material safety, issue orders to licensees, and adjudicate legal matters" (www.nrc.gov/about-nrc/organization/commfuncdesc.html). The NRC maintains a detailed website with information about nuclear power plants and nuclear materials.

Evaluate the website and work with the information to answer the following questions:

1. Is this a reliable information source? Does the organization have a clear and transparent agenda?
   a. Who runs the website? Do this person's/group's credentials make the information reliable or unreliable?
   b. Is the information on the website up to date? When was the website last updated? Explain.

Go to the NRC's Facilities page (www.nrc.gov/info-finder/region-state). Find your state in the list of states and territories.

2. How many nuclear sites are active in your state? How many nuclear sites are being decommissioned? (Select another state if yours does not have any nuclear sites.)

3. Click on a link for an active nuclear site.
   a. Identify the location of this facility, the type of reactor in use, and the intended expiration date. Find the location on a map and record the closest large body of water.
   b. Look at the links provided (e.g., "Severe Accident Inspections" and "Enforcement Actions") to assess the impact of this facility on the environment and describe the environmental issues associated with the site (if any), and the plan of action to correct the issues.

4. Consider the information you have gathered and do a basic risk assessment analysis for one of the sites.
   a. Do you think the facility is a danger to the environment? To human health? Why or why not?
   b. Is the facility located in an area that may have natural disasters? If so, what are they? If a natural disaster were to occur, predict some outcomes.

FUELED BY THE SUN

A tiny island makes big strides with renewable energy

CORE MESSAGE

In order to become a sustainable society, we need to transition to reliable, renewable energy sources with acceptable environmental and social impacts. No single energy source can replace fossil fuels. Instead, a variety of methods, selected to meet the needs of the population, and availability of local energy sources, will help communities shift to sustainable energy use. Fortunately, we have many good options already at our disposal, with other new methods currently in research and development.

AFTER READING THIS CHAPTER, **YOU** SHOULD BE ABLE TO ANSWER THE FOLLOWING **GUIDING QUESTIONS**

1

What are the characteristics of a sustainable energy source, and what role does renewable energy play in terms of global energy production?

2

How do wind and solar power technologies capture energy? What are the advantages and disadvantages of wind and solar power, and how does each compare to fossil fuels in terms of true costs?

Playing soccer under the shade of wind turbines on Samsø Island, Denmark, the first island in the world with 100% renewable energy. Collectively, Samsø's land-based turbines produce about 26 million kilowatt-hours a year, enough to meet the island's demands for electricity. Alessandro Grassani/Invision/Aurora Photos

3

In what ways can we harness geothermal energy and the power of water? What are some of the trade-offs associated with each of these resources?

4

What roles do conservation and energy efficiency play in helping us meet our energy needs sustainably?

5

What combination of actions did Samsø take to become an energy-positive island? What is the take-home message to other communities that might want to reduce their use of fossil fuels?

On a typical cold, misty January day in Denmark in 2003, many of the 4,100 residents of a small island gathered together at the beach. Everyone, including the mayor, strained their eyes to see the faint outline of several structures, each over 30 stories tall, located more than 3 kilometers offshore.

Nestled in the crook of Denmark's mainland, Samsø is home to a small, windswept community of Danish farmers known for their sweet strawberries and tender early potatoes. It is a quiet and serene place. Yet it has been the site of a dramatic revolution—a community transformation that made headlines around the world.

The transformation began a new chapter on that cold day in 2003. The mayor pushed a button, and the offshore structures slowly creaked to life. Through the gray mist and rain, people could gradually see the massive blades begin to rotate, converting the power of wind into energy. It was a landmark day in Samsø's ambitious attempt to become the greenest, cleanest, and most energy-independent place on Earth.

"That was a very big moment," recalls Søren Hermansen, a Samsø resident who was key in getting the community behind the project. "Nobody really thought it would happen when we started."

Sustainable ecosystems and societies rely on renewable energy.

Samsø used to be just like most other communities on Earth: fully dependent on fossil fuels like coal, oil, and gas. As recently as 1998, Samsø imported all of its energy resources from the mainland: Tankers hauled oil into its ports, and the island imported electricity generated from burning coal via cables.

But in 1997, the government of Denmark decided it was important to promote the idea of **renewable energy**, energy from sources that are replenished over short time scales or that are perpetually available. To do so, it announced a competition: Which local area or island could become self-sufficient on renewable energy? The contest invited applicants to describe which renewable resources were available in their community and how they would be used to replace fossil fuels.

The purpose of the contest was to put communities on track to be more sustainable. To qualify as a **sustainable energy** source, the energy must be renewable,

renewable energy Energy from sources that are replenished over short time scales or that are perpetually available.

sustainable energy Energy from sources that are renewable and have a low environmental impact.

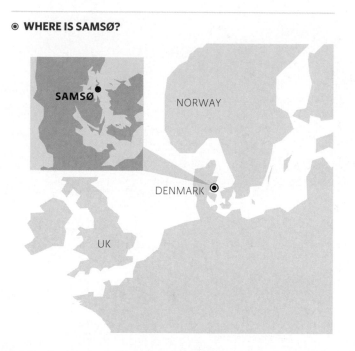

⊙ **WHERE IS SAMSØ?**

SAMSØ

NORWAY

DENMARK

UK

with a low enough environmental impact that it could be used for the long term. As a result, any source of energy that causes environmental damage when it is captured or produced, such as damaging habitat or generating pollution, was not ideal. For example, experts continue to debate the value of "clean coal"; this type of coal does burn more cleanly than typical coal, but it causes environmental damage when it is extracted and requires a lot of energy to process into a cleaner fuel—while still generating toxic waste (see Chapter 18).

KEY CONCEPT 23.1

Sustainable energy sources are those that meet our needs, have acceptable impacts, and are readily replenished.

↑ Samsø's offshore wind turbines provide much of the electricity used by the island. Since 2005, Samsø has produced more electricity than it uses and exports the extra to mainland Denmark and beyond to neighboring countries.

Even though Samsø is small —about 26 kilometers (16 miles) long and only 7 kilometers (4 miles) at its widest—it has plenty of the natural resources that are becoming increasingly important sources of sustainable energy. A small firm put together an application for the contest, and the little island won.

To become sustainable, Samsø turned to one of its most plentiful natural resources.

In an ambitious plan, Samsø's leaders wanted to do more than merely transition to renewable energy sources. Their goal was for the island to become *energy neutral* (produce as much energy as its residents consume) or even *energy positive* (produce more energy than consumed). A major part of Samsø's transformation was the installation of 11 onshore and 10 offshore wind turbines designed to harness the power of **wind energy**, or energy contained in the motion of air across Earth's surface. There is no shortage of wind on Samsø, and the powerful breezes turn huge blades (up to 40 meters [130 feet] in length) that

are connected to a generator, which converts mechanical energy into electricity.

Nine of the onshore turbines on Samsø were purchased collectively by groups of farmers who bought shares in their construction. "People on the island are personally invested in this," says Bernd Garbers, a German engineer who lives on the island and is a consultant for the firm that won the energy contest, Samsø Energy Academy. By 2013 the windmills were producing enough power to supply 100% of the electricity used by the islanders— sometimes more. In just 10 years, Samsø had become an energy-positive island.

Samsø uses about 500 billion kilojoules (kJ) of energy each year. A barrel of oil produces 6.15 million kJ, so Samsø uses 81,300 *barrels of oil equivalents (BOE)* annually. (That's 500 billion divided by 6.15 million.) In 2013, the entire human population used an estimated 90 billion BOE, and only about 9% of that energy came from renewable resources. The United States ranks far below that average, fulfilling only

wind energy Energy contained in the motion of air across Earth's surface.

INFOGRAPHIC 23.1   **RENEWABLE ENERGY USE**

↓ According to the U.S. Energy Information Administration, as of 2010, renewable sources of energy contributed 4.18 trillion kilowatt-hours (kWh) of total electricity energy production worldwide (about 21% of the total). By 2040, electricity generation by renewable energy sources is projected to increase by more than 40% but will account for only about 25% of the total because total usage is expected to double over that time. Of renewable sources, biomass (mostly fuels like wood, charcoal, and animal waste) makes up the largest proportion, followed by energy produced by hydroelectric dams.

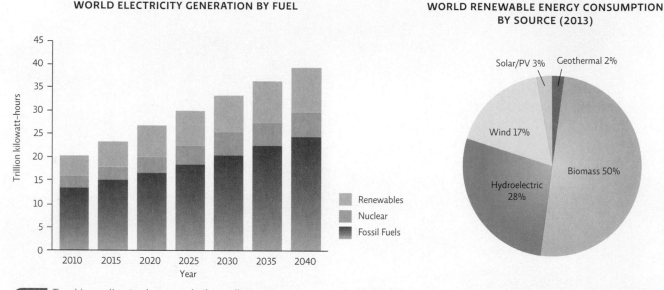

WORLD ELECTRICITY GENERATION BY FUEL

WORLD RENEWABLE ENERGY CONSUMPTION BY SOURCE (2013)

To address climate change and other pollution issues associated with fossil fuels, we should begin to phase out fossil fuels and increase our use of renewables. Are we on track? Summarize the change in total energy use and the usage of renewables and fossil fuels from 2010 to the projected values for 2040.

about 5% of our energy needs through renewable energy in 2013. **INFOGRAPHIC 23.1**

In addition to wind power, Samsø also generates heat that is piped to homes by burning locally grown straw, a **biomass energy** source. Biomass fuels, or *biofuels*, have the advantage of being carbon neutral; they do release CO_2 into the air, but this is CO_2 that the growing plants recently removed from the air. They also can be locally sourced, reducing transportation costs (monetary and environmental). On the other hand, they release air pollutants (particulates), and raising biofuel crops can displace food crops, reducing overall food supplies. (For much more on biofuels, see LaunchPad Chapter 32.)

The straw used in the power plants is the only energy source that is burned on the island other than the gasoline and diesel fuel used in vehicles. Ideally, Samsø residents would have swapped their vehicles for cars that run on hydrogen or electricity, but those technologies were too expensive and not efficient enough to use on the island. Since Samsø would continue to rely on conventional fuels for transportation, engineers decided to install large offshore wind turbines to produce an equivalent amount of clean energy, offsetting uses by motor vehicles and boats. Islanders are working toward a gradual transition to electric vehicles and those powered by biofuels. The island's small size helps make all this feasible since daily "commutes" are short and miles traveled per person per day are lower than in most industrialized nations.

Far off the coast, the offshore turbines are especially efficient at generating energy because wind conditions are better at sea. In some especially windy locations, people have built strategically placed *wind farms*, which can contain dozens of turbines. In 2014, the world's largest was located in California, with a generation capacity of 1,548 megawatts (MW), but Texas is the leading producer of wind energy in the United States. It has 7,772 wind turbines that generate enough electricity—12,355 megawatts (MW)—to power 3.3 million homes per year. **INFOGRAPHIC 23.2**

But wind energy isn't perfect. First, even for a blustery location like Samsø, wind is intermittent—it stops and starts irregularly, not producing a steady stream

biomass energy Energy from biological material such as plants (wood, charcoal, crops) and animal waste.

KEY CONCEPT 23.2

Fossil fuels are the leading fuel for electricity production. The use of both renewables and fossil fuels is rising, due to increased energy demand.

INFOGRAPHIC 23.2 **HOW IT WORKS: WIND TURBINES**

Rotating blades turn a shaft inside the turbine. The shaft is attached to a gear that rotates a higher-speed shaft on the generator.

SHAFT

GEAR BOX

GENERATOR

The spinning generator produces electrical current.

Wind turns the blades.

Some people feel that windmills and wind farms are eyesores, but one must also consider the alternatives. Would you rather live near a wind farm or a mountaintop removal coal mine? Explain.

← Wind turbines can be large or small. All work by the same principle: Spinning blades turn a shaft inside a generator and produce electricity. New designs produce quieter windmills, addressing one of the criticisms of the technology.

© Dang Ngo/ZUMAPRESS.com

Stefan Falke/laif/Redux

↑ Top: Palm Springs, California: Wind turbines generate electricity for southern California. Bottom: A solar and wind energy-powered house in San Francisco, California.

The scalloped edge of this windmill blade is a new design inspired by the fin of a humpback whale—an example of biomimicry. Founders of the company WhalePower have demonstrated that this shape makes the blade more aerodynamic and less likely to stall out at low speeds, allowing it to produce more power than traditional designs.

↑ A solar heating plant near Marup, Samsø Island. Changes in Samsø energy use mean that instead of importing electricity, the island exports it.

KEY CONCEPT 23.3

Wind power is pollution free but is an intermittent energy source. In addition, turbines are dangerous to wildlife, and some consider them eyesores.

---

solar energy Energy harnessed from the Sun in the form of heat or light.

photovoltaic (PV) cells A technology that converts solar energy directly into electricity.

active solar technologies Mechanical equipment for capturing, converting, and sometimes concentrating solar energy into a more usable form.

solar thermal system An active technology that captures solar energy for heating.

passive solar technologies Technologies that allow for capture of solar energy (heat or light) without any electronic or mechanical assistance.

of power. And large wind turbines are not cheap. Each onshore turbine cost the Samsø islanders the equivalent of $1 million; offshore turbines rang up at $5 million apiece. (Less expensive, smaller wind turbines can be purchased by homeowners for personal use.) Beyond cost, wind turbines can create noise, and some people see them as eyesores. They can also have an impact on the local environment, threatening birds and bats, who are unable to nest near turbines or are killed by rotating turbine blades. More than 1 million birds and bats are killed by wind turbines annually in the United States. While that may seem like a large number, it is far less than the number of birds killed by flying into communication towers (40 million) or those killed by domestic cats (hundreds of millions). Still, to decrease the risk from windmills, engineers now avoid placing them in known migratory flight paths or close to areas frequented by birds of prey such as eagles.

The most abundant sustainable energy source is the one that powers the planet—the Sun.

Each year, Earth receives a staggering amount of energy from the Sun, more than 4 million exajoules (1 exajoule is 10^{18}, or a million trillion, joules). We use the Sun for many things—to warm our homes, heat our pools, and provide light—but new technologies allow more effective use, and even storage, of the Sun's power.

In addition to wind power, the islanders on Samsø decided to tap directly into this boundless natural resource.

Solar energy is energy harnessed from the Sun in the form of heat or light. Wind power is actually an indirect form of solar energy. Wind results from the difference in temperature between different regions of Earth, such as the poles and the equator, causing air to move from cooler regions to warmer regions.

Solar energy can be used in two ways: through active or passive technologies. Around the countryside in Samsø, homes are dotted with **photovoltaic (PV) cells**, also called PV cells or solar panels. PV cells are **active solar technologies** that convert solar energy directly into electricity. If just 4% of the world's deserts were covered in PV cells, it would supply all of the world's electricity needs. When the Samsø renewable energy project first began, locals were able to buy PV panels for their homes at a low price subsidized by the government.

The islanders also rely on **solar thermal systems**, another active technology that captures solar energy for heating. In a field at the north of the island, rows and rows of solar collectors face the sky, absorbing the Sun's rays and using that energy to heat a massive tank of water to 160°F (71°C). The hot water is then piped into 178 nearby homes for use in heating systems. Another 200 homes on the island, ones farther away from district heating plants, have individual solar collectors to capture solar energy for heat.

INFOGRAPHIC 23.3 SOLAR ENERGY TECHNOLOGIES TAKE MANY FORMS

2

↓ There are many ways to capture and use solar energy. Passive solar homes are constructed in a way that maximizes solar heating potential.

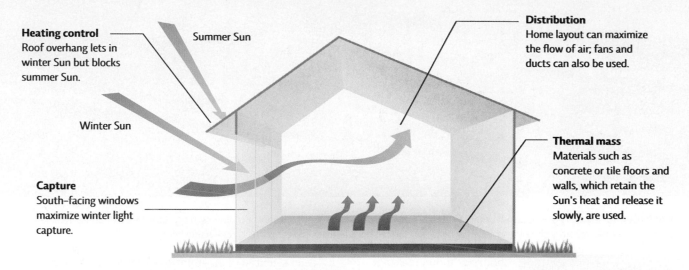

Heating control
Roof overhang lets in winter Sun but blocks summer Sun.

Summer Sun

Distribution
Home layout can maximize the flow of air; fans and ducts can also be used.

Winter Sun

Capture
South–facing windows maximize winter light capture.

Thermal mass
Materials such as concrete or tile floors and walls, which retain the Sun's heat and release it slowly, are used.

↓ Active solar technologies capture and convert solar energy for another use such as heating water or producing electricity.

Gilles ROLLE/REA/Redux

Electrons

Current

Negatively charged silicon

P/N junction

Positively charged silicon

Holes

Current

Current

↑ In a solar water heater (shown here on the railing), sunlight heats a fluid in pipes, which then heats water in a tank. Photovoltaic panels (shown here on the roof) capture sunlight and convert it to electricity that can be used by the homeowner or fed back to the grid and sold to the local utility.

↑ Photovoltaic (PV) cells convert sunlight to electricity. They are made of a negatively charged (n-type) and a positively charged (p-type) layer of semiconductor material, such as silicon. When light strikes the surface, its energy causes electrons to move from one layer to "holes" in the other, creating an electrical current that can be used to power devices or sent out on the power grid.

? In terms of reducing fossil fuel use, why should you make your home as energy efficient as possible before installing active solar technologies?

Less expensive alternatives to PV cells or solar thermal systems are **passive solar technologies**. A greenhouse is a simple example of such a system: It captures heat without any electronic or mechanical assistance. Many energy-conscious homes are designed with passive energy in mind, incorporating strategically oriented windows to maximize sunlight in a room and dark-colored walls or floors to absorb that light and heat the home.

Because solar energy is conceptually simple, safe, and clean, with no noise or moving parts, it is the most popular member of the renewable energy club. Today,

hundreds of thousands of buildings around the world are powered by PV cells. **INFOGRAPHIC 23.3**

But like wind energy, solar energy is plagued by intermittency and start-up costs. PV cells are becoming cheaper every day, but it can still take a homeowner or business owner many years to recoup the cost of installation. Intermittency is a bigger problem. Sunlight is available for only roughly half of each day—and even less in places like Anchorage, Alaska, that see only a few hours a day of pale sunlight in the winter. (However, in those places, summertime production can go on for 18

↑ At least 450 residents of Samsø own shares in the onshore turbines, and a roughly equal number own shares in those offshore. Residents are proud of their accomplishments in creating the first island in the world operating on 100% renewable energy.

KEY CONCEPT 23.4

Solar power can be harnessed in many ways. It is pollution free but is expensive and is less productive at higher latitudes, on cloudy days, and at night.

hours or more, making solar panels popular in Alaskan locations that are "off the grid.") Because of this, it is unlikely that any community will rely solely on solar power for its energy needs.

Diversification is a hallmark of a sustainable energy future: Because sustainable energy sources have strengths and weaknesses, no single source will likely meet the needs of any particular community, and certainly not those of the entire world. But together, each can provide a unique contribution. For example, solar is productive in daylight hours, whereas wind tends to blow harder at night; thus these two renewable sources complement each other in many places.

But in some regions, it makes more sense to tap other sources of renewable energy.

Energy that causes volcanos to erupt and warms hot springs can also heat our homes.

Unlike Samsø, some communities are fortunate enough to be located near sources of **geothermal energy**. A tremendous amount of heat is produced deep in Earth from radioactive decay of isotopes; temperature increases with depth, and Earth's core may be 9,000°F (5,000°C). It is the same heat that bubbles hot springs and causes geysers to erupt from the ground at Yellowstone National Park in Wyoming. Within the past century, technological advances have enabled us to tap into these vast resources for heat and electricity. "There has been a huge surge in the development of geothermal resources," says Wilfred Elders,

co-chief scientist of an ambitious geothermal project known as the Iceland Deep Drilling Project (IDDP). "It is one of the most viable and economically attractive sources of renewable energy."

A wide variety of geothermal systems are currently in use. **Geothermal heat pumps** (also called ground-source heat pumps) are used in more than half a million homes around the world—not to generate electricity but to reduce our use of it. If you have ever been in a cave, you have probably been struck by the cool, constant temperatures there—always around 55°F (12.5°C), whether it's freezing outside or in the middle of a heat wave. No matter what the temperature above, at modest depths underground, the temperature remains a constant 55°F. Engineers bury fluid-filled pipes, and the fluid takes on that ground temperature. The fluid is then pumped into the home, where its temperature is transferred to air ducts and circulated through the home, essentially providing residents with year-round 55°F temperatures. In the winter, the home only needs to be warmed up from 55°F, rather than the outside colder temperatures, and in the summer, this system provides natural cooling. Such pumps are fairly expensive to install ($20,000 to $25,000 on average) but result in lower monthly energy bills than conventional heating and cooling systems. Areas with more extreme climates save enough money in monthly bills to offset the cost of installation—a metric known as **payback time**—in as little as 5 years.

On the other end of the spectrum are **geothermal power plants**. Generators above geothermal wells use steam

Because sustainable energy sources have strengths and weaknesses, no single source will likely meet the needs of any particular community, and certainly not those of the entire world.

released from hydrothermal reservoirs (hot springs) to spin turbines and produce electricity. Hydrothermal reservoirs are areas in Earth's crust that hold heat energy in the form of water, either as steam or liquid. Geysers in northern California power the world's largest geothermal electrical system, generating more than 1,000 MW of energy—enough for a city as large as San Francisco.

Geothermal power plants are reliable and efficient, but their potential is entirely dependent on location. Because drilling is expensive, only sites with enough accessible heat to generate significant electricity are considered for development. These tend to be hot zones, locations like Iceland and the western United States, where the tectonic plates in Earth's crust pull away from or rub against each other, allowing the hot magma to flow upward through cracks in the rock to reach areas closer to the surface. The ambitious IDDP is attempting to drill into a high-temperature geothermal system (the Krafla volcano, in this case), penetrate twice as deep as conventional geothermal wells (which are usually less than 3,000 meters [10,000 feet] deep), and reach

supercritical fluids, substances with temperatures above 700°F (350°C) that are confined by such intense pressure that they exist in a limbo state between liquid and gas. The resulting superheated steam could generate 10 times the electricity of most geothermal wells.

This type of renewable energy is becoming more popular. Between 2005 and 2010, use of geothermal power worldwide increased by 20%, according to the International Geothermal Association, and it is currently growing at an annual rate of 4% to 5%. As of 2013, geothermal power plants were annually producing 30,000 MW of geothermal power online, with another 12,000 MW of projects under construction or proposed.

INFOGRAPHIC 23.4

geothermal energy Heat stored underground, contained in either rocks or fluids.

geothermal heat pump A system that transfers the steady 55°F (12.5°C) underground temperature to a building to help heat or cool it.

payback time The amount of time it takes to save enough money in operation costs to pay for equipment.

geothermal power plants Power plants that use the heat of hydrothermal reservoirs to produce steam and turn turbines to generate electricity.

INFOGRAPHIC 23.4   GEOTHERMAL ENERGY CAN BE HARNESSED IN A VARIETY OF WAYS   3

↓ The high temperatures found underground in some regions can be tapped to generate electricity. Geothermal heat can also be piped directly to communities to provide heat or hot water. Geothermal energy can also be tapped using ground-source heat pumps, reducing the cost to heat and cool individual buildings.

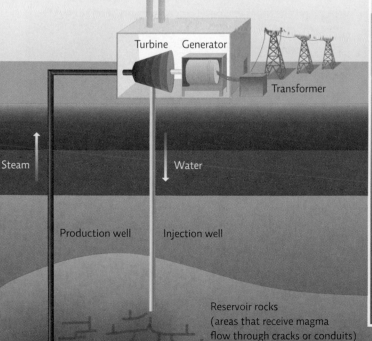

GEOTHERMAL POWER PLANT
In dry steam geothermal power plants, water is injected into deep wells. When the water hits hot rocks deep underground, steam is produced. This steam rises through a separate pipe to turn a turbine at the surface, generating electricity.

Turbine   Generator

Transformer

Steam   Water

Production well   Injection well

Reservoir rocks (areas that receive magma flow through cracks or conduits)

GROUND-SOURCE HEAT PUMP

A pump in the home circulates a fluid similar to antifreeze through a closed-loop system of buried pipes, bringing the ground temperature into the home.

Pump

The EPA considers ground-source heat pumps (a type of geothermal energy) to be one of the most efficient heating and cooling systems. The home only needs to be further heated from the ground temperature of about 55°F (12.5°C) in the winter; in the summer this temperature can cool the home.

The two geothermal technologies are quite different. Which one actually produces electricity, and which one helps you conserve energy?

Geothermal energy can be captured on a small scale to lower home heating and cooling costs or on a large scale to produce electricity or heat.

Harnessing such a powerful heat source can be difficult, however. In early 2007, an earthquake of magnitude 3.4 on the Richter scale shook the town of Basel, Switzerland. The quake was attributed to a geothermal mining project in the area, and the project was halted. The Iceland drilling project also experienced serious problems when the drill repeatedly got stuck 2 kilometers (1.3 miles) deep into a volcanic crater, leading to months of jammed drill bits and broken pipes.

The power of water can be harnessed but comes with trade-offs.

The island of Samsø is shaped curiously like a violin. It is also surrounded by one of the most significant sources of renewable energy worldwide: water.

Humans have harnessed the power of falling water for thousands of years, ever since early civilizations used watermills to grind grain into flour. Today, energy produced from moving water—known as **hydropower**—supplies more electricity to

hydropower Energy produced from moving water.

the population than any other single renewable resource. Approximately 7% of electricity used in the United States and 28% around the world is generated from hydropower. Large-scale hydroelectric power plants at giant dams are the source of most of that power.

Like wind energy, hydropower is an indirect form of solar energy. The downward flow of water from mountaintop to ocean is a consequence of the water cycle, a process of evaporation and condensation driven by the Sun's heat. (See Chapter 14 for more information about the water cycle.) Hydropower is abundant, clean, and does not typically produce greenhouse gases.

There are a variety of ways to harness the energy of moving water—from ocean waves and the tides to capturing energy released from variations in ocean temperatures (ocean thermal energy conversion)—but the most common way to generate hydropower is with dams. The Grand Coulee Dam on the Columbia River in Washington State is one of the largest concrete structures on the planet and the third-largest dam. The 12 million cubic yards of concrete that make up the dam could form a 1.2-meter-wide (4-foot-wide) sidewalk wrapped twice around the equator. The dam is the largest electrical power producer in the United States, with a total generating capacity of 6,809 MW. The reservoir created by the dam, an artificial lake that pools behind the structure (called an impoundment), stretches some 240 kilometers (150 miles) to the Canadian border. Three power plants at the Grand Coulee Dam contain

↓ The geothermal heat produced from the Reykjanes Power Plant in Iceland is mostly used to heat freshwater, which heats 89% of the houses in Iceland.

↓ Installation of a geothermal heating system in a residence. Closed-loop tubing can be installed horizontally in shallow trenches or vertically in much deeper wells. Here, a horizontal loop is being installed.

24 generators, which produce enough electricity to power two cities the size of Seattle.

But hydropower from large dams is far from an ideal resource. It wasn't a real option on Samsø, for instance, because the island lacks high mountain peaks whose runoff would feed large rivers. So geography is a key factor. Importantly, hydroelectric systems also come with some significant drawbacks. Electricity generation capacity can vary from year to year based on rainfall amounts. Also, dams that restrict water flow can curtail water supplies to downstream areas. Further, in hot climates, huge amounts of water evaporate from reservoirs. But more significantly, dams are responsible for the loss of major habitats and the displacement of tens of millions of people around the globe. It is the most debated and contentious of any renewable energy resource.

This is nowhere more apparent than at the Grand Coulee. "The Grand Coulee has had a huge impact for good and for ill," says Michael Garrity, Washington Conservation Director for American Rivers, an organization dedicated to protecting America's rivers. As the final loads of concrete were placed and the first generator at Grand Coulee was switched on in 1941, crowds gathered on the hill above the dam to marvel at the "birth of one of the world's greatest waterfalls," a local paper proclaimed. But nearby, thousands of Native Americans watched in horror as habitats, homes, and livelihoods were irreversibly lost under the pool of water engulfing the land behind the dam.

For nearby Native American communities, the Grand Coulee was a humanmade disaster that sacrificed not only land but a staple of their economy and culture—salmon. Salmon are migratory fish: They hatch in the freshwater tributaries of the Columbia, migrate downstream into the Pacific Ocean to spend most of their lives in salt water, then return to the freshwater environment of their origin to spawn. But the Grand Coulee Dam blocked the way; salmon and trout runs upstream of the dam were completely wiped out.

Not all hydropower systems have the same impact as a giant dam. A *run-of-the-river hydroelectric* system, often a low stone or concrete wall, doesn't block the water; it merely directs some of the flowing water past a turbine and thus generates electricity from the natural flow and elevation drop of a river, which is less disruptive to the river ecosystem. But energy production is dependent on the flow of a river at any particular time, so such a system is suitable only for rivers with dependable flow rates year-round. Such systems are useful in remote areas where electricity can be costly, and they are good candidates to supplement solar power systems, since annual periods of high sunlight often have low water flow and vice versa. **INFOGRAPHIC 23.5**

Garrity and other conservation groups still hope that new technologies will someday make existing dams like the Grand Coulee more eco-friendly. "We are advocates for harnessing resources from our rivers," he says, "but in a way that's sustainable."

The true cost of various energy technologies can be difficult to estimate.

One recurring theme with renewable energy technologies is that they are more expensive than fossil fuel methods. Even though sunlight, wind, geothermal heat, and water are free, constructing and installing the solar cells, wind turbines, underground pipes, and hydroelectric systems needed to harness their energy are not.

But perhaps we should not be asking why renewables are so expensive but instead asking why fossil fuels are so cheap. We need to consider *all* the costs—environmental, social, and economic—of these technologies to fairly compare them. Even if the costs to construct and install a technology are high, those costs might be offset by the lower environmental and health costs of the sustainable technology. This makes comparisons between the energy technologies difficult. Various studies have tried and so far agree on several points: Fossil fuels, led by coal, have the highest external (environmental) costs. Purely from an economic standpoint, solar energy is the most expensive form of energy. Finally, when external costs are added to production costs to estimate a true cost per kilowatt-hour, wind leads the way as the least expensive method. (For a look at the trade-offs of biofuels, see LaunchPad Chapter 32.) **TABLE 23.1**

It was this logic that helped convince the people of Samsø to personally invest in the ambitious project to convert to 100% sustainable energy. Such a project would not be possible everywhere, says Hermansen. Indeed, he wasn't even sure at first it would succeed because it needed the buy-in of every one of the island's 4,000-plus

KEY CONCEPT 23.6

Water power can be harnessed to generate electricity. Large dams are the most common and productive method, but there are other, less destructive technologies in use and development.

INFOGRAPHIC 23.5   **HARNESSING THE POWER OF WATER**

↓ The energy of moving water can be captured in a variety of ways to produce electricity.

HYDROELECTRIC DAM

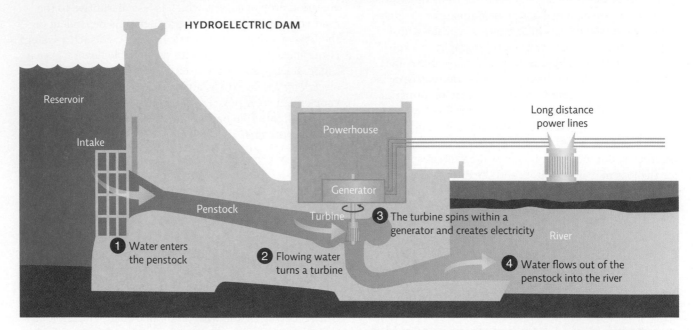

↑ The most common way to harness the power of water is with a hydroelectric dam. When a large river is dammed, a reservoir is formed behind the dam. The buildup of water in the reservoir creates an enormous amount of pressure; water diverted from the top of the dam flows through long pipes, called penstocks, to turbines below. As the water rushes past the curved blades of the turbines, they spin and generate electricity.

↑ A run-of-the-river hydroelectricity station on an estuary of the Hudson River, New York. In such stations, the river still flows freely and is not impounded to produce a lake.

↑ A variety of new technologies are being developed and tested to harness the power of ocean tides and waves. This giant turbine will be fastened to the seafloor to capture energy from tidal currents.

 Do you feel the energy production and recreational/flood control benefits of large hydroelectric dams outweigh the permanent ecological and community destruction they cause? Explain your position.

TABLE 23.1   THE TRADE-OFFS OF RENEWABLE ENERGY SOURCES   ［2］［3］

↓ All energy sources come with trade-offs that must be examined to make the wisest energy choices. When the true costs of fossil fuels are factored in, however, sustainable options are the environmental—and possibly economic—best choice.

Advantages	Disadvantages
Wind • Abundant in some areas • Pollution free • Low environmental impact and carbon footprint • Effective at large and small scales • Lowest-cost renewable energy source • The industry creates jobs • Surplus energy can be stored in batteries	• Wind is unpredictable and intermittent • Not all areas are windy enough to support it • Wildlife (birds and bats) are killed by windmills • Considered an eyesore by some • Usually must be paired with other energy sources to meet needs
Solar • Abundant • Pollution free • Low environmental impact and carbon footprint • Effective at large and small scales • Portable • The industry creates jobs • Surplus energy can be stored in batteries	• Location matters; shady and high-latitude areas get less sunlight • Little or no generation on cloudy days or at night • Expensive compared to other renewable energy sources • Usually must be paired with other energy sources to meet needs
Geothermal • Pollution free • Low environmental impact and carbon footprint • Can be used at large or small scales to produce electricity or to capture heat • Geothermal heat pumps can be installed anywhere and will reduce monthly heating and air conditioning costs	• Geothermal power plants are expensive to build and are suitable only in some areas • Geothermal heat pumps are more expensive to install than traditional HVAC systems • Usually must be paired with other energy sources to meet needs
Hydroelectric • Pollution free • Low environmental impact and carbon footprint (once dams are built) • Large dams offer recreation and flood control advantages • Small run-of-the-river systems can generate electricity without damming a river • Tremendous power potential remains to be harnessed from the oceans	• Large dams are expensive to build and permanently damage habitats and displace communities • Production capacity varies from year to year based on rainfall • Usually must be paired with other energy sources to meet needs • Damming large rivers restricts water flow to downstream areas

? Consider the area where you live. What mix of renewable energy sources would work best for your community?

KEY CONCEPT 23.7

Renewable energy sources may be more expensive to use but only if the true costs of using fossil fuels are ignored.

residents—literally. The only way the project would work, he and others believed, was if it was owned by the people. For years, Hermansen and others had to go practically door to door, trying to convince everyone to become part owners in the island's sustainable energy infrastructure. "If we didn't own these wind turbines and other sustainable energy projects on Samsø, a big power company would," says Hermansen. "Eventually, people will get their money back."

At first, the islanders were skeptical. "People were a little scared in the beginning. They were afraid the changes would disturb the landscape and affect tourism," says Garbers. But attitudes slowly changed, thanks largely to Hermansen's efforts.

Hermansen was born on Samsø and grew up there on a family farm. He was an environmental studies teacher with an enthusiasm for all things renewable when he

Gary Weathers/Tetra images/Getty Images

↑ The Grand Coulee Dam, located on the Columbia River in central Washington, is the largest concrete structure in the United States. In addition to producing up to 6.5 million kilowatts of power, the dam irrigates more than 200,000 hectares of farmland and provides wildlife and recreation areas. However, its construction eradicated salmon runs and inundated Aboriginal villages and ancient burial sites.

heard about the contest, and it wasn't long before he became the project's first employee. Through extensive community meetings and seminars, he rallied the islanders to the cause. "We are lucky to have him," says Garbers. "If it had been some guy from the mainland promoting the project, maybe the islanders wouldn't have believed him. But Søren was local." This constellation of forces—a small, close-knit community open to the project, community members who personally invested in the venture, led by a local individual with a passion for sustainability—helped Samsø achieve its goal.

By now, most of Samsø's residents have invested, starting at around $600 for 1 share in a wind turbine. The logic, says Hermansen, is that all residents share the profits. In a good wind year, the island sells its excess electricity, and people who purchased 10 shares receive 10 shares' worth of profits. He estimates that each Samsø resident has invested an average of $20,000 over the past 10 years in revolutionizing electricity on the island.

conservation Efforts that reduce waste and increase efficient use of resources.

Conservation plays a vital role in a sustainable energy society.

Even though the people of Samsø saw the importance of investing in sustainable energy, they realized that one of the best ways to help achieve energy independence would be to simply use less of it. As energy advisors like to say, the *greenest kilowatt is the one you never use*. Luckily, there are lots of ways to reduce electricity use right now. **TABLE 23.2**

Conservation—making choices that result in less energy consumption—is a vital part of our quest to become sustainable users of energy. Conservation means simply not using energy at certain times, as well as using energy more efficiently when we do use it.

KEY CONCEPT 23.8

Meeting our energy needs with renewable sources becomes more likely when we pair renewables with energy conservation measures.

TABLE 23.2   SAVING ENERGY   4

↓ Energy conservation is about wise use: using less and using energy more efficiently. Many of the steps below can be taken immediately, with no investment; others require money and time to implement. Whatever you can do will not only reduce your energy use but also reduce your energy bill and the pollution generated from producing that energy.

No-Cost Ways to Save Energy	Steps That Cost Money but Have Quick Payback Times	More Expensive Options with Payback Times Greater Than 5 Years
Turn off your computer and monitor and unplug phone chargers when not in use.	The first step is to have an energy audit performed to identify where your home is energy inefficient. (Go to www.energysavers.gov for more information.)	Replace older windows with double-pane windows suitable for your area. A variety of low-emissivity (low-e) window products restrict loss of heating or cooling while allowing in plenty of light.
Plug home electronics, such as TVs and DVD players, into power strips; turn off the power strips when the equipment is not in use.	Install attic and wall insulation to meet recommendations for your area.	Replace your heating, ventilation, and air conditioning (HVAC) system with an energy-efficient model; consider a ground-source heat pump.
Lower the thermostat on your water heater to 120°F (50°C); turn it off if you will be away for several days.	Weatherstrip and caulk doors and window frames; insulate electrical outlets on outside walls with inexpensive foam cutouts. Check heating ducts for leaks and seal if needed.	Install an on-demand (tankless) water heater; this saves energy because water is not constantly being heated and reheated.
In the winter, open curtains on south-facing windows to let in heat and light during the day; close them at night to reduce heat loss.	Insulate hot-water pipes and the water heater, especially if the water heater is located in an unheated area of your home.	
Be sure to close the fireplace damper when not in use to prevent loss of heat.	Install a programmable thermostat; you can program it to automatically adjust the temperature when you are not at home or are sleeping.	
Take short showers instead of baths to reduce hot water use.	Look for the Energy Star label on home appliances and light bulbs. Energy Star products meet strict-efficiency guidelines set by the U.S. Department of Energy and the EPA.	
Wash only full loads of dishes and clothes; wash clothes with cold water.	Upgrade to a high-efficiency washing machine that uses less energy (and less water—which also saves energy).	

? Which of the suggestions for saving energy could be done by a student living in a dormitory or rental apartment?

Much of conservation involves changing behavior. For example, lighting typically accounts for about 25% of the average home's electric bill, and simply turning off the lights when you leave a room can make a big difference. Other options include studying in a room with ample natural light instead of a dark corner that requires artificial light and lighting your home with compact fluorescent light bulbs (CFLs) or LEDs (light-emitting diodes), which consume less electricity than traditional incandescent bulbs. **INFOGRAPHIC 23.6**

For the residents of Samsø, because the island has such a cold climate, one of the first conservation efforts revolved around heating homes, mainly by transitioning away from electric heaters. Home energy audits pinpointed other steps individuals could take, such as installing more

INFOGRAPHIC 23.6 **ENERGY EFFICIENCY**   4

↓ Lighting accounts for about 25% of the electricity use in the average U.S. home, so efficient lighting can significantly reduce your energy use. There are a variety of lighting options for the home, and lighting technology is still improving.

CONSIDER LIGHT BULBS A 60-watt incandescent light bulb generates mostly heat and is a very inefficient way to produce light. On the other hand, a 13-watt compact fluorescent light bulb (CFL) produces less heat but the same amount of light as a traditional bulb. A 10-watt light–emitting diode (LED) is even more energy efficient than a CFL. Here we compare the three bulbs in terms of amount of energy used and the number of light bulbs required to provide 50,000 hours of use (the average life span of an LED bulb). Though they may cost more to purchase, LED and CFL bulbs save money and energy over the course of a bulb's life.

42 incandescent bulbs (3,000 kilowatts)   5 CFL bulbs (700 kilowatts)   1 LED bulb (500 kilowatts)

HAZARDOUS? Though CFLs actually contain mercury, if both an incandescent and a CFL bulb were illuminated using energy from a coal-fired power plant, the incandescent would release much more mercury to the environment (and the CFL will only release its 0.4 mg of mercury if it is broken). LEDs do not contain mercury but may contain other heavy metals such as lead, nickel, and copper; the type and amount of metal contained within depends on the color of the bulb (e.g., some red LEDs contain high amounts of lead). Though LEDs are not considered hazardous waste, as with a CFL, if an LED bulb breaks, you should carefully sweep up the material (gloves are recommended) and place it in a sealable plastic bag before disposing in the trash.

  Though the price is decreasing, LED bulbs are much more expensive than CFLs. In 2013 there was no real cost difference to buy and burn one LED bulb compared to five CFL bulbs. Why is it still enviromentally advantageous to use LEDs bulbs rather then CFLs if your electricity comes from a coal power plant?

KEY CONCEPT 23.9

Samsø became energy positive by using a variety of renewable technologies that fit its locale and reducing overall energy needs through conservation.

home insulation or using straw, wood chips, and other sources of biomass for heat. Older gas stoves were replaced with more efficient electric stoves; people began heating water using electricity generated by the wind turbines or solar energy. Residents were encouraged to apply for grants toward upgrades that made their homes more energy efficient. Similar programs exist in the United States. (See Chapter 25 for more ways consumers can cut energy use.)

In their quest to become an energy-neutral or even an energy-positive island, the people of Samsø focused their efforts on a variety of renewable technologies and conservation measures. In just a few years, the island of Samsø has transformed itself, capturing the attention of the other countries, such as the United States and Japan, that are impressed by how a small farming community became the most energy-efficient place on Earth. By 2005, the island was producing more energy from renewable resources than it was using. This transition has the added benefit of cleaner air: Samsø has reduced its nitrogen and sulfur air pollution by 71% and 41%, respectively, and its CO_2 emissions have fallen to below zero, thanks to the fact that the windmills offset more CO_2 than the islanders' vehicles emit. **INFOGRAPHIC 23.7**

Much of Samsø's success stems from community members working together to solve their own local problems, says Hermansen. As he told *Time* magazine in 2008, "People say: 'Think globally and act locally.' But I say you have to think locally and act locally, and the rest will take care of itself."

Select References:
Birol, F. (2013). *World Energy Outlook.* Paris: International Energy Agency. doi: 10.1787/weo-2013-en.
Geothermal Energy Association. (2014). *Annual U.S. and Global Geothermal Power Production Report.* http://geo-energy.org/reports.aspx.
Smallwood, K. S. (2013). Comparing bird and bat fatality rate estimates among North American wind energy projects. *Wildlife Society Bulletin,* 37(1): 19–33.

INFOGRAPHIC 23.7   **SAMSØ: THE ENERGY-POSITIVE ISLAND**

↘ Samsø is pursuing several methods to help it produce enough renewable energy to meet all its needs.

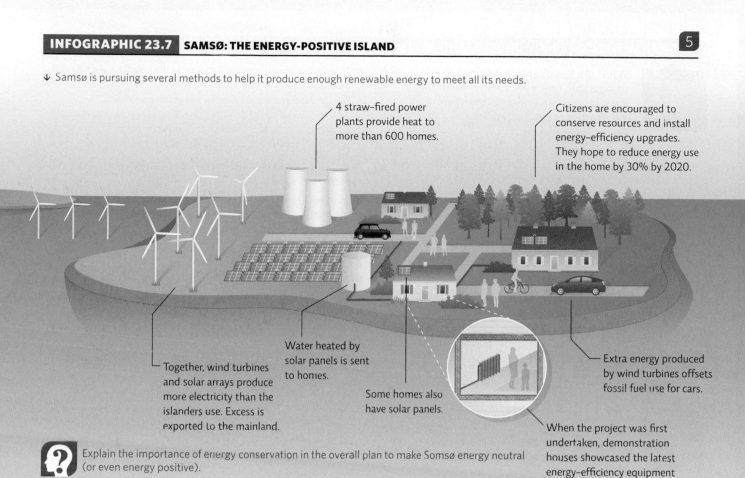

4 straw–fired power plants provide heat to more than 600 homes.

Citizens are encouraged to conserve resources and install energy–efficiency upgrades. They hope to reduce energy use in the home by 30% by 2020.

Together, wind turbines and solar arrays produce more electricity than the islanders use. Excess is exported to the mainland.

Water heated by solar panels is sent to homes.

Some homes also have solar panels.

Extra energy produced by wind turbines offsets fossil fuel use for cars.

When the project was first undertaken, demonstration houses showcased the latest energy–efficiency equipment and insulation options.

? Explain the importance of energy conservation in the overall plan to make Somsø energy neutral (or even energy positive).

BRING IT HOME

PERSONAL CHOICES THAT HELP

A key component to developing a sustainable society is using renewable energy. Use of renewable energy decreases harmful impacts of mining associated with nonrenewable energy, and it also reduces the amount of air and water pollution produced when the fuel is processed, transported, and burned. The efficiency and availability of renewable energy are rapidly increasing because of the development of new trends and technologies.

Individual Steps

• Contact your energy provider to see if you can purchase a percentage of your energy from a renewable source.

• Regardless of your home's energy source, make sure you are using energy efficiently. Review Table 23.2 and visit http://energy.gov/energysaver/energy-saver for more ideas.

• If you have a smart phone, download the PVme app to see how many solar panels you would need to meet your household energy needs.

• A major barrier to using solar energy for many people is the cost. For information on tax incentives, rebates, and other programs that make using renewable energy easier, check out www.dsireusa.org or look into the growing trend of leasing solar panels.

Group Action

• Learn about opportunities to invest in solar energy projects for lower-income communities or start one in your own community from the nonprofit Community Power Network, at www.communitypowernetwork.com.

Policy Change

• Contact your state and federal legislators and ask them to support or sponsor legislation that provides financial incentives for the purchase of renewable technologies.

iStockphoto/Thinkstock

ENVIRONMENTAL LITERACY UNDERSTANDING THE ISSUE

 1 What are the characteristics of a sustainable energy source, and what role does renewable energy play in terms of global energy production?

INFOGRAPHIC 23.1

1. True or False: The use of fossil fuels is projected to decline in the next 50 years as the use of renewable energy sources increases.

2. Which of the following renewable energy sources contributes the largest proportion of energy worldwide?
   a. Wind
   b. Water
   c. Solar
   d. Biomass

3. Which of the following in *not* a characteristic of sustainable energy sources?
   a. They must be renewable.
   b. We must use them at or less than the rate at which they are replenished.
   c. They must have no environmental impact.
   d. All of these are characteristics of sustainable energy.

 2 How do wind and solar power technologies capture energy? What are the advantages and disadvantages of wind and solar power, and how does each compare to fossil fuels in terms of true costs?

INFOGRAPHICS 23.2, 23.3 AND TABLE 23.1

4. Solar technologies that capture heat without any electronic or mechanical assistance are called:
   a. passive solar technologies.
   b. active solar technologies.
   c. solar thermal systems.
   d. photovoltaic solar cells.

5. Which of the following is not an advantage that solar and wind power share?
   a. Pollution free
   b. Effective at large and small scales
   c. The most inexpensive renewable technologies to install
   d. Job creation

6. Compare and contrast the production of electricity by a wind turbine and the production of electricity by heating water to produce steam (thermoelectric production).

 3 In what ways can we harness geothermal energy and the power of water? What are some of the trade-offs associated with each of these resources?

INFOGRAPHICS 23.4 , 23.5 AND TABLE 23.1

7. True or False: While geothermal power plants need to be built near sources of underground heat such as hot springs, geothermal heat pumps can be installed in any homes built on land.

8. Which disadvantage of large dams can the smaller run-of-the-river dams avoid?
   a. The smaller dams do a better job at flood control.
   b. The environmental damage of creating a reservoir is avoided.
   c. The smaller dams don't generate air pollution like the large dams do.
   d. Production capacity will be steady not variable, as with large dams.

9. Explain how a geothermal heat pump works and how it can lower both heating and cooling costs for homeowners.

10. How were Native Americans and wild salmon populations adversely affected by the construction of the Grand Coulee Dam? Do you think this dam should have been built? Justify your answer.

 4 What roles do conservation and energy efficiency play in helping us meet our energy needs sustainably?

INFOGRAPHIC 23.6 AND TABLE 23.2

11. Energy advisors say "the greenest kilowatt is the one:
    a. produced by solar power."
    b. that doesn't harm the Earth."
    c. that's cheapest."
    d. you never use."

12. Which of the following should you do first if you want to lower your electric bill?
    a. Install PV panels on your rooftop
    b. Have an energy audit done to see where your home is most inefficient
    c. Replace your appliances with energy-efficient models
    d. Install more insulation in your attic and exterior walls

13. What is the concept of "payback time," and how can it be useful in deciding what types of conservation measures to pursue?

 5 What combination of actions did Samsø take to become an energy-positive island? What is the take-home message to other communities that might want to reduce their use of fossil fuels?

INFOGRAPHIC 23.7

14. Samsø used _____, _____, and _____ energy sources to meet its energy needs.

15. As a member of the city council in a small town in New Mexico, you must vote on a new source of energy for your growing community. What should be the first step in determining where the energy will come from?
    a. Rely on historical practices and build a new coal-fired power plant.
    b. Pass a new law that requires all buildings to have solar panels installed.
    c. Conduct an assessment of available energy sources, both renewable and nonrenewable.
    d. Provide citizens a tax subsidy for improving the energy efficiency of their homes.

16. Describe the characteristics of the four sustainable energy sources presented here for generating electricity: wind, solar, geothermal, and hydroelectric power. Why are all sources not practical for every location?

SCIENCE LITERACY WORKING WITH DATA

A regional power company has proposed building a wind farm in your community. There is plenty of wind, so many people in the community think this is a good idea. Some, however, have cited the environmental impacts of windmills, especially bird and bat mortality. Examine the following graphs and use them to answer the questions that follow.

Graph A

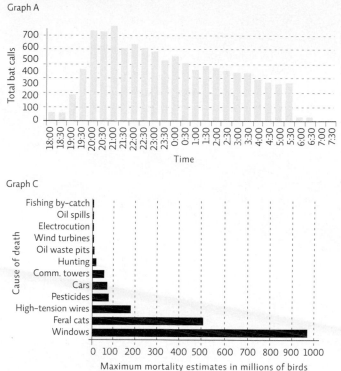

Graph B

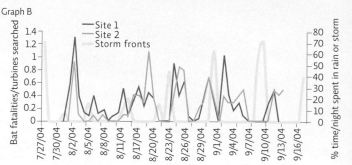

Graph C

Table 1

Date	Buffalo Ridge, MN	Vansycle, OR	Buffalo Mtn., TN	Stateline OR/WA	Foote Creek Rm, WY	Total bat carcasses	%
May 1–15	0	0	0	–	0	0	0.0
May 16–31	1	0	0	–	1	2	0.4
June 1–15	0	0	0	–	1	1	0.2
June 16–30	3	0	0	–	2	5	0.9
July 1–15	9	0	9	0	2	13	2.8
July 16–31	88	0	0	0	26	110	22.2
Aug 1–15	127	0	10	0	19	151	28.2
Aug 16–31	75	4	0	11	33	128	23.9
Sep 1–15	52	4	8	0	21	81	13.1
Sep 16–30	4	2	–	10	0	20	3.7
Oct 1–15	1	0	0	8	2	11	2.1
Oct 16–31	2	0	0	0	0	2	0.4
Nov 1–15	0	0	0	1	0	1	0.2

Interpretation

1. In one or two sentences, summarize the information provided in each graph or table.

2. For each graph or table, provide a title and a summary of the results.

Advance Your Thinking

3. Based on the data and where you live, what do you think the impact of a wind farm in your area would be on bird and bat mortality? If a wind farm were installed, what recommendations could you make to reduce bird and bat mortality? Make sure to discuss your answer in relation to each of the graphs/tables.

INFORMATION LITERACY EVALUATING INFORMATION

Renewable energy resources have become a priority around the world as we grapple with the cost and limited supply of fossil fuels, not to mention the effect on the environment resulting from their use. How do we know which energy source is appropriate to invest in, with both our time and money? Do we have to give up our current standard of living if we switch to an alternative energy source?

Go to the EPA's Clean Energy website (www.epa.gov/cleanenergy/ energy-and-you/how-clean.html). Read the information on the home page and enter your zip code in the box provided. You will then be asked to select the power company that provides energy to your community. Choose your company from the drop-down menu and continue by clicking the link View Report. Look at the Electricity Generation in Your Region graphs and read the information provided in the Make a Difference section.

Evaluate the website and work with the information to answer the following questions:

1. Is this a reliable information source? Does it have a clear and transparent agenda?
   a. Who runs this website? Do this organization's credentials make it reliable or unreliable? Explain.
   b. Who are the authors? What are their credentials? Do they have scientific background and expertise that lend credibility to the website?

2. What fuel provides most of the energy in your region? Were you aware of this, or are you surprised?

3. How does your community differ from the nation as a whole in terms of both fuel sources and emissions?

4. Does your local provider offer green power options? Can you sign up to use green power? If so, describe the program and how you could participate.

Find an additional case study online at http://www.macmillanhighered.com/launchpad/saes2e

CORE MESSAGE

Society seeks to protect the natural environment and public health by establishing environmental policies that define what is acceptable behavior for individuals, groups, or nations with respect to the environment. National and international policies are needed when environmental problems extend across state or national boundaries. Because environmental problems are complex, policies are often compromises between various stakeholders. Getting agreement on national or international policies can be difficult.

AFTER READING THIS CHAPTER, YOU SHOULD BE ABLE TO ANSWER THE FOLLOWING **GUIDING QUESTIONS**

Why are environmental policies sometimes needed at a national or even international level? What are some of the major U.S. environmental laws?

How are policy decisions made? How does lobbying influence policy decisions?

COUNTERFEIT COOLING

In the global efforts to thwart climate change, some lessons are learned after the fact

Smoke rising from the chimneys of a large factory in Beijing, China. TAO Images Limited/Getty Images

3

What policy tools can be used to implement and enforce environmental policy?

4

How are international policies established and enforced? How has the international community responded to the issue of climate change?

5

How successful have efforts to reduce CO_2 emissions been? Why has the Clean Development Mechanism been less effective than expected in reducing greenhouse gases?

In the spring of 2006, when he was still a law student at Stanford University, Michael Wara had a Eureka! moment—a discovery that would eventually confirm scientists' and policy makers' worst fears. Wara, a former climate-change scientist who was now focused on climate law, had set out to assess a particularly controversial global environmental policy—one that had been implemented a few years earlier in an effort to slow the rise of greenhouse gas emissions.

The *emissions trading* policy known broadly as **cap-and-trade** was based on a deceptively simple-sounding idea: A *cap*, or upper limit, is set on the amount of any given pollutant—in this case greenhouse gases—that a country (or an industry within a country if it is a national program) is allowed to emit each year. Greenhouse gas producers within that country (i.e., industry or power companies) are each issued permits for the amount of pollution they are allowed to emit, and all those permits all add up to the preestablished cap. If a company reduces its emissions below its allowance, it can sell, or *trade*, those leftover permits as "carbon credits" in the global marketplace—often for a hefty sum, usually to other companies that have yet to meet their own emission-reduction targets. By purchasing the carbon credits, companies that can't manage to reduce their own emissions can still contribute to reduction initiatives elsewhere—in other industries or other parts of the world (see Chapter 20).

In an effort to reduce carbon emissions in developed countries, the **Clean Development Mechanism (CDM)** was created in 2003. Under this cap-and-trade program, industrialized nations that were required to lower their carbon emissions would fund emissions-curbing technology and projects in developing, unindustrialized nations by purchasing carbon credits (also referred to as carbon offset credits) that these projects earned. In this trading scheme, 1 credit offsets 1 metric ton of carbon; this means a company can release 1 metric ton of carbon over its allowable amount for every carbon credit it purchases. "For example," Wara explained in a 2007 *Nature* article, "rather than build an ineffectual but cheap coal-fired plant, a Chinese utility might instead build a more efficient gas-fired plant that emits less CO_2; the difference in potential carbon emissions between the coal plant and the gas plant can, after monitoring and certification, be converted into CDM [carbon offset] credits that can be sold to an industrialized nation. The revenue from the credits enables the utility to afford the more expensive gas plant." These clean energy projects were seen as the most cost-effective ways to reduce overall carbon emissions worldwide because they helped developing countries that would otherwise resort to using fossil fuel resources as they grew and developed.

Straightforward as it all sounded, though, policy makers and environmentalists had been sharply divided since the program's inception on one major question: Would the policy actually result in emission reductions? With the emissions trading program now well established, it was time to find out. Wara had deployed an army of formulas across a mountain of spreadsheets, over many, many weeks.

The final tallies were more than disheartening: Somehow, this program that had been designed to reduce greenhouse gas emissions was actually contributing to their increase. "It really shocked us," Wara recalls. "Nobody could believe it at first."

It looked like one group in particular was gaming the system: the coolant makers of India and China, whose gaseous products (called HFCs) are used to keep air conditioners and refrigerators nice and cold. Producing this coolant also generates a waste gas as a by-product. Both the coolant and this by-product waste gas are well-known potent greenhouse gases. As Wara discovered, these companies were eliminating the waste gas and earning thousands of credits (that they sold for tens of millions of dollars to developed countries) to do so. And the companies were churning out both gases in spades. "They were producing twice to three times as much coolant gas as they needed to meet market demand," says Wara. "It didn't matter if the coolant sold or not, because they were making a fortune off the credits they were earning by destroying all the by-product."

---

cap-and-trade Regulations that set an upper limit for pollution emissions, issue permits to producers for a portion of that amount, and allow producers that release less than their allotment to sell permits to those who exceeded their allotment.

Clean Development Mechanism (CDM) A UN program that allows a country that is committed to reducing greenhouse gases to implement emission-reduction projects in developing countries.

Because the causes and consequences of environmental problems often transcend state or national borders, national and international policies are needed to address them.

Carbon and greenhouse gas regulation—including the cap-and-trade system created for it—had come amid a protracted and impassioned global conversation. Political leaders, policy makers, scientists, and environmental activists all over the world had converged on the question of how best to curb emissions and thus stave off global warming. (See Chapter 21 for more on climate change.) With all that chatter and debate, how did it come to this?

The answer to that question contains a broad lesson about the challenges of protecting the environment through treaties and legislation.

Public policies aim to improve life in societies.

Environmental policies give us guidelines meant to restore or protect the natural environment—sometimes by repairing damaged ecosystems, other times by reducing or mitigating the impact we humans have on our planet. As discussed in Chapter 1, environmental problems are often "wicked problems," meaning that they tend to be very complex; they have multiple causes and consequences, along with multiple stakeholders. Most of the biggest environmental issues that we now face—like pollution, species endangerment, and climate change—are also **transboundary problems**; they occur across state and national boundaries. Therefore, solving them requires the cooperation of individual states and countries around the world. **INFOGRAPHIC 24.1**

This makes environmental policy tricky. From the start, lawmakers must juggle a handful of potentially daunting factors: effectiveness, or whether the policy can attain the desired goal; the negative trade-offs that might result from the policy; who will absorb the cost burden (external and internal costs) of the policy, for both its enactment and its after-effects; and whether the policy is flexible enough to accommodate changes—the task of **adaptive management**. Even once all those hurdles are cleared, another remains, and it's often the largest of the bunch: public support. Policies that may be effective, affordable, and flexible can still die before they ever have a chance to be enacted if voters and/or lawmakers don't agree that they are necessary.

Within the United States, policy can be set at three basic levels: local, state, and national. Before the 1960s, environmental issues mostly dealt with how best to use resources (see Chapter 1). Addressing pollution or environmental damage was not a key objective. And environmental issues were primarily handled at the state level. To be sure, there were a few environmentally focused federal laws on the books; for example, the Oil Pollution Act of 1924 banned the release of oil into coastal waters. But in general, federal environmental regulations were considered an intrusion on state sovereignty. In fact, most environmental problems were addressed only after the fact, through litigation—an arrangement that too often favored the polluters: It was even more difficult back then than it is today to prove that toxins from a factory or dump that had seeped into the water or permeated the air were killing livestock or causing human illnesses.

Eventually, though, things began to change. Industry grew, and so did pollution. As it did, environmental problems began slipping across state lines, so that water and air pollution from one state affected another. In the 1960s and 1970s, federal legislators, prompted by a massive national outcry, realized that more regulation was needed, and so they devised a new set of policies that could function across state lines: **performance standards**. By determining how much pollution could be released in the first place, environmental regulation shifted from after-the-fact litigation to prevention. The shift proved effective, and air pollution emissions from industry and vehicles began to drop; since 1970, emissions of the six criteria air pollutants have fallen more than 70%. (See Chapter 20 for more on criteria air pollutants.)

By 1969, the era of modern environmental policy had begun. That year, with performance standards gaining a foothold, the **National Environmental Policy Act (NEPA)** was codified into law. NEPA established environmental protection as a guiding policy for the nation, mandating that the federal government take the environment into consideration before taking any action that

environmental policy A course of action adopted by a government or an organization that is intended to improve the natural environment and public health or reduce human impact on the environment.

transboundary problem A problem that extends across state and national boundaries; pollution that is produced in one area but falls in or reaches other states or nations.

adaptive management A plan that allows room for altering strategies as new information becomes available or as the situation itself changes.

performance standards Targets that specify acceptable levels of pollution that can be released or exist in ambient (outdoor) air; industries must act to meet these standards.

National Environmental Policy Act (NEPA) A 1969 U.S. law that established environmental protection as a guiding policy for the nation and required that the federal government take the environment into consideration before taking action that might affect it.

INFOGRAPHIC 24.1   ADDRESSING TRANSBOUNDARY ENVIRONMENTAL PROBLEMS REQUIRES
INTERNATIONAL COOPERATION   1

↓ National and international policies are needed to address global environmental problems because what happens in one area can affect another. Human impact, such as the transport of non-native species, can have far-reaching consequences. In addition, the atmosphere, rivers, and oceans can very effectively deliver pollution from one area to the next. For example, greenhouse gases released in any part of the world impact global climate. Environmental issues that affect the entire planet require a global response, guided by policies we all agree to follow.

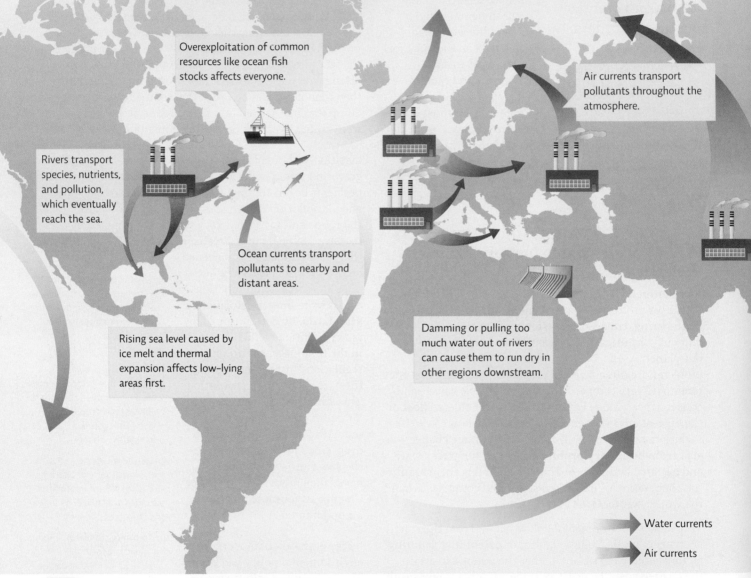

Overexploitation of common resources like ocean fish stocks affects everyone.

Air currents transport pollutants throughout the atmosphere.

Rivers transport species, nutrients, and pollution, which eventually reach the sea.

Ocean currents transport pollutants to nearby and distant areas.

Rising sea level caused by ice melt and thermal expansion affects low-lying areas first.

Damming or pulling too much water out of rivers can cause them to run dry in other regions downstream.

Water currents

Air currents

 What are some of the difficulties in trying to set and manage policy for international environmental problems?

might affect it. It also established a process that remains central to environmental regulation where the need for legislation is determined by available scientific evidence, and where various solutions are analyzed and compared, in excruciating detail, before a decision is made.

NEPA's signature feature has been the **environmental impact statement (EIS)**—a report that details the likely effect of a proposed federal action, such as building a road or upgrading a nuclear facility. The goal of an EIS is to identify problems before they occur so that stakeholders can choose the most acceptable course of action (maybe move that road a few miles to the south to avoid disturbing that forest; maybe don't build the road at all). To keep the process transparent, the findings are made available to everyone—citizens, policy makers, and

INFOGRAPHIC 24.2   POLICY DECISION MAKING—THE NEPA PROCESS

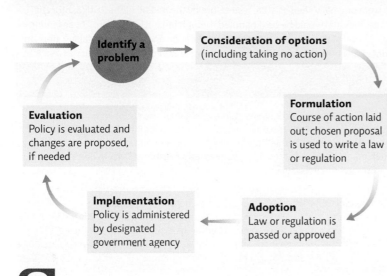

← Policies are created and revised using some basic steps that allow policy makers to systematically evaluate the situation and possible responses. The process starts with identifying the problem, considering available options for responding, and evaluating the costs and benefits. A policy is then drafted, further evaluated by interested parties, and, if found acceptable, formally adopted. In the United States, bills must be passed by the House of Representatives and the Senate and then signed by the president to become law. Regulations based on those laws are proposed and administered by regulatory agencies (like the Environmental Protection Agency) in a similar manner. The process itself is responsive and allows for adaptive management—reentering the policy cycle for revision as new or changing information comes to light.

What problems can arise if the language in a U.S. law that the EPA is supposed to enforce is vague or ambiguous?

KEY CONCEPT 24.2

The NEPA process is a useful guideline for policy making that includes systematically considering all options before setting policy and evaluating policy after it is implemented.

KEY CONCEPT 24.3

Many iconic U.S. environmental laws originated in the 1960s and 1970s. Some of these laws allow citizens to sue violators to ensure that the laws are properly enforced.

special interest groups—and everyone is given a chance to respond (through letters and public hearings). **INFOGRAPHIC 24.2**

In NEPA's wake came a wave of iconic legislation, most of it passed with overwhelming bipartisan support. Many of the environmental laws passed in the 1970s have a mechanism that allows individual citizens (or groups—including state governments) to demand enforcement via the **citizen suit provision**. Violations can be reported; if they aren't dealt with in a satisfactory manner, the citizen or group can file a lawsuit against the violator (an individual, a private company, or the government—even the regulatory agency mandated to enforce regulation) that has allegedly failed to uphold the existing law.

To implement and enforce all these new federal environmental laws,

Congress established the **Environmental Protection Agency (EPA)** in 1970. The EPA is a regulatory agency that establishes rules and regulations to support each environmental law as it is passed. EPA officials set the standards which ensure that the goals of any given law are met. They are also tasked with holding individual states and corporations accountable. If a given entity fails to comply with a given rule, the EPA has the authority to step in and mandate changes. It can, for example, force a power plant to make upgrades that decrease pollution, close down a factory for repeated violations, or fine an individual state for failing to curb its vehicle-generated air pollution. It can also force entities to pay clean-up costs and, in certain cases, can revoke operating permits. **TABLE 24.1**

Of course, the EPA's reach extends only to U.S. borders. And, as the case of the coolant factories shows, most environmental problems tend to stretch way beyond those.

Policy making involves many players.

There are currently more than 500 international environmental agreements in effect. They go by a range of names (conventions, accords, agreements, treaties, and so on), regulate a range of human activities, and protect

environmental impact statement (EIS) A document that outlines the positive and negative impacts of a proposed action (including alternative actions and the option of taking no action); used to help decide whether that action will be approved.

citizen suit provision A provision that allows a private citizen to sue, in federal court, a perceived violator of certain U.S. environmental laws, such as the Clean Air Act, in order to force compliance.

Environmental Protection Agency (EPA) The federal agency responsible for setting policy and enforcing U.S. environmental laws.

TABLE 24.1   NOTABLE U.S. ENVIRONMENTAL LAWS

↓ In the United States, environmental protection and regulation were originally seen as state issues, but by the middle of the 20th century, it became apparent that many environmental problems crossed state lines and would be best handled through federal legislation. These landmark environmental laws were passed, beginning in the 1960s, during a period of tremendous bipartisan cooperation and support for taking steps to ensure a clean and healthy environment. They have been amended many times to deal with changing or new environmental problems.

Law	Description
National Environmental Policy Act (NEPA), 1969	Established environmental protection as a guiding policy for the nation. It mandates that the federal government take the environment into consideration by completing an environmental assessment before pursing any federal action that might have an environmental impact. (Several states have similar laws regarding state actions that might have an environmental impact.)
Clean Air Act (CAA), 1970 (originally passed in 1963)	Regulates air pollutants that are hazardous to human health by setting standards about the amount of pollutants that can be present or released into the air. Greenhouse gases such as CO_2 were not originally covered by the CAA since they are not toxic, but in 2009 the Supreme Court gave the EPA the authority under the CAA to regulate greenhouse gases because they directly impact climate change.
Clean Water Act (CWA), 1972	Regulates water quality by setting standards for the release or presence of specified toxic or hazardous water pollutants.
Endangered Species Act (ESA), 1973	Protects and aids in the recovery of endangered and threatened species of fish, wildlife, and plants in the United States.
Toxic Substances Control Act (TOSCA), 1976	Regulates the production and distribution of designated toxic chemicals.
Comprehensive Environmental Response, Compensation, and Liability Act (CERCLA), 1980	Commonly called "Superfund," CERCLA requires that responsible parties clean up sites contaminated by hazardous materials and holds them liable for the costs and damages. It also provides funding to decontaminate sites when the owners cannot be found or cannot afford the cost of clean-up.

Do you feel that these U.S. environmental laws should be strengthened, weakened, or remain as they are?

a litany of environmental issues—whaling, fishing, endangered species, and the ozone layer, to name a few. The vast majority of them do what they were intended to do: influence human behavior in ways that protect the natural environment. "Overall, it's a very methodical and well-done process," says Stephen Andersen, co-chair of the economic assessment panel for the 1987 Montreal Protocol, which is an international treaty that successfully phased out most ozone-depleting substances (see Chapter 2).

Ideally, environmental policy begins with scientific insights gleaned through careful measurement and observation. Those insights come to policy makers through a range of venues: congressional hearings with expert witnesses; federal advisory committees; federally funded research organizations, like the American Association for the Advancement of Science or the Oak Ridge National Laboratory; and in many cases, international scientific organizations like the Intergovernmental Panel on Climate Change (IPCC), the World Meteorological Organization, and the United Nations Environment Programme.

Once a problem (say, a hole in the ozone layer, or an endangered species, or a rapidly warming planet) is identified, legislators and scientists work together to arrive at a set of policy recommendations. Oftentimes they use statistical analyses to gauge uncertainty—that is, how sure or unsure scientists are about future outcomes—and determine whether the **precautionary principle** should be employed. As various policy

precautionary principle Acting in a way that leaves a safety margin when the data is uncertain or severe consequences are possible.

political lobbying Contacting elected officials in support of a particular position; some professional lobbyists are highly organized, with substantial financial backing.

↑ One of the first CDM projects, a 10.6 MW wind farm at Bada Bagh in India's northern state of Rahasthan generates 19 million kWh of electricity annually. It prevented more than 160,000 metric tons of carbon from being released into the atmosphere in its first ten years of operation.

options are vetted, the judiciary weighs in about the constitutionality of the options.

But policy is determined by much more than science. **Political lobbying** also plays a large role. In the United States and even on the international stage, political lobbying—contacting elected officials in support of a particular position—is part of the democratic process. (We have access to our elected officials and can share our opinions with them.) Citizens and private organizations (e.g., nonprofits, labor unions, and industry groups) lobby for or against specific proposals, based on their own interests—which can range from the health of the environment to the health of the economy (or their own bottom line) to the future of the planet. Critics say that professional lobbying has grown alarmingly sophisticated and well financed, making individual voices harder to hear and potentially interfering with policy makers' judgment. Not only do industries run ad campaigns promulgating ideas that serve their own best interests ("clean" coal, for example, is not as clean as it sounds; it still creates heavy pollution and environmental damage), they also contribute large sums of money to candidates for elected office, hoping to influence those candidates, if elected, to act in ways favorable to the industry.

Nonprofits like the National Resources Defense Council (NRDC) also have professional lobbying divisions and do the same thing—spend money to promote their positions to elected officials and to the general public. Taken together, these environmental nonprofit organizations spend millions of dollars per year in federal lobbying efforts; they spent a record $24.6 million in 2009 in the United States, according to the Center for Responsive Politics. Despite this, the deep pockets of industry often outspend these groups—in that same year, the oil and gas industry alone spent an industry-record $175 million for

KEY CONCEPT 24.4

Policy decision making should be influenced by sound science, but political lobbies, public opinion, and the press also strongly influence the process.

INFOGRAPHIC 24.3 **INFLUENCES ON U.S. ENVIRONMENTAL POLICY DECISION MAKING**   2

↓ Many organizations and individuals influence not only whether we institute a policy to deal with an environmental issue but also the design of that policy—what it covers and how it will be implemented and enforced. The wide variety of voices, many representing differing viewpoints, can make it difficult to create new policies. Though political ideologies might influence how one goes about addressing a problem, policy makers ideally look to the best available science when making decisions about whether a policy is needed to protect the health and well-being of the public and environment.

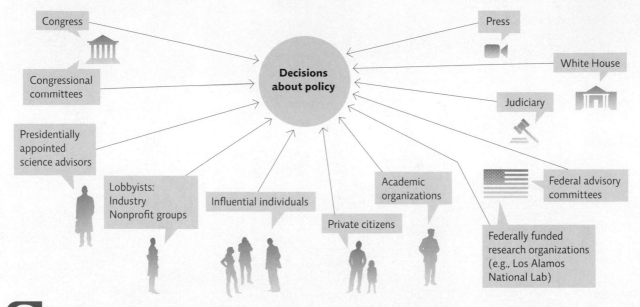

 Based on how you see the process playing out in the United States today, rank the parties that influence U.S. policy making from most influential to least influential. In your opinion, is this ranking as it should be?

INFOGRAPHIC 24.4 **POLICY TOOLS**   3

COMMAND-AND-CONTROL REGULATION

ADVANTAGES
- Simple in concept and may achieve desired goals quickly
- Directly changes the behavior of the regulated industry
- Especially effective policy tool when the potential for severe environmental or health impact is high, such as with extremely toxic substances
- Useful when level of control is known and uniform across all regulated industries

DISADVANTAGES
- Making changes to policy takes time, which makes it hard to keep up with new technologies.
- A one-size-fits-all approach may limit some industries' ability to use the most cost-effective methods to address their impact.
- No incentive for companies to reduce pollution below mandated limits
- Regulatory agencies must have sufficient funding to enforce compliance.

IMPLEMENTATION
Permits: Authorization required for operation; permits specify acceptable actions or environmental releases

Performance and technology standards: Impose emission limits for a given source, identify ambient levels of pollutants that are acceptable, and/or specify the technology and methods that must be used to reduce pollution

Penalties: Lost contracts, jail time, paying for clean-up or damage, fines, liability that requires violators to compensate others for harm or damage

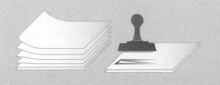

 In general, which approach do you support to reduce our use of fossil fuel and other activities that contribute to climate change: command-and-control regulation or economic incentives?

lobbying; the industry total was more than $400 million that year—often supporting provisions that oppose restrictions or regulation, steps that can potentially lead to increased environmental damage. **INFOGRAPHIC 24.3**

A variety of policy tools are being used to address climate change.

In general, there are a range of **policy tools** that lawmakers use to protect the environment. Governmental **command-and-control regulation** is one approach. This can take several different forms: issuing permits to authorize operation or establishing performance standards that regulate emissions and specifying the practices that must be used to meet those standards.

Alternatively, *market-based approaches* create economic incentives for the private sector to reduce environmentally harmful actions, without dictating exactly how to reach a desired target. For example, governments pursue the *polluter pays* principle when they levy taxes based on the amount of pollution

produced—so-called **green taxes**. A different option is to reduce taxes by offering tax credits to consumers or businesses that pursue environmentally friendly actions—such as buying hybrid vehicles or installing energy-efficient equipment. Cap-and-trade policies represent another market-based approach. First used to successfully curb acid rain pollution in the 1990s, this method sets limits and distributes tradable permits for allowable emissions which businesses can freely trade to meet performance standards. Financial incentives such as **subsidies**, grants, and low-interest loans can also encourage environmentally beneficial actions that might otherwise be hard for individuals or businesses to afford. **INFOGRAPHIC 24.4**

The HFC coolant policy that has proved so troublesome involves cap-and-trade as well as subsidies. It is part of the sluggish, decades-long path that constitutes the international attempt to address a warming climate. Its origins can be traced

KEY CONCEPT 24.5

Policy can be enforced with command-and-control regulation or through economic incentives that favor preferred responses.

policy tools Methods that can be used to enforce or implement regulations or achieve desired outcomes.

command-and-control regulation Regulations that set an upper allowable limit of pollution release which is enforced with fines and/or incarceration.

green tax A tax (a fee paid to the government) assessed on environmentally undesirable activities (e.g., a tax per unit of pollution emitted).

subsidies Financial assistance given by a government or another party in support of actions that are expected to benefit the public good.

ECONOMIC INCENTIVES

ADVANTAGES
- Fund actions that otherwise might be too costly for individuals or business to afford
- Taxes generate revenue that could benefit environmental causes
- Provide incentives to reduce pollution below requirements
- Encourage innovation for reducing environmental impacts
- May be more cost effective than command and control regulation
- Stimulate the economy

DISADVANTAGES
- Citizens may oppose tax dollars being used for endeavors they do not support.
- The cost of green taxes levied against industry may be passed on to the consumer.
- Cap-and-trade programs can create pollution hotspots in areas where most users choose to buy credits rather than reduce emissions.
- May inadvertently support undesirable activities (perverse subsidies)

IMPLEMENTATION
Green taxes: The *polluter pays* principle: taxes are levied on an environmentally harmful action (e.g., per pollution unit emitted).

Subsidies: Financial assistance: Includes things like cash transfers, lower costs for resources, and tax credits—reduction in one's tax because of environmentally favorable actions.

Grants or low-interest loans: Money to support the purchase of environmentally beneficial products such as solar panels

Tradable permits (cap-and-trade): Individual industries can trade or sell their allotment of emissions if they release less than their permit allows; all permits fall under a maximum cap for that pollutant or the pollutant in that area.

Jamey Stillings

↑ The Ivanpah Solar Electric Generating System, which began operation in 2013, is the world's largest solar-thermal power plant. Located in the California Mohave Desert, the 377 MW system uses more than 170,000 mirrors to reflect solar energy onto receivers on top of towers 137 meters (450 feet) tall. This heat is used to generate steam for the production of electricity. This facility will save more than 400,000 metric tons of carbon emissions per year compared to a coal-fired power plant producing the same amount of electricity.

back to the **United Nations Framework Convention on Climate Change (UNFCCC)**—an international treaty that most experts say has been the foundation for global climate change policy. Born of the 1992 Earth Summit in Rio de Janeiro—a multinational, UN-led conference that set the stage for modern-era sustainable development and environmental stewardship—the UNFCCC was the first convention to formally recognize climate change as a serious emerging threat against which precautions should be taken. Its signatories—including the United States and 95 other nations—set themselves the lofty goal of stabilizing greenhouse gas emissions "at a level that would prevent dangerous anthropogenic [human-induced] interference with the climate system."

That treaty was only the beginning of a long, slow, deliberative process that involved myriad entities, and a seemingly endless stream of meetings, debates, negotiations, and votes. Agreeing that something needs to be done

is one thing; coming to a consensus on what to do is another. Policy making, especially at the international level, is a painfully slow business. In fact, it took several years for the nations' representatives participating in drafting the UNFCCC to establish a formal statement, essentially saying "we should do something about climate change."

After that came the **Kyoto Protocol** (1997), an international treaty that laid out exactly what that *something* is: specific greenhouse gas reduction targets, with deadlines. But agreeing on these targets was not easy. Debates raged for months and months, making international headlines each step of the way and frequently pitting the business sector against the environmentalist movement, scientists against policy makers, and countries against countries. At issue was how, exactly, to curb greenhouse gas emissions. **INFOGRAPHIC 24.5**

United Nations Framework Convention on Climate Change (UNFCCC) A 1992 international treaty that formally recognized climate change as an emerging problem and that said precautions should be taken to prevent dangerous anthropogenic interference with Earth's climate system.

Kyoto Protocol The 1997 amendment to the UNFCCC that set legally binding specific goals for reductions in greenhouse gas emissions for certain nations that ratified the treaty.

KEY CONCEPT 24.6

International policy is established though treaties that range from a simple agreement that action is needed to protocols that specify procedures and targets for participants.

INFOGRAPHIC 24.5   **SETTING INTERNATIONAL POLICIES**

↓ Establishing policies at the international level takes time, a lot of work, and compromise. Effective policies are generally those that provide benefits across sectors (not just for one group or region), address the causes as well as the consequences, identify specific targets, and include the flexibility of science-based adaptive management to allow for revision. To succeed, international policies must have buy-in at multiple levels—government, citizens, industry, and among nations—and have effective enforcement.

← Brazilian President Fernando Collor de Mello, center, is applauded as he signs the United Nations Framework Convention on Climate Change (UNFCCC) at the 1992 Earth Summit in Rio de Janeiro, Brazil.

International meetings like this are held to address pressing issues and may result in treaties signed by participating countries that agree to abide by the treaty's content. A *convention* is a treaty that represents a position on a particular issue and may identify the courses of action that signatory parties agree should be pursued. The UNFCCC established that climate change was a major problem that needed to be addressed. Meetings typically include presentations by policy makers, scientists, and even individual citizens, and sessions that assess the situation.

← Chairperson Raúl Estrada-Oyuela shakes hands with an official after the Kyoto Protocol was adopted at the Kyoto International Conference Center in Japan, December 11, 1997.

Broad treaties like the 1992 UNFCCC are just the beginning and require additional agreements that lay out exactly what will be done. In this case the Kyoto Protocol provided that direction. A *protocol* is a treaty that specifically indicates what will be done—precise goals or targets are set. The protocol also indicates how compliance will be assessed and enforced.

← Canadian Environment Minister Peter Kent's notes for his announcement that Canada was withdrawing from the Kyoto Protocol in December 2011.

For international treaties like the Kyoto Protocol, compliance depends on the cooperation of the signatory parties to do what they said they would do when they signed the treaty. Beyond this, there are few international avenues available to enforce compliance. Other nations may put pressure on the violator by imposing economic sanctions (reducing or cutting off trade) but this does not guarantee compliance and is often seen as a last resort.

 There are different types of international agreements. In general, what is the difference between a *convention* treaty and a *protocol* treaty?

INFOGRAPHIC 24.6   INTERNATIONAL EFFORTS TO ADDRESS GLOBAL CLIMATE ISSUES   4

↓ The international community has been grappling with climate change since 1979, when the first World Climate Conference was held. Though progress has been slow, recent meetings and treaties have focused on identifying the appropriate response for countries, setting emission-reduction targets, and establishing policies to help countries that are most vulnerable to the effects of climate change.

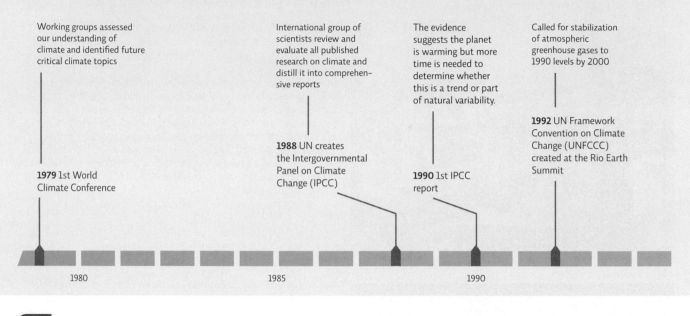

Working groups assessed our understanding of climate and identified future critical climate topics

1979 1st World Climate Conference

International group of scientists review and evaluate all published research on climate and distill it into comprehensive reports

1988 UN creates the Intergovernmental Panel on Climate Change (IPCC)

The evidence suggests the planet is warming but more time is needed to determine whether this is a trend or part of natural variability.

1990 1st IPCC report

Called for stabilization of atmospheric greenhouse gases to 1990 levels by 2000

1992 UN Framework Convention on Climate Change (UNFCCC) created at the Rio Earth Summit

1980   1985   1990

? Do you feel that the international response aimed at addressing climate change has progressed at a reasonable pace, or has it progressed too slowly or too quickly? Explain.

Ultimately, countries like the United States and Great Britain that had historically released the most greenhouse gases were given much bigger reduction targets than other countries; in all, 37 developed nations and the European Union itself (called Annex 1 parties) were given specific reduction targets relative to their emissions in the baseline year, 1990. Because developing countries had not released as much in the past, and did not necessarily have the resources to reduce current emissions, they were not given any reduction targets.

Developed nations—the United States especially— argued that this was unfair: If developing nations could burn more fossil fuels—and China and India were two developing nations that were burning lots of fossil fuels—they would gain an unfair advantage in the global marketplace; lucrative industries would desert us and flock to them. But developing nations were adamant: Emission caps of any kind would stunt their economic growth. Why should they have to bear even a fraction of the burden of reducing pollution that was caused almost exclusively by the developed, industrialized world? To address this divide, the Clean Development Mechanism was created by the Kyoto Protocol in 2003 to broker a compromise that would encourage both developed and developing countries to curb their emissions.

While the United States still declined to ratify Kyoto (the lack of specific reduction targets for all nations proved unacceptable to the members of the U.S. Senate, who rejected the Kyoto Protocol in a 97−0 vote), every other developed nation did sign on. Eventually, the CDM and its cap-and-trade system gave rise to an entire industry built around carbon credits. But it didn't happen overnight. Like most other international environmental policies, Kyoto was designed to allow a gradual shift so that countries would have time to

KEY CONCEPT 24.7

International efforts to address climate change include the passage of a variety of treaties and the establishment of the IPCC to review climate science, but progress has been slow.

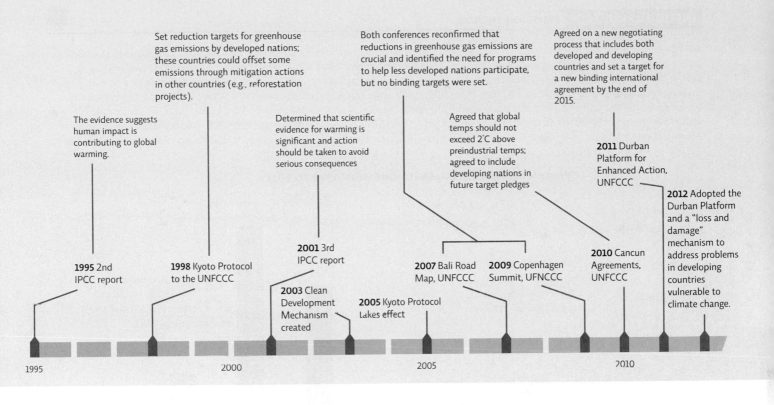

The evidence suggests human impact is contributing to global warming.

Set reduction targets for greenhouse gas emissions by developed nations; these countries could offset some emissions through mitigation actions in other countries (e.g., reforestation projects).

Determined that scientific evidence for warming is significant and action should be taken to avoid serious consequences

Both conferences reconfirmed that reductions in greenhouse gas emissions are crucial and identified the need for programs to help less developed nations participate, but no binding targets were set.

Agreed that global temps should not exceed 2°C above preindustrial temps; agreed to include developing nations in future target pledges

Agreed on a new negotiating process that includes both developed and developing countries and set a target for a new binding international agreement by the end of 2015.

2011 Durban Platform for Enhanced Action, UNFCCC

2012 Adopted the Durban Platform and a "loss and damage" mechanism to address problems in developing countries vulnerable to climate change.

1995 2nd IPCC report

1998 Kyoto Protocol to the UNFCCC

2001 3rd IPCC report

2003 Clean Development Mechanism created

2005 Kyoto Protocol takes effect

2007 Bali Road Map, UNFCCC

2009 Copenhagen Summit, UFNCCC

2010 Cancun Agreements, UNFCCC

1995   2000   2005   2010

---

respond and the economy would not take too big a hit— and also so that adaptive technology could be developed that would make the tasks at hand easier to accomplish. This is a common approach: Treaties and protocols like Kyoto identify a target and then outline a robust time frame, replete with interim targets and deadlines, all in service to the ultimate goal—in this case, zero carbon emissions. **INFOGRAPHIC 24.6**

But as Wara and others know all too well, when policies are slow to move forward, problems are also slow to emerge.

Policies sometimes have unintended consequences.

The numbers on the computer screen in Wara's Stanford office are digital avatars; their real-life counterparts— invisible gaseous molecules that trap heat and thus warm our atmosphere—exist a world away. They can be found in the large, fluffy plumes of smoke that curl up from the pipes of aged coolant factories in regions like the state of Gujarat in western India, then billow and disperse across low-slung, densely packed metropolises. Those factories produce two gases in particular: HCFC-22, used

in refrigeration and air conditioning; and HFC-23, which is merely a by-product of creating HCFC-22. Because both gases are potent global warmers, both are regulated under the Kyoto Protocol.

There are six Kyoto gases in all: carbon dioxide (CO_2), methane, nitrous oxide, hydrofluorocarbons (HFCs), perfluorocarbons (PFCs), and sulfur hexafluoride. For the purposes of crediting, each is converted to its *equivalent in carbon dioxide*, widely agreed to be the most abundant and most massively emitted of the six. So carbon dioxide is assigned a value of 1 (meaning that a single credit represents permission to emit 1 metric ton of CO_2 per year), and the remaining five are valued in relation to that, based on how potent they are as warmers and how long they can be expected to remain in the atmosphere. Methane is valued at 25 (1 ton of it equals 25 tons of CO_2, and thus 21 carbon credits), nitrous oxide at 298, and so on.

To be sure, the Kyoto Protocol has achieved some successes. The United Kingdom, for example, met its 2012 Kyoto target—a 12.5% reduction in greenhouse gas emissions relative to the 1990 baseline—in 2000, more

INFOGRAPHIC 24.7 | **EMISSION TRENDS**

↓ The International Energy Agency tracks CO_2 emissions from fossil fuel combustion, the leading contributor of greenhouse gas emissions. Data comparing 2011 emissions to the baseline year of 1990 are shown here. While some nations did have a significant drop in CO_2 emissions, others actually increased their emissions. Overall, the Annex 1 nations of the Kyoto Protocol (developed nations that committed to emission targets) reduced emissions and exceeded their Kyoto target of a 4.7% reduction by 7.4%. Much of this reduction was due to the economic restructuring of many of the countries of the former Soviet Union and Eastern bloc. The United States, which did not ratify the protocol and was not bound by its target of a 7% reduction, actually increased CO_2 emissions by 8.6%. Developing nations were not given any reduction targets and many had an increase in emissions. Overall, compared to 1990 emission levels, the world saw a net increase of 49.3%.

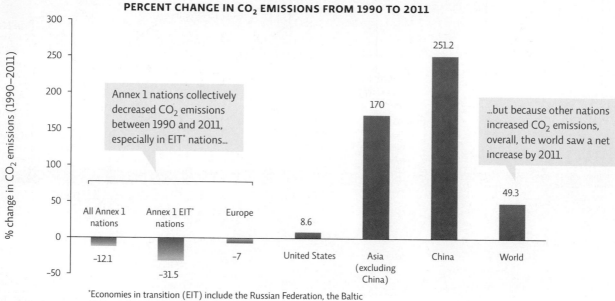

PERCENT CHANGE IN CO_2 EMISSIONS FROM 1990 TO 2011

Annex 1 nations collectively decreased CO_2 emissions between 1990 and 2011, especially in EIT* nations…

…but because other nations increased CO_2 emissions, overall, the world saw a net increase by 2011.

*Economies in transition (EIT) include the Russian Federation, the Baltic states, and several central and eastern European states.

? This graph shows the percentage change in CO_2 emissions between 1990 and 2011. Why might it also be helpful to see the actual CO_2 emissions of each group to determine the contribution of each to the problem and solution?

than a decade early. By the actual deadline, it had cut emissions by a whopping 23%—or double its Kyoto target. Germany also exceeded its Kyoto targets by nearly 10% by the 2012 deadline.

The United States also took steps to monitor and reduce emissions as part of its UNFCCC commitment. Though U.S. greenhouse gas emissions are still higher than 1990 levels, thanks to energy-efficiency improvements and land-use changes (like reforestation projects), they are starting to fall—dropping 10% between 2005 and 2012. Improving fuel efficiency in vehicles is also part of the U.S. approach. New targets for **corporate average fuel efficiency (CAFE) standards** of vehicles have been set; the 2011

corporate average fuel efficiency (CAFE) standards A target of the minimum fuel efficiency (MPG) that manufacturers must meet; evaluated as a weighted average of all the cars and light trucks each manufacturer produces.

average for cars and light trucks of 27.3 miles per gallon (MPG) will incrementally increase to 54.5 MPG by 2025.
INFOGRAPHIC 24.7

The coolant factories, though, stand in stark contrast to these achievements, in large part because of the way gases were assigned value under Kyoto. HFC-23, which is a potent greenhouse gas and has an atmospheric lifetime of 270 years, has been valued at 11,700. This means eliminating just 1 metric ton of this by-product gas earns a company 11,700 carbon credits. And that means the gas—a mere by-product with no commercial value of its own—is very valuable. When

KEY CONCEPT 24.8

Despite efforts to reduce CO_2 emissions, and successes in some countries, worldwide emissions continue to increase.

Once policies are implemented, flaws may emerge or conditions may change, but adaptive management can inform policy changes that address these issues.

approved as a CDM project, industries in developing countries earn carbon credits by destroying these gases; they then sell these credits to Annex 1 parties that are themselves trying to meet Kyoto targets. According to Wara's calculations, in some years—when carbon credits were selling for a lot and coolant was selling for just a little—the companies made more than twice as much from the credits as they did from the coolant itself. "That's a profound distortion of the market," he says. "Once that balance shifts, you're not in the coolant business anymore, you're in the carbon business."

At least some factories resorted to dubious measures to maintain that business: Data from Wara and others revealed that during crediting periods, they would deliberately use less efficient manufacturing processes in order to generate as much HFC-23 as possible, so that they could maximize their profits by destroying the gas. They

also overproduced HCFC-22 "above levels that otherwise would be produced in response to HCFC-22 market demand," according to a report by the NRDC, "simply in order to maximize HFC-23 for destruction."

"It's just such a far cry from what policy makers envisioned happening when they set this whole thing up," says Andersen. "The idea was that the money paid for the credits would fund renewable energy projects in the developing world—things like solar and wind power." Instead, the vast majority of that money— billions of dollars—has been given to the refrigeration industry. The CDM reports that since the UN program began, more than 40% of all credits have been given to just 19 HCFC-22-producing coolant factories—the majority of them in China and India. Countries in Sub-Saharan Africa, which were initially expected to be big beneficiaries of the CDM program, have been left out in the cold. **INFOGRAPHIC 24.8**

> *" Once that balance shifts, you're not in the coolant business anymore, you're in the carbon business. "*
> —Michael Wara

INFOGRAPHIC 24.8   A CARBON CREDITING SYSTEM GETS SIDETRACKED

% OF CDM EMISSION CREDITS ISSUES, BY TYPE (2012)

Other 3%
Fossil fuel switch* 3%
Methane 4%
Landfill gas 4%
Energy efficiency 5%
HFCs 36%
Nitrous oxide 19%
Renewables 26%

* using natural gas instead of coal

← The main intention of the Clean Development Mechanism was to fund low-carbon energy production facilities in developing nations in order to reduce global greenhouse gas emissions. Because it is cheaper to build a new power plant from scratch in a developing country (that would be building one anyway) than to prematurely close down or retrofit an older polluting power plant in a developed country, this was seen as a way to reduce global carbon emissions in a more cost effective way. Early on, however, almost half of CDM money went to HFC projects. Other projects are beginning to catch up with non-HFC projects making up around 64% of the total in 2012.

What could be done to get the CDM system back on track to achieve its main goal of funding low-carbon energy production facilities in developing nations?

Adapting policies is necessary but difficult.

So far, repairing the flawed CDM program has been as challenging as getting it into place was—a lesson, say critics, in the importance of adaptive management. Reforming the system would be much easier if more flexibility or fail-safe options had been built into the original plan.

These days, at any rate, there are no shortages of solutions being proposed. Some experts advocate a more intense focus on short-lived greenhouse gases like methane and black carbon (or soot). For his part, Wara advocates going in the opposite direction and making the global carbon market a market for CO_2 rather than for all six gases. "That's the most important one," he says. "It's emitted in the most prodigious quantities, and has a very long atmospheric life." Most CO_2 comes from the energy sector, so focusing on that gas alone would be a good way to ensure that the carbon market pushes humanity away from fossil fuels toward more sustainable energy sources.

China and India have fiercely resisted such changes to the CDM program, and critics charge that politics and political lobbying have had undue influence on reform efforts. When European delegates at one global climate conference suggested that any payments for incineration of HFC-23 should go into an international fund to help factories retool or phase out both the by-product and its underlying coolant product altogether, the Chinese government blocked the initiative, insisting that the money go directly to its own clean development fund.

Likewise, when several Kyoto countries objected to awarding credits to natural gas–burning plants (proponents argued that natural gas emits less CO_2 than a coal plant would; opponents argued that such projects stray too far from the stated goals of the CDM—to fund and support clean, non-fossil fuel–based energy projects), the Chinese factions within the council overrode them.

In the few short years since the CDM was launched, the coolant manufacturers on the panel have amassed both power and influence. As Martin Hession, a past chair of the CDM, told the *New York Times* recently, even raising the possibility of trimming future payments was "politically hard." "China and India both have representatives on the panel," the paper reported. "And the new chairman, Maosheng Duan, is Chinese." Some policy makers have worried that if the coolant makers

aren't paid to destroy the HFC-23, they will simply release it into the atmosphere. That, says Wara, would be catastrophic.

In 2010, though, the European Union finally put its foot down. And so did the United Nations. Responding to public pressure, both groups began dramatically altering the way they value and pay out HFC-23 emission-reduction credits. In 2013, the European Union stopped accepting HFC-23 credits from companies in its carbon trading system (which happens to be the largest in the world, by a long shot). The United Nations is refusing to award HFC-23 credits to any new factories, and in the fall of 2011 revised downward the percentage of coolant gas that would be eligible for the HFC-23 reduction credit. As Hession told the *New York Times*, the United Nations believed that such measures would eliminate the incentive to overproduce coolant gases. Some fear that the CDM program will collapse altogether before the CDM policy is modified to favor quality projects that actually do what the CDM was intended to do: help implement clean energy programs in developing countries. CDM backers and the UN hope that new, more ambitious, post-Kyoto climate targets will increase demand for offsets (and bring the United States, China, and other large emitters on board) and breathe new life into the CDM project.

Others are not so sure. Only time will tell.

In the meantime, many economists still agree that emissions trading schemes are the best hope we have for curbing greenhouse gas emissions. Even Wara has remained hopeful. "I am still enamored of market-based approaches to these problems," he says. "For better or worse, those are the incentives that people respond to." Besides, he adds, "As messy a story as Kyoto is, it's still one we can draw valuable lessons from. In California we have a cap-and-trade program spinning into action now. They've learned a lot of lessons from the CDM and I think their program is much stronger because of it."

Select References:
Andersen, S. O., & K. M. Sarma. (2010). *Making Climate Change and Ozone Treaties Work Together to Curb HFC-23 and Other "Super Greenhouse Gases."* New York: NRDC Issue Paper. www.nrdc.org/globalwarming/files/hfc23.pdf.
Rosenthal, E. and A. W. Lehren. (2012). Carbon credits gone awry raise output of harmful gas. *New York Times*, August 9, 2012.
Van der Hoeven, M. (2013). *CO2 Emissions from Fuel Combustion; Highlights.* Paris: IEA Statistics.
Wara, M. (2007). Is the global carbon market working? *Nature*, 445(7128): 595–596.

↑ The 28th session of the Intergovernmental Panel on Climate Change (IPCC) in Budapest, Hungary on April 9, 2008. The IPCC evaluates current climate science and makes policy recommendations to the international community.

STR/AFP/Getty Images

BRING IT HOME

PERSONAL CHOICES THAT HELP

The process of writing and revising policy, proposing it, voting on amendments, and finally enacting it as law is a complex and often messy one. The legislative process is often referred to as "sausage making" because of all the steps and input, as well as the fact that the final product often looks much different than the original.

Individual Steps

• Find out who and what is influencing your elected politicians. The website www.opensecrets.org allows you to look up the top individuals and industries that contribute to any candidate's campaign.

Group Action

• When an important issue is not adequately addressed, concerned citizens often form petition drives. Signatures are collected and delivered to politicians, who can propose new legislation. Form a group to petition for an important issue that you feel is being overlooked and present the collected signatures to any politician who can propose new policy.

Policy Change

• Actions speak louder than words. Visit www.votesmart.org and find the voting record of your representative. How does he or she vote on environmental issues like climate change? If you do not feel your representative's record is moving the country forward, volunteer for another candidate whose policies you support during the next election cycle.

ENVIRONMENTAL LITERACY UNDERSTANDING THE ISSUE

1 Why are environmental policies sometimes needed at a national or even international level? What are some of the major U.S. environmental laws?

INFOGRAPHIC 24.1 AND TABLE 24.1

1. Why are international laws and policies necessary to address some environmental issues?
   a. Legislation at the national level does not address important environmental problems.
   b. National legislation cannot address environmental problems that cross national boundaries.
   c. National legislation is not effective because people won't vote for pro-environment laws.
   d. International laws and policies are easier to enforce than national laws and policies.

2. A unique feature of many U.S. environmental laws is the citizen-suit provision. Explain this provision. Why is it a useful part of these laws?

3. What did the landmark environmental laws of the 1960s and 1970s have in common?

2 How are policy decisions made? How does lobbying influence policy decisions?

INFOGRAPHICS 24.2 AND 24.3

4. Modern U.S. environmental policy:
   a. requires that environmental impacts be evaluated before federal action is taken.
   b. allows input only from stakeholders (e.g., land owners or local citizens).
   c. is focused on repair of damage rather than on prevention.
   d. is mainly found at the state and local levels.

5. Explain the steps of the NEPA process for policy making and identify the strengths and weaknesses of this process.

6. How does political lobbying affect national environmental policy? Do you agree with the critics that political lobbies are too powerful? Explain.

3 What policy tools can be used to implement and enforce environmental policy?

INFOGRAPHIC 24.4

7. When there is uncertainty about an environmental problem or severe consequences are possible, policy makers should invoke the _____ _____ when deciding how best to proceed.

8. A consumer who buys an electric car can get a reduction in his or her income tax that year. This is an example of:
   a. a green tax.
   b. command-and-control regulation.
   c. a tax break.
   d. a federal grant.

9. Give an example of a market-driven approach to solving environmental problems. How does this differ from command-and-control regulation of environmentally damaging behavior?

4 How are international policies established and enforced? How has the international community responded to the issue of climate change?

INFOGRAPHICS 24.5 AND 24.6

10. Effective international environmental policies:
    a. are simpler to implement than national policies.
    b. benefit only a small number of nations and interest groups.
    c. allow for revision based on science or changing needs.
    d. focus on the causes, not the consequences, of environmental issues.

11. International policies may be enforced:
    a. through a multitude of international laws.
    b. by international bodies such as the United Nations.
    c. more easily and effectively than national laws and policies.
    d. mainly through voluntary compliance of the nations involved.

12. Which treaty would you identify as the foundation of international policy on climate change: the UN Framework Convention on Climate Change (UNFCCC) or the Kyoto Protocol? Explain your reasoning.

5 How successful have efforts to reduce CO_2 emissions been? Why has the Clean Development Mechanism been less effective than expected in reducing greenhouse gases?

INFOGRAPHICS 24.7 AND 24.8

13. International climate change policy:
    a. has established new binding targets to replace Kyoto targets, which expired in 2012.
    b. requires that developing countries reduce carbon emissions.
    c. has relied heavily on market solutions such as emissions caps and carbon credit trading.
    d. has decreased global carbon emissions to below 1990 levels.

14. Since the 1997 Kyoto Protocol:
    a. some nations have decreased their carbon emissions.
    b. the United States has met its Kyoto target of a 6% reduction below 1990 levels.
    c. global carbon emissions have fallen below 1990 levels.
    d. All of the above.

15. How did HFC producers manipulate the Clean Development Mechanism, and what did this do to the effectiveness of this program? How would you recommend eliminating this abuse?

SCIENCE LITERACY **WORKING WITH DATA**

The U.S. Energy Information Administration (EIA) was established by law to be an independent and impartial energy authority. It provides statistical and analytical information on energy issues, including pollution and climate change, to support public understanding and policy making. The EIA's data on carbon dioxide (CO_2) emissions and atmospheric concentrations are shown in the following graph.

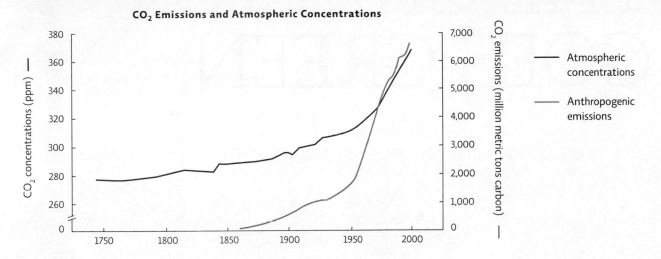

CO_2 Emissions and Atmospheric Concentrations

Interpretation

1. Describe in one or two sentences how atmospheric concentrations of carbon dioxide have changed since about 1750.

2. How have anthropogenic (human-caused) emissions of carbon dioxide changed since about 1860? Why do you think that there are no data on anthropogenic emissions before 1860?

3. What relationship do you see between anthropogenic emissions of carbon dioxide and atmospheric concentrations of carbon dioxide?

Advance Your Thinking

4. How would you predict that the two lines on this graph will change if new, more stringent emission targets are set and countries start meeting their goals? Explain your reasoning.

5. About 75% of anthropogenic emissions of carbon dioxide are created during the burning of fossil fuels, and about one-third of this is from vehicle emissions. Suppose we started using electric cars. Identify the circumstances under which this might or might not lead to decreased atmospheric carbon dioxide.

INFORMATION LITERACY **EVALUATING INFORMATION**

Climate change can feel like an overwhelming problem, far beyond one person's ability to influence. How much does it really affect the environment if you drive a truck instead of a car or keep your air conditioner set at 70°F instead of 72°F? Well, the Environmental Protection Agency (EPA) has created a calculator to answer these questions.

Evaluate the website and work with the information to answer the following questions:

1. Visit the Environmental Protection Agency website (www.epa.gov).
   a. What is the mission of the EPA?
   b. Is the EPA website up to date? Does it appear to be accurate? Reliable? Explain.
   c. How long has the EPA been a part of the U.S. government? Does it appear to have been effective? Explain.

2. Go to the EPA emissions calculator by visiting www.epa.gov/climatechange/ghgemissions/Ind-calculator.html or searching the EPA website for "household carbon footprint calculator."
   a. Complete the Household Carbon Footprint Calculator.
   b. What is your estimated annual level of greenhouse gas emissions (personal or family)? What is the largest source of your greenhouse gas emissions?
   c. Do these estimates seem accurate to you? Why or why not?
   d. What actions did the calculator identify that you could use to decrease your emissions? How much emissions could you reasonably save if you followed the recommendations?
   e. Identify one action that you would be most likely to do. What is it, and how much impact would it have?
   f. What action would you be least likely to do? Why?
   g. If individuals were to follow the EPA recommendations, what impact do you think this would have on global carbon emissions? Explain your reasoning.

Find an additional case study online at http://www.macmillanhighered.com/launchpad/saes2e

THE GHETTO GOES GREEN

Building a better backyard in the Bronx

CORE MESSAGE

Cities can be both an environmental blessing and a curse. Using green strategies to plan or retrofit cities can benefit citizens, businesses, and the environment—not to mention reduce environment-related health problems and degradation of natural resources.

AFTER READING THIS CHAPTER, YOU SHOULD BE ABLE TO ANSWER THE FOLLOWING **GUIDING QUESTIONS**

1
What has been the pattern of global urbanization and megacity growth in recent decades?

2
What trade-offs are associated with cities or urban areas?

Clay Garden, built on the site
of a burned-down home by
a resident across the street,
provides urban farming
opportunities for local
residents. In the background
are the Webster Morrisania
public housing projects,
which provide homes for
some of the poorest people
in the Bronx. Nina Berman/Noor/
Redux

3

What is environmental justice? How
does urban flight contribute to and
result from environmental justice
problems?

4

What environmental problems does
suburban sprawl generate?

5

How can we create cities that are
environmentally sustainable and
promote good quality of life for the
residents?

As she walked her dog Xena—a scruffy puppy she had found tied to a tree in her South Bronx neighborhood—Majora Carter considered her options. The 32-year-old aspiring filmmaker had moved back home to save money while she attended graduate school. Initially, she had wanted as little to do with the decaying neighborhood as possible. But then she'd gotten involved in a local artists' group and taken work at a community development center. Now, a colleague at the city parks department was offering her a $10,000 grant to come up with a waterfront development project for her neighborhood. Carter was balking.

At the moment, she and her neighbors were busy fighting a mammoth waste facility that the city was trying to move from Staten Island to the East River waterfront. With 30 transfer stations in the South Bronx, their tiny parcel of New York already handled 40% of the entire city's commercial waste, not to mention having four power plants, two sludge processing plants, and the largest food distribution center in the world. All told, some 60,000 diesel trucks passed through the neighborhood every week. In exchange for this burden, area residents boasted the highest asthma and obesity rates in the country, along with some of the poorest air quality. Another waste facility would only make matters worse. Consumed with this battle, Carter wasn't sure she had the time or energy to take on a development project.

Besides, the idea of developing waterfront property in the South Bronx seemed a bit naïve to her. Like most of her neighbors, Carter had lived in the neighborhood most of her life; she knew full well how inaccessible the surrounding river was to residents.

The waterfront—all of it—had long been claimed by industry. There was simply nothing left to develop, she thought, as she and Xena made their way along their usual route—past the transfer station, roaring with diesel-powered, garbage-filled 18-wheelers, along a winding string of garages filled with auto glass shops, metal work, and produce shipments.

And then Xena began pulling her toward an abandoned lot, one they had passed a million times without bothering to notice. After a futile effort to resist, Carter allowed herself to be led down a garbage-strewn path, through a ramble of towering weeds—the kind of place one would never venture alone at night. There at the end, sparkling in the early morning light, was the East

urban areas Densely populated regions that include cities and the suburbs that surround them.

urbanization The migration of people to large cities; sometimes also defined as the growth of urban areas.

◉ **WHERE IS THE BRONX, NEW YORK?**

NY

BRONX

NJ

MANHATTAN

NY

◉ **NEW YORK CITY**

River. Carter stood in awe. How many other forgotten patches of waterfront were there, she wondered? Maybe the river wasn't so inaccessible after all.

More people live in cities than ever before.

For the first time in human history, more than half the world's population lives in **urban areas**—densely populated regions that include both cities and the suburbs that invariably surround them. In the United States the proportion is even higher: 80% of Americans are urban dwellers. **Urbanization**, the migration of people to large cities, is happening around the world at an unprecedented rate. As global population swells, rural lands are morphing into urban and suburban ones, and ordinary cities are growing into *megacities*—those with at least 10 million residents. With more than 19 million inhabitants, the New York City metropolitan area qualifies as the largest city in the United States and one of the world's 29 megacities. **INFOGRAPHIC 25.1**

Concrete Plant Park, a 3-hectare park on the site of a former concrete batch mix plant, is one of the milestones in the creation of the Bronx River Greenway, an environmental effort to transform the Bronx River. New York City officials and local activists re-established salt marshes on the riverbank once strewn with trash and tires and opened it up to recreation and water activities. Nina Berman/Noor/ Redux

INFOGRAPHIC 25.1   **URBANIZATION AND THE GROWTH OF MEGACITIES**   1

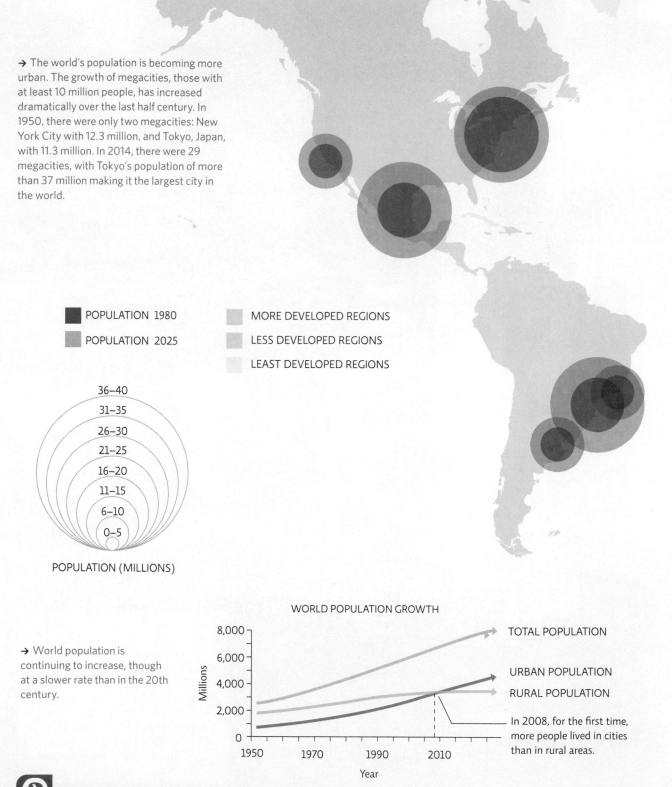

→ The world's population is becoming more urban. The growth of megacities, those with at least 10 million people, has increased dramatically over the last half century. In 1950, there were only two megacities: New York City with 12.3 million, and Tokyo, Japan, with 11.3 million. In 2014, there were 29 megacities, with Tokyo's population of more than 37 million making it the largest city in the world.

POPULATION 1980
POPULATION 2025

MORE DEVELOPED REGIONS
LESS DEVELOPED REGIONS
LEAST DEVELOPED REGIONS

36–40
31–35
26–30
21–25
16–20
11–15
6–10
0–5

POPULATION (MILLIONS)

WORLD POPULATION GROWTH

→ World population is continuing to increase, though at a slower rate than in the 20th century.

TOTAL POPULATION

URBAN POPULATION

RURAL POPULATION

In 2008, for the first time, more people lived in cities than in rural areas.

Millions — 8,000 / 6,000 / 4,000 / 2,000 / 0

Year — 1950 / 1970 / 1990 / 2010

What are the advantages and disadvantages of living in a megacity?

→The United Nations predicts that by 2030 there will be 41 megacities. Most of those cities will be in Asia and Latin America.

FUTURE GLOBAL URBAN GROWTH WILL OCCUR IN CITIES OF ALL SIZES

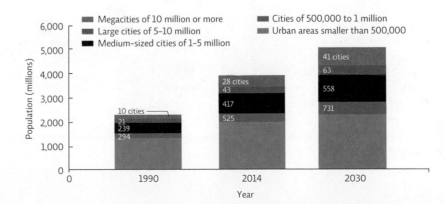

Legend:
- Megacities of 10 million or more
- Large cities of 5–10 million
- Medium-sized cities of 1–5 million
- Cities of 500,000 to 1 million
- Urban areas smaller than 500,000

Y-axis: Population (millions) — 0, 1,000, 2,000, 3,000, 4,000, 5,000, 6,000
X-axis: Year — 1990, 2014, 2030

1990: 10 cities, 21, 239, 294
2014: 28 cities, 43, 417, 525
2030: 41 cities, 63, 558, 731

KEY CONCEPT 25.1

As of 2008, more people live in cities than in rural areas. The number of megacities has greatly increased since 1950 and will continue to do so in the future.

For the first time in human history, more than half the world's population lives in urban areas.

To be sure, cities bring some obvious advantages to their inhabitants: more job opportunities, better access to education and health care, and more cultural amenities, to name a few. But as far as the environment is concerned, urbanization is both a blessing and a curse. On the plus side, concentrating people in smaller areas (building *up* rather than *out*) can reduce the development of outlying agricultural land and wild spaces and thus protect existing farms and ecosystems. Higher population densities also make some environmentally friendly practices more cost-effective. For example, it's easier to implement recycling and mass transit programs in cities because there are more people to share the costs of these services. Living in smaller homes that are closer to needed amenities and having access to mass transit also decreases the energy use—and the carbon footprint—of urban dwellers compared to those who live in suburban areas. **INFOGRAPHIC 25.2**

On the minus side, cities are *locally* unsustainable: They require the import of resources like food and energy and the export of waste. Because they are densely populated, most cities are also hotbeds of traffic congestion (which pollutes the air) and sewage overflow (which pollutes the water).

Another problem stems from the way cities are designed and built—namely, the replacement of vegetation with pavement. Plants absorb water, filter air, and regulate area temperatures; pavement and concrete do not. In fact, the blacktop that covers most cities prevents rainwater from being absorbed into the ground, which in turn diminishes groundwater supplies and can lead to flooding (see Chapter 15). Cities also require an abundance of energy. This trifecta—too few plants, too much pavement, and high energy use—conspires to trap solar heat absorbed, and put off, by buildings, making most cities warmer than their surrounding countrysides. This phenomenon is known as the **urban heat island effect**. **TABLE 25.1**

INFOGRAPHIC 25.2   MANY URBAN AREAS HAVE LOWER PER CAPITA ECOLOGICAL FOOTPRINTS THAN AVERAGE   2

→ Due to higher population densities, less personal vehicle travel, smaller homes, and efficiencies of scale, people living in large urban areas typically have a lower ecological footprint than those in suburban areas. A 2009 study by geographer David Dodman compared the carbon footprints of various countries and large cities within those countries. Almost all of the cities evaluated had lower carbon footprints than their national average.

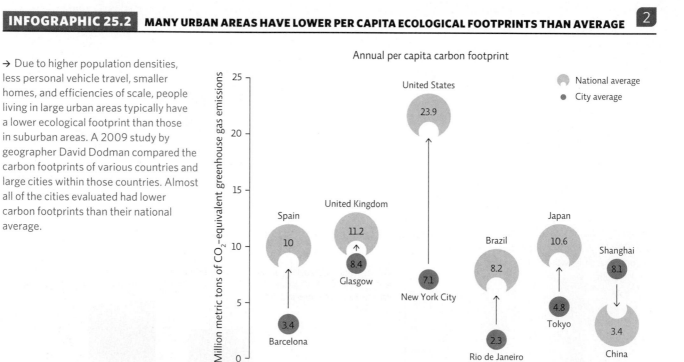

Annual per capita carbon footprint

National average
City average

(y-axis: Million metric tons of CO$_2$-equivalent greenhouse gas emissions)

- United States 23.9 / New York City 7.1
- Spain 10 / Barcelona 3.4
- United Kingdom 11.2 / Glasgow 8.4
- Brazil 8.2 / Rio de Janeiro 2.3
- Japan 10.6 / Tokyo 4.8
- Shanghai 8.1 / China 3.4

According to this data, only Shanghai has a carbon footprint higher than the national average. Why do you think this is true for Shanghai but not the other cities evaluated?

TABLE 25.1	TRADE-OFFS OF URBANIZATION

Advantages	Disadvantages
• Lower impact per person due to smaller homes and less traveling	• Dependence on food and resource inputs from outside the city • Concentrated wastes that have to be transported away
• Higher energy efficiency in stacked housing than in freestanding buildings	• Urban heat island effect, which increases energy needs and can have health consequences
• More transportation options, which lessens the need for personal vehicles • Closer proximity to destinations, which makes walking and mass transit viable options	• Traffic congestion and its associated air pollution due to high population densities
• Easier-to-implement zoning ordinances	• Possibly higher disease and violence in concentrated inner-city areas
• More job opportunities because local collaboration from a diverse community fosters innovation and ingenuity	• Higher cost of living, which limits who can afford to live in the city
• More services for citizens, including more educational and cultural opportunities and better health care options	• Less green space, which leads to stormwater problems

? How might the disadvantages of urbanization be addressed to lessen their impact?

All urban dwellers are vulnerable to the health effects associated with pollutants. But in most cities, the pros and cons of city living are unevenly realized. For example, New York City is one of the wealthiest, most populous cities in the world, but most of the cultural amenities, top-notch health care facilities, and job opportunities are concentrated in Manhattan, while most of the garbage, sewage, and power plants are located in the Bronx. This imbalance has spawned a whole new area of activism known as **environmental justice**, based on the idea that no community should be saddled with more environmental burdens and fewer environmental benefits than any other.

The movement is particularly relevant in the most impoverished cities in the world. In Mumbai, the largest city in India, with 20.5 million people, almost 7 million are slum dwellers who live in horrid, overcrowded conditions—as many as 18,000 people per acre—without adequate sanitation or running water. One report showed only one toilet per 1,440 residents. Globally, more than 1 billion of the world's population live in slums, mostly in large cities in developing countries.

With 90% of future population growth predicted to occur in large cities, urban planners are desperately searching for ways to create cities where the basic needs of residents are met and where the environmental benefits outweigh the environmental costs. The story of how the South Bronx waterfront was lost and then reclaimed provides important lessons about how to do this.

Suburban sprawl consumes open space and wastes resources.

In the late 1940s, when Carter's father, a Pullman porter and the son of a slave, first bought the house Carter would grow up in, the South Bronx was a mostly European-descended white working class suburb of Manhattan. But as more Hispanic

KEY CONCEPT 25.2

Cities offer many services and opportunities and a lower per capita ecological footprint than other areas but have problems with waste, stormwater, and higher rates of crime and disease.

urban heat island effect The phenomenon in which urban areas are warmer than the surrounding countryside due to pavement, dark surfaces, closed-in spaces, and high energy use.

environmental justice The concept that access to a clean, healthy environment is a basic human right.

INFOGRAPHIC 25.3   URBAN FLIGHT CONTRIBUTES TO SUBURBAN SPRAWL

MONTREAL CMA (CENSUS METROPOLITAN AREA)

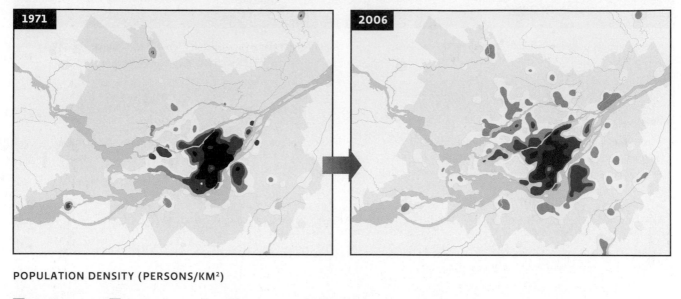

POPULATION DENSITY (PERSONS/KM²)

■ 12,000 and over   ■ 6,000 to 11,999   ■ 3,000 to 5,999   ■ 1,500 to 2,999   ☐ 400 to 1,499

↑ Urban flight, the movement of people out of inner-city areas, is often driven by the decay of urban areas ("flight from blight") and the influx of lower-income groups who are often minorities.

? What kind of urban development might encourage people of different ethnicities, levels of education, or incomes to return to inner city areas?

KEY CONCEPT 25.3

Minorities and low-income populations are more likely than others to be subject to societal choices that damage their environment—an environmental justice issue.

and Black Americans moved to the area seeking their share of the American Dream, whites moved to nearby commuter towns. The process of people leaving a city center for surrounding areas, originally made possible by the automobile and later by mass transit, is known as **urban flight**. While in many cities—especially those in developing nations—immigration exceeds emigration, urban flight today is triggered by a variety of forces, including overcrowding, noise and air pollution, the high cost of city living, and, in some cases, racial tensions. **INFOGRAPHIC 25.3**

No matter what the cause, urban flight results in **suburban sprawl**—a slow conversion of rural areas outside of a city into suburban and exurban ones. **Exurbs** are more sparsely populated towns beyond the immediate suburbs whose residents also commute into the city for work.

As its name suggests, suburban sprawl tends to spread out over long corridors in an unplanned and often inefficient manner. By covering ever-greater swaths of terrain with concrete and pavement, sprawl reduces the amount of land available for farming, wildlife, and ecosystem services. Because of the haphazard way in which they are developed, the resulting communities are heavily dependent on driving; unlike cities, which are densely populated and can accommodate mass transit systems, suburbs and exurbs require residents to drive almost everywhere they need to go. And because suburban homes are typically larger than urban ones (and exurban homes are often even larger than suburban ones), they tend to have a greater ecological footprint. **INFOGRAPHIC 25.4**

Urban planners today are well aware of these perils and often work to mitigate them. But in the 1960s,

INFOGRAPHIC 25.4   SUBURBAN SPRAWL

↓ Urban flight often leads to suburban sprawl—low population density in developments that appear outside a city. Homes typically get larger the farther they are from the city, and residents have a larger ecological footprint (larger homes and more time spent driving). The suburbs now have their own suburbs—the exurbs, which are commuter towns that are beyond the traditional suburbs but whose residents still commute into the city, often an hour or more each way. Both suburbs and exurbs often displace farmland and wildlands.

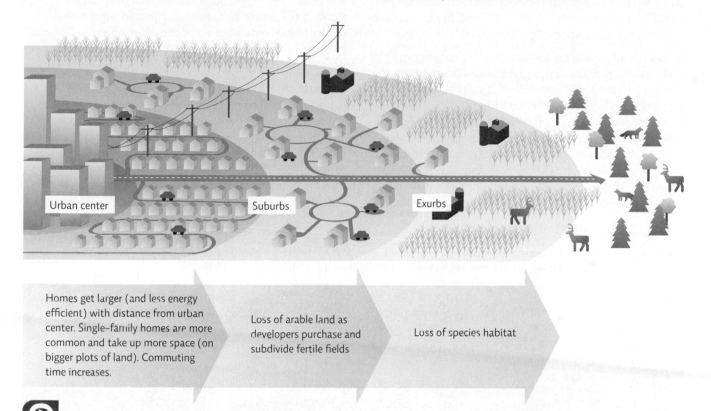

Urban center

Suburbs

Exurbs

Homes get larger (and less energy efficient) with distance from urban center. Single-family homes are more common and take up more space (on bigger plots of land). Commuting time increases.

Loss of arable land as developers purchase and subdivide fertile fields

Loss of species habitat

? If you worked in an urban center, would you rather live there, in the suburbs, or in the exurbs? Explain.

prolific urban planner Robert Moses was all too happy to accommodate urban flight and the sprawl that came with it. It was Moses who commissioned the Cross Bronx Expressway, a highway that enabled commuters from suburban Westchester County north of the city to completely bypass the Bronx as they traveled in and out of Manhattan each day. But the expressway also displaced 600,000 Bronx residents and further segregated the ailing borough from the rest of New York City. "The South Bronx was utterly cut off,"

KEY CONCEPT 25.4

Suburban sprawl displaces farmland and natural areas; residents often have higher per capita ecological footprints than urban dwellers.

says Marta Rodriguez, a lifelong Bronx resident and a colleague of Carter's. "We didn't stand a chance."

To make matters worse, in the Bronx and elsewhere in the 1960s, urban flight led to redlining—the process whereby banks rule certain sections of a city off-limits to any type of investment. Bronx landlords quickly discovered that if their neighborhood was redlined, torching their buildings and collecting the insurance money would yield greater profits than renting or selling. By the 1970s, the burning of tenement houses had spiraled out of control, so

urban flight The process of people leaving an inner-city area to live in surrounding areas.

suburban sprawl Low-population-density developments that are built outside of a city.

exurbs Towns beyond the immediate suburbs whose residents commute into the city for work.

much so that at one Yankees game, an addled sportscaster famously declared, "Ladies and gentlemen, the Bronx is burning." And as the shopping centers and apartment buildings were shuttered or burned, other industries took their place—namely, the garbage disposal operations and auto parts manufacturers that had been shunned by wealthier enclaves. Before long, the South Bronx had been transformed into an industrial wasteland.

In most cities, *zoning laws*—laws that restrict the type of development allowed in a given area—create a buffer between commercial and residential areas so that factories aren't wedged between houses. But in the Bronx, such laws were routinely ignored. Without other options, area residents were forced to accept factories and warehouses built on the ashes of apartment complexes. And because these new industrial neighbors preferentially hired commuters from outside the Bronx, unemployment rates skyrocketed—along with crime, poverty, and asthma.

It turned out that the patch of waterfront Carter and Xena had stumbled upon was a relic of the Moses-era highway expansion, sitting as it did beneath the Sheridan Expressway, a stretch of highway originally meant to cut across the entire northeast Bronx. The Sheridan was abandoned when planners realized it would run through the Bronx Zoo, a popular tourist destination. But by then the damage was done. Hunts Point—Carter's neighborhood—had been isolated and the surrounding waterfront condemned to wasteland.

Carter knew that replacing the abandoned lot with a park would be a big first step toward righting some of the wrongs that her community had endured. The trees and plants would trap pollutants from the air, preventing them from infiltrating people's lungs. The grass and soil would absorb rainwater so that it could no longer carry trash and detritus from the streets into the river. And claiming even a small patch of waterfront for themselves

↓ A volunteer gardener at Finca Del Sur, a garden in the South Bronx, tends the corn stalks while a passenger train goes by in the background. The garden was created on an empty plot of land bordered by a highway exit ramp and a commuter train line.

Nina Berman/Noor/Redux

↑ Majora Carter received a MacArthur "genius" grant for her work with Sustainable South Bronx. She stresses that the environmental movement is not just one of the middle-class majority who can afford to buy organic food, drive hybrid cars, and live in areas with little pollution. Low-income families also deserve a clean and healthy environment.

would give Carter and her neighbors a sense of ownership, not to mention a connection to nature and a place to stretch their legs.

Indeed, studies by urban planning expert Reid Ewing and others have shown that parks improve both the physical and psychological health of people who live near them. And in cities themselves—cities that, like the Bronx, were once plagued by drug trafficking and rampant gun violence—more green space can also mean less crime. In Bogotá, Colombia, in the late 1990s, for example, a particularly environmentally conscious mayor noticed that while his city was

Carter knew that replacing the abandoned lot with a park would be a big first step toward righting some of the wrongs that her community had endured.

designed to accommodate heavy automobile traffic, the vast majority of his electorate did not drive. So he narrowed municipal thoroughfares from five lanes to three, expanded bike lanes and pedestrian walkways, and established a string of parks and public plazas throughout the city. The result? People stopped littering. Crimes rates dropped. And slowly but surely, city residents reclaimed their streets.

Bogotá was not so different from the South Bronx, Carter thought. If that city could go green on a developing country's budget, surely she and her colleagues could raise enough money to do the same. Starting with the $10,000 seed grant from the city parks department, they leveraged a small fortune in additional grants, donations, and private investment, until they had finalized plans to build a $3 million park, complete with gardens, grassy knolls, and East River kayaking. Hunts Point Riverside Park—the spot that Carter stumbled upon—would be the borough's first waterfront park in more than 60 years. But that was just the beginning.

Environmental justice requires engaged citizens.

Energized by their successful riverside park project, Carter and her neighbors formed a nonprofit called the Sustainable South Bronx (SSBx). The group immediately set its sights on an even grander vision: They would create a greenbelt around the entire community—2.5 kilometers (1.5 miles) of waterfront greenway, 5 hectares (12 acres) of new waterfront open space, and 14 kilometers (8.5 miles) of green streets (with landscaped medians)—all connected by an interlinking system of bike and pedestrian pathways that stretched from the Hunts Point Riverside Park, around the South Bronx's winding edges, all the way to the existing 160-hectare (400-acre) park on Randall's Island. They would also disassemble the Sheridan Expressway and turn it into 11 hectares (28 acres) of additional parkland, some of which they would designate as *conservation easements*—tracts of land that the city would agree not to develop.

It was an ambitious agenda indeed—an expensive one, too—and would require the support of administrators and elected officials from the Bronx to Manhattan to the state capital in Albany. "There is a big fear that environmental justice is fiscally irresponsible," says Carter. "People running the city think 'How can we spend money on parks when we're coming up short on schools, and clinics, and job training, and health care?' What they don't realize is that parks can actually help with those things, too." Parks not only increase community pride but also create green jobs and improve health.

Convincing community members of these benefits would prove as difficult as convincing legislators. Getting them to come out and oppose a landfill was one thing; area residents knew all too well what another trash heap would do to their neighborhood. But getting them to support a park? They had more pressing concerns. Theirs was the poorest congressional district in the city; at the time, more than 20% of residents were unemployed. And their neighborhood hadn't had a waterfront park in more than 60 years, let alone an entire greenway. Why bother now? "We'd ask people, 'What would you like to see in your neighborhood?' and they really didn't have an answer," Rodriguez says. "They'd never been asked that question before. It was as if having parks was too far in the future for them."

New Urbanism A movement that promotes the creation of compact, mixed-use communities with all the amenities of day-to-day living close by and accessible.

In fact, the SSBx vision folded readily into a growing movement aimed at making cities more environmentally friendly and socially equitable. **New Urbanism**, as the movement is called, maintains that cities

(both now and in the future) have the capacity to reduce our per-person ecological footprint, even as they improve the quality of life for people, provided they are designed properly. The City University of New York Institute for Research on the City Environment estimates that if a city is designed and built with an eye toward sustainability, the ecological footprint of any given urban dweller could be trimmed to about half that of the average American.

On top of the carbon savings, the consensus emerging from environmentalists, sociologists, and economists is that the future lies in cities—where most people will live and perhaps where most people should live. Cities promote interaction among a diverse group of people. This in turn promotes the exchange of ideas and lessens cultural and economic barriers. Many cities around the world are pursuing sustainable development and paving the way for others to do the same. **INFOGRAPHIC 25.5**

KEY CONCEPT 25.5

Engaged citizens can help revitalize neglected or damaged areas, improving their own community and helping to reduce urban flight and sprawl.

The future depends on making large cities sustainable.

Sustainable cities are cities where the environmental pros outweigh the cons—where sprawl is minimized, walkability is maximized, and the needs of inhabitants are met locally. In recent years, urban planners have come up with a wide range of strategies for accomplishing these goals. To achieve self-sufficiency, for example, a sustainable city might maintain a mixture of open and agricultural land along its outskirts. Such land could provide a large part of the local food, fiber, and fuel crops, along with recreational opportunities and ecological services. Waste and recycling facilities could also be located nearby, along with other enterprises aimed at producing resources needed by area residents. To stave off sprawl, the same city might establish urban growth boundaries—outer city limits beyond which major development would be prohibited. Keeping any outward growth that does occur as close to mass transit as possible minimizes the impacts of transportation, just as building "up" (a parking garage) rather than "out" (an expansive parking lot) minimizes the amount of land used. To encourage more walking and less driving, zoning laws might allow for mixed land uses, where residential areas are located reasonably close to commercial and light industrial ones.

INFOGRAPHIC 25.5 | GREEN CITIES

↓ Many cities of the world are taking steps to develop more sustainably in an effort to improve their local environments and their standards of living. Green cities have many things in common, such as recognizing the importance of civic involvement, having a government commitment to sustainable development, and pursuing a holistic approach that looks at all the ways the city can reduce its ecological footprint while still developing and growing. While there are many others, here are some notable examples.

© Marion Kaplan/Alamy

© PRISMA ARCHIVO/Alamy

South America Curitiba, Brazil, has become a model for sustainable development by focusing on choices that are good for the environment (reducing pollution and waste and increasing energy efficiency) and good for its residents (a walkable city with green spaces, an efficient and low cost bus system, and high-quality, low-cost housing).

Europe Copenhagen, Denmark, has lower-than-average per capita energy use due to efficient heating systems and energy-efficient buildings and a strong commitment to using renewable energy. Public education programs have reduced water consumption and waste generation. High car taxes and road infrastructure that favors cyclists are helping Copenhagen pursue its goal to become the "World's Best Cycle City."

Christophe Testi/Shutterstock

Eye Ubiquitous/UIG via Getty Images

North America San Francisco, California's per capita energy use is low thanks to city programs that encourage or require energy efficiency and conservation. Many city buildings have been retrofitted for energy and water efficiency, and the city is installing solar power systems on many municipal buildings. It has a well-developed public transportation system, and its "Better Streets Plan" is making it a safer and more walkable city for pedestrians.

Asia The prosperous city-state of Singapore has reduced its carbon footprint by transitioning from oil to natural gas and setting high energy-efficiency standards for buildings. In addition, water reclamation facilities purify wastewater to provide one-fifth of the city's water supply. It has a robust public transportation system and strong emission standards for vehicles, giving the city much better air quality than most other large Asian cities.

 What steps would you recommend that your own city or community take in an effort to reduce its ecological footprint? Justify your answer.

The quest for sustainability must include cities. Smart growth allows cities to develop in a way that minimizes environmental impact while enhancing community living.

Of course, building an ideal city from scratch is easy compared with the task of overhauling an existing city, especially when that city is as densely populated and ever expanding as New York. Upgrading decaying infrastructure like roads, public places, and sewage and water lines can be more expensive than new construction, and the process is disruptive to residents. Urban retrofits are certainly possible, but sometimes their very success raises property values to the point that the original residents can no longer afford to live in their own neighborhoods. Even so, there are plenty of ways that American cities can push themselves into the environmental plus column. For example, **infill development**—the development of empty lots within a city—can significantly reduce suburban sprawl. And even the most car-friendly of cities has a range of options for reducing traffic congestion and the air pollution that comes with it: reliable public transportation, car-sharing programs that allow residents to use cars when needed for a monthly fee, and sidewalks and overhead passageways that allow pedestrians to safely cross busy roads. The cumulative effect of strategies like these, which help create walkable communities with lower ecological footprints, is known as **smart growth**. **INFOGRAPHIC 25.6**

Persuading people to support smart growth, as Carter and her colleagues soon discovered, is a matter of showing them that the benefits could be economic as well as environmental. "You need to show them what we call the triple bottom line," says James Chase, vice president of SSBx (and Carter's husband), referring to the economic, social, and environmental impacts of any decision. "Developers, government, and residents all need some tangible, positive return." A major park project would surely be a boon for all three. Developers would be guaranteed millions in waterfront development contracts. Residents could look forward to cleaner air and water, a prettier neighborhood, and better health

as a result. The state and city governments would save a bundle in health care costs. The greenbelt would also spur the local economy: Such a vast stretch of public space would attract street vendors, food stands, bicycle shops, and sporting goods stores.

Smart growth also requires a green workforce. Some of the undeveloped property that Carter and her neighbors hoped to convert into parkland was contaminated with hazardous waste. These sites are called *brownfields*, and they require a special type of clean-up, or remediation, before they can be developed.

The surrounding wetlands, suffering from decades of neglect, would also need to be restored. And maintaining the new trees, plants, and parks SSBx hoped to create would require a workforce trained in urban forestry. Anxious to claim these emerging professions—all of which promised job security and living wages—for their own community, Carter and her neighbors launched the Bronx Environmental Stewardship Training—a green job training program that teaches South Bronx residents the principles of urban forestry, brownfield remediation, and wetland restoration. Program participants—many of them ex-convicts and high school dropouts facing prison time—also learn how to install solar panels and retrofit older buildings to make them energy efficient. So far, 82% of the participants have found jobs in the green economy, and 15% have gone on to college. "Once we figured out the employment factor, we had a win–win–win," says Chase. "There's all these jobs—good jobs—that are going to take off in the next decade, but that not a lot of people know how to do right now. It was a clear opportunity for us."

The green economy includes a movement known as **green building**—that is, the construction of buildings that are better for the environment and the health of those who use them. Buildings that meet a minimum standard are awarded a **Leadership in Energy and Environmental Design (LEED)** certification by the U.S. Green Building Council. Points are awarded based the council's evaluation of how well the building meets certain criteria, such as energy and water efficiency, the use of sustainable resources, and indoor environmental quality. The Bronx Library Center is a silver-certified LEED building. It earned the silver certification by recycling 90% of the waste materials created during the construction of the building, using architectural design and efficient heating

Green building design focuses on efficient use of energy, and water, and on building materials with low environmental and health impacts.

infill development The development of empty lots within a city.

smart growth Strategies that help create walkable communities with lower ecological footprints.

green building Construction and operational designs that promote resource and energy efficiency and provide a better environment for occupants.

LEED (Leadership in Energy and Environmental Design) A certification program that awards a rating (standard, silver, gold, or platinum) to buildings that include environmentally sound design features.

INFOGRAPHIC 25.6   SUSTAINABLE CITIES AND SMART GROWTH

5

↓ Smart growth can be applied to large cities or to smaller communities. It employs strategies that make efficient use of land to create pleasant livable communities with a lower ecological footprint than current suburban areas.

Take advantage of compact building design and incorporate environmentally friendly technologies.

Create a range of housing opportunities and choices.

Renovate and develop existing communities (rather than build outside the city).

Foster distinctive, attractive communities with a strong sense of place.

Encourage community and stakeholder collaboration; make development decisions that are fair and cost-effective.

Mix land uses to place residential and commercial areas together.

Provide urban green space; preserve farmland and critical environmental areas.

COMMUNITY MEETING
June 23rd

Create walkable and bike-friendly neighborhoods.

NATURAL GAS BUS

Provide a variety of "clean" transportation choices into and around the city.

 Which of these smart growth principles would be most appealing to you if you were looking for a place to live in a city?

INFOGRAPHIC 25.7   **GREEN BUILDING**

↓ Many steps can be taken to build or retrofit a building so that it has less environmental impact and is a healthier environment for those who live, work, or go to school there. The nonprofit group Green Building Council certifies buildings through its LEED program (Leadership in Energy and Environmental Design). A building receives a standard, silver, gold, or platinum rating, based on a variety of criteria that include energy efficiency, sustainable building material use, and innovative design.

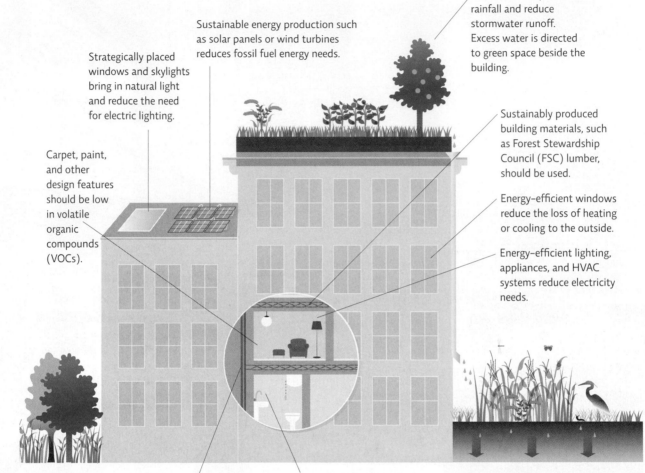

Sustainable energy production such as solar panels or wind turbines reduces fossil fuel energy needs.

Green roofs help absorb rainfall and reduce stormwater runoff. Excess water is directed to green space beside the building.

Strategically placed windows and skylights bring in natural light and reduce the need for electric lighting.

Carpet, paint, and other design features should be low in volatile organic compounds (VOCs).

Sustainably produced building materials, such as Forest Stewardship Council (FSC) lumber, should be used.

Energy-efficient windows reduce the loss of heating or cooling to the outside.

Energy-efficient lighting, appliances, and HVAC systems reduce electricity needs.

Energy-efficient heating systems might recycle heat from equipment (industrial or commercial settings) or use more efficient designs such as radiant floor heat.

Water-saving devices such as waterless urinals and motion-sensitive low-flow faucets are encouraged.

Green space, rain gardens, and permeable or grass pavers allow stormwater infiltration and offer opportunities for building users to enjoy nature.

? How could a LEED-certified building help address urban problems such as the urban heat island effect and stormwater issues?

and cooling systems to save 20% of energy costs, and using sustainably grown wood in 80% of the construction lumber.

Carter and her team also launched Smart Roofs, LLC, a green-roof and green-wall installation company. Green roofs are one type of rain garden—an area seeded with plants suited to local temperature and rainfall conditions. (See Chapter 15 for more on rain gardens.) A 2004 study by the New York City Department of Design and Construction found that consumers could save more than

$5 million in annual cooling costs if green roofs were installed on just 5% of the city's buildings. According to a study by Columbia University, the same amount of green roofing could achieve an annual reduction of 350,000 metric tons of greenhouse gases. And Riverkeep, an environmental nonprofit, found that green roofs can retain 3,000 liters of storm water for every $1,000 of investment—easing pressure on the city's overburdened sewer systems and mitigating water pollution from storm runoff. Green roofs also lessen the urban heat island effect. **INFOGRAPHIC 25.7**

To help the company take off, SSBx secured tax credits from the state legislature. Building owners who install green roofs on at least 50% of their available rooftop now receive a 1-year property tax credit of up to $100,000. Carter offered up her own roof as the first test case.

By the time the Hunts Point Riverside Park opened, dozens of cities across the country—from Madison, Wisconsin, to Miami, Florida—had taken up the mantle of sustainability and smart growth. In 2013 the United States led the world in green building, with more than 44,000 LEED-certified buildings or projects under construction. That same year, the EPA gave out five National Awards for Smart Growth Achievement. The Bronx received an honorable mention award for Via Verde, a LEED Gold mixed-income housing complex built on a former rail yard that incorporates a wide variety of green building features such as solar panels, natural lighting,

and green roofs. It also has easy access to transportation and a neighborhood medical clinic on site. The project was so successful, it prompted New York City to change its green zoning rules to make it easier to implement similar projects throughout the city.

Select References:
Beckett, K., & A. Godoy. (2010). A tale of two cities: A comparative analysis of quality of life initiatives in New York and Bogotá. *Urban Studies,* 47(2): 277–301.
Carter, M. (2006). "Greening the Ghetto," www.ted.com/talks/majora_carter_s_tale_of_urban_renewal.
Dodman, D. (2009). Blaming cities for climate change? An analysis of urban greenhouse gas emissions inventories. *Environment and Urbanization,* 21(1): 185–201.
Ewing, R., et al. (2003). Relationship between urban sprawl and physical activity, obesity, and morbidity. *American Journal of Health Promotion,* 18(1): 47–57.
Heilig, G. K. (2012). *World Urbanization Prospects: The 2011 Revision.* New York: United Nations, Department of Economic and Social Affairs (DESA), Population Division, Population Estimates and Projections Section.

BRING IT HOME

PERSONAL CHOICES THAT HELP

A sustainable community is one that promotes economic and environmental health and social equity. It is one in which the health and well-being of all citizens are considered, while those citizens help implement and maintain the community.

Individual Steps
• Investigate and support sustainable businesses in your area (see www .sustainablobusiness.com). Research products before you purchase them to understand the impact of your consumption choices (see www.goodguide .com).
• If you have a balcony or yard, plant flowers, vegetables, or trees.
• Support local businesses by shopping and dining close to home.

Group Action
• Join neighborhood clean-up days. If you can't find one, organize one.
• Reduce reliance on cars. Start a petition to get more bike lanes in your city. Ride public transit more often.

• Find out how colleges and universities are working toward sustainable practices at www. AASHE.org.

Policy Change
• Attend a meeting of your city council or county commission and ask members to look into smart growth opportunities.

• See how well you can plan for a sustainable community. Play the PC strategy game Fate of the World (www .fateoftheworld.net) and see how policies you put in place impact global climate change, rain forest preservation, and resource use.

Monty Rakusen/Cultura RF/Getty Images

ENVIRONMENTAL LITERACY UNDERSTANDING THE ISSUE

1 What has been the pattern of global urbanization and megacity growth in recent decades?
INFOGRAPHIC 25.1

1. True or False: Currently more people live in rural areas that cities, but that is expected to change soon.

2. In 2025, where will most of the world's megacities be found?
   a. North America
   b. Europe
   c. Africa
   d. Asia

2 What trade-offs are associated with cities or urban areas?
INFOGRAPHIC 25.2 AND TABLE 25.1

3. True or False: Urban dwellers typically have lower carbon footprints than people who live in rural or suburban areas.

4. The urban heat island effect is caused by:
   a. minimal green space.
   b. lots of pavement and buildings.
   c. high energy use in the city as a whole.
   d. all of the above.

5. Point out several reasons why the typical city dweller has a lower ecological footprint than a suburban dweller. What are some problems that might be greater for a large city than for a smaller suburban one?

3 What is environmental justice? How does urban flight contribute to and result from environmental justice problems?
INFOGRAPHIC 25.3

6. Which of the following is an example of environmental injustice?
   a. Locating industries away from where people live
   b. Building garbage dumps in high-poverty, low-income areas
   c. People chaining themselves to trees to prevent the trees from being cut down
   d. Preventing the construction of a dam to save an endangered species of fish

7. In response to redlining in the 1960s, landlords in the Bronx:
   a. increased rents, forcing working-class people to move out of the borough.
   b. subdivided apartments into smaller units to increase the number of renters.
   c. burned their buildings to collect insurance money.
   d. constructed walkways for tenants to access the nearest subway.

8. If urban centers have better health care and job opportunities than outlying areas, why are some city residents exposed to worse, not better, living conditions?

4 What environmental problems does suburban sprawl generate?
INFOGRAPHIC 25.4

9. True or False: As one moves from the city center to the suburbs and exurbs, home size tends to increase.

10. Compared to inner city dwellers, suburban and exurban dwellers tend to:
    a. have lower ecological footprints.
    b. live in lower population density communities.
    c. walk more and drive less.
    d. work close to home and have short commutes.

11. The exurbs are:
    a. discrete communities within the city center.
    b. regions too distant from the city for city commuters to live.
    c. areas outside suburbs with even bigger houses and longer commutes.
    d. suburban areas that are losing residents to the city.

12. Identify several reasons suburban dwellers might have bigger ecological footprints than urban dwellers.

5 How can we create cities that are environmentally sustainable and promote good quality of life for the residents?
INFOGRAPHICS 25.5, 25.6, AND 25.7

13. True or False: To be LEED-certified, a building must be constructed with recycled materials and use sustainable energy sources such as solar panels.

14. A city that promotes smart growth:
    a. encourages development at the city edges.
    b. provides tax incentives for people who own more than one car.
    c. allows vacant lots to accumulate in the city for a more open look.
    d. mixes land uses to place residential and commercial areas together.

15. Explain the triple bottom line, using the Bronx waterfront restoration as an example.

16. What is New Urbanism, and why do its proponents feel that cities may be one of our biggest answers to many of the environmental problems we face today?

SCIENCE LITERACY WORKING WITH DATA

One common measurement of how well a building performs with regard to energy usage is its energy use intensity (EUI). EUI is expressed as energy used (in gigajoules) relative to a building's size so that structures of different sizes can be compared. This graph plots data for 36 LEED-certified office buildings, comparing their actual performances to expected performances, based on their designs. (Each data point represents a different building's EUI.) The line shown is not a trend line for the actual data; it is a line that bisects the graph at a 45-degree angle. This line allows us to see how closely the expected values match the observed values.

Measured versus Expected EUI in LEED–Certified Buildings

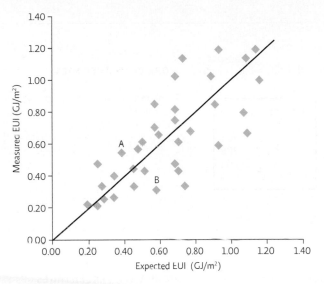

Interpretation

1. Look at the points labeled A and B. For each, what was the expected EUI predicted by the design plan? What was the actual measured EUI for this building?

2. Which of these two buildings, A or B, is performing better than expected?

3. Give the coordinate points for the one building that performed exactly as expected.

4. What does the line in the scatter plot tell you about those buildings above or below it?

Advance Your Thinking

5. About how well do LEED-certified buildings meet or exceed their predicted EUI?

6. The average EUI for non-LEED-certified office buildings is 2.19 GJ/m². How do LEED-certified buildings, even those that do not perform as well as expected, compare to this average?

INFORMATION LITERACY EVALUATING INFORMATION

More and more people around the world live in cities, and the number of megacities is increasing. There are costs and benefits to living in cities, and in an effort to increase the benefits, there is a growing movement toward "greening" cities. Environmental and social scientists have published a number of studies documenting the effects of greener cities on environmental and human health.

Go to the website Green Cities: Good Health (http://depts.washington.edu/hhwb/). Read the introduction to the site.

Evaluate the website and work with the information to answer the following questions:

1. Is this a reliable information source? Does it have a clear and transparent agenda?
   a. Who runs this website? Do the organization's credentials make it reliable or unreliable? Explain.

   b. Who are the authors? What are their credentials? Do they have the scientific background and expertise to lend credibility to the website?

Look under Research Themes to see what research the program pursues. Choose one of the links under Fast Facts, such as Crime & Public Safety, Active Living, or Mental Health & Function.

2. What type of information is provided on the page? What is the source of the information?

3. List a couple of the fast facts provided. Then scroll down the page and read each section. For the topic you chose, what is the primary claim? What data is provided to support the claim?

4. Do you find the data convincing? Why or why not? How does what you read relate to your own life? Give specific examples.

Find an additional case study online at http://www.macmillanhighered.com/launchpad/saes2e

APPENDIX 1

BASIC MATH SKILLS

Math skills are needed to evaluate data and even to understand much of the information in news reports. Here we present a review of some basic skills that will be useful in this class and in other science classes.

AVERAGES (MEANS)

To calculate an average, add all the numbers in the data set and divide by the number of numbers.

Example: The sum of these numbers is 10,547; 10,547 / 8 = 1318.375. You could round this off for an average or mean of 1,318.

Data set	
1,004	766
2,349	988
456	1,203
1,882	1,899

WORK WITH AVERAGES

Problem 1: If these were your grades on exams, what would be your exam average?
84, 73, 93, 95, 79, 86

PERCENTAGES/FREQUENCIES

To **convert a fraction to a percentage**, divide the numerator (top number) by the denominator (bottom number) and multiply by 100.
Example: To express the fraction $^2/_5$ as a decimal, divide 2 by 5, which equals 0.4. Multiply this by 100 for your answer: 40%.
To **convert a decimal to a percentage**, multiply by 100; a shortcut for this is simply to move the decimal over two places to the right.
Example: $0.08 \times 100 = 8\%$

WORK WITH PERCENTAGES

Problem 2: If 8 out of 32 frogs in a pond have deformities, what percentage of frogs have deformities?

Problem 3: In a pond, 25% of the frogs have leg deformities. If there are 100 frogs in the pond, how many have deformities?

Problem 4: If there are 68 frogs in the pond and 25% have deformities, how many have deformities? (First, make an estimate based on your answer to Problem 3—will it be a higher or lower number? This will help you decide if the answer you calculate is reasonable or whether you might need to recalculate.)

SCIENTIFIC NOTATION

In science, we often use very large or very small numbers. To make these easier to present, scientists use scientific notation, which multiplies a number (called the coefficient) by 10 (the base) raised to a given power (the exponent). If the coefficient is 1, we can leave it off and simply show the base and exponent (e.g. $1 \times 10^2 = 10^2$). The exponent tells us how many orders of magnitude larger or smaller to make the number. In other words, the exponent is telling us how many zeros the number will have: $10^2 = 100$; $10^3 = 1,000$, and so on. Negative exponents represent decimals; for example: $10^{-2} = 0.01$; $10^{-3} = 0.001$, and so on.

Here is a simple shorthand way to evaluate numbers given in scientific notation. Move the decimal place to the right if 10 has a positive exponent, and to the left if the exponent is negative. The number of spaces the decimal place is moved is equal to the exponent. For example, 10^2 tells us to move the decimal place 2 spaces to the right; 10^{-2} means we move it 2 spaces to the left.

By convention, we always designate the coefficient as a whole number (2) or a decimal, with the decimal point at the "10" position (2.3). In other words we would write 2.3×10^5, not 23×10^4. Both are technically correct but the first is the preferred format.

Examples:
$2 \times 10^6 = 2,000,000$
$2.36 \times 10^5 = 236,000$
$4.99 \times 10^{-4} = 0.000499$

Some typical values you might run across include:
$10^6 = 1$ million
$10^9 = 1$ billion
$10^{12} = 1$ trillion

MEASUREMENTS AND UNITS OF MEASURE

There are many handy conversion calculators on the Internet, but it is still useful to have a general idea of how large various units of measure are and how metric and English systems of measurement compare.

LENGTH

Metric
1 kilometer (km) = 1,000 meters (10^3)
1 meter (m) = 100 centimeters
1 centimeter (cm) = 10 millimeters
1 millimeter (mm) = 0.000001 meters (10^{-6})
1 micrometer (μm) = 0.000000001 meters (10^{-9})
1 nanometer (nm) = 0.000000000001 meters (10^{-12})

English
1 mile (mi) = 5,280 feet
1 yard (yd) = 36 inches (in) or 3 feet
1 foot (ft) = 12 inches

Conversions
1 km = 0.621 mi
1 m = 39.4 in
1 cm = 0.394 in

1 mi = 1.609 km
1 yd = 0.914 m
1 in = 2.54 cm

MASS

Metric
1 metric ton (mt) = 1,000 kilograms
1 kilogram (kg) = 1,000 grams
1 gram (g) = 1,000 milligrams
1 milligram (mg) = 0.001 grams (10^{-3})

English
1 U.S. ton (t) = 2,000 pounds
1 pound (lb) = 16 ounces (oz)

Conversions
1 mt = 2,200 lb
1 kg = 2.2 lb
1 g = 0.035 oz

1 t = 0.907 mt
1 lb = 4.54 g or 0.454 kg
1 oz = 28.35 g

VOLUME

Metric
1 liter (L) = 1,000 milliliters
1 milliliter (ml) = 0.001 liters

English
1 gallon (gal) = 4 quarts
1 quart (qt) = 2 pints or 4 cups
1 pint (pt) = 16 fluid oz

Conversions
1 L = 0.265 gal or 1.06 qt
1 gal = 3.79 L

AREA

Metric
1 hectare (ha) = 10,000 square km

English
1 acre (ac) = 4,840 square yards (yd)

Conversions
1 ha = 2.47 ac
1 ac = 0.405 ha

CONCENTRATIONS

Metric
1 part per million (ppm) = 1 mg/L
1 part per billion (ppb) = 1 μg/L
1 part per trillion (ppt) = 1 ng/L

TEMPERATURE CONVERSIONS

Fahrenheit (°F) to Celsius (°C): °C = (°F − 32) × $\frac{5}{9}$
Celsius (°C) to Fahrenheit (°F): °F = (°C − $\frac{5}{9}$) + 32

In general:
1°C = 1.8°F
1°F = 0.56°C

Answers to problems:
1. 85   **2.** $\frac{8}{32}$ = 0.25 = 25%   **3.** 100 × 0.25 = 25   **4.** 68 × 0.25 = 17

APPENDIX 2

DATA-HANDLING AND GRAPHING SKILLS

This tutorial offers a quick look at the basics of working with data and graphing.

Scientists gather data to learn about the natural world. Data can be organized into graphs, which are "pictures" or visual representations of the data. Because they can condense and organize large amounts of information, graphs are often easier to interpret than a simple list of numbers. They show relationships between two or more variables that help us determine whether the variables are correlated in any way and allow us to look for trends or patterns that might emerge. To be effective, graphs should be constructed according to conventions, and must be accurately plotted and properly labeled. Certain types of graphs are more suitable than others to show particular types of data, so it is important to choose the correct graph for your data.

The following sections describe variables found in graphs; data tables; and the types of graphs commonly used in environmental science.

VARIABLES

The **independent variable** is the parameter the experimenter manipulates—it could be whether or not a group is exposed to a treatment (given a medicine, exposed to a particular wavelength of light), is part of a distinct group (trees at specified distances from a stream), or is a group followed over a period of time (monitored daily, yearly, etc.). If you were setting up a data table in which to record the data your experiment would produce, you would be able to fill in the values for the independent variable *before beginning* the actual experiment.

The **dependent variable** is the response being measured in the experiment—the responding variable. The experiment is being conducted to see if this variable is "dependent on" the independent variable. In other words, when you change the independent variable, does the dependent variable change as a result? If you were setting up a data table in which to record data, you would be able to include a column heading for the dependent variable, but you would not be able to enter the values until the experiment was complete. There may be more than one dependent variable being tested.

DATA TABLES

Data tables have a conventional format. The independent variable is shown in the left-hand column and the data for the dependent variable or variables are shown in columns to the right of that. To be useful, the data table needs to have a descriptive title (what data are we looking at?) and the units of measure must be included.

Independent variable

Dependent variable

Units of measure are given—here we multiply each value by 1,000 to find the number of tons of catch that year; using the 1,000-fold conversion in the units allows us to use numbers that are easier to interpret.

Annual Atlantic Herring Catch

Year	Herring catch (1,000 tons)
1965	731
1970	580
1975	382
1980	270
1985	180
1990	150
1995	45
2000	20
2005	26

TYPES OF GRAPHS

A. LINE GRAPHS

In science, researchers often test the effect of one variable on another. Line graphs are used when the independent variable is represented by a numerical sequence (1, 2, 3 . . .) rather than discrete categories (red, yellow, blue . . .). The dependent variable is always a numerical sequence.

Steps to producing a line graph

1. **Determine the *x*-axis and the *y*-axis.** The independent variable is usually shown on the *x*-axis (the horizontal axis). In a line graph, this variable is one that changes in a predictable, numerical sequence, such as the passing of time, increasing concentrations of a solution being tested, habitat distance from the seashore, etc. The data being collected (the response being observed) represent the dependent variable, which is shown on the *y*-axis (the vertical axis). The axes require a descriptive title that indicates exactly what each axis represents, along with units of measure, if needed.

2. **Set up the axes.** Set up each axis so that the largest data value for that variable is close to the end of the axis, leaving as much of the available space for your graph as possible. Aim for 5 to 15 "ticks" (the small dividing marks) on any given axis—don't overload it with 50 tiny ticks or have so few that it is hard to place data points. It is essential to evenly space the ticks on a given axis, keeping the increments the same numerical size. On the sample graph, all the x-axis ticks are 10 years apart; all the y-axis ticks are 100 units apart. The increments will depend on your particular data; however, be sure they are presented in the same size for a given axis. Give the graph a descriptive title.

3. **Plot the points.** Plot each point by finding the x-value on the x-axis and moving up until you reach the y-value position across from the y-axis. If you are graphing more than one set of data, draw the data points as different shapes or colors and provide a legend to identify each data set.

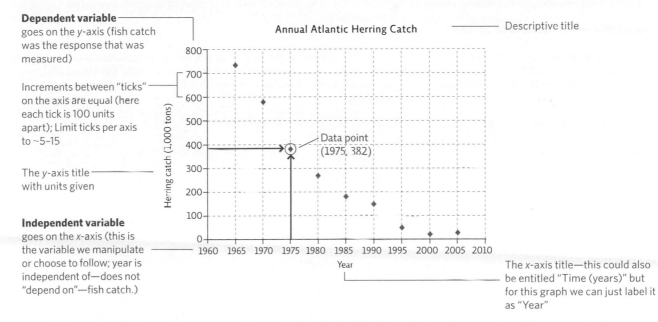

Dependent variable
goes on the y-axis (fish catch was the response that was measured)

Increments between "ticks" on the axis are equal (here each tick is 100 units apart); Limit ticks per axis to ~5–15

The y-axis title with units given

Independent variable
goes on the x-axis (this is the variable we manipulate or choose to follow; year is independent of—does not "depend on"—fish catch.)

Descriptive title

Data point
(1975, 382)

The x-axis title—this could also be entitled "Time (years)" but for this graph we can just label it as "Year"

4. **Draw the line.** Once data points are in place, you can draw your line, but don't simply connect the dots unless they all line up exactly. Step back and visualize what kind of trend the data are showing and draw a line that approximates that trend. These trend lines can be mathematically determined but can also be fairly accurately estimated by simply drawing in a line that goes through the center of the data—about as many points will be above as below the line. You can draw a straight line or you may elect to draw a curve to accommodate shifts in the trend.

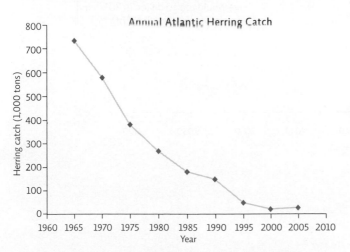

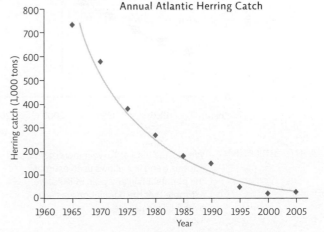

"Connecting the dots" like this implies that each data point is perfectly accurate and that this exactly represents the relationship between the two variables.

Drawing in a "trend line" that floats through the cloud of points is a more accurate estimation of the actual relationship seen between the two variables. We could draw a straight trend line for this data, but since it seems that the rate of decline lessens as times goes by, the curve seen here may better represent the relationship.

Interpreting the data

Once the graph is made, we can evaluate the data and draw conclusions. The first step is to simply describe the relationship seen—this is a statement of the *results* (observations). Here we see that between 1965 and 1980, herring catches dropped off dramatically and thereafter continued to drop, but more slowly. Now that we understand the relationship between the two variables, we can draw *conclusions*—make some inferences: What might have caused this relationship? What else may be true because this relationship exists? We could infer from these data that the herring population size also decreased in this time frame. If we know that the same number of fishers were fishing for herring the same number of days each year, the slower decline after 1980 might represent the fact that the fish are more difficult to catch because the population size is smaller. We might also conclude that it has not been as profitable to fish commercially for herring since 1980 as it was in the 1960s and 1970s. These last three statements are conclusions (inferences) based on the results of the study (observations); they are not observations themselves.

Interpolation and extrapolation (projections)

Plotting line graphs also allows us to estimate values of *y* or *x* within the range of our data set, values that we did not actually measure (*interpolation*). We can also create a projection of data points beyond our data set (*extrapolation*) by extending the line. This assumes the same trend will hold at higher or lower *x*-axis values, which may or may not be true; therefore, extrapolations are not likely to be as accurate as interpolations. Extrapolations, also called projections, are often shown as dashed lines.

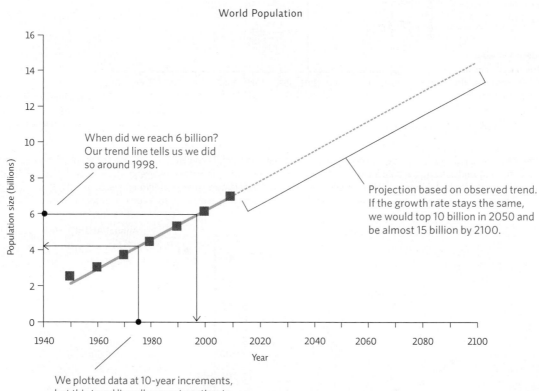

World Population

When did we reach 6 billion? Our trend line tells us we did so around 1998.

Projection based on observed trend. If the growth rate stays the same, we would top 10 billion in 2050 and be almost 15 billion by 2100.

We plotted data at 10-year increments, but this trend line allows us to estimate the population in any year. In 1975, we estimate that the population was just over 4 billion people.

B. SCATTER PLOTS

Scatter plots (with or without a trend line) are used when any x-value could have multiple y-values. For instance, in the second graph shown here, data were collected from various countries. Girls in four of the countries surveyed receive, on average, 4 years of schooling; therefore, there are 4 data points over the *x*-axis value of 4. But each of those countries had different y-values (total fertility rate). Here it would make no sense to "connect the dots." The resulting line would be impossible to follow.

It is more appropriate to construct a line that passes through the cloud of points and shows the "trend," just as we did with the line graph. Data points can be entered into a computer graphing program to calculate a "best fit" line, but you can also estimate the path yourself. To pick the best fit line, draw a line (straight or curved) that passes centrally through the cloud of data points, with about as many points above as below the line—the occasional point far away from the others won't impact the line significantly. When completed, your line may or may not be straight, but it should not connect the dots.

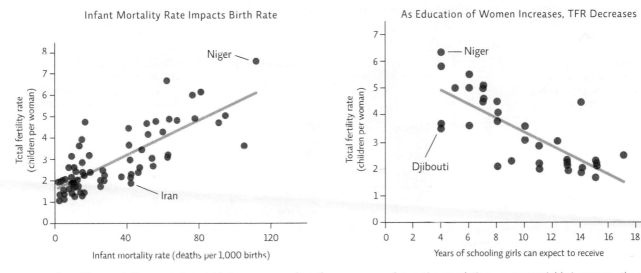

A *positive correlation*—as one variable increases, so does the other—an upward-sloping line

A *negative correlation*—as one variable increases, the other decreases—a downward-sloping line

C. PIE CHARTS

Pie charts are useful when the groups represented by the independent variable are all discrete categories (e.g. red, yellow, blue. . .) rather than a numerical sequence (e.g. 1,2,3. . .). In addition, the categories also represent all the subsets of a whole—all the category values add up to 100%. In other words, you have the entire pie! Data values and/or category titles can be shown either inside each "slice" or outside the pie. The data could also be shown as a bar graph (see Section D on the next page), but showing it as a pie chart instead allows one to more easily compare the size of each group to the other groups and to the whole.

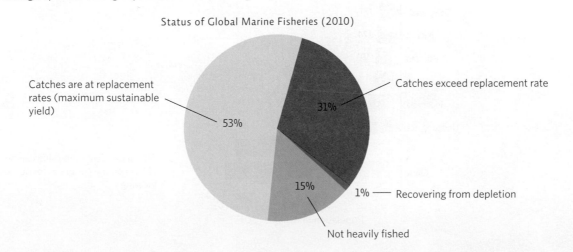

D. BAR GRAPHS

Bar graphs are appropriate in some cases. As with a pie chart, the key consideration is whether the independent variable (the variable you are manipulating in your experiment) is part of a numerical sequence (in which case, a line graph or scatter plot would be used) or represents discrete, separate groups. When you have discrete, separate groups, a bar graph can be used—it would make no sense to connect the data from one group to the next in a line. As with a line graph or scatter plot, the independent variable usually goes on the *x*-axis and the dependent variable is shown on the *y*-axis.

For example, researchers examined the stomach contents of birds from two different colonies. "Colony" is the independent variable because the researchers chose to see if where a bird lived would affect the type of food it ate. The volume of each food type ingested is the dependent variable—it is the data that the researcher set out to find and it may change according to colony location.

Food and Plastic Eaten by Two Albatross Colonies

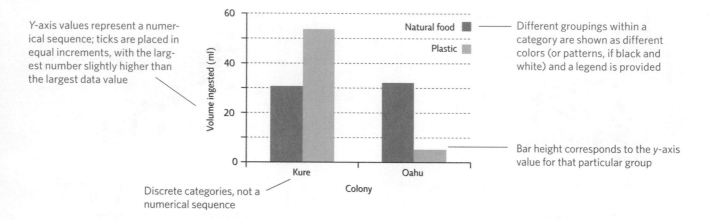

Y-axis values represent a numerical sequence; ticks are placed in equal increments, with the largest number slightly higher than the largest data value

Different groupings within a category are shown as different colors (or patterns, if black and white) and a legend is provided

Bar height corresponds to the *y*-axis value for that particular group

Discrete categories, not a numerical sequence

Sometimes it is easier to place the independent variable on the *y*-axis if the labels themselves are long. This prevents the need to place labels sideways, making them harder to read. The graph below, which shows the population size of the ten most populous countries, displays the independent variable (country) on the *y*-axis and the dependent variable (population size) on the *x*-axis.

Ten Most Populous Countries

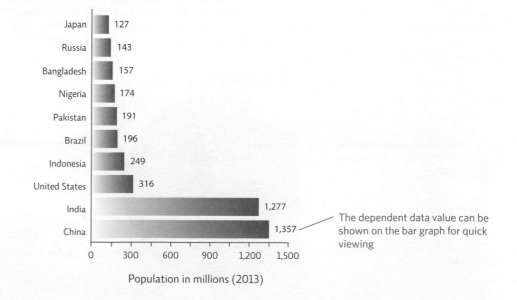

The dependent data value can be shown on the bar graph for quick viewing

Population in millions (2013)

E. AREA GRAPHS

Another useful graph that allows us to view the relative proportion of all the groups being compared is an area graph. It is used when we have a line graph (the independent variable on the x-axis is a numerical sequence) showing multiple lines. Each data set (line) is part of a larger group—here we show total fish catch broken down by type of fish. Each line is graphed "on top" of the other, and the space between the lines is filled in with a different color. This is useful because at any given x-axis point (say, the year 1968) we can see what the total fish catch was as well as how much each type of fish contributed to the total catch. The width of the "ribbon" for each fish type at that point represents its y-axis value—in this case its catch in 1,000 tons. (The y-axis value opposite the ribbon represents the total for all groups.)

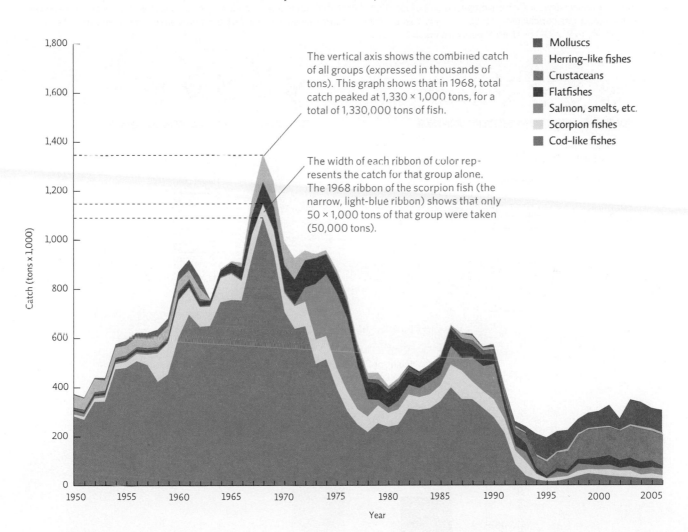

Fish Catch by Commercial Group: Newfoundland–Labrador Shelf

The vertical axis shows the combined catch of all groups (expressed in thousands of tons). This graph shows that in 1968, total catch peaked at 1,330 × 1,000 tons, for a total of 1,330,000 tons of fish.

The width of each ribbon of color represents the catch for that group alone. The 1968 ribbon of the scorpion fish (the narrow, light-blue ribbon) shows that only 50 × 1,000 tons of that group were taken (50,000 tons).

Legend:
- Molluscs
- Herring-like fishes
- Crustaceans
- Flatfishes
- Salmon, smelts, etc.
- Scorpion fishes
- Cod-like fishes

Y-axis: Catch (tons x 1,000)
X-axis: Year

APPENDIX 3

STATISTICAL ANALYSIS

DESCRIPTIVE STATISTICS

In science it is not enough to simply collect data and graph it in order to draw conclusions. Suppose we see a difference between data collected for different groups. How different must the data sets be in order to conclude that the groups are different from each other? And how can we determine whether the treatment we applied—say, growing plants with a new fertilizer—really affected growth? We turn to statistical analysis.

Let's look at an example. We are growing two sets of plants, identical in every way except that one is grown without any fertilizer (the control group) and the other is grown with fertilizer (the test group). To draw conclusions, examine the values in the data table.

We begin with descriptive statistics—what are the characteristics of our data set? We calculate useful statistical values for each data set such as the *mean* (the average), the *range* (the highest value minus the lowest value), and the *sample size* (the number of subjects in each group). We might also calculate other values (with the help of any number of readily available online or calculator-based programs) that help describe the data set, such as *standard deviation* (the average amount of variation of each data value from the mean) and *standard error* (a measure that gives us an idea of how accurate our calculated mean really is, based on the standard deviation). Standard error bars are often shown with data as $\pm$ values (for example, 14.9 $\pm$ 1.4) or as error bars on a graph.

Mean
14.9 cm

Mean
19 cm

CONTROL GROUP—
GROWN WITHOUT FERTILIZER

	Height (cm)
Sample size: 10	
Range: 15	5
Mean: 14.9	11
Std deviation: 4.3	14
Std error: 1.4	15
	15
	16
	17
	18
	18
	20

TEST GROUP—
GROWN WITH FERTILIZER

	Height (cm)
Sample size: 10	
Range: 4	17
Mean: 19	17
Std deviation: 1.2	18
Std error: 0.38	18
	19
	20
	20
	20
	20
	21

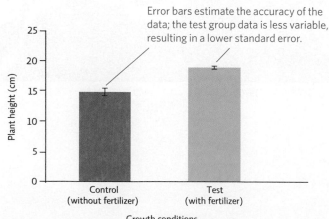

Average Height of Plants Grown With
and Without Fertilizer

Error bars estimate the accuracy of the data; the test group data is less variable, resulting in a lower standard error.

INFERENTIAL STATISTICS

We can take our analysis further and evaluate the data with an inferential statistical test to determine how likely it is that the data we obtained from the two groups in our experiment actually represent different responses or whether our two groups are both just subsets of a single, larger group. The statistical test gives us a *p-value*—a number that tells us how much overlap there is between the data sets. In science, we generally require that there be no more than a 5% overlap between the two data sets. If the high end of one set (the control group here) overlaps just a little with the low end of the other data set (our test group), and this overlap is no more than 5%, we can conclude that the two groups most likely represent two distinct populations, a result of the treatment we applied—in this case, the addition of fertilizer. If the overlap had been more than 5% (a p-value > 0.05), we would not have sufficient evidence to conclude that they were indeed two different groups, but instead we would say they were likely to be a single group that varied widely.

The data showed this:

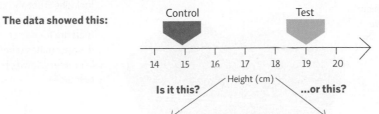

The average height of the plants in the two groups is different—but are they different enough to be considered two different populations?

THE DATA POINTS COULD ALL BE PART OF ONE POPULATION

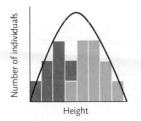

If the two groups are really part of one population, we might have inadvertently put faster-growing individuals in the test group and/or slower-growing ones in the control group. If that is true, retesting or using a larger sample size should produce some short control and some tall test plants.

THE DATA POINTS COULD REPRESENT 2 SEPARATE POPULATIONS

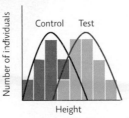

If the two groups really did respond differently to the treatment (fertilizer use), retesting or using a larger sample size should still produce this same trend, with most test plants being larger than control plants.

For this data set, a *t-test* (a simple statistical test) yields a p-value of 0.035—our data sets overlap 3.5%. Therefore, we can conclude that the two groups are different at the 0.05 level. Because our experimental design eliminated other variables that might have affected growth (the only difference between the two groups was whether plants received fertilizer), it is reasonable to conclude that the greater growth was caused by the fertilizer. As you read about studies and evaluate the authors' conclusions, look for the p-value given with the analysis of the data—if the calculated p-value is larger than 0.05, the author will probably conclude that there is not sufficient evidence to conclude that the variable tested had an effect.

OTHER FACTORS AFFECT RESULTS

VARIABILITY

Data sets with a lot of variability are less likely to show significant differences, even if the means are different—there is more of a chance that the two data sets overlap so much that they must be considered part of the same population.

LITTLE VARIABILITY

LOTS OF VARIABILITY

SAMPLE SIZE IS IMPORTANT

Small sample sizes are less reliable because, due to sampling error, we may have inadvertently sampled mostly unusual subjects; this would give us an incorrect picture of the entire population.

Mean

SMALLER CONTROL GROUP

This smaller sample overestimates the mean

Mean

LARGER CONTROL GROUP

This larger sample suggests that larger plants are not the norm

APPENDIX 4
GEOLOGY

EARTH IS A DYNAMIC PLANET THAT IS CONSTANTLY CHANGING

Earth is composed of discrete layers; mineral and fossil fuel deposits are found in the layers of Earth's crust. Powerful geologic forces are constantly but slowly rearranging the face of Earth.

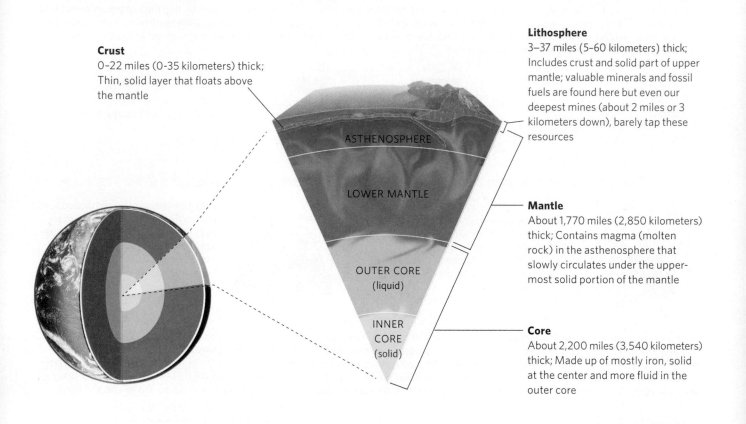

Crust
0–22 miles (0-35 kilometers) thick; Thin, solid layer that floats above the mantle

Lithosphere
3–37 miles (5–60 kilometers) thick; Includes crust and solid part of upper mantle; valuable minerals and fossil fuels are found here but even our deepest mines (about 2 miles or 3 kilometers down), barely tap these resources

Mantle
About 1,770 miles (2,850 kilometers) thick; Contains magma (molten rock) in the asthenosphere that slowly circulates under the uppermost solid portion of the mantle

Core
About 2,200 miles (3,540 kilometers) thick; Made up of mostly iron, solid at the center and more fluid in the outer core

ASTHENOSPHERE

LOWER MANTLE

OUTER CORE
(liquid)

INNER CORE
(solid)

EARTH'S CRUST

The crust is a very thin layer compared to the size of Earth; the continental crust is thicker than the oceanic crust.

The lithosphere (the solid part of Earth's surface) exists as tectonic plates that float above the asthenosphere.

OCEAN CRUST

CONTINENTAL CRUST

MANTLE'S SOLID PORTION

ASTHENOSPHERE

Where plates meet head on, one slides below (subducts) the other. This sometimes causes earthquakes, a sudden release of energy, as one plate slides past the other. Trenches may form in this area.

Magma exits the asthenosphere where plates move apart in volcanic eruptions, solidifying to form ridges or islands.

THE ROCK CYCLE CONSTANTLY FORMS AND REFORMS ROCKS IN EARTH'S CRUST

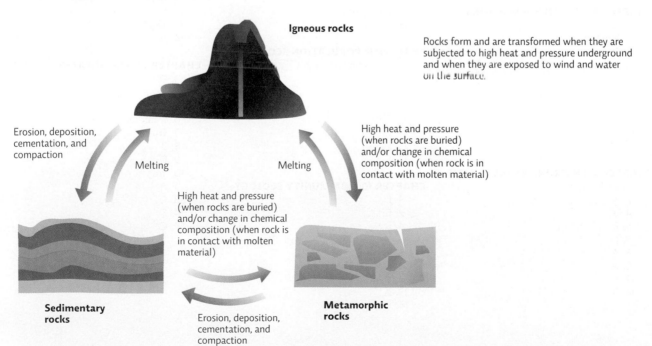

Igneous rocks

Rocks form and are transformed when they are subjected to high heat and pressure underground and when they are exposed to wind and water on the surface.

Erosion, deposition, cementation, and compaction

Melting

Melting

High heat and pressure (when rocks are buried) and/or change in chemical composition (when rock is in contact with molten material)

High heat and pressure (when rocks are buried) and/or change in chemical composition (when rock is in contact with molten material)

Sedimentary rocks

Metamorphic rocks

Erosion, deposition, cementation, and compaction

SELECTED ANSWERS TO END-OF-CHAPTER PROBLEMS

Answers to multiple choice, true/false, and fill-in-the-blank questions are shown here. Go online for the rest of the answers to the short answer and discussion questions.
→ www.macmillanhighered.com/launchpad/saes2e

CHAPTER 1: ENVIRONMENTAL LITERACY
1. living and nonliving
2. B
4. environmental, economic, and social
5. A
7. D
8. C
10. A
12. the tragedy of the commons
13. A

CHAPTER 2: SCIENCE LITERACY AND THE PROCESS OF SCIENCE
2. D
3. B
4. D
6. D
8. observational
9. experimental
11. True

CHAPTER 3: INFORMATION LITERACY
1. C
2. D
4. C
7. D
8. B
10. B
12. A

CHAPTER 4: HUMAN POPULATIONS
1. C
2. 7 billion
4. B
5. C
6. False
8. B
9. C
11. D
12. B
14. B

CHAPTER 5: ENVIRONMENTAL HEALTH
1. pathogens
2. A
4. B
5. B
7. D
8. A
10. A
12. D
13. B

CHAPTER 6: ECONOMICS
1. D
2. A
4. C
5. B
7. False
9. C
10. A
12. D
13. C

CHAPTER 7: SOLID WASTE
1. C
4. False
5. C
7. E
10. False
13. True
14. A
15. D

CHAPTER 8: ECOSYSTEMS AND NUTRIENT CYCLING
1. B
2. D
4. C
6. A
8. C
9. B
11. True
12. phosphorus
13. A

CHAPTER 9: POPULATION ECOLOGY
1. population density
2. D
4. D
5. A
7. False
9. C
12. False
13. C

CHAPTER 10: COMMUNITY ECOLOGY
1. D
2. C
4. B
5. True
7. C
10. False
11. C
13. A
14. B

CHAPTER 11: EVOLUTION AND EXTINCTION
1. gene frequencies
2. A
4. C
6. D
8. B
9. D
12. C

CHAPTER 12: BIODIVERSITY
2. A
3. D
4. True
5. genetic
6. C
8. False
9. C
10. habitat destruction
11. B
13. True
14. D

CHAPTER 13: PRESERVING BIODIVERSITY
1. False
2. D
4. False
5. C
7. A
9. False
10. C
13. True
14. A

CHAPTER 14: FRESHWATER RESOURCES
1. D
3. False
4. C
6. water table
7. True
8. B
9. D
10. True
11. A
13. C
14. E
15. C

CHAPTER 15: WATER POLLUTION
1. D
2. A
4. True
5. Answer: 4, 1, 3, 2
7. C
9. A
11. False
12. C
13. C
14. A

CHAPTER 16: FEEDING THE WORLD
1. False
2. D
3. D
4. A
6. B
7. C
9. False
10. D
12. A

CHAPTER 17: SUSTAINABLE AGRICULTURE: RAISING CROPS
1. True
3. C
4. E
6. A
7. B
8. D
9. C
10. B
11. False
12. D
13. B

CHAPTER 18: COAL
1. False
2. B
3. D
4. A
5. B
7. C
9. True
10. A
11. A
13. False
14. B

CHAPTER 19: OIL AND NATURAL GAS
1. C
2. B
4. False
5. D
7. True
8. C
10. A
11. B
13. False
14. A

CHAPTER 20: AIR POLLUTION
1. 5 million
2. True
3. B
4. C
5. A
7. False
8. D
10. True
11. C
13. B

CHAPTER 21: CLIMATE CHANGE
1. weather; climate
2. True
4. D
5. A
7. False
8. B
11. False
12. A
14. True
15. mitigation, adaptation
16. C

CHAPTER 22: NUCLEAR POWER
1. protons
2. C
3. D
5. True
6. A
8. C
9. D
11. B

CHAPTER 23: SUN, WIND, WATER, AND EARTH ENERGY
1. False
2. D
3. C
4. A
5. C
7. True
8. B
11. D
12. B
14. wind, solar, and biomass
15. C

CHAPTER 24: ENVIRONMENTAL POLICY
1. B
4. A
7. precautionary principle
8. C
10. C
11. D
13. C
14. A

CHAPTER 25: URBANIZATION AND SUSTAINABLE COMMUNITIES
1. False
2. D
3. True
4. D
6. B
7. C
9. True
10. B
11. C
13. False
14. D

GLOSSARY

A

abiotic The nonliving components of an ecosystem, such as rainfall and mineral composition of the soil. (Chapter 8)

acid deposition Precipitation that contains sulfuric or nitric acid; dry particles may also fall and become acidified once they mix with water. (Chapter 20)

acid mine drainage Water flowing past exposed rock in mines, leaching out sulfates. These sulfates react with the water and oxygen to form acids (low-pH solutions). (Chapter 18)

active solar technologies Mechanical equipment for capturing, converting, and sometimes concentrating solar energy into a more usable form. (Chapter 23)

adaptation A trait that helps an individual survive or reproduce. (Chapter 11); Efforts intended to help deal with a problem that exists, such as climate change. (Chapter 21)

adaptive management A plan that allows room for altering strategies as new information becomes available or as the situation itself changes. (Chapter 2, 24)

additive effects Exposure to two or more chemicals that has an effect equivalent to the sum of their individual effects. (Chapter 3)

age structure The percentage of the population that is distributed into various age groups. (Chapter 4)

age structure diagram A graphic that displays the relative sizes of various age groups, with males shown on one side of the graphic and females on the other. (Chapter 4)

agroecology Scientific field that considers the area's ecology and indigenous knowledge, and favors agricultural methods that protect the environment and meet the needs of local people. (Chapter 17)

air pollution Any material added to the atmosphere (naturally or by humans) that harms living organisms, affects the climate, or impacts structures. (Chapter 21)

albedo The ability of a surface to reflect away solar radiation. (Chapter 21)

annual crops Crops that grow, produce seeds, and die in a year and must be replanted each season. (Chapter 17)

antagonistic effects Exposure to two or more chemicals that has a lesser effect than the sum of their individual effects would predict. (Chapter 3)

anthropocentric worldview A human-centered view that assigns intrinsic value only to humans. (Chapter 1)

anthropogenic Caused by or related to human action. (Chapter 1, 21)

applied science Research whose findings are used to help solve practical problems. (Chapter 1)

aquifer An underground, permeable region of soil or rock that is saturated with water. (Chapter 14, 15)

artificial selection A process in which humans decide which individuals breed and which do not in an attempt to produce a population with desired traits. (Chapter 11)

asthma A chronic inflammatory respiratory disorder characterized by "attacks" during which the airways narrow, making it hard to breathe; can be fatal. (Chapter 20)

atmosphere The blanket of gases that surrounds Earth and other planets. (Chapter 2)

B

background rate of extinction The average rate of extinction that occurred before the appearance of humans or that occurs between mass extinction events. (Chapter 11)

benthic macroinvertebrates Easy-to-see (not microscopic) arthropods such as insects that live on the stream bottom. (Chapter 15)

bioaccumulation The buildup of substances in the tissue of an organism over the course of its lifetime. (Chapter 3)

biocentric worldview A life-centered approach that views all life as having intrinsic value, regardless of its usefulness to humans. (Chapter 1)

biochar A form of charcoal, produced when organic matter is partially burned, that can be used to improve soil quality. (Chapter 17)

biodegradable Capable of being broken down by living organisms. (Chapter 7)

biodiversity The variety of life on Earth; it includes species, genetic, and ecological diversity. (Chapter 1, 12)

biodiversity hotspot An area that contains a large number of endangered endemic species. (Chapter 12)

biological assessment The process of sampling an area to see what lives there as a tool to determine how healthy the area is. (Chapter 15)

biomagnification The increased levels of substances in the tissue of predatory (higher trophic level) animals that have consumed organisms that contain bioaccumulated toxic substances. (Chapter 3)

biomass energy Energy from biological material such as plants (wood, charcoal, crops) and animal waste. (Chapter 23)

biome One of many distinctive types of ecosystems determined by climate and identified by the predominant vegetation and organisms that have adapted to live there. (Chapter 8)

biosphere The sum total of all of Earth's ecosystems. (Chapter 8)

biotic The living (organic) components of an ecosystem, such as the plants and animals and their waste (dead leaves, feces). (Chapter 8)

biotic potential (*r*) The maximum rate at which the population can grow due to births if each member of the population survives and reproduces. (Chapter 9)

boom-and-bust cycles Fluctuations in population size that produce a very large population followed by a crash that lowers the population size drastically, followed again by an increase to a large size and a subsequent crash. (Chapter 9)

bottleneck effect The situation that occurs when population size is drastically reduced, leading to the loss of some genetic variants, and resulting in a less diverse population. (Chapter 11)

bottom-up regulation The control of population size by factors that enhance growth and survival (growth factors) such as nutrients, water, sunlight, and habitat. (Chapter 9)

C

cap-and-trade Regulations that set an upper limit for pollution emissions, issue permits to producers for a portion of that amount, and allow producers that release less than their allotment to sell permits to those who exceeded their allotment. (Chapter 20, 24)

carbon capture and sequestration (CCS) Removing carbon from fuel combustion emissions or other sources and storing it to prevent its release into the atmosphere. (Chapter 18)

carbon cycle Movement of carbon through biotic and abiotic parts of an ecosystem. Carbon cycles via photosynthesis and cellular respiration as well as in and out of other reservoirs, such as oceans, soil, rock, and atmosphere. It is also released by human actions such as the burning of fossil fuels. (Chapter 8)

carbon footprint The amount of carbon released to the atmosphere by a person, company, nation, or activity. (Chapter 17)

carbon taxes Governmental fees imposed on activities (such as fossil fuel use) that release CO_2 into the atmosphere. (Chapter 21)

carrying capacity (*K*) The population size that an area can support for the long term; depends on resource availability and the rate of per capita resource use by the population. (Chapters 1, 4, 9)

cash crops Food and fiber crops grown to sell for profit, rather than for use by local families or communities. (Chapter 16)

cause-and-effect relationship An association between two variables that identifies one (the effect) occurring as a result of or in response to the other (the cause). (Chapter 2)

cellular respiration The process in which all organisms break down sugar to release its energy, using oxygen and giving off CO_2 as a waste product. (Chapter 8)

citizen suit provision A provision that allows a private citizen to sue, in federal court, a perceived violator of certain U.S. environmental laws, such as the Clean Air Act, in order to force compliance. (Chapter 24)

Clean Air Act (CAA) First passed in 1963 and amended most recently in 1990, this U.S. law authorizes the EPA to set standards for dangerous air pollutants and enforce those standards. (Chapter 20)

Clean Development Mechanism (CDM) A UN program that allows a country with a greenhouse gas reduction commitment to implement emission-reduction projects in developing countries. (Chapter 24)

Clean Water Act (CWA) U.S. federal legislation that regulates the release of point source pollution into surface waters and sets water quality standards for those waters. It also supports best management practices to reduce nonpoint source pollution. (Chapter 15)

climate Long-term patterns or trends of meteorological conditions. (Chapter 21)

climate change Alteration in the long-term patterns and statistical averages of meteorological events. (Chapter 21)

climax species Species that move into an area at later stages of ecological succession. (Chapter 10)

closed-loop system A production system in which the product is returned to the resource stream when consumers are finished with it or is disposed of in such a way that nature can decompose it. (Chapter 6)

clumped distribution A distribution in which individuals of a population are found in groups or patches within the habitat. (Chapter 9)

coal Fossil fuel formed when plant material is buried in oxygen-poor conditions and subjected to high heat and pressure over a long time. (Chapter 18)

coevolution A special type of natural selection in which two species each provide the selective pressure that determines which traits are favored by natural selection in the other. (Chapter 11)

coliform bacteria Bacteria often found in the intestinal tract of animals; monitored to look for fecal contamination of water. (Chapter 14)

command-and-control regulation Regulations that set an upper allowable limit of pollution release which is enforced with fines and/or incarceration. (Chapter 20, 24)

commensalism A symbiotic relationship between individuals of two species in which one benefits from the presence of the other, but the other is unaffected. (Chapter 10)

community All the populations (plants, animals, and other species) living and interacting in an area. (Chapter 8)

community ecology The study of all the populations (plants, animals, and other species) living and interacting in an area. (Chapter 10)

competition Species interaction in which individuals are vying for limited resources. (Chapter 10)

composting Allowing waste to biologically decompose in the presence of oxygen and water, producing a soil-like mulch. (Chapter 7)

concentrated animal feeding operation (CAFO) A method in which large numbers of meat or dairy animals are reared at high densities in confined spaces and fed a calorie rich diet to maximize growth.

condensation The conversion of water from a gaseous state (water vapor) to a liquid state. (Chapter 14)

conservation Efforts that reduce waste and increase efficient use of resources. (Chapter 23)

conservation biology The science concerned with preserving biodiversity. (Chapter 13)

conservation genetics The scientific field that relies on species' genetics to inform conservation efforts. (Chapter 13)

consumer An organism that obtains energy and nutrients by feeding on another organism. (Chapter 8, 10)

contour farming Farming on hilly land in rows that are planted along the slope, following the lay of the land, rather than oriented downhill. (Chapter 17)

control group The group in an experimental study that the test group's results are compared to; ideally, the control group will differ from the test group in only one way. (Chapter 2)

control rods Rods that absorb neutrons and slow the chain reaction in a fission nuclear reactor. (Chapter 22)

conventional oil reserves Light- or medium-density crude oil deposits; extracted by pumping. (Chapter 19)

Convention on Biological Diversity (CBD) An international treaty that promotes sustainable use of ecosystems and biodiversity. (Chapter 13)

Convention on International Trade in Endangered Species of Wild Fauna and Flora (CITES) An international treaty that regulates the global trade of selected species. (Chapter 13)

core species Species that prefer core areas of a habitat—areas deep within the habitat, away from the edge. (Chapter 10)

corporate average fuel efficiency (CAFE) standards A target of the minimum fuel efficiency (MPG) that manufacturers must meet; evaluated as a weighted average of all the cars and light trucks each manufacturer produces. (Chapter 24)

correlation Two things occurring together but not necessarily having a cause-and-effect relationship. (Chapter 2)

cover crop A crop planted in the off-season to help prevent soil erosion and to return nutrients to the soil. (Chapter 17)

cradle-to-cradle Management of a resource that considers the impact of its use at every stage, from raw material extraction to final disposal or recycling. (Chapter 6)

critical thinking Skills that enable individuals to logically assess information, reflect on that information, and reach their own conclusions. (Chapter 3)

crop rotation Planting different crops on a given plot of land every few years to maintain soil fertility and reduce pest outbreaks. (Chapter 17)

crude birth rate The number of offspring born per 1,000 individuals per year. (Chapter 4)

crude death rate The number of deaths per 1,000 individuals per year. (Chapter 4)

crude oil A mix of hydrocarbons that exists as a liquid underground; can be refined to produce fuels or other products. (Chapter 19)

cultural eutrophication Nutrient enrichment of an aquatic ecosystem that stimulates excess plant growth and disrupts normal energy uptake and matter cycles. (Chapter 15, 17)

D

dam A structure that blocks the flow of water in a river or stream. (Chapter 14)

debt-for-nature-swaps Arrangements in which a wealthy nation forgives the debt of a developing nation in return for a pledge to protect natural areas in that developing nation. (Chapter 13)

decomposers Organisms such as bacteria and fungi that break organic matter all the way down to constituent atoms or molecules in a form that plants can take back up. (Chapter 10)

demographic factors Population characteristics such as birth rate or life expectancy that influence how a population changes in size and composition. (Chapter 4)

demographic transition A theoretical model that describes the expected drop in once-high population growth rates as economic conditions improve the quality of life in a population. (Chapter 4)

density-dependent factors Factors, such as predation or disease, whose impact on a population increases as population size goes up. (Chapter 9)

density-independent factors Factors, such as a storm or an avalanche, whose impact on a population is not related to population size. (Chapter 9)

dependent variable The variable in an experiment that is evaluated to see if it changes due to the conditions of the experiment. (Chapter 2)

desalination The removal of salt and minerals from seawater to make it suitable for consumption. (Chapter 14)

detritivores Consumers (including worms, insects, crabs, etc.) that eat dead organic material. (Chapter 10)

discounting future value Giving more weight to short-term benefits and costs than to long-term ones. (Chapter 6)

dissolved oxygen (DO) The amount of oxygen in the water. (Chapter 15)

dose-response curve A graph of the effects of a substance at different concentrations or levels of exposures. (Chapter 3)

E

ecocentric worldview A system-centered view that values intact ecosystems, not just the individual parts. (Chapter 1)

eco-industrial parks Industrial parks in which industries are physically positioned near each other for "waste-to-feed" exchanges; the waste of one becomes the raw material for another. (Chapter 7)

ecolabeling Providing information about how a product is made and where it comes from. Allows consumers to make more sustainable choices and support sustainable products and the businesses that produce them. (Chapter 6)

ecological diversity The variety within an ecosystem's structure, including many communities, habitats, niches, and trophic levels. (Chapter 12)

ecological economics New theory of economics that considers the long-term impact of our choices on people and the environment. (Chapter 6)

ecological footprint The land area needed to provide the resources for, and assimilate the waste of, a person or population. (Chapter 1, 6)

ecological succession Progressive replacement of plant (and then animal) species in a community over time due to the changing conditions that the plants themselves create (more soil, shade, etc.). (Chapter 10)

economics The social science that deals with the production, distribution, and consumption of goods and services. (Chapter 6)

ecosystem All of the organisms in a given area plus the physical environment in which, and with which, they interact. (Chapter 8, 10)

ecosystem conservation A management strategy that focuses on protecting an ecosystem as a whole in an effort to protect the species that live there. (Chapter 13)

ecosystem restoration The repair of natural habitats back to (or close to) their original state. (Chapter 13)

ecosystem services Benefits that are important to all life, including humans, provided by functional ecosystems; includes such things as nutrient cycles, air and water purification, and ecosystem goods such as food and fuel. (Chapter 6, 12)

ecotones Regions of distinctly different physical areas that serve as boundaries between different communities. (Chapter 10)

ecotourism Low-impact travel to natural areas that contributes to the protection of the environment and respects the local people. (Chapter 13)

edge effects The different physical makeup of an ecotone that creates different conditions that either attract or repel certain species (e.g., it is drier, warmer, and more open at the edge of a forest and field than it is further in the forest). (Chapter 10)

edge species Species that prefer to live close to the edges of two different habitats (ecotone areas). (Chapter 10)

effluent Wastewater discharged into the environment. (Chapter 14)

electricity The flow of electrons (negatively charged subatomic particles) through a conductive material (such as wire). (Chapter 18)

emerging infectious diseases Infectious diseases that are new to humans or that have recently increased significantly in incidence, in some cases by spreading to new ranges. (Chapter 5)

emigration The movement of people out of a given population. (Chapter 4)

empirical evidence Information gathered via observation of physical phenomena. (Chapter 2)

empirical science A scientific approach that investigates the natural world through systematic observation and experimentation. (Chapter 1)

endangered/endangered species Describes a species that faces a very high risk of extinction in the immediate future. (Chapter 11, 12, 13)

Endangered Species Act (ESA) The primary federal law that protects biodiversity in the United States. (Chapter 13)

endemic/ endemic species Describes a species that is native to a particular area and is not naturally found elsewhere. (Chapter 11, 12)

endocrine disruptor A substance that interferes with the endocrine system, typically by mimicking a hormone or preventing a hormone from having an effect. (Chapter 3)

energy The capacity to do work. (Chapter 18)

energy flow The one-way passage of energy through an ecosystem. (Chapter 8)

energy independence Meeting all of one's energy needs without importing any energy. (Chapter 19)

energy return on energy investment (EROEI) A measure of the net energy from an energy source (the energy in the source minus the energy required to get it, process it, ship it, and then use it). (Chapter 18, 19)

energy security Having access to enough reliable and affordable energy sources to meet one's needs. (Chapter 19)

environment The biological and physical surroundings in which any given living organism exists. (Chapter 1)

environmental ethic The personal philosophy that influences how a person interacts with his or her natural environment and thus affects how one responds to environmental problems. (Chapter 1)

environmental health The branch of public health that focuses on factors in the natural world and the human-built environment that impact the health of populations. (Chapter 5)

environmental impact statement (EIS) A document that outlines the positive and negative impacts of a proposed federal action (including alternative actions and the option of taking no action); used to help decide whether or not that action will be approved. (Chapter 18, 24)

environmental justice The concept that access to a clean, healthy environment is a basic human right. (Chapter 20, 25)

environmental literacy A basic understanding of how ecosystems function and of the impact of our choices on the environment. (Chapter 1)

environmental policy A course of action adopted by a government or organization that is intended to improve the natural environment and public health or reduce human impact on the environment. (Chapter 24)

Environmental Protection Agency (EPA) The federal agency responsible for setting policy and enforcing U.S. environmental laws. (Chapter 3, 24)

environmental racism Occurs when minority communities face more exposure to pollution than average for the region. (Chapter 20)

environmental science An interdisciplinary field of research that draws on the natural and social sciences and the humanities in order to understand the natural world and our relationship to it. (Chapter 1)

epidemiologist A scientist who studies the causes and patterns of disease in human populations. (Chapter 3, 5)

eutrophication A process in which excess nutrients in aquatic ecosystems feed biological productivity, ultimately lowering the oxygen content in the water. (Chapter 15, 17)

evaporation The conversion of water from a liquid state to a gaseous state. (Chapter 14)

evolution Differences in the gene frequencies within a population from one generation to the next. (Chapter 11)

e-waste Unwanted computers and other electronic devices such as discarded televisions and cell phones. (Chapter 7)

experimental study Research that manipulates a variable in a test group and compares the response to that of a control group that was not exposed to the same variable. (Chapter 2)

exponential growth The kind of growth in which a population becomes progressively larger each breeding cycle; produces a J curve when plotted over time. (Chapter 9)

external cost A cost associated with a product or service that is not taken into account when a price is assigned to that product or service but rather is passed on to a third party who does not benefit from the transaction. (Chapter 6)

extinction The complete loss of a species from an area; may be local (gone from an area) or global (gone for good). (Chapter 11)

extirpated/extirpation Describes a species that is locally extinct in one or more areas but still has some individual members in other areas. (Chapter 11, 12)

exurbs Towns beyond the immediate suburbs whose residents commute into the city for work. (Chapter 25)

F

falsifiable Being capable of being proved wrong by evidence. (Chapter 2)

famine A severe shortage of food that leads to widespread hunger. (Chapter 16)

fertilizer A natural or synthetic mixture that contains nutrients that is added to soil to boost plant growth. (Chapter 16, 17)

flagship species The focus of public awareness campaigns aimed at generating interest in conservation in general; usually an interesting or charismatic species, such as the giant panda or tiger. (Chapter 13)

food chain A simple, linear path starting with a plant (or other photosynthetic organism) that identifies what each organism in the path eats. (Chapter 10)

food miles The distance a food travels from its site of production to the consumer. (Chapter 17)

food security Having physical, social, and economic access to sufficient, safe, and nutritious food. (Chapter 16)

food web A linkage of all the food chains together that shows the many connections in the community. (Chapter 10)

fossil fuels Nonrenewable resources like coal, oil, and natural gas that were formed over millions of years from the remains of dead organisms. (Chapter 18, 19)

fossil record The total collection of fossils (remains, impressions, traces of ancient organisms) found on Earth. (Chapter 11)

founder effect The situation that occurs when a small group with only a subset of the larger population's genetic diversity becomes isolated and evolves into a different population, missing some of the traits of the original. (Chapter 11)

fracking (hydraulic fracturing) The extraction of oil or natural gas from dense rock formations by creating factures in the rock and then flushing out the oil/gas with pressurized fluid. (Chapter 19)

freshwater Water that has few dissolved ions such as salt. (Chapter 14)

fuel rods Hollow metal cylinders filled with uranium fuel pellets for use in fission reactors. (Chapter 22)

G

gene Stretches of DNA, the hereditary material of cells, that each direct the production of a particular protein and influence an individual's traits. (Chapter 11)

gene frequencies The assortment and abundance of particular variants of genes relative to each other within a population. (Chapter 11)

genetically modified organisms (GMOs) Organisms that have had their genetic information modified to give them desirable characteristics such as pest or drought resistance. (Chapter 16)

genetic diversity The heritable variation among individuals of a single population or within a species as a whole. (Chapter 11, 12)

genetic drift The change in gene frequencies of a population over time due to random mating that results in the loss of some gene variants. (Chapter 11)

geothermal energy The heat stored underground, contained in either rocks or fluids. (Chapter 23)

geothermal heat pump A system that transfers the steady 55°F (12.5°C) underground temperature to a building to help heat or cool it. (Chapter 23)

geothermal power plants Power plants that use the heat of hydrothermal reservoirs to produce steam and turn turbines to generate electricity. (Chapter 23)

global warming The observed and ongoing rise in the Earth's average temperature that is contributing to climate change. (Chapter 21)

green building Construction and operational designs that promote resource and energy efficiency, and provide a better environment for occupants. (Chapter 25)

green business Doing business in a way that is good for people and the environment. (Chapter 6)

Green Revolution Plant-breeding program in the mid-1900s that dramatically increased crop yields and led the way for mechanized, large-scale agriculture. (Chapter 16)

Green Revolution 2.0 Programs that focus on the production of genetically modified organisms (GMOs) to increase crop productivity. (Chapter 16)

green tax A tax (a fee paid to the government) assessed on environmentally undesirable activities (e.g., a tax per unit of pollution emitted). (Chapter 20, 24)

greenhouse effect The warming of the planet that results when heat is trapped by Earth's atmosphere. (Chapter 21)

greenhouse gases Molecules in the atmosphere that absorb heat and reradiate it back to Earth. (Chapter 21)

greenwashing Claiming environmental benefits about a product when the benefits are actually minor or nonexistent. (Chapter 17)

gross primary productivity A measure of the total amount of energy captured via photosynthesis and transferred to organic molecules in an ecosystem. (Chapter 10)

ground-level ozone A secondary pollutant that forms when some of the pollutants released during fossil fuel combustion react with atmospheric oxygen in the presence of sunlight. (Chapter 20)

groundwater Water found underground in aquifers. (Chapter 14)

growth factors Resources individuals need to survive and reproduce that allow a population to grow in number. (Chapter 9)

H

habitat The physical environment in which individuals of a particular species can be found. (Chapter 8, 10)

habitat destruction The alteration of a natural area in a way that makes it unsuitable for the species living there. (Chapter 12)

habitat fragmentation The destruction of part of an area that creates a patchwork of suitable and unsuitable habitat area that may exclude some species altogether. (Chapter 12)

hazardous waste Waste that is toxic, flammable, corrosive, explosive, or radioactive. (Chapter 7)

high-level radioactive waste (HLRW) Spent nuclear reactor fuel or waste from the production of nuclear weapons that is still highly radioactive. (Chapter 22)

high-yield varieties (HYVs) Strains of staple crops selectively bred to produce more grain than their natural counterparts. (Chapter 16)

hormone A chemical released by organisms that directs cellular activity and produces changes in how their bodies function. (Chapter 3)

hydropower Energy produced from moving water. (Chapter 23)

hypothesis A possible explanation for what we have observed that is based on some previous knowledge. (Chapter 2)

hypoxia A situation in which a body of water contains inadequate levels of oxygen, compromising the health of many aquatic organisms. (Chapter 15)

I

immigration The movement of people into a given population. (Chapter 4)

incinerators Facilities that burn trash at high temperatures. (Chapter 7)

independent variable The variable in an experiment that a researcher manipulates or changes to see if the change produces an effect. (Chapter 2)

indicator species A species that is particularly vulnerable to ecosystem perturbations, and that, when we monitor it, can give us advance warning of a problem. (Chapter 10, 13)

industrial agriculture Farming methods that rely on technology, synthetic chemical inputs, and economies of scale to increase productivity and profits. (Chapter 16, 17)

infant mortality rate The number of infants who die in their first year of life per every 1,000 live births in that year. (Chapter 4)

infectious disease An illness caused by an invading pathogen such as a bacterium or virus. (Chapter 5)

inferences Conclusions we draw based on observations. (Chapter 2)

infill development The development of empty lots within a city. (Chapter 25)

infiltration The process of water soaking into the ground. (Chapter 14)

information literacy The ability to find and evaluate the quality of information. (Chapter 3)

integrated pest management (IPM) The use of a variety of methods to control a pest population, with the goal of minimizing or eliminating the use of chemical toxins. (Chapter 17)

Intergovernmental Panel on Climate Change (IPCC) An international group of scientists that evaluates scientific studies related to climate change to thoroughly and objectively assess the data. (Chapter 21)

internal cost A cost—such as for raw materials, manufacturing costs, labor, taxes, utilities, insurance, or rent—that is accounted for when a product or service is evaluated for pricing. (Chapter 6)

instrumental value An object's or species' worth, based on its usefulness to humans. (Chapter 1, 12)

intrinsic value An object's or species' worth, based on its mere existence; it has an inherent right to exist. (Chapter 1, 12)

invasive species A non-native species (a species outside its range) whose introduction causes or is likely to cause economic or environmental harm or harm to human health. (Chapter 11)

in vitro study Research that studies the effects of experimental treatment on cells in culture dishes rather than in intact organisms. (Chapter 3)

in vivo study Research that studies the effects of an experimental treatment in intact organisms. (Chapter 3)

IPAT model An equation ($I = P \times A \times T$) that measures human impact (I), based on three factors: population (P), affluence (A), and technology (T). (Chapter 6)

isotopes Atoms that have different numbers of neutrons in their nucleus but the same number of protons. (Chapter 22)

K

kerogen shale/oil shale Rocks that contain kerogen, a mix of solid organic material that can be converted to a liquid fossil fuel for extraction. (Chapter 19)

keystone species A species that impacts its community more than its mere abundance would predict, often altering ecosystem structure. (Chapter 10, 13)

K-selected species Species that have a low biotic potential and that share characteristics such as long life span, late maturity, and low fecundity; generally show logistic population growth. (Chapter 9)

Kyoto Protocol The 1997 amendment to the UNFCCC that set legally binding specific goals for reductions in greenhouse gas emissions for certain nations that ratified the treaty. (Chapter 24)

L

landscape conservation An ecosystem conservation strategy that specifically identifies a suite of species, chosen because they use all the vital areas within an ecosystem; meeting the needs of these species will keep the ecosystem fully functional, thus meeting the needs of all species that live there. (Chapter 13)

LD$_{50}$ (lethal dose 50%) The dose of a substance that would kill 50% of the test population. (Chapter 3)

leachate Water that carries dissolved substances (often contaminated) that can percolate through soil. (Chapter 7)

LEED (Leadership in Energy and Environmental Design) A certification program that awards a rating (standard, silver, gold, or platinum) to a building that includes environmentally sound design features. (Chapter 25)

less developed country A country that has a lower standard of living than a developed country and has a weak economy; may have high poverty. (Chapter 4)

life expectancy The number of years an individual is expected to live. (Chapter 4)

life-history strategies Biological characteristics of a species (for example, life span, fecundity, maturity rate) that influence how quickly a population can potentially increase in number. (Chapter 9)

limiting factor The critical resource whose supply determines the population size of a given species in a given ecosystem. (Chapter 8)

logical fallacies Arguments that attempt to sway the reader without using reasonable evidence. (Chapter 3)

logistic growth The kind of growth in which population size increases rapidly at first but then slows down as the population becomes larger; produces an S curve when plotted over time. (Chapter 9)

low-level radioactive waste (LLRW) Material that has a low level of radiation for its volume. (Chapter 22)

M

malnutrition A state of poor health that results from inadequate or unbalanced food intake; includes diets that provide too few or too many calories and/or do not provide the proper nutrients (deficient in one or more nutrients). (Chapter 16)

Milankovitch cycles Predictable variations in Earth's position in space relative to the Sun that affect climate. (Chapter 21)

minimum viable population The smallest number of individuals that would still allow a population to be able to persist or grow, ensuring long-term survival. (Chapter 9)

mitigation Efforts intended to minimize the extent or impact of a problem such as climate change. (Chapter 21)

monoculture Farming method in which a single variety of one crop is planted, typically in rows over huge swaths of land, with large inputs of fertilizer, pesticides, and water. (Chapter 17)

Montreal Protocol An international treaty that laid out plans to phase out ozone-depleting chemicals such as CFCs. (Chapter 2)

more developed country A country that has a moderate to high standard of living on average and an established market economy. (Chapter 4)

mountaintop removal Surface mining technique that uses explosives to blast away the top of a mountain to expose the coal seam underneath; the waste rock and rubble is deposited in a nearby valley. (Chapter 18)

municipal solid waste (MSW) Everyday garbage or trash (solid waste) produced by individuals or small businesses. (Chapter 7)

mutualism A symbiotic relationship between individuals of two species in which both parties benefit. (Chapter 10)

N

National Environmental Policy Act (NEPA) A 1969 U.S. law that established environmental protection as a guiding policy for the nation and required that the federal government take the environment into consideration before taking action that might affect it. (Chapter 24)

natural capital The wealth of resources on Earth. (Chapter 6)

natural gas A gaseous fossil fuel composed mainly of simpler hydrocarbons, mostly methane. (Chapter 19)

natural interest Readily produced resources that we could use and still leave enough natural capital behind to replace what we took. (Chapter 6)

natural selection The process by which organisms best adapted to the environment survive to reproduce, leaving more offspring than less well-adapted individuals. (Chapter 11)

negative feedback loop Changes caused by an initial event that trigger events that then reverse the response (for example, warming leads to events that eventually result in cooling). (Chapter 21)

net primary productivity (NPP) A measure of the amount of energy captured via photosynthesis and stored in a photosynthetic organism. (Chapter 10)

New Urbanism A movement that promotes the creation of compact, mixed-use communities with all of the amenities for day-to-day living close by and accessible. (Chapter 25)

niche The role a species plays in its community, including how it gets its energy and nutrients, what habitat requirements it has, and what other species and parts of the ecosystem it interacts with. (Chapter 8, 10)

nitrogen cycle A continuous series of natural processes by which nitrogen passes from the air to the soil, to organisms, and then returns back to the air or soil. (Chapter 8)

nitrogen fixation Conversion of atmospheric nitrogen into a biologically usable form, carried out by bacteria found in soil or via lightning. (Chapter 8)

noncommunicable diseases (NCDs) Illnesses that are not transmissible between people; not infectious. (Chapter 5)

nondegradable Incapable of being broken down under normal conditions. (Chapter 7)

nonpoint source pollution Runoff that enters the water from overland flow and can come from any area in the watershed or enters the air from dispersed or mobile sources. (Chapter 15, 20)

nonrenewable resource A resource that is formed more slowly than it is used, or is present in a finite supply. (Chapter 1, 19)

nuclear energy Energy released when an atom is split (fission) or combines with another to form a new atom (fusion). (Chapter 22)

nuclear fission A nuclear reaction that occurs when a neutron strikes the nucleus of an atom and breaks it into two or more parts. (Chapter 22)

nutrient cycle Movement of life's essential chemicals or nutrients through an ecosystem. (Chapter 8)

O

observational study Research that gathers data in a real-world setting without intentionally manipulating any variable. (Chapter 2)

observations Information detected with the senses—or with equipment that extends our senses. (Chapter 2)

oil A liquid fossil fuel useful as a fuel or as a raw material for industrial products. (Chapter 19)

open dumps Places where trash, both hazardous and nonhazardous, is simply piled up. (Chapter 7)

organic agriculture Farming that does not use synthetic fertilizers, pesticides, or other chemical additives like hormones (for animal rearing). (Chapter 17)

overburden The rock and soil removed to uncover a mineral deposit during surface mining. (Chapter 18)

overpopulation More people living in an area than its natural and human resources can support. (Chapter 4)

ozone A molecule with three oxygen atoms that absorbs UV radiation in the stratosphere. (Chapter 2)

P

parasitism A symbiotic relationship between individuals of two species in which one benefits and the other is negatively affected. (Chapter 10)

particulate matter (PM) Particles or droplets small enough to remain aloft in the air for long periods of time. (Chapter 20)

passive solar technologies Technologies that allow for the capture of solar energy (heat or light) without any electronic or mechanical assistance. (Chapter 23)

pathogen An infectious agent that causes illness or disease. (Chapter 5)

payback time The amount of time it would take to save enough money in operation costs to pay for the equipment. (Chapter 23)

peak oil The moment in time when oil will reach its highest production levels and then steadily and terminally decline. (Chapter 19)

peer review A process whereby researchers submit a report of their work to outside experts who evaluate the study's design and results to determine whether it is of a high enough quality to publish. (Chapter 2, 3)

perennial crops Crops that do not die at the end of the growing season but live for several years, allowing them to be harvested annually without replanting. (Chapter 17)

performance standards Targets set that specify acceptable levels of pollution that can be released or exist in ambient (outdoor) air; industries must act to meet these standards. (Chapter 24)

persistence The ability of a substance to remain in its original form; often expressed as the length of time it takes a substance to break down in the environment. (Chapter 3)

persistent chemicals Chemicals that don't readily degrade over time. (Chapter 3)

pesticide A natural or synthetic chemical that kills or repels plant or animal pests. (Chapter 17)

pesticide resistance The ability of a pest to withstand exposure to a given pesticide, the result of natural selection favoring the survivors of an original population that was exposed to the pesticide. (Chapter 17)

petrochemicals Distillation products from the processing of crude oil such as fuels or industrial raw materials. (Chapter 19)

phosphorus cycle A series of natural processes by which the nutrient phosphorus moves from rock to soil or water, to living organisms, and back to soil. (Chapter 8)

photovoltaic (PV) cells A technology that converts solar energy directly into electricity. (Chapter 23)

pioneer species Plant species that move into an area during early stages of succession; these are often r species and may be annuals—species that live 1 year, leave behind seeds, and then die. (Chapter 10)

point source pollution Pollution from discharge pipes (or smoke stacks) such as that from wastewater treatment plants or industrial sites. (Chapter 15, 20)

policy A formalized plan that addresses a desired outcome or goal. (Chapter 2)

policy tools Methods that can be used to enforce or implement regulations or achieve desired outcomes. (Chapter 24)

political lobbying Contacting elected officials in support of a particular position; some professional lobbyists are highly organized, with substantial financial backing. (Chapter 24)

pollution standards Allowable levels of a pollutant that can be present in environmental waters or released over a certain time period. (Chapter 15)

polyculture Farming method in which a mix of different species are grown together in one area. (Chapter 17)

population All the individuals of a species that live in the same geographic area and are able to interact and interbreed. (Chapter 8, 9)

population density The number of individuals per unit area. (Chapter 4, 9)

population distribution The location and spacing of individuals within their range. (Chapter 9)

population dynamics Changes over time in population size and composition. (Chapter 9)

population growth rate The change in population size over time that takes into account the number of births and deaths as well as immigration and emigration numbers. (Chapter 4, 9)

population momentum The tendency of a young population to continue to grow even after birth rates drop to "replacement fertility" (two children per couple). (Chapter 4)

positive feedback loop Changes caused by an initial event that then accentuate that original event (for example, a warming trend gets even warmer). (Chapter 21)

potable Clean enough for consumption. (Chapter 14)

precautionary principle A principle that encourages acting in a way that leaves a margin of safety when there is a potential for serious harm but uncertainty about the form or magnitude of that harm. (Chapter 2, 3, 21, 24)

precipitation Rain, snow, sleet, or any other form of water falling from the atmosphere. (Chapter 14)

predation Species interaction in which one individual (the predator) feeds on another (the prey). (Chapter 10)

prediction A statement that identifies what is expected to happen in a given situation. (Chapter 2)

primary air pollutants Air pollutants released directly from both mobile sources (such as cars) and stationary sources (such as industrial and power plants). (Chapter 20)

primary sources Sources that present new and original data or information, including novel scientific experiments or observations and firsthand accounts of any given event. (Chapter 3)

primary succession Ecological succession that occurs in an area where no ecosystem existed before (e.g., on bare rock with no soil). (Chapter 10)

producer An organism that captures solar energy directly and uses it to produce its own food (sugar) via photosynthesis. (Chapter 8, 10)

protected areas Geographic spaces on land or at sea that are recognized, dedicated, and managed to achieve long-term conservation of nature. (Chapter 13)

proven reserves A measure of the amount of a fossil fuel that is economically feasible to extract from a known deposit using current technology. (Chapter 19)

public health The science that deals with the health of human populations. (Chapter 5)

R

radioactive Atoms that spontaneously emit subatomic particles and/or energy. (Chapter 22)

radioactive half-life The time it takes for half of the radioactive isotopes in a sample to decay to a new form. (Chapter 22)

radiative forcer Anything that alters the balance of incoming solar radiation relative to the amount of heat that escapes out into space. (Chapter 21)

random distribution A distribution in which individuals of a population are spread out over the environment irregularly, with no discernable pattern. (Chapter 9)

range of tolerance The range, within upper and lower limits, of a limiting factor that allows a species to survive and reproduce. (Chapter 8)

receptor A structure on or inside a cell that binds to a particular molecule, thus allowing the molecule to affect the cell. (Chapter 3)

reclamation Restoring a damaged natural area to a less damaged state. (Chapter 18)

recycle The fourth of the waste-reduction four *Rs:* Return items for reprocessing into new products. (Chapter 7)

reduce The second of the waste-reduction four *Rs:* Make choices that allow you to use less of a resource by, for instance, purchasing durable goods that will last or can be repaired. (Chapter 7)

reduced-tillage cultivation Planting crops in soil that is minimally disturbed and that retains some plant residue from the previous planting. (Chapter 17)

refuse The first of the waste-reduction four *Rs:* Choose not to use or buy a product if you can do without it. (Chapter 7)

remediation Restoration that focuses on the cleanup of pollution in a natural area. (Chapter 13)

renewable energy Energy from sources that are replenished over short time scales or that are perpetually available. (Chapter 1, 23)

replacement fertility rate The rate at which children must be born to replace those dying in the population. (Chapter 4)

reservoirs (or sinks) Abiotic or biotic components of the environment that serve as storage places for cycling nutrients. (Chapter 8)

reservoir Abiotic or biotic components of the environment that serve as storage places for cycling nutrients. (Chapter 8); An artificial lake formed when a river is impounded by a dam. (Chapter 14)

resilience The ability of an ecosystem to recover when it is damaged or perturbed. (Chapter 10)

resistance factors Things that directly (predators, disease) or indirectly (competitors) reduce population size. (Chapter 9)

resource partitioning A strategy in which different species use different parts or aspects of a resource rather than compete directly for exactly the same resource. (Chapter 10)

restoration ecology The science that deals with the repair of damaged or disturbed ecosystems. (Chapter 10)

reuse The third of the waste reduction four *Rs:* Use a product more than once for its original purpose or for another purpose. (Chapter 7)

riparian areas The land areas close enough to a body of water to be affected by the water's presence (for example, areas where water-tolerant plants grow) and that affect the water itself (for example, provide shade). (Chapter 15)

risk assessment The process of weighing the risks and benefits of a particular action in order to decide how to proceed. (Chapter 3)

r-selected species Species that have a high biotic potential and that share other characteristics, such as short life span, early maturity, and high fecundity. (Chapter 9)

_____

S

saltwater intrusion The inflow of ocean (salt) water into a freshwater aquifer that happens when an aquifer has lost some of its freshwater stores. (Chapter 14)

sanitary landfills Disposal sites that seal in trash at the top and bottom to prevent its release into the atmosphere; the sites are lined on the bottom, and trash is dumped in and covered with soil daily. (Chapter 7)

science A body of knowledge (facts and explanations) about the natural world and the process used to get that knowledge. (Chapter 2)

scientific method The procedure scientists use to empirically test a hypothesis. (Chapter 2)

secondary air pollutants Air pollutants formed when primary air pollutants react with one another or with other chemicals in the air. (Chapter 20)

secondary sources Sources that present and interpret information solely from primary sources. (Chapter 3)

secondary succession Ecological succession that occurs in an ecosystem that has been disturbed; occurs more quickly than primary succession because soil is present. (Chapter 10)

seed banks Places where seeds are stored in order to protect the genetic diversity of the world's crops. (Chapter 17)

selective pressure A nonrandom influence that affects who survives or reproduces. (Chapter 11)

service economy A business model whose focus is on leasing and caring for a product in the customer's possession rather than on selling the product itself (that is, selling the *service* that the product provides). (Chapter 6)

sex ratio The relative number of males to females in a population; calculated by dividing the number of males by the number of females. (Chapter 4)

single-species conservation A management strategy that focuses on protecting one particular species. (Chapter 13)

sliding reinforcer Actions that are beneficial at first but that change conditions such that their benefit declines over time. (Chapter 1)

smart growth Strategies that help create walkable communities with lower ecological footprints. (Chapter 25)

smog Hazy air pollution that contains a variety of pollutants including sulfur dioxide, nitrogen oxides, tropospheric ozone, and particulates. (Chapter 20)

social trap Decisions by individuals or groups that seem good at the time and produce a short-term benefit but that hurt society in the long run. (Chapter 1)

solar energy Energy harnessed from the Sun in the form of heat or light. (Chapter 23)

solar thermal system An active technology that captures solar energy for heating. (Chapter 23)

solubility The ability of a substance to dissolve in a liquid or gas. (Chapter 3)

species A group of plants or animals that have a high degree of similarity and can generally only interbreed among themselves. (Chapter 8)

species diversity The variety of species, including how many are present (richness) and their abundance relative to each other (evenness). (Chapter 10, 12)

species evenness The relative abundance of each species in a community. (Chapter 10)

species richness The total number of different species in a community. (Chapter 10)

statistics The mathematical evaluation of experimental data to determine how likely it is that any difference observed is due to the variable being tested. (Chapter 2)

stormwater runoff Water from precipitation that flows over the surface of the land. (Chapter 15)

stratosphere The region of the atmosphere that starts at the top of the troposphere and extends up to about 31 miles; contains the ozone layer. (Chapter 2)

strip cropping Alternating different crops in adjacent strips, several rows wide; helps keep pest populations low. (Chapter 17)

subsidies Financial assistance given by the government or other party in support of actions that are expected to benefit the public good. (Chapter 20, 24)

subsurface mines Sites where tunnels are dug underground to access mineral resources. (Chapter 18)

suburban sprawl Low-population-density developments that are built outside of a city. (Chapter 25)

surface mining A form of mining that involves removing soil and rock that overlays a mineral deposit close to the surface in order to access that deposit.

surface water Any body of water found above ground, such as oceans, rivers, and lakes. (Chapter 14)

sustainable Capable of being continued without degrading the environment. (Chapter 1, 6)

sustainable agriculture Farming methods that can be used indefinitely because they do not deplete resources, such as soil and water, faster than they are replaced. (Chapter 17)

sustainable development Economic and social development that meets present needs without compromising the ability of future generations to do the same. (Chapter 1, 6)

sustainable energy Energy from sources that are renewable and have a low environmental impact. (Chapter 23)

symbiosis A close biological or ecological relationship between two species. (Chapter 10)

synergistic effects Exposure to two or more chemicals that has a greater effect than the sum of their individual effects would predict. (Chapter 3)

T

tar/oil sands Sand or clay formations that contain a heavy-density crude oil (crude bitumen); extracted by surface mining. (Chapter 19)

tax credit A reduction in the tax one has to pay in exchange for some desirable action. (Chapter 20)

terracing On steep slopes, land is leveled into steps; reduces soil erosion and runoff down the hillside. (Chapter 17)

tertiary sources Sources that present and interpret information from secondary sources. (Chapter 3)

testable Having a possible explanation that generates predictions for which empirical evidence can be collected to verify or refute the hypothesis. (Chapter 2)

test group The group in an experimental study that is manipulated such that it differs from the control group in only one way. (Chapter 2)

theory A widely accepted explanation of a natural phenomenon that has been extensively and rigorously tested scientifically. (Chapter 2)

threatened Describes a species that is at risk for extinction. (Chapter 11)

threatened species Species that are at risk for extinction; various threat levels have been identified, ranging from "least concern" to "extinct." (Chapter 13)

tight oil Light (low density) oil in shale rock deposits of very low permeability; extracted by fracking. (Chapter 19)

time delay Actions that produce a benefit today and set into motion events that cause problems later on. (Chapter 1)

top-down regulation The control of population size by factors that reduce population size (resistance factors) such as predation, competition, or disease. (Chapter 9)

total fertility rate (TFR) The number of children the average woman has in her lifetime. (Chapter 4)

toxic substances/toxics Chemicals that cause damage to living organisms through immediate or long-term exposure. (Chapter 3)

toxicologist A scientist who studies the specific properties of potentially toxic substances. (Chapter 3)

trade-offs The imperfect and sometimes problematic responses that we must at times choose between when addressing complex problems. (Chapter 1)

tragedy of the commons The tendency of an individual to abuse commonly held resources in order to maximize his or her own personal interest. (Chapter 1)

transboundary problem A problem that extends across state and national boundaries; pollution that is produced in one area but falls in or reaches other states or nations. (Chapter 20, 24)

transgenic organism An organism that contains genes from another species. (Chapter 16)

transpiration The loss of water vapor from plants. (Chapter 14)

triple bottom line The combination of the environmental, social, and economic impacts of our choices. (Chapter 1, 6)

trophic levels Feeding levels in a food chain. (Chapter 10)

troposphere The region of the atmosphere that starts at ground level and extends upward about 7 miles. (Chapter 2)

true cost The sum of both external and internal costs of a good or service. (Chapter 6)

U

ultraviolet (UV) radiation Short-wavelength electromagnetic energy emitted by the Sun. (Chapter 2)

unconventional reserves Deposits of oil or natural gas that cannot be recovered with traditional oil/gas wells but may be recoverable using alternate techniques. (Chapter 19)

uniform distribution A population distribution in which individuals are spaced evenly, perhaps due to territorial behavior or mechanisms for suppressing the growth of nearby individuals. (Chapter 9)

United Nations Framework Convention on Climate Change (UNFCCC) A 1992 international treaty that formally recognized climate change as an emerging problem and that precautions should be taken to prevent dangerous anthropogenic interference with Earth's climate system. (Chapter 24)

urban areas Densely populated regions that include cities and the suburbs that surround them. (Chapter 25)

urban flight The process of people leaving an inner-city area to live in surrounding areas. (Chapter 25)

urban heat island effect The phenomenon in which urban areas are warmer than the surrounding countryside due to pavement, dark surfaces, closed-in spaces, and high energy use. (Chapter 25)

urbanization The migration of people to large cities; sometimes also defined as the growth of urban areas. (Chapter 25)

U.S. Nuclear Waste Policy Act (1982) The federal law which mandated that the federal government build and operate a long-term repository for the disposal of high level radioactive waste. (Chapter 22)

V

vector-borne disease An infectious disease acquired from organisms that transmit a pathogen from one host to another. (Chapter 5)

W

waste Any material that humans discard as unwanted. (Chapter 7)

wastewater Used and contaminated water that is released after use by households, industry, or agriculture. (Chapter 14)

wastewater treatment The process of removing contaminants from wastewater to make it safe enough to release into the environment. (Chapter 14)

waterborne disease An infectious disease acquired through contact with contaminated water. (Chapter 5)

water cycle The movement of water through various water compartments such as surface waters, atmosphere, soil, and living organisms. (Chapter 14)

water footprint The water appropriated by industry to produce products or energy; this includes the water actually used and water that is polluted in the production process. (Chapter 14)

water pollution The addition of any substance to a body of water that might degrade its quality. (Chapter 15)

water scarcity Not having access to enough clean water. (Chapter 14)

watershed The land area surrounding a body of water over which water such as rain can flow and potentially enter that body of water. (Chapter 15)

watershed management Management of what goes on in an area around streams and rivers. (Chapter 15)

water table The uppermost water level of the saturated zone of an aquifer. (Chapter 14)

weather The meteorological conditions in a given place on a given day. (Chapter 21)

wetland An ecosystem that is permanently or seasonally flooded. (Chapter 14)

wind energy Energy contained in the motion of air across Earth's surface. (Chapter 23)

worldview The window through which one views one's world and existence. (Chapter 1)

Z

zero population growth The absence of population growth; occurs when birth rates equal death rates. (Chapter 4)

zoonotic disease A disease that is spread to humans from infected animals (not merely a vector that transmits the pathogen but another host that harbors the pathogen through its life cycle). (Chapter 5)

CREDITS/SOURCES

INFOGRAPHIC SOURCES AND REFERENCES

CHAPTER 1:
IG 1.7 Data Source: UN Human Development Indices, 2008 (http://is.gd/7UqClm).

CHAPTER 2:
IG 2.2 Data Source: Refrigerant Reclaim Australia. (2010). Annual Report 2009/2010.
IG 2.6 Research from: Abarca, J.F., & C.C. Casiccia. (2002). *Photodermatology, Photoimmunology & Photomedicine*, 18(6): 294–302 and Menzies, S. W., *et al.* (1991). *Cancer Research*, 51(11): 2773–2779.
IG 2.7 Data Source: UN Vital Ozone Graphics, 2007.

CHAPTER 3:
IG 3.3 Research from: Nagel, S. C., *et al.* (1997). *Environmental Health Perspectives*, 105(1): 70–76; Ishido, M. & J. Suzuki. (2010). *Journal of Health Science*, 56(2): 175–181; Lang, I. A., *et al.* (2008). *Journal of the American Medical Association*, 300: 1303–1310.

CHAPTER 4:
IG 4.1 Data Source: *UN World Population Prospects, the 2010 Revision.*
IG 4.2 Data Source: Population Reference Bureau, *2013 World Population Data Sheet.*
IG 4.3 Information from: Population Action International
IG 4.4 Data Sources: Table: Population Reference Bureau, *2013 World Population Data Sheet*; Graph: United Nations Department of Economic and Social Affairs
IG 4.6 Data from: U.S. Census Bureau
IG 4.7 Data Sources: Infant Mortality: CIA Factbook, 2011 estimates; Desired Fertility: Pritchett, L. (1994). *Population and Development Review*, 20: 1-55; Education: Pew Research Center. (2011). *The Future of the Global Muslim-Population*; Family Planning: Sedgh, G., *et al.* (2007). *Women With an Unmet Need for Contraception in Developing Countries and Their Reasons for Not Using a Method*, Guttmacher Institute.
IG 4.8 Data Source: Global Footprint Network. (2010). *The Ecological Footprint Atlas 2010.*

CHAPTER 5:
IG 5.3 Data Source: World Health Organization (WHO). (2006). *Preventing Disease Through Healthy Environments.*
IG 5.5 Information from the CDC (www.dpd.cdc.gov/dpdx/HTML/Dracunculiasis.htm) and CNN (http://www.cnn.com/2010/HEALTH/04/05/guinea.worm.lifecycle/index.html)
IG 5.6 Data Sources: Map: WHO. (2006). *Preventing Disease Through Healthy Environments* (page 8) and WHO. (2004). Global Health Observatory Data: Total Environment—Total Burden of Disease Top Ten Causes of Death: World Health Organization, *Deaths across the globe: An overview.*
IG 5.7 Information from Emory University. (2011). *End game for Guinea worm disease is near.* (http://www.emory.edu/EMORY_REPORT/stories/2011/03/report_from_carter_center_guinea_worm_disease.html)

CHAPTER 6:
IG 6.1 Research from: Costanza, R., *et al.* (1997). *Nature.* 387: 253-260.
IG 6.2 Data from: Global Footprint Network. (2012). *Living Planet Report 2012 and Footprint Interactive Graph,* http://wwf.panda.org/about_our_earth/all_publications/living_planet_report/living_planet_report_graphics/footprint_interactive/

CHAPTER 7:
IG 7.1 Data Source: EPA
IG 7.2 Information from EPA
IG 7.4 Research from: Young, L.C., *et al.* (2009). *PLoS ONE*, 4(10): e7623.
IG 7.7 Research from: Gertler, N. (1995). *Industrial Ecosystems: Developing Sustainable Industrial Structures.* Dissertation. Massachusetts Institute of Technology.

CHAPTER 8:
IG 8.4 Information from: Map: The Nature Education Knowledge Project; Graph: Ricklefs, R. (2000). *The Economy of Nature,* W.H. Freeman, New York.

CHAPTER 9:
IG 9.6 Showshoe hare/lynx graph: Research from: Krebs, C.J. (2011). *Proceedings of the Royal Society B.* 278(1705): 481-489.

CHAPTER 10:
IG 10.7 Information from: Comprehensive Everglades Restoration Plan website (http://is.gd/l6XXjK)

CHAPTER 11:
IG 11.2 Research from: Hoekstra, H. E., J. G. Krenz, & M. W. Nachman. (2005). *Heredity*, 94(2): 217-228.
IG 11.4 Research from: Savidge, J.A. (1987). *Ecology*, 68: 660-668.

CHAPTER 12:
IG 12.1 Data Source: Chapman, A.D. (2009). *Number of Living Species in Australia and the World.*
IG 12.4 Information from Conservation International

CHAPTER 13:
IG 13.1 Data Source: Baillie, J.E.M, *et al.* (2010). *Evolution Lost: Status and Trends of the World's Vertebrates,* Zoological Society of London.
IG 13.2 Information from the International Union for Conservation of Nature
IG 13.6 Data Sources: Map: UN Environmental Programme. (2012). *Global Environmental Outlook-5 Report*; Graph: Statistics Sweden.

CHAPTER 14:
IG 14.3 Data Sources: Pie Chart: UN World Water Assessment Programme; Map: UN Food and Agriculture Organization; Improved Sanitation graph: Population Reference Bureau; Water Use Graph:

Hoekstra, A. Y. & A. K. Chapagain. (2007). *Water Resources Management*, 21: 35–48.
IG 14.6 Data Sources: Pie Chart: University of Georgia Cooperative Extension; Water Use for Electricity Production: Institute of Electrical and Electronics Engineers; Gallons of Water per Food or Product: *National Geographic* (http://is.gd/MCeBGJ)

CHAPTER 15:
IG 15.1 Information from EPA
IG 15.4 Data Sources: Nitrogen fertilizer use and Gulf of Mexico hypoxia graph: The Fertilizer Institute and the EPA: Nitrogen Losses Graph: Oquist, K. A., *et al.* (2007). *Journal of Environmental Quality*, 36(4): 1194 -1204.
IG 15.5 Information from U.S. Department of Agriculture.
IG 15.7 Information from EPA. (2011). *Gulf of Mexico Regional Ecosystem Restoration Strategy.*

CHAPTER 16:
IG 16.1 Data Sources: Map: UN Food and Agriculture Organization (http://www.fao.org/hunger/en/); Bar Graph from UN Food and Agriculture Organization. (2010). *State of Food Insecurity in the World.*
IG 16.4 Data Source: Department of Plant and Soil Sciences, Oklahoma State University, http://www.nue.okstate.edu/Crop_Information/World_Wheat_Production.htm

CHAPTER 17:
IG 17.4 Information from: Furuno, Takao. (2001). *The Power of the Duck*. Tasmania: Takari Publications.
IG 17.7 Information from Environmental Working Group (www.ewg.org)

CHAPTER 18:
IG 18.1 Data Source for Pie Chart: U.S. Energy Information Administration,
IG 18.3 Information from Britannica Encyclopedia (http://is.gd/MUgevK)
IG 18.5 Information from Kentucky Geological Survey (http://is.gd/nyyEbh)
IG 18.7 Information from World Coal Association

CHAPTER 19:
IG 19.3 Data Source: U.S. Energy Information Association (www.eia.gov/countries/)
IG 19.3 Information from The Coming Oil Crisis: Colin Campbell's research (http://www.oilcrisis.com/campbell/)
IG 19.7 Information from FracFocus Chemical Disclosure Registry (www.fracfocus.ca)
IG 19.8 Information from USGS (www.eia.gov/oil_gas/rpd/northamer_gas.pdf), Alberta Geological Survey (www.ags.gov.ab.ca/energy/oilsands/), and PacWest Consulting Partners (http://pacwestcp.com/)

CHAPTER 20:
IG 20.1 Data Source: the World Health Organization
IG 20.2 Data Source: EPA
IG 20.3 Research from Laden, Francine, *et al.* (2006). *American Journal of Respiratory and Critical Care Medicine*. 173: 667-672.
IG 20.4 Information from National Atmospheric Deposition Program
IG 20.5 Information from EPA
IG 20.6 Information from: Smokestack Scrubber Diagram: Encyclopedia Britannica; Cap-and-Trade Diagram: *Washington Post*

CHAPTER 21:
IG 21.1 Information from Intergovernmental Panel on Climate Change (IPCC). (2007). 4th Assessment Report, Working Group 1 Report: *The Physical Science Basis*
IG 21.2 Data Sources: Surface Temperature, Temperature Anomalies, and Precipitation Changes graphs: IPCC (2013), 5th Assessment Report, Working Group I Report: *The Physical Science Basis*; Cumulative Loss of Glacier Ice Graph: UN Environmental Programme; sea level change over time source: National Climatic Data Center/NOAA; Tropical cyclones: Chylek, P., & G. Lesins. (2008). Journal of Geophysical Research: Atmospheres, 113(D22) doi:10.1029/2008JD010036.
IG 21.3 Data Source: Bar graph: IPCC. (2013). 5th Assessment Report, Working Group 1 Report: *The Physical Science Basis*
IG 21.5 Data Source: IPCC. (2013). 5th Assessment Report, Working Group 1 Report: *The Physical Science Basis*

IG 21.7 Data Sources: Upper Graphs: NASA; Lower graph: United States Global Change Research Program
IG 21.8 Data Source: IPCC. (2013). 5th Assessment Report, Working Group 1 Report: *The Physical Science Basis.*
IG 21.10 Information from: Future Scenarios Graph: IPCC. (2013). 5th Assessment Report, Working Group I Report: *The Physical Science Basis*; Stabilization Wedges Diagram: Pacala, S. & R. Socolow. (2004). *Science*, 305: 968-972.
IG 21.11 Information from IPCC. (2013). 5th Assessment Report, Working Group II Report: *Impacts, Adaptation and Vulnerability*

CHAPTER 22:
IG 22.1 Information from www.nuclearconnect.org
IG 22.2 Decay Chain Diagram: Information from U.S. Department of Energy
IG 22.3 Mining Flowchart: Information from World Information Service on Energy Uranium Project; Fuel assembly: Information from www.climateandfuel.com
IG 22.4 Information from www.atomicarchive.com
IG 22.5 Information from the Nuclear Regulatory Commission and the Union of Concerned Scientists
IG 22.6 Information from U.S. Office of Environmental Management (http://gtcceis.anl.gov/guide/rad/index.cfm)

CHAPTER 23:
IG 23.1 Data Source: U.S. Energy Information Administration
IG 23.2 Information from U.S. Department of the Interior, *Wind Energy Development Programmatic Environmental Impact Statement*
IG 23.3 Information from the U.S. Department of Energy and Wave Green Energy (www.wavege.com/solar-panel-diagram.html)
IG 23.4 Information from U.S. Department of Energy and Tennessee Valley Authority
IG 23.5 Information from Tennessee Valley Authority
IG 22.6 Information from U.S. Office of Energy Efficiency and Renewable Energy (www.energy.gov) and eartheasy.com

CHAPTER 24:

IG 24.7 Information from UN Environmental Programme. (2005). *Vital Climate Change Graphics Update* at http://www.grida.no/graphicslib/detail/kyoto-protocol-timeline-and-history_76cf#

IG 24.8 Data Source: International Energy Agency. (2011). *CO_2 Emissions from Fuel Combustion: Highlights.*

IG 24.9 Data Source: UN Environmental Programme Risoe Centre, *CDM/JI Pipeline*, accessed August 2014 from http://cdmpipeline.org/cdm-projects-type.htm

CHAPTER 25:

IG 25.1 Data Sources: Map: UN Population Division; CIA World Factbook; Urban vs. Rural Graph: United Nations, Population Division; Megacities Bar Graph: United Nations, *World Urbanization Prospects: The 2014 Revision, Highlights*

IG 25.2 Information from *National Geographic*

IG 25.3 Information from Stats Canada

IG 25.5 Information from European Green City Index (www.siemens.com/entry/cc/en/greencityindex.htm)

IG 25.6 Information from the U.S. EPA

SCIENCE LITERACY: WORKING WITH DATA

CHAPTER 1:

Data Source: Brown, L.R. *World on the edge—food and agriculture data—livestock and fish.* pp. 7, 15. Earth Policy Institute. www.earthpolicy.org/datacenter/pdf/book_wote_livestock.pdf.

CHAPTER 2:

Data Source: http://undsci.berkeley.edu/article/ozone_depletion_01. Figure 28 - A Plot of Chlorine Monoxide and Ozone Concentrations from Ozone Depletion: Uncovering the hidden hazard of hairspray. Originally from: Anderson, J.G., *et al.* (1989). Ozone destruction by chlorine radicals within the Antarctic vortex: The spatial and temporal evolution of ClO–O3 anticorrelation based on in situ ER-2 data. *Journal of Geophysical Research* 94: 11465–11479.

CHAPTER 3:

Data Source: Schreder, E. (2006). *Pollution in People.: A Study of Toxic Chemicals in Washingtonians.* Toxic-Free Legacy Coalition, Seattle, WA. (http://pollutioninpeople.org/). Mercury Levels Graph: Figure 3; DDT Levels Graph: Figure 7.

CHAPTER 4:

Data Source: United Nations, Department of Economic and Social Affairs, Population Division. (2011). *World Population Prospects: The 2010 Revision.* New York. (Updated: 15 April 2011). http://esa.un.org/unpd/wpp/Analytical-Figures/htm/fig_1.htm (Graph A); (Updated: 5 July 2011). http://esa.un.org/unpd/wpp/Analytical-Figures/htm/fig_13.htm (Graph B).

CHAPTER 5:

Data Source: Fan V. (2012). *Malaria Estimate Sausages by WHO and IHME.* Center for Global development. http://blogs.cgdev.org/globalhealth/2012/02/malaria-estimate-sausages-by-who-and-ihme.php

CHAPTER 6:

Data Source: Brown, L.R. (2011). *Learning from China: Why the existing economic model will fail.* Earth Policy Institute. http://www.earthpolicy.org/data_highlights/2011/highlights18

CHAPTER 7:

Data Source: Greenpeace. (2006). *Plastic Debris in the World's Oceans.* Table 2.1 Number and Percentage of Marine Species Worldwide with Documented Entanglement and Ingestion Records. Greenpeace. http://www.unep.org/regionalseas/marinelitter/publications/docs/plastic_ocean_report.pdf

CHAPTER 8:

Data Source: http://www.globalchange.umich.edu/globalchange1/current/lectures/kling/rainforest/rainforest_table.html. Originally from: Terborgh, J. (1992). Diversity and the tropical rain forest. *Scientific American Library*, W. H. Freeman, New York, xii + 242 pages.

CHAPTER 9:

Data Source: Kaibab Plateau Deer Population: 1907–1940, http://www.hhh.umn.edu/centers/stpp/pdf/KaibabPlateauExercise.pdf (Redrawn from Leopold, A.)

CHAPTER 10:

Data Source: Rogers, J.D. (2008). *Journal of Geotechnical and Geoenvironmental Engineering,* http://web.mst.edu/~rogersda/levees/History%20New%20Orleans%20Flood%20Control-Rogers.pdf. Page 6, Fig. 7 (adapted from Kolb and Saucier 1982).

CHAPTER 11:

Data Sources: Human Population Growth and Extinction. Center for Biological Diversity. http://www.biologicaldiversity.org/programs/population_and_sustainability/extinction/. Originally from: Scott, J.M. (2008). *Threats to Biological Diversity: Global, Continental, Local.* U.S. Geological Survey, Idaho Cooperative Fish and Wildlife, Research Unit, University Of Idaho.

CHAPTER 12:

Data Source: Gibson, L., *et al.* (2013). Near-complete extinction of native small mammal fauna 25 years after forest fragmentation. *Science*, 341(6153): 1508-1510.

CHAPTER 13:

Data Source: IUCN; Figure 3 from *Summary Statistics*, www.iucnredlist.org/about/summary-statistics. Accessed September 4, 2012.

CHAPTER 14:

Data Source: Public Health Information and Geographic Information Systems (GIS). (2011). DALYs attributable to water, sanitation and hygiene (diarrhea), 2004. World Health Organization. http://gamapserver.who.int/mapLibrary/Files/Maps/Global_wsh_daly_2004.png

CHAPTER 15:
Data Source: Collins, S.J. & R.W. Russell. (2009). Toxicity of road salt to Nova Scotia amphibians. *Environmental Pollution*, 157(1): 320-324.

CHAPTER 16:
Data Source: UN Food and Agriculture Organization. (2010). *The State of Food Insecurity in the World, 2010*, figures 1 and 2 on page 9. http://www.fao.org/docrep/013/i1683e/i1683e.pdf

CHAPTER 17:
Data Source: Rodale Institute. (2011). *The Farming Systems Trial. Figure Comparison of FST Organic and Conventional Systems.* http://rodaleinstitute.org/assets/FSTbooklet.pdf

CHAPTER 18:
Data Source: National Institute for Occupational Safety and Health, 2007 World Report. http://www.cdc.gov/niosh/programs/mining/risks.html

CHAPTER 19:
Data Source: Jackson, R. B., *et al.* (2013). Increased stray gas abundance in a subset of drinking water wells near Marcellus shale gas extraction. *Proceedings of the National Academy of Sciences*, 110(28): 11250-11255.

CHAPTER 20:
Data Source: Environment Canada, National Air Pollution Surveillance (NAPS) Network, Ottawa, 2004.

CHAPTER 21:
Data Source: Woodall, C. W., *et al.* (2009). An indicator of tree migration in forests of the eastern United States. *Forest Ecology and Management*, 257(5): 1434–1444.

CHAPTER 22:
Data Source: Clean Energy Insight. http://www.cleanenergyinsight.org/wp-content/uploads/2009/08/comparingindustrysafety_graph.jpg. Original source: U.S. Bureau of Labor Statistics, 2007.

CHAPTER 23:
Data Source: Graphs 1 and 2: Arnett, E., *et al.* (2005). Relationships between bats and wind turbines in Pennsylvania and West Virginia: An assessment of fatality search protocols, patterns of fatality, and behavioral interactions with wind turbines. Report prepared for Bats and Wind Energy Cooperative.
Graph 3: Sibley, D. (2010). Causes of bird mortality, *Sibley Guides: Identification of North American Birds and Trees* (http://www.sibleyguides.com/conservation/causes-of-bird-mortality/);

Table 1: Erickson, W., *et al.* (2001). *Avian collisions with wind turbines: A summary of existing studies and comparisons to other sources of avian collision mortality in the United States*. West Inc., prepared for the National Wind Coordinating Committee. http://www.duke.edu/web/nicholas/bio217/ptb4/batdata.html

CHAPTER 24:
Data Source: Energy Information Agency, *What are Greenhouse Gases?*, Figure 1. (www.eia.gov/oiaf/1605/ggccebro/chapter1.html)

CHAPTER 25:
Data Sources: New Buildings Institute. (2008). *Energy Performance of LEED® for New Construction Buildings: Final Report.* (http://newbuildings.org/sites/default/files/Energy_Performance_of_LEED-NC_Buildings-Final_3-4-08b.pdf) and the EPA (http://www.energystar.gov/index.cfm?fuseaction=buildingcontest.eui)

INDEX

atoms, 432f
author bias in studies, 57
authority, appeal to, 56
averages, A–2
average temperature, change in, 406, 406f, 408
azolla, 318–319, 324

B

Bacillus thuringiensis, 308
background rate of extinction, 210–211
bacteria, 88f, 282, 284
 coliform, 266
Bakken formation, 360f, 362, 362f
 upsides and downsides of, 372
Bangladesh, 68f
 carrying capacity, 76
bar graphs, A–8
Barlow, Lisa, 7–8
barrels of oil equivalents (BOE), 453
benthic macroinvertebrates, 288
Benyus, Janine, *Biomimicry*, 12
beta radiation, 439, 439f
Big Sandy River, 350
bioaccumulation, 48–49, 48f, 128
biochar, 330, 330f
biodegradable, defined, 122
biodiesel, 112f
biodiversity, 218–237, 226f
 climate change and, 424
 definition, 11, 221
 on Earth, 222f
 ecological diversity and, 225–226
 on Hawaiian Islands, 233f
 hotspots, 226, 227f, 228
 levels of, 225
 local, 11
 preserving, 238–257
 protecting, 234t
 of rainforest, 220–221
 threats to, 228–233, 241, 242f
biofuels, 454
biological assessment, 288
biological hazards, 84, 85f
 addressing, 88–89
biomagnification, 49, 49f
biomass energy, 143, 182, 454, 454f
biomes, 142f
 in Biosphere 2, 146–148
 definition, 143

distribution of, 145f
 environmental conditions in, 146–149
 freshwater, 143
 marine, 143
 resource use in, 157
 terrestrial, 143, 144f, 145f
biomimicry, 12, 112
Biomimicry (Benyus), 12
biosphere, 142, 142f, 143
Biosphere 1, 143
Biosphere 2, 138f, 147f
 biomes in, 146–148
 biospherians, 140, 141f, 149
 breeches in, 154
 carbon cycle, 150–151
 desert biome, 141f
 design of, 146–148
 goals of, 140
 location, 140f
 Mission 1, 148–149
 nitrogen cycle, 152
 nutrient cycling in, 149
 oxygen levels in, 150
 phosphorus cycle in, 152–153
 Pinon Pine Tree Drought Experiment, 154f
 as research tool, 146
biotic, definition, 149
biotic factors, communities and, 183–186
biotic potential, 164, 165f
birth control, 73
 one-child policy and, 70
 TFR and, 74f
birth rate, 164
 death rate and, 71f
Biscayne National Park, Florida, 184f. See also Florida Everglades
bisphenol A (BPA), 44
 in aquatic life, 128
 ban on, 58
 bioaccumulation of, 128
 cancer and, 56
 concentrations of, 47
 dose, 54
 epidemiological studies of, 51
 exposure limits, 58
 medical conditions associated with, 44, 51
 persistence, 49
 safe exposure levels, 58
 in vitro studies of, 50
 in vivo studies of, 50
bisphenol S (BPS), 58
Blake, Dan, 114f

Blake, Steven, 243, 254–255
Bloomberg, Michael, 228
Boaz, Noah, 277
BOE. *See* barrels of oil equivalents
Bogotá, Colombia, 501
Boiga irregularis. See brown tree snake
boiling water reactor (BWR), 438
boom-and-bust cycles, 169, 169f
boreal forest, 144f
 distribution of, 145f
bottled water, 133
bottleneck effect, 208, 209f
bottom-up regulation, 170, 171f
BPA. *See* bisphenol A
BPS. *See* bisphenol S
Brattahlid, Greenland, 13f
Brazil, population control in, 74
British Antarctic Survey, 25
Brix, Hans, 187
Brody, Aaron L., 58
Bronx, New York City, 492f
 air quality in, 492
 asthma rates in, 388f, 391
 Bronx Library Center, 504, 506
 Clay Garden, 490f
 Concrete Plant Park, 493f
 Cross Bronx Expressway, 499
 Finca Del Sur garden, 500f
 Hunts Point Riverside Park, 501, 507
 industry in, 492, 500
 tenement houses, 499–500
 waterfront, 500–501
Bronx Environmental Stewardship Training, 504
Bronx River Greenway, 493f
brown tree snake (*Boiga irregularis*), 198f
 birds on Guam and, 205–206
 population control of, 213–214
 spread of, 207f
Bruno, David, 133
Bt crops, 308
 pesticide use and, 309–310
Bullard, Perry, 35f
Burgess, Tony, 146–147, 149
Burkina Faso, Africa, 300, 300f
 agricultural technology in, 310–311
bushmeat trade, 87
business
 environmental impact of, 101–105
 green, 111–112
 sustainable, 111–112
BWR. *See* boiling water reactor

WORLD MAP

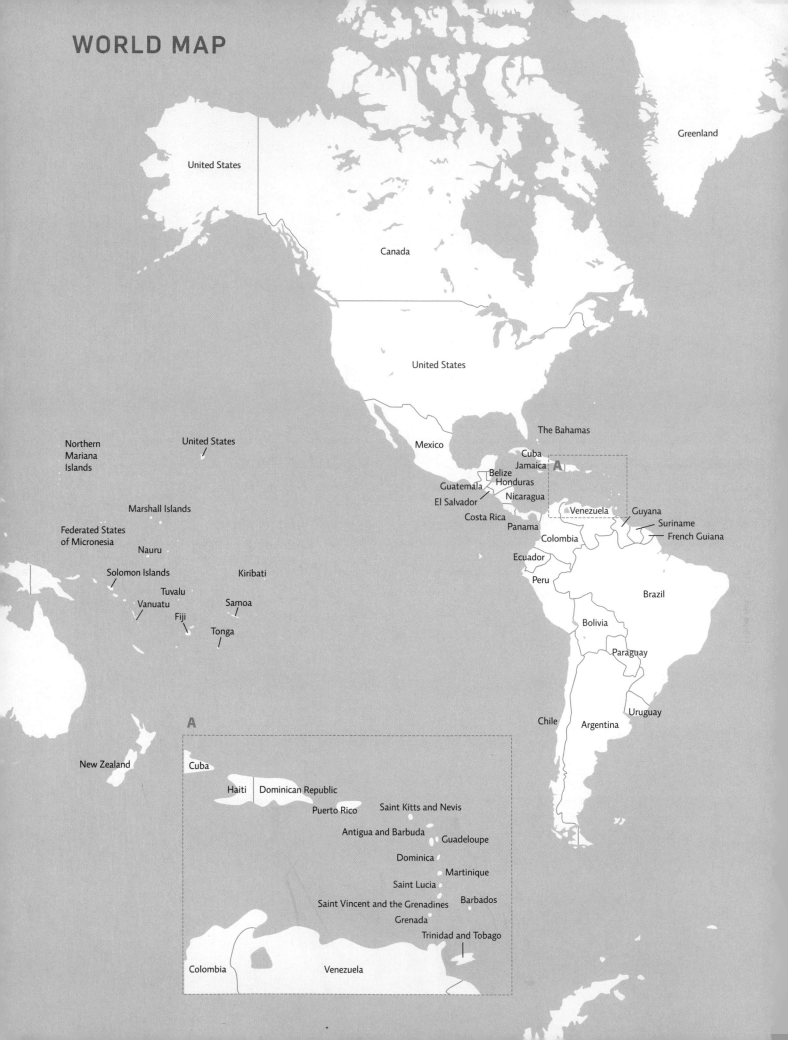

Greenland

United States

Canada

United States

Northern Mariana Islands

United States

Mexico

The Bahamas

Cuba
Jamaica

A

Belize
Guatemala
Honduras
El Salvador
Nicaragua
Costa Rica
Panama
Venezuela
Guyana
Suriname
French Guiana

Marshall Islands

Colombia

Federated States of Micronesia

Nauru

Ecuador

Peru

Brazil

Solomon Islands

Kiribati

Tuvalu

Vanuatu
Fiji
Samoa
Tonga

Bolivia

Paraguay

New Zealand

Uruguay

Chile

Argentina

A

Cuba

Haiti
Dominican Republic

Puerto Rico
Saint Kitts and Nevis

Antigua and Barbuda
Guadeloupe

Dominica
Martinique

Saint Lucia

Saint Vincent and the Grenadines
Barbados

Grenada

Trinidad and Tobago

Colombia
Venezuela